CW00602126

# AutoCAD® 2002:
# The Complete Reference

David S. Cohn

**McGraw-Hill**/Osborne

New York  Chicago  San Francisco
Lisbon  London  Madrid  Mexico City
Milan  New Delhi  San Juan
Seoul  Singapore  Sydney  Toronto

**McGraw-Hill**/Osborne
2600 Tenth Street
Berkeley, California 94710
U.S.A.

To arrange bulk purchase discounts for sales promotions, premiums, or fund-raisers, please contact **McGraw-Hill**/Osborne at the above address. For information on translations or book distributors outside the U.S.A., please see the International Contact Information page immediately following the index of this book.

## AutoCAD® 2002: The Complete Reference

Copyright © 2002 by The McGraw-Hill Companies. All rights reserved. Printed in the United States of America. Except as permitted under the Copyright Act of 1976, no part of this publication may be reproduced or distributed in any form or by any means, or stored in a database or retrieval system, without the prior written permission of publisher, with the exception that the program listings may be entered, stored, and executed in a computer system, but they may not be reproduced for publication.

1234567890 DOC DOC 0198765432

ISBN 0-07-222429-0

**Publisher**
  Brandon A. Nordin

**Vice President & Associate Publisher**
  Scott Rogers

**Acquisitions Editor**
  Megg Morin

**Project Editor**
  Katie Conley

**Acquisitions Coordinator**
  Tana Allen

**Technical Editor**
  Christal Elliot

**Copy Editor**
  Bill McManus

**Proofreader**
  Marian Selig

**Indexer**
  Irv Hershman

**Computer Designers**
  Apollo Publishing Services
  Mickey Galicia

**Illustrators**
  Michael Mueller, Lyssa Sieben-Wald

**Series Design**
  Peter F. Hancik

This book was composed with Corel VENTURA™ Publisher.

Information has been obtained by **McGraw-Hill**/Osborne from sources believed to be reliable. However, because of the possibility of human or mechanical error by our sources, **McGraw-Hill**/Osborne, or others, **McGraw-Hill**/Osborne does not guarantee the accuracy, adequacy, or completeness of any information and is not responsible for any errors or omissions or the results obtained from the use of such information.

To my family
Genevieve
Clarice
Emma
Andrea
and Bruce

To my sisters
Bobbi
and Lori

And to my father
Barney

## About the Author

David S. Cohn is a computer consultant and technical writer specializing in helping architects, engineers, and designers leverage computer technology. He is also a licensed architect with over 20 years of experience, including 18 years using AutoCAD. The editor-in-chief of *Engineering Automation Report*, he is also a contributing editor to *Desktop Engineering* magazine, and for over 7 years served as the senior editor of *CADalyst*, the AutoCAD magazine. In addition to this book and over a dozen other books on AutoCAD, David has written books on Internet web design and his many articles also appear in *Cadence, Computer Graphics World,* and *PC Magazine.*

Working from his home in Bellingham, Washington—which includes its own LAN and intranet—David also consults to numerous software companies and has written the user manuals and other documentation for several popular CAD and engineering programs. When he's not writing or consulting, David can be found skiing, sailing, or working on model railroads, in the company of his wife and four children. You may contact him at **david@dscohn.com**.

# Contents

**Part I**

**AutoCAD Basics**

**Part II**

**Moving Beyond the Basics**

## 13   Working with Text ............................... 391

## 14   Dimensioning Your Drawing ..................... 433

**Part III**

**Becoming an Expert**

**Part IV**

**Advanced Topics**

**Part V**

# Acknowledgments

This book is an example of what happens when you work with professionals. I am indebted to everyone whose efforts helped make this book possible. My appreciation and thanks to my editors, Megg Morin, Gretchen Campbell, Katie Conley, Tana Allen, and Gretchen Ganser for their constant encouragement and support during all phases of the creation of this book. Thanks also to Joanne Cutherbertson, Stephane Thomas, Cynthia Douglas, and Carolyn Welch, for their work on previous editions of this book. Their contributions are still very apparent. Working with all of them was a real delight. Thanks also to Bill McManus whose thorough but gentle copy editing was once again extremely appreciated. Thanks also to Marian Selig for her meticulous proofing of the final text, Irv Hershman for his very complete index, and to illustrators Lyssa Seiben-Wald and Michael Mueller.

An incredibly huge thank you once again to my wonderful technical editor, Christal Elliott. After using and writing about AutoCAD continuously for over 18 years, I thought I knew just about everything. Yet, she pointed out features that I knew nothing of, clarified technical issues that I didn't fully understand, and called my attention to issues that affect a wide range of AutoCAD users. Her comments and suggestions have made this a much more valuable book.

I also owe a debt of gratitude to numerous people at Autodesk. While there are too many people to mention individually, I must single out my good friend Lynn Allen, who was always able to put me in contact with the right person to answer some of the more trying questions. Thanks also to my compatriots at Cyon Research, Brad Holtz, Dr. Joel Orr, Rick Stavanja (who was also instrumental in setting up the companion web site) and Evan Yares, for all of their help and encouragement.

Finally, a special thank you to my wife Genevieve, and my children Clarice, Emma, Andrea, and Bruce for putting up with me yet again during the trials and tribulations that go into the creation of a new book, and for encouraging me to come out from behind the computer monitor occasionally to share in the life of my wonderful family. I've always said that a book consumes as much time as the author can possibly give it, and this book, in particular, consumed more than its fair share of my life. More than anything else, the love and support of my family is what made this book possible.

# Introduction

AutoCAD is the leading computer-aided design and drafting (CAD) program in the world. Since its original introduction in November 1982, AutoCAD has grown in sales and functionality to become the standard PC-based CAD program against which all other similar programs compete and against which they are judged. Over the years, AutoCAD has kept pace with developments in the computer industry. The program has grown from its original command line driven DOS-based roots to become a fully compatible Windows application.

While previous releases were available for DOS and Windows, as well as for a number of UNIX-based computer systems (and for a few years even the Apple Macintosh), AutoCAD 2002, the current version and the version detailed in this book, runs only under

Windows XP, Windows 2000, Windows 98, Windows Millennium Edition, or Windows NT 4 (service pack 5 or later). By concentrating on these modern Microsoft operating systems running on Intel-compatible PCs, Autodesk has been able to improve AutoCAD far beyond the capabilities found in earlier versions. By any estimation, AutoCAD 2002 is the most powerful AutoCAD ever.

There are numerous reasons for AutoCAD's incredible success. When first introduced, AutoCAD provided its users with 80 percent of the capabilities found in other mainframe-based CAD programs of the time, while selling for less than 20 percent of the cost of those programs. Whereas other PC-based CAD programs that followed imposed limits on drawing size, layer names, or the level of drawing accuracy, AutoCAD enabled users to draw almost anything they could imagine. In addition, from the outset AutoCAD presented numerous ways in which users could accomplish similar tasks, and provided many opportunities through which the program could be customized to suit the specific needs of its customers. Users could do anything from modifying the program's menus to creating custom applications using AutoCAD's AutoLISP programming language.

Those capabilities are even more enhanced in AutoCAD 2002. This latest release incorporates numerous improvements over the previous version of AutoCAD, including the ability to leverage the Internet for collaboration with others, true associative dimensioning, new text features, the ability to define and use drafting standards, the incorporation of DesignXML, and powerful new tools to help you create and use attributes.

Numerous other new features, such as a Publish to Web wizard, double-click editing, and improvements to numerous commands, make this new version of AutoCAD more powerful and yet easier to use than any previous release. The fact that AutoCAD 2002 accomplishes all of this while also performing faster than the previous release is even more amazing.

If you're new to AutoCAD, the improved dialog boxes, dockable toolbars, and pull-down menus make AutoCAD 2002 the ideal version to learn to use. Of course, if you've used earlier versions of AutoCAD, you'll still be able to use AutoCAD's familiar command line and screen menus. All of these improvements and more are explained in detail in the pages that follow.

# Who Should Read This Book?

*AutoCAD 2002: The Complete Reference* is your guide to all aspects of AutoCAD. It is written for anyone who wants to create dimensionally accurate drawings and three-dimensional CAD models. Whether you're new to CAD and AutoCAD or a long-time user, you'll find what you need in this book. New users will find clear, easily understood explanations of all of AutoCAD's commands. Experienced users will learn about AutoCAD 2002's new features. Regardless of your experience level, this book will help you work faster and smarter, leveraging all of AutoCAD's incredible power.

# What Is in This Book?

*AutoCAD 2002: The Complete Reference* is designed as an easy-to-use guide to every aspect of AutoCAD. Throughout you will find explanations of the purpose and use of every command and feature. While each chapter builds on the material already presented, this book is organized to help you quickly look up the information you need. To locate instructions for using specific AutoCAD commands, use the table of contents or the index. To help you better understand its organization, the following sections explain how the book is organized.

## Part I: AutoCAD Basics

Part I introduces you to AutoCAD. In Chapter 1, you learn about the AutoCAD interface, how to start commands, and how to open and save drawings. Chapter 2 builds on this knowledge, explaining how to start a new drawing and control drawing settings such as layers, colors, and linetypes. You'll also learn how to use features such as object snap and grid to ensure the accuracy of your drawings. Beginning with simple objects and progressing on to more complex objects, Chapters 3 and 4 explain how to draw two-dimensional objects. Chapter 5 then explains how to control the onscreen appearance of your drawings, teaching you how to use functions such as pan, zoom, multiple viewports, and AutoCAD's lineweight capabilities. In Chapter 6, you'll learn how to work with coordinate systems to further increase drawing accuracy. Much of the power of computer-aided design comes from the reuse of objects you have already drawn and the ease with which you can make changes. Chapter 7 completes the discussion of AutoCAD basics by teaching you how to modify objects in your drawings.

## Part II: Moving Beyond the Basics

In Part II, you move beyond the basic tasks of creating and modifying objects in your AutoCAD drawings. In Chapter 8, you learn how to organize the information in your drawings, while Chapter 9 explains how to get information such as area and distance from your drawings. Chapter 10 extends upon what you already know about modifying objects, describing how to edit complex drawing objects such as polylines. Chapter 11 continues this theme with a discussion of changing object properties. In Chapters 12, 13, and 14, you learn how to add cross hatching, text, and dimensions to your drawings, while Chapter 15 describes how to combine objects into blocks and how you can then use those blocks and associated attribute information to extract data from your drawing. You also learn how to combine several drawings using external references. In Chapter 16, you learn about AutoCAD DesignCenter, while in Chapter 17 you learn how to create plot layouts in preparation for printing your drawings. Finally, in Chapter 18, you learn everything you need to know to print your AutoCAD drawings.

## Part III: Becoming an Expert

Although the skills you learn in the first two parts of the book apply to everything you are likely to do using AutoCAD, these sections deal primarily with two-dimensional drawings. In Part III, you learn how to work in three-dimensional space. Chapter 19 explains how to draw in three dimensions. You also learn the difference between wireframe, mesh, and solid objects, and how to manipulate 3-D views interactively using AutoCAD's 3DORBIT command. In Chapter 20, you learn how to edit those three-dimensional objects, while Chapter 21 expands upon what you have learned by specifically showing you how to edit three-dimensional solid models. In Chapter 22, you learn how to create shaded and rendered images of your three-dimensional models, while Chapter 23 explains how to include raster images in your AutoCAD drawings. This section of the book concludes in Chapter 24 with a discussion of ways in which you can use AutoCAD in conjunction with other programs.

## Part IV: Advanced Topics

Part IV introduces a number of advanced AutoCAD topics. Chapter 25 shows you how to save your drawings in Web-enabled formats and add hyperlinks to your drawings. Chapter 26 continues this discussion, explaining how to make your drawings accessible via the Internet and how to collaborate with others via the Web. In Chapter 27, you learn how to link information in AutoCAD drawings to external databases. Chapters 28 and 29 conclude the discussion of advanced topics with a look at how you can customize AutoCAD.

## Part V: About the Appendixes Online

Part V rounds out the book with a guide to the appendixes available online. Appendix A provides a handy reference to all of AutoCAD's commands, while Appendix B provides a similar list of all the program's system variables. Appendix C explains how to install and configure AutoCAD. Appendix D explains how to use the AutoCAD Migration Assistance toolkit, a collection of tools that help you convert your files from earlier versions of AutoCAD. Appendix E provides a list of the various file types used in AutoCAD, while Appendix F contains a glossary of CAD terms. Finally, Appendix G describes how to access and use the companion web site at **www.dscohn.com/acad2002**.

## Conventions Used in This Book

This book is written in plain English and provides a no-nonsense approach that is meant to be easy to follow and understand. Throughout the book, you will also find several conventions used to help you quickly find and understand the material presented.

### Icons

Occasionally, when we want to alert you to important information, advice, a shortcut, or a potential pitfall, you'll see a Note, Tip, or Caution icon along with additional information.

*Notes like this are included when there is something that needs to be taken into consideration, or when there is a special feature that is worth giving extra attention to.*

When there is something in particular that can save you time or provide an easier way to get something done, it appears as a Tip, set off from the rest of the text.

 Sometimes a particular command or instruction can present problems. When this is likely, the information contained in a Caution tells you in advance what to expect and how to proceed without running into trouble.

In addition to Notes, Tips, and Cautions, you will also see Learn by Example icons.

 **LEARN BY EXAMPLE**
The Learn by Example icon tells you when a practice file is available that you can use to test an AutoCAD 2002 feature. The text that accompanies these icons provides a description of the sample file and its name.

---

## Sidebars

When there is something interesting that is somewhat tangential to the main discussion, it is presented in a sidebar like this one.

---

# Typefaces

For clarity, a number of different typefaces are used throughout this book to identify specific types of information. Although neither Windows nor AutoCAD makes a distinction between upper- and lowercase text (except when typing literal text that will appear in the drawing), all commands appear in CAPITAL LETTERS. When the text includes instructions for you to type something in response to a command, the actual text is shown in **boldface**. Actual keypresses, such as pressing ENTER, are shown in small capital letters. Similarly, when you need to press several keys simultaneously, such as CTRL-C to copy something to the Windows Clipboard, the key combination is shown in small caps separated by a hyphen. In that case, you should hold down the CTRL key and press the letter C simultaneously. If the text lists the keystrokes as ^C, this means to actually type the caret and then the letter C.

There are instances where we show the text that AutoCAD actually displays on its command line. When this occurs, all the actual AutoCAD prompts appear. (Occasionally, I have omitted responses such as the *1 selected, 1 found* message displayed during object selection.) The AutoCAD prompts appear unaltered. Literal user response is always in **bold** and is meant to be typed exactly as shown. For example,

 Command: **LINE**

means that you should type **L I N E** (the LINE command). Although not always shown, when typing commands at AutoCAD's command prompt, you must press ENTER to complete the command sequence. If the command sequence includes an instruction to make a selection of your choice, the response is shown in italics and enclosed in parentheses. For example,

```
Specify first point: (select a point)
```

means that you should select any point you wish. If you enter a coordinate value, you will need to complete your entry by pressing ENTER. Only when you must press ENTER without making any other entry is the ENTER keystroke actually included in the command sequence.

When AutoCAD prompts you to respond, a default value is often displayed. This default value appears on the command line enclosed in angle brackets:

```
Rotate arrayed objects? [Yes/No] <Y>:
```

You select the default value by pressing ENTER. In such instances, the actual text appearing on AutoCAD's command line also appears in the examples in this book. When the actual default value will necessarily vary depending on the particular situation, a sample value or an explanation of the value will appear in italics within the angle brackets:

```
Object snap target height (1-50 pixels) <current value>:
```

## Command Names

For simplicity and consistency, when instructing you to start a command, the sequence of steps is presented in the order in which you would actually make your selection. For example, you may be instructed "From the File menu, choose New" or "On the Standard toolbar, click New." After all, you must click the File menu first before you can click the New command. When you must make a selection from a cascading submenu, the sequence may be shortened. For example, the instruction "From the View menu, choose Zoom | Previous" means to click the View menu, click Zoom to display the View pull-down menu, and then click Previous in the submenu that appears.

## Let's Get Started

Now it's time to get to work. I hope that this book proves to be both a teaching guide and a useful reference. I've included lots of tips and tricks that I've learned from my 18 years of working with AutoCAD. However, if I've missed something or there is a shortcut or something that you'd like to share with other readers, please send it to me at **david@dscohn.com** so I can post it to the companion web site and include it in a future edition of this book. And thank you for choosing *AutoCAD 2002: The Complete Reference*.

# The Complete Reference

# Part I

## AutoCAD Basics

# The Complete Reference

# Chapter 1

## Getting Started with AutoCAD

This chapter takes you on a tour of AutoCAD and provides an overview of the different components of the AutoCAD screen. It teaches you about AutoCAD's drawing environment, how to start commands, and how to open existing drawings. As you will see, AutoCAD provides many different ways to accomplish the same task. For example, to activate a command, you can select it from a menu or a toolbar, or type the command name at AutoCAD's command prompt.

There is no right or wrong way to do something in AutoCAD. As you become more familiar with the program, you will develop techniques that you find most comfortable for the way you work and the types of drawings you create. Once you become proficient, you may want to customize AutoCAD for your particular needs. There are also many common drafting tasks that can be automated by using scripts and macros, or by writing or purchasing add-on applications that run inside AutoCAD, tailoring the program for specific drafting and design disciplines. While this book will not teach you how to write custom programs, you will learn how to customize many AutoCAD features and utilize third-party add-ons.

AutoCAD's *open architecture* enables it to be customized to suit your individual needs, and once you have used it for a while, you'll be able to make the most of its amazing capabilities. But we're getting a bit ahead of ourselves. First, let's learn the basics. This chapter explains the following concepts:

- Starting AutoCAD
- Understanding the AutoCAD interface
- Selecting commands
- Correcting mistakes
- Getting online help
- Opening existing drawings
- Working with multiple drawings
- Using Partial Open and Partial Load
- Saving your work
- Exiting from AutoCAD

## Starting AutoCAD

This chapter assumes that you have already installed AutoCAD. If you are going to work in AutoCAD as you follow along in this book, you should install AutoCAD now, before proceeding. Installing AutoCAD 2002 is quite simple, particularly compared to earlier versions of the program. An easy-to-use Setup program guides you through the AutoCAD installation process, transferring the files from the CD-ROM to a folder it creates on your hard disk. The Setup program also creates a menu item on the Windows Start menu, and a shortcut icon on your desktop. If you need additional help installing AutoCAD, see Appendix C.

You can start AutoCAD by choosing it in the Start menu or by double-clicking the AutoCAD 2002 icon on the Windows desktop. To start AutoCAD from the Start menu, choose Start | Programs | AutoCAD 2002 | AutoCAD 2002.

**Note**  *The first time you start AutoCAD, the program displays the Authorization wizard, in which you provide the authorization code to unlock your copy of AutoCAD. You register your copy of AutoCAD and obtain this authorization code from Autodesk, either via the Web or by e-mail, phone, fax, or mail. If you elect to authorize AutoCAD at this time, the wizard guides you through the process, offering options such as connecting to Autodesk's registration web site, automatically generating an e-mail message, displaying the proper phone numbers, or printing a registration form that you can fax or mail to Autodesk. If you decide to defer this process until a later time, you can begin using AutoCAD now. You have 15 days from the first time you start AutoCAD in which to register and authorize your copy. The Authorization wizard appears every time you start AutoCAD until you have registered your copy and obtained your authorization code. Once you obtain the code, write it down and save it along with your AutoCAD 2002 CD-ROM, in case you ever need to reinstall the software.*

When you start AutoCAD, the program displays the AutoCAD 2002 Today window. This window provides tools to help you start a new drawing, load symbol libraries, access an online bulletin board for design collaboration within your company, and use the Autodesk Point A design portal. You'll learn more about this window later in this chapter (see Figure 1-17). For now, click the Close button to dismiss the AutoCAD 2002 Today window.

**Note**  *The AutoCAD 2002 Today window serves as the default startup dialog box whenever you start AutoCAD or begin a new drawing. If you prefer, you can reconfigure AutoCAD to use a more traditional style startup dialog box, similar to the one used in earlier versions, by changing the Startup setting on the System tab of the Options dialog box.*

## Understanding the AutoCAD Interface

The AutoCAD screen is divided into six distinct areas:

- Title bar
- Menu bar
- Toolbars
- Document window or drawing area
- Command window
- Status bar

Figure 1-1 shows the typical layout of the AutoCAD screen. Most of these components are standard Windows features. For example, the *title bar* along the

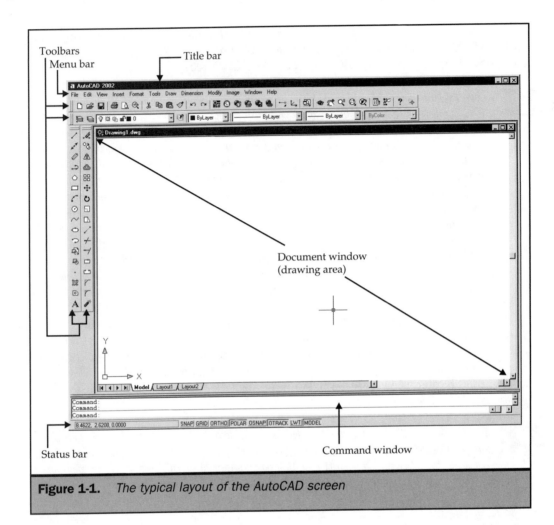

**Figure 1-1.** *The typical layout of the AutoCAD screen*

top of the window shows the name of the program, AutoCAD 2002. The name of the current drawing (or "Drawing1," if the current drawing has not been saved) appears in the title bar of the *document window*. Each open drawing has its own document window. If the document window has been maximized, the name of the current drawing appears in the main AutoCAD window title bar, enclosed within square brackets. The *menu bar*, located directly below the title bar, provides pull-down menus from which you can choose commands. You can also activate commands by clicking the buttons on the various *toolbars*. The *status bar* along the bottom of the screen shows the coordinates of the screen cursor as well as the current setting of various AutoCAD program modes. You'll learn more about the status bar later in the chapter.

**Note**   *By default, AutoCAD now displays the full drawing filename, including the drive and full path, in the title bar. This option is controlled in the File Open area of the Open and Save tab of the Options dialog box.*

The *command window* is one component of AutoCAD that does not have an equivalent in most other Windows programs. You can start any AutoCAD command by typing the command and then pressing ENTER. Some of the components always appear in the same location. Others, such as the toolbars and command window, can be turned off or relocated anywhere on your Windows desktop. Figure 1-2 shows an AutoCAD screen in which some of these components have been rearranged.

The *document window,* or drawing area, occupies most of the screen. This is the area in which you actually create your drawing. (Remember that you can have more than one drawing open at a time. Each has its own document window.) Notice that there are

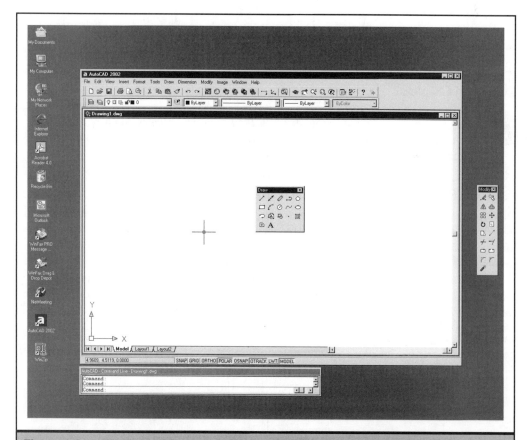

**Figure 1-2.**   *The AutoCAD screen after rearranging several of the components*

two other elements within this window: an icon with two arrows pointing at 90-degree angles, and an icon that looks like a small plus sign (+) with a box at its center. These are the *User Coordinate System (UCS)* icon and the drawing cursor, respectively.

## UCS Icon

The UCS icon helps you understand how your drawing is oriented. The icon consists of two arrows, one pointing to the right and one pointing to the top of the drawing area.

Notice that one arrow is labeled X and the other Y. These labels indicate the current orientation of the drawing's X and Y axes. Notice, too, the inclusion of a square where the two arrows intersect. This indicates that the UCS corresponds to the *World Coordinate System (WCS)*. You will learn more about coordinates and coordinate systems in Chapter 6.

   *The UCS icon looks considerably different than it did in earlier versions of AutoCAD. Although the new icon is much more intuitive, you can change the appearance of the icon to that used in earlier versions, as described in Chapter 6.*

## Crosshairs, Pickbox, and Cursor

Notice that the drawing cursor moves around the screen as you move the mouse. You use the cursor for selecting points or objects within the drawing area. The appearance of the cursor changes depending on which AutoCAD command is currently active or where you move the cursor within the AutoCAD screen.

By default, the cursor appears as a small plus sign with a box at its center. The point at which the *crosshairs* meet is the actual cursor position and corresponds to a specific point within the AutoCAD drawing. The box, called a *pickbox*, is used to select objects within the drawing.

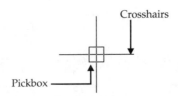

When you activate an AutoCAD command used to create a new object, such as the LINE command (used to draw lines), the pickbox disappears, leaving just the crosshairs. Move the cursor to a start position in the drawing area and click to select that point. Then, move the cursor to a different position corresponding to the end point of the line and click again to select the end point of the line. The line is drawn.

**Note**    *Throughout this book, the term click is used to mean pressing the left-mouse button (or the pick button on a multibutton digitizer puck) one time. Double-click means to click the left-mouse button twice in quick succession. Right-click means to click the right-mouse button (or button two on a multibutton puck), and SHIFT-right-click means that you should press the SHIFT key while simultaneously clicking the right-mouse button. Drag means that you should press and hold down the left-mouse button while moving the mouse.*

If you activate a command to modify an existing object, such as the ERASE command, the crosshairs disappear, leaving just the pickbox. You can then select the object to be erased by moving the cursor so that the pickbox is over the object, and clicking to select the object.

To ensure accuracy when selecting points with the drawing cursor, you can use AutoCAD's *object snap* modes to snap the crosshairs to a specific point on an existing object, such as the end point of a line or the center of a circle. When an object snap mode is active, the cursor appears with both the crosshairs and a slightly different pickbox, called an *aperture box.* When prompted to select a point, move the cursor so that the aperture box falls over a line. As you click the cursor, it automatically snaps to the end point of the line.

If you move the cursor outside the drawing area, the cursor changes to one of several standard Windows pointers. For example, when moving the cursor over a toolbar or the status bar, the cursor changes to a Windows arrow. You can then select a command by clicking the toolbar button or menu command.

## Status Bar

The status bar at the bottom of the AutoCAD screen displays both the current cursor position and the status of various AutoCAD modes (see Figure 1-3). The cursor position displays as either X, Y, Z coordinates or, when certain drawing commands are active, as a distance and angle relative to the last point selected. As you move the cursor, the coordinates update automatically. You can toggle the automatic coordinate display on and off by clicking within the coordinate display area, or by pressing the F6 function key.

The other option buttons on the status bar indicate the current Snap mode, grid display, Ortho mode, polar tracking, object snap, and object snap tracking settings, whether lineweights are visible, and the current drawing space (model space or paper space). You can toggle these modes on and off by clicking the appropriate button. You'll learn more about these modes in Chapter 2.

When you move the cursor over a toolbar or menu command, the status bar changes to display information about the selected command (see Figure 1-4).

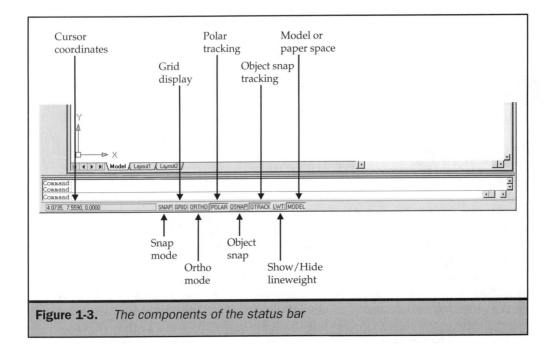

**Figure 1-3.** *The components of the status bar*

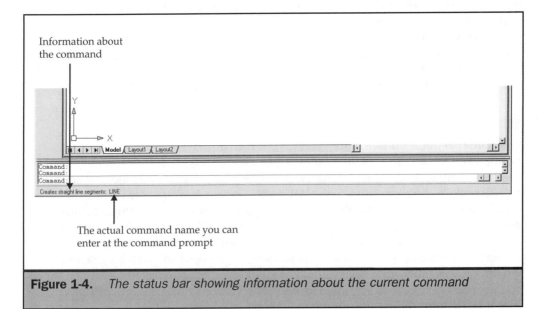

**Figure 1-4.** *The status bar showing information about the current command*

*If you look carefully at the command description on the status bar, you'll see a single word to the right of the colon at the end of the description. This is the actual AutoCAD command. You can activate the command by typing this command name at the AutoCAD command prompt, described later in this chapter. The names used for commands in the menus may be different from the actual command names.*

## Pull-Down Menus

Most AutoCAD commands, as well as numerous standard Windows functions, are available from pull-down menus on the menu bar. These menus are arranged in a hierarchical fashion. For example, all commands for opening, saving, and printing drawings (which are standard Windows functions) are available in the File pull-down menu. Commands for drawing new AutoCAD objects are found in the Draw pull-down menu.

Figure 1-5 shows a typical pull-down menu. Notice that some menu items display a small black arrow to the right of the command name. Clicking such a command or item expands the menu to display a cascading submenu containing additional options to the command or a collection of related commands. Other menu items have an ellipsis (three dots) immediately following the command name. This indicates that by selecting the command, a dialog box will be displayed.

Pull-down menus often contain other components. An underlined letter on a menu corresponds to the access key, which you can type from the keyboard to start the command. To display a pull-down menu, press the ALT key in combination with the access key that is shown for the menu name. Shortcut keys indicate a keyboard key or a key combination that invokes a particular command without requiring the use of a menu, such as F2 to display AutoCAD's Text window, or CTRL-C to copy objects to the Windows Clipboard. Although you might not use access keys and shortcut keys at first, as you become more familiar with AutoCAD, you will likely use these faster alternatives for starting AutoCAD commands.

Notice that the pull-down menu also contains *separators,* lines that help divide the commands into logical groups of related commands. For example, the Zoom, Pan, and Aerial View commands all can be used to change the way the drawing is displayed, and thus are grouped together. Although not shown in Figure 1-5, sometimes a command cannot be used. For example, the REDO command can be used only immediately after using the UNDO command. When a command is unavailable, it appears dimmed or grayed. Also, notice that in the UCS Icon submenu, the On and Origin selections appear with a check mark. A command or option preceded by a check mark indicates that the command or option can be turned on or off.

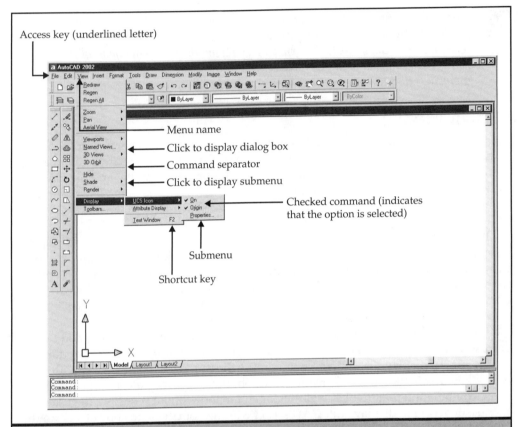

**Figure 1-5.** *A typical pull-down menu, expanded to show the UCS Icon submenu*

## Shortcut Menus

The *shortcut menus* are special menus that display at the cursor position when you press the right-mouse button. Shortcut menus are completely context-sensitive. The functions displayed in the menu vary depending on the location of the cursor when you right-click, the type of object selected, and whether an AutoCAD command is active. If you press the SHIFT key and right-click at the same time, AutoCAD displays the Object Snap shortcut menu.

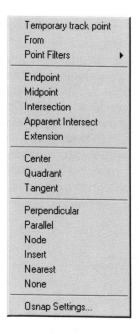

 *If you are using an IntelliMouse or a three-button mouse, pressing the middle button either displays the Object Snap shortcut menu or activates real-time panning, depending on the current value of the MBUTTONPAN system variable.*

## Toolbars

When you start AutoCAD for the first time, the Standard, Object Properties, Draw, and Modify toolbars are displayed. AutoCAD's standard menu provides 26 toolbars, each of which contains a group of related commands. You can have any of these toolbars visible at any time and control where they are placed on the desktop. All of these toolbars can be customized by adding and deleting buttons. You can also move and resize the toolbars, and create new toolbars. Toolbars are probably the easiest and fastest way to start AutoCAD commands. As you become more proficient, you will probably want to modify the default toolbars or create your own so that the commands you use most often are always conveniently available.

Since you want the most frequently used commands readily available, it makes sense that two toolbars in particular appear by default across the top of the AutoCAD window. The Standard toolbar contains buttons for standard Windows functions, such as opening, saving, and printing files, cutting-and-pasting objects to and from the Windows Clipboard, and undoing or redoing your previous actions. It also contains many standard AutoCAD functions, such as panning and zooming the display of the drawing area. The Object Properties toolbar, as its name implies, contains buttons and drop-down list boxes for controlling the properties of AutoCAD objects, such as the current layer, color, and linetype. These toolbars are shown in Figure 1-6.

The TOOLBAR command displays the Toolbars tab of the Customize dialog box, where you choose which toolbars are displayed (see Figure 1-7). To display a toolbar, click the box adjacent to its name so that an X appears in the box. To close a toolbar, click the box adjacent to its name so that the X disappears. You can also set other options, such as whether toolbars are displayed with large or small buttons, and whether to display or hide ToolTips. When ToolTips are enabled, a brief description of the command appears as you momentarily pause the arrow cursor over a toolbar button.

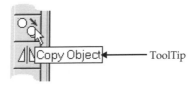

ToolTips make it easier to learn the purpose of each toolbar button until you have memorized what each icon represents. The Customize dialog box also contains the tools you use to customize toolbars.

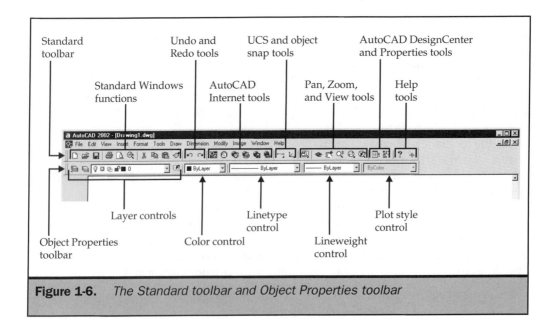

**Figure 1-6.**   *The Standard toolbar and Object Properties toolbar*

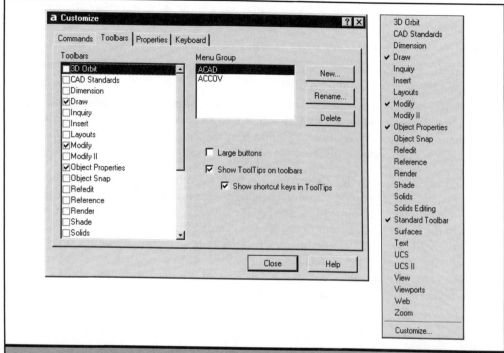

**Figure 1-7.**   *You control which toolbars are visible by using (from left to right) the Toolbars tab of the Customize dialog box or the Toolbars shortcut menu.*

You can also simply right-click while the arrow cursor is over any toolbar to display the Toolbars shortcut menu containing a list of all the toolbars (see Figure 1-7). You can then toggle selected toolbars on and off by clicking their name in the menu. A check mark indicates that the toolbar is currently visible. Clicking Customize displays the Toolbars tab of the Customize dialog box.

Toolbars are either *docked* (attached to any of the edges of the drawing area) or *floating* (freestanding elements anywhere on the desktop). When you start AutoCAD for the first time, the displayed toolbars are all docked. A floating toolbar has a Title bar and a Close box, and can be resized. Here are some techniques for controlling the display and placement of toolbars:

■ To undock (or float) a toolbar, click anywhere inside the toolbar, except on a button, and drag the toolbar away from the perimeter of the drawing area.

■ To dock a toolbar, drag it to the perimeter of the drawing area (also called the *docking area*).

- To position a toolbar in a docking area without docking it, press CTRL while you drag it.
- To move a toolbar, drag it to a new location.
- To resize a toolbar, move your cursor to the edge of the toolbar until it changes to a resize arrow, and then drag.
- To close a toolbar, right-click to display the Toolbars shortcut menu, and then click the toolbar name within that menu to remove the check mark. You can also close a toolbar by floating it and then clicking the Close button in the upper-right corner of the toolbar, or by using the Toolbars dialog box.

Some buttons, such as the Zoom Window button on the Standard toolbar, contain flyouts. *Flyouts* either provide options for using the command with different methods or contain other, related commands (see Figure 1-8). Flyouts are indicated by a small triangle in the lower-right corner of the button. To display a flyout, click the toolbar button and hold down the mouse button. To select a button from a flyout, continue to hold down the mouse button while pointing to the button you want, and then release the mouse button. In addition to starting the command you specified, the button you selected on the flyout becomes the default button on the toolbar.

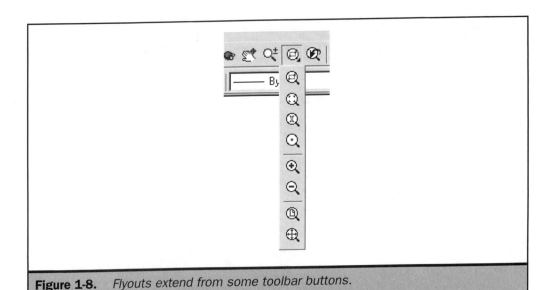

**Figure 1-8.**    *Flyouts extend from some toolbar buttons.*

# Model Tab and Layout Tabs

If you've used previous versions of AutoCAD, you'll notice something new at the bottom of the drawing area, between the document window and the command window. The Model tab and Layout tabs (see Figure 1-9) enable you to switch your drawing between model space and paper space. You generally create your drawings in model space, and then switch to paper space to create layouts and print your drawing.

In AutoCAD R14 and earlier, AutoCAD used a single model space and a single paper space. Beginning with AutoCAD 2000, model space has become much more visual, showing you exactly what your drawing will look like when printed. Paper space is also much more flexible than in the past. You can now create multiple paper space layouts from the same drawing. This enables you to consolidate multiple drawing sheets into a single drawing file. For example, you can create individual layouts for the floor plan, electrical plan, and plumbing plan of a building from a single drawing file, without having to constantly turn layers on and off. Buttons to the left of the Model and Layout tabs enable you to scroll through these tabs when your drawing has more tabs than will display across the screen. You'll learn about creating paper space layouts in Chapter 17.

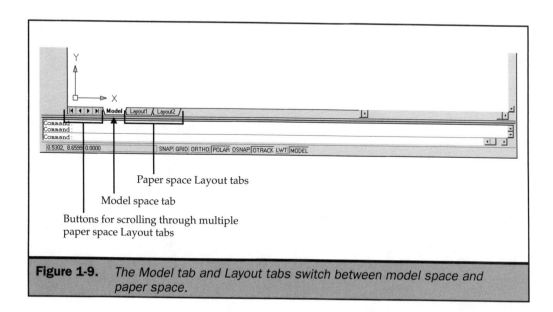

**Figure 1-9.**    *The Model tab and Layout tabs switch between model space and paper space.*

## Command Window

The Command window is where you type AutoCAD commands and view prompts
and messages. When initially displayed, the Command window is docked at the
bottom of the screen, between the drawing area and the status bar. The Command
window initially displays the three most recent lines of prompts, but you can change
the number of lines displayed. Scroll bars on the right side of this window let you
scroll back to see previous prompts. You can undock and move this window by
dragging it, and also dock it at the top of the drawing area.

The Command window can be resized to change the number of lines of text it
displays, by dragging the split bar, which divides the Command window from
the drawing area (see Figure 1-10). When the Command window is floating, you
can also resize its width. When docked, it always extends across the width of the
AutoCAD window.

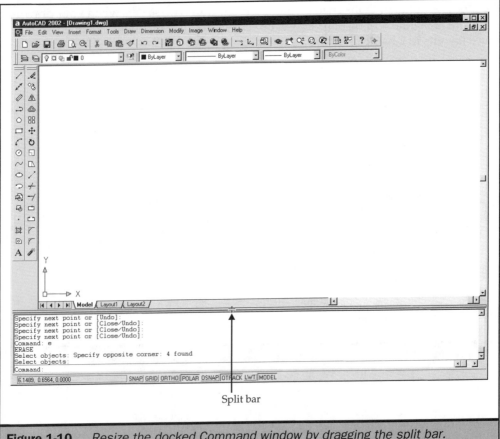

Split bar

**Figure 1-10.**   *Resize the docked Command window by dragging the split bar.*

## The Options Dialog Box

Throughout this chapter and in other areas throughout this book, you will find references to the Options dialog box. This dialog box contains controls for customizing many of AutoCAD's settings. A detailed description of this dialog box can be found in Appendix C.

The Options dialog box contains nine separate pages, or tabs, each controlling a different aspect of AutoCAD:

- **Files**    Specifies the directories that AutoCAD searches to find special files, such as menus, drivers, and support files. It also specifies optional user-defined settings, such as the dictionary used for spell checking.

- **Display**    Contains controls for customizing AutoCAD's display, such as screen colors and the number of lines in the Command window.

- **Open and Save**    Controls how often AutoCAD automatically saves your drawings and whether it creates a backup copy.

- **Plotting**    Controls options related to plotting, such as the default plot settings for new drawings.

- **System**    Controls general AutoCAD system settings, such as the current graphics and pointing device drivers and whether you can open multiple drawings.

- **User Preferences**    Contains controls that let you optimize the way you work with AutoCAD, such as the default drawing units and the use of accelerator keys and shortcut menus.

- **Drafting**    Enables you to control various editing options, such as the use of AutoSnap and Auto Tracking.

- **Selection**    Controls AutoCAD's selection modes and the use of grips.

- **Profiles**    Allows you to create user-defined configurations.

**Note**    *You can also change the number of lines of text displayed in the docked Command window from the Display tab of the Options dialog box.*

## Text Window

In addition to the AutoCAD drawing environment (sometimes called the *Graphics window*) described thus far, the Text window is another important element to the interface. The *Text window* is a second window in which you can type AutoCAD commands and view prompts and messages (see Figure 1-11). Initially, the Text window is not visible, although it becomes visible when you use certain AutoCAD commands. For example, the LIST command automatically activates the Text window.

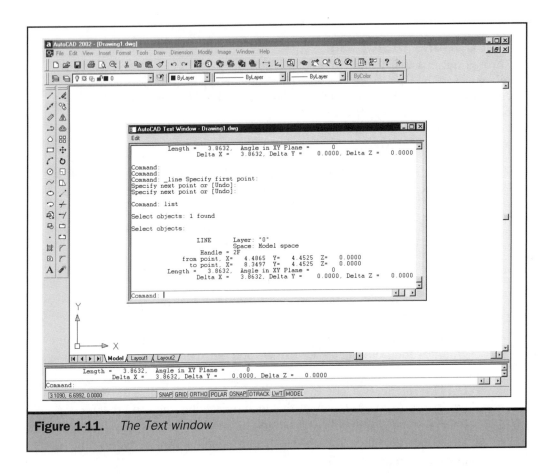

**Figure 1-11.** *The Text window*

Unlike the Command window, the Text window cannot be docked. It always appears in its own window and, when activated, has its own task button on the Windows taskbar. The Text window contains many more lines of prompts, along with its own scroll bars. This window can be resized, minimized, or closed entirely when not needed, without affecting the main AutoCAD window.

**Tip** *To switch between the main AutoCAD drawing window and the Text window, press F2. If the Text window is not active, pressing F2 activates and immediately displays the Text window. Pressing F2 subsequently toggles back and forth between the Graphics window and the Text window.*

You can cut-and-paste text between the Text window and the Windows Clipboard. Most standard Windows CTRL key combinations and cursor keys can also be used in the Text window.

## Screen Menu

Although it does not initially appear, AutoCAD has an additional screen component that you can display and use for starting commands. The *screen menu* (sometimes called the *side menu*) consists of an area on the right edge of the drawing area. When you move the cursor over a menu item, the item becomes highlighted. You then click to select a menu item. Clicking an item shown in all uppercase letters switches the menu to display a related selection of commands. Clicking an item shown in mixed uppercase and lowercase letters starts that command. You turn the screen menu on or off from the Display tab of the Options dialog box. The screen menu behaves like a toolbar. Once displayed, you can undock it and reposition it as you would any toolbar.

**Note**    *The screen menu is similar to menu systems used by older CAD systems. In earlier versions of AutoCAD, the screen menu was the primary method of starting commands, other than typing them at AutoCAD's command prompt. Because the screen menu is always visible when it is active, it consumes a considerable portion of the drawing area. For that reason, most users choose not to display this screen component.*

## Digitizer Template

While not part of the AutoCAD screen, the digitizer template represents an additional interface that you can use when working with AutoCAD. A *digitizer* is a large, flat pad with an associated pen or puck. The *puck* may look similar to a mouse, but often has many more buttons. Digitizers vary in size from a few inches square to tablets measuring 48 inches by 60 inches or more.

As you move the puck, wires below the surface of the digitizer track the puck's movement. AutoCAD's cursor follows these movements. Thus, as a pointing device, a digitizer or tablet is somewhat similar to a mouse, but there are a few significant differences. A mouse is a *relative pointing device,* meaning that the position of the cursor onscreen is only relative to its previous position. If you roll the mouse to the edge of your mouse pad, you can lift it up and reposition it without changing the position of the cursor. With a digitizer, the position of the puck is directly related to the position of the cursor. Thus, a digitizer is an *absolute pointing device.* You can accurately trace paper drawings by using a digitizer, which is something you can't do with a mouse.

Digitizers provide another capability in AutoCAD in addition to tracing paper drawings. AutoCAD comes with a sample drawing called Tablet 2000.dwg. You can plot a copy of this drawing and attach it to your digitizer. This digitizer *template* includes a screen pointing area and numerous rectangular areas filled with icons. The standard AutoCAD menu maps AutoCAD commands to specific areas of the template. Once properly configured, you can start a command by moving the puck over the specific area and clicking. Refer to Appendix C for additional information about configuring a digitizer template.

*Although the trend today seems increasingly to be toward the use of toolbars rather than digitizer template menus to select commands, digitizers are still an essential tool for accurate tracing.*

# Selecting Commands

Now that you've learned about the AutoCAD interface, you have a better understanding of the many ways in which you can select commands. You can select AutoCAD commands by using any of these methods:

- Click a button in a toolbar.
- Choose a command from a pull-down menu.
- Type the command in the Command window or Text window.
- Choose a command from a shortcut menu.
- Click a command on the screen menu.
- Click a command on the digitizer template.

Some commands remain active until you end them, so that you can repeat an action without having to select a command repeatedly. You can end such a command by pressing ENTER or ESC, or by right-clicking and choosing Enter or Cancel from the shortcut menu.

*The discussions that follow do not discuss command selection from the screen menu or digitizer template, since these methods are used less often. Feel free to experiment with using all six methods for selecting commands, and use whichever ones you find most comfortable or efficient.*

## Starting Commands from Menus

To start a command from a menu, choose it from the list of available menu options. You can also use the ALT key in combination with the appropriate access key to open a specific pull-down menu and then press the access key for the desired command.

To use a command on a menu to draw a line:

1. Move the cursor to the menu bar and choose the Drawing menu by clicking Draw. The menu displays the various drawing commands.

2. Click Line. The menu disappears and AutoCAD displays the prompt:

   ```
   Specify first point:
   ```

   in the Command window, asking you to select the starting point of the line.

3. Move the cursor within the drawing area and then click to select the line's starting point. AutoCAD now prompts:

   ```
   Specify next point or [Undo]:
   ```

   asking you to select the end point of the line.

4. Move the cursor again and then click to select the end point of the line. Again, AutoCAD prompts you to select an end point.

5. Move the cursor and click to select another point.

6. Right-click to display the shortcut menu, and then choose Enter to end the command.

 *To close a pull-down menu without choosing a command, press* ESC. *You can also click anywhere in the drawing area or choose another pull-down menu item.*

## Starting Commands from Toolbars

To start a command from a toolbar, click a button and then respond to the prompts that appear in the Command window.

To use a button from a toolbar to draw a circle:

1. From the Draw toolbar, click the Circle button. In the Command window, AutoCAD displays the prompt:

   ```
   Specify center point for circle or [3P/2P/Ttr (tan tan radius)]:
   ```

   This prompt displays several options of the CIRCLE command. The default option is to select the center point of the circle.

2. Move the cursor within the drawing area and then click to select the circle's center point. AutoCAD now prompts:

   ```
   Specify radius of circle or [Diameter]:
   ```

   Again, AutoCAD displays several options. The default option is to specify the radius of the circle.

3. Notice that as you move the cursor, you can change the size of the circle by dragging a radius line from the circle's center point. To complete the circle, select a point or type a value for the radius and then press ENTER.

# Starting Commands from the Command Line

To start a command from the command line, type the command and then press ENTER.

As you already saw, when an AutoCAD command has several options, these options are displayed on the command line. The default option is displayed as part of the command prompt, and the other options are enclosed within square brackets and separated by forward slash characters.

You don't need to type anything to select the default option. For example, when you drew the circle in the previous example, you simply selected the center point when AutoCAD displayed it as the default option. Similarly, to specify the radius of the circle, you simply selected a point on the radius of the circle or typed the radius value.

If you want to specify the size of the circle by using its diameter, however, you select the diameter option. To select a different option, type the letter or letters (shown capitalized) to select the option and press ENTER. For example, to specify the diameter of the circle, type **D** and press ENTER. AutoCAD then prompts you to specify the diameter of the circle. You can also select the command option from the shortcut menu that is displayed by right-clicking. You'll learn more about the shortcut menus in the next section.

**Note**    *All AutoCAD commands can be started by typing the complete command name. Some commands also have abbreviated names, called aliases. You can start these commands by typing the command alias. For example, you can start the LINE command by typing L and pressing ENTER. In the chapters that follow, whenever an alias is available, the instructions for starting a command by typing show the command aliases along with the command name. Command aliases are defined in the ACAD.PGP file. You'll learn more about this file in Chapter 28.*

# Using Shortcut Menus

When you right-click the mouse, AutoCAD displays a shortcut menu from which you can choose various options. The shortcut menu is *context-sensitive*, which means that the options presented in the menu vary depending on the cursor position when you right-click, any objects already selected, and the current status of AutoCAD, such as whether a command is already active.

Right-clicking in the drawing area displays one of the following six shortcut menus. The menu that displays depends on AutoCAD's current status, and whether you right-click a blank area or an object within the drawing.

- **Default**   Displayed when no command is active and no objects are selected. This menu contains common options, such as Copy, Paste, Pan, and Zoom.

- **Edit**   Displayed when no command is active, by right-clicking when a drawing object is selected. This menu contains editing commands specific to the type of object selected.

■ **Command**    Displayed when a command is active. This menu contains the command options, as well as other commands, such as Pan and Zoom.

Enter
Cancel

Arc
Halfwidth
Length
Undo
Width

Pan
Zoom

■ **Object Snap**    Displayed by pressing SHIFT or CTRL while you right-click. This menu contains all the object snaps, object snap settings, and point filters. You'll learn about object snaps in Chapter 2.

■ **Hot Grips**    Displayed by right-clicking a hot grip. This menu contains grip editing commands. You'll learn about grips in Chapter 7.

■ **OLE**    Displayed by right-clicking an OLE object. This menu contains options for editing OLE objects. You'll learn about OLE objects in Chapter 24.

■ **Hyperlinks**    Displayed by right-clicking when a graphical object with an attached hyperlink is selected. This menu contains options for opening, copying, and editing the hyperlink, as well as for adding it to your Favorites list. You'll learn more about hyperlinks in Chapter 25.

**Note**    *In R14 and earlier versions of AutoCAD, instead of displaying a shortcut menu, right-clicking was equivalent to pressing ENTER. Since some users may wish to retain that behavior, AutoCAD provides several settings that control the use of shortcut menus. The Object Snap, Hot Grips, and OLE shortcut menus are always available. Within the User Preferences tab of the Options dialog box, however, you can individually control whether AutoCAD displays the Default, Edit, Command, and Hyperlink shortcut menus or behaves as it did in these previous versions. In addition to turning these menus on and off, you can customize the options displayed in the menus as you would customize any other aspect of AutoCAD's menus. You'll learn about customizing AutoCAD's menus in Chapter 28.*

In addition to clicking in the drawing area, you can right-click other AutoCAD window areas to display one of the following additional shortcut menus. Again, the menu that displays depends on where you click.

- **Toolbar**  Right-clicking any portion of any toolbar displays a shortcut menu you can use to display, hide, or customize any available toolbar.
- **Command line**  Right-clicking the command line or text window accesses the six most recently used commands, as well as the Copy and Paste commands.
- **Dialog box or window**  Right-clicking within any dialog box or window displays a shortcut menu containing options appropriate to the particular function.
- **Status bar**  Right-clicking the coordinate display or any of the buttons on the status bar displays a shortcut menu for controlling the selected setting.
- **Model/Layout**  Right-clicking the Model tab or any of the Layout tabs displays a shortcut menu containing the plotting, page setup, and layout setup commands. You'll learn more about these functions in Chapters 17 and 18.

## Using Pointing Devices

In addition to using your mouse or digitizer to select commands and points within the drawing, you can use the other buttons on the pointing device to activate other AutoCAD functions. For example, as you've just learned, you can right-click to display shortcut menus. If you are using a three-button mouse, the middle button either activates real-time panning or displays the Object Snap shortcut menu, depending on how AutoCAD is currently configured.

If you are using an IntelliMouse, the small wheel between the left and right buttons provides the following additional functions:

- Rotate the wheel forward to zoom in or rotate it backward to zoom out. By default, each increment changes the zoom factor by 10 percent, but you can change this increment by changing the ZOOMFACTOR system variable.
- Double-click the wheel to zoom to the drawing extents.
- Press the wheel and drag the mouse to pan the drawing.

You'll learn about panning and zooming to move around within a drawing in Chapter 5.

## Repeating a Command

You can repeat the command you just used, without having to reselect it, by pressing either the SPACEBAR or ENTER, or by right-clicking and choosing the Repeat *command* item at the top of the shortcut menu, where *command* represents the name of the most recently used command.

| Repeat Line |
|---|
| Cut |
| Copy |
| Copy with Base Point |
| Paste |
| Paste as Block |
| Paste to Original Coordinates |
| Undo |
| Redo |
| Pan |
| Zoom |
| Quick Select... |
| Find... |
| Options... |

When entering commands from the command line, you can type **MULTIPLE** before starting some commands (such as CIRCLE or ARC) to repeat a command indefinitely. When you are done with the command, press ESC.

*You can repeat any command you have used during the current AutoCAD session by using the* UP ARROW *and* DOWN ARROW *keys to navigate to a previous command. When the command name is displayed on the command line, press* ENTER. *You can also locate a previously used command in the command history (in the Text window), cut the command to the Windows Clipboard, and paste it to the command line. Remember that you can also right-click in the command window to access the six most recently used commands.*

## Using Commands Transparently

Certain commands can be used while another command is active. For example, while drawing a line, you may want to use the PAN command to move the drawing across the screen to select the end point of the line. You can also change the settings of drawing aids, such as snap or grid, while other commands are active. Commands that can be used while another command is active are called *transparent* commands.

When you start a command from a toolbar or menu and another command is already active, AutoCAD automatically starts the new command transparently, if possible. If the command can't be used transparently, the active command is canceled and the new command is started.

To use a command transparently from the command line, type an apostrophe (') before the name of the command. AutoCAD displays double angle brackets preceding the prompt on the command line to indicate when a command is being used transparently. For example, to change the snap spacing to ten units while you are drawing a line, enter the following:

```
Command: LINE
Specify first point: 'SNAP
>>Specify snap spacing or [ON/OFF/Aspect/Rotate/Style/Type] <0.5000>: 10
Resuming LINE command.
Specify first point:
```

*Most commands that do not select objects, create new objects, or cause the drawing to be regenerated can be used transparently. Changes made by using dialog boxes that have been opened transparently do not take effect until the original command has been completed.*

## Correcting Mistakes

AutoCAD keeps track of all the commands you use and the changes you make. If you change your mind or make a mistake, you can undo, or reverse, the last action or several previous actions. You can also redo the last action that you reversed.

The Undo and Redo buttons on the Standard toolbar provide the easiest means to undo or redo the previous action.

Undo ⟶   ⬅ Redo

To undo the most recent action, use one of the following methods:

- On the Standard toolbar, click Undo.
- From the Edit menu, choose Undo.
- At the command line, type **U** and then press ENTER.
- Press the CTRL-Z shortcut key combination.
- Right-click to display the shortcut menu, and then choose Undo.

You can also use the UNDO command to reverse several actions at once. To undo a specific number of actions:

1. At the command prompt, type **UNDO**.
2. On the command line, enter the number of actions to undo, and then press ENTER. For example, to reverse the last five actions, type **5**.

*The UNDO command provides other options that let you mark actions as you work. You can then restore the drawing to its condition at that point by undoing back to a previously established mark. You can also use the Begin and End options of the UNDO command to group several actions and later reverse the actions of the entire group.*

*If you erase one or more objects by mistake, you can use the OOPS command to restore them to the drawing.*

To redo an action, do one of the following:

- On the Standard toolbar, click Redo.
- From the Edit menu, choose Redo.
- At the command line, type **REDO** and then press ENTER.
- Press the CTRL-Y shortcut key combination.
- Right-click to display the shortcut menu, and then choose Redo.

*The REDO command reverses the action of the last U or UNDO command. To redo something, you must use the REDO command immediately after using the U or UNDO command.*

## Getting Online Help

AutoCAD includes an online Help system containing all the printed documentation that comes with AutoCAD as well as various additional documents supplied only in electronic form. You can get help about any AutoCAD command or topic by using the online Help system.

To display AutoCAD's online Help system, do one of the following:

- On the Standard toolbar, click the Help button.
- From the Help menu, choose AutoCAD Help.
- At the command line, type **HELP** and then press ENTER.
- Press the F1 key.

The first time you access the Help system when no command is active, AutoCAD displays the AutoCAD Help Contents tab (see Figure 1-12). Once you have used Help, the last tab you used is recalled the next time you bring up Help. When a command is active, accessing the Help system displays information about that command. For example, if you start the LINE command and then access the Help system by using any of the methods just listed, the program displays information about the LINE command (see Figure 1-13).

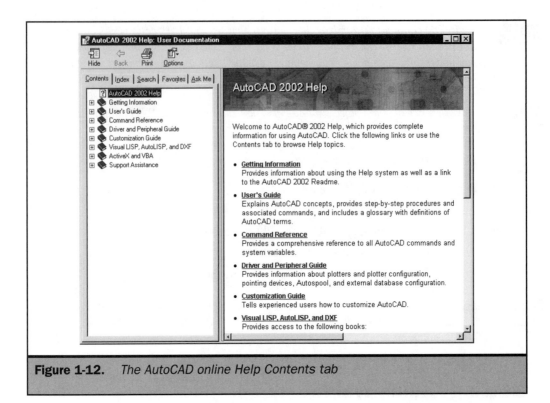

**Figure 1-12.** *The AutoCAD online Help Contents tab*

AutoCAD also includes an Active Assistance window. The contents of this window change dynamically as you work, displaying information about the current command or dialog box. You can manually hide or show this window and also control how and when the Active Assistance window is automatically opened. For example, you may only want to display the window when you're working in a dialog box, or to only display information when you use a new command for the first time.

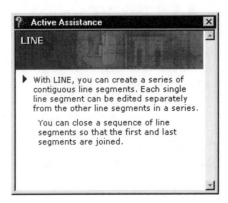

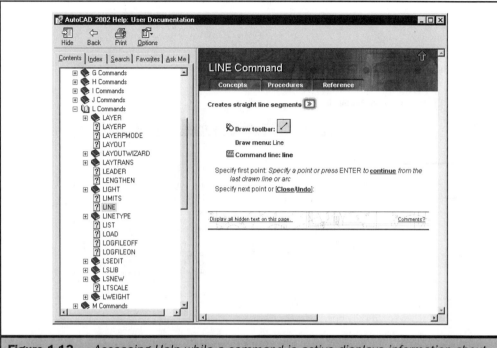

**Figure 1-13.**    *Accessing Help while a command is active displays information about that command.*

To display the Active Assistance window, do one of the following:

- On the Standard toolbar, click the Active Assistance button.
- From the Help menu, choose Active Assistance.
- At the command line, type **ASSIST** and then press ENTER.

You can also control the type of information displayed in the Active Assistance window. For example, you can configure the window to display general information about a dialog box, or specific information about the dialog box option beneath the arrow cursor. To change any of the Active Assistance settings, right-click in the Active Assistance window and then select Settings from the shortcut menu to display the Active Assistance Settings dialog box (see Figure 1-14).

*If you change the Activation setting to On Demand, Active Assistance remains active even when it is not visible, and you may hear a clicking sound when moving the cursor within a dialog box. To disable Active Assistance so that it no longer monitors your actions, right-click the Active Assistance icon located in your system tray in the bottom right-hand corner of your display to display the shortcut menu, and then choose Exit. To reactivate Active Assistance, simply display the Active Assistance window again.*

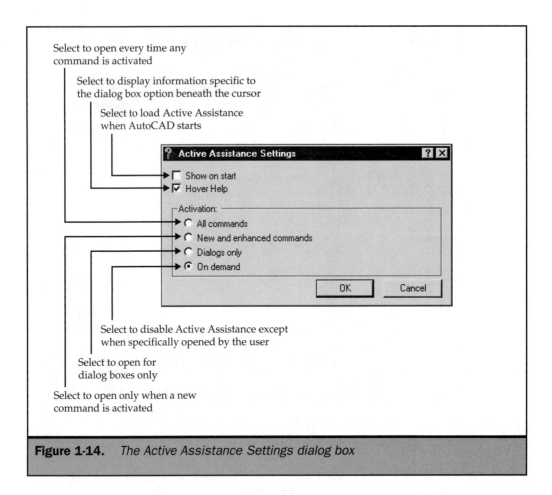

Select to open every time any
command is activated

Select to display information specific to
the dialog box option beneath the cursor

Select to load Active Assistance
when AutoCAD starts

Select to disable Active Assistance except
when specifically opened by the user

Select to open for
dialog boxes only

Select to open only when a new
command is activated

**Figure 1-14.**    *The Active Assistance Settings dialog box*

In addition to AutoCAD's Help system, the Help pull-down menu provides a link
to the Product Support page on Autodesk's Point A web site. There you will find links
to technical support, a knowledge base of Autodesk technical support information,
copies of the latest patches and software updates, discussion groups, and other on-line
resources. The Help menu also provides links to the Autodesk Users Group
International web site and the Autodesk Learning Assistance.

**Note**    *AutoCAD comes with a Learning Assistance CD that provides interactive lessons to
help you learn how to use AutoCAD. In order to access Learning Assistance from within
AutoCAD, you must first install the Learning Assistance software by running the
Setup program on the Learning Assistance CD. If you prefer to run this software from
the CD rather than installing it onto your system, run the Setup program and choose
Run rather than Install. You'll be able to access all of the lessons, but Learning
Assistance will not be accessible from the Help menu.*

# Opening Existing Drawings

Now that you're more familiar with the AutoCAD environment, you are ready to learn how to open an existing drawing. Since it often takes several days to complete a detailed drawing, you will probably open existing drawings more often than you create new ones. In the next chapter, you'll learn how to create new drawings.

To open an existing drawing, use one of the following methods:

- On the Standard toolbar, click the Open button.
- From the File menu, choose Open.
- At the command line, type **OPEN** and then press ENTER.
- Press the CTRL-O shortcut key combination.

*AutoCAD remembers the names of the most recent drawings that you worked on. To quickly open a drawing file that you recently used, choose its name from the list in the lower portion of the File pull-down menu.*

AutoCAD displays the Select File dialog box. You can preview the drawing by clicking it to highlight the filename in the list of files (see Figure 1-15). To open a drawing, either double-click the filename in the list of files or highlight the file and then click Open. You can also type its name (or a partial name and wildcards) in the File Name edit box and then click Open.

*The icons on the left side of the dialog box provide quick access to commonly used files and file locations. Click any one of these icons to display the files at that location. You can reorder these icons by dragging them to a new location in the list. To add or modify an icon, right-click the icon to display a shortcut menu. You cannot remove the Point A, Buzzsaw, or Red Spark icons.*

AutoCAD can also help you find drawings and other AutoCAD files located in other folders on your hard drive. You can instruct AutoCAD to search in one or more directories and on one or more drives, even across a local area network.

To use the file search utility, click Tools | Find in the Select File dialog box. AutoCAD displays the Find dialog box (see Figure 1-16). Notice that the dialog box has two tabs. The Name & Location tab provides tools to help filter your search based on the file type, filename, and location. The Date Modified tab lets you filter your search based on the date the file was created or modified. For example, you can search for all drawing files on the C drive that were created after a specific date. To start the actual search, click Find Now. After several seconds, AutoCAD displays a list of all the drawing files that match your search criteria. To open a drawing, select it in the list, click OK, and then click Open in the Select File dialog box.

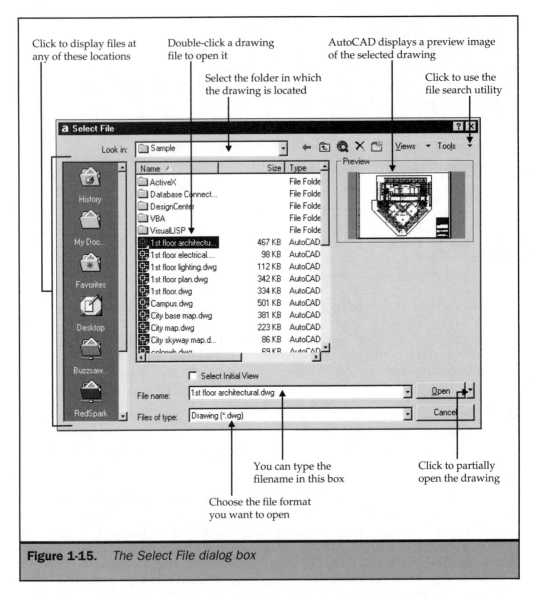

Click to display files at
any of these locations

Double-click a drawing
file to open it

AutoCAD displays a preview image
of the selected drawing

Select the folder in which
the drawing is located

Click to use the
file search utility

You can type the
filename in this box

Click to partially
open the drawing

Choose the file format
you want to open

**Figure 1-15.**    *The Select File dialog box*

You can also open an existing drawing using the AutoCAD 2002 Today window (see Figure 1-17). To open the AutoCAD 2002 Today window (if it is not already open), do one of the following:

- On the Standard toolbar, click the Today button.
- From the Tools menu, choose Today.
- At the command line, type **TODAY** and then press ENTER.

Specify the file type.

Click to limit the search to files saved
before or after a specific date and time

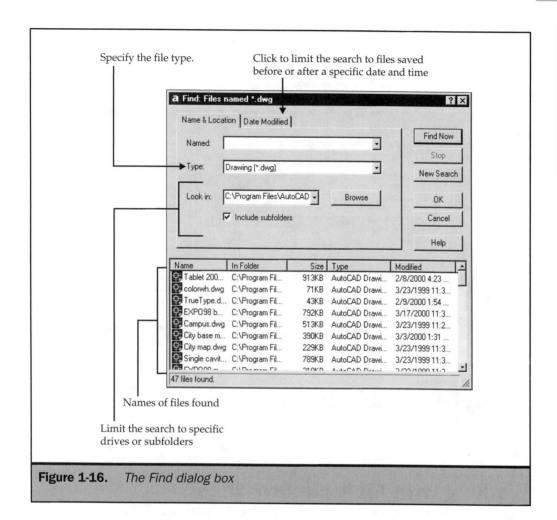

**Figure 1-16.**    *The Find dialog box*

Names of files found

Limit the search to specific
drives or subfolders

The upper portion of this window includes an area called My Workplace.
On the left side of this area is an area called My Drawings, which contains three
tabs. If you select the Open Drawings tab, AutoCAD displays a list of recently
opened drawings. By default, this list is sorted by date. You can also sort the list
alphabetically by filename or by location (drive letter and folder name). When
you move the cursor over a drawing name, AutoCAD displays a preview image
of the drawing. If you pause the cursor over a drawing name, AutoCAD displays
its complete path and the date and time it was last opened. To open a drawing,
simply click the drawing name.

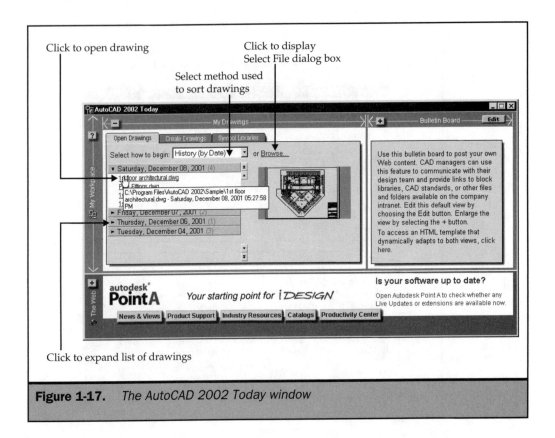

**Figure 1-17.**   *The AutoCAD 2002 Today window*

# Working with Multiple Drawings

You can open more than one drawing in a single AutoCAD session. Each open drawing appears within its own document window with the drawing name displayed in the title bar of each window. Although you may have multiple drawings open at once, only one drawing is active (or current) at a time. AutoCAD commands only affect the active drawing.

When multiple drawings are open, you can simply click anywhere in a drawing's document window to make it the active drawing. You can also press CTRL-F6 or CTRL-TAB to switch between open drawings, or select the active drawing from the Window pull-down menu. Individual document windows can be moved, resized, minimized, and maximized just like any other Windows document window, and you can use the other commands on the Windows pull-down menu to tile or cascade all of the open drawings.

Having multiple drawings open simultaneously also offers the ability to copy-and-paste objects, settings, and properties between drawings.

# Using Partial Open and Partial Load

You can use the Partial Open option to open just a portion of a drawing file, thus improving AutoCAD's performance when working with a large drawing file. When you use the Partial Open option to open a drawing, you load only the geometry contained within a previously saved view and on specific layers. For example, if you load geometry from the WEST-WING view and the WALLS layer, AutoCAD loads into the drawing everything from the WALLS layer that falls within the WEST-WING view. Once loaded, you can edit only the geometry that has been loaded, but you can subsequently use the PARTIALOAD command to load additional geometry from another view or on other layers.

To partially open a drawing, follow these steps:

1. Do one of the following:

   ■ On the Standard toolbar, click the Open button.

   ■ From the File menu, choose Open.

   ■ At the command line, type **OPEN** and then press ENTER.

   ■ Press the CTRL-O shortcut key combination.

2. In the Select File dialog box, select the drawing you want to open, and then click the arrow next to the Open button (see Figure 1-15) and choose Partial Open from the drop-down list to display the Partial Open dialog box (see Figure 1-18).

3. In the Partial Open dialog box, under View Geometry To Load, select the name of the view you want to load. The default view *Extents* loads the entire drawing.

4. Under Layer Geometry To Load, select one or more layers.

5. Click Open.

**Caution** *If you don't select any layers to load, no geometry will be visible, but all layers still exist in the drawing. If you draw on a layer whose geometry has not been loaded, you may be drawing on top of existing geometry that isn't loaded.*

When a drawing has been partially loaded, you can edit only the geometry that has been loaded. If necessary, you can load additional geometry using the Partial Load dialog box. This dialog box is almost identical to the Partial Open dialog box, except that in addition to specifying a previously named view, you can click the Pick A Window button and then specify an area within the drawing in which to load the additional geometry. To load additional geometry into a partially open drawing, follow these steps:

1. Do one of the following:

   ■ From the File menu, choose Partial Load.

   ■ At the command line, type **PARTIALOAD** and then press ENTER.

2. In the Partial Load dialog box, select the view name or click the Pick A Window button to define a rectangular view area.

3. Select one or more layers whose geometry you want to load.

4. Click OK.

**Note**  *The PARTIALOAD command is available only if the current drawing is a partially open drawing. When a drawing is partially open, the designation "(Partially loaded)" appears on the title bar of its document window, adjacent to the drawing name.*

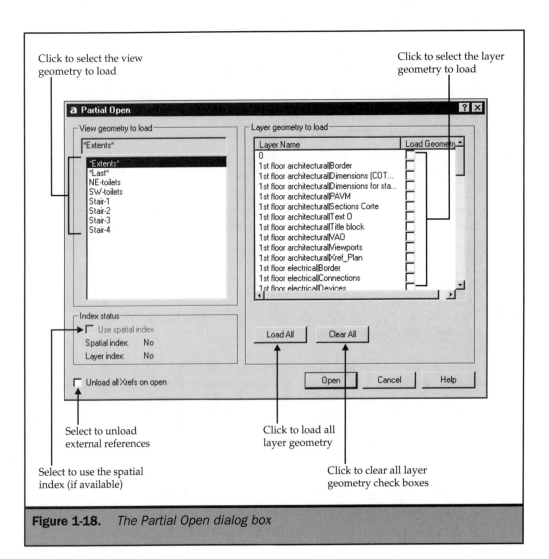

Click to select the view geometry to load

Click to select the layer geometry to load

Select to unload external references

Click to load all layer geometry

Select to use the spatial index (if available)

Click to clear all layer geometry check boxes

**Figure 1-18.** *The Partial Open dialog box*

# Saving Your Work

As you work, you should save your drawing periodically. This ensures that you don't accidentally lose important work in the event of a power failure or other mishap. When you first start a new drawing, the file has no name. Thus, the first time you save it, AutoCAD displays a dialog box and prompts you to name the file. Once the file has been saved, or if you open an existing drawing, the SAVE command saves the drawing immediately, without bringing up a dialog box, and includes any changes you may have made to the current file. If you want to save a copy of the drawing without saving changes to the original file, you can use the SAVEAS command to save the drawing under another name.

To save a drawing, do one of the following:

- On the Standard toolbar, click the Save button.
- From the File menu, choose Save.
- At the command line, type **SAVE** and then press ENTER.
- Press the CTRL-S shortcut key combination.

> **Tip**   *When you save a drawing by using the SAVE command, AutoCAD also saves the previous version of the file as a backup, using the same filename but appending the file extension .BAK. You can recover the previous version of the drawing by renaming the backup copy so that it has a .DWG file extension and then loading it into AutoCAD. These backup copies can consume a considerable amount of disk space, however. If you decide that you will not need to maintain backup copies of your drawings, you can turn off this feature from the Open and Save tab of the Options dialog box.*

If the drawing has not previously been saved, AutoCAD displays the Save Drawing As dialog box. Type the drawing name in the File Name text box and then click Save.

To save a drawing using a different name, do one of the following:

- From the File menu, choose Save As.
- On the command line, type **SAVEAS** and then press ENTER.

In the Save Drawing As dialog box, type the new drawing name in the File Name text box, and then click Save.

> **Note**   *As you work, AutoCAD periodically saves your drawing. However, it doesn't save the drawing to the current file. Instead, AutoCAD saves the drawing to the file specified by the SAVEFILEPATH and SAVEFILE system variables, appending the file extension .SV$. Initially, AutoCAD updates this saved file every 120 minutes. You can change the name of the file and the interval between periodic updates from within the Options dialog box. The time interval is controlled from the Open and Save tab. You can also turn off the automatic save feature from this tab. The location of this saved file is controlled from the Files tab (under Automatic Save File Location).*

# Viewing and Updating Drawing Properties

You can save custom property information, such as the title, author, subject, keywords, and hyperlink addresses, when you save your drawings. The new Drawing Properties dialog box lets you save, view, and update this information at any time, making it much easier to keep track of your drawings.

To display the Drawing Properties dialog box, do one of the following:

- From the File menu, choose Drawing Properties.
- At the command line, type **DWGPROPS** and then press ENTER.

The Drawing Properties dialog box contains four tabs:

- **General**   Displays the filename, drawing type, location, size, and other information. Since the information on this tab is derived from the operating system, all the information is read-only.

- **Summary**   Enables you to enter the drawing title, subject, author's name, keywords, comments, and a hyperlink base (see Figure 1-19).

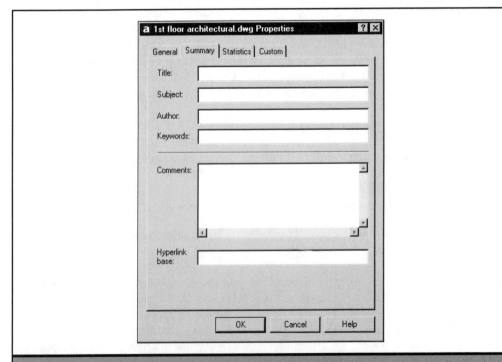

**Figure 1-19.**   *You can store the drawing title, subject, author, and other information in the Summary tab of the Drawing Properties dialog box.*

- **Statistics**   Displays data such as the date the drawing was created and last modified and how long the drawing has been edited.
- **Custom**   Enables you to enter up to ten custom properties. You specify the names of the fields in the Name column, and specify the value for each field in the Value column.

You can then use the drawing property information when searching for drawing files, using AutoCAD DesignCenter or Windows Explorer.

*Properties entered in the Drawing Properties dialog box are not associated with the drawing until you save the drawing.*

# Exiting from AutoCAD

When you are finished working in AutoCAD, you should exit from the program. To exit from AutoCAD, do one of the following:

- On the main AutoCAD title bar, either click the Close button or double-click the program button.
- From the File menu, choose Exit.
- At the command line, type **EXIT** or **QUIT** and then press ENTER.
- Press the ALT-F4 shortcut key combination.

If you haven't saved your most recent changes for each open drawing, AutoCAD displays a dialog box for each one, asking whether you want to save the changes to the current drawing.

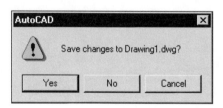

Click Yes to save the changes, No to exit from AutoCAD without saving the changes, or Cancel to remain in AutoCAD.

**Note** *If a command is active in any open drawing window, AutoCAD displays an alert dialog box, informing you that you must first complete the command and then try again. Switch to the drawing in which the command is active, complete or cancel the command, and then try to exit from AutoCAD again.*

**Tip** *You can use the CLOSEALL command to close all open drawings without exiting from AutoCAD.*

# The Complete Reference

# Chapter 2

## Drawing Basics

AutoCAD helps you to organize information in your drawings for greater efficiency. For example, you can draw objects representing different types of information on various layers, and then use those layers to control the color, display or plot visibility, linetype, and lineweight of those objects. You can create new drawings more quickly by reusing information from your existing drawings. And, you can maintain standards across all the drawings you create by using the same drawing sheet layout, text styles, and symbols. You generally consider these issues whenever you start a new drawing, and AutoCAD can help by automatically establishing common settings whenever you begin a new drawing.

When you start a new drawing, you also need to decide what type of dimensional units you will use (such as feet and inches or millimeters) and the number of decimal places required. You may also want to think about the scale at which you will later print copies of your drawing.

In this chapter, you will learn how to do the following:

- Start a new drawing.

- Use drawing aids, such as grid, snap, and orthogonal settings, as well as polar tracking and polar snap tracking to draw accurately.

## Starting a New Drawing

When you start a new drawing, you can base that drawing on a template that contains standard settings that you generally use. A *template* is simply a normal AutoCAD drawing that has been saved as a drawing template file (with a .DWT file extension). AutoCAD comes with numerous templates representing different standard bordered drawing sheets. These templates also have preestablished layers, linetypes, and other settings. You can use one of these templates as is, modify a template to suit your specific needs, or create your own template. You can also start a new drawing from scratch, without the use of a template.

**Note**    *Although templates have virtually the identical format as drawing files, they are differentiated from drawing files by their .DWT file extension. Templates also contain a short description that appears in the Create New Drawing dialog box. You can save any drawing as a template by using the SAVEAS command.*

AutoCAD also comes with two wizards that help you to start a new drawing. These wizards let you set the type of linear and angular units to be used, and predetermine the size of the drawing area.

**Tip**    *Because the Setup wizards step you through the decision process of setting up a new drawing, you may find it easier to start new drawings by using the wizards, particularly if you're new to AutoCAD. If most of your drawings use the same basic settings, you may find it faster to use a template to start a new drawing. Because of the extra steps involved when using a wizard, starting from a template is much faster.*

To start a new drawing, do one of the following:

- On the Standard toolbar, click the New button.
- From the File menu, choose New.
- At the command line, type **NEW** and then press ENTER.
- Press the CTRL-N shortcut key combination.

If AutoCAD is configured to use the AutoCAD Today window, whenever you start a new drawing, AutoCAD displays the Create Drawings tab in the My Drawings area of the AutoCAD 2002 Today window (see Figure 2-1). From this tab, you can choose to start a new drawing using one of the following methods:

- Use one of the wizards to be led through the steps of setting up a new drawing.
- Use an existing template as the basis of a new drawing.
- Start a new drawing from scratch, using either the default English (feet and inches) or metric settings.

You can select the method you want to use by choosing it from the drop-down list. The default method initially displayed is the one used the last time the AutoCAD 2002 Today window was displayed.

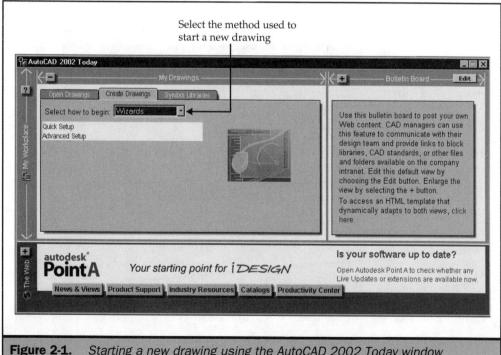

**Figure 2-1.**    *Starting a new drawing using the AutoCAD 2002 Today window*

If AutoCAD is configured to use a traditional startup dialog, whenever you start a new drawing, AutoCAD displays the Create New Drawing dialog box (see Figure 2-2). From this dialog box, you can also choose to start a new drawing using any one of the three methods available from the AutoCAD 2002 Today window. Other than starting the process from either the AutoCAD Today window or the Create New Drawing dialog box, the actual process of starting a new drawing is the same.

# Using the Drawing Wizards

If you decide to start a new drawing using a wizard, AutoCAD offers two different wizards you can use:

- **Quick Setup wizard**   Sets the unit of measurement and the drawing area
- **Advanced Setup wizard**   An expanded version of the Quick Setup wizard, offering additional control over angle measurements

Both the Quick Setup and Advanced Setup wizards use the template file ACAD.DWT.

## Quick Setup Wizard

The Quick Setup wizard has only two steps: Units and Area. When you start the Quick Setup wizard, AutoCAD displays the Units page (similar to Figure 2-3). Select the unit of measurement you want to use by selecting the appropriate radio button. The dimension

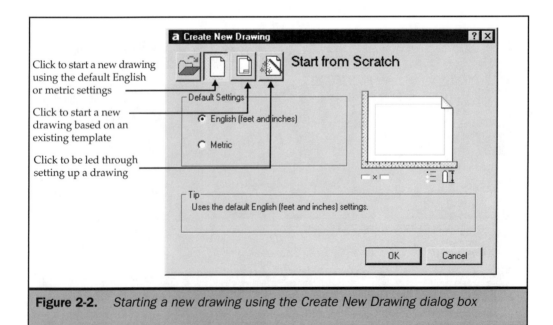

**Figure 2-2.**   *Starting a new drawing using the Create New Drawing dialog box*

adjacent to the image of the window changes to show the selected units of measurement, which may be any one of the following:

- **Decimal**    Decimal units (15.5000)
- **Engineering**    Feet and decimal inches (1'-3.5000")
- **Architectural**    Feet, inches, and fractional inches (1'-3 1/2")
- **Fractional**    Fractional units (15 1/2)
- **Scientific**    Scientific notation (1.5500E+01)

After you select the units of measurement, click Next. AutoCAD displays the Area page (similar to Figure 2-7). Specify the width and length (expressed in full-scale units) of the area in which you plan to draw. This action limits the area of the drawing covered by grid dots when the grid is turned on. You can also adjust this setting individually after creating the drawing.

After you specify the width and length, click Finish. AutoCAD immediately creates the drawing, using the settings that you specified.

## Advanced Setup Wizard

The Advanced Setup wizard contains the same two steps as the Quick Setup wizard, but adds three additional steps that let you specify the type of angle measurement you want to use.

When you start the Advanced Setup wizard, AutoCAD displays the Units page (see Figure 2-3). This step is similar to the first step in the Quick Setup wizard, but also enables you to specify the number of decimal places used for linear distances. Select the unit of measurement you want to use by selecting the appropriate radio button. The dimension adjacent to the image of the window changes to show the selected units of measurement, which may be any one of the following:

- **Decimal**    Decimal units (15.5000)
- **Engineering**    Feet and decimal inches (1'-3.5000")
- **Architectural**    Feet, inches, and fractional inches (1'-3 1/2")
- **Fractional**    Fractional units (15 1/2)
- **Scientific**    Scientific notation (1.5500E+01)

Specify the number of decimal places or the fractional size to which you want linear measurement displayed by selecting from the Precision drop-down list. The values presented in this list vary according to the type of units selected.

After you select the units of measurement and precision, click Next. AutoCAD displays the next step (see Figure 2-4). Select the angle of measurement you want to use

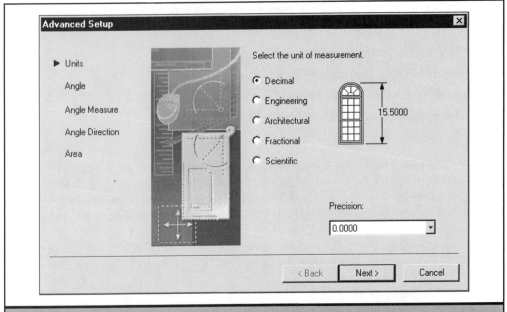

**Figure 2-3.** *The Units page of the Advanced Setup wizard*

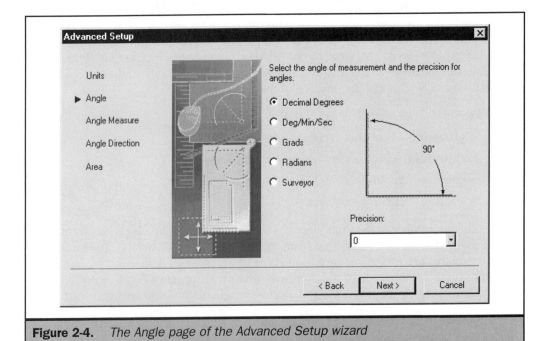

**Figure 2-4.** *The Angle page of the Advanced Setup wizard*

by selecting the appropriate radio button. The image of the angular dimension changes to show the selected units of measurement, which may be any one of the following:

- **Decimal Degrees**   Display partial degrees as decimals (90.0000°)
- **Deg/Min/Sec**   Display partial degrees as minutes and seconds of an arc (90d0'0")
- **Grads**   Display angles as grads (100.0000g)
- **Radians**   Display angles as radians (1.5708r)
- **Surveyor**   Display angles in surveyor units (N)

Specify the number of decimal places or the fractional size to which you want angular measurement displayed by selecting from the Precision drop-down list. The values presented in this list vary based on the type of units selected.

After you select the units of measurement and the precision for angles, click Next. AutoCAD displays the next step (see Figure 2-5). Select the direction of the zero angle by selecting the appropriate radio button. The angles shown on the compass rose change to reflect the direction you select, which may be any one of the following:

- **East**   Sets the east compass point as the zero angle
- **North**   Sets the north compass point as the zero angle
- **West**   Sets the west compass point as the zero angle
- **South**   Sets the south compass point as the zero angle
- **Other**   Sets the zero angle to the compass point that you specify as the angle value in the adjacent edit box

After you select the zero angle direction, click Next. AutoCAD displays the Angle Direction page (see Figure 2-6). Select the direction in which you want positive angle values to increase, by selecting the appropriate radio button. The direction arrow in the image changes to show the direction you have selected, which may be either of the following:

- **Counter-Clockwise**   Angles increase counterclockwise (default, based on the right-hand rule)
- **Clockwise**   Angles increase clockwise

After you select the direction, click Next. AutoCAD displays the Area page (see Figure 2-7). This step is identical to the second step in the Quick Setup wizard. Specify the width and length (expressed in full-scale units) of the area in which you plan to draw. This action limits the area of the drawing that is covered by grid dots when the grid is turned on.

After you specify the width and length, click Finish. AutoCAD immediately creates the drawing, using the settings that you specified. You can also adjust these settings individually after creating the drawing.

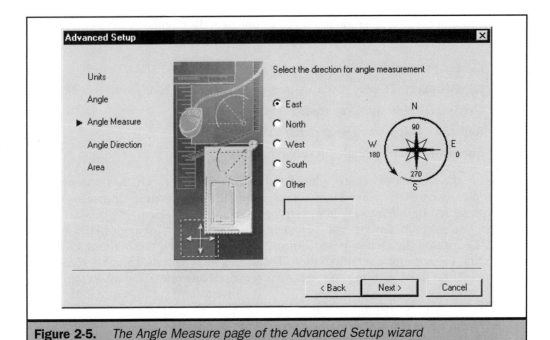

**Figure 2-5.**    *The Angle Measure page of the Advanced Setup wizard*

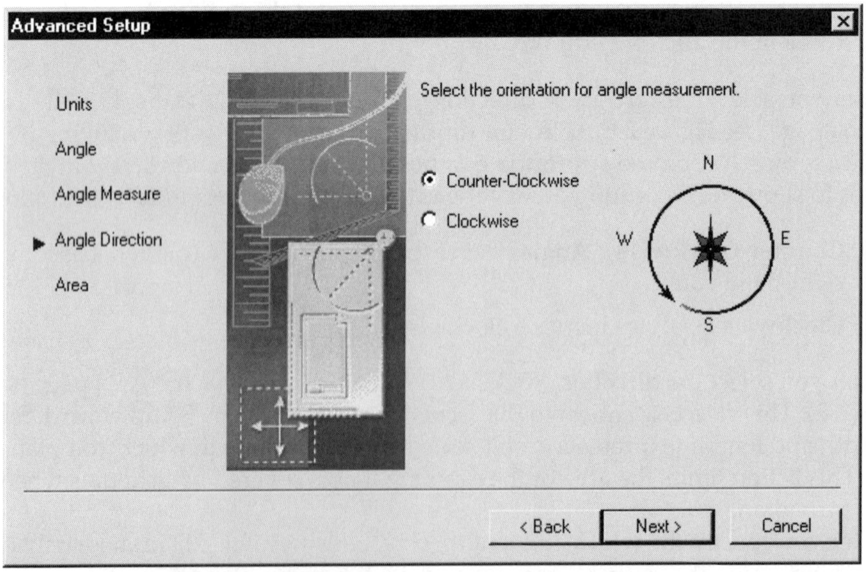

**Figure 2-6.**    *The Angle Direction page of the Advanced Setup wizard*

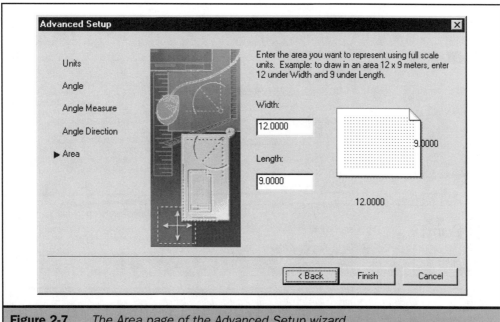

**Figure 2-7.**     *The Area page of the Advanced Setup wizard*

# Using Template Drawings

When you choose to use a template to start a new drawing, AutoCAD displays a list of 66 available templates (see Figure 2-8). If you're using the AutoCAD 2002 Today window, when you move the cursor over a template name, AutoCAD displays a preview image of the template. To start a drawing using a template, simply click the template name. When using the traditional startup dialog, AutoCAD displays a similar list.

The list contains the actual template filenames. These predefined template files are stored in AutoCAD's Template directory. This list actually includes 32 different template designs, each available in two different formats (one with a color-dependent plot style and one with a named plot style), plus two additional templates (collections of metric and ANSI template layouts). Each template file contains predefined settings for layers and drawing limits, a predrawn title border and title block, and a determination as to whether measurements will be in English or metric units. Template files can also contain predefined dimension styles, units, and views. You'll learn about plot styles in Chapter 17.

To start a new drawing based on a template, either double-click the template or select the template and click OK.

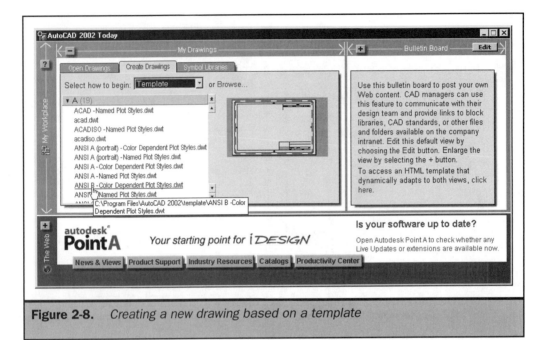

**Figure 2-8.**   *Creating a new drawing based on a template*

*If you use the traditional startup dialog and select one of the templates in the list, the dialog box displays a brief description and a preview image of the template. This description is not available when using the AutoCAD 2002 Today window.*

When you use one of the predefined templates containing a border, AutoCAD creates the border and title block in a layout (referred to as *paper space* in earlier versions of AutoCAD) and creates a floating model-space viewport within the border.

If the file you want to use as a template isn't listed, click the Browse button. AutoCAD displays a standard file dialog box labeled Select a Template File. You can then select the template or drawing file from the list of files.

*Unlike the Setup wizards, after you start a new drawing based on a template, it is up to you to make any necessary changes to the drawing setup. For example, you may need to change the type of linear or angular units. If you regularly start new drawings with several preestablished settings, you should create a custom template file containing these settings and use it when starting new drawings. You can save any drawing as a template by specifying the file type as Drawing Template File (*.DWT) when you save a new drawing using the Save Drawing As dialog box.*

## Starting a Drawing from Scratch

When you click Start From Scratch, AutoCAD offers just two setting selections. You can set the measurement to English or metric units. If you select English units, AutoCAD

AUTOCAD BASICS

creates a new drawing based on the ACAD.DWT template file. If you select metric units, the new drawing is based on the ACADISO.DWT template file.

*If you accidentally change the ACAD.DWT or ACADISO.DWT templates from their original default values, you can restore them by starting a new drawing from scratch using either English or metric units, respectively, and then using the SAVEAS command to save that drawing as a template, using the filename ACAD.DWT or ACADISO.DWT.*

# Setting Up a Drawing

Regardless of the method you use to start a new drawing, you may find it necessary to alter some of the basic settings established at the start. Of course, if you find that you regularly change the same settings, you can save yourself a great deal of time in the future by creating a custom template containing all of those settings.

*The types of settings normally determined in template files are layers, colors, linetypes, scale factors, drawing units, text heights, and drawing aids such as snap and grid spacing. However, any setting that you might save as part of the drawing file can be predetermined by using a template file, including drawing objects.*

## Setting the Current Layer

In traditional drafting, you often separate elements, such as walls, dimensions, structural steel members, and electrical plans, onto separate translucent overlays. Layers in an AutoCAD drawing are like the overlays that you use in manual drafting. When you want to print the working drawings, you can create several different drawings by choosing different layers for each plot.

The layers feature in AutoCAD offers numerous advantages over physical transparencies. The number of overlays you can combine to print a manually drafted drawing is limited by the printing process. No such limitation exists in AutoCAD. AutoCAD lets you define an unlimited number of layers, any of which can be visible or invisible at any time. You can name each layer and assign to each its own color, linetype, lineweight, and plot style. You can also lock individual layers to ensure that information on those layers is not altered accidentally, and set any layer so that it is visible onscreen but does not print.

Every drawing has at least one layer, the default layer, named "0." You can add to the drawing any number of additional layers. One layer of your choice must be selected as the *current*, or active, layer. When you create an object, it is created on the current layer.

Since layers can control the linetype, lineweight, and color assigned to objects, you can use layers to organize drawing objects to a much finer degree than you could by using manual overlay drafting. For example, you can create centerlines on a centerline layer, and assign a centerline linetype to that layer. Then, whenever you draw a centerline, you switch to that layer (by making it the current layer) and then draw the line.

To set the current layer, on the Drawing Properties toolbar, choose the current layer from the Layer Control drop-down list.

With careful planning, you can divide your drawing objects into discrete categories, with objects that represent different drawing elements—such as part outlines, cross-hatching, dimensions, centerlines, and notes—each on its own respective layer.

*AutoCAD sorts layer names alphabetically. Therefore, when organizing a layer-naming scheme, use common prefixes when naming layers, so that AutoCAD displays them in some sort of hierarchical fashion. Careful organization also makes it easier to use filters to display specific groups of layers. For example, if you are creating floor plans for a multistory building, using a prefix representing the floor level (such as 02 for the second floor) makes it easy to display only those layers representing a specific floor. You'll learn more about layers, including how to set layer filters, in Chapter 8.*

## Creating New Layers

You create and name layers by using the Layer Properties Manager dialog box. Display this dialog box by doing one of the following:

- On the Object Properties toolbar, click the Layers button.
- From the Format menu, choose Layer.
- At the command line, type **LAYER** (or **LA**) and press ENTER.

AutoCAD displays the Layer Properties Manager dialog box (see Figure 2-9).
To create a new layer, follow these steps:

1. In the Layer Properties Manager dialog box, click New. AutoCAD creates a new layer called Layer1.

2. Type a name for the new layer over the highlighted default name, and then press ENTER. The layer name can contain up to 255 characters, including spaces.

3. Click OK to complete the command and return to your drawing.

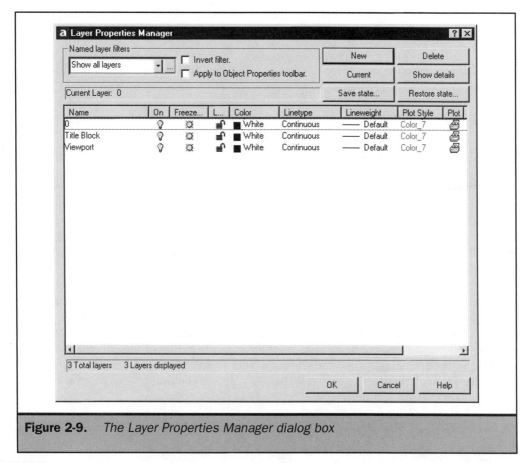

**Figure 2-9.**    *The Layer Properties Manager dialog box*

**Note**    *While the Layer Properties Manager dialog box is displayed, you can also change the color, linetype, or lineweight associated with any layer, turn layers on and off, rename layers, or select a different layer as the current layer. You'll learn more about working with layers in Chapter 8.*

## Setting the Current Object Color

The color assigned to an object determines how the object is displayed and printed. Depending on whether your drawing uses a color-dependent or named plot style, the object's color can determine its print color or can be used to assign different pen numbers and widths to specific colors. You'll learn about plot styles in Chapters 17 and 18.

Objects are created in the current color. There are 255 regular colors, and 2 additional color properties that are often referred to as colors. You can use the first 7 of the 255 colors by name.

1            red

2            yellow

| 3 | green |
| 4 | cyan |
| 5 | blue |
| 6 | magenta |
| 7 | white |

*If the background of your graphics area is white, color 7 actually appears black.*

Each color has a unique number from 1 to 255. The two additional color properties are BYLAYER and BYBLOCK. These color properties cause an object to adopt the color either of the layer on which the object is drawn or of the block into which it is grouped.

**Note**    *When assigning color numbers, AutoCAD assigns color number 256 to the BYLAYER property and assigns color number 0 to the BYBLOCK property.*

When you start a new drawing, objects are created in the color BYLAYER, which means that all objects adopt the color of the current layer (the layer on which they are drawn). Initially, layer 0 is both the only layer and the current layer. Its default color is white. When you create a new layer, its color defaults to that of the current layer.

To set the current color, you can choose a color from the Color Control drop-down list on the Drawing Properties toolbar. This list displays the first seven colors and the special colors BYLAYER and BYBLOCK.

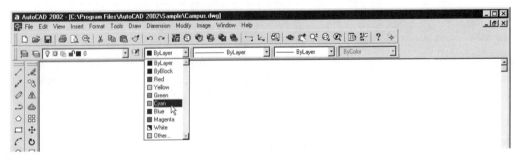

**Note**    *If you have set the current color to any other color number during the current editing session, that color number also appears in the Color Control drop-down list.*

Or, you can use the Select Color dialog box to set the current color to something other than one of the colors included in the Color Control drop-down list. Display the Select Color dialog box by doing one of the following:

■ On the Object Properties toolbar, choose Other from the Color Control drop-down list.

- From the Format menu, choose Color.
- At the command line, type **COLOR** (or **COL**) and then press ENTER.

AutoCAD displays the Select Color dialog box (see Figure 2-10).
To set the current color, follow these steps:

1. In the Select Color dialog box, click BYLAYER, BYBLOCK, or the color of your choice. You can also type the color number or one of the standard color names in the Color edit box.

2. Click OK.

*Assigning colors by layer makes managing colors and layers easier, because you can quickly change the color of all objects drawn on a layer. Using specific colors for different layers also provides a visual aid to assist in determining the layer on which a particular object is drawn. Assigning color on a per-object basis is useful, however, in certain circumstances. For example, you may assign color on a per-object basis to avoid having to create a new layer just to draw a few objects using a different color, particularly when those objects are more logically grouped on an existing layer.*

## Setting the Current Linetype

Linetypes help convey information about the meaning of objects in a drawing. You use different linetypes to differentiate the purpose of one line from another. A linetype

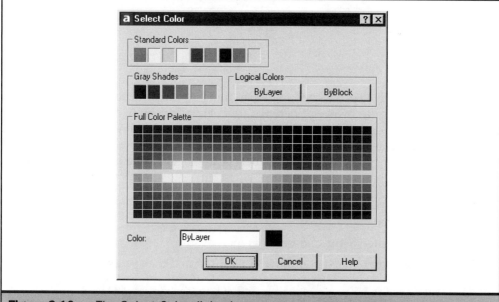

**Figure 2-10.**    *The Select Color dialog box*

definition consists of a repeating pattern of dots, dashes, and blank spaces. It can also include a repeating pattern of text and shapes. The definition determines both the sequence and relative length of the pattern. Linetypes determine the appearance of objects both onscreen and when printed. By default, every drawing has at least these three linetypes: CONTINUOUS, BYLAYER, and BYBLOCK. Your drawing may also contain an unlimited number of additional linetypes.

When you create an object, it is created using the current linetype. By default, the current linetype is BYLAYER, meaning that an object's actual linetype is determined by the linetype assigned to the layer on which it is drawn. With a BYLAYER setting, if you change the linetype assigned to the layer, all the objects created on that layer change to reflect the new linetype.

You can also select a specific linetype as the current linetype. Doing so overrides the layer's linetype setting. Objects are then created using that linetype, and changing the layer linetype has no effect on them.

As a third option, you can use the special linetype BYBLOCK. When BYBLOCK is selected, all objects are initially drawn using the CONTINUOUS linetype. Once the objects have been grouped into a block, however, they will inherit the linetype setting of the current layer when you subsequently insert the block into a drawing.

To set the current linetype, on the Drawing Properties toolbar, choose the current linetype from the Linetype Control drop-down list.

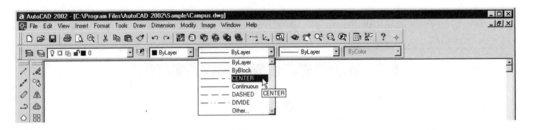

AutoCAD comes with 45 predefined linetypes, and you can also create your own linetypes. Unlike colors, however, linetypes are not available until after they have been loaded into the drawing. Linetype definitions are stored in linetype library files (having a .LIN file extension). You can use the Linetype Manager dialog box to load linetypes from a library file. Display the Linetype Manager dialog box by doing one of the following:

- On the Object Properties toolbar, choose Other from the Linetype Control drop-down list.
- From the Format menu, choose Linetype.
- At the command line, type **LINETYPE** (or **LT**) and then press ENTER.

AutoCAD displays the Linetype Manager dialog box (see Figure 2-11).

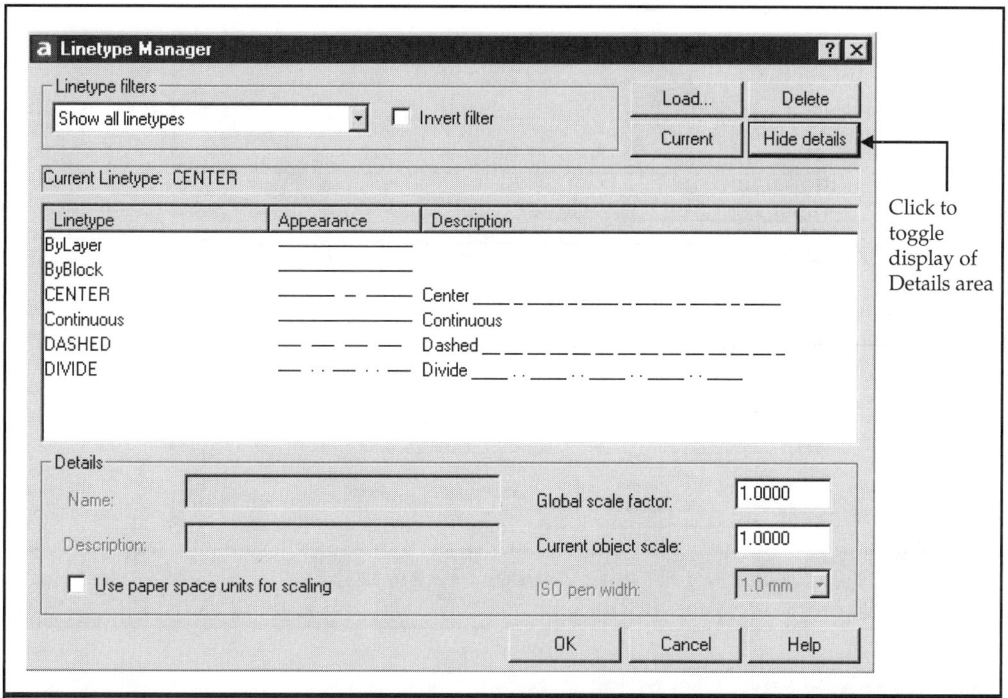

**Figure 2-11.** *The Linetype Manager dialog box with the Details area expanded*

**Note**    *Linetypes control the appearance of all AutoCAD objects except points, text, hatch patterns, 3-D polylines, and the borders of viewports. These objects always display and print with a continuous line.*

To load a linetype, follow these steps:

1. In the Linetype Manager dialog box, click Load. AutoCAD displays the Load or Reload Linetypes dialog box (see Figure 2-12). AutoCAD normally uses one of its default linetype library files (ACAD.LIN for English measurement, and ACADISO.LIN for metric). You can click File to load linetype definitions from a different linetype library file.

2. In the Load or Reload Linetypes dialog box, select one or more linetypes to load from the Available Linetypes list and then click OK. AutoCAD adds these linetypes to the list displayed in the Linetype Manager dialog box. The new linetypes also will be listed in the Linetype Control drop-down list on the Object Properties toolbar.

3. Click OK to complete the command and return to your drawing.

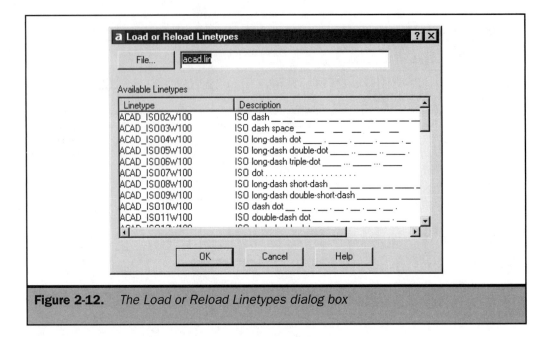

**Figure 2-12.**  *The Load or Reload Linetypes dialog box*

  *When selecting linetypes in the Load or Reload Linetypes dialog box, you can use the standard Windows mouse procedures* SHIFT-*click,* CTRL-*click, and* CTRL-A *to select several adjacent linetypes, several individual linetypes, or all the linetypes, respectively. You can also right-click to display a shortcut menu, as you can in any other dialog box.*

## Setting the Current Linetype Scale

While linetype definitions describe the relative lengths of dashes and blank spaces, you can further control the appearance of linetypes by adjusting the linetype scale factor. The smaller the scale, the more repetitions of the linetype pattern that are generated per drawing unit. For example, if a linetype pattern is defined as a sequence of dashed lines and open spaces, each 0.25-unit long, the sequence of dashes and spaces will actually measure 0.25 unit when applied using a linetype scale factor of 1. Changing the linetype scale factor to 0.5 would reduce the length of each line and space to 0.125 unit; a scale factor of 2 would increase the length of each to 0.5 unit.

  *If a line segment is too short to hold even one dash sequence, AutoCAD draws a continuous line. Similarly, setting the linetype scale too large may cause AutoCAD to draw a continuous line.*

AutoCAD initially uses a global linetype scale factor of 1.0. This scale factor is applied to all objects drawn with a noncontinuous linetype.

You can also control the linetype scale on a per-object basis. The current linetype scale factor is applied relative to the global linetype scale factor. Therefore, a line created with an object linetype scale factor of 0.5 in a drawing with a global linetype scale factor of 2 would appear the same as a line with an object linetype scale factor of 1 in a drawing with a global linetype scale factor of 1.

LTSCALE = 2

LTSCALE = 1

You control both the global and object linetype scale factors from the Linetype Manager dialog box. To change the linetype scale factor, follow these steps:

1. From the Format menu, choose Linetype.

2. In the Linetype Manager dialog box, click Show Details. AutoCAD expands the dialog box, as shown in Figure 2-11.

3. In the Details area, enter the global and current object scales. The Global Scale Factor changes the scale factor of all new and existing lines in the drawing. The Current Object Scale affects only subsequently drawn objects.

4. Click OK.

# Setting the Current Lineweight

You can also assign lineweight on a per-layer or per-object basis and then see the actual lineweight as you work on your drawing. Like linetypes, the use of different lineweights helps to convey information about the meaning of objects in a drawing. For example, you can use a heavy lineweight to show the outline of a cross-section and use a thin lineweight for the cross-hatching within that cross-section.

**Note**    *Lineweight is created as a solid fill in the object's assigned color, and affects the appearance of all AutoCAD objects except TrueType fonts, raster images, points, and 2-D solids and solid fills. Although polylines can be drawn using lineweights, the actual polyline width generally overrides its lineweight.*

AutoCAD has 23 valid lineweight values, ranging from 0.05 to 2.11 millimeters (0.002 to 0.083 inch), plus values of BYLAYER, BYBLOCK, DEFAULT, and 0. These values are shown in Table 2-1. A value of 0 is always displayed as one pixel in model space and plots with the thinnest line possible on your printer. The DEFAULT lineweight is initially set to a value of 0.25 millimeter (0.01 inch), although that value can be changed to one of the other valid lineweight values. Any value equal to or less than the default value displays as one pixel in model space, but plots using the assigned weight.

| Millimeter(s) | Inch | Point(s) | Pen Size | ISO | DIN | JIS | ANSI |
|---|---|---|---|---|---|---|---|
| 0.05 | 0.002 | | | | | | |
| 0.09 | 0.003 | 1/4 | | | | | |
| 0.13 | 0.005 | | | | ✓ | | |
| 0.15 | 0.006 | | | | | | |
| 0.18 | 0.007 | 1/2 | 0000 | ✓ | ✓ | ✓ | |
| 0.20 | 0.008 | | | | | | |
| 0.25 | 0.010 | 3/4 | 000 | ✓ | ✓ | ✓ | |
| 0.30 | 0.012 | | 00 | | | | 2H or H |
| 0.35 | 0.014 | 1 | 0 | ✓ | ✓ | ✓ | |
| 0.40 | 0.016 | | | | | | |
| 0.50 | 0.020 | | 1 | ✓ | ✓ | ✓ | |
| 0.53 | 0.021 | 1-1/2 | | | | | |
| 0.60 | 0.024 | | 2 | | | | H, F, or B |
| 0.70 | 0.028 | 2-1/4 | 2-1/2 | ✓ | ✓ | ✓ | |
| 0.80 | 0.031 | | 3 | | | | |
| 0.90 | 0.035 | | | | | | |
| 1.00 | 0.039 | | 3-1/2 | ✓ | ✓ | ✓ | |
| 1.06 | 0.42 | 3 | | | | | |
| 1.20 | 0.47 | | 4 | | | | |
| 1.40 | 0.056 | | | ✓ | ✓ | ✓ | |
| 1.58 | 0.062 | 4-1/4 | | | | | |
| 2.0 | 0.078 | | | ✓ | ✓ | | |
| 2.11 | 0.083 | 6 | | | | | |

**Table 2-1.** *Valid Lineweight Values*

When you create an object, it is created using the current lineweight. By default, the current lineweight is BYLAYER, meaning that an object's actual lineweight is determined by the lineweight assigned to the layer on which it is drawn. With a BYLAYER setting, if you change the lineweight assigned to the layer, all the objects created on that layer change to reflect the new lineweight. By default, layers use the DEFAULT lineweight.

You can also select a specific lineweight as the current lineweight. Doing so overrides the layer's lineweight setting. Objects are then created using that lineweight, and changing the layer lineweight has no effect on them.

As with linetypes, you can also use the special lineweight setting BYBLOCK. When BYBLOCK is selected, all objects are initially drawn using the DEFAULT lineweight. Once the objects have been grouped into a block, however, they will inherit the lineweight setting of the current layer when you subsequently insert the block into a drawing.

To set the current lineweight, on the Drawing Properties toolbar, choose the current lineweight from the Lineweight Control drop-down list.

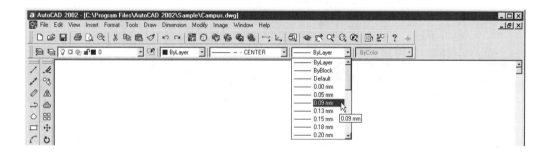

When you first draw objects using a wide lineweight, you may not actually see the lineweight in your drawing. By default, AutoCAD is configured to *not* display lineweights, because the display of lines using more than one pixel slows AutoCAD's performance. To actually see various lineweights, you must explicitly turn on the display of lineweight. You can easily turn lineweight display on and off by doing one of the following:

■ Toggle the LWT button on the status bar.

■ At the command line, type **LWDISPLAY** and press ENTER, and then change the system variable value to 0 or 1 to toggle lineweight display off or on.

You can also control lineweight display, as well as control other lineweight settings, by using the Lineweight Settings dialog box, shown in Figure 2-13. To display this dialog box, do one of the following:

■ From the Format menu, choose Lineweight.

■ At the command line, type **LWEIGHT** and press ENTER.

■ Right-click the LWT button on the status bar and then select Settings from the shortcut menu.

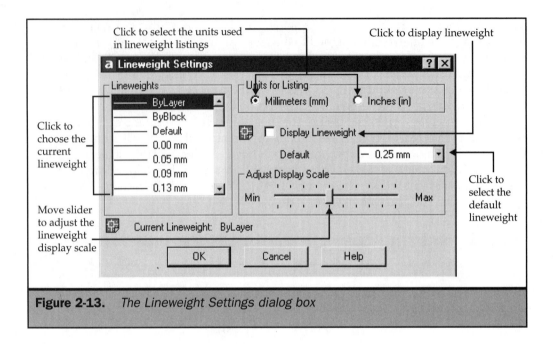

**Figure 2-13.** *The Lineweight Settings dialog box*

To turn on the display of lineweight, select the Display Lineweight check box. Within the Lineweight Settings dialog box, you can also choose the current lineweight, control whether AutoCAD uses millimeters or inches when listing lineweights, set the default lineweight, and adjust the lineweight display scale.

**Note** *The lineweight display scale determines how AutoCAD displays lineweights in model space. In Paper Space, AutoCAD always displays the actual lineweight; thus, the lineweight changes as you change the zoom scale factor. In model space, the lineweight display doesn't change as you zoom in and out. Instead, AutoCAD adjusts the display of lineweight based on the lineweight display scale setting. By default, all lineweights less than or equal to the default lineweight value are displayed at one pixel. If you move the Adjust Display Scale slider to the right, lineweights in model space are displayed at a larger scale. If you move the slider to the left, lineweights are displayed at a smaller scale. Moving the slider all the way to the left turns off lineweight display altogether. As you move the slider, the Lineweights list shows how the various lineweights will display at the current display scale. You'll learn more about working with lineweights in Chapter 8. In Chapters 17 and 18, you'll learn about paper space and how to plot with lineweights.*

## Setting the Drawing Units

When you draw using AutoCAD, you typically draw at full-size (1:1 scale), and then set a scale factor when you print or plot your drawing. Before you begin drawing, however, you need to determine the relationship between drawing units and real-world units.

For example, you can decide whether one linear drawing unit represents an inch, a foot, a meter, or a mile. In addition, you can specify the way the program measures angles. For both linear and angular units, you can also set the degree of display precision, such as the number of decimal places or the smallest denominator used when displaying fractions. The precision settings affect only the display of distances, angles, and coordinates. AutoCAD always stores distances, angles, and coordinates using floating-point accuracy.

When you start a new drawing by using a Setup wizard, some or all of these drawing units are set up for you based on the selections you make. If you start a new drawing from scratch, or decide later that you need to change the drawing units, you can easily change any of the units settings.

To set the units type and precision, follow these steps:

1. Do either of the following:

   ■ From the Format menu, choose Units.

   ■ At the command line, type **UNITS** (or **UN**) and then press ENTER.

   AutoCAD displays the Drawing Units dialog box (see Figure 2-14).

2. In the Drawing Units dialog box, under Length, select a unit type and precision for linear measurements. The Precision drop-down list shows selections based on the type of units selected.

3. Under Angle, select an angle type and precision for angular measurements. The Precision drop-down list shows selections based on the type of angular units selected.

4. Select the Clockwise check box if you want positive angles to increase in a clockwise direction. (Note that, by default, positive angles increase counterclockwise, and this check box is not selected.)

5. To specify the direction of the zero angle, click Direction. AutoCAD displays the Direction Control dialog box (see Figure 2-15). By default, the zero angle is at the "three o'clock" or East position.

6. Select or specify the direction of the zero angle.

7. Click OK to close each dialog box.

**Note**    *The drop-down list in the Drawing Units For DesignCenter Blocks area of the Drawing Units dialog box controls the unit of measure used when inserting blocks using AutoCAD DesignCenter. You'll learn more about this component of AutoCAD in Chapter 16.*

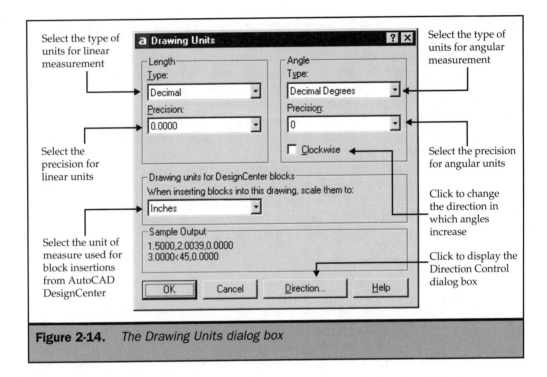

Select the type of units for linear measurement

Select the type of units for angular measurement

Select the precision for linear units

Select the precision for angular units

Click to change the direction in which angles increase

Select the unit of measure used for block insertions from AutoCAD DesignCenter

Click to display the Direction Control dialog box

**Figure 2-14.**    *The Drawing Units dialog box*

## Understanding Scale Factors

Although it's a good idea to keep your scale factor in mind when setting up a drawing, you don't need to set the scale until you print the drawing. For example, when you draw a mechanical part that is 40 inches in length, you actually draw it 40 inches in

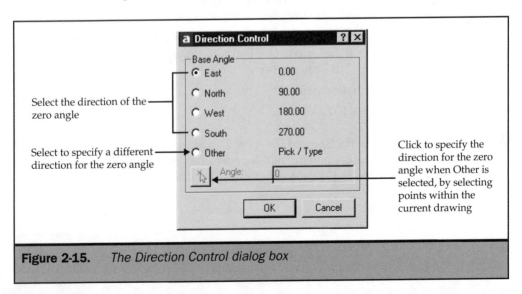

Select the direction of the zero angle

Select to specify a different direction for the zero angle

Click to specify the direction for the zero angle when Other is selected, by selecting points within the current drawing

**Figure 2-15.**    *The Direction Control dialog box*

AutoCAD, rather than applying a scale factor as you draw. When you print your drawing, you can assign the scale at which you want the drawing to print.

Scale, however, does affect the way a few elements, such as text, arrows, or linetypes, appear when you print or plot your drawing. For these elements, you can make adjustments when you first set up your drawing so that they print or plot at the correct size. For example, when you draw text, you need to determine the text size so that when you plot it later at a particular scale, the text height is correct.

After you determine the eventual scale of your finished drawing, you can calculate the scale factor for the drawing as a ratio of one drawing unit to the actual scale unit represented by each drawing unit. For example, if you plan to print your drawing at 1/8"=1'-0", your scale factor ratio is 1:96 (1/8"=12" is the same as 1=96). If you want your printed scale to be 1"=100', your scale factor ratio is 1:1200. You can then use this scale factor to calculate the size of objects, such as text, in the drawing.

Table 2-2 shows some standard architectural and engineering scale ratios and equivalent text heights required to create text that measures 1/8-inch high when you plot the drawing at the specified scale.

| Scale | Scale Factor | Text Height |
|---|---|---|
| 1/16"=1'-0" | 192 | 24" |
| 1/8"=1'-0" | 96 | 12" |
| 3/16"=1'-0" | 64 | 8" |
| 1/4"=1'-0" | 48 | 6" |
| 3/8"=1'-0" | 32 | 4" |
| 1/2"=1'-0" | 24 | 3" |
| 3/4"=1'-0" | 16 | 2" |
| 1"=1'-0" | 12 | 1.5" |
| 1 1/2"=1'-0" | 8 | 1" |
| 3"=1'-0" | 4 | 0.5" |
| 1"=10' | 120 | 15" |
| 1"=20' | 240 | 30" |
| 1"=30' | 360 | 45" |
| 1"=40' | 480 | 60" |
| 1"=50' | 600 | 75" |
| 1"=60' | 720 | 90" |
| 1"=100' | 1200 | 150" |

**Table 2-2.**   *Standard Scale Ratios and Equivalent Text Heights*

## Setting the Drawing Limits

The *drawing limits* represent an invisible boundary around your drawing. You can use drawing limits settings to make sure that you do not create a drawing larger than what can fit on a specific sheet of paper when it is printed at a specific scale.

You can use the scale factors shown in Table 2-2 to predetermine the size of your drawing, to make sure that it fits on a specific paper size when you print it. You control the size of your drawing by setting the drawing limits. To calculate the drawing limits to match the size of your paper, multiply the dimensions of your paper size by your scale factor.

For example, if the paper you use to print measures 36×24 inches, and you plot your drawing at 1/8"=1'-0" (in other words, using a scale factor of 96), the size of your drawing is measured in drawing units as 36×96 (or 3,456 units) wide and 24×96 (or 2,304 units) high. You therefore set the X,Y coordinate of the upper-right corner to 3456,2304.

*You might want to set the drawing limits slightly smaller than the actual paper size, to provide a margin around the edges of the printed image. If you use predrawn templates, the margin is already calculated.*

Keep in mind that you can print the finished drawing at any scale, regardless of the scale factor you calculate. You can also print on paper of a different size or use Layouts to create different views of your drawing, with different positions and scales for each of those views. The scale factor is not related to the size of the objects you draw; it simply provides a preliminary guide to help you establish the text height and drawing limits when you begin your drawing. You can change the text height, drawing limits, and any other scale-related factor (such as linetype scale) at any time.

To set the drawing limits, follow these steps:

1. Do one of the following:

   ■ From the Format menu, choose Drawing Limits

   ■ At the command line, type **LIMITS** and then press ENTER

   AutoCAD prompts:

   ```
   Specify lower left corner or [ON/OFF] <0.0000,0.0000>:
   ```

2. Specify the X,Y coordinate of the lower-left corner of the limits of your drawing by either entering a coordinate value or selecting a point in the drawing; or, press ENTER to use the default (0,0). AutoCAD prompts for the upper-right corner.

3. Specify the X,Y coordinate of the upper-right corner of the limits of your drawing by either entering a coordinate value or selecting a point in the drawing corresponding to the upper-right corner. For example, to set the upper-right corner for a 36×24-inch drawing that will later be printed at 1/8"=1'-0", type **3456,2304**.

**Note**    *Entering the coordinates for the lower-left and upper-right limits only sets their values. To configure AutoCAD so that it prevents you from drawing outside the limits, you must turn on the limits. To do so, start the LIMITS command again and then type **ON** in response to the first prompt.*

# Using Grid and Snap

AutoCAD's grid and snap settings are effective tools to use in your drawing to ensure accuracy. The *grid* is a pattern of dots that appears onscreen but that does not print. *Snap* restricts the movement of the crosshairs to predetermined intervals. Although you can set the spacing of grid and snap intervals independently, their values are often related. For example, you might set the snap spacing to 1-inch increments, but display the grid at 12-inch intervals.

## Setting a Reference Grid

AutoCAD's reference grid consists of a regular pattern of dots. Using this grid is similar to drawing on a sheet of nonreproducing grid paper. Although the reference grid is visible on the screen, it does not print as part of the drawing, nor does it affect where you can draw.

The reference grid extends only to the limits of the drawing, helping you to visualize the boundary of your drawing, align objects, and visualize distances between objects. You can turn the grid on and off, as needed. You can also change the spacing of the grid at any time.

To turn on the grid and set the grid spacing, follow these steps:

1. Do one of the following:

   ■ From the Tools menu, choose Drafting Settings.

   ■ At the command line, type **DSETTINGS** (or **DS**, **RM**, **SE**, or **DDRMODES**) and then press ENTER.

   ■ Right-click the SNAP or GRID button on the status bar, and then select Settings from the shortcut menu.

   If not already displayed, select the Snap and Grid tab of the Drafting Settings dialog box (see Figure 2-16).

2. Select the Grid On check box to display the grid.

3. In the Grid X Spacing box, enter the horizontal distance between grid dots.

4. To use the same spacing for vertical and horizontal grid dots, press TAB. Otherwise, in the Grid Y Spacing box, enter the vertical distance between grid dots.

5. Click OK.

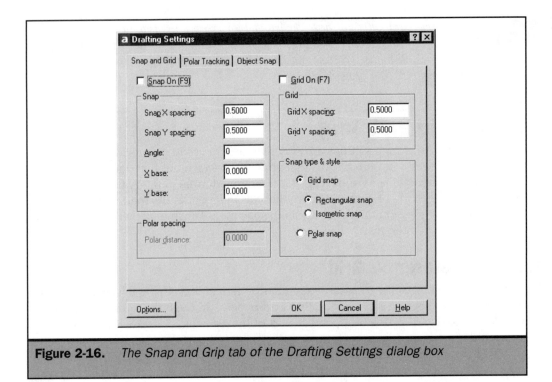

**Figure 2-16.** *The Snap and Grip tab of the Drafting Settings dialog box*

 *You can also turn the grid on and off by either clicking the GRID button on the status bar, pressing* CTRL-G *or* F7, *or typing the GRID command. Setting the grid spacing to 0,0 automatically matches the grid spacing to that of the snap spacing.*

## Setting Snap Spacing

Another way to ensure drawing accuracy is to use AutoCAD's snap feature. When snap is turned on, the crosshairs "snap" to a predetermined snap interval. Although matching the grid spacing to the snap spacing is often helpful, the two settings do not have to match.

To turn on snap and set the snap spacing, follow these steps:

1. Do one of the following:

   ■ From the Tools menu, choose Drafting Settings.

   ■ At the command line, type **DSETTINGS** (or **DS**, **RM**, **SE**, or **DDRMODES**) and then press ENTER.

   ■ Right-click the SNAP or GRID button on the status bar and then select Settings from the shortcut menu.

If not already displayed, select the Snap and Grid tab of the Drafting Settings dialog box (refer to Figure 2-16).

2. Select the Snap On check box to turn on snap.

3. In the Snap X Spacing box, enter the horizontal distance between snap points.

4. To use the same spacing for vertical and horizontal snap points, press TAB. Otherwise, in the Snap Y Spacing box, enter the vertical distance between snap points.

5. Click OK.

*You can also turn snap on and off by clicking the SNAP button on the status bar, pressing CTRL-B or F9, or typing the SNAP command.*

# Changing the Snap Angle and Base Point

The snap and grid normally are both based on the *drawing origin,* the 0,0 coordinate in the World Coordinate System. You can relocate the snap and grid origin, however, to help you draw objects in relation to a different location. You can also rotate the grid to a different angle, to realign the crosshairs to the new grid angle.

To change the snap angle and base point, follow these steps:

1. Display the Snap and Grid tab of the Drafting Settings dialog box.

2. Select the Snap On check box to turn on snap.

3. Under the Snap area, type the desired snap rotation angle in the Angle box.

4. In the X Base and Y Base boxes, type the X and Y coordinates, respectively, of the new base point.

5. Click OK.

*Although the grid spacing does not necessarily match the snap interval, it will always match the snap angle.*

# Using Isometric Snap and Grid

You can use the isometric snap and grid options to create 2-D isometric drawings. With the isometric option, you are simply drawing a simulated 3-D view on a 2-D plane, much the same as you might draw on a piece of paper. Do not confuse isometric drawings with 3-D drawings. You create 3-D drawings in 3-D space.

The isometric option always uses three preset planes, which are denoted as the left, right, and top planes. You cannot alter the arrangement of these planes. If the snap angle is 0, the three isometric axes are 30 degrees, 90 degrees, and 150 degrees.

*You can cycle through the isometric planes by pressing* CTRL-E *or* F5. *As you change planes, the crosshairs change to indicate the current isometric plane: left (90- and 150-degree axes), right (90- and 30-degree axes), and top (30- and 150-degree axes).*

Left          Right          Top

When you turn on the isometric snap and grid options and select an isometric plane, the snap intervals, grid, and crosshairs align with the current plane. The grid is always shown as isometric and uses Y coordinates to calculate the grid spacing.

To turn on the isometric snap and grid options, follow these steps:

1. Display the Snap and Grid tab of the Drafting Settings dialog box.

2. Select the Grid On check box.

3. Under Snap Type & Style, select the Isometric Snap radio button.

4. Click OK.

**Note**  *If you also turn on AutoCAD's Ortho mode, AutoCAD restricts the drawing of objects to the current isometric plane.*

## Using Polar Snap

You can also use a special snap control called *polar snap*, which causes the cursor to snap to increments along polar alignment angles. The increment is specified on the Snap and Grid tab of the Drafting Settings dialog box, and the polar alignment angle increments are specified on the Polar Tracking tab. You'll learn more about polar tracking later in this chapter.

  *Polar snap has the advantage of snapping to increments relative to a starting point rather than to the snap spacing. You'll learn more about polar snap later.*

To set the polar snap distance and activate polar snap:

1. Display the Snap and Grid tab of the Drafting Settings dialog box.

2. Under Snap Type & Style, select the Polar Snap radio button.

3. Under Polar Spacing, type the polar snap distance in the Polar Distance box.

4. Click OK.

5. Click the POLAR button on the status bar.

**Note** *To use polar snap, you must turn on both snap and polar tracking. You can turn polar tracking on and off by clicking the POLAR button on the status bar, pressing F10, or selecting the Polar Tracking check box on the Polar Tracking tab of the Drafting Settings dialog box. When polar tracking is on, AutoCAD displays an alignment path and ToolTip display of the distance and angle whenever you move the cursor so that it crosses a polar alignment angle, even if snap is currently turned off. When both snap and polar tracking are on, the cursor snaps to polar distance increments along the polar alignment angle.*

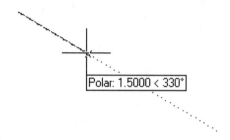

# Using Ortho Mode

Sometimes, it helps to restrict cursor movement to the current horizontal and vertical axes, so that you can draw at right angles, or *orthogonally*. For example, with the default 0-degree orientation (angle 0 at the "three o'clock" or East position), when Ortho mode is enabled, lines are restricted to 0, 90, 180, or 270 degrees. As you draw lines, the rubber-band line follows either the horizontal or vertical axis, depending on the direction in which you move the cursor. When you enable the isometric snap and grid, cursor movement is restricted to orthogonal equivalents within the current isometric plane.

To turn Ortho mode on or off, do one of the following:

- Click the ORTHO button on the status bar.
- Press CTRL-L or F8.
- Type the **ORTHO** command.

**Note** *AutoCAD ignores the Ortho setting when you type coordinates at the command line or use object snaps.*

# Using Object Snaps

*Object snaps* enable you to quickly select exact geometric points on existing objects without having to know the exact coordinates of those points. With object snaps, you can select the end point of a line or arc, the center point of a circle, the intersection of any two objects, or any other geometrically significant position. You can also use object snaps to draw objects that are tangent or perpendicular to an existing object.

You can use object snaps any time AutoCAD prompts you to specify a point—for example, if you are drawing a line or other object. You can work with object snaps in one of two ways:

- Enable a running object snap that remains in effect until you turn it off.

- Enable a one-time object snap for a single selection by choosing an object snap when another command is active. A one-time object snap can also be used to override a running object snap.

When using object snaps, AutoCAD recognizes only visible objects or visible portions of objects. You cannot snap to objects on layers that have been turned off, or to the blank portions of dashed lines.

You can set object snaps by using a number of different methods. The method you use depends on whether you are setting a running object snap or a one-time object snap. For example, you can quickly set a one-time object snap by selecting the object snap either from the Object Snap flyout on the Standard toolbar or from the Object Snap toolbar (see Figure 2-17). The actual methods are described later in this chapter.

When you specify one or more object snaps and then select an object, AutoCAD snaps to the object snap point that is closest to the intersection of the crosshairs. You can also display a target box, called the AutoSnap Aperture box, which is added to the crosshairs when an object snap is active. AutoCAD can then snap to any object snap point that falls within this aperture. If AutoSnap is also enabled, AutoCAD displays

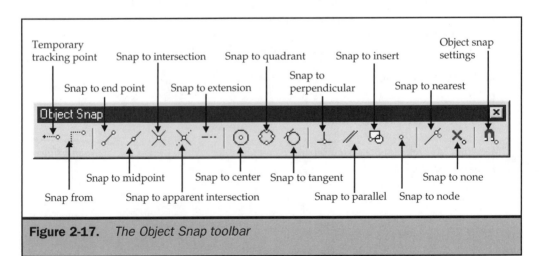

**Figure 2-17.** *The Object Snap toolbar*

an AutoSnap marker when you move the crosshairs over an object that geometrically matches the currently active object snap. You can control the presence and size of this aperture box and turn the AutoSnap marker on and off from the Drafting tab of the Options dialog box.

## Endpoint

The *endpoint object snap* finds the end point of an object. You can snap to the closest end point of an arc, elliptical arc, mline, line, polyline segment, or ray, or to the closest corner of a trace, solid, or 3-D face. If an object has thickness, the endpoint object snap also snaps to the end points of the edges of objects and to the end points of the edges of 3-D solids and regions. As shown in Figure 2-18, to snap to the end point of an object, click anywhere on the object near its end point.

## Midpoint

The *midpoint object snap* finds the middle point of an object. You can snap to the midpoint of an arc, elliptical arc, mline segment, line, polyline segment, solid, spline, or xline. In the case of xlines, the midpoint snaps to the first defined point. For splines and elliptical arcs, the midpoint snaps to a point on the object midway between the start point and the end point. If an object has thickness, the midpoint object snap will snap to the midpoints of any edge of the object, and to the midpoints of any edge of 3-D solids and regions. As shown in Figure 2-19, to snap to the midpoint of an object, pick anywhere on the object.

## Intersection

The *intersection object snap* finds the intersection of any combination of objects. You can snap to the intersection of arcs, circles, ellipses, elliptical arcs, mlines, lines, polylines, rays, splines, xlines, and any combination of these objects. The intersection object snap also snaps to the intersection of the edges of regions and curves, but it doesn't snap

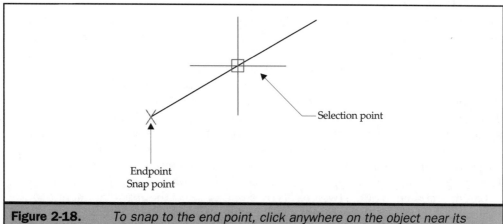

Selection point

Endpoint
Snap point

**Figure 2-18.** *To snap to the end point, click anywhere on the object near its end point.*

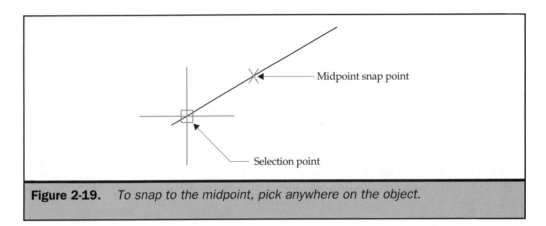

**Figure 2-19.** *To snap to the midpoint, pick anywhere on the object.*

to the edges or corners of 3-D solids. You can snap to the corners of objects that have thickness. If you have thickened two objects in the same direction and they have intersecting bases, you can snap to the intersection of their edges. If the thickness of the two objects isn't the same, however, the lesser thickness defines the intersection point.

You can snap to the intersections of arcs and circles that are part of a block if you have uniformly scaled the block.

The *extended intersection mode* snaps to the imaginary intersection of two objects that would intersect if the objects were extended along their natural paths. To use the extended intersection mode, you must explicitly select a one-time intersection object snap and then click one of the objects. AutoCAD then prompts you to select a second object. As soon as you click the second object, the program snaps to the imaginary intersection formed by extending those objects. As shown in Figure 2-20, to snap to the intersection of two objects, either pick near the intersection of the two objects or explicitly select the intersection object snap, select one object, and then select the other object.

## Apparent Intersection

The *apparent intersection object snap* finds the apparent intersection of any combination of two objects that don't actually intersect in 3-D space but appear to intersect onscreen. You can snap to the apparent intersection of two objects consisting of arcs, circles, ellipses, elliptical arcs, mlines, lines, polylines, rays, splines, or xlines. The *extended apparent intersection object snap mode* snaps to the imaginary intersection of two objects that would appear to intersect if the objects were extended along their natural paths. To use the extended apparent intersection mode, you must explicitly select a one-time apparent intersection object snap and then click one of the objects. AutoCAD then prompts you to select a second object. As soon as you click the second object, the program snaps to the apparent intersection formed by extending those objects. As shown in Figure 2-21, to snap to the apparent intersection of two objects, either pick near the apparent intersection of the two objects or explicitly select the apparent intersection object snap, select one object, and then select the other object.

AUTOCAD BASICS

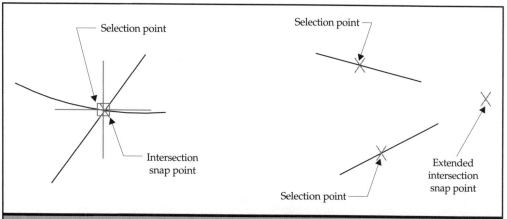

**Figure 2-20.** *To snap to the intersection of two objects, pick near the intersection. To select the point at which two objects would intersect if extended, select one object and then the other.*

## Extension

The *extension object snap* can be used to find a point along the natural extension of a line or arc. To use the extension object snap, pause the cursor over the end of a line or arc. AutoCAD adds a small plus sign (+) at the cursor location to indicate that the line or arc has been selected for extension. As you then move the cursor along the natural extension path of the line or arc, AutoCAD displays the temporary extension path. This is referred to as *object snap tracking,* which is a specific form of *AutoTrack.* You'll learn more about AutoTrack and object snap tracking later in this chapter.

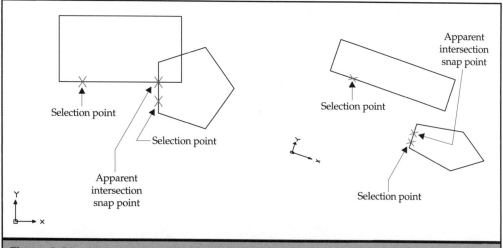

**Figure 2-21.** *To snap to the apparent intersection of two objects, pick near their apparent intersection. To select the points at which the two objects would appear to intersect if extended, select one object and then the other.*

You can use the extension object snap in conjunction with other object snaps, such as the intersection or apparent intersection object snap, to find the extended intersection of a line or arc with another object, or you can select more than one line or arc for extension to locate the extended intersection of the two objects (see Figure 2-22).

## Center

The *center object snap* finds the center point of curved objects. You can snap to the center of an arc, circle, ellipse, elliptical arc, or polyline arc segment. To snap to the center, you must select a point on the visible part of the object, as shown in Figure 2-23.

## Quadrant

The *quadrant object snap* finds the quadrant points of curved objects. You can snap to the closest quadrant (the 0-, 90-, 180-, and 270-degree points) of an arc, circle, ellipse, elliptical arc, or polyline arc segment. As shown in Figure 2-24, to snap to a quadrant, pick the object near the desired quadrant point.

## Tangent

The *tangent object snap* finds tangent points on objects. You can snap to the point on an arc, circle, ellipse, elliptical arc, or polyline arc segment that, when connected to the previous point, forms a line tangent to that object. The *deferred tangent snap mode* is automatically enabled when you select an arc, circle, or polyline arc as the start point for a tangent line. If AutoSnap is enabled, a Snaptip and marker are displayed when you pass the cursor over a deferred tangent snap point. You'll learn more about AutoSnap and Snaptips later in this chapter. The deferred tangent snap mode doesn't

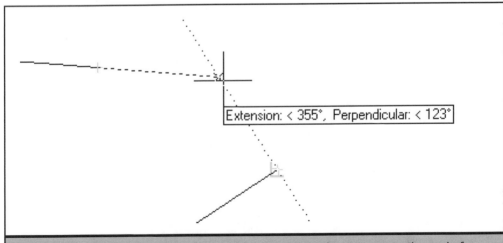

**Figure 2-22.**  *To use the extension object snap, pause the cursor over the end of a line or arc, and then move the cursor along the temporary extension path.*

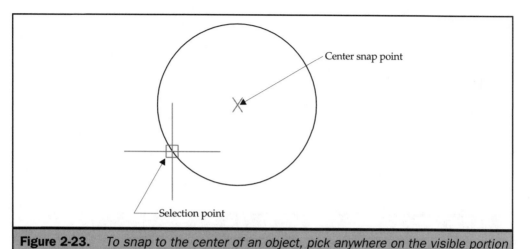

**Figure 2-23.** *To snap to the center of an object, pick anywhere on the visible portion of the object.*

work with ellipses or splines. As shown in Figure 2-25, to snap to a tangent pick the object near the tangent point.

## Perpendicular

The *perpendicular object snap* finds points that form a perpendicular condition with another object. You can snap to the point on an arc, circle, ellipse, elliptical arc, mline, line, polyline, ray, solid, spline, or xline that forms a perpendicular alignment with another object or with an imaginary extension of that object. The *deferred perpendicular snap mode* is automatically enabled when you use an arc, circle, mline, line, polyline, xline, or 3-D solid edge as the first snap point from which to draw a perpendicular line.

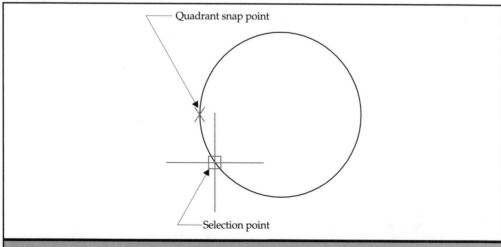

**Figure 2-24.** *To snap to a quadrant, pick the object near the quadrant point.*

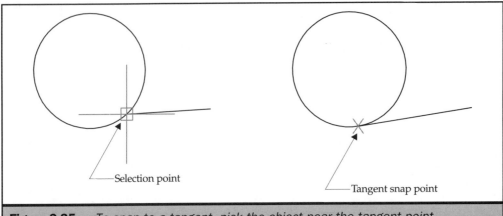

**Figure 2-25.**    *To snap to a tangent, pick the object near the tangent point.*

If AutoSnap is enabled, a Snaptip and marker are displayed when the cursor passes over a deferred perpendicular snap point. As shown in Figure 2-26, to snap to a point perpendicular to an object, pick anywhere on the object.

## Parallel

The *parallel object snap* can be used to create a straight-line segment (such as a line or polyline) parallel to an existing straight-line segment. To use the parallel object snap, begin drawing a straight-line segment, and then pause the cursor over an existing straight-line segment. As you move the cursor so that the rubber-band line extending from the previous point is approximately parallel to the existing straight-line segment, AutoCAD adds a parallel-line symbol over the existing segment and displays the temporary parallel-line path. You can then specify the end point of the straight-line segment anywhere along the temporary line.

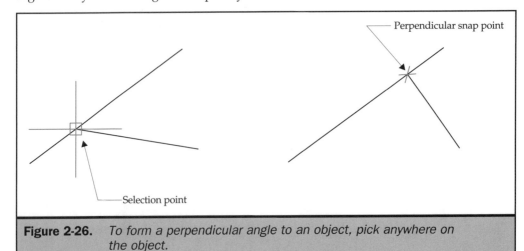

**Figure 2-26.**    *To form a perpendicular angle to an object, pick anywhere on the object.*

You can use the parallel object snap in conjunction with other object snaps, such as intersection, apparent intersection, or extension, to find the point where the parallel line meets another object or the extension of an object (see Figure 2-27).

## Insertion

The *insertion object snap* finds the insertion point of an attribute, attribute definition, block, shape, or text object. As shown in Figure 2-28, to snap to the insertion point of an object, pick anywhere on the object.

 *The insertion object snap sometimes appears as Insert in the ToolTips and cursor menu.*

## Node

The *node object snap* finds and snaps to a point object.

 *The node object snap is particularly useful for snapping to point objects inserted by the DIVIDE and MEASURE commands. You'll learn about these commands in Chapter 9.*

## Nearest

The *nearest object snap* finds the nearest point on another object that is visually closest to the cursor. You can snap to the nearest point on an arc, circle, ellipse, elliptical arc, mline, line, point, polyline, ray, spline, or xline.

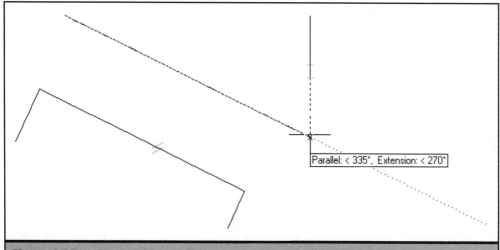

**Figure 2-27.** *To use the parallel object snap, start a straight-line segment, pause the cursor over the existing straight-line segment to which you want it drawn perpendicular, and then move the cursor along the temporary parallel-line path.*

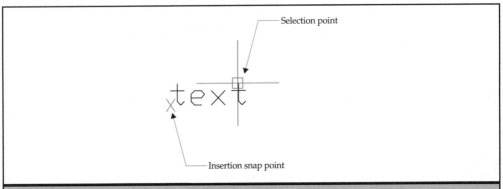

**Figure 2-28.**   *To snap to the insertion point of an object, pick anywhere on the object.*

## None

The *none object snap* instructs AutoCAD not to use any object snap modes.

*Use the none object snap as a one-time object snap to temporarily turn off running object snaps for a single selection. After making the selection, the running object snaps become active again. Don't confuse this snap mode with the Clear All button in the Osnap Settings dialog box, which turns off all the running object snaps.*

## Setting a One-Time Object Snap

You can enable a one-time object snap for a single selection by choosing an object snap when another command is active. For example, while drawing lines, if you want to snap to the midpoint of an existing line, you can activate the midpoint object snap. The one-time object snap is active only for the current selection. Once you select a point in the drawing, the object snap is turned off.

*If you enable a one-time object snap, and AutoCAD cannot locate a geometric point matching the object snap condition, the program displays a message indicating that no point was found. Since the one-time object snap mode is no longer active, you'll need to reselect the object snap mode.*

You can set a one-time object snap by using any of the following methods:

- On the Standard toolbar, select the Object Snap flyout and then choose the object snap buttons.
- On the Object Snap toolbar, click one of the object snap buttons.

■ At the command line, when you are prompted to select a point or an object, type the object snap mode and then press ENTER.

■ Hold down the SHIFT key and right-click to display the object snap shortcut menu; then, select the object snap you want to use.

*When typing object snap modes at the command line, you can combine any two or more modes by entering the modes, separated by commas. For example, you could have AutoCAD locate the end point or intersection of two objects by typing **ENDP,INT**. In this case, AutoCAD scans the entities in the aperture box and locks onto the end point or intersection of two entities, depending on which condition it finds closest to the crosshairs. When you type the name of the object snap mode at the command line, you need to type only the first three letters.*

## Setting a Running Object Snap

When you enable a running object snap, the snap modes you select remain in effect until you turn them off. You can turn on running object snaps, and change the size of the object snap aperture, from the Object Snap tab of the Drafting Settings dialog box (see Figure 2-29) by using any of the following methods:

■ On either the Object Snap flyout or the Object Snap toolbar, click the Object Snap Settings button.

■ From the Tools menu, choose Drafting Settings, and then choose the Object Snap tab.

■ At the command line, type **OSNAP** and then press ENTER.

■ Right-click the OSNAP button on the status bar and then select Settings from the shortcut menu, or use SHIFT-right-click to display the object snap shortcut menu and choose Osnap Settings.

*When one or more running object snaps have been set, you can quickly turn them all off or on by either clicking the OSNAP button on the status bar, pressing CTRL-F, or pressing F3. If no object snap mode is currently active, you can also display the Object Snap tab of the Drafting Settings dialog box by clicking the OSNAP button on the status bar. Doing this when an object snap mode is active, however, simply disables the running object snap.*

## Controlling AutoSnap

AutoSnap is a powerful object snap enhancement first introduced in AutoCAD Release 14. AutoSnap lets you visually preview possible snap points before actually making your selection. When enabled, AutoSnap is triggered automatically whenever you use

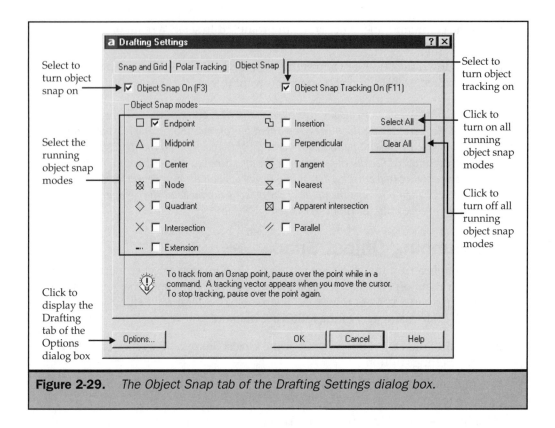

**Figure 2-29.** *The Object Snap tab of the Drafting Settings dialog box.*

either a one-time or running object snap mode. AutoSnap displays a special marker and an AutoSnap ToolTip whenever you move the cursor over an object snap point.

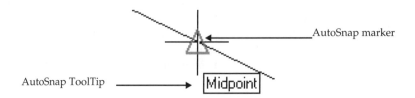

An additional Magnet option attracts the cursor to the object snap point as if it were magnetic.

You can control AutoSnap options from the Drafting tab of the Options dialog box (see Figure 2-30). You can display this dialog box by doing any of the following:

■ From the Tools menu, choose Options.

■ At the command line, type **OPTIONS** and then press ENTER.

■ Right-click in the command window, or right-click in the drawing area when no commands are active, and then choose Options from the shortcut menu.

■ Click the Options button in the Drafting Settings dialog box.

**Note** *You can't display the Options dialog box while another command is active.*

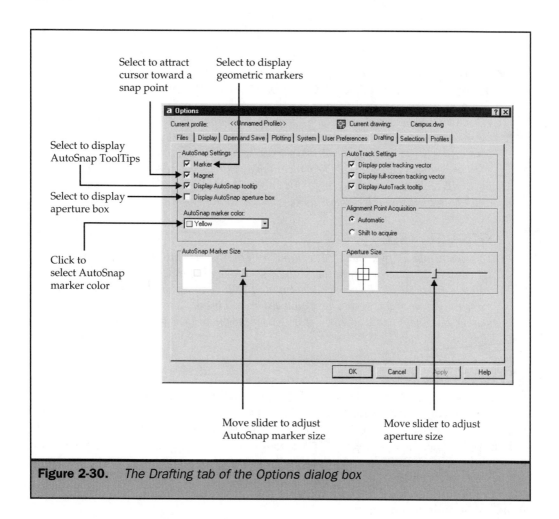

**Figure 2-30.** *The Drafting tab of the Options dialog box*

The AutoSnap area of this dialog box contains six controls:

- **Marker** Displays a geometric symbol indicating the type of object snap at a given location as you move the cursor over the snap points of an object.

- **Magnet** Causes the cursor to automatically lock onto a snap point when you move the cursor close enough to the object.

- **Display AutoSnap ToolTip** Displays a small text box containing the name of the snap mode when you move the cursor over a snap point.

- **Display AutoSnap Aperture Box** Displays an aperture box at the center of the crosshairs whenever an object snap mode is active.

- **AutoSnap Marker Color** Determines the AutoSnap marker color by selecting one of the first seven standard colors from the drop-down list.

- **AutoSnap Marker Size** Adjusts the size of the AutoSnap marker. As you move the slider, the adjacent display changes to indicate the new marker size.

*When working on complex drawings, instead of zooming in to ensure that you select the correct point, you can use AutoSnap to cycle through all the viable points matching the currently active object snap modes. To use the cycling feature, activate one or more object snap modes, begin an AutoCAD command that requires you to select a point (such as the LINE command), and then move the cursor over an object. You can then press the TAB key to cycle through the possible object snap locations.*

*LEARN BY EXAMPLE*
*To see how to use AutoSnap to cycle through all the viable points matching the currently active object snap modes, open the AutoCAD drawing Figure 2-31 from the companion web site. Start the LINE command by typing **LINE** and then pressing ENTER. Then, move the cursor over a particularly crowded area of the drawing. An AutoSnap marker and Snaptip will appear. Press TAB repeatedly to cycle through the other viable object snaps.*

# Using AutoTrack

AutoTrack helps you draw objects at specific angles or in specific relationships to other objects. When you turn on AutoTrack, AutoCAD displays temporary alignment paths to help you create objects at precise positions and angles. There are actually two AutoTrack options: polar tracking and object snap tracking. You've already learned about some of the power of AutoTrack's polar tracking in the discussion of polar snap earlier in this chapter, and you've learned about object snap tracking in the explanation of the extension and parallel object snaps. The polar tracking and object snap tracking options can be turned on or off independently.

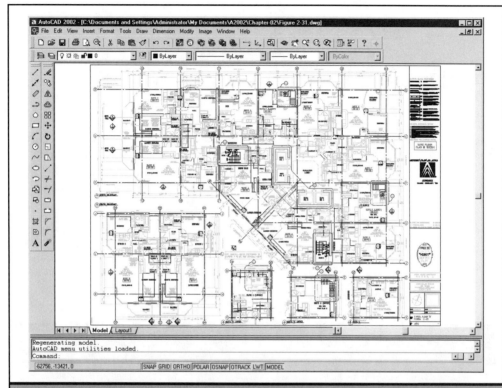

**Figure 2-31.**    *When working in a particularly crowded area of a drawing, press* TAB *repeatedly to cycle through the viable object snaps.*

Polar tracking tracks the cursor position along temporary alignment paths defined by polar angles relative to a point selected during a command. Object snap tracking works in conjunction with object snaps based on points acquired by pausing the cursor over specific geometry. The size of the AutoSnap aperture determines how close the cursor must be to the geometry or alignment path before AutoCAD acquires the geometry or displays the temporary alignment path.

## Using Polar Tracking

When you draw or edit objects, polar tracking helps you to select positions at specific distance and angle increments relative to the last point. You control the angle increments from the Polar Tracking tab of the Drafting Settings dialog box, shown in Figure 2-32, and control the distance increments from the Snap and Grid tab of that same dialog box (refer to Figure 2-16). When you use polar tracking, AutoCAD displays a temporary alignment path and a ToolTip displaying the distance and angle from the last point.

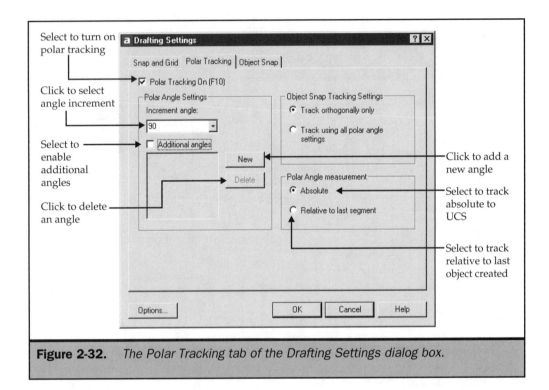

**Figure 2-32.**   *The Polar Tracking tab of the Drafting Settings dialog box.*

For example, if the angle increment is set to 45 degrees, when you start drawing a line and specify its starting point, AutoCAD displays a temporary alignment path whenever your cursor passes over a 45-degree increment. In addition, if the distance increment is set to 1 drawing unit and snap is turned on, as you move the cursor along a temporary alignment path, the cursor snaps to one-unit increments along the path.

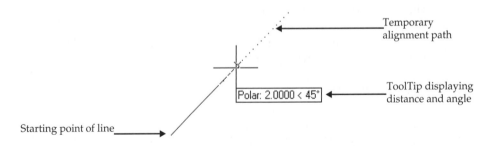

You can use polar tracking to track along any polar angle increment of 90, 45, 30, 22.5, 18, 15, 10, or 5 degrees. You can also specify additional angles and control whether AutoCAD measures polar angles based on the current User Coordinate System (UCS)

or relative to an existing line segment. For example, if the last line was drawn at a 66-degree angle, when you track relative to the line, if the increment angle is set to 45 degrees, you can snap to increments at 45-degree angles to that line. You'll learn more about coordinates and coordinate systems in Chapter 6.

To turn on polar tracking and set the polar tracking increment angle, follow these steps:

1. Do one of the following:

   ■ From the Tools menu, choose Drafting Settings.

   ■ At the command line, type **DSETTINGS** (or **DS**, **RM**, **SE**, or **DDRMODES**) and then press ENTER.

   ■ Right-click the POLAR button on the status bar and then select Settings from the shortcut menu.

   If not already displayed, select the Polar Tracking tab of the Drafting Settings dialog box (see Figure 2-32).

2. Select the Polar Tracking On check box to enable polar tracking.

3. Under Polar Angle Settings, select the angle increment from the Increment Angle drop-down list.

4. Click OK.

**Tip**    *You can also turn polar tracking on and off by either clicking the POLAR button on the status bar or pressing F10. You can't use Ortho mode and polar tracking at the same time. If you turn on polar tracking, AutoCAD turns off Ortho mode. If you turn Ortho mode back on, AutoCAD turns off polar tracking.*

The angle that you select from the Increment Angle drop-down list results in incremental angles as you move the cursor. For example, if you select 30 degrees, the temporary alignment path appears at every 30-degree increment (30, 60, 90 degrees, and so on). You can also add specific additional angles. For example, if you also want to track along a 55-degree angle, you can add that angle as an additional angle, by doing the following:

1. Display the Polar Tracking tab of the Drafting Settings dialog box.

2. Select the Additional Angles check box.

3. Click New and then type the additional angle.

4. Click OK.

You can add as many additional angles as you want. Note, however, that these are distinct angles. In other words, AutoCAD will only track along the specific additional angle, not at additional increments of that angle. For example, if you want to track along both a 55-degree and a 110-degree angle, you must specify both angles in the Additional Angles list.

AUTOCAD BASICS

**Note** *If you enter a negative angle, it is converted into a positive angle in the range from 0 to 360 degrees by adding 360 degrees to the negative angle.*

The angles you add remain in the list and are available whenever you select the Additional Angles check box. If you no longer want to track along a particular angle, you must select it in the list and then click the Delete button.

If you need to track along a specific angle only one time, it is generally more convenient to simply specify that angle as a polar angle override. You can do this at the command line by typing the angle preceded by a left-angle bracket (<). For example, to draw a line at a 55-degree angle, you could use the following command sequence:

```
Command: LINE
Specify first point: (select starting point)
Specify next point or [Undo]: <55
Angle Override: 55
Specify next point or [Undo]:
```

Notice that once you specify the angle override, the cursor is locked along the 55-degree angle. You can then specify the distance as any point along that angle. After you specify the next point, the angle override disappears and the cursor is free to move normally.

## Using Object Snap Tracking

When you draw or edit objects, object snap tracking helps you to select positions along alignment paths based on object snap points. For example, you can select a point along a path based on the end point or midpoint of an object.

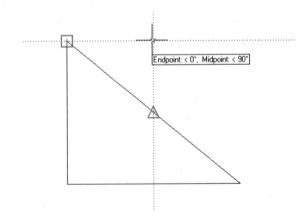

To use object snap tracking, you must first turn on object snap tracking and set one or more running object snaps. You can turn on object snap tracking from the Object

Snap tab of the Drafting Settings dialog box, by clicking the OTRACK button on the status bar or by pressing F11.

Once object snap tracking is enabled and one or more object snaps have been set, when a command prompts you to specify a point, move the cursor over the object point you want to track on, and briefly pause the cursor over that point. Don't click the point. As soon as AutoCAD acquires the point, a small plus sign (+) appears adjacent to the point. If AutoSnap markers are enabled, a marker also appears at the acquired point. Then, as you move away from the point, AutoCAD displays a temporary alignment path.

You can acquire more than one point and use them to specify the next point. For example, in the previous illustration, AutoCAD has located a point based on alignment paths from the end point of one corner of the triangle and the midpoint of the hypotenuse.

If you acquire a point that you don't want to use, simply move the cursor back over that point to clear the acquisition mark. AutoCAD also clears acquired points with each new command prompt and every time that you toggle object snap tracking on or off.

By default, AutoCAD acquires object snap tracking points automatically when you pause the cursor over a significant point. You can change this behavior, however, by changing the Alignment Point Acquisition setting on the Drafting tab of the Options dialog box (refer to Figure 2-30). If you select the Shift To Acquire radio button, you must press the SHIFT key while the cursor is over an object snap point to acquire that point.

**Tip**    *You can also use the controls under AutoTrack Settings to control other aspects of the AutoTrack behavior. For example, when the Display Polar Tracking check box is cleared, no polar tracking path is displayed. When the Display Full-Screen Tracking Vector check box is selected, the alignment path extends from the previous point to the cursor and beyond to the edge of the drawing window. When this check box is cleared, the alignment path ends at the cursor position. The Display AutoTrack ToolTip check box determines whether AutoCAD displays the ToolTip containing the alignment angle and distance information.*

You can also determine whether object snap tracking works in conjunction with the polar angle settings. By default, AutoCAD displays alignment paths only at orthogonal angles (0, 90, 180, and 270 degrees) when using object snap tracking. If you select the Track Using All Polar Angle Settings radio button under Object Snap Tracking Settings on the Polar Tracking tab of the Drafting Settings dialog box (refer to Figure 2-32), AutoCAD displays alignment paths at all the current polar angle settings.

### LEARN BY EXAMPLE

*To see how the power of AutoTracking can eliminate the need to create construction lines, complete the side view of the mechanical part shown in Figure 2-33. Open the AutoCAD drawing Figure 2-33 from the companion web site.*

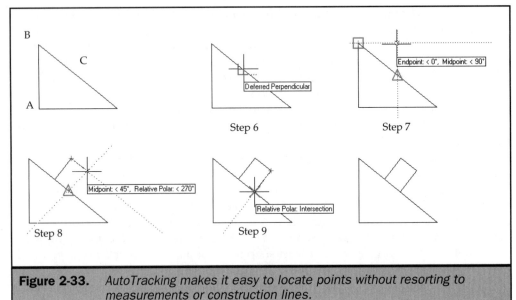

Step 6

Step 7

Step 8

Step 9

**Figure 2-33.** *AutoTracking makes it easy to locate points without resorting to measurements or construction lines.*

To draw the remainder of the mechanical part, use the following instructions:

1. Display the Drafting Settings dialog box.

2. On the Object Snap tab, select the Endpoint, Midpoint, and Intersection check boxes in the Object Snap Modes area. Turn on both object snap and object snap tracking by selecting both the Object Snap On and Object Snap Tracking On check boxes.

3. On the Polar Tracking tab, select an Increment Angle of 45 degrees. Under Object Snap Tracking Settings, select Track Using All Polar Angle Settings. Under Polar Angle Measurement, select Relative To Last Segment. Turn on polar tracking.

4. Click OK to close the Drafting Settings dialog box.

5. Start the LINE command. (You can select the command from the Draw toolbar, or type **LINE** and press ENTER.)

6. When AutoCAD prompts you to specify the first point, acquire point A, select a one-time perpendicular object snap, move the cursor over the hypotenuse (so that AutoCAD displays a Deferred Perpendicular ToolTip), and then click the hypotenuse of the triangle.

7. When AutoCAD prompts you to specify the next point, acquire the end point at point B and the midpoint at point C. Then, move the cursor until AutoCAD displays the intersection of the two orthogonal alignment paths connecting these points. Click to select the point.

8. AutoCAD draws the first line and then prompts you to specify the next point. Again, acquire the midpoint at point B and move the cursor along the 270-degree relative polar alignment path until you also see an alignment path extending from the midpoint at point B intersecting the 270-degree relative polar alignment path. Click to select the point.

9. Move the cursor down, perpendicular to the hypotenuse, until AutoCAD displays the Relative Polar: Intersection ToolTip. Click to select the point and then press ENTER to end the command.

# The Complete Reference

# Chapter 3

## Creating Simple
## 2-D Objects

E very AutoCAD drawing is composed of objects, most of them simple two-dimensional objects. For purposes of organization, this book classifies these 2-D objects into two groups: simple objects and more complex objects. The simple objects include lines (both finite and infinite), circles, arcs, ellipses, elliptical arcs, points, and rays. In addition, AutoCAD includes a freehand sketch command. For purposes of this discussion, freehand sketches are also classified as simple objects.

This chapter explains how to create simple objects. As with other AutoCAD commands, you can start the drawing command for each of these objects by using any of the following methods:

- On the Draw toolbar, click the button to start the drawing command.
- From the Draw menu, choose the drawing command.
- At the command line, type the drawing command and press ENTER.

Most drawing commands present several different ways to create an object. For example, when drawing arcs, AutoCAD offers 11 different options. Although all the possible options are explained, in most instances, examples of only one or two methods are provided.

When you use a drawing command, AutoCAD prompts you to enter coordinate points, such as the end points of a line or the center point of a circle, or distances, such as the radius of a circle. You generally enter the points or distances either by using your mouse or by typing actual coordinate values at the command line.

After you create objects, you can modify them by using any of AutoCAD's object modification commands. You'll learn about these commands in Chapter 7.

## Drawing Lines

Lines are probably the most commonly used AutoCAD object. Over 50 percent of typical drawings consist of lines. A line object itself consists of two points: a start point and an end point. You can connect a series of lines, but each line segment is considered a separate line object.

To draw a line:

1. Do one of the following:

- On the Draw toolbar, click the Line button.
- From the Draw menu, choose Line.
- At the command line, type **LINE** (or **L**) and press ENTER.

AutoCAD prompts:

```
Specify first point:
```

2. Specify the start point. Notice that a rubber-band line extends from the start point to the cursor position and changes size and location as you move the cursor. AutoCAD prompts:

```
Specify next point or [Undo]:
```

3. Specify the end point. As soon as you select the end point, the line segment is drawn and AutoCAD repeats the previous prompt. You can then draw another line segment.

4. Press ENTER to end the command.

While the LINE command is active, you can type **U** to undo the previous line segment. Repeatedly using the Undo option removes each previous line segment. After you draw two or more line segments, you can type **C** (which stands for Close) to create a segment that returns to the original start point, thus ending the LINE command.

 *You can start a new line at the end point of the previous line by restarting the LINE command and pressing ENTER when prompted to select the start point.*

Figure 3-1 shows a simple drawing created using lines. The following command sequence was used to create this drawing:

```
Command: LINE
Specify first point: 6,4
Specify next point or [Undo]: 0,4
Specify next point or [Undo]: 0,0
Specify next point or [Close/Undo]: 6,0
Specify next point or [Close/Undo]: ENTER
```

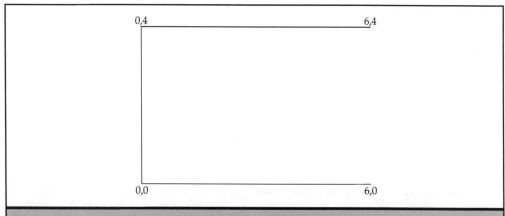

**Figure 3-1.** *A simple drawing created using the LINE command*

*This example uses coordinates to specify the end points of each line segment. You will learn about coordinates in Chapter 6. You could also set the snap interval to one-inch increments and draw this object by pointing while watching the coordinate display on the status bar.*

## Drawing Circles

Circles represent another common AutoCAD object. The default method used to create circles is to specify the center point and the radius. Of course, you can use other methods, as well. For example, you can draw circles by specifying the center point and the diameter. Or you can select two points that define the end points of the circle's diameter, or select three points on the circumference of the circle. In addition, you can generate a circle with a specified radius tangent to two objects already in the drawing, or three points on the circumference that are tangent to three existing objects. In each case, you can select points within the drawing, enter coordinates, or use combinations of the two methods. Figure 3-2 shows several different methods for drawing circles.

To draw a circle by specifying its center and radius:

1. Do one of the following

   - On the Draw toolbar, click the Circle button.

   - From the Draw menu, choose Circle | Center, Radius.

   - At the command line, type **CIRCLE** (or **C**) and press ENTER.

   AutoCAD prompts:

   ```
   Specify center point for circle or [3P/2P/Ttr (tan tan radius)]:
   ```

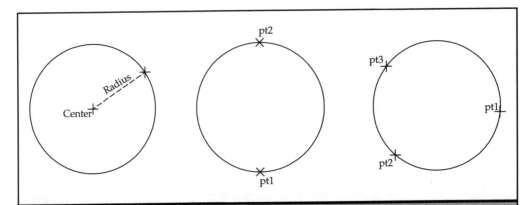

**Figure 3-2.** *Methods for drawing circles (from left to right): by specifying the center point and radius; by specifying two points on the circle; and by specifying three points on the circle*

2. Specify the center point. Notice that a rubber-band line extends from the center point to the cursor position. You also see a circle. The size of the circle changes as you move the cursor. AutoCAD prompts:

```
Specify radius of circle or [Diameter]:
```

3. Specify the radius by either selecting the end point of the radius or typing the actual radius and then pressing ENTER. As soon as you specify the radius, the circle is drawn and the command ends.

To draw a circle tangent to two existing objects (see Figure 3-3):

1. Do one of the following:

   ■ On the Circle toolbar, click the Circle button.

   ■ From the Draw menu, choose Circle | Tan, Tan, Radius. Go to Step 3.

   ■ At the command line, type **CIRCLE** (or **C**) and press ENTER.

   AutoCAD prompts:

```
Specify center point for circle or [3P/2P/Ttr (tan tan radius)]:
```

2. Right-click and then select Ttr (tan tan radius) from the shortcut menu, or type **TTR** and then press ENTER. AutoCAD prompts:

```
Specify point on object for first tangent of circle:
```

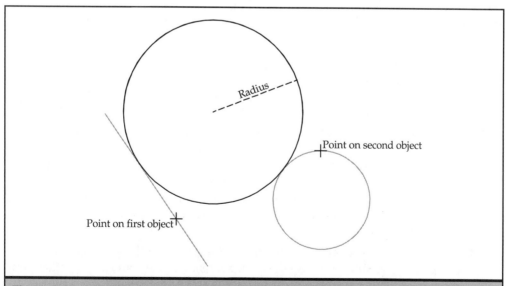

**Figure 3-3.**  *A circle drawn tangent to two existing objects*

3. Select the first object to which the circle will be tangent. AutoCAD prompts:

   ```
   Specify point on object for second tangent of circle:
   ```

4. Select the second object to which the circle will be tangent. AutoCAD prompts:

   ```
   Specify radius of circle <default>:
   ```

5. Specify the radius. You can type the actual radius value and then press ENTER, specify the distance by selecting two points in the drawing (the radius being the distance between those points), or press ENTER to accept the default radius value. As soon as you specify the radius, the circle is drawn and the command ends.

**Tip**   *As you learned in the previous example, when you start the CIRCLE command from the menu, you can choose the method you want to use to draw the circle prior to starting the command. This eliminates several keystrokes, making it faster than starting the command from the toolbar or typing it at the command line. As you will learn, this is true of several other drawing commands, as well.*

## Drawing Arcs

An arc is a portion of a circle. The default method used to create arcs is to specify three points—the start point, a second point, and the end point—with the resulting arc passing through each of those points. In addition, AutoCAD provides seven other methods for defining arcs, enabling you to draw arcs based on such parameters as the center point, radius, chord length, included angle, or direction, in various combinations. In total, AutoCAD offers 11 different ways to draw arcs. These methods break down into the following five groups:

- **3 Points**   AutoCAD draws the arc passing through three points that you specify. The resulting arc starts at the first point, passes through the second point, and ends at the third point. The arc can be drawn in a clockwise or a counterclockwise direction.

- **Start, Center**   You specify the start point and center point of the arc. You can then complete the arc by specifying its end point, the angle subtended by the arc, or the length of its chord. Specifying a positive angle causes the arc to be drawn counterclockwise. Specifying a negative angle results in the arc being drawn in a clockwise direction. Similarly, a positive chord length causes the arc to be drawn counterclockwise, while a negative chord length causes the arc to be drawn clockwise.

- **Start, End**   You specify the start point and end point of the arc. You can then complete the arc by specifying the angle subtended by the arc, the direction from the start point to the end point, or the radius of the arc. Specifying a positive angle causes the arc to be drawn counterclockwise. Specifying a negative angle results in the arc being drawn in a clockwise direction. When specifying the radius, AutoCAD always draws the arc in a counterclockwise direction.

■ **Center, Start** You specify the center point and start point of the arc. You can then complete the arc by specifying the end point, the angle subtended by the arc, or the length of its chord. Specifying a negative angle results in the arc being drawn in a clockwise direction. Similarly, a positive chord length causes the arc to be drawn counterclockwise, while a negative chord length causes the arc to be drawn clockwise.

■ **Continue** This option draws an arc that is tangent to and extending from the last object you drew.

To draw an arc by specifying three points (see Figure 3-4):

1. Do one of the following:

   ■ On the Draw toolbar, click the Arc button.

   ■ From the Draw menu, choose Arc | 3 Points.

   ■ At the command line, type **ARC** (or **A**) and press ENTER.

   AutoCAD prompts:

   ```
   Specify start point of arc or [Center]:
   ```

2. Specify the start point. Notice that a rubber-band line extends from the start point to the cursor position. AutoCAD prompts:

   ```
   Specify second point of arc or [Center/End]:
   ```

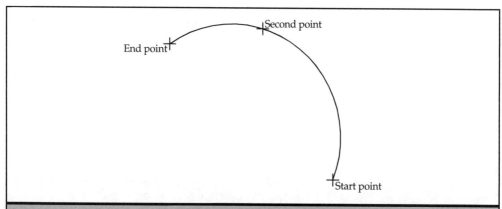

**Figure 3-4.** *An arc drawn by specifying the start point, second point, and end point*

3. Specify a second point. Notice that once you specify the second point, the rubber-band line turns into an arc that extends from the start point, through the second point, to the cursor position. AutoCAD prompts:

```
Specify end point of arc:
```

4. Specify the end point. As soon as you specify the end point, the arc is drawn and the command ends.

To draw an arc by specifying its start point, center point, and end point (see Figure 3-5):

1. Do one of the following:

   ■ On the Draw toolbar, click the Arc button.

   ■ From the Draw menu, choose Arc | Start, Center, End.

   ■ At the command line, type **ARC** (or **A**) and press ENTER.

   AutoCAD prompts:

```
Specify start point of arc or [Center]:
```

2. Specify the start point. AutoCAD prompts:

```
Specify second point of arc or [Center/End]:
```

3. If you started the command by using the menu, AutoCAD automatically prompts you for the center point. Otherwise, type **C** and press ENTER, or right-click and then select Center from the shortcut menu. AutoCAD prompts:

```
Specify center point of arc:
```

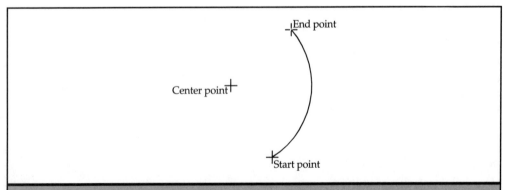

**Figure 3-5.** *An arc drawn by specifying the start point, center point, and end point*

4. Specify the center point. Notice that once you specify the center point, the rubber-band line turns into an arc that extends from the start point through a line extending from the center point to the cursor position. AutoCAD prompts:

```
Specify end point of arc or [Angle/chord Length]:
```

5. Specify the end point. As soon as you specify the end point, the arc is drawn and the command ends.

**Note**   *The end point you specify may not necessarily be the end point of the arc. Rather, it defines the end point of the line extending from the center of the arc. The actual end point of the resulting arc will be on this line.*

To draw an arc by specifying the start point, end point, and included angle (see Figure 3-6):

1. Do one of the following:
   - On the Draw toolbar, click the Arc button.
   - From the Draw menu, choose Arc | Start, End, Angle.
   - At the command line, type **ARC** (or **A**) and press ENTER.

   AutoCAD prompts:

   ```
   Specify start point of arc or [Center]:
   ```

2. Specify the start point. AutoCAD prompts:

   ```
   Specify second point of arc or [Center/End]:
   ```

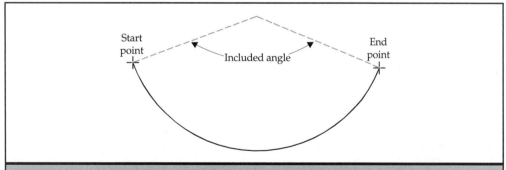

**Figure 3-6.**   *An arc drawn by specifying the start point, end point, and included angle*

3. If you started the command by using the menu, AutoCAD automatically prompts you for the end point. Otherwise, type **E** and press ENTER, or right-click and select End from the shortcut menu. AutoCAD prompts:

```
Specify end point of arc:
```

4. Specify the end point. AutoCAD prompts:

```
Specify center point of arc or [Angle/Direction/Radius]:
```

5. If you started the command by using the menu, AutoCAD automatically prompts you for the included angle. Otherwise, type **A** and press ENTER, or right-click and select Angle from the shortcut menu. Notice that once you specify the end point, the rubber-band line turns into an arc that extends from the start point to the end point, along with a line extending from the start point to the cursor position. AutoCAD prompts:

```
Specify included angle:
```

6. Specify the included angle either by typing the actual angle and then pressing ENTER or by dragging the rubber-band line and selecting a point in the drawing (the resulting angle being measured from the zero-angle direction). As soon as you specify the angle, the arc is drawn and the command ends.

If the last object you drew was an arc, a line, or an open 2-D polyline, you can draw an arc that is tangent to and starting from the end point of that object. To draw an arc tangent to the last object drawn (see Figure 3-7):

1. Do one of the following:

   - On the Draw toolbar, click the Arc button.
   - From the Draw menu, choose Arc | Continue.
   - At the command prompt, type **ARC** (or **A**) and press ENTER.

   AutoCAD prompts:

```
Specify start point of arc or [Center]:
```

2. If you started the command by using the menu, AutoCAD automatically prompts you for the end point. Otherwise, press ENTER. AutoCAD prompts:

```
Specify end point of arc:
```

   Notice that AutoCAD displays a rubber-band arc extending from the end point of the last object you drew, at an angle tangent to the object.

3. Specify the end point. As soon as you specify the end point, the arc is created and the command ends.

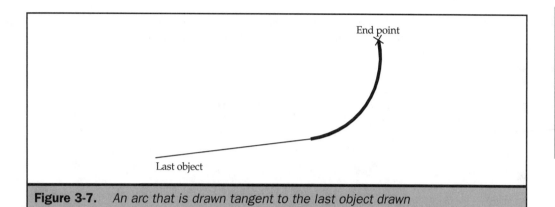

**Figure 3-7.** *An arc that is drawn tangent to the last object drawn*

Online

**EXAMPLES**

*LEARN BY EXAMPLE*
*Practice using the LINE, CIRCLE, and ARC commands to draw the plan view of a small mechanical part. Start a new drawing by using the Figure 3-8 template file on the companion web site.*

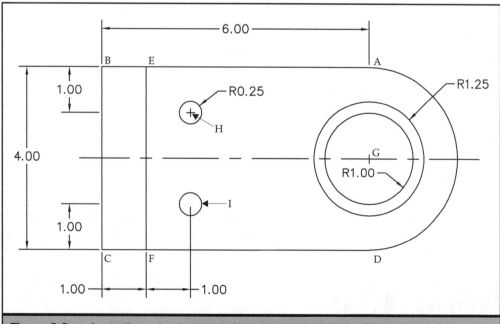

**Figure 3-8.** *A small mechanical part drawn using the LINE, ARC, and CIRCLE commands*

To draw the mechanical part shown in Figure 3-8, use the following instructions and command sequence. The dimensions and point labels are for information only.

```
Command: LINE
Specify first point: (select point A)
Specify next point or [Undo]: (select point B)
Specify next point or [Undo]: (select point C)
Specify next point or [Close/Undo]: (select point D)
Specify next point or [Close/Undo]: ENTER
Command: ARC
Specify start point of arc or [Center]: ENTER
Specify end point of arc: (select point A)
Command: LINE
Specify first point: (select point E)
Specify next point or [Undo]: (select point F)
Specify next point or [Undo]: ENTER
Command: CIRCLE
Specify center point for circle or [3P/2P/Ttr (tan tan radius)]:
(select point G)
Specify radius of circle or [Diameter]: 1.25
Command: ENTER
Specify center point for circle or [3P/2P/Ttr (tan tan radius)]:
(select point G)
Specify radius of circle or [Diameter] <1.2500>: 1
Command: ENTER
Specify center point for circle or [3P/2P/Ttr (tan tan radius)]:
(select point H)
Specify radius of circle or [Diameter] <1.0000>: .25
Command: ENTER
Specify center point for circle or [3P/2P/Ttr (tan tan radius)]:
(select point I)
Specify radius of circle or [Diameter] <0.2500>: ENTER
```

When you finish drawing the mechanical part, use the Layer Control drop-down list to make the Centerline layer the current layer. Then, use the LINE command along with the midpoint object snap to draw the centerline through the part.

# Drawing Ellipses

Geometrically, an ellipse is defined by two axes. The default method for drawing an ellipse is to specify the end points of one axis of the ellipse, and then specify a distance representing half the length of the second axis. The end points of the axis determine the orientation of the ellipse. The longer axis of the ellipse is called the *major axis,* and the shorter one is the *minor axis.* The order in which you define the axes does not matter. AutoCAD determines the major and minor axes based on their relative lengths.

When drawing an ellipse, you can use either of the following methods:

- Specify the center of the ellipse and the two axes.
- Specify the end points of one of the axes of the ellipse, and then either specify the other axis or determine the ellipse by rotating a circle about the first axis.

Both of these options create complete ellipses. You can also create an elliptical arc with the ELLIPSE command.

To draw an ellipse by specifying its axis end points (see Figure 3-9):

1. Do one of the following:

   - On the Draw toolbar, click the Ellipse button.
   - From the Draw menu, choose Ellipse | Axis, End.
   - At the command line, type **ELLIPSE** (or **EL**) and press ENTER.

   AutoCAD prompts:

   ```
   Specify axis endpoint of ellipse or [Arc/Center]:
   ```

2. Specify the first end point. Notice that a rubber-band line extends from the end point to the cursor position. AutoCAD prompts:

   ```
   Specify other endpoint of axis:
   ```

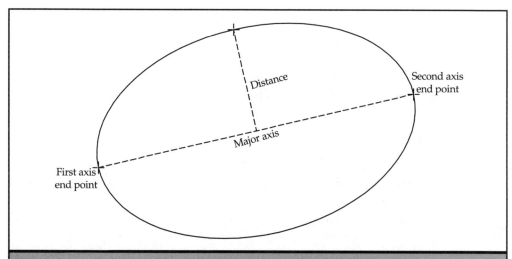

**Figure 3-9.**    *An ellipse drawn by specifying its axis end points*

3. Specify the second end point. Notice that the rubber-band line now extends from the midpoint of the axis you just defined. You also see an ellipse that changes as you move the cursor. AutoCAD prompts:

```
Specify distance to other axis or [Rotation]:
```

4. Specify the half-length of the other axis by selecting a point in the drawing or by typing its length and pressing ENTER. As soon as you specify the length, the ellipse is drawn and the command ends.

**Note**   *If Isometric mode is on, the prompt appears as follows:*

```
Specify axis endpoint of ellipse or [Arc/Center/Isocircle]:
```

In that case, you are actually using AutoCAD's ISOPLANE command to draw in one of the three isometric planes. Although you can use the CIRCLE command to draw a circle in any User Coordinate System (UCS) and then view it from any point in space, you must use the ELLIPSE command when drawing 2-D representations of circles viewed from an isometric angle.

## Drawing Elliptical Arcs

An elliptical arc is a portion of an ellipse. You create elliptical arcs by using the Arc option of the ELLIPSE command. The default method for drawing elliptical arcs is to specify the end points of one axis of the ellipse, and then specify a distance representing half the length of the second axis. Then, you specify the start and end angles for the arc, measured from the center of the ellipse in relation to the direction of its major axis. You can also specify the start angle and an included angle. If the start and end angles are the same, you create a full ellipse.

To draw an elliptical arc by specifying its axis end points (see Figure 3-10):

1. Do one of the following:

   ■ On the Draw toolbar, click the Ellipse Arc button. Go to Step 3.

   ■ From the Draw menu, choose Ellipse | Arc. Go to Step 3.

   ■ At the command line, type **ELLIPSE** (or **EL**) and press ENTER.

   AutoCAD prompts:

```
Specify axis endpoint of ellipse or [Arc/Center]:
```

2. Type **A** and then press ENTER, or right-click and select Arc from the shortcut menu. AutoCAD prompts:

```
Specify axis endpoint of elliptical arc or [Center]:
```

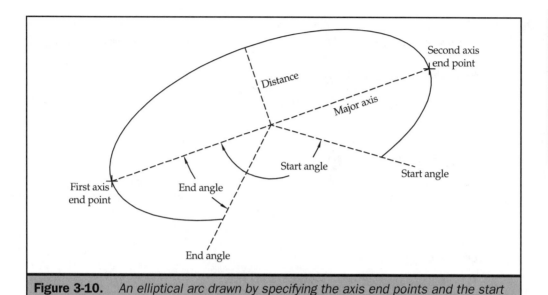

**Figure 3-10.**    *An elliptical arc drawn by specifying the axis end points and the start and end angles*

3. Specify the first end point. AutoCAD prompts:

```
Specify other endpoint of axis:
```

4. Specify the second end point. AutoCAD prompts:

```
Specify distance to other axis or [Rotation]:
```

5. Specify the half-length of the other axis. Notice that AutoCAD shows you a complete ellipse. You also see a rubber-band line extending from the center of the ellipse to the cursor position. The start angle is measured counterclockwise from the angle of the major axis of the ellipse. AutoCAD prompts:

```
Specify start angle or [Parameter]:
```

6. Specify the start angle of the arc by either typing an angle or selecting a point. Notice that the rubber-band line now extends from the center of the ellipse to the cursor. You also see an elliptical arc extending from the point defined by the start angle to the rubber-band line. AutoCAD prompts:

```
Specify end angle or [Parameter/Included angle]:
```

7. Specify the end angle. Again, the angle is measured counterclockwise from the angle of the major axis. As soon as you specify the end angle, the elliptical arc is drawn and the command ends.

*Versions of AutoCAD before Release 14 used polylines to approximate an ellipse. Newer versions create true ellipse objects. The PELLIPSE system variable determines the type of ellipse created. If PELLIPSE equals 0, AutoCAD creates a true ellipse object. If PELLIPSE equals 1, AutoCAD creates the ellipse by using a polyline that approximates the ellipse. You can then use the PEDIT command to edit the ellipse. You'll learn about this command in Chapter 10.*

# Creating Point Objects

Point objects, which consist of a single dot, or one of 19 other possible display styles, are useful as nodes or reference points. For example, you may want to use point objects to mark station points along a roadway centerline.

To draw a point:

1. Do one of the following:

   - On the Draw toolbar, click the Point button.
   - From the Draw menu, choose Point | Single Point (or Multiple Point).
   - At the command line, type **POINT** (or **PO**) and press ENTER.

   AutoCAD prompts:

   ```
   Specify a point:
   ```

2. Specify the location of the point.

 *Depending on how you start the command, AutoCAD may repeat the prompt. In that case, to end the command, press ESC.*

## Changing Point Styles

Changing the size and appearance of point objects affects all point objects already in the drawing, as well as all points that you subsequently draw. You can control the size and appearance of point objects by using the Point Style dialog box (see Figure 3-11). To display this dialog box, do one of the following:

- From the Format menu, choose Point Style.
- At the command line, type **DDPTYPE** and press ENTER.

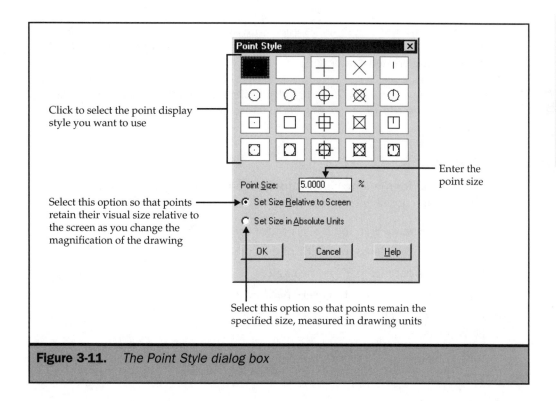

Click to select the point display style you want to use

Enter the point size

Select this option so that points retain their visual size relative to the screen as you change the magnification of the drawing

Select this option so that points remain the specified size, measured in drawing units

**Figure 3-11.**   *The Point Style dialog box*

# Drawing Construction Lines

This section digresses from the discussion of simple drawing objects for a moment to discuss a special type of object. While you can certainly use AutoCAD's other tools to draw most objects without ever drawing the types of construction lines commonly used in manual drafting, sometimes it is still helpful to create construction lines. When the need arises, AutoCAD provides two objects specifically for use as construction lines: xlines (or infinite lines) and rays. An *xline* is a construction line that extends to infinity in both directions. A *ray* is a construction line that starts at a specific point and extends to infinity in one direction.

Because both xlines and rays extend to infinity, they are not calculated as part of the overall area or extents of the drawing. Yet, you can modify construction lines by moving, copying, and rotating them, as you can with any other drawing object. And, if visible onscreen and drawn on a layer whose plotting setting is turned on, construction lines appear in your printed drawings.

*When creating construction lines, you should create them on their own layer, so that this layer can be either turned off or frozen before plotting, or have its plotting setting turned off.*

## Drawing Xlines

The default method for drawing xlines is to select a point along the line and then specify the direction of the line. You can draw an xline in any of the following ways:

- Draw xlines parallel to the X axis of the current UCS.
- Draw xlines parallel to the Y axis of the current UCS.
- Draw xlines parallel to a specific angle.
- Draw an xline that bisects a user-specified angle.
- Draw xlines parallel to and offset a specified distance from existing linear objects.

To draw an xline by specifying a point on the line and its direction (see Figure 3-12):

1. Do one of the following:

- On the Draw toolbar, click the Construction Line button.
- From the Draw menu, choose Construction Line.
- At the command line, type **XLINE** (or **XL**) and press ENTER.

   AutoCAD prompts:

   ```
   Specify a point or [Hor/Ver/Ang/Bisect/Offset]:
   ```

2. Specify a point through which the xline will pass. Notice that as soon as you specify a point, AutoCAD displays an infinite line passing through the point. As you move the cursor, the alignment of the line changes. AutoCAD prompts:

   ```
   Specify through point:
   ```

3. Specify the direction by specifying another point. As soon as you specify the direction, AutoCAD draws the xline and then repeats the preceding prompt. You can draw additional xlines passing through the first point.

4. To end the command, press ENTER or ESC.

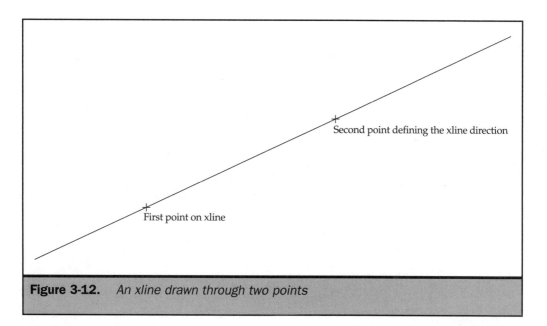

**Figure 3-12.**   *An xline drawn through two points*

## Drawing Rays

Unlike xlines, AutoCAD provides only one method for drawing rays. You select the start point of the ray and then specify its direction. To create a ray (see Figure 3-13):

1. Do one of the following:

   ■ From the Draw menu, choose Ray.

   ■ At the command line, type **RAY** and press ENTER.

   AutoCAD prompts:

   ```
   Specify start point:
   ```

2. Specify the start point for the ray. Notice that as soon as you specify a point, AutoCAD displays a ray starting at that point and extending to infinity in the direction of the cursor. As you move the cursor, the alignment of the ray changes. AutoCAD prompts:

   ```
   Specify through point:
   ```

3. Specify a point through which the ray will pass. As soon as you specify the direction, AutoCAD draws the ray and then repeats the preceding prompt, so that you can create additional rays. Each subsequent ray begins at the same start point.

4. To end the command, press ENTER or ESC.

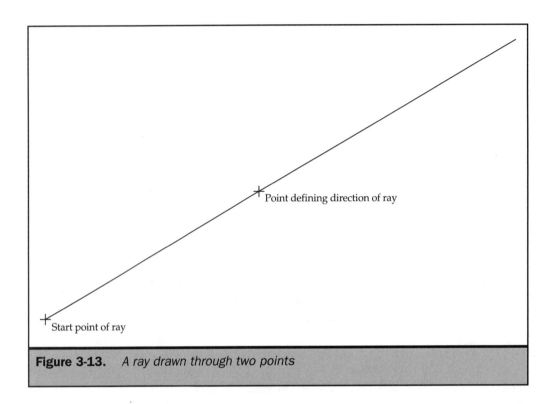

**Figure 3-13.** *A ray drawn through two points*

# Creating Freehand Sketches

A freehand sketch consists of many straight-line segments, created either as individual line objects or as a polyline. Before you begin creating a freehand sketch, you must set the length, or increment, of each segment. The smaller the segments, the more accurate your sketch is, but segments that are too small can greatly increase the drawing file size.

Sketches are useful for creating very irregularly shaped objects, such as irregular boundaries or topographic contour lines, or for tracing existing paper drawings by using a digitizer. Sketches otherwise are usually an impractical method of drawing. As mentioned, the small segment increment can quickly increase the drawing file size. Sketches are also more difficult to modify, should the need arise. Still, used sparingly and only when appropriate, sketching is a useful tool in your AutoCAD arsenal.

**Note** *Because the SKETCH command is not often used, it does not initially appear on the Draw toolbar or in the Draw menu. If you use the command often enough, you may want to modify the menu or toolbar to include this command. You'll learn about modifying menus and toolbars in Chapter 28.*

The SKETCH command operates differently than other AutoCAD drawing commands. You must use the mouse like a pen, first clicking to put the "pen" down to draw, and then clicking again to lift it up. If you are satisfied with the results, you can then record them to the drawing. Otherwise, you can erase part or all of the sketch and start over.

To create a freehand sketch:

1. At the command line, type **SKETCH** and press ENTER.

2. Specify the length of the sketch segments.

3. Click the mouse button to place the command in the pen-down mode.

4. Move the mouse to draw a temporary freehand sketch.

5. Click the mouse button again to lift up the pen to stop sketching. You can then move the cursor to another location in the drawing without sketching.

6. Type **R** at any time to record (save) the temporary sketch into the drawing. If the pen is down, you can simply continue sketching after recording. If the pen is up, click the mouse button again to resume sketching. You can then begin sketching at any location in the drawing. When the pen is up, to resume sketching at the end point of the last sketch segment, type **C** and then move the cursor over the end point.

7. Press ENTER to complete the sketch and record all unrecorded sketch segments.

*While sketching, when the pen is in the up position, you can draw a straight line from the end of the last sketched line to the current cursor position by typing a period.*

## Erasing Freehand Sketch Lines

You can use the Erase option of the SKETCH command to erase temporary freehand sketch lines that have not yet been recorded to the drawing. When in erase mode, you can erase portions of the freehand sketch by moving the cursor over the temporary sketch lines. Everything from the cursor location to the end of the sketched line is erased.

To erase unrecorded freehand lines:

1. While the pen is up or down, type **E**.

2. Move the cursor to the end of the freehand sketch line that you just drew and then move it as far back along the line as you want to erase.

3. To complete the erasure, type **P** or click the mouse button. To restore the sketched line, type **E**. Note that after you complete the erasure, the erased segments cannot be restored.

**Note**

*Once you record a freehand sketch into the drawing, you cannot use the erase mode. You must instead use AutoCAD's ERASE command to remove the sketch from the drawing.*

# Setting the Sketch Method

Using polylines for freehand sketches makes it easier to go back and edit or erase sketches. Unfortunately, by default, AutoCAD creates sketches by using individual line segments. To control whether AutoCAD creates freehand sketches by using individual line segments or polylines, you change the value of the SKPOLY system variable.

To specify lines or polylines when sketching:

1. At the command line, type **SKPOLY** and press ENTER. AutoCAD displays the current SKPOLY value.

2. To create freehand sketches by using polylines, type **1** and press ENTER.
   To create freehand sketches by using lines, type **0** and press ENTER.

The
Complete
Reference

# Chapter 4

## Creating More Complex 2-D Objects

In the last chapter, you learned how to create simple 2-D objects. In this chapter, you'll learn about AutoCAD's more complex 2-D objects, including rectangles, polygons, multilines (mlines), polylines, spline curves, solids, and regions.

The basic objects covered in the previous chapter had one thing in common: they consisted of individual objects. Whether straight or curved, each line, arc, or circle exists in the AutoCAD drawing as an individual object. For example, if you draw a rectangle by using the LINE command, each side of the rectangle consists of an individual line object.

The objects you learn about in this chapter are different. While each of these objects is composed of multiple segments, all the segments are treated as belonging to a single object. When you draw a rectangle by using the RECTANGLE command, each side of the rectangle is created as a polyline segment, but selecting any side of the rectangle for editing causes the entire rectangle to be selected.

After you create complex objects, you can modify them by using any of AutoCAD's object modification commands. Commands for moving, rotating, or copying these objects affect the entire object. Because of the unique nature of these complex objects, however, most of them have special editing commands for modifying the objects themselves.

## Drawing Rectangles

The RECTANGLE command creates rectangles as closed polylines with four sides. You draw a rectangle by specifying its opposite corners. The resulting rectangle is aligned parallel to the current User Coordinate System (UCS).

To draw a rectangle (see Figure 4-1):

1. Do one of the following:

   ■ On the Draw toolbar, click Rectangle.

   ■ From the Draw menu, choose Rectangle.

   ■ At the command line, type **RECTANGLE** (or **REC**) and press ENTER.

   AutoCAD prompts:

   ```
   Specify first corner point or [Chamfer/Elevation/Fillet/Thickness/Width]:
   ```

2. Specify one corner of the rectangle. Notice that as soon as you specify the first corner, a rubber-band rectangle extends from that point to the cursor position; its size changes as you move the cursor. AutoCAD prompts:

   ```
   Specify other corner points or [Dimensions]:
   ```

3. Specify the opposite corner of the rectangle.

As soon as you specify the other corner, the rectangle is drawn and the command ends.

Although the default method of drawing rectangles is simply to specify the opposite corners, AutoCAD provides several options. For example, you can create the rectangle with chamfered (beveled) or filleted (rounded) corners, or with wide lines.

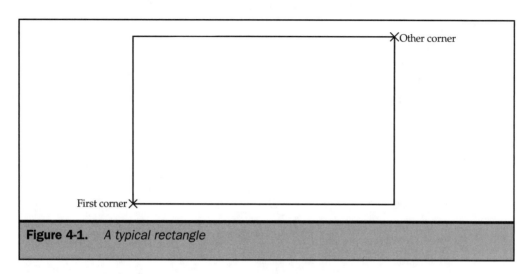

**Figure 4-1.** *A typical rectangle*

You can also specify the dimensions (the length or width), change the elevation (the height above the XY plane) or the thickness of the rectangle. You will learn about these concepts in later chapters.

To draw a rectangle with filleted corners (see Figure 4-2):

1. Do one of the following:

   ■ On the Draw toolbar, click Rectangle.

   ■ From the Draw menu, choose Rectangle.

   ■ At the command line, type **RECTANGLE** (or **REC**) and press ENTER.

   AutoCAD prompts:

   ```
   Specify first corner point or [Chamfer/Elevation/Fillet/Thickness/Width]:
   ```

2. Type **F** and press ENTER, or right-click and then select Fillet from the shortcut menu. AutoCAD prompts:

   ```
   Specify fillet radius for rectangles <0.0000>:
   ```

3. Specify the fillet radius either by entering a value and pressing ENTER or by selecting two points in the drawing (the fillet radius being the distance between the points). AutoCAD repeats the prompt:

   ```
   Specify first corner point or [Chamfer/Elevation/Fillet/Thickness/Width]:
   ```

4. Specify one corner of the rectangle. Notice that as soon as you specify the first corner, a rubber-band rectangle with rounded corners extends from that point to the cursor position; its size changes as you move the cursor. AutoCAD prompts:

   ```
   Specify other corner point or [Dimensions]:
   ```

5. Specify the opposite corner of the rectangle. As soon as you specify the other corner, the rectangle is drawn and the command ends.

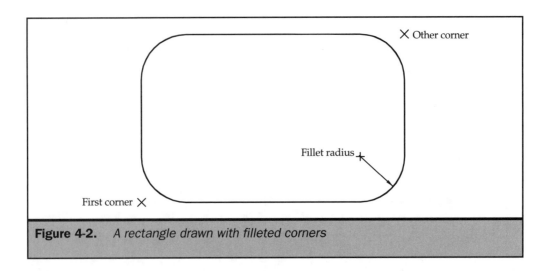

**Figure 4-2.** *A rectangle drawn with filleted corners*

 **Caution** *If the length of the shortest side of the rectangle is less than twice the fillet radius, the corners of the resulting rectangle will not be rounded.*

**Note** *If you change any value, such as the width or elevation, the new value remains in effect until you change it again.*

**Tip** *You can edit the sides of a rectangle individually by using the PEDIT command. You can convert the sides into individual line objects by using the EXPLODE command.*

# Drawing Polygons

Polygons are closed polylines comprised of a minimum of 3 and a maximum of 1,024 equal-length sides. The default method for drawing a polygon is to specify the center of the polygon and the distance from the center to each vertex, so that the entire polygon falls on an imaginary circle (referred to as *inscribed*). Alternately, you can draw a polygon so that the midpoints of each side of the polygon fall on an imaginary circle (referred to as *circumscribed*), or you can specify the start point and end point (and therefore the length) of one side.

## Drawing Inscribed Polygons

An *inscribed polygon* consists of an equal-sided polygon defined by its center point and the distance to its vertices. Thus, the entire polygon is contained or inscribed within a circle of a specified radius. You specify the number of sides, the center point, and either the radius or the location of one vertex, which determines both the size and orientation of the polygon.

To draw an inscribed polygon (see Figure 4-3):

1. Do one of the following:
   - On the Draw toolbar, click Polygon.
   - From the Draw menu, choose Polygon.
   - At the command line, type **POLYGON** (or **POL**) and press ENTER.
   
   AutoCAD prompts:
   
   ```
   Enter number of sides <4>:
   ```

2. Specify the number of sides by typing a value (from 3 to 1,024) and then pressing ENTER. AutoCAD prompts:
   
   ```
   Specify center of polygon or [Edge]:
   ```

3. Specify the center of the polygon. AutoCAD prompts:
   
   ```
   Enter an option [Inscribed in circle/Circumscribed about circle] <C>:
   ```

4. Type **I** and then press ENTER, or right-click and then select Inscribed In Circle from the shortcut menu. Notice that a rubber-band polygon appears with a line extending from the center of the polygon to the cursor position at a vertex. The size of the polygon changes as you move the cursor. AutoCAD prompts:
   
   ```
   Specify radius of circle:
   ```

5. Specify the radius of the circle by either typing a value or selecting a point in the drawing (the radius being the distance from the center of the polygon to the vertex point). As soon as you specify the radius of the circle, the polygon is drawn and the command ends.

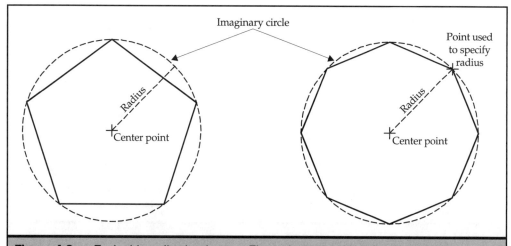

**Figure 4-3.**  *Typical inscribed polygons. The polygon on the left was drawn by specifying the radius; the one on the right was drawn by specifying the location of one of the vertices.*

**Note**    *If you specify the radius by entering a value, the resulting polygon is drawn so that the bottom side is aligned parallel to the X axis of the current UCS. If you specify the radius by specifying a point in the drawing, however, you can also control the alignment of the polygon. These differences are shown in Figure 4-3.*

## Drawing Circumscribed Polygons

A *circumscribed polygon* consists of an equal-sided polygon defined by its center point and the distance to the midpoint of its sides. Thus, the entire polygon is outside, or circumscribed around, a circle of a specified radius. You specify the number of sides, the center point, and either the radius or the location of the midpoint of one side, which determines both the size and orientation of the polygon.

To draw a circumscribed polygon (see Figure 4-4):

1. Do one of the following:

   ■ On the Draw toolbar, click Polygon.

   ■ From the Draw menu, choose Polygon.

   ■ At the command line, type **POLYGON** (or **POL**) and press ENTER.

   AutoCAD prompts:

   ```
   Enter number of sides <default>:
   ```

2. Specify the number of sides by typing a value (from 3 to 1,024) and pressing ENTER. AutoCAD prompts:

   ```
   Specify center of polygon or [Edge]:
   ```

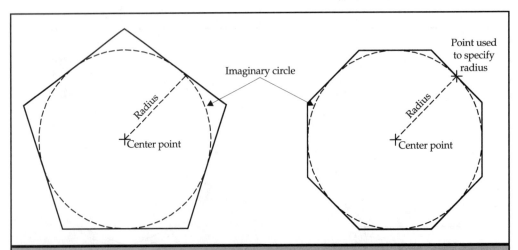

**Figure 4-4.**    *Typical circumscribed polygons. The polygon on the left was drawn by specifying the radius; the one on the right was drawn by specifying the location of the midpoint of one of the sides.*

3. Specify the center of the polygon. AutoCAD prompts:

```
Enter an option [Inscribed in circle/Circumscribed about circle] <I>:
```

4. Type **C** and then press ENTER, or right-click and then select Circumscribed About Circle from the shortcut menu. Notice that a rubber-band polygon appears with a line extending from the center of the polygon to the cursor position at the midpoint of one side. The size of the polygon changes as you move the cursor. AutoCAD prompts:

```
Specify radius of circle:
```

5. Specify the radius of the circle by either typing a value or selecting a point in the drawing (the radius being the distance from the center of the polygon to the midpoint of one of the sides of the polyline). As soon as you specify the radius of the circle, the polygon is drawn and the command ends.

**Note**  *If you specify the radius by entering a value, the resulting polygon is drawn so that the bottom side is aligned parallel to the X axis of the current UCS. If you specify the radius by specifying a point in the drawing, however, you can also control the alignment of the polygon. These differences are shown in Figure 4-4.*

# Drawing Multilines

You will often need to draw parallel lines. For example, architects use parallel lines to represent walls in plan view. Although you could draw each individual parallel line object, AutoCAD provides a special object type specifically for this purpose. This object, called a *multiline* (or mline), consists of from 1 to 16 parallel lines (referred to as *elements*). You draw multilines similar to the way you draw lines: by specifying a start point and an end point. Unlike lines, however, a single multiline can consist of one or more segments of parallel lines.

Each multiline is based on a predefined multiline style that determines the number of elements in the multiline, as well as their color, linetype, and spacing. Each style is given a unique name. The style can also determine the appearance at the terminating ends of the multiline and the visibility of its intermediate joints.

Unfortunately, multilines have not proven to be a very effective AutoCAD object. They're difficult to define, and although they can be copied or moved using standard AutoCAD editing commands, they require a special editing command in order to modify their intersections and vertices, or to create gaps (such as when representing doors and windows). For that reason, many experienced AutoCAD users choose not to use multilines, relying instead on parallel lines or polylines.

Although the MLINE command, used to create multilines, appears on the Draw toolbar and in the Draw menu, it is not covered here. Those wishing to learn more about multilines will find detailed information on the companion web site.

*Architects who depend upon parallel lines to represent walls should consider upgrading to Autodesk's Architectural Desktop, a program based on AutoCAD but tailored specifically to the needs of architects and other building design professionals.*

# Drawing Polylines

A *polyline* is a connected sequence of arcs and lines that is treated as a single object. As you've already learned, a number of AutoCAD objects, such as rectangles and polygons, are actually created as polyline objects. You can draw a polyline using any linetype style. Unlike other objects, such as individual lines, arcs, and circles, polylines can also have a width that either remains constant or tapers over the length of any segment. Figure 4-5 shows some typical polylines.

Since the entire polyline is a single object, the standard AutoCAD editing commands affect the entire polyline. AutoCAD also provides a special command that allows you to edit individual polyline segments. You'll learn more about this command in Chapter 10.

When you start to draw a polyline, AutoCAD prompts you for the start point, similar to drawing a line. After you select the start point, however, AutoCAD displays the current line width and prompts:

```
Specify next point or [Arc/Close/Halfwidth/Length/Undo/Width]:
```

Initially, AutoCAD assumes that you will be drawing a straight-line segment using the current width. You can draw the segment by specifying its end point (just like when drawing lines), or you can select one of the other polyline options. If you select an end point, AutoCAD draws the segment and then prompts for another end point, displaying all the other options. To end the command, press ENTER.

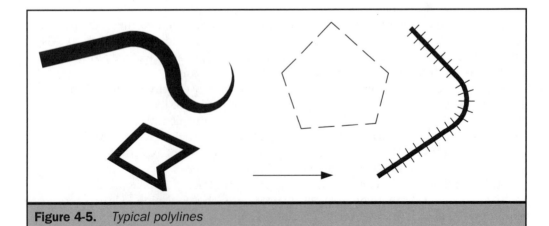

**Figure 4-5.**  *Typical polylines*

# Straight-Line Polylines

When drawing straight-line polyline segments, AutoCAD prompts you for the end point of the current segment and presents the following command options:

- **Arc**   Switches to the polyline arc mode, allowing you to draw arc segments. The command presents a different series of options, similar to the ARC command.

- **Close**   Closes the polyline by drawing a line segment from the current point to the start point of the first segment you drew.

- **Halfwidth**   Lets you specify the width of the next polyline segment by prompting you to specify the distance from the polyline's center line to one edge (half the width). You can set the starting width and ending width separately to create a tapered polyline segment. AutoCAD then draws subsequent segments with the same ending width as the previous segment, unless you change the width again.

- **Length**   Draws a polyline segment of a specified length, continuing the polyline at the same angle as that of the previous segment.

- **Undo**   Removes the previous polyline segment.

- **Width**   Lets you specify the overall width of the next polyline segment. You can set the starting width and ending width separately to create a tapered polyline segment. AutoCAD then draws subsequent segments with the same ending width as the previous segment, unless you change the width again.

To draw a polyline with straight segments (see Figure 4-6):

1. Do one of the following:

    - On the Draw toolbar, click Polyline.
    - From the Draw menu, choose Polyline.
    - At the command line, type **PLINE** (or **PL**) and press ENTER.

    AutoCAD prompts:

    ```
    Specify start point:
    ```

2. Specify the start point. Notice that a rubber-band line extends from the start point to the cursor position and changes as you move the cursor. AutoCAD prompts:

    ```
    Current line-width is 0.0000
    Specify next point or [Arc/Halfwidth/Length/Undo/Width]:
    ```

3. Specify the end point of the polyline segment. When you specify the end point, AutoCAD draws the polyline segment and then repeats the previous prompt. You can then select an option or draw another segment.

4. To complete the command, press ENTER or, if you have drawn two or more segments, type **C** and press ENTER to close the polyline and end the command.

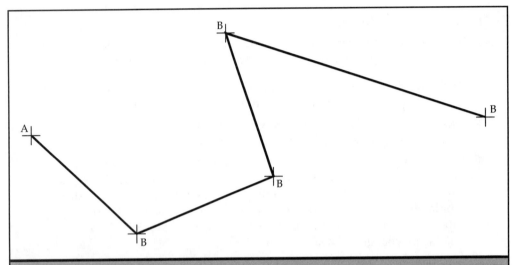

**Figure 4-6.** *A typical polyline drawn from start point (A) to each segment end point (B)*

# Polyline Arcs

The Arc option switches to the polyline arc mode. Once in this mode, new polyline segments are created as arc segments, until you end the command or switch back to line mode. When you draw polyline arc segments, the first point of the arc is the end point of the previous segment. By default, you draw arc segments by specifying the end point of each arc segment. Each successive arc segment is drawn tangent to the previous arc or line segment.

When you switch to arc mode, AutoCAD prompts you for the end point of the current arc segment and presents a different series of prompts:

```
Specify endpoint of arc or
[Angle/CEnter/CLose/Direction/Halfwidth/Line/Radius/Second
pt/Undo/Width]:
```

■ **Angle**  Prompts you for the included angle, or the angle the arc is to span. A positive number specifies a counterclockwise angle. A negative number results in a clockwise included angle. You then specify the center, radius, or end point of the arc.

■ **CEnter**  Prompts you to specify the center point of the arc segment. You then specify the angle, length, or end point of the arc.

■ **CLose**   Closes the polyline by drawing an arc segment from the current point and tangent to the previous segment to the start point of the first segment you drew.

■ **Direction**   Prompts you to specify the starting direction for the arc segment tangent to the previous segment. You then specify the end point.

■ **Halfwidth**   Prompts for the halfwidth. This is the same option as in line mode.

■ **Line**   Switches to the polyline line mode, allowing you to draw straight-line polyline segments.

■ **Radius**   Prompts you for the radius of the arc segment. Then, you can either select the end point (the default) or provide the angle that the arc subtends. The arc is drawn tangent to the previous segment.

■ **Second pt**   Prompts you for two additional points through which the arc will pass.

■ **Undo**   Removes the previous polyline segment.

■ **Width**   Prompts for the overall width of the next polyline segment. This is the same option as in line mode.

If you simply specify the end point of the current arc segment (the default option), the arc is drawn tangent to the previous polyline segment.

To draw a polyline with arc segments (see Figure 4-7):

1. Do one of the following:

   ■ On the Draw toolbar, click Polyline.

   ■ From the Draw menu, choose Polyline.

   ■ At the command line, type **PLINE** (or **PL**) and press ENTER.

   AutoCAD prompts:

   ```
   Specify start point:
   ```

2. Specify the start point. AutoCAD prompts:

   ```
   Current line-width is 0.0000
   Specify next point or [Arc/Halfwidth/Length/Undo/Width]:
   ```

3. Type **A** and press ENTER, or right-click and then select Arc from the shortcut menu. Notice that a rubber-band arc segment extends from the start point to the cursor position and changes as you move the cursor. AutoCAD prompts:

   ```
   Specify endpoint of arc or
   [Angle/CEnter/CLose/Direction/Halfwidth/Line/Radius/Second
   pt/Undo/Width]:
   ```

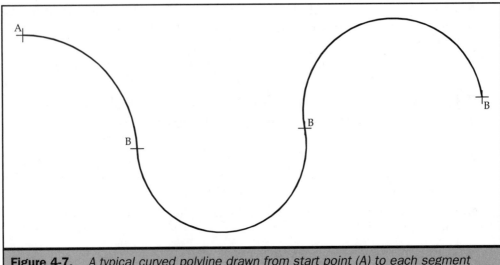

**Figure 4-7.** *A typical curved polyline drawn from start point (A) to each segment end point (B)*

4. Specify the end point of the polyline arc segment. When you specify the end point, AutoCAD draws the polyline segment and then repeats the previous prompt. You can then select an option or draw another segment.

5. To complete the command, press ENTER or type **CL** and press ENTER (or select CLose from the shortcut menu) to close the polyline and end the command.

*LEARN BY EXAMPLE*

*Practice using the PLINE command to draw a left-turn arrow for use in a typical parking lot layout. Start a new drawing by using the Figure 4-8 template file on the companion web site.*

To draw the left-turn arrow shown in Figure 4-8, use the following command sequence:

```
Command: PLINE
Specify start point: (select a point)
Current line-width is 0.0000
Specify next point or [Arc/Halfwidth/Length/Undo/Width]: W
Specify starting width <0.0000>: ENTER
Specify ending width <0.0000>: 18
```

```
Specify next point or [Arc/Halfwidth/Length/Undo/Width]: @18,0
Specify next point or [Arc/Close/Halfwidth/Length/Undo/Width]: W
Specify starting width <18.0000>: 8
Specify ending width <8.0000>: ENTER
Specify next point or [Arc/Close/Halfwidth/Length/Undo/Width]: L
Specify length of line: 12
Specify next point or [Arc/Close/Halfwidth/Length/Undo/Width]: A
Specify endpoint of arc or
[Angle/CEnter/CLose/Direction/Halfwidth/Line/Radius/Second
pt/Undo/Width]: A
Specify included angle: -90
Specify endpoint of arc or [CEnter/Radius]: R
Specify radius of arc: 12
Specify direction of chord for arc <0>: -45
Specify endpoint of arc or
Angle/CEnter/CLose/Direction/Halfwidth/Line/Radius/Second
pt/Undo/Width]: L
Specify next point or [Arc/Close/Halfwidth/Length/Undo/Width]: L
Specify length of line: 12
Specify next point or [Arc/Close/Halfwidth/Length/Undo/Width]: ENTER
```

When you are finished, be sure to save your drawing.

**Figure 4-8.**    *A left-turn arrow drawn using the PLINE command*

## Lightweight Polylines

Beginning in Release 14, AutoCAD now creates 2-D polylines using an optimized format, referred to as a *lightweight polyline,* that conserves memory and disk space. Each polyline object can contain a large amount of vertex data. The lightweight polyline object stores all the vertex data in a single array. Prior to Release 14, polylines were created with an older format in which the data for each vertex was stored separately.

The use of the lightweight format is accomplished automatically. In addition, when you load a drawing that was created using an earlier version, AutoCAD normally converts the older-style polyline objects into lightweight polylines. AutoCAD's editing commands make no distinction between polyline formats. The formats affect only the amount of data stored in the AutoCAD drawing file. If you save an AutoCAD 2000 format drawing back into a pre-R14 format, the lightweight polylines are automatically converted into the older format.

In some instances, however, AutoCAD either cannot convert old-style polylines or saves new polylines by using the older style. You can also override the creation of lightweight polylines. The PLINETYPE (Polyline Type) system variable determines whether AutoCAD uses the new lightweight polylines or the older-style polylines, and also determines whether AutoCAD converts polylines from older drawings to the new polyline format. The following possible PLINETYPE values determine how AutoCAD deals with polylines:

| VALUE | Meaning |
|---|---|
| 0 | Polylines are created with the older format. Polylines in existing drawings are not converted. |
| 1 | Polylines are created with the new format. Polylines in existing drawings are not converted. |
| 2 | Polylines are created with the new format. Polylines in existing drawings are automatically converted to the new format when the drawing is opened. |

AutoCAD 2002 also provides a command that explicitly converts older-style polylines into the new lightweight format. The CONVERT command also updates pre-Release 14 associative hatch objects into a new optimized format. You start the command by typing **CONVERT** and pressing ENTER. The command asks whether you want to convert hatch patterns, polylines, or both. You can then select these objects individually or convert all the objects in the current drawing. The command prompts appear as follows:

```
Command: CONVERT
Enter type of objects to convert [Hatch/Polyline/All] <All>:
Enter object selection preference [Select/All] <All>:
```

AutoCAD displays a dialog box warning you that hatch objects may change appearance when converted, and asking you if you still want to proceed. After the conversion is complete, AutoCAD displays one or both of the following messages:

```
number hatch objects converted
number 2D polyline objects converted
```

where *number* is the actual number of objects converted.

Polylines in existing drawings that contain curve-fit or splined segments remain in the old format when opened in AutoCAD 2002. Polylines that have extended entity data stored on their vertices also remain in the old format. Note that sometimes polylines need to be converted into the older format to be compatible with other programs. AutoCAD 2002 includes an undocumented command, CONVERTPOLY, that accomplishes this function. When you start the command, AutoCAD prompts:

```
Enter polyline conversion option [Heavy/Light] <Light>:
```

Specify Heavy to convert selected polylines into the old-style polyline format, or Light to convert them into the new lightweight format. The command then prompts you to select the polylines you want to convert.

# Drawing Spline Curves

A *spline* is a smooth curve defined by a set of points. Splines are often used to accurately represent the sculpted shapes of objects as diverse as boat hulls, turbine blades, or cellular phones. Whereas these types of curves would be approximated in conventional drafting by using a French curve, they can be mathematically and accurately produced using AutoCAD. Actually, several types of splines (called *B-splines*) exist, all of which share these three similarities:

■ The curve is divided into segments, called *knots*. When these knots are unevenly spaced along the curve, the curve is referred to as a *non-uniform* B-spline.

■ *Control points* are used to pull or shape the curve, as if weight were being applied to the curve at the control point. These control points do not necessarily fall on the curve itself. When the weight is not equal at each control point, the curve is referred to as a *rational* B-spline.

■ The *order* of the spline indicates the number of times the curve segment can change curvature. A curve with an order of one is a straight line; an order of three has a constant curvature (an arc). Third-order curves are also called *quadratic* curves. Fourth-order curves, also called *cubic* curves, can have one curvature change, fifth-order curves can have two changes, and so on.

AutoCAD creates non-uniform rational B-splines, or NURBS curves, which produce a smooth curve between a set of control points. Although these splines have a default order of 4, you can increase this as high as 26. Generally, fourth-order splines work fine, producing smooth curves. Higher-order splines require more control points, often resulting in weird fluctuations.

**Note** *You can also create approximations of true spline curves by using the PEDIT command to smooth a polyline. If desired, you can then use the SPLINE command to convert these smoothed polylines into true splines.*

True splines have several advantages over splined polylines (polylines that were smoothed and then converted to splines):

- True splines are more accurate than splined polylines.
- True splines retain their definitions when edited; splined polylines lose their definitions.
- True splines consume less disk space, resulting in smaller drawing files.

Splines can be created using any linetype. Unlike splined polylines, however, you cannot apply a width value to splines.

You create a spline by specifying its control points, the tangency of the spline at its end points, and its *fit tolerance*, which defines how closely the spline fits its control points. Since the entire spline is a single object, the standard AutoCAD editing commands affect the entire spline. AutoCAD also provides a special command that allows you to edit the spline by changing features, such as its control points and tolerance. You'll learn more about this command in Chapter 10.

To draw a spline (see Figure 4-9):

1. Do one of the following:

- On the Draw toolbar, click Spline.
- From the Draw menu, choose Spline.
- At the command line, type **SPLINE** (or **SPL**) and press ENTER.

    AutoCAD prompts:

    ```
    Specify first point or [Object]:
    ```

2. Specify the first control point. Notice that a rubber-band line extends from the first point to the cursor position and changes as you move the cursor. AutoCAD prompts:

    ```
    Specify next point:
    ```

3. Specify the second control point. When you specify the second control point, AutoCAD draws a portion of the spline and displays a rubber-band line

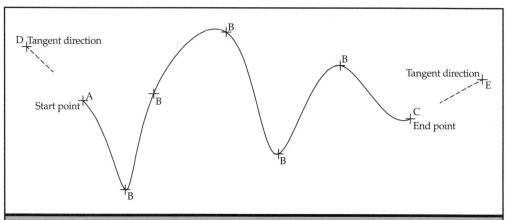

**Figure 4-9.**    *A typical spline curve drawn from the first point (A), through intermediate points (B), to the last point (C), with tangent directions specified by selecting points D and E*

extending from the second control point to the cursor position. AutoCAD prompts:

```
Specify next point or [Close/Fit Tolerance] <start tangent>:
```

4. Specify the next control point. AutoCAD repeats the previous prompt. You can continue specifying as many more control points as you want.

5. When you are done specifying control points, press ENTER. Notice that a rubber-band line now extends from the first control point to the cursor position. The tangency of the spline at the start point changes as you move the cursor.

AutoCAD prompts:

```
Specify start tangent:
```

6. Specify the tangent direction for the start point of the spline. A similar rubber-band line extends from the last control point, and AutoCAD prompts:

```
Specify end tangent:
```

7. Specify the tangent direction for the end point of the spline. As soon as you specify the end tangent, the spline is drawn and the command ends.

## SPLINE Command Options

The SPLINE command offers three options:

- Create a closed spline
- Adjust the spline's fit tolerance
- Convert an existing splined polyline into a spline object

## Creating Closed Splines

When you create a closed spline, the first control point is also the last control point, and both points share the same tangent information. The following command sequence was used to create the closed spline shown in Figure 4-10:

```
Command: SPLINE
Specify first point or [Object]: (select point A)
Specify next point: (select point B)
Specify next point or [Close/Fit Tolerance] <start tangent>: C
Specify tangent: (select point C)
```

*LEARN BY EXAMPLE*
*Practice using the SPLINE command to draw the outline of a gasket. Start a new drawing by using the Figure 4-11 template file on the companion web site.*

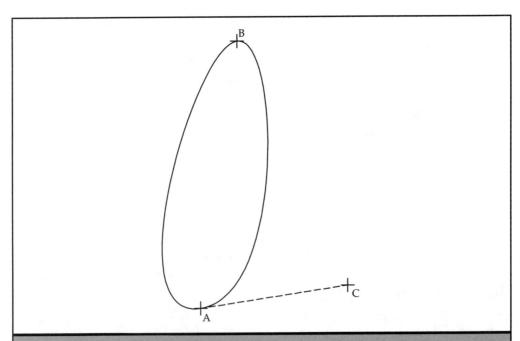

**Figure 4-10.** *A typical closed spline*

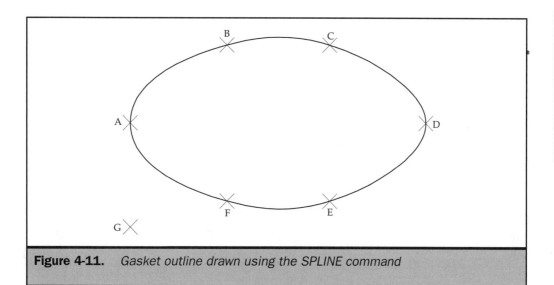

**Figure 4-11.** *Gasket outline drawn using the SPLINE command*

To draw the outline of the gasket shown in Figure 4-11, use the following command sequence. You can use the node object snap to snap to the labeled points.

```
Command: SPLINE
Specify first point or [Object]: (select point A)
Specify next point: (select point B)
Specify next point or [Close/Fit Tolerance] <start tangent>: (select point C)
Specify next point or [Close/Fit Tolerance] <start tangent>: (select point D)
Specify next point or [Close/Fit Tolerance] <start tangent>: (select point E)
Specify next point or [Close/Fit Tolerance] <start tangent>: (select point F)
Specify next point or [Close/Fit Tolerance] <start tangent>: C
Specify tangent: (select point G)
```

When you are finished, be sure to save your drawing.

## Changing the Fit Tolerance

After you specify two or more control points, you can change the tolerance for fitting the current spline curve. AutoCAD prompts:

```
Specify fit tolerance <0.0000>:
```

The spline curve is redefined from the tolerance value. A value of 0 results in a spline curve that passes through all the control points. A tolerance value greater than 0 results in a spline curve that passes through fit points that fall within the specified tolerance. Figure 4-12 illustrates the effect of changing the fit tolerance.

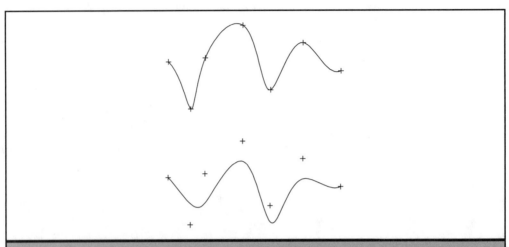

**Figure 4-12.** *A spline created with a fit tolerance greater than 0 passes through the first and last control points, but not necessarily through any of the intermediate points.*

## Converting Objects into Splines

The Object option converts an existing 2-D or 3-D quadratic or cubic spline-fit polyline into an equivalent spline. AutoCAD prompts you to select objects. You can use any object selection method. Press ENTER after you finish selecting objects. AutoCAD immediately converts the spline-fit polylines into splines and ignores any objects you select that are not spline-fit polylines:

```
Command: SPLINE
Specify first point or [Object]: O
Select objects to convert to splines ..
Select objects: (select the objects)
Select objects: ENTER
```

**Caution** *When you convert a spline-fit polyline into a spline, any width information is lost. The DELOBJ system variable controls whether the original polyline is retained or deleted from the drawing. Spline objects are less flexible than polylines. For example, you can't extrude or explode splines or join two splines together. If you convert a spline-fit polyline into a spline and later discover that it should have remained a polyline (but can't reverse your actions in the current drawing), the only way to recover the old polyline is to save the drawing in AutoCAD R12 format and then open it again. Be aware, however, that when saving the drawing in R12 format, you may lose other AutoCAD 2000 drawing information that has no equivalent in the earlier version.*

# Drawing Donuts

Donuts are solid, filled circles or rings that are created as closed, wide polylines. Donuts are often used to create filled dots or to indicate pads on an electronic circuit diagram. You draw a donut by first specifying the inside and outside diameters of the donut and then specifying its center. You can then create multiple copies of the same donut by specifying additional center points, until you press ENTER to complete the command.

To draw a donut (see Figure 4-13):

1. Do one of the following:

   ■ From the Draw menu, choose Donut.

   ■ At the command line, type **DONUT** (or **DO**) and press ENTER.

   AutoCAD prompts:

   ```
   Specify inside diameter of donut <0.5000>:
   ```

2. Specify the inside diameter of the donut. AutoCAD prompts:

   ```
   Specify outside diameter of donut <1.0000>:
   ```

3. Specify the outside diameter of the donut. Notice that as soon as you specify both diameters, you see a ghost image of the donut centered at the cursor position. AutoCAD prompts:

   ```
   Specify center of donut or <exit>:
   ```

4. Specify the center of the donut. AutoCAD draws the donut and then repeats the previous prompt.

5. Specify the center point to draw another donut, or press either ENTER or ESC to complete the command.

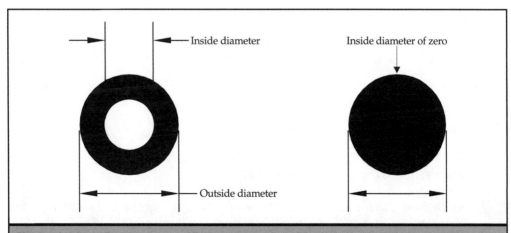

**Figure 4-13.**    *Typical donuts. To create a solid, filled circle, specify an inside diameter of 0.*

# Creating Solid-Filled Areas

You can draw rectangular, triangular, or quadrilateral areas filled with a solid color. AutoCAD prompts you for the first, second, third, and fourth points, and then, if you continue to enter points, alternates back and forth from third to fourth to third points, and so on. You specify the points in a triangular manner. The first and second points establish a starting edge. After that, you enter points diagonally from each other at opposite corners of the object. You can continue to specify points in this fashion, building up a complex solid object.

 *Although you could draw a boundary and then fill the enclosed area with a solid hatch pattern, using solids for simple shapes, such as triangles or rectangular columns on a floor plan, results in a small drawing-file size. Creating simple solids also requires fewer steps than drawing a boundary and then filling it with a solid hatch.*

To draw a rectangular solid-filled area (see Figure 4-14):

1. Do one of the following:

   - On the Surfaces toolbar, click 2D Solid. (*Hint:* You can display this toolbar by using the Toolbars shortcut menu. To display this menu, right-click any visible toolbar.)

   - From the Draw menu, choose Surfaces | 2D Solid.

   - At the command line, type **SOLID** (or **SO**) and press ENTER.

   AutoCAD prompts:

   ```
   Specify first point:
   ```

2. Specify the first point. AutoCAD prompts:

   ```
   Specify second point:
   ```

3. Specify the second point. AutoCAD prompts:

   ```
   Specify third point:
   ```

4. Specify the third point. AutoCAD prompts:

   ```
   Specify fourth point or <exit>:
   ```

5. Specify the fourth point. AutoCAD prompts:

   ```
   Specify third point:
   ```

6. Press ENTER to complete the command.

**Note** *Solids, donuts, wide polylines, and other solid-filled objects appear solid only when Fill mode is turned on. You'll learn more about turning Fill mode on and off in Chapter 5.*

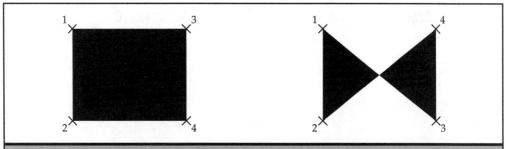

**Figure 4-14.** *Typical solid-filled areas. The order in which you specify points determines the shape of the resulting 2-D solid.*

# Creating Regions

Regions are two-dimensional enclosed areas created from closed shapes called *loops*. A loop consists of a connected sequence of arcs, circles, ellipses, elliptical arcs, lines, polylines, solids, splines, traces, and 3Dfaces that form one or more closed boundaries on a single plane. The boundaries of these loops must form closed areas by being closed objects (such as a closed polyline or spline) or by consisting of a series of objects that share end points with adjacent objects, forming a closed area. As shown in Figure 4-15, these boundaries can't intersect themselves. As explained later in this chapter, you can use Boolean operations to combine regions, creating shapes that would be difficult to draw using other commands. You can also quickly determine the area of a region. In addition, regions are often used to create three-dimensional solid objects.

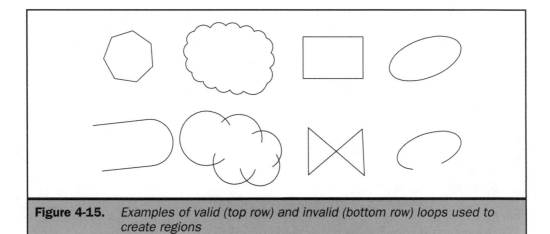

**Figure 4-15.** *Examples of valid (top row) and invalid (bottom row) loops used to create regions*

# Creating Regions with the REGION Command

The REGION command creates a region object from a selection set of existing objects. The command prompts you to select the objects to be converted to regions. When you've finished selecting objects, press ENTER. AutoCAD immediately converts the valid objects selected into regions. If more than one valid boundary is selected, each becomes a separate region.

To create a region from an existing object:

1. Do one of the following:

- On the Draw toolbar, click Region.
- From the Draw menu, choose Region.
- At the command line, type **REGION** (or **REG**) and press ENTER.

    AutoCAD prompts:

    ```
    Select objects:
    ```

2. Select the object or objects that you want to convert into regions. After you select each object, AutoCAD repeats the Select objects prompt, so that you can select additional objects.

3. After you finish selecting objects, press ENTER. AutoCAD immediately converts the valid objects selected into regions, reporting the number of loops extracted and the number of regions created.

---

*LEARN BY EXAMPLE*
*Practice using the REGION command to convert objects into regions. Start a new drawing by using the Figure 4-16 template file on the companion web site.*

To convert the objects into regions, use the following command sequence and instructions. (*Hint:* When selecting the objects, use the *window* object selection method. To do this, when prompted to select objects, click in the drawing to the lower-left of the objects, and then move the cursor to the upper-right and click again. When you move the cursor the first time, you should see a rectangle. Clicking the second time selects all the objects that fall entirely within the rectangle. You'll learn more about selecting objects in Chapter 7.)

```
Command: REGION
Select objects: (select all the objects)
Select objects: ENTER
16 loops extracted.
16 Regions created.
```

When you are finished, save the drawing.

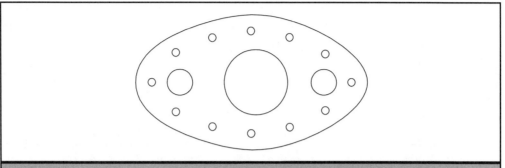

**Figure 4-16.** *A gasket consisting of 15 circles and a closed spline, converted into regions*

  *When you convert objects into regions by using the REGION command, the DELOBJ system variable controls whether the original objects are retained or deleted from the drawing.*

## Creating Regions with the BOUNDARY Command

The BOUNDARY command creates either a polyline boundary or a region from any enclosed area. Unlike the REGION command, the BOUNDARY command works regardless of whether the objects share end points or intersect themselves. When you use the BOUNDARY command, AutoCAD displays the Boundary Creation dialog box, shown in Figure 4-17.

The BOUNDARY command analyzes the objects making up the *boundary set.* When you click the Pick Points button, the command prompts you to select a point in the drawing. It then determines a boundary from existing objects that form an enclosed area. As shown in Figure 4-18, the resulting region is comprised only of the enclosed area formed by the boundary.

To create a region with the BOUNDARY command:

1. Do one of the following:

   ■ From the Draw menu, choose Boundary.

   ■ At the command line, type **BOUNDARY** (or **BO**) and press ENTER.

2. In the Boundary Creation dialog box, select Region from the Object Type drop-down list.

3. Click the Pick Points button. The dialog box temporarily closes and AutoCAD prompts:

   ```
   Select internal point:
   ```

4. Specify a point inside the enclosed area. AutoCAD analyzes the boundary set and then highlights the proposed boundary.

5. Press ENTER to complete the command.

**Note**    *Unlike the REGION command, the BOUNDARY command does not delete the original objects when creating a boundary, regardless of the DELOBJ value.*

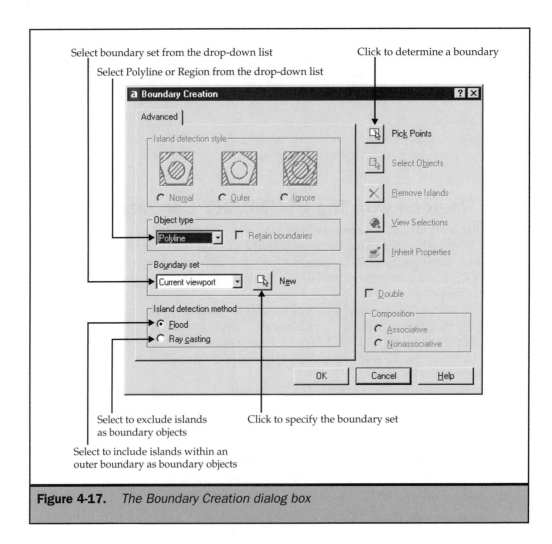

**Figure 4-17.**    *The Boundary Creation dialog box*

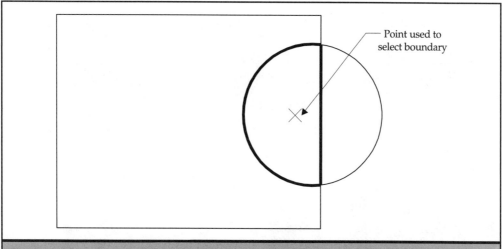

Point used to
select boundary

**Figure 4-18.**   *Boundary region (dark line) created from enclosed area formed by the rectangle and the circle*

## Understanding Boolean Operations

Regions offer interesting drawing possibilities. Although they are only two-dimensional, they are very similar to a special group of AutoCAD objects called *solids*. Solids provide a simple yet sophisticated way to model three-dimensional objects. A solid object has edges, surfaces, and volume. You'll learn more about solids and creating three-dimensional solid models in Chapter 19.

Regions behave like two-dimensional solids. Although regions have no volume, you can perform *Boolean* operations on regions, as well as on solids. Boolean logic, which was originally developed by the British mathematician George Boole, defines algebraic operations of *and, or, not,* and *exclusive or.* In AutoCAD, these Boolean operations become the commands UNION, SUBTRACT, and INTERSECT. Results typical of these commands are shown in Figure 4-19.7

- ■ **UNION**   Combines two or more regions into a single region comprised of their combined areas. The two regions do not need to intersect.
- ■ **SUBTRACT**   Removes one or more regions from another region.
- ■ **INTERSECT**   Creates a new composite region consisting of the area common to two or more overlapping regions.

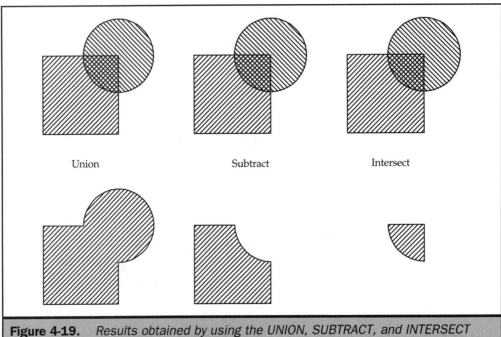

**Figure 4-19.**   *Results obtained by using the UNION, SUBTRACT, and INTERSECT commands on regions*

To create a composite region by union:

1. Do one of the following:

   ■ On the Solids Editing toolbar, click Union.

   ■ From the Modify menu, choose Solids Editing | Union.

   ■ At the command line, type **UNION** (or **UNI**) and press ENTER.

2. Select the objects to be joined, and then press ENTER.

To create a composite region by subtraction:

1. Do one of the following:

   ■ On the Solids Editing toolbar, click Subtract.

   ■ From the Modify menu, choose Solids Editing | Subtract.

   ■ At the command line, type **SUBTRACT** (or **SU**) and press ENTER.

2. Select the region from which you want to subtract, and then press ENTER.

3. Select the region to subtract.

To create a composite region by intersection:

1. Do one of the following:

- On the Solids Editing toolbar, click Intersect.
- From the Modify menu, choose Solids Editing | Intersect.
- At the command line, type **INTERSECT** (or **IN**) and press ENTER.

2. Select the objects to be intersected, and then press ENTER.

*If you use the INTERSECT command on regions that do not actually overlap, AutoCAD deletes the regions and creates a null region. Use the UNDO command to restore the original regions to the drawing.*

**LEARN BY EXAMPLE**
*Practice using the SUBTRACT command to remove the holes from the gasket you drew previously (refer to Figure 4-16). Either load your drawing or load Figure 4-16.DWG on the companion web site.*

To subtract the holes from the gasket, use the following command sequence and instructions:

```
Command: SUBTRACT
Select solids and regions to subtract from ..
Select objects: (select the outline of the gasket)
Select objects: ENTER
Select solids and regions to subtract ..
Select objects: (select all the holes)
Select objects: ENTER
```

Although the resulting drawing does not look any different than before, it has been converted into a single region object. You can see the difference by creating a shaded or rendered view of the gasket, as shown in Figure 4-20.

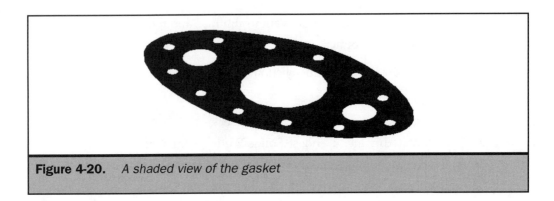

**Figure 4-20.**    *A shaded view of the gasket*

---

**Tip**    *Because regions behave like solids, you can change their shape only by using the three Boolean commands. The EXPLODE command reduces a region back into its constituent parts, at which point you can modify its individual objects.*

The
Complete
Reference

AutoCAD 2002

# Chapter 5

## Viewing Your Drawing

AutoCAD provides many ways to display and view your drawing. You can also change various display settings, to speed up the display or printing of a drawing. This chapter explains how to do the following:

- Navigate within a drawing by scrolling and panning
- Change the magnification of a drawing by zooming in and out
- Use the Aerial View to zoom and pan the drawing
- Work with multiple windows or views of a drawing
- Control the display of elements, to optimize performance when working with large or complex drawings

# Redrawing and Regenerating a Drawing

As you work on a drawing, visual elements may remain on the screen after the completion of a command. For example, when you specify points on the screen, AutoCAD optionally places small marks, called *blips,* which remain on the screen. You can remove these elements by refreshing or redrawing the display.

As you will learn later in this chapter, the AutoCAD graphics area can be divided into multiple viewports, with each viewport displaying a different view of your drawing. When you refresh the display, you can redraw just the current viewport or all the viewports of the active drawing.

To redraw the current viewport of the active drawing:

- From the View menu, choose Redraw.
- At the command line, type **REDRAW** (or **R**) and press ENTER.

To redraw all the viewports of the active drawing, at the command line, type **REDRAWALL** (or **RA**) and press ENTER.

Information about drawing objects is stored in a database as floating-point values, ensuring a high level of precision. Sometimes, a drawing must be recalculated, or regenerated, from the floating-point database, to convert the floating-point values into the appropriate screen coordinates. Some commands automatically regenerate the entire drawing and recalculate the screen coordinates. You can also manually initiate a regeneration. When the drawing is regenerated, it is also redrawn. Regeneration takes longer than redrawing, but sometimes you may need to regenerate the drawing. For example, after turning Fill mode on or off, you must regenerate the drawing to see the change.

As with redrawing, you can regenerate just the current viewport or all the viewports of the active drawing.

To regenerate the current viewport of the active drawing, do one of the following:

- From the View menu, choose Regen.
- At the command line, type **REGEN** (or **RE**) and press ENTER.

To regenerate all the viewports of the active drawing, do one of the following:

- From the View menu, choose Regen All.
- At the command line, type **REGENALL** (or **REA**) and press ENTER.

**Note**
*Some commands automatically force AutoCAD to regenerate the drawing under certain conditions. Because regenerating the drawing can be a lengthy process for large drawings, the command REGENAUTO allows you to control this automatic regeneration. To disable automatic regeneration of a drawing, type **REGENAUTO** at the command line and set automatic regeneration to Off.*

# Moving Around Within a Drawing

You can move the view of a drawing displayed in the current viewport by scrolling or panning. Doing so changes the portion of the drawing that you are viewing, without changing the current magnification. *Scrolling* lets you move around in the drawing horizontally and vertically. *Panning* lets you move the drawing in any direction.

## Using Scroll Bars

Each AutoCAD drawing window contains vertical and horizontal scroll bars that you can use to navigate within the drawing. The position of the scroll box in relation to the scroll bar indicates the location of the drawing in relation to the virtual screen extents of the drawing (the extents of the current regenerated view of the drawing).

**Note**
*When you are working with multiple viewports, the scroll bars affect only the current viewport. If you prefer, you can turn off the scroll bars from the Display tab of the Options dialog box to make more room for viewing your drawing. You'll still be able to use AutoCAD's other methods, such as panning, to move around within your drawings.*

## Using the PAN Command

The PAN command lets you move about the drawing in any direction. Panning shifts or slides the drawing horizontally, vertically, or diagonally. The magnification of the drawing remains the same; only the portion of the drawing that is displayed changes.

When you activate the PAN command, the cursor changes to a hand. Click and hold down the pick button (the left-mouse button) on your pointing device to lock the cursor into its current position. Then, drag the drawing to move or pan it over to the desired location. Release the pick button to stop panning. You can then press the pick button again to pan to a different location.

If you pan to the edge of the virtual screen extents, AutoCAD displays a bar adjacent to the hand cursor on the side where the extent has been reached.

Left extent      Right extent      Top extent    Bottom extent

AutoCAD also displays a message in the status bar, indicating that you have panned to the extents of the drawing and can't pan any farther in that direction.

To pan the drawing:

1. Do one of the following:

   - On the Standard toolbar, click Pan Realtime.
   - From the View menu, choose Pan | Real Time.
   - Right-click in the drawing window and select Pan from the shortcut menu.
   - At the command line, type **PAN** (or **P**) and press ENTER.

2. Press and hold down the pick button.

3. Drag the drawing.

4. Release the pick button.

5. Press ENTER or ESC to end the PAN command.

*If you are using a Microsoft Intellimouse, you can pan by depressing the mouse wheel while dragging the mouse. Pressing the SHIFT key while depressing the mouse wheel locks the pan motion to the horizontal or vertical direction. You can also pan when the Intellimouse is set to joystick mode by holding down CTRL and the wheel button while you move the mouse. Joystick panning is controlled by the direction and speed with which you drag your mouse. The system variable MBUTTONPAN must be set to 1 to pan using the mouse wheel.*

# Changing the Magnification of Your Drawing

You can change the magnification of your drawing at any time by zooming. Zoom out to reduce the magnification, so that you can see more of the drawing, or zoom in to increase the magnification, so that you can see a portion of the drawing in greater detail. Changing the magnification of the drawing affects only the way the drawing is displayed; it has no effect on the dimensions of the objects in your drawing.

When you start the ZOOM command, AutoCAD prompts:

```
Specify corner of window, enter a scale factor (nX or nXP), or
[All/Center/Dynamic/Extents/Previous/Scale/Window] <real time>:
```

Specify the option that you want to use by typing the capital letter corresponding to that option, or right-click and then select the option from the shortcut menu. In most cases, starting the command from a toolbar or menu automatically selects the option that you want.

The Previous and Realtime options of the ZOOM command are available on the Standard toolbar. The other ZOOM command options are available both from the Zoom flyout on the Standard toolbar and the Zoom toolbar. You can also select the individual options from the View menu or by typing the command at the command line.

## Realtime Pan and Zoom

Realtime panning and zooming is a feature first available in AutoCAD Release 14. This feature enables you to shift the position of the drawing, or change its magnification, interactively. In prior versions of AutoCAD, if you wanted to pan the drawing, you had to either pick points within the drawing or enter coordinates in response to command prompts. Similarly, to change the magnification of the drawing, you had to use one of the ZOOM command options. Although you can still use the older methods, in AutoCAD 2002, the default method to pan or zoom your drawing is to use the Realtime feature.

When you use the Realtime Pan feature, the cursor changes to a hand. When you use the Realtime Zoom feature, the cursor changes into a magnifying glass with plus (+) and minus (-) signs. Whenever the Realtime Pan or Zoom feature is active, you can also right-click to display this shortcut menu:

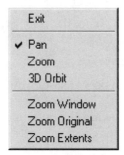

You can use the shortcut menu to exit from the command, switch between the Pan and Zoom modes, zoom in to a designated window, redisplay the previous zoomed or panned view, or zoom to the drawing's extents. You can also activate the 3DORBIT command. You'll learn about 3DORBIT in Chapter 19.

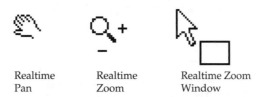

| Realtime Pan | Realtime Zoom | Realtime Zoom Window |

**Note**  *As you learned in Chapter 1, flyouts are indicated by a small triangle in the lower-right corner of the button. To display a flyout, click the toolbar button and hold down the mouse button. To select a button from a flyout, continue to hold down the mouse button while pointing to the toolbar button that you want, and then release the mouse button. In addition to starting the command that you specified, the button you selected on the flyout becomes the default button on the toolbar.*

The most often used options of the ZOOM command are described in the following sections.

## Zooming in Realtime

The default method of using the ZOOM command is to use the Realtime Zoom feature. Like panning in Realtime, zooming in Realtime lets you interactively change the magnification of your drawing.

To zoom in Realtime:

1. Do one of the following:

   ■ On the Standard toolbar, click Zoom Realtime.

   ■ From the View menu, choose Zoom | Realtime.

   ■ Right-click in the drawing window and select Zoom from the shortcut menu.

   ■ At the command line, type **ZOOM** (or **Z**) and then press ENTER twice.

   When you activate Realtime Zoom, the cursor changes into a magnifying glass with plus (+) and minus (-) signs, and AutoCAD prompts:

   ```
   Press Esc or Enter to exit, or right-click to display shortcut menu.
   ```

2. Click (press the pick button on your pointing device) and then drag the magnifying glass toward the top of the screen to zoom in, or drag the magnifying glass toward the bottom of the screen to zoom out.

3. Release the pick button to stop zooming.

4. Press ENTER or ESC to end the ZOOM command.

Moving half the distance from the bottom toward the top of the screen is equivalent to doubling the magnification; moving half the distance from the top toward the bottom of the screen is equivalent to reducing the magnification by one-half.

If you reach the limit to which AutoCAD can zoom in, the plus sign disappears and AutoCAD displays the following warning on the status line:

```
Already zoomed in as far as possible.
```

Similarly, if you reach the limit to which AutoCAD can zoom out, the minus sign disappears and AutoCAD displays the following warning on the status line:

```
Already zoomed out as far as possible.
```

 *If you are using a Microsoft Intellimouse, you can zoom by rolling the mouse wheel. Simply roll the wheel toward you to zoom in, and away from you to zoom out. To adjust the zoom percentage that the wheel uses, change the value of the ZOOMFACTOR system variable.*

## Controlling the Virtual Display

AutoCAD stores information about the drawing in two different ways. The actual drawing database contains detailed information about every object, including its exact coordinate points. (You'll learn more about coordinates in the next chapter.) This information is saved to the DWG file when you save your drawing. To improve performance, AutoCAD also maintains information about what is displayed on the screen. This virtual display or display list contains only the information that you need to manipulate the drawing on the screen. When you use the REDRAW or REDRAWALL commands, AutoCAD refreshes the display from this display list. When you regenerate the drawing, you actually rebuild this virtual display from the drawing database.

As you've already seen, Realtime Pan and Zoom do not work beyond the limits of the virtual display. To pan or zoom beyond these limits, you must regenerate the drawing. You can use VIEWRES to control the way that AutoCAD displays curved objects, by setting the circle zoom percent value in the range from 1 to 20000. The lower the number, the faster the drawing is redrawn, but at the expense of fewer segments being used to display circles and arcs (resulting in a coarser appearance). The higher the number, the more segments that are used to display curves, but at the expense of longer redrawing time. Regardless of the way that curves are displayed on the graphics screen, they are always plotted as accurate curves. The default value is 100. You can also change this value by changing the Arc and Circle Smoothness value in the Display tab of the Options dialog box.

Note that if you use the VIEWRES command, AutoCAD first prompts you to specify whether or not you want fast zooms. Your response has no effect, however. This prompt simply remains from previous versions of AutoCAD to maintain compatibility with automated scripts that expect this prompt to be present.

The WHIPARC system variable determines whether AutoCAD displays circles and arcs as true curves or as a series of vectors. By default, curved objects are displayed as a series of vectors, because doing so results in faster redraws, object snaps, and entity selections. Displaying the true curves, however, results in a higher-quality display and uses less display list memory, thus improving drawing regeneration.

# Using a Zoom Window

You can quickly magnify a specific area of a drawing by specifying the corners of a rectangular window around it. The lower-left corner of the window becomes the lower-left corner of the new area displayed.

To zoom in to an area by using a window (see Figure 5-1):

1. Do one of the following:

- On the Standard toolbar or the Zoom toolbar, click Zoom Window.
- From the View menu, choose Zoom | Window.
- At the command line, type **ZOOM** (or **Z**) and press ENTER; then, either type **W** and press ENTER, or right-click and choose Window from the shortcut menu.

2. Specify one corner of the area that you want to view.

3. Specify the opposite corner of the area that you want to view.

*If you are using the Realtime Zoom or Pan feature, you can also right-click to display the shortcut menu and then choose Zoom Window. The cursor changes to an arrow with an adjacent rectangle. To zoom in to a specified area, click and hold down the pick button to select one corner, drag the cursor toward the opposite corner, and then release the pick button to complete the zoom window operation.*

*Although the lower-left corner becomes the lower-left corner of the new display area, the top and right edges depend on your display window and may not correspond exactly to the area that you specify.*

## Displaying the Previous View of a Drawing

After you zoom in or use pan to view a portion of your drawing in greater detail, you may want to zoom back out to see the entire drawing, or shift the drawing back to its previous position. AutoCAD enables you to do so, by letting you repeatedly step back through each previous view of the drawing.

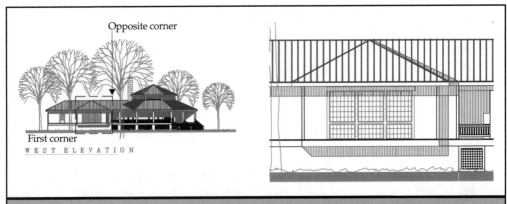

**Figure 5-1.**    *To zoom by using a window, specify the opposite corners of the area that you want to view.*

To restore the previous view, do one of the following:

- On the Standard toolbar, click Zoom Previous.
- From the View menu, choose Zoom | Previous.
- At the command line, type **ZOOM** (or **Z**) and press ENTER; then, either type **P** and press ENTER, or right-click and choose Previous from the shortcut menu.

## Zooming to a Specific Scale

You can increase or decrease the magnification of your view by a precise scale factor, measured in one of three ways:

- Relative to the overall size of the drawing (the drawing limits)
- Relative to the current view
- Relative to paper space units

When you change the magnification factor, the portion of the drawing located at the center of the current viewport remains centered on the screen.

To change the magnification of the view relative to the overall size of the drawing, start the ZOOM command and type a number that represents the magnification scale factor. For example, if you type a scale factor of 2, the drawing appears at twice its original size. If you type a magnification factor of .5, the drawing appears at half its original size.

You can change the magnification of the drawing relative to its current magnification (the current view) by adding an X after the magnification scale factor. For example, if you type a scale factor of 2X, the drawing changes to twice its current size. If you type a magnification factor of .5X, the drawing changes to half its current size.

### Zoom Relative to Paper Space

When you are working in paper space, you can also change the magnification of the current model space view, relative to paper space, by entering a scale factor followed by XP. This method is similar to zooming the drawing relative to the drawing limits, but it affects only the current viewport and scales it relative to the paper space units rather than to the absolute drawing size. For example, suppose that you're designing a widget. If the widget measures three inches across its side, and you want the side view to be drawn at half scale, you can use the XP option to scale the side-view viewport, using a scale factor of .5XP. If you later plot the paper space view at full size, the resulting side view would be half size. You'll learn more about paper space in Chapter 17.

To zoom by using a specific scale:

1. Do one of the following:

- On the Standard toolbar or the Zoom toolbar, click Zoom Scale.
- From the View menu, choose Zoom | Scale.
- At the command line, type **ZOOM** (or **Z**) and press ENTER; then, either type **S** and press ENTER, or right-click and choose Scale from the shortcut menu.

AutoCAD prompts:

```
Enter scale factor (nX or nXP):
```

2. Type the scale factor relative to the drawing limits, current view, or paper space view, and then press ENTER.

*You don't actually need to type S before you enter the scale factor. You can simply type the scale factor in response to the list of ZOOM command options, and then press* ENTER, *saving an intermediate step. Using these step-saving shortcuts is one way to work faster with AutoCAD.*

AutoCAD provides another easy way to quickly change the magnification relative to the current view. The Zoom In and Zoom Out tools on the Zoom flyout, and the Zoom In and Zoom Out selections on the View | Zoom menu, perform the same function as zooming by using a factor of 2X and .5X, respectively.

Zoom In          Zoom Out

## Centering the Zoom Area

When you change the zoom scale factor, the portion of the drawing that is located at the center of the current viewport remains centered after changing the magnification. You can use the Center option, however, to specify the point that you want at the center of the view when you change the drawing magnification. When you use this option, AutoCAD first prompts you to select the center point of the resulting zoom area. You can then specify the magnification relative to the drawing limits, current view, or paper space.

To change the center of the zoom area:

1. Do one of the following:

- On the Standard toolbar or the Zoom toolbar, click Zoom Center.
- From the View menu, choose Zoom | Center.

■ At the command line, type **ZOOM** (or **Z**) and press ENTER; then, either type **C** and press ENTER, or right-click and choose Center from the shortcut menu.

AutoCAD prompts:

```
Specify center point:
```

2. Select the point that you want located at the center of the new view. AutoCAD prompts:

```
Enter magnification or height <current>:
```

3. Type the scale factor relative to the drawing limits, current view, or paper space view, and then press ENTER.

## Displaying the Entire Drawing

To display the entire drawing, you can zoom out to either its limits or *extents* (the area actually containing drawing objects) by using either the All or Extents options, respectively, of the ZOOM command. If all objects are drawn within the drawing's limits, Zoom All displays the drawing to its limits. If objects extend outside the limits, this option displays the drawing extents, instead. Zoom Extents displays the drawing only to its extents, even if that is less than the limits.

To zoom to the drawing limits, do one of the following:

■ On the Standard toolbar or the Zoom toolbar, click Zoom All.

■ From the View menu, choose Zoom | All.

■ At the command line, type **ZOOM** (or **Z**) and press ENTER; then, either type **A** and press ENTER, or right-click and choose All from the shortcut menu.

To zoom to the drawing extents, do one of the following:

■ On the Standard toolbar or the Zoom toolbar, click Zoom Extents.

■ From the View menu, choose Zoom | Extents.

■ At the command line, type **ZOOM** (or **Z**) and press ENTER; then, either type **E** and press ENTER, or right-click and choose Extents from the shortcut menu.

    *Infinite construction lines (xlines) and rays have no effect on the drawing extents.*

    *If you've changed the drawing extents since the last time that you zoomed the drawing to its extents, Zoom Extents may require the regeneration of the drawing. If REGENAUTO is turned off, AutoCAD warns you that a regeneration is required.*

## Using the Aerial View

The Aerial View is a navigation tool that lets you see the entire drawing in a separate window. You can use the controls within the Aerial View window to move quickly within your drawing. By keeping the Aerial View window open as you work, you can zoom and pan without ever starting a PAN or ZOOM command.

To display the Aerial View window (see Figure 5-2), do one of the following:

- From the View menu, choose Aerial View.

- At the command line, type **DSVIEWER** (or **AV**) and press ENTER.

> **Note**    *Repeating these steps when the Aerial View window is open closes the window.*

Initially, the Aerial View window displays the entire drawing. A dark rectangle in the Aerial View window (the current view box) surrounds the area corresponding to the portion of the drawing currently displayed in the main AutoCAD drawing window.

The Aerial View window has three buttons that control panning and zooming of the image within the Aerial View. You can also control the operation of the Aerial View window by using its View and Options menus.

When you click inside the Aerial View, AutoCAD displays a second rectangle, called a *pan and zoom box.* When the box has an X in its center, the Aerial View is in Pan mode, and moving the box pans the drawing. You can move this box to the desired location and then click to switch to Zoom mode. In Zoom mode, the box has an arrow along its right edge. You can then change the size of the box by moving your pointing device. Click to toggle back and forth between Pan and Zoom modes. After you select

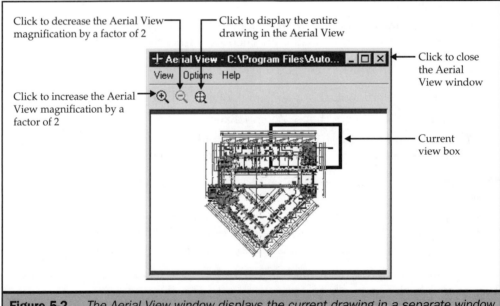

**Figure 5-2.** *The Aerial View window displays the current drawing in a separate window.*

the area that you want displayed in the drawing area, right-click or press ENTER to lock the current view box in position.

*If the Aerial View's Realtime Zoom mode is active, the drawing area updates continuously as you pan and zoom in the Aerial View.*

To pan and zoom using the Aerial View:

1. Click in the Aerial View window to display the pan and zoom box.
2. While in Pan mode (indicated by an X in the center of the box), move the box to pan to the desired area of the drawing.
3. In the Aerial View window, click to switch to Zoom mode.
4. Move the cursor to change the size of the zoom box.
5. When you are satisfied with the magnification, click to switch back to Pan mode; or, either right-click or press ENTER to lock the current view box in position.

## Resizing the Aerial View Image

You can change the size of the image appearing in the Aerial View window without affecting the actual AutoCAD drawing. The Zoom In button increases the Aerial View magnification incrementally by a factor of 2, while the Zoom Out button decreases the magnification by the same factor. The Global button redisplays the entire drawing in the Aerial View window. These functions are duplicated by the Zoom In, Zoom Out, and Global commands in the Aerial View's View menu, and they also appear in the shortcut menu that is displayed when you right-click in the Aerial View window.

*When the entire drawing is displayed in the Aerial View window, the Zoom Out item in the View menu and the Zoom Out button are grayed out (unavailable). When the current view nearly fills the Aerial View window, the Zoom In item in the View menu and the Zoom In button are grayed out (unavailable). Both conditions can occur at the same time, such as immediately after you enter a Zoom Extents command.*

## Changing the Aerial View Options

AutoCAD normally updates the Aerial View window automatically whenever you make any changes to the drawing. If you work with multiple viewports, the Aerial View window also changes when you select different viewports. When you are working on large or complex drawings, this may slow down AutoCAD's performance. To improve performance, you may want to turn off these Aerial View options.

To turn an Aerial View option on or off:

1. In the Aerial View window, select the Options menu; or right-click in the Aerial View window to display the shortcut menu.

2. Choose the desired option to toggle it on or off. A check mark indicates that the option is turned on; the absence of a check mark indicates that the option is turned off.

The Aerial View options are summarized as follows:

■ **Auto Viewport**   When on, AutoCAD automatically updates the Aerial View to match the active viewport.

■ **Dynamic Update**   When on, any changes that you make to the drawing are immediately visible in the Aerial View. When off, changes don't appear in the Aerial View until you perform a zoom or pan function in the Aerial View.

■ **Realtime Zoom**   When on and the Aerial View window is in Zoom mode, once you select the first corner of the view window, the drawing updates in real time as you move the cursor to select the second corner. When off, the drawing doesn't update until you select the second corner.

**Note**   *Although the Aerial View works in both paper space and model space, in paper space, the Aerial View window shows only paper space entities, including viewport borders. In addition, the Aerial View window doesn't update in real time when in paper space. The Aerial View window is also not available in nonzoomable viewports, such as viewports displaying perspective or shaded views. When you switch back to a conforming viewport, the Aerial View window immediately returns to the screen. In addition, AutoCAD has only one Aerial View window. When multiple drawings are open, only the current drawing is displayed in the Aerial View window. As you switch between the open drawings, the Aerial View window updates to show the current drawing.*

## Using Named Views

As you work on a drawing, you may find that you frequently switch among different portions of it. For example, if you are drawing a floor plan of a house, you may zoom in to particular rooms of the house and then zoom out to display the entire house. Although you can use the PAN and ZOOM commands or the Aerial View to do this, it is much easier to save various views of the drawing as named views. You can then quickly switch between these views by restoring the named views.

You can save, restore, and—if no longer needed—delete named views from the View dialog box, shown in Figure 5-3. To display this dialog box, do one of the following:

■ On the Standard or View toolbar, click Named Views.

■ From the View menu, choose Named Views.

■ At the command line, type **VIEW** (or **V**) and press ENTER.

When you save a named view, AutoCAD saves the view's center point, viewing direction, zoom factor, perspective setting, and whether the view was created in model

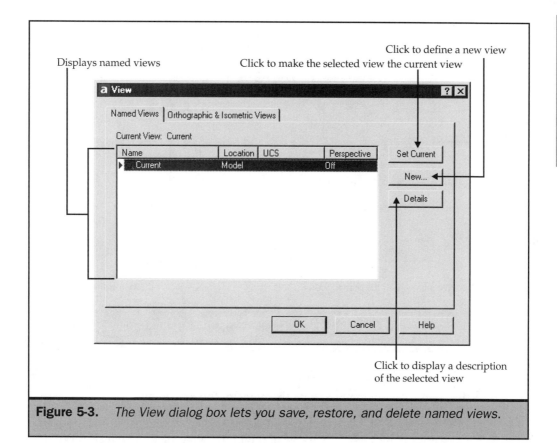

Click to define a new view

Click to make the selected view the current view

Displays named views

Click to display a description of the selected view

**Figure 5-3.**    *The View dialog box lets you save, restore, and delete named views.*

space or a layout. You can also save the current UCS with the view so that when you restore the view, you also restore the UCS. You'll learn about the UCS in the next chapter. In Chapter 19, you'll learn about perspective views and changing the viewing direction.

## Saving Named Views

You can save as a named view either the current view (everything displayed in the current viewport) or a windowed area. After you save a named view, you can restore that view in the current window at any time. You save a new view by using the New View dialog box, shown in Figure 5-4, which is displayed when you click the New button in the View dialog box.

To save the current view as a named view:

1. Do one of the following:

   ■ On the Viewpoint toolbar, click Named Views.

   ■ From the View menu, choose Named Views.

   ■ At the command line, type **VIEW** (or **V**) and press ENTER.

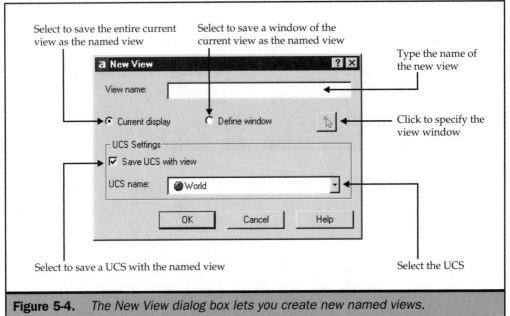

Select to save the entire current
view as the named view

Select to save a window of the
current view as the named view

Type the name of
the new view

Click to specify the
view window

Select to save a UCS with the named view

Select the UCS

**Figure 5-4.**    *The New View dialog box lets you create new named views.*

2. In the View dialog box, click New.

3. In the New View dialog box, type the name of the new view.

4. Choose Current display.

5. If you want to save a coordinate system with the view, select Save UCS With View, and then select the UCS name.

6. Click OK to close the New View dialog box.

7. Click OK to complete the command.

To save a windowed area of the current viewport as a named view:

1. Display the View dialog box.

2. In the View dialog box, click New.

3. In the New View dialog box, type the name of the new view.

4. Choose Define Window.

5. Click the Define View Window button. The dialog boxes temporarily disappear, and AutoCAD prompts:

   ```
   Specify first corner:
   ```

6. Specify the first corner of the view window. AutoCAD prompts:

   ```
   Specify opposite corner:
   ```

7. Specify the opposite corner of the view window.

8. If you want to save a coordinate system with the view, select Save UCS With View, and then select the UCS name.

9. Click OK to close the New View dialog box.

10. Click OK to complete the command.

## Restoring Named Views

After you save one or more named views, you can restore any of those views in the current viewport.

To restore a named view in the current viewport:

1. Do one of the following:

   ■ On the View toolbar, click Named Views.

   ■ From the View menu, choose Named Views.

   ■ At the command line, type **VIEW** (or **V**) and press ENTER.

2. In the View dialog box, select the named view that you want to restore.

3. Click Set Current, or right-click and select Set Current from the shortcut menu.

4. Click OK to complete the command.

 *To delete a named view, display the View dialog box, select the view you want to delete, and then either press the* DELETE *key or right-click and select Delete from the shortcut menu.*

## Using Multiple Viewports

The Model tab can be split into multiple viewports. When you begin a new drawing, it is usually displayed in a single viewport that fills the entire drawing area in the Model tab. You can split the drawing area into multiple viewports, each of which can display a different portion of the drawing. For example, you could display a view of the entire drawing in one viewport and a detailed view in another. Or, you could display the top, front, and side views of an object, each in its own viewport. Figure 5-5 shows a typical drawing with several viewports.

After you divide the screen into multiple viewports, you can control each viewport separately. For example, you can zoom or pan in one viewport without affecting the display in any of the other viewports. You can control the grid, snap, view orientation, and UCS separately for each viewport. You can restore views in individual viewports, draw by selecting points or objects from one viewport to another, and name viewport configurations individually, so that you can reuse them later.

As you draw, any changes that you make in one viewport are immediately visible in the others. You can switch from one viewport to another at any time, even in the middle of a command. To switch viewports, you simply click inside the new viewport to make it the current viewport. For example, you can begin a line in one viewport,

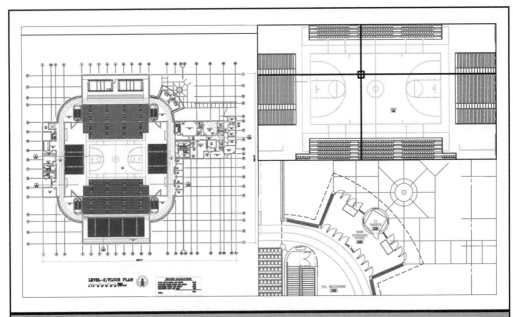

**Figure 5-5.** *A drawing with three viewports. Notice the crosshairs in the upper-right viewport, indicating its status as the current viewport.*

click inside another viewport, and then specify the end point of the line. When you click inside a viewport to make it the current viewport, the crosshairs appear in that viewport, and the viewport border appears darker.

Viewports divided in this fashion are referred to as *tiled viewports,* because they fill the drawing area completely and cannot overlap. Tiled viewports are created in model space and differ from the floating viewports created in paper space on a Layout tab. Paper space viewports are used to create the final layout prior to printing a copy of a drawing. You'll learn more about floating paper space viewports in Chapter 17.

When you are working with tiled viewports, you can also select objects for modification in any viewport. The objects become highlighted only in the viewport in which they are selected. If you start dragging the selected objects, all the selected objects become highlighted in the current viewport. The other viewports show highlighting only for those objects that are actually selected in those viewports. When you finish modifying the objects, the changes appear in all the viewports.

## Dividing the Current Viewport into Multiple Viewports

You create and manipulate tiled viewports by using the VPORTS command. Either the display or the current viewport can be divided into two, three, or four viewports. You control the number of viewports created and the arrangement of the viewports. Figure 5-6 shows the 12 standard viewport configurations.

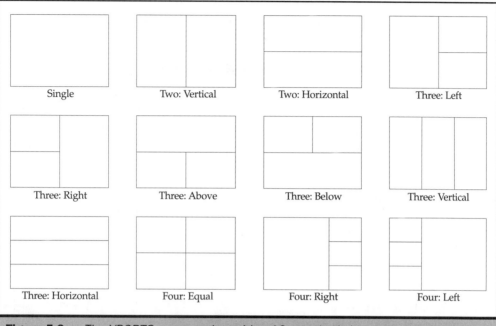

**Figure 5-6.** *The VPORTS command provides 12 standard viewport configurations when working in model space.*

**Note**    *The Single viewport selection always restores the screen to a single viewport. The Four: Right and Four: Left selections can be applied only to the entire display and are available only when working in model space. The other nine configurations can be applied to the entire display or can also be used to subdivide the current viewport.*

To start the command, do one of the following:

- On the Viewports toolbar, click Display Viewports Dialog.
- From the View menu, choose Viewports and then select the desired option from the submenu.
- At the command line, type **VPORTS** and press ENTER.

If you select the 2, 3, or 4 Viewports configuration from the submenu, AutoCAD subdivides the current viewport. In the case of the 2 Viewports or 3 Viewports configurations, AutoCAD prompts you on the command line to select which of the possible two or three viewport configurations you want to use. When you start the command from the command line or select New Viewports from the Viewports submenu, AutoCAD displays the New Viewports tab of the Viewports dialog box, shown in Figure 5-7. The left side of the dialog box contains a list of standard viewport configurations, while the Preview image on the right side shows how the display will

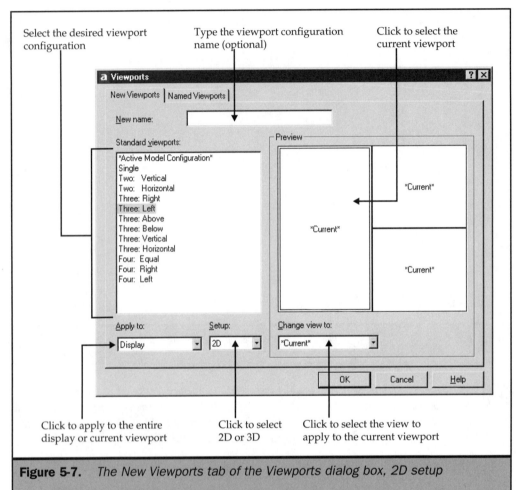

**Figure 5-7.**   *The New Viewports tab of the Viewports dialog box, 2D setup*

be divided based on the selections you make. The current viewport is indicated by a double rectangle.

In the Apply To drop-down list, you can choose whether the viewport configuration is applied to the entire display or only within the current viewport. Applying a configuration to the current viewport subdivides that viewport without affecting the others.

If you select 2D in the Setup drop-down list, the new viewport configuration is created with the current view in all of the viewports. If you select 3D, however, a set of standard orthogonal views (such as top, front, and side) are applied to the viewports. By typing a name in the New Name box, you can also save the resulting viewport arrangement as a named configuration.

To create multiple tiled viewports:

1. Do one of the following:

   ■ On the Viewports toolbar, click Display Viewports Dialog.

   ■ From the View menu, choose Viewports | New Viewports.

   ■ At the command line, type **VPORTS** and press ENTER.

2. In the Viewports dialog box, under Standard Viewports, choose the desired viewport configuration (such as Three: Right).

3. Under Apply To, do one of the following:

   ■ Select Display to discard any current viewports and apply the selected viewport configuration to the entire Model tab.

   ■ Select Current Viewport to apply the selected viewport configuration to the current viewport.

   If you are subdividing the current viewport, make sure the viewport you want to divide is selected in the Preview area.

4. Click OK.

You can also join adjacent viewports to create nonstandard viewport arrangements, as long as the resulting joined viewport forms a rectangle.

To join two viewports:

1. From the View menu, choose Viewports | Join. AutoCAD prompts:

   ```
   Select dominant viewport <current viewport>:
   ```

2. Click inside the viewport containing the view that you want to keep. AutoCAD prompts:

   ```
   Select viewport to join:
   ```

3. Click inside the adjacent viewport to join it to the first viewport.

**Note**   *You can also use the VPORTS command to create tiled paper space viewports in a Layout tab. Although you can't name paper space viewports, you can restore viewports that were created and saved previously in model space. You'll learn about paper space viewports in Chapter 17.*

## Saving and Restoring Viewport Configurations

If you specify a name when creating a viewport arrangement, AutoCAD saves that arrangement as a named configuration, so that you can recall it to the screen later. The number and placement of the viewports are saved exactly as they are currently displayed. AutoCAD also saves the zoom factor, grid and snap settings, viewing direction, and coordinate system settings for each viewport.

To name and save a viewport configuration in model space:

1. Display the New Viewports tab of the Viewports dialog box.

2. Type the name of the viewport configuration in the New Name box and then press ENTER.

3. Configure the viewports as desired.

4. Click OK.

You restore previously saved named viewports from the Named Viewports tab of the Viewports dialog box, shown in Figure 5-8. The left side of the dialog box contains a list of previously saved viewport configurations, while the Preview image on the right side displays the viewport arrangement of the configuration you select from the list.

To restore a named viewport configuration:

1. Do one of the following:

   ■ On the Viewports toolbar, click Display Viewports Dialog.

   ■ From the View menu, choose Viewports | Named Viewports.

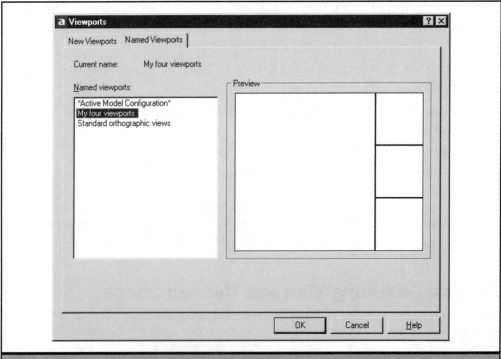

**Figure 5-8.** *The Named Viewports tab of the Viewports dialog box*

■ At the command line, type **VPORTS** and press ENTER; then, select the Named Viewports tab.

2. Under Named Viewports, select the viewport configuration you want to restore.

3. Click OK.

*To delete a named viewport configuration, display the Named Viewports tab of the Viewports dialog box, select the named viewport that you want to delete, and then either press the DELETE key or right-click and select Delete from the shortcut menu.*

### LEARN BY EXAMPLE
*To see how to use the Viewports dialog box to create standard orthogonal views, open the AutoCAD drawing Figure 5-9 from the companion web site.*

Figure 5-9 shows the standard orthogonal views—top, front, right, and isometric. When you first open this drawing, you see only an isometric view. Use the following steps to automatically divide the Model tab into the four orthogonal views:

1. Display the New Viewports tab of the Viewports dialog box.

2. In the Setup drop-down list, choose 3D.

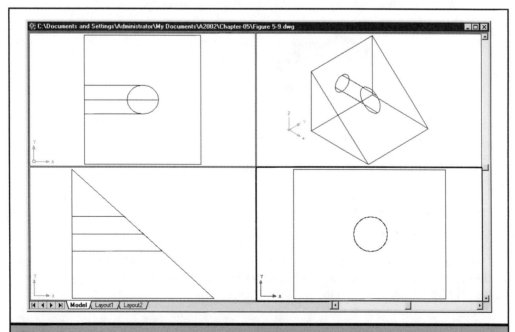

**Figure 5-9.** *Standard orthogonal views created using the Viewports dialog box*

3. Under Standard Viewports, select Four: Equal.

4. Click OK.

# Controlling Visual Elements

The number of objects in your drawing and the complexity of the drawing affect how quickly AutoCAD can process commands and display your drawing. You can improve overall program performance by turning off the display of certain visual elements, such as solid fills and text, while you work on the drawing. When you are ready to plot your drawing, you can then turn on these elements so that your drawing plots the way that you want.

You can also improve performance by turning off features, such as object selection highlighting and the display of marker blips created when you select locations in the drawing. Many of these options are available from the Display tab of the Options dialog box (see Figure 5-10).

## Turning Fill On and Off

You can reduce the time that it takes to display or print a drawing by turning off the display of solid fill. When solid fill is turned off, all filled objects, such as wide polylines, hatched areas, and solids, display and print as outlines. When you turn solid fill on or off, you must regenerate the drawing before the change becomes visible.

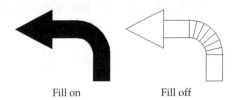

Fill on          Fill off

To turn solid fill on or off:

1. At the command line, type **FILL** and press ENTER. AutoCAD prompts:

   ```
   Enter mode [ON/OFF] <current>:
   ```

2. Type either **ON** or **OFF** and then press ENTER, or use the shortcut menu.

3. Regenerate the drawing to display the change.

You can also turn solid fill on and off by toggling the Apply Solid Fill check box on the Display tab of the Options dialog box (refer to Figure 5-10) or by changing the value of the FILLMODE system variable.

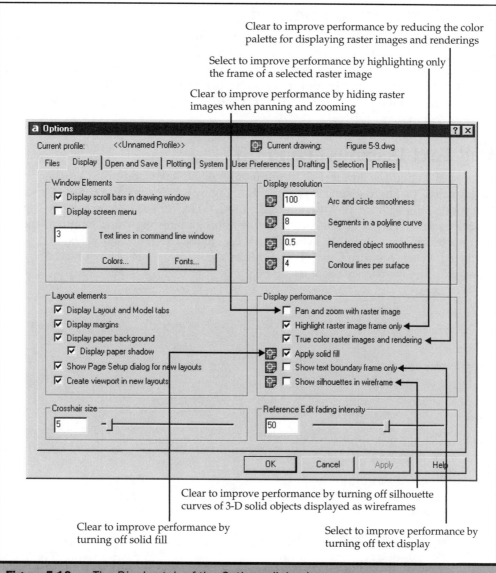

Clear to improve performance by reducing the color palette for displaying raster images and renderings

Select to improve performance by highlighting only the frame of a selected raster image

Clear to improve performance by hiding raster images when panning and zooming

Clear to improve performance by turning off silhouette curves of 3-D solid objects displayed as wireframes

Clear to improve performance by turning off solid fill

Select to improve performance by turning off text display

**Figure 5-10.**    *The Display tab of the Options dialog box*

# Turning Lineweight On and Off

You can also improve AutoCAD's performance by turning off the display of lineweights. Lineweights add width to objects. Any lineweight width that displays as more than 1 pixel in width may slow down AutoCAD's performance. You can toggle lineweights on and off by clicking the LWT button on the status bar, by selecting the Display Lineweight check box in the Lineweight Settings dialog box, or by changing the value of the LWDISPLAY system variable.

# Turning Text Display On and Off

Text objects require a considerable amount of time to display and print. You can reduce the time that it takes to display or print a drawing by turning on AutoCAD's Quick Text mode. When Quick Text mode is on, AutoCAD indicates only the location of text entities by displaying a rectangle of the same approximate height and length as the string of text that it represents. When Quick Text mode is off, the actual text is displayed. When you turn Quick Text mode on or off, you must regenerate the drawing before the change becomes visible.

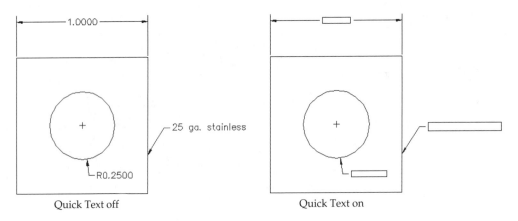

Quick Text off                    Quick Text on

To turn Quick Text mode on or off:

1. At the command line, type **QTEXT** and press ENTER. AutoCAD prompts:

   ```
   Enter mode [ON/OFF] <current>:
   ```

2. Type either **ON** or **OFF** and then press ENTER.

3. Regenerate the drawing to display the change.

You can also turn Quick Text mode on and off by toggling the Show Text Boundary Frame Only check box on the Display tab of the Options dialog box or by changing the value of the QTEXTMODE system variable.

## Turning Highlighting On and Off

When you select objects to modify, AutoCAD highlights them by using a dashed-line type. This highlight disappears when you finish modifying the objects. Sometimes, highlighting objects can take a considerable amount of time. You can improve overall program performance by turning highlighting off.

To turn highlighting on and off:

1. At the command line, type **HIGHLIGHT** and press ENTER. AutoCAD prompts:

   ```
   Enter new value for HIGHLIGHT <current>:
   ```

2. Type **0** to turn highlighting off, or **1** to turn it on.

## Turning Blips On and Off

*Blips* are temporary markers that appear on the screen when you select an object or location in the drawing. Blips are visible only until you redraw the drawing. You can't select blips; they are used only for reference and never print. Nonetheless, blips can sometimes clutter the drawing while you're working.

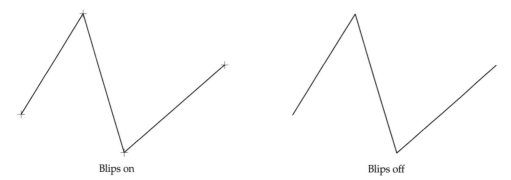

Blips on                                                                      Blips off

To turn blips on and off:

1. At the command line, type **BLIPMODE** and press ENTER. AutoCAD prompts:

   ```
   Enter mode [ON/OFF] <current>:
   ```

2. Type **ON** to turn blips on, or **OFF** to turn them off.

## Improving AutoCAD Performance

In addition to the methods that you've already learned, such as changing the VIEWRES value or turning off Fill mode, the following are several other things that you can do to improve AutoCAD's performance:

- Set your drawing limits to a minimum area, encompassing only that area needed for your drawing.
- Set a reasonable grid density and turn off the grid when it is not needed.
- Make sure that the drawing origin falls within the drawing limits.
- Freeze unneeded layers.
- Unload unneeded images or external references.
- Use FONTMAP to convert complex fonts to simpler fonts that display faster.

You'll learn more about these and other methods of improving AutoCAD's performance in later chapters. Also, consider using the Partial Open option to select which view and what layer geometry you want to display when opening large drawings. You learned about using Partial Open and Partial Load in Chapter 1.

# Chapter 6

## Working with Coordinates

To create accurate drawings in AutoCAD, you can locate points within a drawing by entering coordinates as you draw or modify objects. When you are working on two-dimensional drawings, you enter two-dimensional coordinates; for three-dimensional objects, you specify three-dimensional coordinates.

You can also specify coordinates in relation to other known locations or objects in a drawing. In particular, when you are working on three-dimensional drawings, it is often easier to specify coordinates in relation to a two-dimensional working plane, called a User Coordinate System (UCS).

This chapter explains how to work with coordinates, including how to do the following:

- Use 2-D and 3-D coordinate systems
- Specify absolute and relative coordinates
- Specify polar, spherical, and cylindrical coordinates
- Define and manipulate User Coordinate Systems

# Using Cartesian Coordinate Systems

Many commands in AutoCAD require that you specify points as you draw or modify objects. You can specify points either by selecting points with the mouse or by typing coordinate values at the command line. AutoCAD locates points within a drawing by using a Cartesian coordinate system. This coordinate system underlies every drawing that you create.

## Understanding How Coordinate Systems Work

The Cartesian coordinate system uses three perpendicular axes, referred to as X, Y, and Z, to specify points in three-dimensional space. Every location in a drawing can be represented as a point relative to a 0,0,0 coordinate point, which is referred to as the *origin*. When you create two-dimensional objects, you need to specify the horizontal coordinate positions along the X axis and specify the vertical coordinate positions along the Y axis. Thus, every point on the plane can be represented as a coordinate pair, comprised of an X coordinate and a Y coordinate. Positive coordinates are located above and to the right of the origin; negative values are located to the left and below the origin. Figure 6-1 illustrates a typical two-dimensional coordinate system.

When you work in 2-D, you need to enter only the X and Y coordinates. AutoCAD assumes that the Z-axis value is always the current elevation, which is zero by default. When you work in 3-D, however, you also specify the Z-axis value. As you look at a plan view of your drawing (a view from above, looking down), the Z axis extends straight up out of the screen at a 90-degree angle to the XY plane. Positive coordinates are located above the XY plane, and negative coordinates are located below the plane. Figure 6-2 shows a three-dimensional coordinate system.

Every AutoCAD drawing uses a fixed coordinate system, referred to as the *World Coordinate System* (*WCS*), and any point in the drawing always has a specific X,Y,Z coordinate in the WCS. You can also define arbitrary coordinate systems, located anywhere

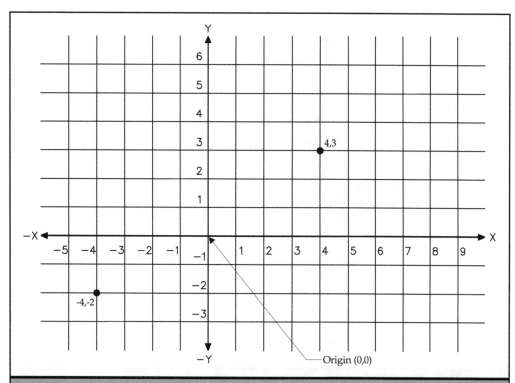

**Figure 6-1.** *A two-dimensional coordinate system. Every point is identified by its X,Y coordinate.*

in three-dimensional space and oriented in any direction. This type of coordinate system is called a *User Coordinate System* (UCS). You can create as many UCSs as you want, saving or moving them to help you construct 3-D objects. By defining a UCS within the WCS, you can simplify the creation of most 3-D objects into combinations of 2-D objects.

## Understanding How Coordinates Are Displayed

As you move the crosshairs, the current cursor position is displayed on the status bar as X,Y,Z coordinates. By default, this display updates dynamically as you move the cursor. You can toggle the coordinate display to static mode by doing one of the following:

3.4830,1.9803,0.0000 ◀——————— Current cursor location (X,Y,Z)

■ Click the coordinate display area on the status bar.

■ Press F6.

■ Press CTRL-D.

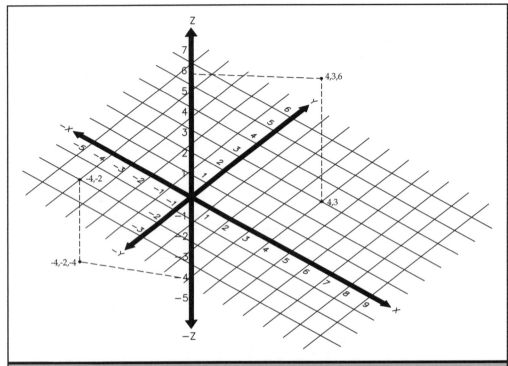

**Figure 6-2.** *A three-dimensional coordinate system. Every point is identified by its X,Y,Z coordinate.*

When a command is active and AutoCAD prompts for a distance or displacement, the coordinate display changes to a different dynamic mode that shows the distance and angle from the previous point, rather than X,Y,Z coordinates.

$$3.5198<31\ \ ,0.0000 \longleftarrow \text{Distance and angle from previous point}$$

 *When AutoCAD displays the coordinates as a distance and angle, toggling the coordinate display cycles through all three display modes: dynamic distance and angle, dynamic X,Y,Z coordinates, and static coordinates.*

## Finding the Coordinates of a Point

You can use the ID command to find the X,Y,Z coordinates of any point. When identifying a point on an object, you should use an appropriate object snap to ensure that you select the desired point. For example, to find the coordinates of the end point of a line:

1. Do one of the following:

- On the Inquiry toolbar, click Locate Point.
- From the Tools menu, choose Inquiry | ID Point.
- At the command line, type **ID** and then press ENTER.

AutoCAD prompts:

```
Specify point:
```

2. Activate the endpoint object snap (if it isn't already active) as a running object snap. (*Hint:* Type **END** and press ENTER, or press SHIFT-right-click and select Endpoint from the shortcut menu.)

3. Click the line near its end point.

AutoCAD displays the coordinates of the point on the command line.

*You can also select the object by using grips, small boxes that appear at specific locations on an object, such as at the end points and midpoints of lines. When you click a grip, the coordinate area on the status bar displays the coordinates of the grip. The coordinate reported is rounded according to the UNITS precision setting, but is stored by AutoCAD as a floating-point number.*

# Using Two-Dimensional Coordinates

When you are working in two dimensions, you specify points on the XY plane. You can specify any point either as an *absolute coordinate* (or Cartesian coordinate), by using the exact X-coordinate and Y-coordinate locations in relation to the origin (the 0,0 coordinate point at which the two axes intersect), or as a *relative coordinate,* relative to the previous point. You can also specify points by using *relative* or *absolute polar coordinates,* which locate a point by using a distance and an angle.

## Entering Absolute Cartesian Coordinates

To enter absolute Cartesian coordinates, type the coordinate location of the point at the command line. For example, to use absolute Cartesian coordinates to draw a line from the origin (0,0) to a point 4 units to the right and 3 units above the origin, as shown in Figure 6-3, start the LINE command and respond to the prompts as follows:

```
Specify first point: 0,0
Specify next point or [Undo]: 4,3
```

When you use absolute Cartesian coordinates, you need to know the exact point locations for anything that you draw. For instance, to use absolute Cartesian coordinates to draw an 8.5-unit square with its lower corner at 4,5, you must determine that the upper-left corner is at coordinate 4,13.5, the upper-right corner is at 12.5,13.5, and the lower-right corner is at 12.5,5.

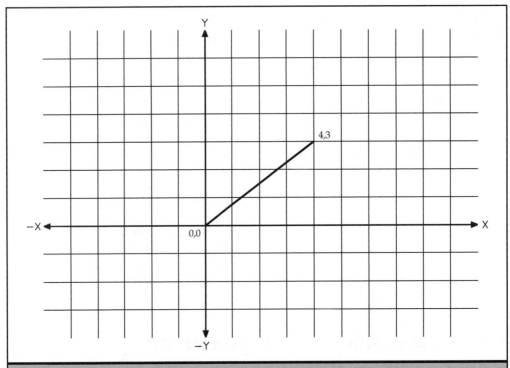

**Figure 6-3.** *Drawing a line by using absolute Cartesian coordinates*

## Entering Relative Cartesian Coordinates

Using relative Cartesian coordinates often is simpler than using absolute Cartesian coordinates. When you use relative coordinates, you specify a location in the drawing by determining its position relative to the last coordinate that you specified. To use relative Cartesian coordinates, type the coordinate values at the command line, preceded by the @ symbol. The coordinate pair following the @ symbol represents the distance along the X axis and the Y axis to the next point relative to the previous point.

For example, to draw an 8.5-unit square with its lower-left corner at 4,5 by using relative coordinates, as shown in Figure 6-4, you don't need to know the coordinates of the other corners. Simply start the LINE command and then respond to the prompts as follows:

```
Specify first point: 4,5
Specify next point or [Undo]: @8.5,0
Specify next point or [Undo]: @0,8.5
Specify next point or [Close/Undo]: @-8.5,0
Specify next point or [Close/Undo]: C
```

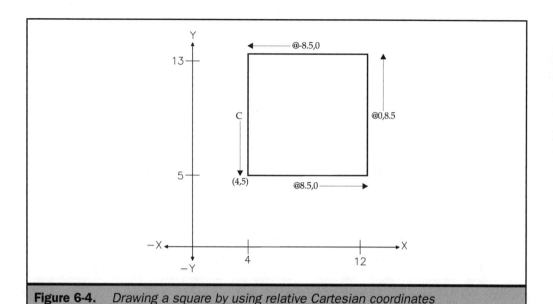

**Figure 6-4.**   *Drawing a square by using relative Cartesian coordinates*

The first relative coordinate (@8.5,0) locates the new point 8.5 units to the right (along the X axis) from the previous point of 4,5. The second relative coordinate (@0,8.5) locates the next point 8.5 units above (along the Y axis) the previous point, and so on. Typing **C** (for Close) draws the final line segment back to the first point that you specified when you started the LINE command.

## Entering Polar Coordinates

Drawing objects becomes more complicated when the object is drawn at an angle. The previous example would be difficult to complete if you wanted to draw the square so that it is tilted at a 45-degree angle. Polar coordinates simplify the task of drawing at an angle. *Polar coordinates* base the location of a point at a distance and angle from either the origin (absolute coordinate) or the previous point (relative coordinate).

To specify polar coordinates, enter a distance and an angle, separated by the angle bracket symbol (<). For example, to use relative polar coordinates to specify a point 1 unit away from the previous point and at an angle of 45 degrees, you would type @1<45.

To draw the square from the previous example, but tilted at a 45-degree angle, as shown in Figure 6-5, start the LINE command and then respond to the prompts as follows:

```
Specify first point: 4,5
Specify next point or [Undo]: @8.5<45
Specify next point or [Undo]: @8.5<315
Specify next point or [Close/Undo]: @8.5<225
Specify next point or [Close/Undo]: C
```

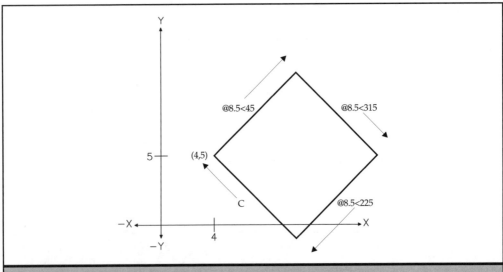

**Figure 6-5.** *Drawing a tilted square by using relative polar coordinates*

| | |
|---|---|
| **Note** | *This example, like all the other examples in this book, assumes AutoCAD's default units settings: Angles increase in a counterclockwise direction and decrease in a clockwise direction, and the 0 angle points to the east (or 3 o'clock) compass point. Thus, an angle of 315 degrees points to the southeast and is the same as -45 degrees. When specifying polar coordinates, you can enter positive and negative angles interchangeably.* |

## Using Direct Distance Entry

Direct distance entry is similar to using relative coordinates, except that you don't specify the angle. Instead, you move the cursor in a particular direction from the previous point and then type the distance. For example, when you use the LINE command to draw a line that is 4.3 units to the right, move the cursor to the right, type **4.3**, and then press ENTER.

To draw a line by using direct distance entry:

1. Start the LINE command.

2. Specify the start point of the line.

3. Move the cursor until the rubber-band line extends at the desired angle.

4. At the command line, type the distance and then press ENTER.

| | |
|---|---|
| **Note** | *Direct distance entry is a good method for specifying the line length, but since the direction of the line is based on the position of the cursor in reference to the previous point, it is only accurate enough for precision drawing when either Ortho mode is turned on or Polar Tracking is turned on and the cursor is aligned with one of the angle increments.* |

*Remember that you can use the polar angle override method described in Chapter 2 to lock the cursor to a specific angle for a single drawing operation. For example, you can start a line, type **<55** to lock the line to a 55-degree angle, and then type the length of the line using direct distance entry. A line of the specified length will be drawn at the specified angle.*

# Using Three-Dimensional Coordinates

Specifying coordinates in three-dimensional space is similar to working in two dimensions, except that you also use the Z axis to locate coordinates. Three-dimensional coordinates are represented in the format X,Y,Z (for example, 4,3,6).

## Using the Right-Hand Rule

To visualize how AutoCAD works with three-dimensional space, you can use a technique known as the *right-hand rule*. Looking at your right hand, imagine that your thumb represents the X axis, your index finger is the Y axis, and your middle finger is the Z axis. Extend your thumb and index finger at right angles and your middle finger perpendicular to your palm. These three fingers are now pointing in the positive X, Y, and Z directions, respectively, as shown in Figure 6-6.

You can also use the right-hand rule to determine the positive rotation direction. Point your thumb in the positive direction of the axis about which you want to rotate. Now, curl your middle finger, ring finger, and little finger in toward your palm. These fingers are now curling in the positive rotation direction, as shown in Figure 6-6.

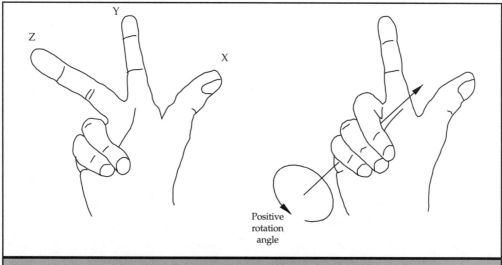

**Figure 6-6.** *The right-hand rule helps you to determine the positive direction of the X, Y, and Z axes, and the positive rotation direction.*

## Entering X,Y,Z Coordinates

When you work in three dimensions, you can specify the X,Y,Z coordinates as absolute distances in relation to the origin (the 0,0,0 coordinate point at which the three axes intersect) or as relative coordinates based on the last point selected. For example, to specify a point 3 units along the positive X axis, 4 units along the positive Y axis, and 2 units along the positive Z axis, you would specify the coordinate as 3,4,2.

## Entering Spherical Coordinates

When you work in three-dimensional space, you can use *spherical coordinates* to specify a three-dimensional point, by entering its distance from either the origin (absolute distance) or the last point (relative distance), along with its angle in the XY plane, and its angle up from the XY plane (in the Z direction). In spherical format, you separate each angle with the open angle bracket symbol (<).

Thus, if you want to draw a line from the origin to a point 10.3923 drawing units away, at an angle of 45 degrees from the X axis of the current UCS, and 35 degrees up from the current XY plane, you would enter **10.3923<45<35**, as shown in the following code and illustrated in Figure 6-7:

```
Specify first point: 0,0
Specify next point or [Undo]: 10.3923<45<35
```

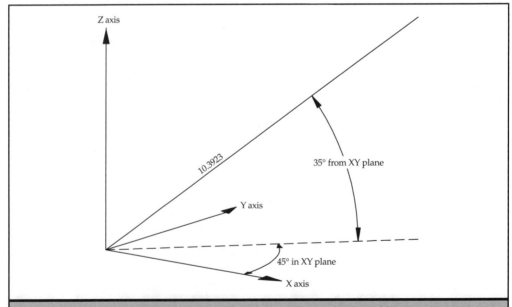

**Figure 6-7.**   *Drawing a line in three-dimensional space by using absolute spherical coordinates*

## Entering Cylindrical Coordinates

When working in three-dimensional space, you can also use *cylindrical coordinates* to specify a three-dimensional point. You specify a point by entering its distance from either the origin (absolute distance) or the last point (relative distance), along with its angle in the XY plane, and its Z-coordinate value.

In cylindrical format, you separate the distance and angle with the open angle bracket symbol (<) and separate the angle and the Z-coordinate value with a comma. For example, to draw a line from the last point to a point that is 7.3485 units away, at an angle of 27 degrees from the X axis of the current UCS and 3 units up in the Z direction (as shown in Figure 6-8), start the LINE command and then respond to the prompts as follows:

```
Specify first point: (specify the start point)
Specify next point or [Undo]: @7.3485<27,3
```

# Using Point Filters and Tracking

Point filters and tracking provide two additional methods of locating a point in a drawing that is relative to another point, without specifying the entire coordinate. Using point filters, you can enter partial coordinates and then have AutoCAD prompt you for the remaining coordinate information. You can use X,Y,Z point filters whenever AutoCAD prompts you for a coordinate, by responding to the prompt with a filter in the following form:

*.coordinate*

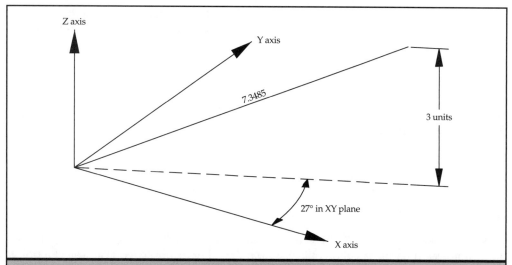

**Figure 6-8.**    *Drawing a line in three-dimensional space, using relative cylindrical coordinates*

in which *coordinate* is one or more of the letters X, Y, and Z. AutoCAD then indicates that you are using a filter by prompting you for the filtered coordinate. For example, if you type **.XY**, AutoCAD stores only the X,Y coordinate pair of the next point that you specify. The program then prompts you for the Z coordinate. The filters .X, .Y, .Z, .XY, .XZ, and .YZ are all valid filters.

Tracking is similar to using point filters, except that instead of specifying the filter, you visually locate points relative to other points in the drawing.

## Using Point Filters in 2-D

You can use point filters when working in two dimensions to locate points in relation to existing objects. For example, to draw a circle centered in a rectangle, as shown in Figure 6-9, start the CIRCLE command and then respond to the prompts as follows:

```
Command: CIRCLE
Specify center point for circle or [3P/2P/Ttr (tan tan radius)]: .X
of MIDPOINT
of (select bottom of rectangle)
of (need YZ): MIDPOINT
of (select right side of rectangle)
Specify radius of circle or [Diameter]: (specify radius)
```

## Using Point Filters in 3-D

You can use point filters when working in three-dimensional space to locate points in two dimensions and then specify the Z coordinate as the elevation above the XY plane. For example, to begin drawing a line from a point with a Z coordinate that is 3 units above the center of a circle, as shown in Figure 6-10, start the LINE command and then respond to the prompts as follows:

```
Command: LINE
Specify first point: .XY
of CENTER
of (select a point on the circle)
of (need Z): 3
Specify next point or [Undo]: @3,3
Specify next point or [Undo]: ENTER
```

## Using Tracking

As you learned in Chapter 2, you can use AutoCAD's AutoTracking to select positions along alignment paths based on object snap points. For example, you can select a point along a path based on the end point or midpoint of an object.

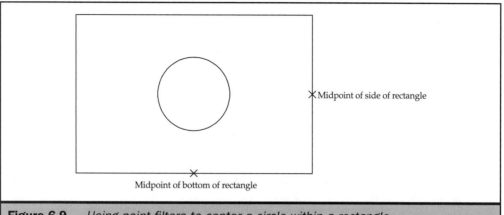

**Figure 6-9.** *Using point filters to center a circle within a rectangle*

Once object snap tracking is enabled and one or more object snaps have been set, when a command prompts you to specify a point, move the cursor over the object point that you want to track on, and briefly pause the cursor over that point. Don't click the point. As soon as AutoCAD acquires the point, a small plus sign (+) appears adjacent to the point. If AutoSnap markers are enabled, a marker also appears at the acquired point. Then, as you move away from the point, AutoCAD displays a temporary alignment path. By acquiring multiple points and moving the cursor so that their alignment paths intersect, you can eliminate the need to create construction lines, and save several keyboard steps. (Refer to Chapter 2 for additional information about AutoTracking.)

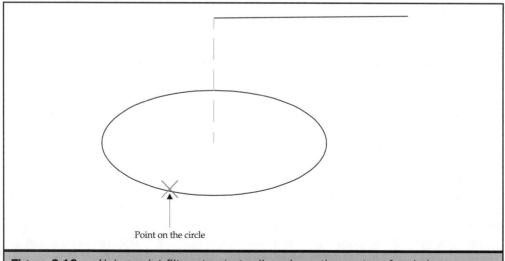

**Figure 6-10.** *Using point filters to start a line above the center of a circle*

*LEARN BY EXAMPLE*

*Practice using tracking to draw a circle centered in a rectangle. Start a new drawing by using the Figure 6-9 template file on the companion web site.*

To use tracking to draw a circle centered in a rectangle, as shown in Figure 6-9, use the following instructions:

1. Display the Drafting Settings dialog box. (*Hint:* Press SHIFT-right-click and select Osnap Settings from the shortcut menu.)

2. In the Object Snap tab, make sure that the Midpoint object snap mode is selected and that both object snap and object snap tracking are enabled. (*Note*: Other object snap modes can remain selected.)

3. Click OK.

4. Start the CIRCLE command.

5. When AutoCAD prompts you to specify the center point for the circle, acquire the midpoint of the bottom of the rectangle. AutoCAD adds a plus sign at the midpoint to indicate that it has been acquired by object snap tracking. Move the cursor directly above the bottom to display the vertical alignment path.

6. Move the cursor and acquire the midpoint of the right side of the rectangle. AutoCAD adds a plus sign at the midpoint of the side of the rectangle.

7. Move the cursor to the left along the horizontal alignment path until you also see the vertical alignment path. Click at the intersection of these two alignment paths to specify the center point of the circle.

8. Specify the radius of the circle.

## Defining User Coordinate Systems

When you work in two-dimensional or three-dimensional space, you can define a UCS, with its own origin and orientation that are separate from the WCS. You can create as many UCSs as you want, saving and recalling them as needed to simplify your construction of 2-D and 3-D objects.

For example, when working in 2-D, you can relocate the UCS, aligning it with the wing of a building oriented at an angle, making it easier to draw the plan view of that wing. When you work in 3-D, you can create a separate UCS for each side of a building. Then, by switching to the UCS for the east side of the building, you can draw the windows on that side by specifying just their X and Y coordinates. When you create one or more UCSs, coordinate entry is based on the current UCS.

Imagine how difficult drawing a window on the side of a house would be if you had to calculate the X,Y,Z coordinates of each corner of the window. The following command sequence illustrates how this task is simplified by first relocating the UCS to the corner of the house (see Figure 6-11). In this case, a UCS corresponding to the north side of the house has previously been saved.

```
Command: UCS
Enter an option [New/Move/orthoGraphic/Prev/Restore/Save/Del/Apply/?/World] <World>: R
Enter name of UCS to restore or [?]: NORTH
Command: LINE
Specify first point: FROM
Base point: 0,0
<Offset>: 3',3'
Specify next point or [Undo]: @30,0
Specify next point or [Undo]: @0,5'
Specify next point or [Close/Undo]: @-30,0
Specify next point or [Close/Undo]: C
```

## Understanding the User Coordinate System Icon

To help you keep your bearings in the current coordinate system, AutoCAD displays a coordinate system icon. Figure 6-11 shows the UCS icon, aligned with the lower-left corner of the house in this example.

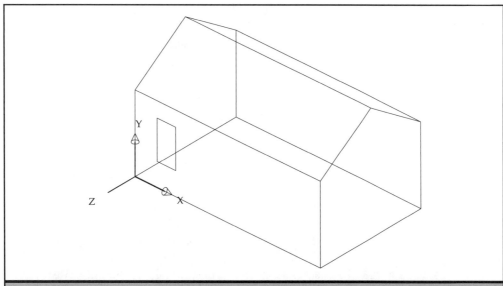

**Figure 6-11.**   *Aligning the UCS to the side of the house makes drawing the windows on that side easy.*

The appearance of the UCS icon changes depending on the orientation of your drawing in 3-D space and also based on whether you're looking at a wireframe or shaded view. The icon itself has also changed considerably from earlier versions of AutoCAD. AutoCAD 2000 and earlier versions displayed a 2-D UCS icon. Newer versions display a 3-D UCS icon, although you can use the UCSICON command to change the appearance of the icon. Figure 6-12 illustrates the various appearances of the UCS icon under different circumstances.

If your view of the current UCS is from within 1 degree of the current XY plane and you are using the older 2-D UCS icon, the UCS icon is replaced with a broken-pencil icon, to warn you that pointing to locations on the screen may yield confusing results.

The coordinate system icon is displayed only when you are working in model space. When you enter paper space, that icon is replaced by a special paper-space icon.

To turn the UCS icon on and off, do one of the following:

■ From the View menu, choose Display | UCS Icon | On. (The check mark indicates that the icon is visible.)

■ At the command line, type **UCSICON**, press ENTER, type either **ON** or **OFF**, and then press ENTER again.

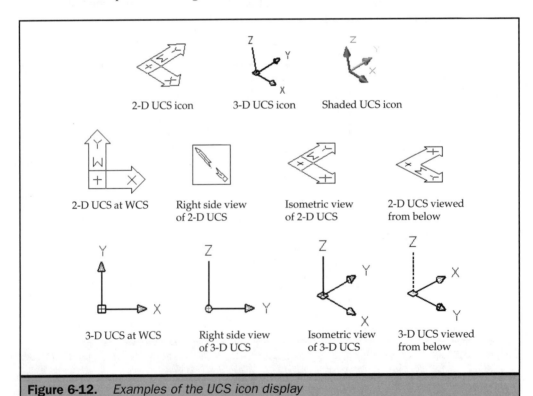

**Figure 6-12.**    *Examples of the UCS icon display*

To display the UCS icon at the UCS origin, do one of the following:

■ From the View menu, choose Display | UCS Icon | Origin. (The check mark indicates that the icon is at the origin.)

■ At the command line, type **UCSICON**, press ENTER, type **OR**, and then press ENTER again.

 *You can also control the UCS icon from the Settings tab of the UCS dialog box. You'll learn more about this dialog box later in this chapter.*

## Changing the Appearance of the UCS Icon

As previously mentioned, you can change the appearance of the UCS icon. The UCS Icon dialog box lets you control whether AutoCAD uses the 3-D UCS icon or the older 2-D icon. You can also change the size of the icon, the line width when using a 3-D icon, and the color of the icon in both model space and paper space. The UCS Icon dialog box is shown in Figure 6-13.

To display the UCS Icon dialog box, do one of the following:

■ From the View menu, choose Display | UCS Icon | Properties.

■ At the command line, type **UCSICON**, press ENTER, type **P**, and then press ENTER again.

## Changing the User Coordinate System

As you saw in the previous example, the UCS command is used to define a new UCS. You can also save, restore, or delete UCSs. You can relocate the current UCS by using any of the following methods:

■ Specify a new origin, without changing the current orientation of the X, Y, and Z axes.

■ Specify a new origin and a point on the positive Z axis.

■ Specify a new origin and points on the positive X and Y axes.

■ Align the UCS with an existing object.

■ Align the UCS with its Z axis parallel to the current viewing direction.

■ Rotate the current UCS around any of its axes.

■ Restore a previously saved UCS.

When you define a new UCS, the UCS icon changes to indicate the origin and orientation of the new UCS.

The various UCS command options are available from both the UCS flyout on the Standard toolbar and the UCS toolbar. You can also select the individual options from the Tools menu, or by typing the command at the command line.

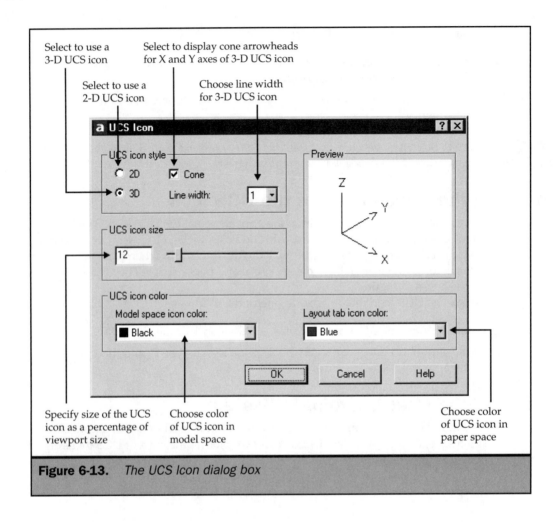

**Figure 6-13.** *The UCS Icon dialog box*

To change the location of the UCS by specifying a new origin (see Figure 6-14):

1. Do one of the following:

- On the Standard toolbar or the UCS toolbar, click Origin UCS.

- On the UCS II toolbar, click Move UCS Origin.

- From the Tools menu, choose New UCS | Origin.

- At the command line, type **UCS** and press ENTER; then, type **O** and press ENTER.

AutoCAD prompts:

```
Specify new origin point <0,0,0>:
```

2. Specify a point for the new origin.

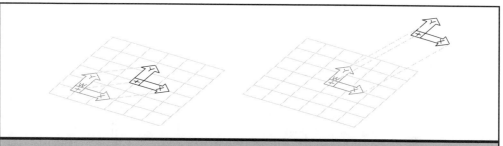

**Figure 6-14.**    *You can move the UCS within the current plane (by changing just the X,Y coordinate) or to a parallel plane (by changing the X,Y,Z coordinate).*

**Tip**    *AutoCAD does not actually display the Origin option on the command line unless you first select the New or Move option. Typing **O** (which is effectively what happens when you select the command from the toolbar or menu) enables you to skip this intermediate step. You can perform similar tricks by typing any of the other options (such as typing **3** to select the 3 Point option, as shown in the next example) at the first UCS command prompt, even though these options don't actually appear until you select the New option.*

To change the UCS by specifying a new origin and X-Y orientation (see Figure 6-15):

1. Do one of the following:

   ■ On the Standard toolbar or the UCS toolbar, click 3 Point UCS.

   ■ From the Tools menu, choose New UCS | 3 Point.

   ■ At the command line, type **UCS** and press ENTER; then, type **3** and press ENTER.

   AutoCAD prompts:

   ```
   Specify new origin point <current>:
   ```

2. Specify a point for the new origin.

   AutoCAD then prompts:

   ```
   Specify point on positive portion of the X-axis <current>:
   ```

3. Specify a point on the positive X axis.

   AutoCAD then prompts:

   ```
   Specify point on the positive-Y portion of the UCS XY plane
   <current>:
   ```

4. Specify a point in the positive Y direction.

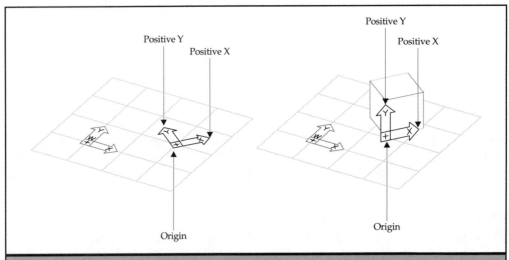

**Figure 6-15.** *You can change the origin and orientation within either the current plane or in any location in three-dimensional space.*

To restore the UCS to the WCS, do one of the following:

■ On the Standard toolbar or the UCS toolbar, click World UCS.

■ On the UCS II toolbar, choose World from the drop-down list.

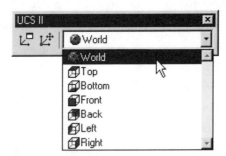

■ From the Tools menu, choose New UCS | World.

■ At the command line, type **UCS** and then press ENTER twice.

## Saving and Restoring a UCS

After you define a UCS, you can save it and later restore it when you need to use it again, similar to the way in which you can save and restore named views. You can save and restore a UCS either by using the UCS command or from the Named UCSs tab of the UCS dialog box (see Figure 6-16). This tab displays a list of named UCSs. Whenever you create a new UCS, AutoCAD adds the UCS to this list. If the UCS is not yet named, it appears in the list as Unnamed. Notice that the current UCS has a small arrow to its left. AutoCAD also adds the previous UCS to the list of previous UCSs, which also appears in the list. You can click the previous UCS to step back through this Previous list, restoring each previous UCS.

*When you use the New option to move the UCS or create a new UCS, AutoCAD adds the previous UCS to the list of UCS names. When you use the Move option, however, AutoCAD does not add a UCS to the Previous UCS list.*

To save a UCS:

1. Do one of the following:
   - On the Standard toolbar, UCS toolbar, or UCS II toolbar, click Display UCS Dialog.
   - From the Tools menu, choose Named UCS.
   - At the command line, type **UCSMAN** and press ENTER.
2. In the UCS dialog box, right-click the current UCS and select Rename from the shortcut menu.
3. Type a new name for the UCS and press ENTER.
4. Click OK to complete the command.

*You can also save the current UCS from the command line by using the Save option of the UCS command.*

You can also restore the previous UCS, or any previously saved UCS, or even delete a previously saved UCS from the Named UCS's tab of the UCS dialog box.

To restore a named UCS:

1. Display the Named UCS's tab of the UCS dialog box.
2. Select the UCS that you want to restore (so that it is highlighted in the list) and then click Set Current.
3. Click OK to complete the command.

## Using a Predefined Orthographic UCS

AutoCAD also provides six standard orthographic UCSs: top, bottom, front, back, left, and right. You can align the UCS to one of these predefined orientations from the Orthographic UCS's tab of the UCS dialog box, shown in Figure 6-17. In addition, you can determine whether the new UCS will be based on the WCS or a named UCS.

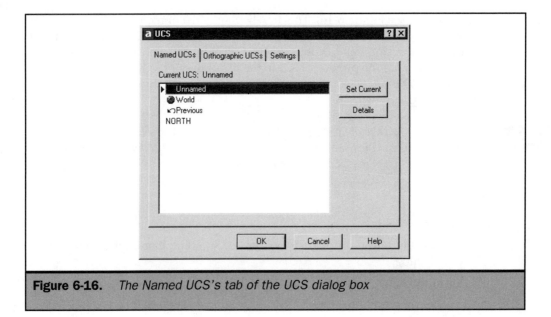

**Figure 6-16.** *The Named UCS's tab of the UCS dialog box*

To switch to a preset UCS:

1. Display the Orthographic UCS's tab of the UCS dialog box.

2. Select the orientation that you want to switch to (so that it is highlighted in the list) and then click Set Current.

3. If desired, do one of the following:

   ■ If you want to change the depth of the orthographic UCS (for example, to move the origin to a different Z coordinate), either double-click the Depth field list of the orthographic UCS you want to change, or highlight the UCS in the list, right-click, and choose Depth from the shortcut menu. You can then specify the new depth either by typing the value in the Top Depth field or by clicking the Select New Origin button and then selecting the origin point in the drawing.

   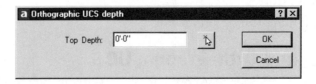

   ■ If you want to calculate an orthographic UCS relative to a named UCS, select the named UCS from the Relative To drop-down list.

4. Click OK.

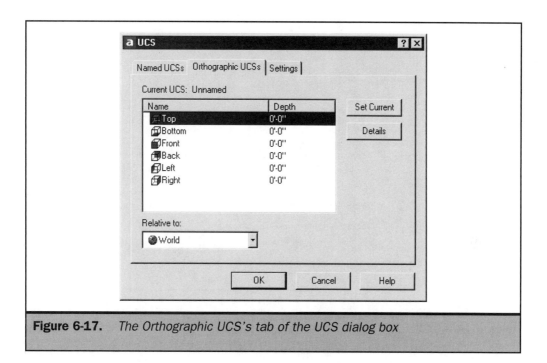

**Figure 6-17.**   *The Orthographic UCS's tab of the UCS dialog box*

**Tip**   *Although the UCS icon and cursor orientation change to reflect the new UCS orientation, the display doesn't change unless the UCSFOLLOW system variable is on, in which case the display will always be a plan view of the current UCS.*

## Assigning a UCS to a Viewport

As you learned in the previous chapter, you can divide the screen into multiple viewports. When working with 3-D models, it is often useful to divide the screen into multiple viewports and then display a different orthographic orientation—such as the top, front, and right side view—each in its own viewport. AutoCAD provides a useful feature that enables you to define and save a different UCS in each viewport, so that when you make a viewport current, it has the same UCS that you used last time it was current.

The UCS in each viewport is controlled by the UCSVP system variable. AutoCAD saves this system variable separately for each viewport, and this system variable, in turn, can be controlled from the Settings tab of the UCS dialog box, shown in Figure 6-18. When the Save UCS With Viewport check box is selected (UCSVP = 1), the UCS is locked to the UCS last used in the current viewport, and does not change to match the UCS of the current viewport. When the check box is not selected (UCSVP=0), the UCS in that viewport changes to match the UCS in the current viewport. By default, AutoCAD 2002 saves the UCS with each viewport.

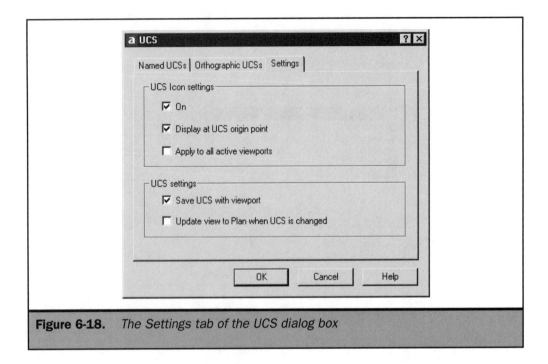

**Figure 6-18.** *The Settings tab of the UCS dialog box*

The other check boxes within this tab also affect the UCS icon and the way that AutoCAD treats the views displayed in each viewport. These settings can be controlled separately for each viewport. The three check boxes under UCS Icon Settings control the following:

- **On** Toggles the UCS icon on and off.
- **Display at UCS Origin Point** Displays the UCS icon at the UCS origin.
- **Apply to All Active Viewports** Specifies whether the UCS icon settings in the current viewport are reflected in all the active viewports.

The two check boxes under UCS Settings control the following:

- **Save UCS with Viewport** Saves the coordinate system settings with the viewport.
- **Update View to Plan when UCS Is Changed** Controls whether the view is restored to plan view when the UCS changes in the viewport. (You'll learn about 3-D views and displaying a plan view in Chapter 19.)

The
Complete
Reference

# Chapter 7

## Modifying Objects

Creating objects is only part of the process of creating a CAD drawing, and for complex drawings, creating objects can take almost as much time as creating paper drawings. When it comes time to make changes to a drawing, however, CAD is much more efficient. AutoCAD provides many editing tools to modify a drawing. You can easily move, rotate, stretch, or change the scale of drawing objects. When you want to remove an object, you can erase it with a few clicks of the mouse. You can also make multiple copies of any object.

You can modify most objects by using general-purpose editing commands. Most of these commands, which are covered in this chapter, are located on the Modify toolbar and the Modify menu. The more complex objects that you learned about in Chapter 4 require special commands to modify specific properties of the objects. You'll learn about those commands in Chapter 10. Although many of the commands presented in this chapter can also be used to modify objects in 3-D space (for example, moving objects to a different elevation), our discussion is confined to working in 2-D. You'll learn about editing in 3-D in Chapter 20. This chapter explains how to do the following:

- Select objects by using object selection methods and grips
- Erase objects from the drawing
- Create copies of existing objects
- Rearrange objects by moving or rotating
- Resize objects by stretching, scaling, extending, trimming, or lengthening
- Break objects

## Selecting Objects

When you modify an object, you must include it in a *selection set,* a collection of objects to be modified. You can use any of the following methods to create a selection set:

- Choose a command first and then select the objects to be modified. With this method, AutoCAD prompts you to select the objects.
- Select the objects first and then choose an object modification command. AutoCAD stores the objects in the *previous* selection set.
- Select objects by pointing to them, and then use grips to modify them.

If highlighting is turned on, AutoCAD highlights objects as you select them.

**Note** *You can also combine several objects into a group and then modify the group. You'll learn more about groups in Chapter 10.*

### Object Selection Methods

After you start a command that prompts you to select objects, you can use any of the following methods to select those objects:

- **Pointing**   Selects the object that you pick directly by using the pickbox or by typing coordinates
- **Window**   Selects objects contained entirely within a rectangular selection window
- **Crossing**   Selects objects contained within or crossing the boundary of a rectangular selection window
- **WPolygon**   Selects objects contained entirely within a polygon selection window
- **CPolygon**   Selects objects contained within or crossing the boundary of a polygon selection window
- **Fence**   Selects objects crossing a multisegment fence line
- **All**   Selects all objects in the drawing
- **Last**   Selects the object most recently added to the drawing
- **Previous**   Selects the objects included in the previous selection set, if one exists

In addition to these methods, you can select objects that match a particular set of properties—for example, all objects on a particular layer or drawn in a specific color.

You can also use a few selection methods automatically without first specifying them. For example, you can simply click to select objects, or use the Window or Crossing selection method by defining the opposite corners of a rectangular selection window. The direction in which you define the corners of the rectangle (left-to-right or right-to-left) determines which type of selection you are using. This is known as *Implied Windowing* mode. When Implied Windowing mode is enabled, if you pick a point in an empty area of the screen, AutoCAD automatically goes into a window selection mode. If you move the cursor to the right from the first point, only objects falling completely within the window are selected. If you move the cursor to the left from the first point, objects within or crossing the window boundaries are selected.

AutoCAD further indicates that you are using a window selection by displaying the rectangle with a solid line. When using a crossing window, the rectangle is displayed using a dashed line. Implied windowing is controlled either by the PICKAUTO system variable or from the Selection tab of the Options dialog box. You'll learn more about this dialog box and its controls in a few pages.

**Tip**   *Normally, you can create a selection window by picking one time to define the first corner and picking a second time to define the diagonally opposite corner. But, you can also instruct AutoCAD to require that you press and hold down the pick button as you drag the cursor diagonally, to draw a selection window or crossing window with your pointing device, and then release the pick button to complete the window. This method, which is similar to the way in which other Windows programs operate, is known as the Press-and-Drag mode and is controlled either from the Selection tab of the Options dialog box or by the PICKDRAG system variable.*

## Pointing to the Object

You can select an object simply by pointing to it. Move the pickbox over an object and click. AutoCAD immediately scans the drawing for an object that crosses the pickbox.

 *When implied windowing is active, if you don't click an object, the selected point becomes the first corner of a window or crossing box. If this becomes annoying, instead of disabling implied windowing, try also enabling Press-and-Drag mode. When this mode is enabled, you must press and hold down the pick button as you drag the cursor diagonally, to draw a selection window or crossing window. In that case, simply clicking does not activate implied windowing. Press-and-Drag mode is controlled either from the Selection tab of the Options dialog box or by changing the value of the PICKDRAG system variable.*

## Object Cycling

In a crowded drawing, it is often difficult to select objects, because other objects are so close or, at times, lie directly on top of one another. When pointing to individual objects, you can cycle through the objects under the pickbox until AutoCAD highlights the one that you want to select. You can turn on object cycling whenever AutoCAD prompts you to select objects. To do so, move the pickbox over the desired object, as close to the object as possible, and then press CTRL-click (hold down the CTRL key while clicking the left-mouse button). AutoCAD displays the following message:

```
Select objects: <Cycle on>
```

After object cycling is activated, each time that you click the left-mouse button, AutoCAD highlights a different object. When the desired object is highlighted, press ENTER, press the SPACEBAR, or right-click. The object is added to the selection set, and AutoCAD turns off object cycling. You can then carry out the current command or continue adding objects to the selection set.

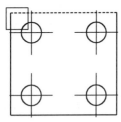 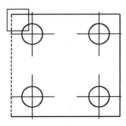 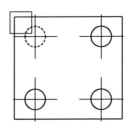

**NOTE:**   *After object cycling has been turned on, the actual position of the cursor has no effect. The position of the pickbox at the time object cycling is turned on determines which objects are considered.*

## Using a Window

Often, using the window selection method explicitly is more convenient than using implied windowing. For example, when you are working in a crowded drawing, you don't have to worry about picking a point in an empty area of the drawing, as you do when relying on implied windowing. You also don't have to consider whether the second corner of the window is to the right or left of the first. When prompted to select objects, type **W** and then press ENTER. AutoCAD prompts for the first corner and other corner, displaying a solid rubber-band rectangle that expands and contracts as you move the cursor. When you pick the second corner, AutoCAD immediately scans the drawing and selects those objects entirely within the rectangle (see Figure 7-1). AutoCAD doesn't select any objects that fall outside the window or that cross its boundaries.

## Using a Crossing Window

The crossing window is similar to the window selection method and often is more convenient than relying on implied windowing, for the same reasons just discussed. To use this selection method, when prompted to select objects, type **C** and press ENTER. AutoCAD prompts for the first corner and then the other corner, displaying a dashed rubber-band rectangle. When you pick the second corner, the program immediately scans the drawing and selects those objects entirely within or crossing the rectangle (see Figure 7-2). AutoCAD doesn't select any objects that fall outside the window.

## Using a Window Polygon

The window polygon method is similar to the window option, but rather than using a rectangular window, you can define an irregularly shaped selection area by picking points. To use this method, you type **WP** and press ENTER. A polygon is formed as you select points. After you select two or more points, AutoCAD shows a closing segment

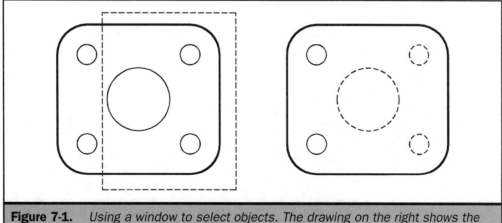

**Figure 7-1.**   *Using a window to select objects. The drawing on the right shows the objects that were selected.*

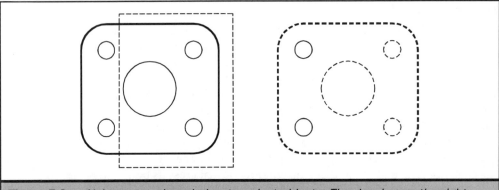

**Figure 7-2.** *Using a crossing window to select objects. The drawing on the right shows the objects that were selected.*

that extends from the cursor back to the first point selected. To complete the window polygon selection, press ENTER. AutoCAD immediately scans the drawing and selects those objects entirely within the polygon (see Figure 7-3). Any objects that fall outside the polygon or cross its boundaries are not selected.

## Using a Crossing Polygon

The crossing polygon method is similar to the crossing option, but rather than using a rectangular window, you can define an irregularly shaped selection area by picking points. To use this method, type **CP** and then press ENTER. The polygon is formed in the same way as when using the window polygon method. To complete the crossing polygon selection, press ENTER. AutoCAD immediately scans the drawing and selects

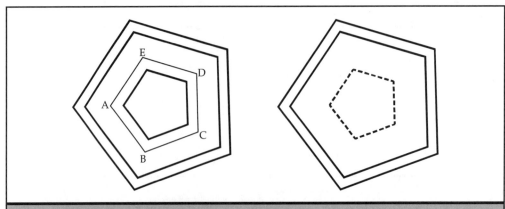

**Figure 7-3.** *Using a window polygon to select objects. The drawing on the right shows the objects that were selected by the polygon that was created by selecting points A, B, C, D, and E.*

those objects entirely within or crossing the polygon (see Figure 7-4). Any objects that fall outside the polygon are not selected.

## Using a Fence

The fence method lets you select objects by drawing a temporary fence line through them. To use this method, type **F** and then press ENTER. A fence line is formed as you select points. To complete the fence selection, press ENTER. AutoCAD immediately scans the drawing and selects those objects through which the fence passes (see Figure 7-5).

## Selecting All Objects

This selection method selects every object in the drawing, except those on frozen or locked layers, even if the objects are not actually visible on the screen. To use this method, type **ALL** and then press ENTER. Because some objects may not be currently visible (they could be offscreen or on layers that are turned off, for example), you should be particularly careful when using this option.

## Selecting the Last Object

This method selects only the last object that was added to the drawing that is currently visible onscreen. To use this method, make sure that the object is in view, and when prompted to select objects, type **L** and then press ENTER.

## Selecting the Previous Selection Set

Sometimes, you will want to reselect the same objects that you have just selected for the previous editing command—for example, to rotate the objects that you've just

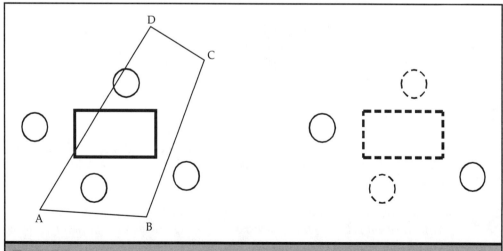

**Figure 7-4.**   *Using a crossing polygon to select objects. The drawing on the right shows the objects that were selected by the polygon that was created by selecting points A, B, C, and D.*

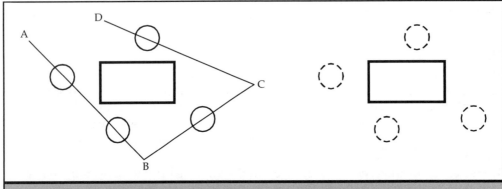

**Figure 7-5.** *Using a fence to select objects. The drawing on the right shows the objects that were selected by the fence that was created by selecting points A, B, C, and D.*

moved. To do so, type **P** and then press ENTER. AutoCAD immediately highlights the previous selection set again.

## Adding and Removing Objects from the Selection Set

As you build a selection set, you may inadvertently pick an object that you don't want included in the selection set. When that happens, simply type **R** and then press ENTER. AutoCAD prompts:

```
Remove objects:
```

You can now select any of the highlighted objects that you want to remove from the current selection set, again by using any of the methods described previously. To switch back to adding objects to the selection set, type **A** and then press ENTER. AutoCAD then prompts:

```
Select objects:
```

Any objects that you subsequently select will again be highlighted, indicating that they have been added to the set. You can toggle back and forth between the Add and Remove modes as often as you want.

*You can also remove objects from the selection set without having to switch into Remove mode. To do so, press the SHIFT key as you select highlighted objects by pointing or using implied windowing. Any objects that you select while pressing the SHIFT key are removed from the selection set.*

## Using Shift-to-Add Mode

Most Windows programs require that you press the SHIFT key when selecting multiple objects. You can make AutoCAD behave this way, as well, by turning on Shift-to-Add mode. When enabled, instead of adding each selected object to the selection set, only the most recently selected objects become part of the current selection set. To add objects to the set, you must press the SHIFT key while selecting them. Similarly, to remove objects from the set, you must reselect them while continuing to press the SHIFT key. Since this requires additional steps, most AutoCAD users prefer to leave Shift-to-Add mode turned off.

Shift-to-Add mode is controlled by the PICKADD system variable. This mode, and other object selection settings, can be controlled from the Selection tab of the Options dialog box, shown in Figure 7-6.

The check boxes in the Selection Modes area of the dialog box toggle AutoCAD's various selection modes on and off.

- ■ **Noun/Verb Selection**   Lets you choose objects first and then start the command that you want to use to edit them

- ■ **Use Shift to Add to Selection**   Adds only the most recent object to the selection set unless you press the SHIFT key while selecting objects

- ■ **Press and Drag**   Requires you to press and hold down the pick button as you drag the cursor diagonally, to draw a selection window or crossing window

- ■ **Implied Windowing**   Automatically activates the Window or Crossing selection mode when you click an empty area of the drawing

- ■ **Object Grouping**   Automatically selects the entire group when you select any object that is a member of that group

- ■ **Associative Hatch**   Causes the boundary of an associative hatch object to also be selected whenever you choose an associative hatch object

# Using Selection Filters

When you select objects, you can limit which objects are selected, by applying a filter to the creation of the selection set. A selection filter lets you select objects based on properties such as color, linetype, object type, or combinations of these properties. For example, you can create a selection filter so that you select only blue circles on a specific layer.

You create selection filters by using either the Quick Select dialog box or the Object Selection Filters dialog box (which is described a bit later in this chapter). As its name implies, the Quick Select dialog box enables you to quickly define a selection set based on filtering criteria that you define within this dialog box. You first create a selection

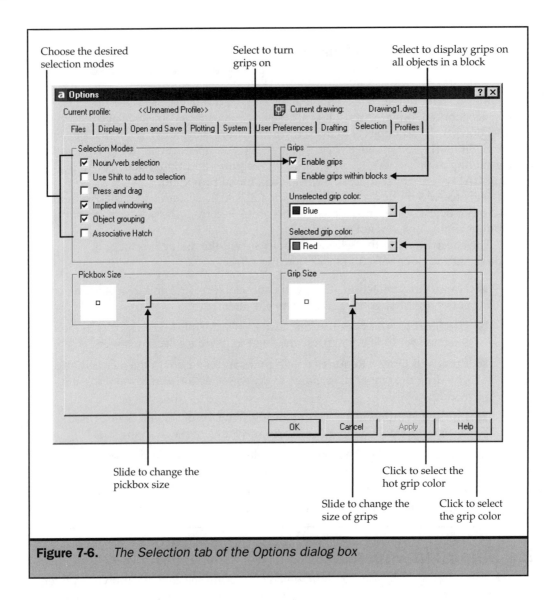

**Figure 7-6.** *The Selection tab of the Options dialog box*

set, and then start the editing command and use the previous selection set. The Object Selection Filters dialog box lets you define more complex filters and also enables you to save and restore named filters. In addition, you can use it to establish the selection set either before starting an editing command or transparently when AutoCAD prompts you to select objects. Regardless of the method you choose, only those selected objects that match the filter are added to the selection set.

*When applying a color or linetype filter, AutoCAD recognizes only colors or linetypes that are explicitly assigned to objects, not colors or linetypes inherited by layer. To select an object whose color is set to BYLAYER (such as a circle that is blue because its layer color is blue), you must set a filter that will match the particular layer or set the color filter to BYLAYER.*

## Using Quick Select

To display the Quick Select dialog box, shown in Figure 7-7, do one of the following:

- From the Tools menu, choose Quick Select.
- At the command line, type **QSELECT** and press ENTER.

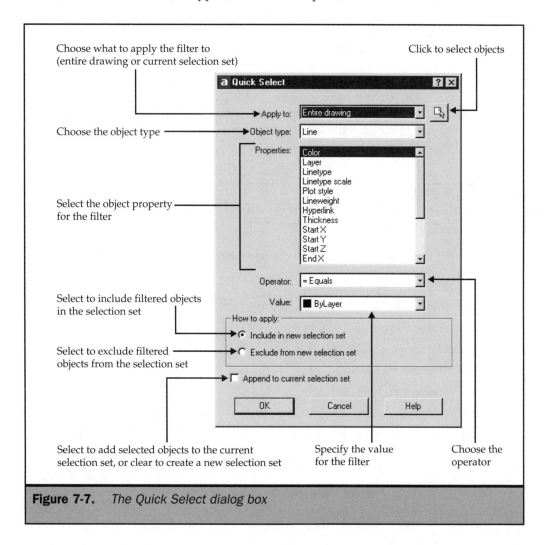

**Figure 7-7.**   *The Quick Select dialog box*

*You can also display the Quick Select dialog box when no commands are active by right-clicking in the drawing area and choosing Quick Select from the shortcut menu. In addition, you can click the Quick Select button in the Properties window. You'll learn about this tool in Chapter 11.*

The Apply To drop-down list enables you to choose what objects to apply the filter to: either the entire drawing or the current selection set. You can also create a new selection set by clicking the Select Objects button and then selecting objects within the drawing. In the Object Type drop-down list, you can either specify the type of object that you want, or select Multiple to filter for more than one object type (such as all objects on a particular layer). The Properties list includes all searchable properties for the object type specified. Select the property on which you want to filter, and then choose the operator from the Operator drop-down list. The content of this list varies depending on the property you choose, and may include Equals, Not Equal, Greater Than, Less Than, and Wildcard Match. For example, Wildcard Match is available only for text fields that can be edited.

Depending on the property and operator you choose, either select the value from the Value drop-down list (if the property has known values) or type the value in this box. Under How To Apply, specify whether you want to include or exclude from the selection set objects that match the specified filtering criteria. When you are ready to apply the filter, click OK.

**LEARN BY EXAMPLE**
*Practice using the Quick Select dialog box to limit the selection of objects. Open the drawing Figure 7-8 on the companion web site.*

To erase the three circles surrounding the text labels in the drawing, do the following:

1. Display the Quick Select dialog box.

2. Under Apply To, choose Entire Drawing.

3. Under Object Type, choose Circle.

4. Under Properties, choose Radius.

5. Under Operator, choose > Greater Than.

6. In the Value box, type **0.09**.

7. Under How To Apply, select Include In New Selection Set.

8. Click OK. AutoCAD selects the three circles surrounding the text labels.

9. In the Modify toolbar, click Erase. AutoCAD immediately erases the three circles.

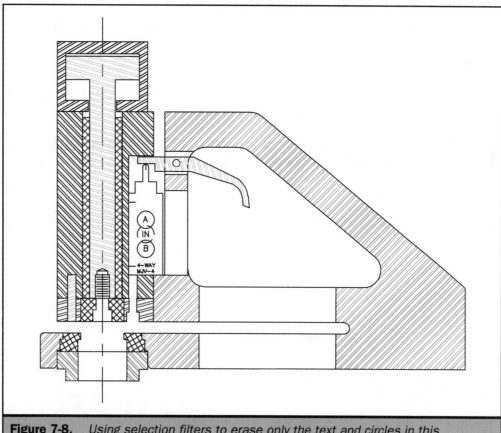

**Figure 7-8.**   *Using selection filters to erase only the text and circles in this mechanical part cross-section*

## Using the FILTER Command

To display the Object Selection Filters dialog box, shown in Figure 7-9, type **FILTER** (or **FI**). This dialog box contains three separate areas. The list at the top of the dialog box shows the current filters that are being used to restrict a selection set. To add a filter to the list, select it from the drop-down list in the Select Filters area and then click the Add To List button. Depending on the type of filter that you select, you can also specify other parameters, such as the start point of a line or the name of a hatch pattern.

The list of filters remains in the dialog box until you clear it or close AutoCAD. If you want, you can assign a name to the current selection filter and save it for reuse at a later time. AutoCAD saves named filters in the FILTER.NFL file.

You can apply a selection filter either before starting a command or while a command is active. If you apply a selection filter when no command is active, the objects can then be edited by specifying the previous selection set, similar to using the

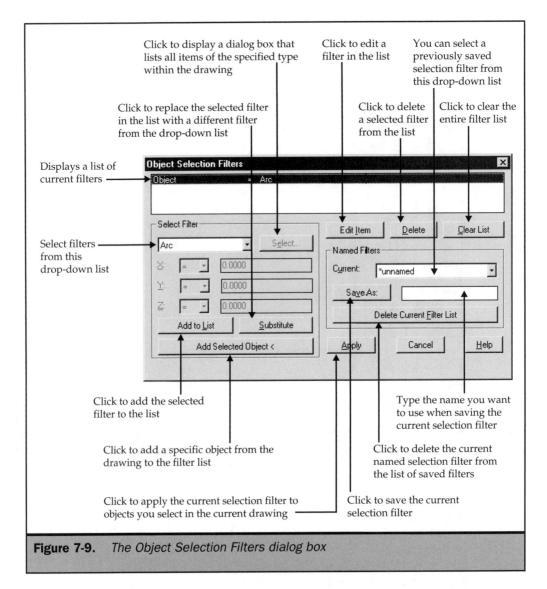

Click to display a dialog box that lists all items of the specified type within the drawing

Click to edit a filter in the list

You can select a previously saved selection filter from this drop-down list

Click to replace the selected filter in the list with a different filter from the drop-down list

Click to delete a selected filter from the list

Click to clear the entire filter list

Displays a list of current filters

Select filters from this drop-down list

Type the name you want to use when saving the current selection filter

Click to add the selected filter to the list

Click to add a specific object from the drawing to the filter list

Click to delete the current named selection filter from the list of saved filters

Click to apply the current selection filter to objects you select in the current drawing

Click to save the current selection filter

**Figure 7-9.** *The Object Selection Filters dialog box*

Quick Select dialog box. To use a selection filter while a command is already active, start the FILTER command transparently by preceding the command with an apostrophe. For example, type **'FILTER** (or **'FI**) and then press ENTER.

*LEARN BY EXAMPLE*
*Practice using the Object Selection Filters dialog box to limit the selection of objects. Continue using the same Figure 7-8 drawing from the previous example.*

To erase all the text objects in the drawing, use the following command sequence and instructions:

```
Command: ERASE
Select objects: 'FILTER
```

1. In the Object Selection Filters dialog box, click Clear List, if any filters are already in the filter list.
2. In the drop-down list in the Select Filter area, choose Text.
3. Click the Add To List button.
4. Click the Apply button and then respond to the remaining command prompts as follows:

```
Select objects: ALL
41 found
36 were filtered out.
Select objects: ENTER
Exiting filtered selection. 5 found
Select objects: ENTER
```

## Defining Complex Selection Filters

Thus far, the selection filters that you've defined have consisted of just a single criteria, such as circles or text. You can also use the FILTER command to define more complex selections, such as all circles with a radius less than 1 and all dimensions on the layer named NOTES. To do this, combine the filter selections with logical operators. You'll find the logical operators at the bottom of the filters drop-down list. These selections mark the beginning and end of each logical operator and must be paired and balanced correctly in the filter list. For example, each Begin OR must have a matching End OR operator. The number of filters that each operator can enclose depends on the operation, as shown in the following table:

| Starting Operator | Encloses | Ending Operator |
|---|---|---|
| Begin AND | One or more filters | End AND |
| Begin OR | One or more filters | End OR |
| Begin XOR | Two filters | End XOR |
| Begin NOT | One filter | End NOT |

*LEARN BY EXAMPLE*
*Practice using complex selection filters to modify the drawing of a mounting plate.*
*Open the drawing Figure 7-10 on the companion web site.*

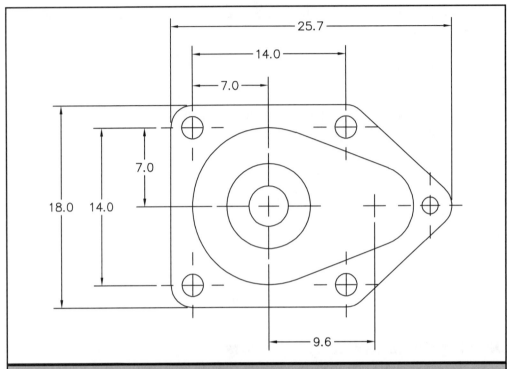

**Figure 7-10.** *Using complex selection filters to erase objects in this mounting plate drawing*

To erase all the circles with a radius of less than 1 drawing unit and all dimensions on the layer named NOTES, use the following command sequence and instructions:

```
Command: ERASE
Select objects: 'FILTER
```

1. In the Object Selection Filters dialog box, click Clear List to clear the previous filter list.

2. In the drop-down list in the Select Filter area, choose **Begin OR and then click the Add To List button. (*Hint:* The logical operators are at the very bottom of the drop-down list.)

3. In the drop-down list, choose **Begin AND, and then click the Add To List button.

4. In the drop-down list, choose Circle and then click the Add To List button.

5. In the drop-down list, choose Circle Radius.

6. In the relationship drop-down list adjacent to the X label, choose <.

7. In the edit box adjacent to the relationship drop-down list, type **1.0**.

8. Click the Add To List button.

9. In the filter drop-down list, choose **End AND and then click the Add To List button.

10. In the filter drop-down list, choose **Begin AND and then click the Add To List button.

11. In the filter drop-down list, choose Dimension and then click the Add To List button.

12. In the filter drop-down list, choose Layer and then click the Select button.

13. In the Select Layer(s) dialog box, choose NOTES and then click OK.

14. Click the Add To List button.

15. In the filter drop-down list, choose **End AND and then click the Add To List button.

16. In the filter drop-down list, choose **End OR and then click the Add To List button.

The list box should now contain these filters:

```
** Begin OR
** Begin AND
Object = Circle
Circle Radius < 1.00
** End AND
** Begin AND
Object = Dimension
Layer = NOTES
** End AND
** End OR
```

17. Click the Apply button, and then respond to the remaining command prompts as follows:

```
Applying filter to selection.
Select objects: ALL
66 found.
62 were filtered out.
Select objects: ENTER
Exiting filtered selection. 4 found
Select objects: ENTER
```

## Loading and Using the Express Tools

The AutoCAD Express Tools consist of a collection of productivity tools—actually, customized add-on software provided by Autodesk—designed to extend the power of AutoCAD. These tools were provided with previous versions of AutoCAD, but are no longer included in AutoCAD 2002. Several of the Express Tools have been enhanced and incorporated into AutoCAD 2002 as standard commands. The others are available free to Autodesk Subscription Program members or may be purchased online from Autodesk's online store. When installing AutoCAD 2002, if you upgraded from a previous version that already had the Express Tools installed, they will also be available for use in AutoCAD 2002. If you performed a clean installation, instructions for installing the Express Tools for use with AutoCAD 2002 are available at Autodesk's support web site. Because the Express Tools are so useful, information about many of the tools is included throughout this book. You'll find additional information about these tools on the companion web site.

When the Express Tools are installed, you can select any of them as you would any other AutoCAD command. You can select most of the commands from one of the Express Tools toolbars, from the Express menu, or by typing the actual command name.

If the Express Tools have been installed and you don't see the Express menu on the menu bar, you can add it to AutoCAD's pull-down menus by using the EXPRESSMENU command. Simply type **EXPRESSMENU** at the command line.

To display the various Express Tools toolbars, display the Toolbars dialog box. Remember that you can display this dialog box by right-clicking any toolbar button and selecting Customize from the shortcut menu. You can also select Toolbars from the View menu or simply type **TOOLBAR** and press ENTER. Under Menu Group, select EXPRESS from the drop-down list, and then select the check box adjacent to the Express Tools toolbar or toolbars that you want to display. Those toolbars can be floated or docked just like any other toolbar.

---

**Tip**  *The Get Selection Set Express tool (GETSEL) creates a temporary selection set, which can then be selected as the previous selection set. You can also use the SELECT command to create a selection set that can then be used as the previous selection set. The SSTOOLS Express tools enable you to create an exclusionary selection set, a selection set containing all objects except those you specifically select. For example, you could select all the objects except those within a selection window.*

# Choosing the Modification Command First

When you choose an object modification command, AutoCAD prompts you to select objects. You can select individual objects or use any of the other object selection methods, such as a selection window or crossing window, to select multiple objects. This is referred to as *verb/noun object selection,* because you specify the action (the verb) before selecting the objects (the nouns) to be acted upon.

When you select objects, you add them to the current selection set. After you select at least one object, you can remove objects from the selection set. To finish adding objects to the selection set, press ENTER to continue with the command. Most object modification commands then act on the entire selection set.

**Tip**
*If you ever forget the available object selection methods, when AutoCAD prompts you to select an object, you can type a question mark (?) and then press ENTER. AutoCAD then displays a list of all the object selection methods. If you display this list, you will likely note several additional selection methods that have not been discussed yet. Box performs like either the window or crossing window, depending on whether you specify the corner points from left to right or right to left, respectively. Auto is actually the default mode; it selects objects, or switches to the box method if you don't point to an object. Single causes object selection to end as soon as you select the first object or objects. Multiple lets you select multiple objects without highlighting each object as you select it, thus speeding up the selection process for complex objects.*

# Selecting Objects First

You can select objects first and then choose how to modify them. This is referred to as *noun/verb object selection,* because you select the object (the noun) before you specify the action (the verb). When you select objects, you can either select individual objects by pointing or use implied windowing to select objects, using either a selection window or a crossing window. As you select each object, it is highlighted.

**Note**
*To use noun/verb object selection, Noun/Verb Selection mode must be enabled on the Selection tab of the Options dialog box. This mode is controlled by the PICKFIRST system variable and, by default, is normally enabled.*

In addition, small squares, called *grips,* appear at strategic points on each object. The locations of the grips depend on the type of object selected. For example, grips appear at the end points and midpoint of a line, at the quadrant points and center point of a circle, and at the end points, midpoint, and center point of an arc.

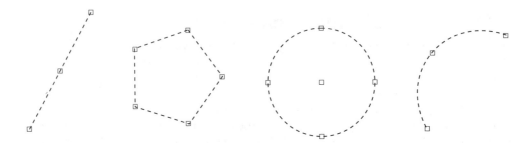

After you select one or more objects, you can choose an object modification command, such as Copy or Move, from the Modify menu or toolbar. You can also use the grips to mirror, move, rotate, scale, stretch, or copy selected objects.

To remove an object from the selection set, press SHIFT and select the object again. When you remove objects from the selection set, the objects are no longer highlighted, but their grips remain, so that you can use the grip as a base point for other operations. To clear grips, press ESC. This also clears the selection set.

> **Note** *When you select objects and then start a command, AutoCAD immediately acts on the objects that you've selected. Therefore, you must select all the objects to be modified before starting the command.*

> **Caution** *You can use noun/verb selection with the following commands: ARRAY, BLOCK, CHANGE, CHPROP, COPY, DVIEW, ERASE, EXPLODE, HATCH, LIST, MIRROR, MOVE, PROPERTIES, ROTATE, SCALE, STRETCH, and WBLOCK. You can't use the following commands with preselected objects; these commands ignore any selection set that is created prior to starting the command: BREAK, CHAMFER, DIVIDE, EDGESURF, EXTEND, FILLET, MEASURE, OFFSET, PEDIT, REVSURF, RULESURF, TABSURF, and TRIM. You can use the PEDIT command with noun/verb selection, as long as only a single polyline is preselected.*

## Editing with Grips

You can edit one or more objects simply by selecting them and then switching into Grip Editing mode. In Grip Editing mode, you can stretch, move, rotate, scale, mirror, or copy the objects. To edit using grips, you first select the object or objects that you want to modify, so that their grips are displayed. To begin editing, click one of the selected grips. The selected grip, referred to as the *hot grip* or *base grip*, changes color.

Selection of a base grip causes AutoCAD to go into Editing mode. Once in Editing mode, you can cycle through the five grip editing commands by pressing the SPACEBAR or ENTER, or by typing a keyboard shortcut (such as **M** for move).

> **Tip** *You can also choose a grip editing command from a shortcut menu. After you select a hot grip, right-click to display a shortcut menu containing grip editing commands.*

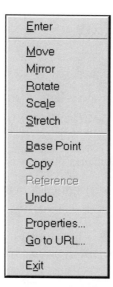

Depending on the type of object and which grip you select as the base grip, you can move the object just by picking the base grip and pointing to a new location. If you press the SHIFT key when you select a new location, you can make multiple copies of the selected objects. You can also switch into Multiple Copy mode by typing **C**. When Multiple Copy mode is active, if you also hold down the SHIFT key while you pick points on the screen, the cursor automatically snaps to an offset position, based on the first two pick points that you selected. For example, if the first copy is placed 2 units away, the cursor then snaps at 2-unit increments. When you release the SHIFT key, you can resume picking points anywhere on the screen.

 *You can use multiple grips as the hot grips to maintain the relationship between the selected grips. To select multiple hot grips, hold down the SHIFT key as you select the grips. Grip Editing mode doesn't become active until you select one grip as a base grip without holding down the SHIFT key.*

### Controlling Grips

You can control the action and appearance of grips by using the controls in the Grips area on the Selection tab of the Options dialog box, shown in Figure 7-6.

Now that you know how to select objects, the remainder of this chapter covers the various ways in which objects can be modified.

## Erasing Objects

You can erase objects from the drawing by using any object selection method. You can use either verb/noun or noun/verb object selection.

To erase one or more objects:

1. Do one of the following:

- On the Modify toolbar, click Erase.
- From the Modify menu, choose Erase.
- At the command line, type **ERASE** (or **E**) and press ENTER.

AutoCAD prompts you to select objects.

2. Select the objects to be erased, and then press ENTER.

You can also select the object to erase, right-click in the drawing area, and then choose Erase from the shortcut menu.

*The OOPS command restores the most recently erased selection set. If you have made additional changes to the drawing since erasing some objects, use OOPS instead of UNDO to restore the erased objects without reversing your other changes. To start the OOPS command, type OOPS and press ENTER.*

## Duplicating Objects

After you draw something once, it is much faster and easier to make copies of it than to draw it again. You can copy one or more objects, and make one copy or multiple copies of those objects. Objects can also be copied to the Windows Clipboard. Use any of the following methods to copy objects within the current drawing:

- Create a copy at a location referenced from the original location by using the COPY command or grips.
- Create a copy that is aligned parallel to the original by using the OFFSET command.
- Create a copy as a mirror image of the original by using the MIRROR command.
- Create several copies in a rectangular or circular pattern by using the ARRAY command.

These commands are discussed in the following sections.

### Copying Objects Within a Drawing

You can duplicate objects within the current drawing. The default method is to create a selection set and then specify a starting point, or *base point*, and a second point, or *displacement*, for the copy, as shown in Figure 7-11. You can also make multiple copies. You can use either verb/noun or noun/verb object selection.

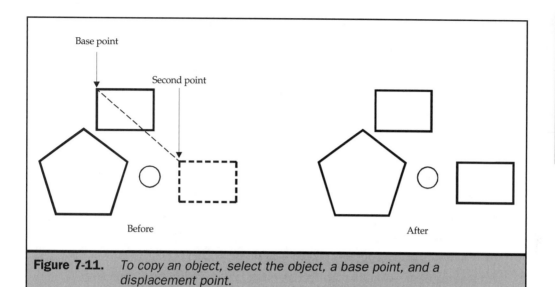

**Figure 7-11.** *To copy an object, select the object, a base point, and a displacement point.*

To copy a selection set once:

1. Do one of the following:

- On the Modify toolbar, click Copy.
- From the Modify menu, choose Copy.
- At the command line, type **COPY** (or **CO** or **CP**) and press ENTER.

AutoCAD prompts you to select objects.

2. Select the objects to be copied and then press ENTER. AutoCAD prompts:

   `Specify base point or displacement, or [Multiple]:`

3. Specify the base point. AutoCAD prompts:

   `Specify second point of displacement or <use first point as displacement>:`

4. Specify the point of displacement.

**Note**    *To copy objects by using the displacement method, when prompted for the base point or displacement, enter a distance instead of specifying a base point, and then press ENTER. When prompted for the second point of displacement, press ENTER again.*

To make multiple copies of a selection set, as shown in Figure 7-12:

1. Do one of the following:

- On the Modify toolbar, click Copy.
- From the Modify menu, choose Copy.

■ At the command line, type **COPY** (or **CO** or **CP**) and then press ENTER. AutoCAD prompts you to select objects.

2. Select the objects to be copied and then press ENTER. AutoCAD prompts:

   ```
   Specify base point or displacement, or [Multiple]:
   ```

3. Type **M** (for Multiple) and then press ENTER, or right-click and select Multiple from the shortcut menu. AutoCAD prompts:

   ```
   Specify base point:
   ```

4. Specify the base point. AutoCAD prompts:

   ```
   Specify second point of displacement or <use first point as displacement>:
   ```

5. Specify the second point of displacement for the first copy. AutoCAD repeats the previous prompt.

6. Specify the second point of displacement for the next copy.

7. Continue specifying displacement points to place additional copies.

8. To complete the command, press ENTER.

**Tip** *You can also copy objects by selecting the objects you want to copy, right-clicking in the drawing area, and then choosing Copy Selection from the shortcut menu.*

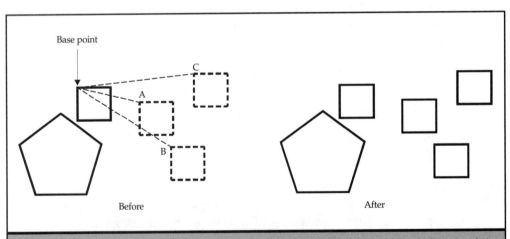

**Figure 7-12.** *To make multiple copies of an object, select the object, a base point, and multiple displacement points (A, B, and C).*

# Copying Using Grips

You can create multiple copies of objects while in any of the Grip Editing modes. For example, you can move objects, leaving a copy of the objects at each point that you specify.

To make multiple copies:

1. Select the objects to be copied.

2. Select a base grip on one of the selected objects.

3. Type **MO** to switch to Move mode.

4. Type **C** to begin making copies.

5. Drag the selected objects to a new location and click.

6. Repeat Step 5 to create multiple copies.

7. To complete the command, press ENTER.

You can also create multiple copies at regularly spaced intervals by creating an offset snap. The offset snap is determined by the distance between the base grip and the first copy.

To make multiple copies by using an offset snap:

1. Select the objects to be copied.

2. Select a base grip on one of the selected objects.

3. Type **MO** to switch to Move mode.

4. Type **C** to begin making copies.

5. Drag the selected object to a new location and click.

6. Hold down the SHIFT key as you place additional copies.

7. To complete the command, press ENTER.

# Copying and Pasting Using the Clipboard

You can use the Windows Clipboard to cut or copy objects from one drawing to another, from paper space to model space and vice versa, or between AutoCAD and another application. Cutting removes the selected objects from the drawing and stores them on the Clipboard. Copying duplicates the selected objects from the drawing and places them on the Clipboard.

To cut objects to the Clipboard:

1. Select the objects that you want to cut.

2. Do one of the following:

- On the Standard toolbar, click Cut To Clipboard.
- From the Edit menu, choose Cut.
- At the command line, type **CUTCLIP** and then press ENTER.
- Press the shortcut keys CTRL-X.
- Right-click in the drawing area and choose Cut from the shortcut menu.

*You can also start the CUTCLIP command first, and then select the objects that you want to cut to the Clipboard.*

To copy objects to the Clipboard:

1. Select the objects that you want to copy.

2. Do one of the following:

- On the Standard toolbar, click Copy To Clipboard.
- From the Edit menu, choose Copy. (*Note:* This is different than the COPY command on the Modify menu.)
- At the command line, type **COPYCLIP** and then press ENTER.
- Press the shortcut keys CTRL-C.
- Right-click in the drawing area and choose Copy from the shortcut menu.

When you copy objects to the Clipboard using COPYCLIP or CUTCLIP, you have very little control over the point used as the base point when later pasting those objects into another drawing. You can use the COPYBASE command, however, to specify the base point when copying objects to the Clipboard.

To specify the base point when copying objects to the Clipboard:

1. Do one of the following:

- From the Edit menu, choose Copy With Base Point.
- At the command line, type **COPYBASE** and then press ENTER.
- Right-click in the drawing area and choose Copy With Base Point from the shortcut menu.

2. Specify the base point.

3. Select the objects you want to copy.

Rather than copying selected objects, you can copy an entire view to the Clipboard. If more than one viewport is visible, AutoCAD copies the current viewport.

To copy a viewport to the Clipboard, do one of the following:

- From the Edit menu, choose Copy Link.
- At the command line, type **COPYLINK** and then press ENTER.

**Note** *The COPYLINK command links the view back to the original source AutoCAD drawing, using Windows' Object Linking and Embedding (OLE). After you paste a copy of the view into another document, if you later update that source drawing, you can update the copies simply by updating the link.*

Anything that you can copy to the Clipboard can also be pasted into a drawing. The format in which the program adds the Clipboard contents to the drawing depends on the type of information in the Clipboard. If you copy AutoCAD drawing objects to the Clipboard, the program pastes them into the drawing as AutoCAD objects. If you copy objects to the Clipboard from other programs, however, they are pasted into the current drawing by using whatever format retains the most information. For example, if the Clipboard contains ASCII text, AutoCAD inserts the text in the current drawing as paragraph text (using the MTEXT command). If the Clipboard contains a graphics object, AutoCAD inserts the object as either an embedded or linked OLE object (depending on how the object was copied to the Clipboard).

To paste objects from the Clipboard, do one of the following:

- On the Standard toolbar, click Paste From The Clipboard.
- From the Edit menu, choose Paste.
- At the command line, type **PASTECLIP** and then press ENTER.
- Press the shortcut keys CTRL-V.
- Right-click in the drawing area and choose Paste from the shortcut menu.

AutoCAD's response when pasting objects from the Clipboard depends on the type of object being pasted. If the Clipboard contains an AutoCAD object, AutoCAD prompts you to specify an insertion point. If the Clipboard contains ASCII text, AutoCAD inserts the text in the upper-left corner of the drawing area, using the MTEXT defaults, and the ASCII text becomes an MTEXT object. You'll learn about the MTEXT command in Chapter 13.

If the Clipboard contains an OLE object, AutoCAD inserts the object in the upper-left corner of the drawing and displays the OLE Properties dialog box. You can then control the size or scale of the object, as well as its plot quality. If the object is an OLE text object, you can also control the font, point size, and text height. With the exception of AutoCAD objects, all objects are inserted as embedded or linked objects. You can edit these embedded or linked objects by double-clicking them in the AutoCAD drawing, which opens the application in which they were created. You'll learn about linking and embedding and working with other applications in Chapter 24.

When pasting an AutoCAD object from one drawing into another, you can use the PASTEORIG command to paste the objects to the same coordinates as in the original drawing from which they were cut or copied.

To paste AutoCAD objects to their original coordinates, do one of the following:

- From the Edit menu, choose Paste To Original Coordinates.
- At the command line, type **PASTEORIG** and then press ENTER.
- Right-click in the drawing area and then choose Paste To Original Coordinates from the shortcut menu.

When you paste copies of AutoCAD objects into a drawing, AutoCAD pastes the individual objects. You can use the PASTEBLOCK command, however, to insert the objects as a single block.

To paste several AutoCAD objects as a single block, do one of the following:

- From the Edit menu, choose Paste As Block.
- At the command line, type **PASTEBLOCK** and then press ENTER.
- Right-click in the drawing area and then choose Paste As Block from the shortcut menu.

AUTOCAD BASICS

*If you subsequently need to edit the individual objects pasted as a single block, use the EXPLODE command to reduce the block to its constituent objects. You'll learn about exploding objects in Chapter 10.*

When you paste objects from other applications into an AutoCAD drawing, you can also convert those objects into AutoCAD objects.

To convert objects into AutoCAD format while pasting from the Clipboard:

1. Do one of the following:

   ■ From the Edit menu, choose Paste Special.

   ■ At the command line, type **PASTESPEC** (or **SP**) and then press ENTER.

   AutoCAD displays the Paste Special dialog box, shown in Figure 7-13.

2. In the Paste Special dialog box, select Paste.

3. In the list of formats, choose the object type that you want to use.

4. Click OK.

Depending on the format that you choose, AutoCAD may issue additional prompts.

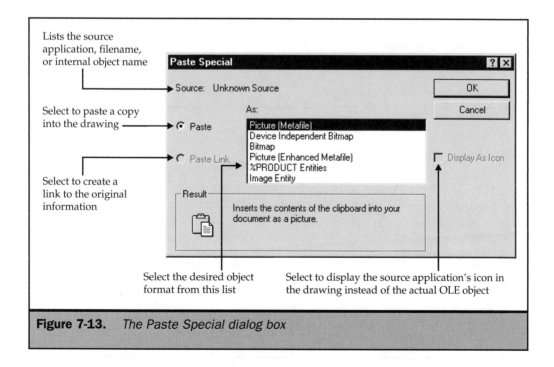

**Figure 7-13.**    *The Paste Special dialog box*

## Offsetting Parallel Copies

The OFFSET command copies a selected object and aligns it parallel to the original object at a specified distance in the current UCS. You can make parallel copies of arcs, elliptical arcs, lines, 2-D polylines, rays, and xlines, as well as concentric circles and ellipses. You *can't* use noun/verb object selection with the OFFSET command.

 *When offsetting ellipses and elliptical arcs, AutoCAD actually creates the new curve as a spline object, because generating an ellipse by offsetting from an existing ellipse isn't mathematically feasible.*

Offsetting curved objects creates larger or smaller curves, depending on which side of the original object you place the copy. For example, placing a parallel copy of a circle outside the circle creates a larger concentric circle; offsetting the copy to the inside of the circle creates a smaller concentric circle.

To make a parallel copy by specifying the distance, as shown in Figure 7-14:

1. Do one of the following:

   ■ From the Modify toolbar, click Offset.

   ■ From the Modify menu, choose Offset.

   ■ At the command line, type **OFFSET** (or **O**) and then press ENTER.

   AutoCAD prompts:

   ```
   Specify offset distance or [Through] <current>:
   ```

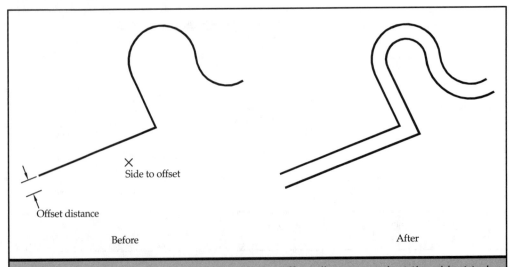

**Figure 7-14.** *To offset an object, specify the offset distance, select the object to be offset, and then specify the side on which to place the parallel copy.*

2. Specify the distance by selecting two points or typing a distance. AutoCAD prompts:

```
Select object to offset or <exit>:
```

3. Select the object to offset. AutoCAD prompts:

```
Specify point on side to offset:
```

4. Specify on which side of the object to place the parallel copy, by pointing. AutoCAD then prompts you to select another object to offset.

5. Repeat Steps 3 and 4, or press ENTER to complete the command.

To make a parallel copy that passes through a point:

1. Do one of the following:

   ■ From the Modify toolbar, click Offset.

   ■ From the Modify menu, choose Offset.

   ■ At the command line, type **OFFSET** (or **O**) and then press ENTER.

   AutoCAD prompts:

```
Specify offset distance or [Through] <current>:
```

2. Type **T** (for Through) and then press ENTER, or right-click and select Through from the shortcut menu. AutoCAD prompts:

```
Select object to offset or <exit>:
```

3. Select the object to offset. AutoCAD prompts:

```
Specify through point:
```

4. Specify a point through which the parallel copy will pass. AutoCAD then prompts you to select another object to offset.

5. Repeat Steps 3 and 4, or press ENTER to complete the command.

---

## Offsetting Polylines

When offsetting polylines, the OFFSETGAPTYPE system variable determines how AutoCAD deals with gaps that occur within the polyline as a result of offsetting the individual polyline segments. This system variable can have the following values:

| | |
|---|---|
| 0 | Extends the segments to fill the gap |
| 1 | Fills the gaps with a filleted arc segment, with the radius of the arc segment equal to the offset distance |
| 2 | Fills the gaps with a chamfered line segment |

# Mirroring Objects

You can create a mirror image of selected objects. You mirror the objects across a mirror line, which you define by specifying two points in the drawing, as shown in Figure 7-15. When mirroring objects, you can retain or delete the original objects. You can use either verb/noun or noun/verb object selection.

To mirror objects:

1. Do one of the following:

    ■ On the Modify toolbar, click Mirror.

    ■ From the Modify menu, choose Mirror.

    ■ At the command line, type **MIRROR** (or **MI**) and then press ENTER.

    AutoCAD prompts you to select objects.

2. Select the objects to be mirrored and then press ENTER. AutoCAD prompts:

    ```
    Specify first point of mirror line:
    ```

3. Specify the first point of the mirror line. AutoCAD prompts:

    ```
    Specify second point of mirror line:
    ```

4. Specify the second point of the mirror line. AutoCAD prompts:

    ```
    Delete source objects? [Yes/No] <N>:
    ```

5. Press ENTER to retain the original objects.

*When mirroring objects, it is often helpful to turn on Ortho mode, so that copies are mirrored vertically or horizontally.*

## Mirroring Objects by Using Grips

To mirror objects by using grips, you first select the objects to display their grips, and then select a grip to make it the hot grip. This becomes the first point of the mirror line. You then switch into Mirror mode and specify the second point of the mirror line.

To mirror objects by using grips:

1. Select the objects to be mirrored.

2. Select a base grip on one of the selected objects.

3. Type **MI** to switch to Mirror mode.

4. Hold down the SHIFT key and specify the second point of the mirror line.

5. Press ENTER to complete the command.

*When you mirror by using grips, the original objects are deleted, unless you also use the Copy option.*

AUTOCAD BASICS

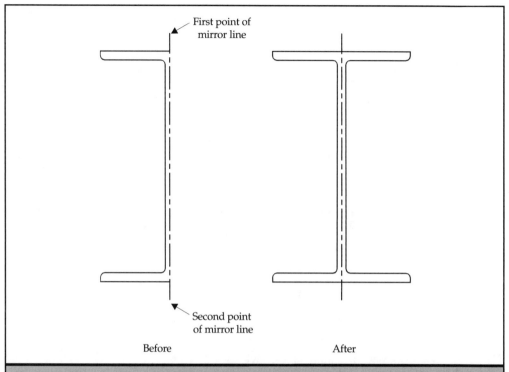

First point of
mirror line

Second point
of mirror line

Before                                After

**Figure 7-15.**   *To mirror an object, select the object to mirror and then specify the
first and second points of the mirror line.*

## Mirroring Text

When you mirror text, AutoCAD normally makes a mirror image of the text. You can
prevent text from being reversed or turned upside down, by changing the MIRRTEXT
system variable. When MIRRTEXT is set to 0, text maintains its original direction, even
when mirrored; a value of 1 causes text to be mirrored, as shown in Figure 7-16.

   *The MIRRTEXT variable affects single-line and multiline text, attributes, and attribute
definitions. It has no effect on text or constant attributes that are contained in blocks.*

# Arraying Objects

You can copy objects in a rectangular or polar (circular) pattern, creating an array. When
you create a *rectangular* array, you control the number of copies in the array by specifying
the number of rows and columns and the distance between them. When you create a
*polar* array, you control the number of copies that comprise the array, and whether to
rotate the copies. You can use either verb/noun or noun/verb object selection.

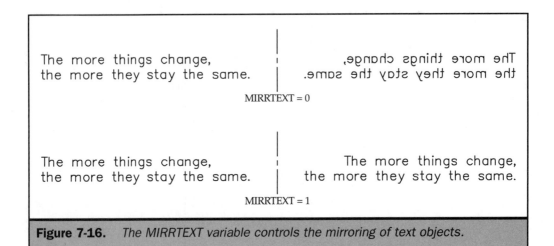

**Figure 7-16.**    *The MIRRTEXT variable controls the mirroring of text objects.*

To create a polar array, as shown in Figure 7-17:

1. Do one of the following:

   - On the Modify toolbar, click Array.

   - From the Modify menu, choose Array.

   - At the command line, type **ARRAY** (or **AR**) and then press ENTER.

2. In the Array dialog box, choose Polar Array. AutoCAD displays the polar array controls (as shown in Figure 7-18).

3. Specify the center point of the array by either entering its X and Y coordinates or by clicking the Pick Center Point button and then selecting the center point within the drawing.

4. Click the Select Objects button. The dialog box temporarily disappears and AutoCAD prompts you to select objects. Select the objects to be arrayed. When you have finished selecting objects, right-click or press ENTER to return to the Array dialog box.

5. In the Method And Values area, select the method you want to use to define the array by selecting one of the methods from the drop-down list, and then specify the required information. (Note that you will need to specify two of these values: the number of items to array, including the original objects; the angle that the array is to fill, from 0 to 360 degrees; and the angle between items. The pair of values required depends on the method you select from the drop-down list. You can specify the angle values by either typing a value or clicking the adjacent button and then specifying the angles by picking points within the drawing.)

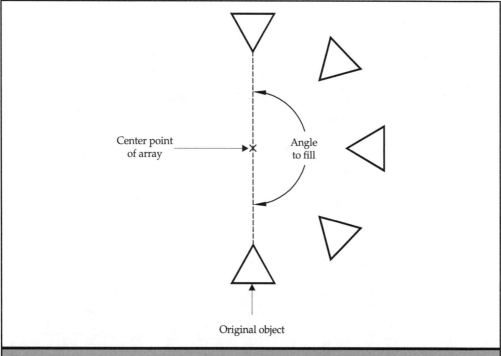

**Figure 7-17.**   *To create a polar array, select the object to be copied, specify the
center point of the array, and then specify the number of items to
array and the angle the array is to fill.*

6. Select the Rotate Items As Copied check box to rotate the objects as they are
   arrayed, or clear this check box to retain the original orientation of each copy as
   it is arrayed.

7. Click OK to create the array.

**Tip**   *You can click More to expand the Array dialog box so that you can specify the base point of
the objects to be arrayed to rotate them about a point other than their default base point.*

After selecting the objects to array and specifying the method and values, but
before completing the array, you can click the Preview button to view a preview of the
resulting array within your drawing. AutoCAD displays a small dialog box. You can
then click Accept to create the array using the current values, or click Modify to return
to the Array dialog box so that you can change one or more values.

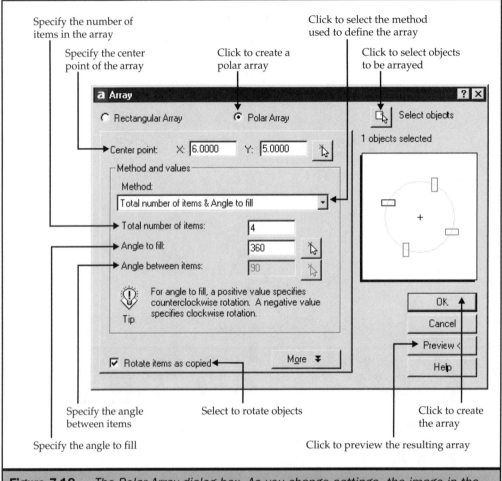

Specify the number of
items in the array

Specify the center
point of the array

Click to create a
polar array

Click to select the method
used to define the array

Click to select objects
to be arrayed

Specify the angle
between items

Select to rotate objects

Click to create
the array

Specify the angle to fill

Click to preview the resulting array

**Figure 7-18.**    *The Polar Array dialog box. As you change settings, the image in the
preview area changes to reflect the current settings in the dialog box.*

To create a rectangular array, as shown in Figure 7-19:

1. Do one of the following:

   ■ On the Modify toolbar, click Array.

   ■ From the Modify menu, choose Array.

   ■ At the command line, type **ARRAY** (or **AR**) and then press ENTER.

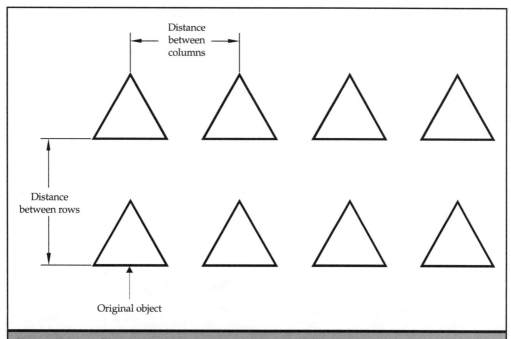

**Figure 7-19.** *To create a rectangular array, select the object to be copied, specify the number of rows and columns, and then specify the distance between rows and columns.*

2. In the Array dialog box, choose Rectangular Array. AutoCAD displays the rectangular array controls (as shown in Figure 7-20).

3. Click the Select Objects button. The dialog box temporarily disappears and AutoCAD prompts you to select objects. Select the objects to be arrayed. When you have finished selecting objects, right-click or press ENTER to return to the Array dialog box.

4. In the Rows and Columns boxes, specify the number of rows and columns in the array.

5. In the Offset Distance And Direction area, specify the row and column offset values by either typing the values or clicking the adjacent buttons and then specifying the distances by picking points within the drawing. (Note that you can also specify the distance between rows and columns in a single step by clicking Pick Both Offsets and then picking two points. The vertical distance between the points defines the distance between rows, and the horizontal distance between the points defines the distance between columns.)

6. To change the rotation angle of the array, enter the new angle in the Angle Of Array box. (Note that this angle is based on the current snap rotation angle and type of angular units.)

7. Click OK to create the array.

As with the polar array, before completing the array, you can click the Preview button to preview the result of the array within your drawing, and then either accept the array or return to the Array dialog box to make any necessary changes.

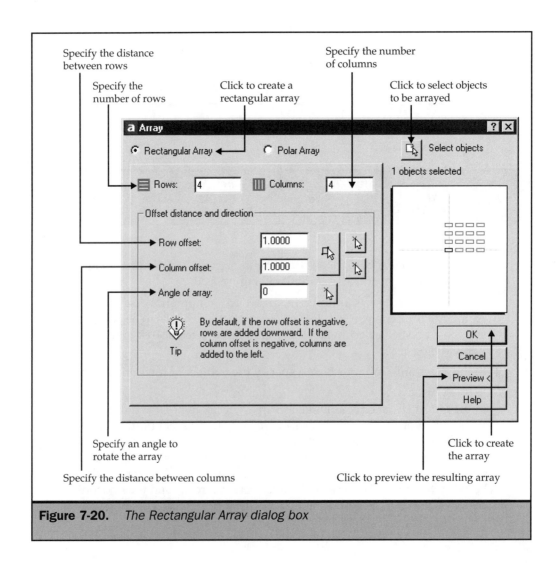

**Figure 7-20.** *The Rectangular Array dialog box*

### Controlling the Maximum Number of Entities in an Array

AutoCAD includes a variable that controls the maximum number of entities that can be created in an array. Unlike most other variables, MaxArray is stored in the Windows System Registry and can't be accessed using the SETVAR command. Instead, you must use AutoLISP. To view the current value, type **(getenv "MaxArray")** at the command line. Note that the variable name is case-sensitive. The initial value is 100000.

To change the MaxArray value, you must use the following command structure:

```
(setenv "MaxArray" "value")
```

where *value* is the actual value. For example, to change the MaxArray value to 500000, you would type **(setenv "MaxArray" "500000")** at the command line. Note that setting MaxArray to a value higher than its default may degrade the performance of AutoCAD.

## Rearranging Objects

You can move one or more objects and also rotate objects about a specified point. When rearranging objects, you can use verb/noun or noun/verb object selection. You can also move and rotate objects by using grips.

## Moving Objects

You can move objects within the drawing. The default method is to create a selection set and then specify a base point and a second point of displacement, as shown in Figure 7-21. When you move objects, their orientation and size remain the same. You can use either verb/noun or noun/verb object selection.

To move objects:

1. Do one of the following:

   - On the Modify toolbar, click Move.
   - From the Modify menu, choose Move.
   - At the command line, type **MOVE** (or **M**) and then press ENTER.

   AutoCAD prompts you to select objects.

2. Select the objects to be moved and then press ENTER. AutoCAD prompts:

   ```
   Specify base point or displacement:
   ```

3. Specify the base point. AutoCAD prompts:

   ```
   Specify second point of displacement or <use first point as displacement>:
   ```

4. Specify the point of displacement.

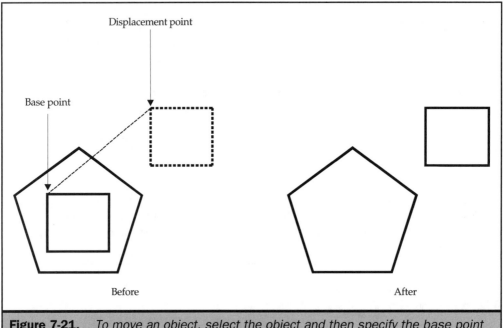

Displacement point

Base point

Before                    After

**Figure 7-21.** *To move an object, select the object and then specify the base point and the second point of displacement.*

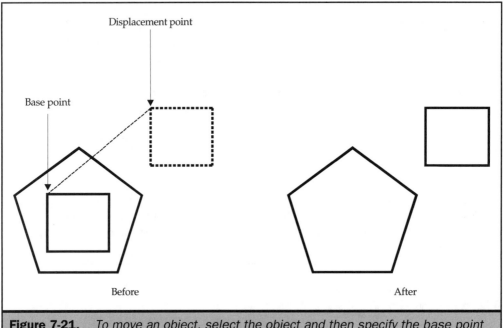 **Note** *To move objects by using the displacement method, when prompted for the base point or displacement, enter a distance instead of specifying a base point and then press ENTER. When prompted for the second point of displacement, press ENTER again.*

You can also move objects by using grips. To move an object by using grips, select the object to display its grips, and then click a grip to make it the hot grip. This becomes the base point (although you can then use the base point option to choose a different base point). You can then switch into Move mode and specify the second point of displacement.

To move an object by using grips:

1. Select the objects to be moved.

2. Select a base grip on one of the selected objects.

3. Type **MO** to switch to Move mode.

4. Drag the object where you want to relocate it and then click to release.

## Rotating Objects

You can rotate objects about a specified point at either a specified rotation angle or an angle relative to a base reference angle. The default method rotates the objects about a specified base point by using a relative rotation angle from their current orientation, as shown in Figure 7-22. When you rotate objects, their size remains the same. You can use either verb/noun or noun/verb object selection.

To rotate objects:

1. Do one of the following:

   ■ On the Modify toolbar, click Rotate.

   ■ From the Modify menu, choose Rotate.

   ■ At the command line, type **ROTATE** (or **ROT**) and then press ENTER.

   AutoCAD prompts you to select objects.

2. Select the objects to be rotated and then press ENTER. AutoCAD prompts:

   ```
   Specify base point:
   ```

3. Specify the base point. AutoCAD prompts:

   ```
   Specify rotation angle or [Reference]:
   ```

4. Specify the rotation angle.

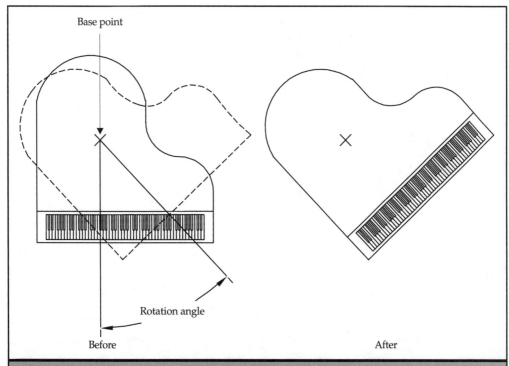

**Figure 7-22.**    *To rotate an object, select the object and then specify the rotation base point and the rotation angle.*

AUTOCAD BASICS

Sometimes, rotating objects in reference to another object is easier than rotating them about a specified base point; for example, when you want to rotate an object so that it aligns with another object. The Reference option lets you select the object that you want to rotate and the object with which you want to align it.

*LEARN BY EXAMPLE*

*Use the Reference angle option of the ROTATE command to align the door symbol with the wall in a simple floor plan. Open the drawing Figure 7-23 on the companion web site.*

To align the door symbol with the wall, as shown in Figure 7-23, use the following command sequence and instructions. Use the endpoint object snap to select the labeled points.

```
Command: ROTATE
Select objects: (select the door)
Select objects: ENTER
Specify base point: (select point A)
Specify rotation angle or [Reference]: R
Specify the reference angle <0>: (select point B)
Specify second point: (select point A)
Specify the new angle: (select point C)
```

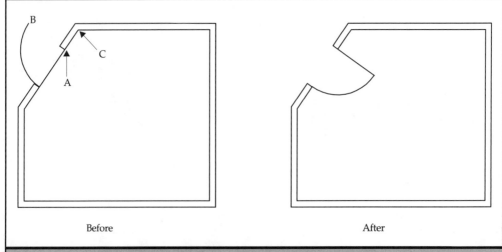

**Figure 7-23.**   *Use the Reference option of the ROTATE command to rotate the door so that it aligns with the wall.*

You can also rotate objects by using grips. To rotate an object by using grips, select the object to display its grips, and then click a grip to make it the hot grip. This becomes the base point. You can then switch into Rotate mode and specify the rotation angle.

To rotate an object by using grips:

1. Select the objects to be rotated.

2. Select a base grip on one of the selected objects.

3. Type **RO** to switch to Rotate mode.

4. Drag the object to rotate it and then click to release. (Note that you can also enter the rotation angle at the command line.)

*One advantage to rotating objects by using grips is that you can also use the grip edit Copy option, so that you can make copies of the objects as you rotate them.*

# Aligning Objects

You can also use the ALIGN command to move, rotate, and optionally resize an object, so that it aligns with another object. Although the ALIGN command often is used to align objects in 3-D, it is equally adept at aligning objects in 2-D. You *can't* use noun/verb object selection with the ALIGN command.

The command first prompts you to select the objects that you want to move. It then prompts you for as many as three pairs of points. These points are referred to as *source points* and *destination points*. The resulting realignment is based on the correlation of these points.

When aligning two-dimensional objects, you work only with two pairs of points. The first pair of points defines the movement of the object. The object is moved so that the first source point matches the first destination point. If you press ENTER without specifying the second source point, the command ends. The object is moved but retains its current alignment.

The second pair of points defines the rotation of the source object. The object is aligned so that a line drawn from the first and second source points aligns with an imaginary line drawn between the first and second destination points. If you then press ENTER without specifying the third source point, the command asks whether you want to scale the objects to match the alignment points. AutoCAD uses the distance between the first and second destination points to determine the scaling factor. Scaling is available only when you align objects by using two pairs of points.

To start the ALIGN command, do one of the following:

■ From the Modify menu, choose 3D Operation | Align.

■ At the command line, type **ALIGN** (or **AL**) and then press ENTER.

*LEARN BY EXAMPLE*
*Use the ALIGN command to align the pipe extension on a drawing of a piping assembly. Open the drawing Figure 7-24 on the companion web site.*

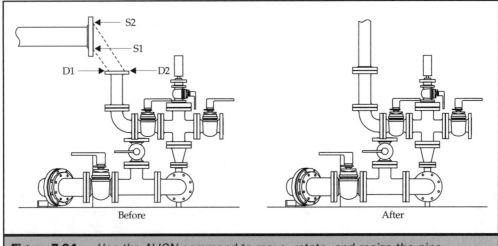

**Figure 7-24.** *Use the ALIGN command to move, rotate, and resize the pipe extension in a single operation.*

To align the pipe extension with the piping assembly shown in Figure 7-24, use the following command sequence and instructions. Use the node object snap to select the labeled points.

```
Command: ALIGN
Select objects: (select the pipe extension)
Select objects: ENTER
Specify first source point: (select point S1)
Specify first destination point: (select point D1)
Specify second source point: (select point S2)
Specify second destination point: (select point D2)
Specify third source point or <continue>: ENTER
Scale objects to alignment points? [Yes/No] <No>: Y
```

*If you need more flexibility when moving, copying, rotating, or scaling objects, you can use the MOCORO Express tool, which combines all four commands within a single command.*

# Resizing Objects

Objects can be resized by stretching, scaling, extending, trimming, or editing their lengths.

## Stretching Objects

You can change the size of objects by stretching them. When you stretch objects, you must select the objects by using either a crossing window or a crossing polygon. You

AUTOCAD BASICS

then either specify a displacement distance or select a base point and a displacement point. Objects that cross the window or polygon boundary are stretched; those completely within the boundary are simply moved, as shown in Figure 7-25. You can use either verb/noun or noun/verb object selection.

To stretch an object:

1. Do one of the following:

■ On the Modify toolbar, click Stretch.

■ From the Modify menu, choose Stretch.

■ At the command line, type **STRETCH** (or **S**) and then press ENTER.

AutoCAD prompts:

```
Select objects to stretch by crossing-window or
crossing-polygon...
Select objects:
```

2. Select the objects and then press ENTER. AutoCAD prompts:

```
Specify base point or displacement:
```

3. Specify the base point. AutoCAD prompts:

```
Specify second point of displacement:
```

4. Specify the second point of displacement.

**Note** *To stretch objects by using the displacement method, when prompted for the base point or displacement point, enter a distance instead of specifying a base point, and then press ENTER. When prompted for the second point of displacement, press ENTER again.*

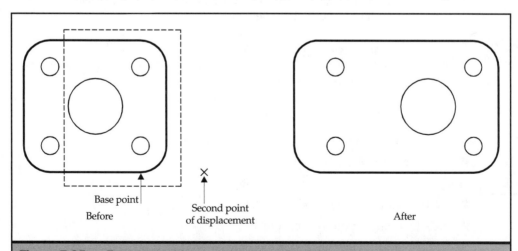

Base point
Before

Second point
of displacement

After

**Figure 7-25.** *To stretch objects, select them by using a crossing window or crossing polygon, and then specify the base point and displacement point.*

You can also stretch objects by using grips. To stretch an object by using grips, select the object to display its grips, and then click a grip to make it the hot grip. This becomes the base point. Because stretch is the default Grip Editing mode, you can immediately stretch the object.

To stretch an object by using grips:

1. Select the objects to be stretched.

2. Select a base grip on one of the selected objects.

3. Drag the grip to stretch the object, and then click to release.

*You will often want to stretch more than one object (or more than one grip) at a time. For example, you may want to stretch a rectangle. You can select more than one hot grip by pressing the* SHIFT *key as you select the grips.*

Some grips, such as the midpoints of lines and the center points of circles and arcs, move the object rather than stretching it.

*The MSTRETCH Express tool lets you stretch multiple objects, by allowing multiple crossing windows and crossing polygons to be specified for a single stretch operation.*

## Scaling Objects

You can change the size of selected objects by scaling them in relation to a base point. AutoCAD lets you change the size of objects either by specifying a scale factor or by specifying a base point and a length, which is used as a scale factor that is based on the current drawing units. You can also use a scale factor referenced to a base scale factor, for example, by specifying the current length and a new length for the object. You can use either verb/noun or noun/verb object selection.

*The SCALE command scales objects uniformly. To scale them nonuniformly, you can first convert them into a block and then insert them with different X and Y scale factors. You'll learn about blocks in Chapter 15.*

To scale an object by specifying a scale factor, as shown in Figure 7-26:

1. Do one of the following:

   ■ On the Modify toolbar, click Scale.

   ■ From the Modify menu, choose Scale.

   ■ At the command line, type **SCALE** (or **SC**) and then press ENTER.

   AutoCAD prompts you to select objects.

2. Select the objects to be scaled and then press ENTER. AutoCAD prompts:

   ```
   Specify base point:
   ```

3. Specify the base point. AutoCAD prompts:

```
Specify scale factor or [Reference]:
```

4. Specify the scale factor.

Sometimes, scaling an object in reference to another object is easier than specifying a scale factor; for example, when you want to resize an object so that it matches a dimension on another object. The Reference option lets you select the object that you want to scale and the reference points that you want to use to resize it.

*LEARN BY EXAMPLE*
*Use the Reference option of the SCALE command to adjust the size of the window so that it fits in the wall on a simple floor plan. Open the drawing Figure 7-27 on the companion web site.*

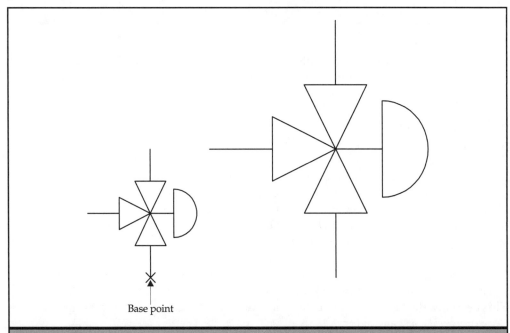

**Figure 7-26.**    *To scale an object by a scale factor, select the object and then specify the base point and scale factor. The object on the right was scaled by a factor of 2.*

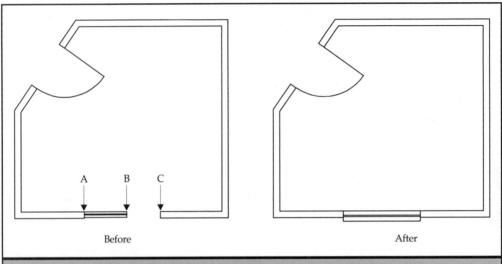

**Figure 7-27.** *Use the Reference option of the SCALE command to scale the window so that it fits within the wall opening.*

To scale the window so that it fits the opening shown in Figure 7-27, use the following command sequence and instructions. Use the endpoint object snap to select the labeled points.

```
Command: SCALE
Select objects: (select the window)
Select objects: ENTER
Specify base point: (select point A)
Specify scale factor or [Reference]: R
Specify reference length <1>: (select point A)
Specify second point: (select point B)
Specify new length: (select point C)
```

You can also scale objects by using grips. To scale an object by using grips, select the object to display its grips, and then click a grip to make it the hot grip. This becomes the base point. You can then switch into Scale mode and either drag or specify a scale factor.

To scale an object by using grips:

1. Select the objects to be scaled.

2. Select a base grip on one of the selected objects.

3. Type **SC** to switch to Scale mode.

4. Drag the object to scale it and then click to release.

**Note** *After switching to Scale mode, you can also type a scale factor and then press* ENTER.

**Tip** *One advantage to scaling objects by using grips is that you can also use the grip edit Copy option, so that you can make copies of the objects as you scale them.*

## Extending Objects

You can extend objects so that they end at a boundary defined by other objects. You can also extend objects to the point at which they would intersect an *implied boundary edge* (an edge that would be intersected if the boundary object was extended). When you use the EXTEND command, you first select the boundary edges and then specify the objects to extend, selecting them either one at a time or by using the fence selection method. You *can't* use noun/verb object selection with the EXTEND command.

Only arcs, elliptical arcs, lines, open 2-D and 3-D polylines, and rays can be extended. Valid boundary objects include arcs, blocks, circles, ellipses, elliptical arcs, floating viewport boundaries, lines, 2-D and 3-D polylines, rays, regions, splines, text, and xlines.

After you start the EXTEND command, you can also trim objects to the selected boundary edges by holding down the SHIFT key and then selecting the objects to be trimmed.

To extend an object, as shown in Figure 7-28:

1. Do one of the following:

   - On the Modify toolbar, click Extend.
   - From the Modify menu, choose Extend.
   - At the command line, type **EXTEND** (or **EX**) and then press ENTER.

   AutoCAD prompts:

   ```
   Current settings: Projection=UCS Edge=None
   Select boundary edges ...
   Select objects:
   ```

2. Select the objects to serve as boundary edges. AutoCAD prompts:

   ```
   Select object to extend or shift-select to trim
     or [Project/Edge/Undo]:
   ```

3. Select an object to extend. AutoCAD then repeats the previous prompt.

4. Select another object to extend, or press ENTER to complete the command.

**Note** *If you select multiple boundary edges, an object lengthens only until it intersects the nearest boundary edge. You can continue to extend the object until it hits the next boundary edge by selecting that object again. If an object could be extended in more than one direction, AutoCAD extends the object in the direction closest to the point that you used to select the object.*

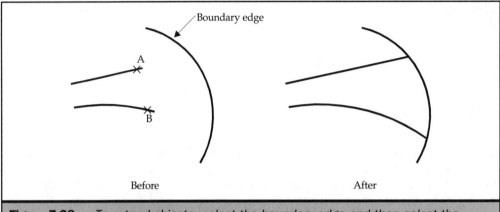

**Figure 7-28.** *To extend objects, select the boundary edge and then select the objects to extend (A and B).*

The Edgemode option lets you control whether objects are extended to an actual boundary or an implied boundary. To extend an object to an implied boundary:

1. Do one of the following:

   ■ On the Modify toolbar, click Extend.

   ■ From the Modify menu, choose Extend.

   ■ At the command line, type **EXTEND** (or **EX**) and then press ENTER.

2. Select the objects to serve as boundary edges and then press ENTER. AutoCAD prompts:

   ```
   Select object to extend or shift-select to trim
     or [Project/Edge/Undo]:
   ```

3. Type **E** (for Edge mode) and then press ENTER, or right-click and select Edge from the shortcut menu. AutoCAD prompts:

   ```
   Enter an implied edge extension mode [Extend/No extend] <current>:
   ```

4. Type **E** (for Extend mode) and then press ENTER, or right-click and select Extend from the shortcut menu.

5. Select an object to extend.

**Note** *The Projmode option lets you determine how AutoCAD locates boundary edges. Normally, objects to be extended must actually intersect the boundary edge in 3-D space. You can also extend objects to a boundary edge that would be intersected if projected into either the XY plane of the current UCS or the current view.*

AUTOCAD BASICS

Online
EXAMPLES

*LEARN BY EXAMPLE*
*Use the fence object selection method to extend several objects simultaneously. Open the drawing Figure 7-29 on the companion web site.*

To extend several objects at once, as shown in Figure 7-29, use the following command sequence and instructions:

```
Command: EXTEND
Current settings: Projection=UCS Edge=None
Select boundary edges ...
Select objects: (select the boundary edge)
Select objects: ENTER
Select object to extend or shift-select to trim
 or [Project/Edge/Undo]: F
First fence point: (select point A)
Specify endpoint of line or [Undo]: (select point B)
Specify endpoint of line or [Undo]: (select point C)
Specify endpoint of line or [Undo]: ENTER
Select object to extend or shift-select to trim
 or [Project/Edge/Undo]: ENTER
```

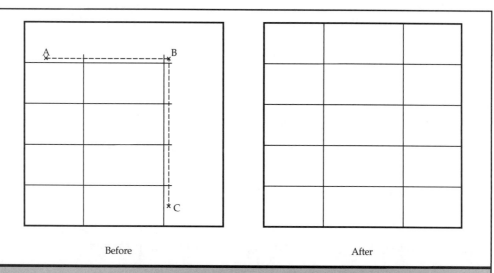

Before

After

**Figure 7-29.** *Use the fence object selection method to extend several objects simultaneously.*

*When you extend a wide polyline, its center line intersects the boundary edge. Because the end of the polyline is always cut at a 90-degree angle, part of the polyline may extend past the boundary edge. A tapered polyline continues to taper until it intersects the boundary edge. If this would result in a negative polyline width, the ending width changes to 0.*

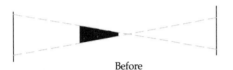

Before                                    After

## Trimming Objects

You can clip, or trim, objects so that they end at a cutting edge that is defined by other objects. You can also trim objects to the point at which they would intersect an *implied cutting edge* (an edge that would be intersected if the cutting edge was extended). When you use the TRIM command, you first select the cutting edges and then specify the objects to be trimmed, selecting them either one at a time or by using the fence selection method. You *can't* use noun/verb object selection with the TRIM command.

Only arcs, circles, ellipses, elliptical arcs, lines, 2-D and 3-D polylines, rays, splines, and xlines can be trimmed. Valid boundary objects include arcs, blocks, circles, ellipses, elliptical arcs, floating viewport boundaries, lines, 2-D and 3-D polylines, rays, regions, splines, text, and xlines.

After you start the TRIM command, you can also extend objects to the selected boundary edges by holding down the SHIFT key and then selecting the objects to be extended.

To trim an object, as shown in Figure 7-30:

1. Do one of the following:

■ On the Modify toolbar, click Trim.

■ From the Modify menu, choose Trim.

■ At the command line, type **TRIM** (or **TR**) and then press ENTER.

   AutoCAD prompts:

   ```
   Current settings: Projection=UCS Edge=None
   Select cutting edges ...
   Select objects:
   ```

2. Select the objects to serve as cutting edges. AutoCAD prompts:

   ```
   Select object to trim or shift-select to extend
     or [Project/Edge/Undo]:
   ```

3. Select an object to trim. AutoCAD then repeats the previous prompt.

4. Select another object to trim, or press ENTER to complete the command.

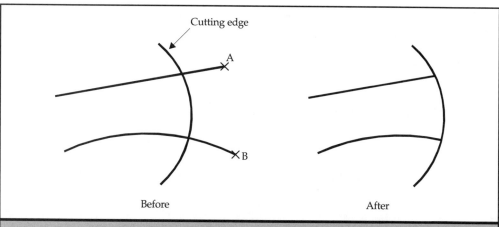

**Figure 7-30.** *To trim objects, select the cutting edge and then select the objects to be trimmed (A and B).*

**Note**    *If you select multiple cutting edges, the object being trimmed is cut where it intersects the first cutting edge that it encounters. If you pick a point on an object lying between two cutting edges, only the portion between those cutting edges is deleted. If the object that you pick to be trimmed is also a cutting edge, the deleted portion disappears from the screen and that cutting edge is no longer highlighted. The visible portion of the object can still serve as a cutting edge, however.*

The Edgemode option lets you control whether objects are trimmed to an actual cutting edge or an implied cutting edge. To trim an object to an implied cutting edge:

1. Do one of the following:

   ■ On the Modify toolbar, click Trim.

   ■ From the Modify menu, choose Trim.

   ■ At the command line, type **TRIM** (or **TR**) and then press ENTER.

2. Select the objects to serve as cutting edges and then press ENTER. AutoCAD prompts:

   ```
   Select object to trim or shift-select to extend
     or [Project/Edge/Undo]:
   ```

3. Type **E** (for Edge mode) and then press ENTER, or right-click and select Edge from the shortcut menu. AutoCAD prompts:

   ```
   Enter an implied edge extension mode [Extend/No extend] <current>:
   ```

4. Type **E** (for Extend mode) and then press ENTER, or right-click and select Extend from the shortcut menu.

5. Select an object to extend.

*The Projmode option lets you determine how AutoCAD locates cutting edges. Normally, objects to be trimmed must actually intersect the cutting edge in 3-D space. You can also trim objects to a cutting edge that would be intersected if projected into either the XY plane of the current UCS or the current view.*

### LEARN BY EXAMPLE

*Use the fence object selection method to trim several objects simultaneously. Open the drawing Figure 7-31 on the companion web site.*

To trim several objects simultaneously, as shown in Figure 7-31, use the following command sequence and instructions:

```
Command: TRIM
Current settings: Projection=UCS Edge=None
Select cutting edges ...
Select objects: (select the cutting edge)
Select objects: ENTER
Select object to trim or shift-select to extend
 or [Project/Edge/Undo]: F
First fence point: (select point A)
Specify endpoint of line or [Undo]: (select point B)
Specify endpoint of line or [Undo]: (select point C)
Specify endpoint of line or [Undo]: ENTER
Select object to trim or shift-select to extend
 or [Project/Edge/Undo]: ENTER
```

*The EXTRIM Express tool trims all objects falling on one side or the other of a cutting edge that is defined by an arc, circle, line, or polyline.*

## Lengthening Objects

You can change the length of objects or the included angle of arcs by using any of the following methods:

- Dynamically drag the end point or angle
- Specify an incremental length or angle measured from an end point
- Specify the new length as a percentage of the total length or angle
- Specify a new length or included angle

You can use the LENGTHEN command to change the length of arcs, lines, elliptical arcs, open polylines, and open splines. You *can't* use noun/verb object selection with the LENGTHEN command.

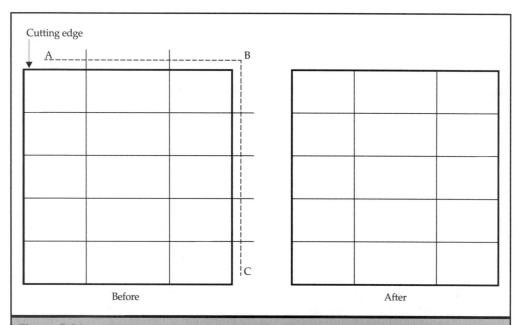

**Figure 7-31.** *Use the fence object selection method to trim several objects at once.*

To change the length of an object by dragging:

1. Do one of the following:

   - On the Modify toolbar, click Lengthen.
   - From the Modify menu, choose Lengthen.
   - At the command line, type **LENGTHEN** (or **LEN**) and then press ENTER.

   AutoCAD prompts:

   ```
   Select an object or [DElta/Percent/Total/Dynamic]:
   ```

2. Type **DY** (for Dynamic mode) and then press ENTER, or right-click and select DYnamic from the shortcut menu. AutoCAD prompts:

   ```
   Select an object to change or [Undo]:
   ```

3. Select the object that you want to lengthen. AutoCAD prompts:

   ```
   Specify new end point:
   ```

   Notice that the cursor snaps to the end point that is closest to the end of the object that you select. The length of the object changes dynamically as you move the cursor.

4. Move the cursor and then click to specify a new end point. AutoCAD then prompts you again to select an object to change.

5. Select another object to lengthen, or press ENTER to complete the command.

*As you've learned, AutoCAD provides many different ways to accomplish the same task. In this case, although the LENGTHEN and EXTEND commands are similar, they operate quite differently. The LENGTHEN command operates on objects one at a time. The EXTEND command requires a real or implied boundary.*

## Breaking Objects

You can break an object into two parts, removing a portion of the object in the process. You specify the two points for the break. By default, the point that you use to select the object is also used as the first break point; however, you can use the First option to select a separate break point from the one used to select the object. You *can't* use noun/verb object selection with the BREAK command.

To break an object by using two specified break points, as shown in Figure 7-32:

1. Do one of the following:

   ■ On the Modify toolbar, click Break.

   ■ From the Modify menu, choose Break.

   ■ At the command line, type **BREAK** (or **BR**) and then press ENTER.

   AutoCAD prompts you to select an object.

2. Select the object. AutoCAD prompts:

   ```
   Specify second break point or [First point]:
   ```

3. Type **F** (for First point) and then press ENTER, or right-click and select First Point from the shortcut menu. AutoCAD prompts:

   ```
   Specify first break point:
   ```

4. Specify the first break point. AutoCAD prompts:

   ```
   Specify second break point:
   ```

5. Specify the second break point.

As soon as you specify the second break point, the object is broken and the command ends.

AutoCAD also provides a tool to simply break an object in two at a selected point. This tool is actually a macro that automatically selects the First option as part of the command sequence. You'll learn more about creating your own macros in Chapter 28.

AUTOCAD BASICS

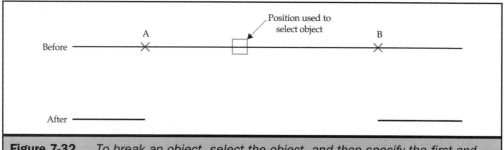

**Figure 7-32.**   *To break an object, select the object, and then specify the first and second break points (A and B, respectively).*

To break an object in two at a selected point:

1. On the Modify toolbar, click Break At Point.

2. Select the object to be broken.

3. Specify the point at which the object is to be broken.

AutoCAD immediately breaks the object in two at the specified point and the command ends. Although the break is not visible, if you select the object, you will immediately see that only the portion you selected is highlighted, clearly indicating that the object has been broken into two separate objects.

# The
# Complete
# Reference

AutoCAD 2002

# Part II

## Moving Beyond the Basics

The
Complete
Reference

# Chapter 8

## Organizing Drawing Information

As you add objects to your drawings, the drawings can quickly become quite complex. AutoCAD offers several tools that help you organize the information represented in your drawings, the primary tool being the use of layers. As you've already learned, with careful planning, you can divide your drawings into different categories of objects—such as walls, doors, and windows—and draw each on its own respective layer. Layers help you to organize objects physically.

Two other drawing components that help you organize drawing information visually are the use of color and linetypes. You are probably familiar with using different linetypes in manual drafting to convey different types of information. For example, you can use a dashed linetype to indicate something hidden by another object, or use a line consisting of alternating long and short dashes to denote the centerline of an object. These same conventions can be applied to AutoCAD drawings. In addition, you can assign any of 256 possible colors to the objects in your drawing. Although you may never print your drawings in color, you can use these color assignments to help you better organize the information in your drawings. Color can also be used to determine pen-width or lineweight variation when you print your drawings.

You already learned the basics of using layers, linetypes, and color in Chapter 2. This chapter explains in greater detail how to do the following:

■ Create and control layers

■ Create, load, and use linetypes

■ Work with lineweights

■ Define and enforce drawing standards

# Organizing Information on Layers

Layers in AutoCAD are much like the transparent overlays you used in manual drafting. You use layers to organize different types of drawing information. In AutoCAD, each object in a drawing exists on a layer. When you draw an object, it is created on the current layer. Each AutoCAD drawing can contain an unlimited number of layers, each with its own distinct name. In addition to deciding how to organize the drawing information with layers, careful thought should be given to the names that you give to those layers.

You can control the visibility of layers. When you turn a layer on or off, the objects drawn on that layer are no longer visible and they don't print or plot. Although a layer may be invisible, you can still select it as the current layer, in which case any new objects that you create are also invisible until you turn the layer back on. Objects on invisible layers can also affect the display and printing of objects that are on other layers. For example, when you work in 3-D, objects on invisible layers can hide other objects when you use the HIDE command to remove hidden lines. You'll learn more about this in Chapter 22.

You can also freeze and thaw layers. Objects on frozen layers do not display, do not print, and do not regenerate. When you freeze a layer, its objects do not affect the display or printing of other objects. For example, objects on frozen layers don't hide other objects when you use the HIDE command to remove hidden lines. In addition, you can't draw on a frozen layer until you thaw it. In a layout (paper space), you can specify the frozen/thawed status of layers individually in each floating viewport.

Layers can be locked or unlocked. The objects on a *locked* layer are still visible and will print, but you can't edit them. Locking a layer prevents you from accidentally modifying objects.

Layers can also be specified as plotted or not plotted. Objects drawn on a layer specified as not plotted can remain visible but do not plot. This feature is useful for storing reference information or construction lines.

As you've already learned, each layer has its own color, linetype, and lineweight. Objects that you draw on a particular layer are displayed in the color, linetype, and lineweight that is associated with that layer, unless you specifically override these settings. You can also assign an object property called *plot style*. Plot styles control how your drawing is plotted. For example, you can use a plot style to plot all the walls in a floor plan with a 50 percent "screened" value. You'll learn about setting the layer plot style in this chapter. In Chapter 18, you will learn about plot styles in greater detail.

You control all the settings associated with layers from the Layer Properties Manager dialog box, shown in Figure 8-1. This dialog box is displayed whenever you start the LAYER command.

To display this dialog box, do one of the following:

- On the Object Properties toolbar, click Layers.
- From the Format menu, choose Layer.
- At the command line, type **LAYER** (or **LA**) and press ENTER.

The layer list contains 11 columns of information. In addition to the layer name displayed in the first column, icons and labels in the other columns indicate the layer's visibility, freeze/thaw state, color, linetype, lineweight, and so on. Note that the Current VP Freeze and New VP Freeze columns appear in the dialog box only when working in a layout. This information is organized from left to right as follows:

- Layer name
- Layer on or off
- Layer frozen or thawed in all viewports
- Layer locked or unlocked
- Layer color
- Linetype assigned to the layer
- Lineweight assigned to the layer
- Plot style assigned to the layer
- Whether the layer plots or not
- Layer frozen only in the current active floating (paper space) viewport
- Layer frozen in any new floating (paper space) viewports

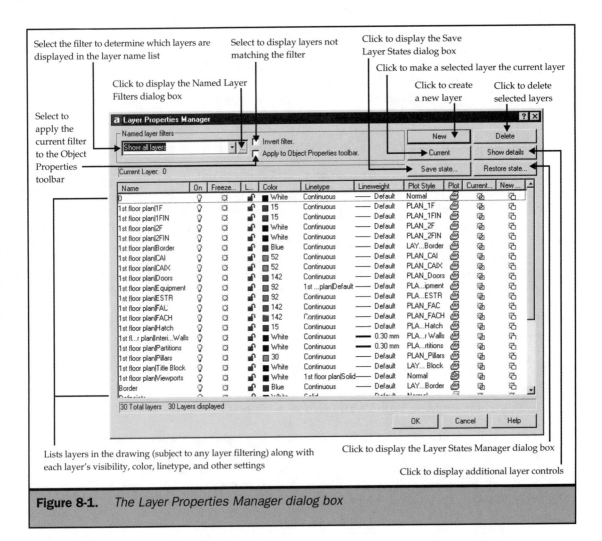

Select the filter to determine which layers are displayed in the layer name list

Select to display layers not matching the filter

Click to display the Save Layer States dialog box

Click to make a selected layer the current layer

Click to display the Named Layer Filters dialog box

Click to create a new layer

Click to delete selected layers

Select to apply the current filter to the Object Properties toolbar

Lists layers in the drawing (subject to any layer filtering) along with each layer's visibility, color, linetype, and other settings

Click to display the Layer States Manager dialog box

Click to display additional layer controls

**Figure 8-1.** *The Layer Properties Manager dialog box*

You can modify any of these properties by clicking the appropriate icon for a specific layer. These columns are identified in greater detail in Figure 8-2.

**Note** *Element lists, such as the list of layer names, normally are sorted alphabetically. The MAXSORT system variable determines the number of elements to be sorted. Its default value is 200. If the total number of elements to be sorted exceeds this value, the list won't be sorted. If you encounter an unsorted list of layers, increase the value of MAXSORT.*

**Tip** *You can sort the list of layers for any layer property by clicking the property heading. For example, to sort the list based on linetype, click the heading at the top of the Linetype column.*

The Details section of the Layer Properties Manager dialog box, shown in Figure 8-3, provides an alternative way to control layer properties. Click the Show Details button to

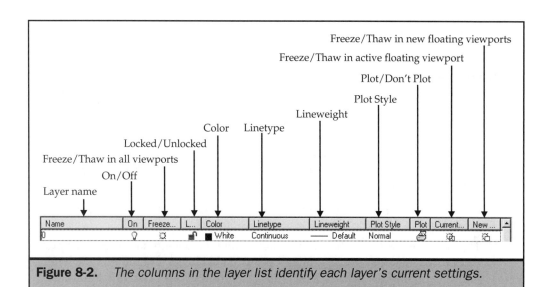

**Figure 8-2.**   *The columns in the layer list identify each layer's current settings.*

enlarge the dialog box to display these additional controls. Clicking the Hide Details button reduces the size of the dialog box, thus hiding this area. You can use the controls in the Details area to change the name of a layer or to change the color, lineweight, linetype, plot style, visibility, or other properties associated with one or more selected layers.

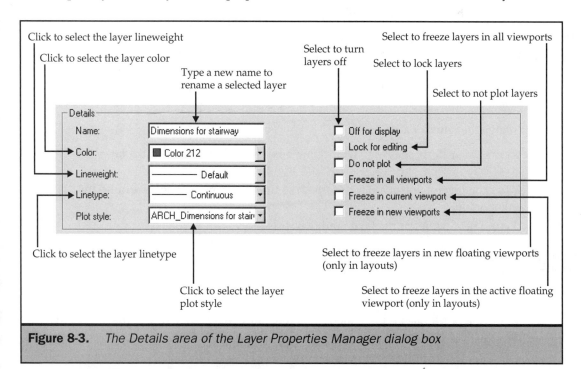

**Figure 8-3.**   *The Details area of the Layer Properties Manager dialog box*

MOVING BEYOND
THE BASICS

## Creating and Naming Layers

You can create an unlimited number of layers in every drawing and use those layers to organize information. When you create a new layer, it is initially assigned the color white (or black, depending on your current drawing background color), the linetype CONTINUOUS, the DEFAULT lineweight, and the Normal plot style. By default, a new layer is also visible (on and thawed). After you create and name a layer, you can change its color, linetype, lineweight, visibility, and other properties.

As you learned in Chapter 2, creating a new layer takes just three simple steps. Since you will probably repeat these steps often, however, it is worth restating them. To create a new layer:

1. In the Layer Properties Manager dialog box, click New. AutoCAD creates a new layer, called Layer1.

2. Type a name for the new layer over the highlighted default name and then press ENTER. The layer name can contain up to 255 characters and can include letters, numbers, and blank spaces. Also note that although layer names are not case-sensitive, AutoCAD retains whatever capitalization you use when specifying layer names.

3. Click OK to complete the command and return to your drawing.

 *When adding a new layer, you can create multiple layers at one time by typing the layer names, separated by commas. You can also right-click in the layer list in the Layer Properties Manager dialog box, and then choose New Layer from the shortcut menu.*

 *If you select a layer just prior to using the New button to create a new layer, the new layer inherits the properties of the selected layer. To create layers with the default properties, be sure that no layers are selected before you pick the New button. You can clear all the selected layers by right-clicking in the layer list and then choosing Clear All from the shortcut menu.*

To change the name of a layer:

1. In the Layer Properties Manager dialog box, click a layer name in the layer list to select it.

2. Click the layer name again (or press F2) so that a rectangle appears around the layer name.

3. Type a new name for the layer and then press ENTER.

 *You can also select the layer and then rename it by typing a new name in the Name box in the Details area of the dialog box.*

To delete a layer:

1. In the Layer Properties Manager dialog box, select the layer name in the layer list.

2. Click the Delete button (or press DELETE).

### Reusing Layer Names

Most AutoCAD users find that they frequently reuse the same standard layer names along with their associated color, linetype, and lineweight settings in many of their drawings. These standard layer names become part of their office standards. For example, architects often reuse standard layer names and settings for creating and organizing specific types of drawing objects, such as walls, doors, windows, and so on. There are several ways to streamline the reuse of standard layer names. The easiest way is to create a template drawing containing all of your predefined layer names. Then, when starting a new drawing, you can base it on this standard template.

You can also use the Layer States Manager to export layer names and layer settings to a file, and then import that file (and thus its layers) into other drawings. You'll learn more about this new feature of AutoCAD 2002 later in this chapter.

AutoCAD 2002 also provides powerful new tools to define and enforce drawing standards for layers and for other named objects such as linetypes, text styles, and dimension styles. You'll also learn about these tools later in this chapter.

You can also reuse layers from existing drawings by using AutoCAD DesignCenter to simply drag existing layer names into your drawing from any other drawing. You'll learn about AutoCAD DesignCenter in Chapter 16.

**Note**    *Layer 0 can never be removed. Layers that have objects on them cannot be deleted until those objects are reassigned to a different layer. You will learn how to change the properties of objects, including their layer assignment, in Chapter 11. You can also use the PURGE command to remove unused layers, as well as other named elements such as linetypes and plot styles that aren't actually used in the drawing. You will learn about the PURGE command in Chapter 11 as well.*

## Setting the Current Layer

When you create new objects, they are drawn on the current layer. To draw new objects on a different layer, you must first make that layer the current layer. Because this is something that you will do quite often, AutoCAD provides three different methods for selecting the current layer:

- Use the Layer Properties Manager dialog box
- Use the Layer Control drop-down list
- Make a selected object's layer the current layer

To set the current layer by using the Layer Properties Manager dialog box:

1. Display the dialog box by using one of the methods that you've already learned.

2. In the layer list, select the layer that you want to make current.

3. Click the Current button, double-click the layer name, or right-click the layer name and then choose Make Current from the shortcut menu.

4. Click OK to close the dialog box and return to your drawing.

Although the Layer Properties Manager dialog box is very useful when you need to create new layers or change the settings of several layers at a time, using it simply to set the current layer generally requires too many steps. For that reason, AutoCAD provides a Layer Control drop-down list. This list is located on the Object Properties toolbar. Using this drop-down list reduces to just two steps the task of setting the current layer.

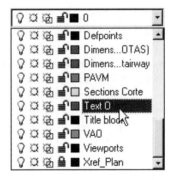

To set the current layer by using the Layer Control drop-down list:

1. On the Object Properties toolbar, click the Layer Control drop-down list.

2. In the list, click the layer name to set it as the current layer. (Note that, as confirmation, the current layer appears in the Layer Control box when the drop-down list is not open.)

*Although the drop-down list may not be wide enough to display very long layer names, if you pause the cursor over a layer name, a ToolTip appears showing the complete layer name.*

Another particularly powerful way to control the current layer is to set it by selecting an existing drawing object. AutoCAD immediately matches the current layer to that of the selected object. Although this method also requires two steps, it is often more convenient than using the Layer Control drop-down list, because you don't need to think at all about the layer name.

To match the current layer to that of a selected object:

1. On the Object Properties toolbar, click the Make Object's Layer Current button. AutoCAD prompts:

```
Select object whose layer will become current:
```

2. Select an object by pointing and clicking.

# Controlling Layer Visibility

Any layer can be visible or invisible. Objects on invisible layers are not displayed and do not print. By controlling the layer visibility, you can turn off unnecessary information, such as construction lines or notes. By changing layer visibility, you can put the same drawing file to multiple uses.

For example, if you are drawing a floor plan, you can draw the walls on one layer, the layout of light fixtures on a second layer, and the location of plumbing fixtures and pipes on a third layer. By selectively turning layers on and off, you can print the basic floor plan, the electrical engineering drawings, and the plumbing drawings all from the same drawing file.

Again, since controlling layer visibility is something that you will do quite often, AutoCAD provides several different methods for turning layers on and off:

- Use the Layer Properties Manager dialog box
- Use the Layer Control drop-down list
- Use one of the special layer-control Express tools

To turn layers on and off by using the Layer Properties Manager dialog box:

1. Display the dialog box by using one of the methods that you've already learned.

2. In the layer list, select one or more layers.

3. Click the light bulb icon adjacent to one of the selected layers to toggle the on/off setting for all the selected layers.

4. Click OK to close the dialog box and return to your drawing.

**Tip**
*When selecting layers, to select several listed layers, you can use the standard Windows mouse procedures:* SHIFT-*click,* CTRL-*click, and* CTRL-A *to select several adjacent layers, several individual layers, or all the layers, respectively. You can also right-click within the layer list to display a shortcut menu from which you can select all the layers, clear all the layers from the layer selection, select all but the current layer, or invert the selection (select all layers not currently selected). You can also use the Off For Display check box in the Details area of the Layer Properties Manager dialog box to turn selected layers on and off.*

The Layer Properties Manager dialog box is most useful for simultaneously turning on or off multiple layers. You can also use the Layer Control drop-down list to turn layers on and off.

To turn layers on or off by using the Layer Control drop-down list:

1. On the Object Properties toolbar, click the Layer Control drop-down list.

2. In the list, click the light bulb icon adjacent to the name of the layer to toggle it on or off.

3. Repeat Step 2, or click anywhere outside the drop-down list to close the list and return to your drawing.

*Because you must turn each layer on or off individually, the Layer Control drop-down list is convenient only for changing the visibility of one or two layers. Use the Layer Properties Manager dialog box when you need to change several layers at once.*

You can also freeze layers to improve the performance of operations, such as removing hidden lines in 3-D drawings or creating shaded images. When a layer is frozen, objects drawn on that layer are no longer visible. Use either the Layer Properties Manager dialog box or the Layer Control drop-down list to freeze and thaw layers, in much the same way as you use them to turn layers on and off.

You can also turn plotting on and off individually for visible layers. This feature enables you to create reference information, such as notes or construction lines, on specific layers. Although those layers can remain visible, the objects on those layers will not plot. Again, you can use either the Layer Properties Manager dialog box or the Layer Control drop-down list to control whether a layer plots or not.

In addition to these standard tools, several Express tools enable you to control the visibility of layers. These tools are described in more detail on the Chapter 8 page of the companion web site.

## Locking and Unlocking Layers

Locking a layer makes it easy to refer to information contained on the layer, but prevents you from accidentally modifying objects drawn on that layer. When a layer is locked, its objects remain visible as long as the layer is not also turned off or frozen. Although you can't edit the objects on a locked layer, you can still set that layer as the current layer and add new objects to it. You can also change the linetype and color associated with that layer. Unlocking a layer restores full editing capabilities.

### Freezing Layers in Specific Viewports

There are several different ways to control the frozen state of layers. By freezing a layer globally, the layer becomes frozen in every viewport and in both model space and all paper space layouts. When you work with floating viewports in layouts, however, there may be instances when you want to freeze specific layers in one viewport but not in others. For example, when you work with three-dimensional drawings, you can use floating viewports to display different orthographic projections—such as a top view in one viewport, a front view in another, and so on. If you add dimensions to these views, you probably don't want the dimensions that you add to the top view to be visible in the front view, and vice versa. Therefore, you can create separate layers for these dimensions and specify that the top-view dimension layer be frozen in all but its viewport, and do the same thing for the dimension layers associated with the other orthographic viewports. You'll learn more about controlling the visibility in viewports in Chapter 17.

The tools that you use for locking and unlocking layers are the same as those tools that you've already learned to use to control layer visibility:

- The Layer Properties Manager dialog box
- The Layer Control drop-down list

To lock or unlock a layer, click the padlock icon adjacent to the layer name in either the dialog box or drop-down list. The current setting toggles between locked and unlocked whenever you click the icon. Again, as you've already learned, use the dialog box when you want to change the settings of several layers at one time. Using the drop-down list is most convenient when you only need to lock or unlock individual layers.

Two Express tools also enable you to lock and unlock layers by selecting an object on the layer that you want to lock or unlock. These tools often are faster and more convenient than using the Layer Properties Manager dialog box or the Layer Control drop-down list, because you don't need to know the layer name.

## Setting the Layer Color

Each layer in a drawing is assigned a color. As long as the current AutoCAD color (shown in the Color Control drop-down list on the Object Properties toolbar) is ByLayer, all new objects that you draw appear in the color that is assigned to the layer on which they are drawn. For example, if you draw a circle and the current layer is assigned the color red, the circle will be red. If you later change the color associated with that layer to green, the circle and all the other objects drawn on that layer will change to green.

As you already learned, when you create a new layer, it is initially either black or white, depending on the background color. You can change the color assigned to any layer at any time by using the Layer Properties Manager dialog box.

To change the color associated with a layer:

1. In the Layer Properties Manager dialog box, click the color icon adjacent to the layer whose color you want to change. AutoCAD displays the Select Color dialog box, shown in Figure 8-4.

2. In the Select Color dialog box, click the color of your choice, or type the color number or one of the standard color names in the Color edit box and then click OK.

**Tip**

*You can select several layers and then click the color icon adjacent to any one of the selected layers to change all the layers to the new color. You can also use the Color drop-down list in the Details area of the dialog box.*

**Caution**

*If you set the current color in the drawing (shown in the Color Control drop-down list on the Object Properties toolbar) to a specific color rather than to ByLayer or ByBlock, new objects will be drawn using the current color rather than the color associated with the current layer. You learned about controlling the current color in Chapter 2.*

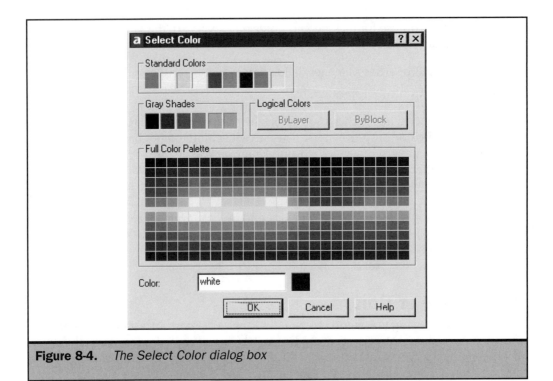

**Figure 8-4.** *The Select Color dialog box*

# Setting the Layer Linetype

In addition to a color, each layer has an associated *linetype* (a repeating pattern of dashes, dots, blank spaces, and symbols). The linetype determines the appearance of objects both on the screen and when printed or plotted. As long as the current AutoCAD linetype (shown in the Linetype Control drop-down list on the Object Properties toolbar) is ByLayer, all new objects that you draw appear using the linetype associated with the layer on which they are drawn. For example, if you draw a line and the current layer has been assigned a dashed linetype, the line will be dashed. If you later change the linetype associated with that layer to a continuous linetype, the line and all the other objects drawn on that layer will be drawn using the continuous linetype.

As you already learned, when you create a new layer, it initially is assigned the continuous linetype. You can change the linetype assigned to any layer at any time by using the Layer Properties Manager dialog box.

To change the linetype associated with a layer:

1. In the Layer Properties Manager dialog box, click the linetype name adjacent to the layer whose linetype you want to change. AutoCAD displays the Select Linetype dialog box, shown in Figure 8-5.

2. In the Select Linetype dialog box, click the linetype that you want to assign to the layer and then click OK.

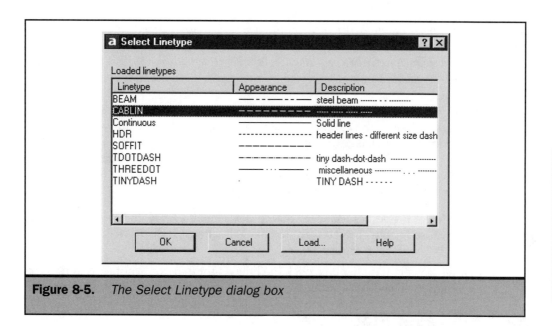

**Figure 8-5.**   *The Select Linetype dialog box*

**Note**   *Only those linetypes that are already loaded in the current drawing appear in the Select Linetype dialog box. To use a linetype that does not appear in the dialog box, you must first load the linetype by clicking the Load button. This is described later in this chapter. As soon as the linetype has been loaded, you can assign it to the layer.*

**Tip**   *When you use the Select Linetype dialog box, you can select several layers and then click the linetype name adjacent to any one of the selected layers to change all the layers to the new linetype. You can also use the Linetype drop-down list in the Details area of the dialog box. Refer to Chapter 2 for a discussion of linetype scale.*

## Setting the Layer Lineweight

As you've already learned in Chapter 2, AutoCAD provides the ability to assign lineweight to the objects in your drawing. Like linetypes, the use of different lineweights helps convey information about the meaning of objects in a drawing. For example, you can use a heavy lineweight to show the outline of a cross-section, and a thin lineweight for the cross-hatching within that cross-section.

Each layer in a drawing is assigned a lineweight. As long as the current AutoCAD lineweight (shown in the Lineweight Control drop-down list on the Object Properties toolbar) is ByLayer, all new objects that you draw appear in the lineweight that is assigned to the layer on which they are drawn. For example, if you draw a line and the current layer is assigned the lineweight 0.80mm, the line will have that lineweight. If you later change the lineweight associated with that layer to 0.25mm, the line and all the other objects drawn on that layer will change to a lineweight of 0.25mm.

*You may not actually see the lineweight in your drawing. By default, AutoCAD is configured to not display lineweights, because the display of lines using more than one pixel slows AutoCAD's performance. In order to actually see various lineweights, you must explicitly turn on the display of lineweights, as described in Chapter 2.*

As you already learned, when you create a new layer, it is initially assigned the Default lineweight (normally 0.01 inch or 0.25mm, although the default value can be changed by using the Lineweight Settings dialog box or by changing the LWDEFAULT system variable value, as described in Chapter 2). You can change the lineweight assigned to any layer at any time by using the Layer Properties Manager dialog box.

To change the lineweight associated with a layer:

1. In the Layer Properties Manager dialog box, click the lineweight adjacent to the layer whose lineweight you want to change. AutoCAD displays the Lineweight dialog box, shown in Figure 8-6.

2. In the Lineweight dialog box, click the lineweight that you want to assign to the layer, and then click OK.

## Setting the Layer Plot Style

Plot styles enable you to change the appearance of a plotted drawing. Plot styles are available only in drawings that use named plot styles (as opposed to those that use the color-dependent plotting mode, which was the only mode available in versions of

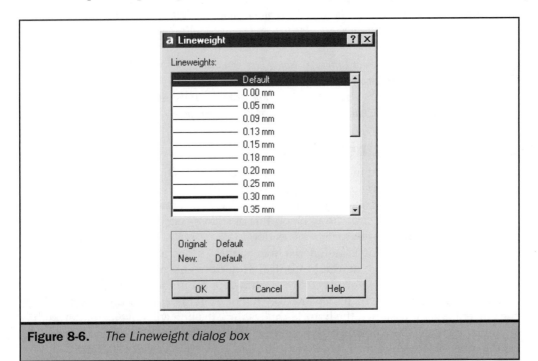

**Figure 8-6.**   *The Lineweight dialog box*

AutoCAD prior to AutoCAD 2000). Named plot styles enable you to control aspects of the plotted drawing—such as the color, screening, linetype, and lineweight, as well as the end style, join style, and fill style of lines—independent of the way they appear in AutoCAD. This feature is particularly useful for plotting the same drawing in different ways. For example, you can create separate layouts of a floor plan for the architectural plan and electrical plan, and assign different plot styles to layers within each layout so that they plot differently in each layout—such as plotting the walls using a 50 percent screen in the electrical plan.

Plot styles are defined in plot style tables. Once created, you can assign plot styles to individual objects or to layers. When working in a drawing that uses named plot styles, each layer in the drawing is assigned a plot style. As long as the current plot style (shown in the Plot Style Control drop-down list on the Object Properties toolbar) is ByLayer, all new objects that you draw are created using the plot style assigned to the layer on which they are drawn. If you later change the plot style associated with that layer, all the objects drawn on that layer will plot using the new plot style. You'll learn more about plot styles in Chapter 18.

When you create a new layer, it is assigned the Normal plot style, which means that the other properties assigned to that layer (color, linetype, and so forth) control the appearance of the objects on that layer when plotted. You can change the plot style assigned to any layer at any time by using the Layer Properties Manager dialog box.

To change the plot style associated with a layer:

1. In the Layer Properties Manager dialog box, click the plot style adjacent to the layer whose plot style you want to change. AutoCAD displays the Select Plot Style dialog box, shown in Figure 8-7.

2. In the Select Plot Style dialog box, click the plot style you want to assign to the layer, and then click OK.

# Undoing Layer Changes

As you create layers and change layer properties, you may find that you want to reverse what you've just done. Although you could use the UNDO command, that would also reverse any other changes you made to the drawing. For example, if you freeze several layers and create some new objects in the drawing, and then decide to thaw the frozen layers, using the UNDO command would thaw the layers but would also remove the new objects. Instead, you could use the Layer Properties Manager dialog box to thaw the layers, but that would require several steps. Or suppose you changed the color and linetype of several layers and then decided later that you wanted to restore their previous settings. In this situation, the UNDO command would be even less useful and it would be even more time consuming to go back into the Layer Properties Manager dialog box to make the changes.

AutoCAD 2002 provides a new command that undoes the changes you made to the layer settings without affecting the rest of the drawing. The Layer Previous command undoes the most recent change or set of changes made using either the Layer Properties Manager dialog box or the Layer Control drop-down list. You can use the Layer Previous command to step back, reversing each previous change to layer settings in succession.

MOVING BEYOND THE BASICS

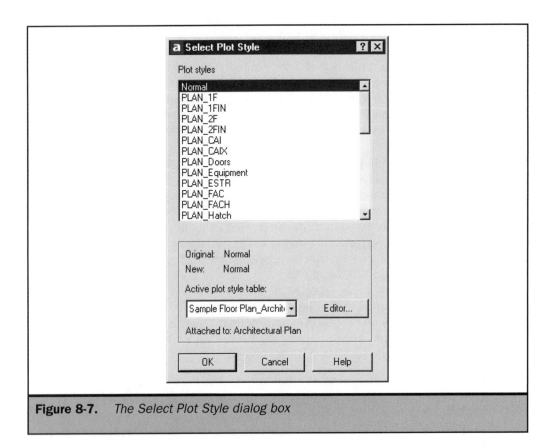

**Figure 8-7.** *The Select Plot Style dialog box*

To undo changes to layer settings, do one of the following:

■ On the Object Properties toolbar, click Layer Previous.

■ At the command line, type **LAYERP** and then press ENTER.

The Layer Previous command does not undo all the changes you can make to layers. Layer Previous will not remove added layers, restore deleted layers, or change layer names. For example, if you rename a layer and change its properties, the command will restore the layer's original properties but will not change its name. If you add a new layer, the Layer Previous command will not remove it from the list of layers.

In order for the Layer Previous command to operate, AutoCAD tracks all changes you make to layer settings. Tracking this information causes a slight decrease in overall program performance. If you wish, you can turn off the tracking of layer property changes. This might be useful prior to running a large script.

To turn off layer property tracking:

1. At the command line, type **LAYERPMODE** and then press ENTER.
   AutoCAD prompts:

   ```
   Enter LAYERP mode [ON/OFF] <current>:
   ```

2. Type **ON** to turn on layer property tracking, or **OFF** to turn tracking off.

# Applying Layer Filters

Although relatively simple AutoCAD drawings may contain just a few layers, complex drawings can easily contain hundreds of layers. By default, all the layers in the drawing appear in the Layer Properties Manager dialog box and the Layer Control drop-down list. Seeing all these layers is often overwhelming, particularly when you really need to deal only with smaller subsets of these layers at any given time. For example, when you work on a drawing of a multistory building, the drawing may contain sets of layers for each individual floor. When you work on just one floor, it is more convenient to see only the layers that are related to that floor. You can accomplish this by applying a layer filter.

In the Layer Properties Manager dialog box, the Named Layer Filters drop-down list enables you to apply layer filters. This drop-down list provides the following options:

- **Show All Layers**   Displays all the layers in the drawing (the default option).
- **Show All Used Layers**   Displays only those layers on which objects have already been drawn. This setting is particularly useful when working with template drawings that contain many predefined layers that haven't been used yet.
- **Show All Xref Dependent Layers**   Displays only those layers added to the drawing from external references.

In addition to these three selections, the drop-down list contains any named layer filters already created using the Named Layer Filters dialog box (shown in Figure 8-8), as well as a filter for each xref that is attached (so you can quickly filter just the layers in a particular attached drawing). You can create named layer filters that filter layers based on any of the following properties:

- Layer name
- Visibility (on or off)
- Freeze or thaw status (globally, in an active viewport, or in all new floating viewports)
- Locked or unlocked status
- Plot or don't plot status

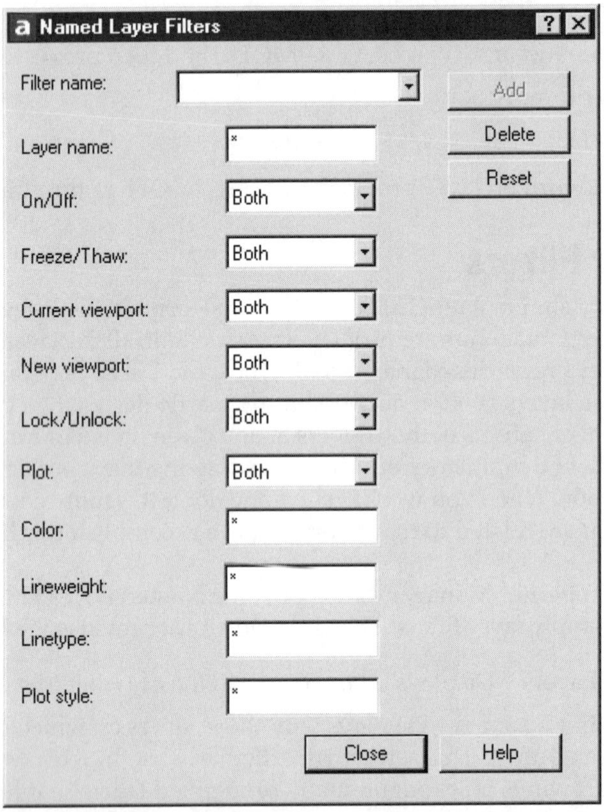

**Figure 8-8.** *The Named Layer Filters dialog box lets you establish a layer filter.*

- Color
- Lineweight
- Linetype
- Plot style

**Tip** *You can also filter for all layers that don't match the specified filter (such as all unused layers or all non-xref-dependent layers), by selecting the Invert Filter check box in the Layer Properties Manager dialog box.*

To set and apply a layer filter:

1. In the Layer Properties Manager dialog box, click the button immediately adjacent to the Named Layer Filters drop-down list to display the Named Layer Filters dialog box.

2. In the Named Layer Filters dialog box, select or type the filter values in the appropriate boxes.

3. Type the name you want to assign to the layer filter in the Filter Name box, and click Add.

4. Repeat Steps 2 and 3 to create another filter. Click Close when done creating layer filters.

5. In the Layer Properties Manager dialog box, in the Named Layer Filters drop-down list, select the name of the filter you want to apply. AutoCAD immediately applies the filter conditions that you specify.

You can filter the layer names based on any of the layer properties. For example, to display just those layers whose names begin with the letter *M*, type **M*** in the Layer Names field. To filter out all the frozen layers, select Frozen from the Freeze/Thaw drop-down list. Other selections let you filter based on visibility, locked or unlocked status, color, and linetype. You can also combine filters; for example, by displaying all layers beginning with the letter *A* that currently are on and whose linetype is dashed.

*The Apply To Object Properties Toolbar check box in the Named Layer Filters area of the Layer Properties Manager dialog box lets you determine whether the filter that you set also affects the Layer Control drop-down list on the Object Properties toolbar. By default, this drop-down list displays all layer names so that you can turn individual layers on and off without constantly having to reset the layer filters. When this check box is selected, however, any layer filter that you establish affects both the Layer dialog box and the Layer Control drop-down list.*

**LEARN BY EXAMPLE**
*Practice working with layers. Open the drawing Figure 8-9 on the companion web site.*

To set a layer filter to turn off quickly all the mechanical ductwork shown in Figure 8-9, use the following instructions:

1. Display the Layer Properties Manager dialog box.

2. Display the Named Layer Filters dialog box.

3. In the Layer Name box, type **H*,M***.

4. In the Filter Name box, type **Mechanical** and then click Add.

5. Click Close.

6. In the Layer Properties Manager dialog box, select Mechanical from the Named Layer Filters drop-down list. AutoCAD immediately filters the layer list, displaying only five layers.

7. In the layer list, select all five layers so that their names are highlighted. (Hint: Right-click and choose Select All from the shortcut menu.) Click the light bulb icon adjacent to any one of the layer names (so that the light bulbs turn blue).

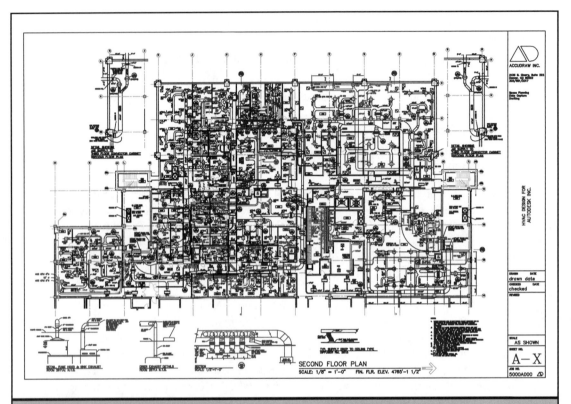

**Figure 8-9.** Practice working with layers by using this complex floor plan.

8. Click OK. AutoCAD immediately turns off all the ductwork.

To change the color of the room number tags, use the following instructions:

1. Display the Layer Properties Manager dialog box.
2. In the Named Layer Filters drop-down list, choose Show All Layers.
3. Click the layer color icon adjacent to the Artag layer name.
4. In the Select Color dialog box, click the red box in the Standard Colors area, and then click OK.
5. Click OK to close the Layer Properties Manager dialog box.

## Saving Layer States

AutoCAD 2002 introduces another powerful feature that lets you save your current layer state (all the current layer settings, such as on, off, freeze, thaw, color, linetype, and so on) so that you can quickly restore those settings at any time. You can then restore any of these previously saved states. When saving a layer state, you can choose which layer properties you want to save. For example, you can choose to save only the Locked/Unlocked settings of the layers, ignoring all the other settings. When you later restore that saved layer state, all the current layer settings remain as they are currently set except whether each layer is locked or unlocked. Layer states are saved as part of the drawing. However, you can save a layer state to a special LAS file and then import that file into another drawing so that you can reuse a previously saved layer state.

**Note** *Saved layer state files contain the names and states of all layers in the drawing from which they are exported except those that are part of external references. When you import an LAS file into a drawing, all of the layer names contained in the LAS file are created in the drawing and their states and properties are set to the settings saved in the LAS file.*

To save the current layer settings as a layer state:

1. Display the Layer Properties Manager dialog box.

2. Set the various layer settings as desired.

3. Click Save State to display the Save Layer States dialog box, shown in Figure 8-10.

4. Specify a name for the new layer state and select those layer states and properties you want to save.

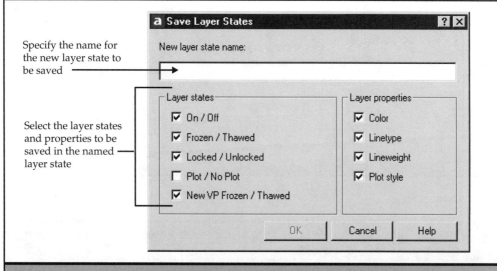

**Figure 8-10.** *The Save Layer States dialog box lets you save selected states and properties of the current layer settings as a saved layer state.*

MOVING BEYOND THE BASICS

5. Click OK to close the Save Layer States dialog box. You can then repeat Steps 2 through 5 to save another layer state.

To restore a named layer state:

1. Display the Layer Properties Manager dialog box.
2. Click Restore State to display the Layer States Manager dialog box, shown in Figure 8-11.
3. In the Layer States list, select the previously saved layer state you want to restore, and then click Restore.
4. Click Close to close the Layer States Manager dialog box.

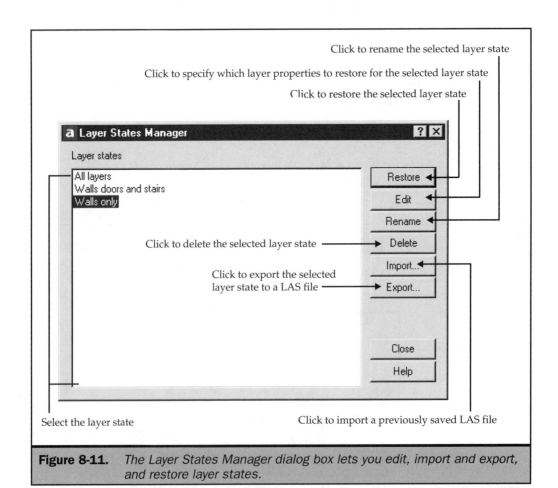

**Figure 8-11.** *The Layer States Manager dialog box lets you edit, import and export, and restore layer states.*

**Note**  *The Express Tools also include a powerful Layer Manager command that lets you save and restore layer states and save those states to a separate LAY file. The AutoCAD Migration Assistance tools, described in Appendix D, include two commands that convert between layer states created by this Express tool and the new saved layer states in AutoCAD 2002.*

# Working with Linetypes

A linetype consists of a repeating pattern of dots, dashes, symbols, or blank spaces. You can use different linetypes to represent specific kinds of information. For example, if you are drawing a site plan, you can draw roads by using a continuous linetype and draw property lines by using a border linetype that consists of dashes and dots. The sewer line that is servicing the site could then be drawn by using a line with the word SEWER repeatedly inserted into a gap in the line.

As you already learned, every drawing, by default, has at least three predefined linetypes: CONTINUOUS, BYLAYER, and BYBLOCK. You cannot rename or delete these linetypes. Your drawings may also contain an unlimited number of additional linetypes. You can load more linetypes into the drawing from a linetype library file or from another drawing, or you can create and save new linetypes.

In Chapter 2, you also learned about setting the current linetype and the linetype scale. When you create an object, it is created using the current linetype, which is normally the linetype associated with the current layer. If that linetype defines a repeating pattern of dots, dashes, symbols, and gaps, the length and spacing of those symbols is multiplied by the linetype scale factor. You control all the settings associated with linetypes in the Linetype Manager dialog box, shown in Figure 8-12. This dialog box is displayed whenever you start the LINETYPE command.

To display this dialog box, do one of the following:

- On the Object Properties toolbar, select Other from the Linetype Control drop-down list.
- From the Format menu, choose Linetype.
- At the command line, type **LINETYPE** (or **LT**) and then press ENTER.

The Details section of the dialog box provides alternative ways to control linetype properties. Click the Show Details button to enlarge the dialog box to display these additional controls. Clicking the Hide Details button hides this area. You can use the controls in the Details area to change the name or description of a linetype, the global or current linetype scale factor, or other properties.

**Note**  *Some objects, such as text and 3-D polylines, always display and plot with a continuous linetype regardless of their linetype setting.*

MOVING BEYOND
THE BASICS

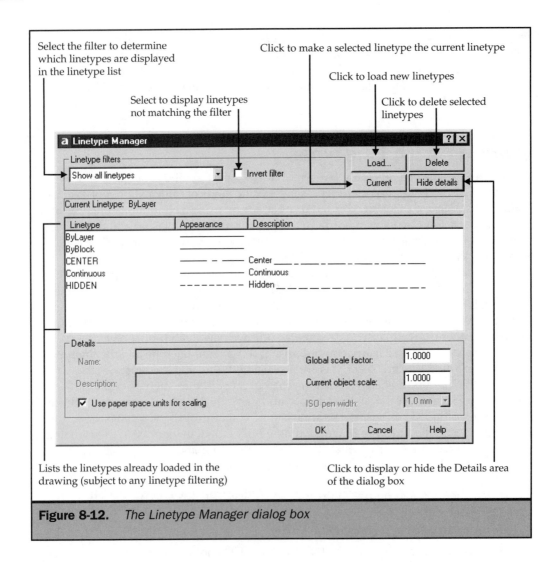

Select the filter to determine which linetypes are displayed in the linetype list

Click to make a selected linetype the current linetype

Click to load new linetypes

Select to display linetypes not matching the filter

Click to delete selected linetypes

Lists the linetypes already loaded in the drawing (subject to any linetype filtering)

Click to display or hide the Details area of the dialog box

**Figure 8-12.** *The Linetype Manager dialog box*

## Loading Additional Linetypes

Before you can select a new linetype to use in a drawing, you must either create the linetype definition, load a predefined linetype from a linetype library file (*.LIN), or drag the linetype from an existing drawing by using AutoCAD DesignCenter. AutoCAD includes two linetype library files (ACAD.LIN and ACADISO.LIN), which contain 45 predefined linetypes.

As you already learned, you can load new linetypes either by using the Linetype Manager dialog box or by clicking the Load button in the Select Linetype dialog box (which is displayed when you click a linetype adjacent to a layer name in the Layer

## Controlling Polyline Linetypes

By default, when you assign a noncontinuous linetype to a 2-D polyline, the linetype pattern is centered on each segment of the polyline so that the linetype starts and ends with a dash at each vertex. If you prefer, you can change the PLINEGEN system variable value so that the linetype is generated continuously throughout the length of the 2-D polyline. When PLINEGEN equals 0, the linetype is centered on each polyline segment. When PLINEGEN equals 1, the linetype is continuous throughout the length of the polyline without regard to the individual vertices.

The setting for PLINEGEN affects any new polylines drawn. To change the linetype display of existing polylines:

1. On the Standard toolbar, click Properties.

2. Select the polyline whose linetype display you want to change.

3. In the Properties window, click Linetype Generation and select Enabled or Disabled.

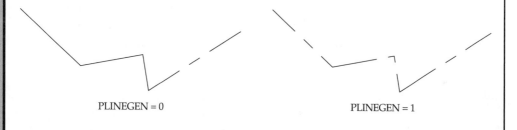

PLINEGEN = 0                          PLINEGEN = 1

Properties Manager dialog box). Either of these actions displays the Load or Reload Linetypes dialog box, shown in Figure 8-13. By default, this dialog box shows all the linetypes available in the ACAD.LIN linetype library file. You can also use it to load linetypes from other custom linetype library files.

Notice in the list of available linetypes that several of the linetype names appear three times, but with slightly different suffixes appended to their names. These are identical linetypes with different internal scaling factors. For example, Hidden2 has dashes and gaps at 1/2 the scale as Hidden, and Hiddenx2 is twice the scale as Hidden.

To load a linetype from a linetype library file other than the default:

1. In the Load or Reload Linetypes dialog box, click the File button.

2. In the Select Linetype File dialog box, select the linetype library file that you want to load, and then click Open. AutoCAD displays the linetypes that are available in the library file that you selected.

3. Select one or more of the available linetypes (so that they are highlighted) and then click OK.

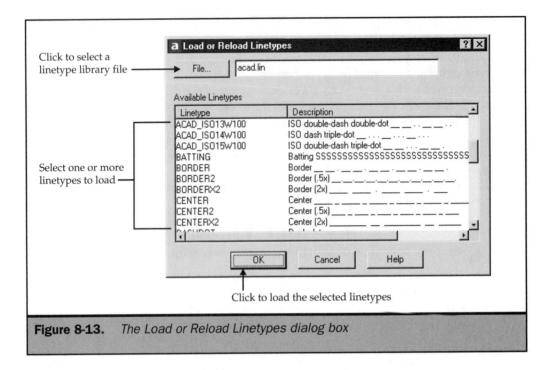

Click to select a linetype library file

Select one or more linetypes to load

Click to load the selected linetypes

**Figure 8-13.** *The Load or Reload Linetypes dialog box*

**Note** *The ISO (International Standards Organization) linetypes are designed for use in metric drawings with an appropriate ISO pen-width setting. For example, if the linetype scale factor is set to 1, you should plot ISO linetypes by using a 1mm pen width. Similarly, if an ISO line is drawn using a scale factor of 0.5, the line should later be plotted using a pen width of 0.5mm.*

## Creating New Linetypes

In addition to loading predefined linetypes from a linetype library file, you can create new linetypes and save them to a linetype library file for use in other drawings. There are three different ways to create linetypes:

- From the command line
- Using a text editor
- Using the Linetype Maker Express tool (described on the companion web site)

There are actually two different classifications of linetypes: simple and complex. A *simple* linetype consists only of line segments (dashes), dots, and open spaces. A *complex* linetype contains these same elements, but can also include shapes or text interspersed with the regular line symbols. You can create simple linetypes by using any of the three methods. Complex linetypes can be created only by using a text editor or the Linetype Maker tool.

*LEARN BY EXAMPLE*
*One of the best ways to learn about creating linetypes is to look at the actual linetype definitions in a linetype library file. You can open this file by using a simple ASCII text editor, such as Windows Notepad. To ensure that you don't accidentally damage the linetype library files that come with AutoCAD, use the file MYLINES.LIN on the companion web site. Since you will be saving your changes to this file, you need to copy it onto your hard drive.*

To open MYLINES.LIN by using Windows Notepad:

1. From the Start menu, choose Programs | Accessories | Notepad.
2. In Windows Notepad, choose File | Open.
3. In the File Name edit box of the Open dialog box, type **\*.lin**, select the file MYLINES.LIN, and then click Open. A portion of the linetype file is shown in Figure 8-14.

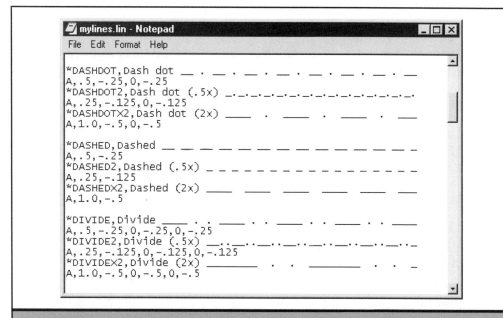

**Figure 8-14.**   *A portion of the linetype library file in Windows Notepad*

Leave this linetype library file open in Windows Notepad so that you can refer to it while you learn about the structure of a simple linetype.

## Creating a Simple Linetype

All linetype definitions consist of two lines of data. The first line of each definition begins with an asterisk, followed by the name of the linetype, a comma, and a text description of the linetype. The description is optional—it usually consists of a series of dots, spaces, and dashes, although it can be any text that you want (such as "This is my special linetype"). This description appears in the Linetype Manager and Select Linetype dialog boxes, in the Description column.

The actual linetype definition code appears on the second line. The definition code line always begins with the letter *A* (an alignment character) and a comma. The remainder of the code line consists of a series of pen down (positive numbers) and pen up (negative numbers) sequences, each separated by a comma. Definitions must always start with a pen down sequence. The following shows two linetype definitions from the standard AutoCAD linetype library file (ACAD.LIN):

```
*CENTERX2,Center (2x) _____  _  _____  __  _____
A,2.5,-.5,.5,-.5
*DASHDOT,Dash dot  __  .  __  .  __  .  __  .  __  .  __  .  __
A,.5,-.25,0,-.25
```

The Centerx2 linetype starts with a line segment 2.5 units long, followed by a .5 unit gap (–.5), a .5 unit dash, and another .5 unit gap. The Dashdot linetype consists of pen down for .5 drawing unit, pen up for .25 unit, a dot (the 0), and then pen up for .25 unit again.

*LEARN BY EXAMPLE*

*Add a new, simple linetype to the linetype library file that is already open in Windows Notepad.*

To add a new, simple linetype:

1. In Windows Notepad, scroll to the bottom of the linetype file and position the text cursor on the bottom line.

2. Add the following new linetype name and description (note the spaces between the dashes and dots) and then press ENTER:

   ```
   *MYLINE,-- - . - -- - . - --
   ```

3. Add the following linetype definition code (note the absence of spaces) and then press ENTER:

   ```
   A,.75,-.25,.25,-.25,0,-.25,.25,-.25,.75
   ```

4. From the Windows Notepad File menu, choose Save.

Now you can try out your new linetype. To load the new linetype, make it the current linetype, and then draw a line with it:

1. Start the LINETYPE command. (Hint: From the Format menu, choose Linetype.)

2. In the Linetype Manager dialog box, click Load.

3. In the Load or Reload Linetypes dialog box, click File and then load the MYLINES.LIN library file.

4. Select MYLINE in the list of available linetypes and then click OK. (Note: Although you added the new linetype to the bottom of the library file, the list is sorted alphabetically in the dialog box.)

5. In the Linetype Manager dialog box, select MYLINE in the linetype list, click Current, and then click OK.

6. Start the LINE command and draw some lines that use your new linetype.

**Note**   *To create a new linetype from the AutoCAD command line, you must start the LINETYPE command from the command line (so that AutoCAD does not display the Linetype Manager dialog box). To do this, type **-LINETYPE** (or **-LT**). Then, use the Create option and follow the command prompts. AutoCAD will prompt you for the name of the linetype library file, the name of the linetype, a description, and then the actual linetype pattern. Creating new linetypes from the command line is less flexible, since you must retype everything if you make a mistake or want to change your linetype definition.*

## Creating a Complex Linetype

Complex linetypes differ from simple linetypes only in that they can incorporate text and shapes in the linetype definition code (the second line of the linetype definition). The first line of the definition, which contains the linetype name and description, is identical to that of the simple linetype. The second line begins with the letter *A*. In addition to the pen up and pen down values, separated by commas, the second line can contain additional definition data, enclosed in square brackets. This data, which is also separated by commas and must contain no spaces, takes the following form:

```
[shapename,shxfilename,R=xx,A=xx,S=xx,X=xx,Y=xx]
```

or

```
["string",stylename,R=xx,S=xx,X=xx,Y=xx]
```

in which *xx* is a positive or negative number.

The components of this data are as follows:

- **shapename**   The name of the shape that you want included in the linetype definition.

- **shxfilename**   The name of the shape file (including the .SHX file extension) in which the shape is saved. AutoCAD must be able to locate the shape file in its search path (or you can include the path as part of the filename). If AutoCAD can't locate the file, the linetype appears without the included shape.

- **string**   The actual text to be used in the linetype. The actual string is enclosed in quotation marks.

- **style**   The text style to be used for the text string. The style must be defined in the current drawing. If you omit this component, the text appears using the current text style.

- **R=xx**   An optional rotation value (expressed in degrees, unless you append *r* for radians or *g* for grads). If you express the value as R=xx, the shape or text is rotated relative to the angle of the line. If you express the value as A=xx, the rotation will be fixed in relation to the current UCS, regardless of the orientation of the line.

- **S=xx**   The scale factor for the shape or text. If the specified text style has a fixed height, it is multiplied by this scale factor. If the text style has a height of 0, this value defines the actual height of the text. This value is multiplied by the linetype scale factor.

- **X=xx**   The distance that the shape or text is to be offset along the length of the line from the end of the previous line segment. If this component is omitted or is 0, the text or shape is not offset from the line segment.

- **Y=xx**   The distance that the shape or text is to be shifted vertically in relation to the line. For example, to align the middle of the text with the line, use a negative value to shift the text down in relation to the line. If this component is omitted or is 0, the text or shape is not shifted.

**Note**   *To make the shape or text fit into a gap in the line, you must follow the bracketed data with a pen up distance that is sufficiently long to create the gap.*

The following shows two complex linetype definitions from the standard AutoCAD linetype library file (ACAD.LIN):

```
*FENCELINE1,Fenceline circle ----0-----0----0-----0----0-----0--
A,.25,-.1,[CIRC1,ltypeshp.shx,x=-.1,s=.1],-.1,1
*GAS_LINE,Gas line ----GAS----GAS----GAS----GAS----GAS----GAS--
A,.5,-.2,["GAS",STANDARD,S=.1,R=0.0,X=-0.1,Y=-.05],-.25
```

The Fenceline1 linetype starts with a line segment .25 unit long, followed by a .1 unit gap. This is followed by the CIRC1 shape located in the LTYPSHP.SHX file. This shape is shifted .1 unit from the end of the line segment and appears at a scale factor of .1. The shape information is followed by a .1 unit gap and a 1 unit line segment.

The Gas_line linetype starts with a line segment .5 unit long, followed by a .2 unit gap into which the word GAS is inserted using the STANDARD text style (which, by default, is always predefined in every AutoCAD drawing). This text is applied at a scale factor of .1 and a relative rotation angle of 0 (thus, aligning the text with the line). The text is offset .1 unit from the end of the previous line segment and −.05 unit down from the line. The text information is followed by a .25 unit gap to ensure that the word GAS fits into the gap. Both of these linetypes are shown in Figure 8-15.

 *You can use some of the special characters available in TrueType fonts as part of a complex linetype. While creating the new linetype, use the Character Map application (usually found by selecting Start | Programs | Accessories | Character Map) to select the character you want to use, click Select to add the character to the Characters To Copy box, and then click Copy to copy the character to the Windows Clipboard. Then, switch back to the program you're using to create the custom linetype, and press* CTRL-V *to paste the character into the correct location in the linetype definition.*

### LEARN BY EXAMPLE
*Add a complex linetype to the linetype library file already open in Windows Notepad.*

To add a new complex linetype:

1. In Windows Notepad, scroll to the bottom of the linetype file and position the text cursor on the bottom line.

2. Add the following new linetype name and description (note the spaces between the dashes and text) and then press ENTER:

   `*SEWER,----SEWER----SEWER----SEWER----`

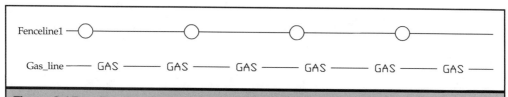

**Figure 8-15.** *The Fenceline1 and Gas_line linetypes*

MOVING BEYOND THE BASICS

3. Add the following linetype definition code (note that no spaces exist) and then press ENTER:

```
A,1,-.05["SEWER",STANDARD,S=.1,R=45,X=.1,Y=-.25],-.5
```

4. From the Windows Notepad File menu, choose Save. Then exit Windows Notepad.

Figure 8-16 shows how the definition code values affect the linetype definition. Now you can try out your new linetype. Follow the same steps that you used previously to load and then make the Sewer linetype the current linetype.

**Tip**   *There are several things to consider when deciding whether to use text or shapes in complex linetypes. Shapes can be hidden behind 3-D objects and need to be used when the linetype requires an object that is not available in a text font. But, if you include shapes in complex linetypes, you must be sure to include the shape file when sharing drawings. This is not necessary if you use standard text fonts.*

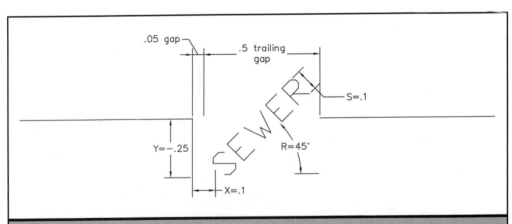

**Figure 8-16.**   *The SEWER linetype showing how each component is applied to the overall linetype definition*

# Applying Drawing Standards

As mentioned earlier, layers provide a useful way to maintain standards within your drawings. By drawing specific types of objects on certain layers, your drawings can be more easily managed and understood by others. For example, if you always draw walls on the WALLS layer and dimensions on the DIMENSIONS layer, the electrical engineer you consult with can quickly turn off the dimensions, turn on the walls, and use your drawing as the basis for her lighting and electrical plans.

In the past, it has often been difficult, however, to enforce these types of standards. Unless you were particularly careful, you could easily draw doors on the WALLS layer or dimensions on the DOORS layer. AutoCAD 2002 provides several new tools to help you create and then enforce drawing standards. You can create standards for layers, linetypes, text styles, and dimension styles.

In order to apply drawing standards, you must first define those standards and save them as a standards file. A standards file is similar to a normal drawing file, but has a .DWS file extension. Once created, you can associate a standards file with an AutoCAD drawing. You can then periodically audit the drawing to make sure that it conforms to its associated standard.

When working with others, you may receive files that utilize layer names different from those you normally use. AutoCAD 2002 also provides a useful tool for mapping and translating layer names so that they match your standards.

## Defining Standards

To establish your standards, you need only create a file that contains the properties for layers, dimension styles, linetypes, and text styles that you want to use as your standard settings and then save that file as a standards file (with a .DWS file extension).

To save a file as a standards file:

1. Start the SAVEAS command to display the Save Drawing As dialog box.

2. In the File Name box, specify the name for the new file.

3. In the Files Of Type drop-down list, select AutoCAD 2000 Drawing Standard (*.dws).

4. Click Save.

Once you have established one or more standards files, the next step is to associate one or more of those standards files with each drawing to which you want the standards to be applied. You will then be able to check those drawings against the standards to make sure that they conform.

To associate a standards file with the current drawing:

1. Do one of the following:

   - On the CAD Standards toolbar, click Configure Standards.
   - From the Tools menu, choose CAD Standards | Configure.
   - At the command line, type **STANDARDS** and then press ENTER.

   AutoCAD displays the Configure Standards dialog box, shown in Figure 8-17.

2. Click the button with the plus sign (+), the Add Standards File button (or press F3).

3. In the Select Standards File dialog box, select the standards file you want to associate with the drawing, and click Open.

4. Repeat Steps 2 and 3, if desired, to associate additional standards files with the current drawing.

5. Click OK.

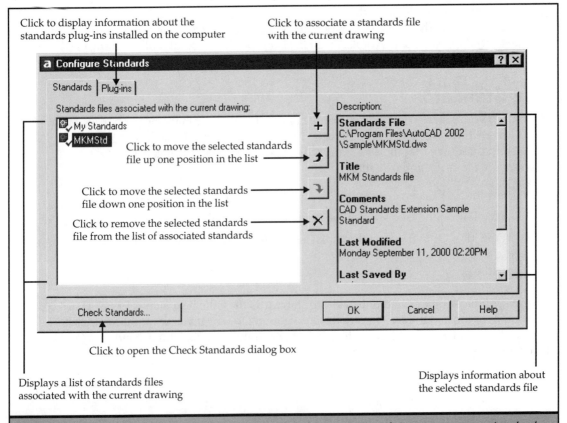

**Figure 8-17.** *The Configure Standards dialog box lets you associate one or more standards files with the current drawing.*

If you associate more than one standards file with a drawing, when you later check for standards violations, if there are any conflicts between the two standards, the first standards file associated with the drawing takes precedence. For that reason, you can use the Move Up and Move Down buttons in the Configure Standards dialog box to change the order in which the standards will be applied.

| Note | *The auditing process uses special applications called plug-ins that define the rules for the properties that are checked. AutoCAD comes with plug-ins for checking dimension styles, layers, linetypes, and text styles. Those plug-ins all appear on the Plug-ins tab of the Configure Standards dialog box. In the future, Autodesk or third-party developers may add standards plug-ins for checking other drawing properties.* |
|---|---|

**MOVING BEYOND THE BASICS**

## Checking for Standards Violations

After you associate a standards file with a drawing, you should periodically check the drawing to make sure that it conforms to the standards. This is particularly important if more than one person is working on the drawing. For example, if you work with a team of consultants, one of them may create new layers that don't comply with the standards you have defined. When that happens, you need to be able to identify the nonstandard layers and fix them.

When you check a drawing for standards violations, each named object is checked against the standards file associated with the drawing. For example, each layer is checked against the layers in the standards file. Such a standards audit can identify two types of problems:

- An object with a nonstandard name is present in the drawing, such as a layer name that does not match any of the layer names in the standards file.

- An object with a standard name but nonstandard properties is present in the drawing, such as a layer name that matches the layer name in the standards file but has a color or linetype different from that assigned to the layer in the standards file.

AutoCAD provides two methods to check for standards violations: a Check Standards command to check the current drawing, and a Batch Standards Checker to audit multiple drawings. When you use the Check Standards command, AutoCAD reports each standards violation and gives you the opportunity to immediately fix the problem. When you use the Batch Standards Checker, AutoCAD checks all the drawings you specify and then produces a web-based report listing all the violations detected.

### Checking the Current Drawing

When you use the Check Standards command to check the current drawing against its associated standards file, AutoCAD displays the Check Standards dialog box, shown in Figure 8-18. If any violations are detected, they are displayed in this dialog box.

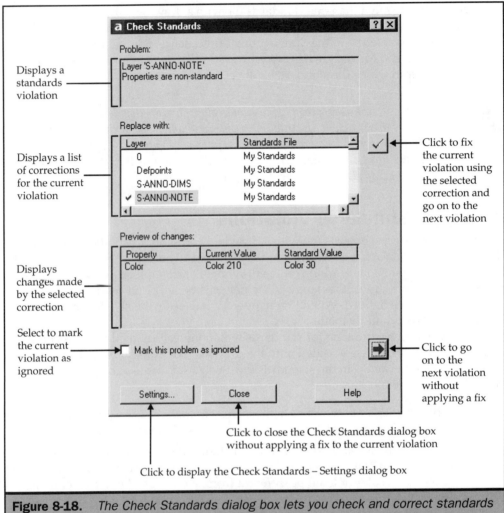

Displays a standards violation

Displays a list of corrections for the current violation

Displays changes made by the selected correction

Select to mark the current violation as ignored

Click to fix the current violation using the selected correction and go on to the next violation

Click to go on to the next violation without applying a fix

Click to close the Check Standards dialog box without applying a fix to the current violation

Click to display the Check Standards – Settings dialog box

**Figure 8-18.**  *The Check Standards dialog box lets you check and correct standards violations in the current drawing.*

To audit the current drawing for standards violations, do one of the following:

- On the CAD Standards toolbar, click Check Standards.
- From the Tools menu, choose CAD Standards | Check.
- At the command line, type **CHECKSTANDARDS** and then press ENTER.
- In the Configure Standards dialog box, on the Standards tab, click Check Standards.

A description of the first violation detected appears in the Problem area of the dialog box along with a list of one or more possible corrections, shown in the Replace With area. If there is a recommended correction, it appears in the list preceded by a check mark. The Preview Of Changes area displays the properties of the nonstandard object along with the fix that will be applied if the correction currently selected in the Replace With list is applied. For example, if the audit indicates that a layer has a nonstandard color value, the Preview Of Changes area will indicate the current nonstandard color value and the standard value that will be applied if the current fix is selected. To fix the nonstandard AutoCAD object using the currently selected correction, click the Fix button (or press F4). AutoCAD then moves on to the next problem detected in the current drawing.

If you wish, you can skip the detected problem by clicking the Next button (or pressing F5). AutoCAD immediately advances to the next nonstandard object in the drawing without applying any fix. If AutoCAD detects a problem that you always want to skip, such as a nonstandard dimension style that you created specifically to accommodate a nonstandard condition, you can ignore the problem in the future by selecting the Mark This Problem As Ignored check box before clicking the Next button.

When you fix objects with nonstandard properties, the standard property is simply applied to the named objects. For example, when a layer has the wrong color assigned to it, AutoCAD changes the color to the one specified in the standards file. When you fix an object with a nonstandard name, however, all drawing objects associated with the nonstandard name are transferred to the replacement standard object you specify, and then the nonstandard object is purged from the drawing. For example, if the audit process detects a nonstandard layer name, when you fix the problem, all objects drawn on that layer will be transferred to the standard layer you specify, and then the nonstandard layer will be purged from the drawing.

**Caution**   *If AutoCAD does not display a recommended fix for a standards violation, none of the possible corrections may be appropriate. You should carefully review the information displayed in the Preview Of Changes area before accepting the fix.*

The Check Standards dialog box does not close when you have cycled through all the detected violations. You must click the Close button to end the command. If any problems remain uncorrected, they continue to display in the dialog box and you can cycle through them again.

If you click the Settings button in the Check Standards dialog box, AutoCAD displays the Check Standards – Settings dialog box, shown in Figure 8-19. You can use the settings in this dialog box to control the operation of the Check Standards command. For example, you can have AutoCAD automatically correct all nonstandard properties when a recommended fix is available. If you've flagged some violations to be ignored, you can also control whether these violations are displayed or not the next time you audit the drawing.

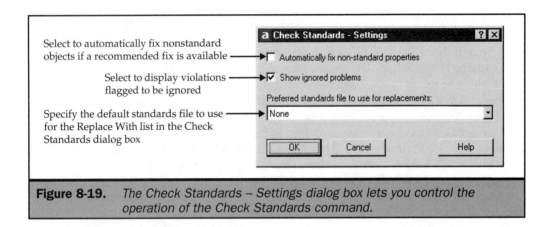

Select to automatically fix nonstandard objects if a recommended fix is available ──►

Select to display violations flagged to be ignored ──►

Specify the default standards file to use ──► for the Replace With list in the Check Standards dialog box

**Figure 8-19.**    *The Check Standards – Settings dialog box lets you control the operation of the Check Standards command.*

*If you are checking a DWG file against a CAD Standards DWS file that has the opposite type of plot style specified (for example, your drawing uses named plot styles but the standards file uses color-dependent plot styles), the plot style property errors of the layer plug-in cannot be fixed. Use the CONVERTPSTYLES command (described in Chapter 18) to first convert the drawing's plot style type, and then run the CAD Standards CHECKSTANDARDS command again.*

## Auditing Multiple Drawings Using the Batch Standards Checker

AutoCAD 2002 also provides a Batch Standards Checker. As its name implies, you can use this function to analyze multiple drawings in a single operation. The Batch Standards Checker is not an AutoCAD command. Rather, it is a separate program that you run independently from AutoCAD. The Batch Standards Checker only checks the drawings and then creates an XML-based report of all the standards violations it detects. It does not provide any tools to fix violations. You will need to load the drawings into AutoCAD in order to correct any violations.

To start the Batch Standards Checker, click Start | Programs | AutoCAD 2002 | Batch Standards Checker. Using this program is a five-step process:

1. Create a list of drawings to be checked.

2. Specify whether you want to check each drawing against its associated standards file or against a standards file that you specify.

3. Save the list of drawings and associated standards to a special file called a standards check file.

4. Run the standards audit.

5. View the resulting standards audit report.

When you start the Batch Standards Checker, it displays the Batch Standards Checker window, shown in Figure 8-20. This window contains its own toolbar and the following five tabs, which provide the tools you'll use to perform the audit:

- **Drawings**   Provides tools to create the list of drawings to be audited
- **Standards**   Lets you control whether each drawing is checked against its associated standards files or against a list of standards files that you specify
- **Plug-ins**   Displays information about the plug-ins used to audit the drawings
- **Notes**   Lets you add notes to be included in the audit report
- **Progress**   Displays summary information about the current audit once you begin the batch audit process

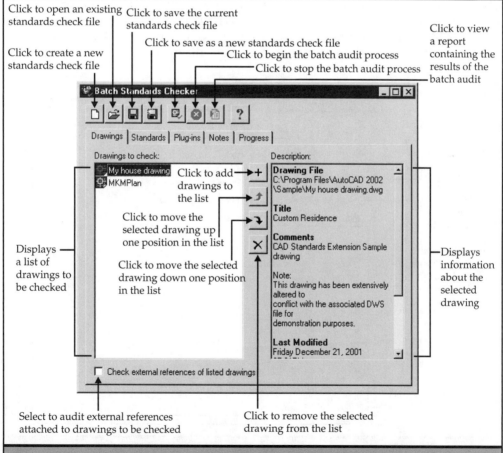

**Figure 8-20.**   *The Drawings tab of the Batch Standards Checker lets you create the list of drawings to be checked for standards violations.*

> **Note**    *The buttons along the top of the dialog box are available at all times, regardless of the tab you select. If you wish to include notes, be sure to add them before you begin the audit.*

The first tab displayed is the Drawings tab, shown in Figure 8-20. To create a list of drawings to be checked, click the Add Drawing button (or press F3), and then select the drawings you want to audit. You can select multiple drawings. The drawings are normally audited in the order in which you add them to the list, but the Drawings tab provides tools so that you can change the order, add more drawings, or remove selected drawings from the list. You can also specify whether you want to audit only the drawings you added to the list, or also check any external references attached to those drawings.

Normally, the Batch Standards Checker audits each drawing using the standards files that are associated with the drawing itself. You can use the settings on the Standards tab, however, to specify that all the drawings be audited instead using one or more standards files that you specify. If you select this option, you must then specify the standards files to be used. The Standards tab is shown in Figure 8-21 and its controls are very similar to those in the Configure Standards dialog box.

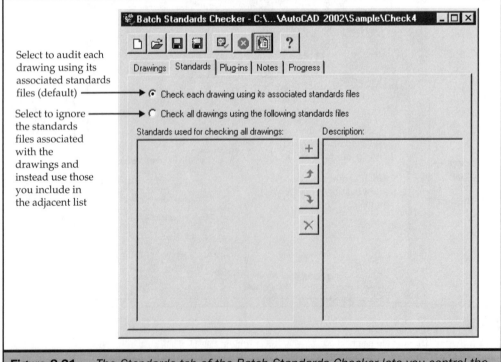

**Figure 8-21.**    *The Standards tab of the Batch Standards Checker lets you control the standards files used to audit the drawings.*

You are now almost ready to run the batch audit process. Before you do, however, you must save your settings to a standards check file (a special file with a .CHX file extension). You can actually save this file at any time after you've created the list of drawings to be checked. If you make any changes to the batch audit settings, however, you will be prompted to save your changes before the audit can actually be performed. Once you have saved a standards check file, you can also open that file again later, either to rerun the audit or to view a previously generated audit report.

To begin the batch audit process, click the Start Check button. The Batch Standards Checker displays its progress in the Progress tab. You can interrupt this process by clicking the Stop Check button. As soon as the audit is complete, the Batch Standards Checker opens a new browser window and displays a complete Standards Audit Report, similar to the one shown in Figure 8-22. You can display report information

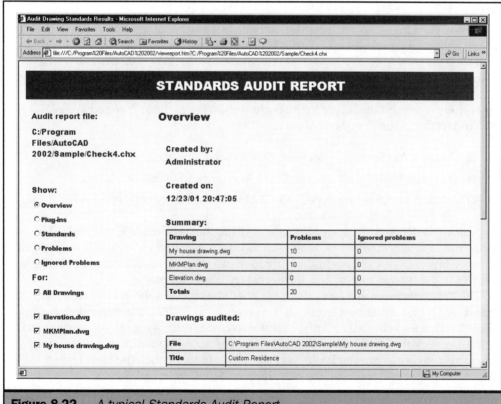

**Figure 8-22.** *A typical Standards Audit Report*

for all the drawings that were audited or for individual drawings, and also select any of the following options under Show to filter the data displayed in the report:

- **Overview**   Displays a summary of the problems encountered in each drawing that was audited

- **Plug-ins**   Displays information about the plug-ins used to run the batch audit

- **Standards**   Displays information about the standards files used to check the drawings in the batch audit

- **Problems**   Displays a detailed list of all the problems encountered for each drawing displayed

- **Ignored Problems**   Displays a detailed list of all problems encountered that were flagged as ignored

## Translating Layer Names and Properties

When you exchange drawing files with others, it is not uncommon to receive files that do not adhere to your layer standards. For example, you may work with a consultant who uses his own in-house layering conventions. AutoCAD 2002 provides a Layer Translator that enables you to map layers in the drawing you're currently working on to different layers in another drawing or in a drawing template or a standards file. You can then convert the layers in the current drawing to those mapped layers.

To start the Layer Translator, do one of the following:

- On the CAD Standards toolbar, click Layer Translate.
- From the Tools menu, choose CAD Standards | Layer Translator.
- At the command line, type **LAYTRANS** and then press ENTER.

AutoCAD displays the Layer Translator dialog box, shown in Figure 8-23. This dialog box contains three areas, each with its own associated buttons. Before you can translate layers from the current drawing to match your standards, you must use the controls in these areas to establish the necessary layer mappings.

The Translate From area contains a list of all the layers in the current drawing. The Translate To area contains a list of all the layers you can map the current drawing's layers to. This list is initially empty. In order to create your layer mappings, you must first either click the Load button to load a drawing, drawing template, or drawing standards file containing the layers you want to map to, or click the New button to define a new layer. If you click the New button, AutoCAD displays a New Layer

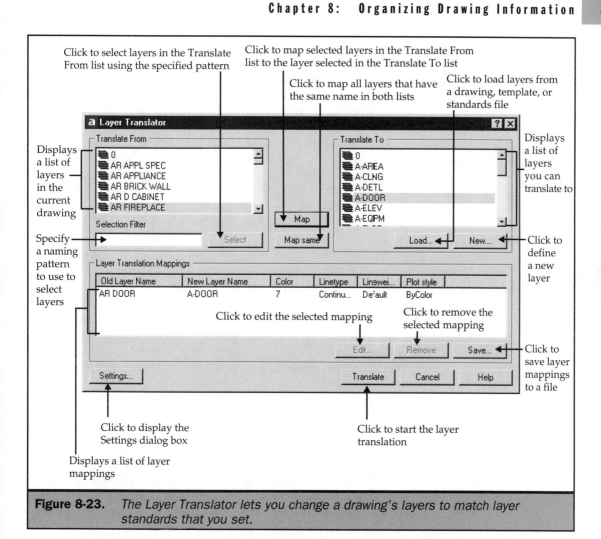

Click to select layers in the Translate
From list using the specified pattern

Click to map selected layers in the Translate From
list to the layer selected in the Translate To list

Click to map all layers that have
the same name in both lists

Click to load layers from
a drawing, template, or
standards file

Displays
a list of
layers
in the
current
drawing

Displays
a list of
layers
you can
translate to

Specify
a naming
pattern
to use to
select
layers

Click to
define
a new
layer

Click to edit the selected mapping

Click to remove the
selected mapping

Click to
save layer
mappings
to a file

Click to display the
Settings dialog box

Click to start the layer
translation

Displays a list of layer
mappings

MOVING BEYOND
THE BASICS

**Figure 8-23.**   *The Layer Translator lets you change a drawing's layers to match layer
standards that you set.*

dialog box, in which you can enter a name for the new layer and select its properties.
You can repeat this process as many times as you want.

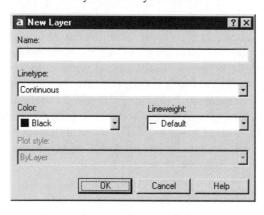

**Note**    *If you load a file containing layer names that are the same as those in a drawing, template, or standards file that you've already loaded, the duplicate layer names and their associated properties are ignored.*

To map layers in the current drawing to those in your standards, select one or more layers in the Translate From list. In the Translate To list, select the layer whose properties you want to map those layers to, and then click the Map button to define the mapping. To map all identically named layers from one list to the other, click the Map Same button. You can also select layers to be mapped by specifying a pattern in the Selection Filter field, and then clicking the Select button. For example, to select all the layers whose names begin with the letters *AR*, type **AR\*** in the Selection Filter field, and then click the Select button.

**Tip**    *You can also purge unused layers in the current drawing by right-clicking in the Translate From list and choosing Purge Layers from the shortcut menu.*

As you add mappings, the mapped layers appear in the list in the Layer Translation Mappings area. Each mapping shows the old layer name, the new layer name, and the properties to which the layer will be mapped. To remove any of these mappings, select the layer in the Layer Translation Mappings list, and then click Remove. If you wish, you can click Edit to keep the layer mapping but change the properties to which the layer will be mapped. You can also right-click in this area of the dialog box and select Edit, Remove, or Remove All from the shortcut menu.

You can also save the layer mappings to a file. When you click the Save button, AutoCAD saves all the layer mappings—the names of the referenced layers and the layers mapped to those layers—to a drawing file. All the linetypes used by those layers are also saved in the drawing file. If you often receive files from a consultant that uses a standard different from yours, once you establish the layer mappings one time, you can save them using this function. Then, the next time you receive a drawing file from the same consultant, you can start the Layer Translator and load the drawing file to which you saved those mappings to quickly map all the layers in the new drawing to your previously established mappings.

The Layer Translator also provides some tools for customizing the translation process. To display these tools, click the Settings button. AutoCAD displays the Settings dialog box, shown in Figure 8-24. The controls in this dialog box let you control whether each translated object takes on the color and linetype assigned to its layer or retains its original color and linetype, whether objects nested within blocks are translated, and whether AutoCAD creates a log file (a text file having the same name as the translated drawing, with a .LOG file extension) detailing the results of the translation. The final check box in this dialog box, Show Layer Contents When Selected, can be particularly helpful when establishing layer mappings. If you select this check box, when you select one or more layers in the Translate From list, AutoCAD displays only those objects in the drawing that are on the selected layers.

After you have established all of the desired layer mappings, click Translate to actually perform the layer translation. If you haven't already saved the layer mappings to a file, AutoCAD asks you if you want to do so. The Layer Translator then performs the layer translation based on the mappings you've established. All layers mapped from an old layer name to a new layer name will be translated. Any unmapped layers will remain unchanged.

*If you are not happy with the results, you can use the UNDO command to restore the drawing to its original condition and then correct any of the mappings as necessary.*

### LEARN BY EXAMPLE
*The companion web site contains several drawings and drawing standards files, along with instructions, so that you can practice defining standards, checking for standards violations, and using the Layer Translator.*

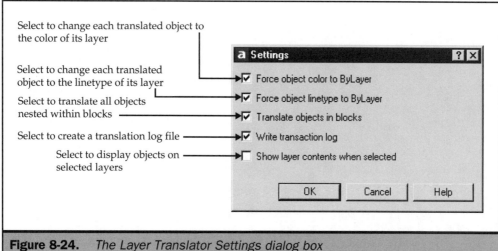

**Figure 8-24.** *The Layer Translator Settings dialog box*

The
Complete
Reference

# Chapter 9

## Getting Drawing Information

A s you work on your drawings, AutoCAD is busy storing accurate, detailed information about all the objects in your drawing. You can utilize this information at any time to get details about an existing drawing and its objects by using the commands that measure distances and calculate areas. You can also track the amount of time that you spend editing a drawing.

In Chapter 1, you learned how to save and display custom property information—such as the title, author, subject, keywords, and hyperlink addresses—along with your drawings. In Chapter 6, you learned how to find the coordinates of any point in the drawing. In Chapter 19, you will learn how to find the volumes of three-dimensional solids. This chapter explains how to do the following:

■ Measure distances along an object

■ Measure distances and angles

■ Divide an object into a number of equal segments

■ Calculate areas

■ Display information about objects in a drawing

■ Track the amount of time spent editing a drawing

■ Display drawing properties

## Specifying Measurements and Divisions

You can place division marks along an arc, circle, ellipse, elliptical arc, line, polyline, or spline, marking off a number of equal segments or marking off intervals of a specific length along any of these objects. For example, you may want to place station-point markers every 50 feet along the centerline of a roadway, or divide the plan view of a window into three equal-width sections of glass, placing a mullion at each division point.

AutoCAD provides two commands for placing various types of markers along the length of objects:

■ The MEASURE command places markers at a specified distance along the length of objects.

■ The DIVIDE command divides objects into a specified number of equal segments, placing markers at each division point.

With either command, you can mark the segments by placing either a block or a point object at the end of each segment. If you use point objects, you can snap to the ends of each segment by using the node object snap. The appearance of the point object is determined by the current point display type, which you control in the Point Style dialog box. You learned about setting point styles in Chapter 3.

To use a block as the marker, the block must already be defined in the current drawing. As shown in Figure 9-1, you can further indicate whether to rotate the block

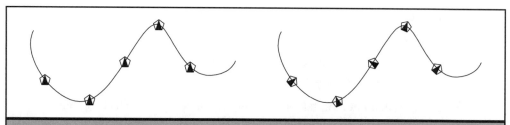

**Figure 9-1.** *When marking segments by using a block, the block can be aligned perpendicular to the object (as shown on the right).*

to align it perpendicular to the object that you are measuring or dividing. You will learn more about blocks in Chapter 15.

When you use either the MEASURE or DIVIDE command, you are prompted to select the object to measure or divide. You must select the object by pointing; no other object selection method is permitted. AutoCAD begins measuring or dividing based on the point at which you select the object and the type of object with which you are working. For most objects, measuring starts from the end point closest to the point that you used to select the object.

## Measuring Segments on Objects

Use the MEASURE command to mark segments of a specified length along a selected object, using either point objects or blocks. To measure segments along an object and mark them by using point objects (see Figure 9-2):

1. Do one of the following:

   ■ From the Draw menu, choose Point | Measure.

   ■ At the command line, type **MEASURE** (or **ME**) and then press ENTER.

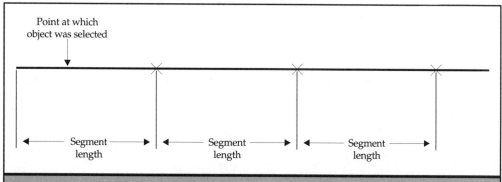

**Figure 9-2.** *When you select the object to be measured, segments are measured from the end closest to the point at which you select the object.*

AutoCAD prompts:

```
Select object to measure:
```

2. Select the object by pointing. AutoCAD prompts:

```
Specify length of segment or [Block]:
```

3. Specify the segment length by either typing the length or selecting two points in the drawing (the distance between these points defines the segment length). AutoCAD immediately places point objects at the specified interval along the object.

To measure segments along an object and mark them by using block insertions:

1. Do one of the following:

   ■ From the Draw menu, choose Point | Measure.

   ■ At the command line, type **MEASURE** (or **ME**) and then press ENTER.

   AutoCAD prompts:

```
Select object to measure:
```

2. Select the object by pointing. AutoCAD prompts:

```
Specify length of segment or [Block]:
```

3. Type **B** (for Block) and then press ENTER, or right-click and select Block from the shortcut menu. AutoCAD prompts:

```
Enter name of block to insert:
```

4. Type the name of the block that you want to use as the marker, and then press ENTER. (*Remember that the block must already exist in the drawing.*) AutoCAD prompts:

```
Align block with object? [Y/N] <Y>:
```

5. Type **Y** (or select Yes from the shortcut menu) to rotate each insertion of the block so that its vertical alignment is always perpendicular to the object, or type **N** (or select No from the shortcut menu) to insert each instance of the block with a zero rotation angle. AutoCAD then prompts:

```
Specify length of segment:
```

6. Specify the segment length by either typing the length or selecting two points in the drawing (the distance between them defining the segment length). AutoCAD immediately places instances of the block at the specified interval along the object.

*LEARN BY EXAMPLE*
*Use the MEASURE command to add striping to a parking lot layout. Open the drawing Figure 9-3 on the companion web site.*

Adding the stripes for a parking lot layout is simple when the parking lot layout is rectilinear. But what about a lot that has curved curbs? This is where the MEASURE command shows its power. Since you can measure segments along a polyline, it's easy to place blocks representing the stripes every nine feet along a curb that was drawn using a polyline. To add the stripes shown in Figure 9-3, use the following command sequence and instructions:

```
Command: MEASURE
Select object to measure: (select point A on polyline)
Specify length of segment or [Block]: B
Enter name of block to insert: STRIPE1
Align block with object? [Y/N] <Y>: ENTER
Specify length of segment: 9'
```

## Dividing Objects into Segments

Use the DIVIDE command to place markers along a selected object, visually dividing it into a specified number of equal-length segments. You can use either a point object or a block to mark the segments.

To divide an object into segments and mark them by using point objects:

1. Do one of the following:

   ■ From the Draw menu, choose Point | Divide.

   ■ At the command line, type **DIVIDE** (or **DIV**) and then press ENTER.

   AutoCAD prompts:

   ```
   Select object to divide:
   ```

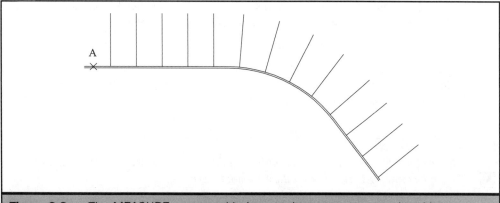

**Figure 9-3.**  *The MEASURE command being used to create a curved parking lot layout*

2. Select the object by pointing. AutoCAD prompts:

```
Enter number of segments or [Block]:
```

3. Specify the number of segments and then press ENTER. AutoCAD immediately places point objects along the object, marking it off into the specified number of equal divisions.

To divide an object into segments and mark them by using block insertions:

1. Do one of the following:

   ■ From the Draw menu, choose Point | Divide.

   ■ At the command line, type **DIVIDE** (or **DIV**) and press ENTER.

   AutoCAD prompts:

```
Select object to divide:
```

2. Select the object by pointing. AutoCAD prompts:

```
Enter number of segments or [Block]:
```

3. Type **B** (for Block) and then press ENTER, or right-click and select Block from the shortcut menu. AutoCAD prompts:

```
Enter name of block to insert:
```

4. Type the name of the block that you want to use as the marker, and then press ENTER. (*Remember that the block must already exist in the drawing.*) AutoCAD prompts:

```
Align block with object? [Y/N] <Y>:
```

5. Type **Y** (or select Yes from the shortcut menu) to rotate each insertion of the block so that its vertical alignment is always perpendicular to the object, or type **N** (or select No from the shortcut menu) to insert each instance of the block with a zero rotation angle. AutoCAD then prompts:

```
Enter the number of segments:
```

6. Specify the number of segments and then press ENTER. AutoCAD immediately places instances of the block along the object, marking it off into the specified number of equal divisions.

*LEARN BY EXAMPLE*
*Use the DIVIDE command to add mullions to the plan view of a storefront window.*
*Open the drawing Figure 9-4 on the companion web site.*

Dividing a large window into equal sections of glass and placing a mullion at each division point illustrates a powerful use of the DIVIDE command. In this case, one instance

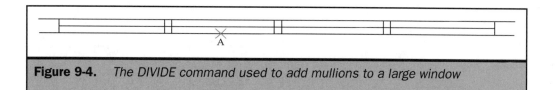

**Figure 9-4.**    *The DIVIDE command used to add mullions to a large window*

of the mullion was already created and saved as a block named MULL. To add the
mullions shown in Figure 9-4, use the following command sequence and instructions:

```
Command: DIVIDE
Select object to divide: (select point A)
Enter number of segments or [Block]: B
Enter name of block to insert: MULL
Align block with object? [Y/N] <Y>: ENTER
Enter the number of segments: 4
```

# Calculating Areas

Because AutoCAD stores detailed information about all the objects in the drawing and
accurately tracks their location, you can easily calculate the area and perimeter of any
closed object, or the area enclosed by a polygon that is formed by a series of points that
you specify. As you calculate these areas, AutoCAD remembers them and can quickly
add or subtract one or more areas from a running total.

    *The area and perimeter are displayed in the current units format. For example, if the
current units type is architectural, the area and perimeter are displayed in feet and
inches. You learned how to set the current drawing units in Chapter 2.*

## Calculating Areas Defined by Points

You can find the area and perimeter of any portion of a drawing by specifying a series
of points, forming a closed polygonal boundary. AutoCAD calculates the area and
perimeter enclosed by this imaginary polygon.

To calculate the area defined by points that you specify (see Figure 9-5):

1. Do one of the following:

   ■  On the Inquiry toolbar, click Area.
   ■  From the Tools menu, choose Inquiry | Area.
   ■  At the command line, type **AREA** (or **AA**) and then press ENTER.

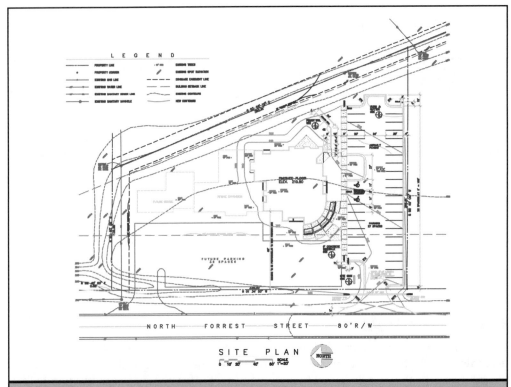

**Figure 9-5.** *Calculating the area and perimeter of a building site by specifying the property corner points*

AutoCAD prompts:

```
Specify first corner point or [Object/Add/Subtract]:
```

2. Specify the first point. AutoCAD then prompts:

```
Specify next corner point or press ENTER for total:
```

3. Specify the second point. AutoCAD then repeats the prompt.

4. Continue specifying points, in sequence, to define the perimeter of the area that you want to measure. To complete the command and display the area, press ENTER. AutoCAD immediately displays the area and perimeter enclosed by the polygon formed by the points that you specified. (Note that you don't need to respecify the first point to close the polygon, but doing so doesn't change the calculation in any way.)

## Calculating the Area Enclosed by an Object

You can find the area of any closed object. In addition, AutoCAD calculates either the circumference or perimeter of the object, depending on the type of object that you select.

To calculate the area of a closed object:

1. Do one of the following:

   ■ On the Inquiry toolbar, click Area.

   ■ From the Tools menu, choose Inquiry | Area.

   ■ At the command line, type **AREA** (or **AA**) and then press ENTER.

   AutoCAD prompts:

   ```
   Specify first corner point or [Object/Add/Subtract]:
   ```

2. Type **O** (for Object) and press ENTER, or right-click and select Object from the shortcut menu. AutoCAD then prompts:

   ```
   Select objects:
   ```

3. Select the object by pointing. As soon as you select the object, AutoCAD displays the area and perimeter or circumference.

## Calculating Combined Areas

You can use the Add and Subtract options to combine areas, thus calculating a running total. When adding or subtracting areas from the total, you can specify the areas being added or subtracted by either selecting objects or specifying points to form polygonal areas.

Before you can calculate combined areas, you must switch the AREA command into either Add or Subtract mode. Once in Add mode, any areas that you calculate are added to the running total. In Subtract mode, any areas that you calculate are subtracted from the running total. You can switch back and forth between these modes by typing either **A** or **S** at the AREA command prompt, or by right-clicking and selecting Add or Subtract from the shortcut menu.

*LEARN BY EXAMPLE*

*Use the Add and Subtract modes of the AREA command to calculate the area of a gasket. Open the drawing Figure 9-6 on the companion web site.*

To calculate the area of the gasket shown in Figure 9-6, use the following command sequence and instructions:

```
Command: AREA
Specify first corner point or [Object/Add/Subtract]: A
Specify first corner point or [Object/Subtract]: O
(ADD mode) Select objects: (select the outline of the gasket)
Area = 3.10 square in. (0.0216 square ft.), Perimeter = 0'-7 5/16"
Total area = 3.10 square in. (0.0216 square ft.)
```

```
(ADD mode) Select objects: ENTER
Specify first corner point or [Object/Subtract]: S
Specify first corner point or [Object/Add]: O
(SUBTRACT mode) Select objects: (select one of the circles)
Area = 0.44 square in. (0.0031 square ft.), Circumference = 0'-2 3/8"
Total area = 2.66 square in. (0.0185 square ft.)
(SUBTRACT mode) Select objects: (select the other circle)
Area = 0.44 square in. (0.0031 square ft.), Circumference = 0'-2 3/8"
Total area = 2.22 square in. (0.0154 square ft.)
(SUBTRACT mode) Select objects: ENTER
Specify first corner point or [Object/Add]: ENTER
```

> **Tip**    *As you learned in Chapter 4, you can also convert the gasket outline into a region. If you used the SUBTRACT command to remove the holes from the gasket, the AREA command immediately displays the actual area of the gasket (with the holes already removed).*

## Calculating Distances and Angles

Because AutoCAD accurately tracks the location of every position in the drawing in relation to the underlying Cartesian coordinate system, the program can very quickly calculate the distance between any two points that you select. AutoCAD displays the following information (see Figure 9-7):

- The distance between the two points, measured in the current drawing units
- The angle of an imaginary line between the two points, measured in the XY plane

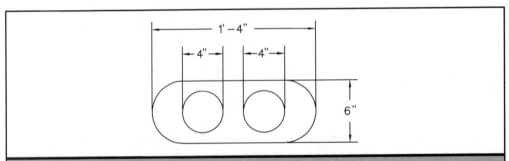

**Figure 9-6.**    *The Add and Subtract modes of the AREA command make it easy to find the area of the gasket.*

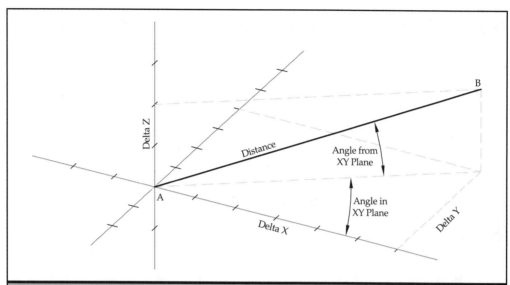

**Figure 9-7.**    *The DIST command displays information about the distance between two specified points (A and B).*

- The angle of an imaginary line between the two points, measured from the XY plane

- The change (delta) in the X, Y, and Z distances between the two points

To calculate the distance between two points:

1. Do one of the following:

- On the Inquiry toolbar, click Distance.

- From the Tools menu, choose Inquiry | Distance.

- At the command line, type **DIST** (or **DI**) and then press ENTER.

    AutoCAD prompts you for the first point.

2. Specify the first point. AutoCAD then prompts you for the second point.

3. Specify the second point. AutoCAD immediately displays the distance information in the following format:

```
Distance = 7.4833,  Angle in XY Plane = 17,  Angle from XY Plane = 358
Delta X = 7.1703,  Delta Y = 2.1254,   Delta Z = -0.2646
```

**Tip**    *You can also use dimensions to measure distances and angles and add those dimensions to your drawings. You'll learn about using dimensions in Chapter 14.*

# Displaying Information About Your Drawing

AutoCAD tracks a great deal of useful information about your drawing and the objects in the drawing. At any given time, you can display the following types of information:

- Information about selected objects
- The status of the current drawing
- The time spent working on the current drawing

This information is displayed in the AutoCAD Text window.

## Displaying Information About Objects

The LIST command displays information about selected objects. The information displayed varies, depending on the type of objects that you select. Regardless of the objects selected, AutoCAD always displays the following information:

- The object type
- The layer on which the object is drawn
- The space in which the object is drawn (model space or paper space)
- The object handle (a unique numeric identifier assigned to every object in the drawing)
- The location of the object (its X, Y, and Z coordinates, relative to the current UCS)
- The size of the object (the exact nature of which varies, depending on the object type)

To display information about an object:

1. Do one of the following:

- On the Inquiry toolbar, click List.
- From the Tools menu, choose Inquiry | List.
- At the command line, type **LIST** (or **LI** or **LS**) and then press ENTER.

AutoCAD prompts you to select objects.

2. Select one or more objects and then press ENTER.

AutoCAD immediately displays information about the selected objects. For example, if you selected a circle, the information displayed might look like Figure 9-8.

*The drawing object information is displayed in the AutoCAD Text window. To switch back to the drawing area, press F2.*

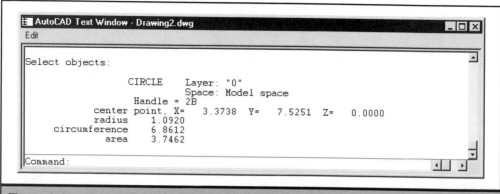

**Figure 9-8.**   *The LIST command displays information about selected objects.*

## Displaying the Drawing Status

It is often useful to determine how much memory a drawing uses or how much free space remains on your computer's hard disk. If you're working on a drawing that has been edited by someone else, you may also need to check the status of various modes and settings, to see whether you need to change them to accommodate the way that you prefer to work with AutoCAD. You can do all of these things in one step by using the STATUS command.

The STATUS command displays lots of information about the drawing, including:

- Drawing name
- Number of objects in the drawing
- Drawing limits
- Insertion base point
- Snap and grid settings
- Current space
- Current layer, color, and linetype
- Current elevation and thickness
- Current settings for various modes (Fill, Grid, Ortho, Snap, and so on)
- Current object snap modes
- The amount of free disk space and physical memory

To display the drawing status, do one of the following:

- From the Tools menu, choose Inquiry | Status.
- At the command line, type **STATUS** and then press ENTER.

AutoCAD immediately displays the drawing status, which looks similar to Figure 9-9.

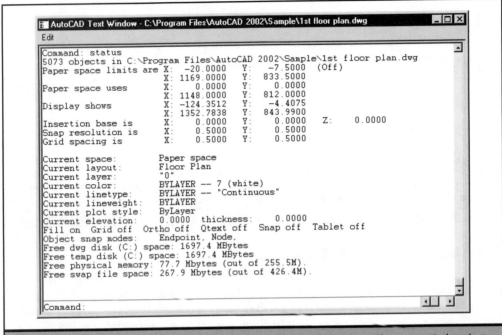

**Figure 9-9.**   *The STATUS command displays information about the current drawing.*

**Note**   *The drawing status information is displayed in the AutoCAD Text window. To switch back to the drawing area, press* F2.

## Tracking Time Spent Working on a Drawing

AutoCAD keeps track of the amount of time that you spend working on a drawing. The program also provides an Elapsed timer that you can turn on and off, and even reset to zero. In addition, AutoCAD keeps track of the time that has elapsed since the drawing was last saved, and can be configured to save your drawing automatically to a backup file at regular intervals, to prevent the accidental loss of data.

The TIME command displays the following information:

- The date and time the drawing was originally created
- The date and time the drawing was most recently saved
- The total amount of time spent working on the drawing
- The status of the Elapsed timer as well as the time elapsed since the timer was last reset
- The time remaining until the next automatic backup

**Tip**   *You can also view much of this information—such as the date and time the drawing was originally created, the date and time the drawing was most recently saved, and the total editing time—on the Statistics tab of the Drawing Properties dialog box. To display this dialog box, from the File menu, select Drawing Properties. You learned about drawing properties in Chapter 1.*

To display the timer information, do one of the following:

■ From the Tools menu, choose Inquiry | Time.

■ At the command line, type **TIME** and then press ENTER.

AutoCAD immediately displays the timer information, which appears similar to Figure 9-10, and then prompts:

```
Enter option [Display/ON/OFF/Reset]:
```

Either press ENTER or ESC to end the command, type the capital letters corresponding to one of the four options, or right-click and select one of the following options from the shortcut menu:

■ **Display**   Redisplay the timer information

■ **ON**   Turn on the Elapsed timer

■ **OFF**   Turn off the Elapsed timer

■ **Reset**   Reset the Elapsed timer to 0

**Note**   *The timer information is displayed in the AutoCAD Text window. To switch back to the drawing area, press F2.*

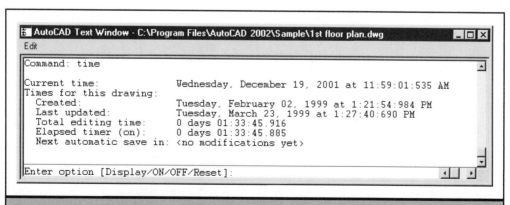

**Figure 9-10.**   *The TIME command displays information about the time spent working on the current drawing.*

# Chapter 10

## Editing
## Complex Objects

Although you can modify most objects by using AutoCAD's general-purpose editing tools to erase, copy, rearrange, resize, or break objects, some of the commands that you learned about in Chapter 7 can't be used to modify AutoCAD's more complex objects, such as multilines, polylines, and splines. AutoCAD provides a number of special-purpose commands for modifying these objects. For example, special-purpose commands let you change the appearance of multilines, convert straight polyline segments into smooth flowing curves, and modify the curvature of splines.

In addition to these special-purpose commands, AutoCAD provides several commands for modifying both simple and complex objects in more sophisticated ways, such as adding fillets or chamfers at the intersections of objects. Sometimes, you may need to reduce complex objects into more basic components, such as converting a polyline into individual line segments. Finally, to help you better manage the objects in your drawing, you can combine individual objects into groups. You can then select all the objects in the group by clicking any one of the objects. This chapter explains how to do the following:

- Combine objects into groups
- Edit polylines, multilines, and splines
- Explode complex objects into more basic components
- Chamfer and fillet both simple and complex objects

## Grouping Objects

In Chapter 7, you learned how to modify individual objects. You may often find, however, that a number of individual objects naturally combine to create a single visual element or group of elements. For example, you may use several rectangles to represent a workstation consisting of a desk and chair. Although you could select each rectangle individually when you want to move the workstation or create a copy, you can cut down on the number of steps by combining the rectangles into a group. Once combined in this way, you can select all the objects in the group by clicking any one of the objects, but you can still modify any of the individual objects, if necessary. This capability is distinctly different from that of blocks, which you will learn about in Chapter 15.

A *group* is a named selection set of objects that is saved as part of the drawing. You can combine objects into a group, which you can then select as a single element. When you create a group, you specify whether or not it is selectable. If a group is selectable, you need to select only one object in the group to select the entire group. Unlike other selection sets, groups are saved with the drawing.

An object can be part of more than one group. For example, a combination of objects representing a chair could be part of a "Chair" group and could also be part of a "Workstation" group. You can list all the groups to which an object belongs, or highlight all the objects belonging to a selected group. Objects in a group are numerically ordered within the group. You can reorder the grouping and control other features of a group by using the Object Grouping dialog box, shown in Figure 10-1.

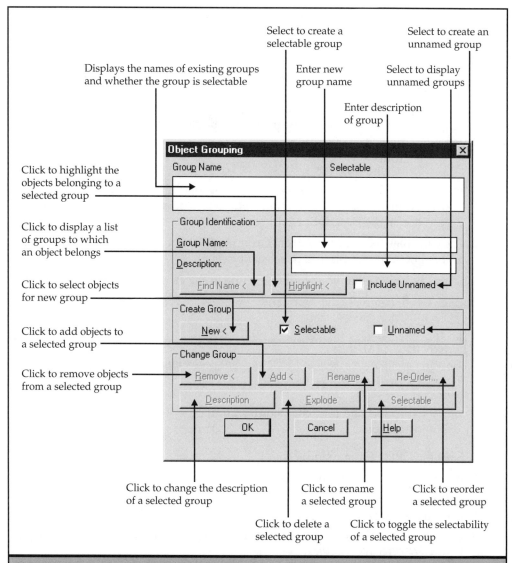

**Figure 10-1.**    *Use the Object Grouping dialog box to create and control groups.*

MOVING BEYOND THE BASICS

*Use groups when you need to combine several objects into a single visual element for easy selection. Although groups function similar to blocks in this regard, you can still edit the individual objects contained in the group. To modify the individual objects in an inserted block, however, you must first explode the block. Blocks offer other advantages, such as the ability to update all instances of a block by redefining the original block definition. You'll learn more about blocks in Chapter 15.*

## Creating Groups

When you create a group, you can assign it a group name and definition. Unnamed groups are not listed in the Group dialog box unless you select the Include Unnamed check box. Groups can be nested within other groups. If you choose a member of a selectable group for inclusion in a new group, all objects in that group are included in the new group. To create a new group:

1. At the command line, type **GROUP** (or **G**) and then press ENTER.

2. In the Object Grouping dialog box, under Group Identification, type the name of the new group in the Group Name edit box. You can also type an optional description in the Description edit box.

3. Click the New button. The Object Grouping dialog box temporarily disappears and AutoCAD prompts you to select objects.

4. Select the objects that you want to include in the group and then press ENTER.

5. Click OK to complete the command.

*LEARN BY EXAMPLE*
*Use object grouping to combine a table and four chairs into a group. Open the drawing Figure 10-2 on the companion web site.*

To combine the table and four chairs into a group, use the following command sequence and instructions:

1. Start the GROUP command.

2. In the Object Grouping dialog box, under Group Identification, type **4CHAIR** in the Group Name edit box.

3. Click the New button.

4. Use the window object selection to select the table and four chairs, as shown in Figure 10-2. Then press ENTER.

5. Click OK to complete the creation of the group.

By default, new groups are both selectable and named. This means that you can later select all the objects in the group either by selecting any one of the objects in the

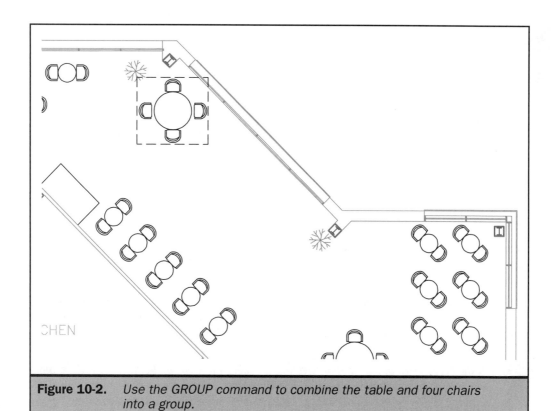

**Figure 10-2.**    *Use the GROUP command to combine the table and four chairs into a group.*

group or by specifying the group by name. In almost all cases, you will want to create selectable groups. Even when a group is selectable, you can still select individual objects within the group, if necessary. Although you won't usually create unselectable groups, it is more common to create unnamed groups. You might create an unnamed group if you simply don't want to take the time to name the group or if you don't think you'll ever need to select the group by name. If you copy a group, AutoCAD creates a new group and assigns it the default name *A*n*, where *n* is a number. Such groups are also considered to be unnamed and don't appear in the group list unless you select the Include Unnamed check box.

# Selecting Groups

Whenever AutoCAD prompts you to select objects, such as during any editing command, you can select a group either by clicking any object within the group or by specifying the name of the group. To select the group by name, type **G** and then press ENTER when AutoCAD prompts you to select objects. AutoCAD then prompts you to enter the group name. Type the group name and then press ENTER. You can then continue to select additional objects or press ENTER to continue with the command.

**LEARN BY EXAMPLE**

*Use the COPY command to copy the 4CHAIR group that you created in the previous exercise. Open the drawing Figure 10-3 on the companion web site.*

To copy the table and chair grouping, as shown in Figure 10-3, use the following command sequence and instructions. Use the center object snap to select the center of the table, and the node object snap to place the copies at the labeled points.

```
Command: COPY
Select objects: G
Enter group name: 4CHAIR
5 found
Select objects: ENTER
Specify base point or displacement, or [Multiple]: M
Specify base point: (select point A using center object snap)
Specify second point of displacement or <use first point as displacement>: (select point B)
Specify second point of displacement or <use first point as displacement>: (select point C)
Specify second point of displacement or <use first point as displacement>: (select point D)
Specify second point of displacement or <use first point as displacement>: ENTER
```

**Note** *In the previous example, simply selecting the table and chair by pointing at the grouping would have been faster than typing the group name. When you work in a very crowded drawing, however, specifying the group name may be easier than trying to select a group by pointing.*

The ability to select the entire group or just one object within the group is controlled by the PICKSTYLE system variable, which, in turn, can be controlled from the Selection tab of the Options dialog box. When object grouping is enabled, if you select any member of the group, the entire group is selected; when disabled, you can select an individual object within the group. You can also toggle object grouping on and off at any time by pressing CTRL-A. (In order to do this, you must first disable Windows standard accelerator keys by clearing the check box on the User Preferences tab of the Options dialog box.)

**LEARN BY EXAMPLE**

*Toggle off object grouping and then use the ERASE command to delete one of the chairs from one of the copies of the 4CHAIR group that you created in the previous exercise.*

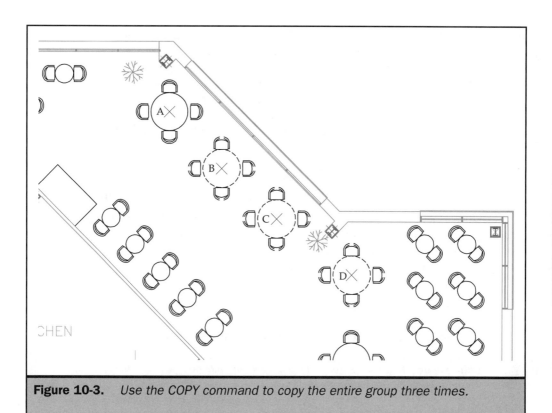

**Figure 10-3.**   *Use the COPY command to copy the entire group three times.*

To erase one of the chairs from the grouping, use the following command sequence and instructions:

```
Command: ERASE
Select objects: CTRL-A
<Group off> (select one of the chairs)
Select objects: ENTER
```

## Editing Groups

As you've just learned, when a group is selectable, you can select the entire group by selecting any object in the group. You can then use any of the object modification commands to edit the entire group. When the group is not selectable or when you turn off object grouping, however, selecting a single object in the group selects only that object. The command then acts only on the selected object.

Some modification commands remove objects from the group. In the previous exercise, for example, when you erased the chair, it was removed from both the drawing and the group. You can also edit the group itself. You can add or remove group members and rename groups. If removing an object from the group leaves the group empty, however, the group definition remains. Exploding a group removes the group definition.

To remove an object from a group:

1. At the command line, type **GROUP** (or **G**) and then press ENTER.

2. In the Object Grouping dialog box, select the group in the list of group names.

3. Under Change Group, click the Remove button. The Object Grouping dialog box temporarily disappears, the members of the selected group are highlighted in the drawing, and AutoCAD prompts you to select objects.

4. Select the highlighted object that you want to remove from the group and then press ENTER.

5. Click OK to complete the command.

Adding a new object to an existing group follows a similar process.

*Adding or deleting objects from an existing group has no effect on any copies of the group that you have already made.*

To delete a named group:

1. At the command line, type **GROUP** (or **G**) and then press ENTER.

2. In the Object Grouping dialog box, select the group in the list of group names.

3. Under Change Group, click Explode.

4. Click OK to complete the command.

*When you insert a drawing that contains groups, the group definition is retained. Drawings inserted in this way initially appear as blocks. You need to explode the block before you can select the groups. If you insert the drawing as an exploded block, however, the group definitions are lost. You'll learn more about blocks in Chapter 15.*

# Editing Polylines

You can modify any type of two-dimensional or three-dimensional polyline. Objects such as rectangles, polygons, and donuts, as well as three-dimensional mesh objects, are all variations of polylines that you can edit. This chapter concentrates on two-dimensional polylines. You'll learn more about editing three-dimensional mesh objects in Chapter 20.

You can edit a polyline by opening or closing it, by changing its overall width or the widths of individual segments, and by converting a polyline that consists of straight line segments into a flowing curve or an approximation of a spline. In addition, you can edit individual polyline vertices by adding, removing, or moving vertices. You can also add new segments to an existing polyline or convert an arc or line object into a polyline.

To edit a polyline, you must first start the PEDIT command and then select the polyline. To start the PEDIT command, do one of the following:

- On the Modify II toolbar, click Edit Polyline.
- From the Modify menu, choose Object | Polyline.
- At the command line, type **PEDIT** (of **PE**) and then press ENTER.

When you start the PEDIT command, AutoCAD prompts you to select a polyline or choose the Multiple option. When PEDIT is the active command, you can't use grips when editing polylines. Although you can use any object selection method to select the polyline, you can edit only one polyline at a time unless you first choose the Multiple option. If you select an object that is not a polyline, the program asks whether you want to turn the object into a polyline. Only arcs and lines can be converted into polylines.

 *If you choose the Multiple option, any changes you make will apply to all the polylines you select. When you use this option, the Edit Vertex option of the PEDIT command is not available.*

To convert an object into a polyline:

1. Start the PEDIT command. AutoCAD prompts you to select a polyline.

2. Select the object. AutoCAD prompts:

   ```
   Object selected is not a polyline
   Do you want to turn it into one? <Y>
   ```

3. Press ENTER. AutoCAD then displays the PEDIT command options.

4. Either type the capital letter that corresponds to the desired option, right-click and select the option from the shortcut menu, or press ENTER to complete the command.

After you select the polyline or convert a selected object into a polyline, AutoCAD displays a list of options. The available options depend on the type and status of the polyline object that you select. For example, if you select an open polyline, the list includes the Close option, whereas if the polyline is already closed, the list includes the Open option, instead. Some of the other options are not available when editing 3-D polylines or meshes. The following options affect the entire polyline:

- **Close**   Draws a polyline segment from the first vertex of the polyline to the last vertex. This option appears only if the polyline is currently open.

MOVING BEYOND THE BASICS

■ **Open**    Removes the closing segment of the polyline. This option appears only if the polyline is currently closed.

■ **Join**    Joins polylines, arcs, and lines to form one continuous 2-D polyline. To join any object to a polyline, that object must already share an end point with an end vertex of the selected polyline, unless you first choose the Multiple option. This option is available only if the selected polyline is open, and it can't be used to join 3-D polylines or to join a 2-D polyline with a 3-D polyline.

■ **Width**    Changes the widths of all the segments of the selected 2-D polyline to one constant width.

■ **Edit Vertex**    Presents a separate set of options for editing the individual polyline vertices (explained later in this chapter).

■ **Fit**    Changes a 2-D polyline into a smooth curve that connects all the vertices of a 2-D polyline.

■ **Spline**    Changes a 2-D or 3-D polyline into a spline-fit curve, which is a smooth curve that is pulled toward the vertices but passes through only the first and last vertices in the case of open polylines. The more vertices in a particular area of a polyline, the more pull that is exerted.

■ **Decurve**    Changes a fit curve or spline curve polyline back into one that is made up of straight line segments.

■ **Ltype Gen**    Controls how linetype patterns are generated around the vertices of 2-D polylines. When Ltype Gen is off, AutoCAD draws the polyline so that the linetype starts and ends with a dash at each vertex. When Ltype Gen is on, AutoCAD draws the polyline with the linetype generated continuously around the vertices.

■ **Undo**    Reverses the previous PEDIT option.

■ **Exit**    Ends the PEDIT command.

## Opening and Closing Polylines

When you close a polyline, the program draws a polyline segment from the first vertex of the polyline to the last vertex. Opening a polyline removes the closing segment. When you select a polyline for editing, the command line prompt displays either the Open or Close option, depending on whether the polyline you select is open or closed.

To close an open polyline:

1. Start the PEDIT command.

2. Select the polyline.

3. Type **C** (for Close) and then press ENTER, or right-click and then select Close from the shortcut menu. AutoCAD immediately adds the closing segment and then repeats the PEDIT options.

4. Choose another option, or press ENTER to complete the command.

*If the last polyline segment before the closing segment is straight, the closing segment is also drawn as a straight segment. If the last polyline segment before the closing segment is curved, however, the closing segment is drawn as a curved segment, drawn tangent to the previous segment.*

## Curving and Decurving Polylines

You can convert a multisegment polyline into a smooth curve by using either the Fit or Spline option. The Fit option creates a smooth curve that connects all the vertices. The Spline option computes a smooth curve that is pulled toward the vertices but passes only through the first and last vertex. The polyline is used as a frame, which can be displayed by using the SPLFRAME system variable. The actual number of segments used to approximate the spline curve is controlled by the SPLINESEGS system variable, and the type of spline curve generated (quadratic or cubic B-spline) is controlled by the SPLINETYPE system variable. The Decurve option returns the polyline to its previous state, before you applied either the Fit or Spline option.

To fit a curve to a polyline:

1. Start the PEDIT command.

2. Select the polyline.

3. Type **F** (for Fit) and then press ENTER, or right-click and then select Fit from the shortcut menu. AutoCAD immediately applies curve fitting and then repeats the PEDIT options.

4. Choose another option, or press ENTER to complete the command.

*LEARN BY EXAMPLE*
*Use the Spline option of the PEDIT command to create smooth, flowing contour lines. Open the drawing Figure 10-4 on the companion web site.*

To convert the rough polyline contours shown in Figure 10-4 into spline curves, use the following command sequence and instructions:

```
Command: PEDIT
Select polyline or [Multiple]: (select one of the polylines)
Enter an option [Close/Join/Width/Editvertex/Fit/Spline/Decurve/Ltype
gen/Undo]: S
Enter an option [Close/Join/Width/Editvertex/Fit/Spline/Decurve/Ltype
gen/Undo]: ENTER
```

Repeat this procedure for the other contour lines.

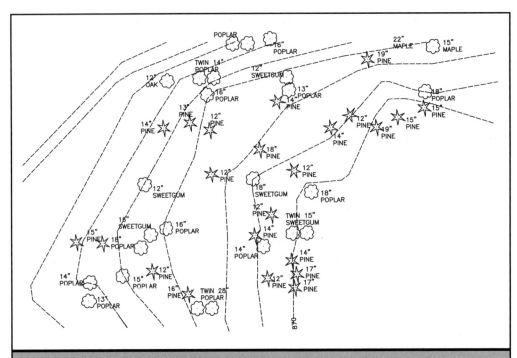

**Figure 10-4.** *Curving polylines create realistic contour lines.*

# Joining Polylines

You can add an arc, line, or polyline object to an existing open polyline, forming one continuous polyline object. To join an object to a polyline, that object must already share an end point with an end vertex of the selected polyline, unless you first choose the Multiple option.

When you join an object to a polyline, the width of the new polyline segment depends on the width of the original polyline and the type of object that you are joining to it:

- A line or an arc assumes the same width as the polyline segment for the end vertex to which it is joined.
- A polyline joined to a tapered polyline retains its own width values.
- A polyline joined to a uniform-width polyline assumes the width of the polyline to which it is joined.

To join an arc, line, or polyline to an existing polyline:

1. Start the PEDIT command.
2. Select the polyline.

3. Type **J** (for Join) and then press ENTER, or right-click and then select Join from the shortcut menu. AutoCAD prompts you to select objects.

4. Select the arc, line, or polyline to join. When you finish selecting objects, press ENTER. AutoCAD immediately joins the objects to the polyline and then repeats the PEDIT options.

5. Choose another option, or press ENTER to complete the command.

*When you add objects to a polyline, you can select multiple objects by using any object selection method. Only those objects that can be joined to form one continuous polyline will be added to the original polyline.*

## Joining Polylines That Don't Exactly Meet

It can be very frustrating when you can't join segments to a polyline because they don't exactly meet. When you use the Multiple option of the PEDIT command, however, you can join two or more polylines whose ends do not meet exactly. In this case, when you select the two polylines and then choose the Join option, you can specify a *fuzz distance,* the maximum distance two end points of individual polylines can be separated and still be joined. The individual polylines are then joined either by inserting a new segment or by trimming or extending the two segments so that they meet at a common point.

When you start the command and choose the Multiple option, AutoCAD prompts you to select objects. You can use any object selection method and may select lines and arcs as well as polylines. After you finish selecting objects, choose the Join option. AutoCAD prompts:

```
Join Type = Extend
Enter fuzz distance or [Jointype] <0.0000>:
```

If you specify the fuzz distance, and the end points of the polylines to be joined fall within that distance from each other, AutoCAD immediately joins the polylines. You can also use the Jointype setting to control how the end points of separated polylines are joined. Three different methods are available for filling the gap between these end points. If you select the Jointype option, AutoCAD prompts:

```
Enter join type [Fillet/Add/Both] <current>:
```

■ **Add**   Inserts a polyline segment between the two end points and then joins the polyline.

■ **Fillet**   Performs a fillet-like operation between the two segments, with an effective fillet radius of 0.0, thus extending or trimming the segments to their resulting intersection point.

■ **Both**   Attempts to join the segments using the Fillet method. If that method can't be used because the resulting intersection point is not within the fuzz distance, then the Add method is used to add a segment between the end points.

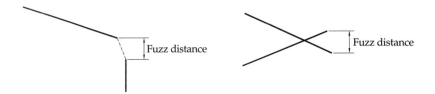

## Changing the Width of Polylines

You can change the width of an entire polyline, applying a uniform width to the entire polyline. To apply a uniform width to an entire polyline:

1. Start the PEDIT command.

2. Select the polyline.

3. Type **W** (for Width) and then press ENTER, or right-click and then select Width from the shortcut menu. AutoCAD prompts you to enter the new width for all segments.

4. Enter the new width and then press ENTER. AutoCAD immediately applies the new width along the entire polyline and then repeats the PEDIT options.

5. Choose another option, or press ENTER to complete the command.

## Editing Polyline Vertices

You can use the Edit Vertex option to modify individual polyline vertices. When you select this option, the PEDIT command switches into a special vertex editing mode and places an X marker on the first vertex. The X marker indicates the vertex that you are editing. When in vertex editing mode, you can edit the polyline one vertex at a time by using the following options:

■ **Next**   Moves the X marker to the next vertex.

■ **Previous**   Moves the X marker to the previous vertex.

■ **Break**   Breaks a polyline into two separate polylines. The first break point is the current vertex when you select the Break option. You can then use the Next and Previous options to move the X marker to another vertex (or leave it at the break point to break the polyline into only two pieces). The Go option then completes the break operation.

■ **Insert**   Inserts a new vertex into the polyline at a location you choose. The new vertex is added after the current vertex, in the direction that you would move when going to the next vertex. AutoCAD prompts you to select the location for the new vertex.

■ **Move** Moves the current vertex to a different location.

■ **Regen** Regenerates the polyline. This option affects only the selected polyline; it doesn't regenerate the entire drawing.

■ **Straighten** Removes all vertices between two specified vertices and substitutes one straight segment for them. The first vertex is the current vertex when you select the Straighten option. You can then use the Next and Previous options to move the X marker to another vertex. The Go option then completes the Straighten operation.

■ **Tangent** Specifies a direction for the current vertex. Later, when you use either the Fit Curve or Spline option to edit the entire polyline, the tangent direction determines the curve that is generated.

■ **Width** Changes the starting and ending widths of the polyline segment immediately following the current vertex. AutoCAD prompts for the starting and ending widths. This option affects only the selected segment, and AutoCAD doesn't immediately redraw the polyline after you change the width. You must use the Regen option to see the changes.

■ **Undo** Reverses the previous vertex editing option.

■ **Exit** Ends the Edit Vertex option and returns to the main PEDIT options prompt.

To move a polyline vertex:

1. Start the PEDIT command.

2. Select the polyline.

3. Type **E** (for Edit Vertex) and then press ENTER, or right-click and then select Edit Vertex from the shortcut menu. AutoCAD switches into vertex editing mode and displays an X marker on the first vertex.

4. Type **N** (for Next) and then press ENTER, or right-click and then select Next from the shortcut menu to move the X marker until it reaches the vertex that you want to move.

5. Type **M** (for Move) and then press ENTER, or right-click and then select Move from the shortcut menu. AutoCAD prompts you to enter the new vertex location.

6. Specify the new location for the vertex. AutoCAD immediately moves the vertex, redraws the polyline, and repeats the vertex editing options.

7. Choose another option, or type **X** (for Exit) or select eXit from the shortcut menu to leave vertex editing mode. When you leave vertex editing mode, AutoCAD repeats the main PEDIT options.

8. Choose another option, or press ENTER to complete the command.

*You can move polyline vertices more easily by using grips. You need to use the PEDIT command, however, to perform other polyline editing.*

To taper the width of an individual polyline segment:

1. Start the PEDIT command.

2. Select the polyline.

3. Type **E** (for Edit Vertex) and then press ENTER, or right-click and then select Edit Vertex from the shortcut menu. AutoCAD switches into vertex editing mode and displays an X marker on the first vertex.

4. Type **N** (for Next) and then press ENTER, or right-click and then select Next from the shortcut menu to move the X marker until it reaches the first vertex of the segment that you want to taper.

5. Type **W** (for Width) and then press ENTER, or right-click and then select Width from the shortcut menu. AutoCAD prompts you to enter the new starting width.

6. Specify the starting width. AutoCAD then prompts you to enter the new ending width.

7. Specify the ending width. AutoCAD immediately redraws the polyline and repeats the vertex editing options.

8. Choose another option, or type **X** (for Exit) or select eXit from the shortcut menu to leave vertex editing mode. When you leave vertex editing mode, AutoCAD repeats the main PEDIT options.

9. Choose another option, or press ENTER to complete the command.

# Editing Multilines

As you learned in Chapter 4, multilines are unique objects in that they can combine several linetypes and colors into a single object. Although you can use standard object modification commands to copy, rotate, stretch, or scale multilines, other commands, such as TRIM, EXTEND, and BREAK, can't be applied to multilines. Because of the special nature of multilines, AutoCAD provides a special command specifically for editing multilines. The MLEDIT command lets you edit the intersections between multilines, change multiline vertices, and cut or weld breaks in multilines.

As noted in Chapter 4, multilines have not proven to be a very effective AutoCAD object, and many experienced AutoCAD users choose not to use multilines, relying instead on parallel lines or polylines. For that reason, although the MLEDIT command appears on the Modify II toolbar and in the Modify menu, it is not covered here. Those wishing to learn more about multilines will find detailed information on the companion web site.

# Editing Splines

You can use most of the standard object modification commands to copy, rotate, stretch, scale, break, or trim splines. To modify the shape of an existing spline, however, you need to use the special SPLINEDIT command. The SPLINEDIT command lets you add, delete, or move control points, change the spline's start and end tangent directions, and open or close an existing spline. You can also change the fit tolerance and refine a spline by changing its order. You learned about fit tolerance and order in Chapter 4.

To start the SPLINEDIT command, do one of the following:

■ On the Modify II toolbar, click Edit Spline.

■ From the Modify menu, choose Object | Spline.

■ At the command prompt, type **SPLINEDIT** (or **SPE**) and then press ENTER.

When you start the SPLINEDIT command, AutoCAD prompts you to select a spline. You can't use noun/verb selection when editing splines, and although you can use any object selection method to select the spline, you can edit only one spline at a time. As soon as you select a spline, AutoCAD displays a list of options. The available options depend on the type and status of the spline object that you select. For example, if you select an open spline, the list includes the Close option, whereas if the spline is already closed, the list includes the Open option, instead.

When you select a spline, AutoCAD displays grips at its control points and presents the following options on the command line:

■ **Fit Data**  Presents a separate set of options for editing the fit data for the selected spline (explained later in this chapter).

■ **Close**  Closes an open spline. If the spline does not have the same start and end points, AutoCAD adds a curve that is tangent at those points. If the start and end points are the same, AutoCAD makes the tangent continuous at that point.

■ **Open**  Opens a closed spline. If the start and end points are the same, AutoCAD removes the tangent information from that point. If a closing curve has formerly been added to close an open spline, that curve is removed along with the tangent information at the start and end points.

■ **Move Vertex**  Moves a selected control point to a new location. AutoCAD highlights the first control point. You can then select the control point that you want to move, and specify its new location.

■ **Refine**  Presents a separate set of options for fine-tuning the selected spline. The Add Control Point option lets you add individual control points that affect a portion of the spline. The Elevate Order option increases the order of the spline; AutoCAD adds a number of control points, uniformly distributed along the spline. The Weight option lets you change the weight at a selected control point.

■ **Reverse**  Reverses the direction of the spline.

- **Undo**   Reverses the last spline-editing operation.
- **Exit**   Ends the SPLINEDIT command.

*You can also modify spline-fit polylines by using the SPLINEDIT command, which converts the polyline into a spline. Because the resulting spline has no fit data, AutoCAD doesn't present the Fit Data option. This option is also omitted for splines that have either been refined, had a control point moved, or had their fit data purged.*

If you select the Fit Data option, AutoCAD displays grips at the fit points rather than at the control points, and the SPLINEDIT command presents the following separate set of options for editing the control points:

- **Add**   Adds a new fit point to the spline. AutoCAD prompts you to select a point, highlights both the fit point that you select and the next point, and then prompts you to enter a new point. The new point that you specify is added between the two highlighted points, with the spline refitting through the fit points.
- **Close**   Closes an open spline. This option is identical to the control point Close option.
- **Open**   Opens a closed spline. This option is identical to the control point Open option.
- **Delete**   Deletes a selected fit point, refitting the spline through the remaining fit points.
- **Move**   Moves a selected fit point to a new location. AutoCAD highlights the first fit point. You can then select the fit point that you want to move, and specify its new location.
- **Purge**   Purges the spline's fit data from the drawing. After removing the fit data, AutoCAD redisplays the main SPLINEDIT prompt without the Fit Data option.
- **Tangents**   Edits the start and end tangents of the spline.
- **Tolerance**   Changes the tolerance fitting of the spline. The spline is redefined based on the new tolerance value.
- **Exit**   Ends the Fit Data option and returns to the main SPLINEDIT options prompt.

*LEARN BY EXAMPLE*
*Use the SPLINEDIT command to move a contour line. Open the drawing Figure 10-5 on the companion web site.*

To move the contour line shown in Figure 10-5, use the following command sequence and instructions:

```
Command: SPLINEDIT
Select spline: (click the spline)
Enter an option [Fit data/Close/Move vertex/Refine/rEverse/Undo]: F
Enter a fit data option
[Add/Close/Delete/Move/Purge/Tangents/toLerance/eXit] <eXit>: M
Specify new location or [Next/Previous/Select point/eXit] <N>: S
Specify control point <exit>: (select fit point A)
Specify new location or [Next/Previous/Select point/eXit] <N>: (select point B)
Specify new location or [Next/Previous/Select point/eXit] <N>: P
Specify new location or [Next/Previous/Select point/eXit] <P>: (select point C)
Specify new location or [Next/Previous/Select point/eXit] <P>: ENTER
Specify new location or [Next/Previous/Select point/eXit] <P>: (select point D)
Specify new location or [Next/Previous/Select point/eXit] <P>: ENTER
Specify new location or [Next/Previous/Select point/eXit] <P>: (select point E)
Specify new location or [Next/Previous/Select point/eXit] <P>: X
Enter a fit data option
[Add/Close/Delete/Move/Purge/Tangents/toLerance/eXit] <eXit>: ENTER
Enter an option [Fit data/Close/Move vertex/Refine/rEverse/Undo]: ENTER
```

MOVING BEYOND
THE BASICS

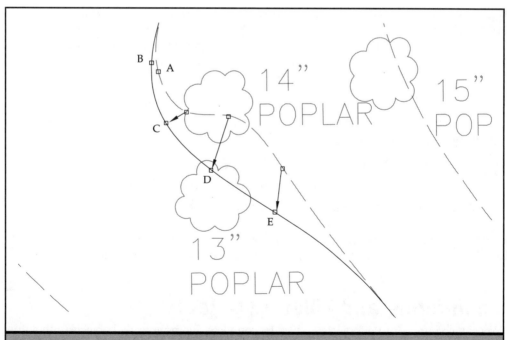

**Figure 10-5.**    *Using the SPLINEDIT command to move a contour line*

# Exploding Objects

You can convert a complex object, such as a polyline, from a single object into its component parts. Exploding a polyline, rectangle, donut, or polygon reduces it to a collection of individual line and arc objects that you can then modify individually.

With the following exceptions, exploding an object usually has no visible effect on the drawing:

- If the original polyline has a width, the width information is lost when you explode it. The resulting lines and arcs follow the center line of the original polyline.

- If you explode a block that contains attributes, the attributes are lost, but the original attribute definitions remain. You'll learn more about blocks in Chapter 15.

- Colors and linetypes that are assigned ByBlock may change in appearance after exploding an object, because block components may revert to the property settings that they had originally, before you assigned the ByBlock setting.

- If you explode a hatch pattern, the objects that make up the hatch pattern subsequently reside on the 0 layer, with color and linetype assigned ByBlock.

**Note**   *Some objects, such as text, external references, and blocks inserted by using the MINSERT command, can't be exploded. Other objects, such as blocks, dimensions, and hatch patterns, are reduced to their individual component parts. You'll learn about each of these types of objects in later chapters.*

To explode an object:

1. Do one of the following:

- On the Modify toolbar, click Explode.
- From the Modify menu, choose Explode.
- At the command line, type **EXPLODE** (or **X**) and then press ENTER.

2. Select one or more objects and then press ENTER.

**Tip**   *The XPLODE command provides a more powerful version of the EXPLODE command, providing control over the color, layer, and linetype of the resulting component objects.*

# Chamfering and Filleting Objects

You can chamfer or fillet objects. A *chamfer* connects two nonparallel objects with a line to create a beveled edge. A *fillet* connects two objects with an arc of a specified radius to create a rounded edge. If both objects that you are working with are on the same layer, the chamfer or fillet is drawn on that layer. If they are on different layers, the chamfer or fillet is drawn on the current layer. In addition to the 2-D objects described in this chapter, you can chamfer and fillet 3-D solids. You'll learn about editing in three dimensions in Chapter 20.

When you chamfer or fillet objects other than polylines, if the objects do not intersect, AutoCAD first extends them until they do intersect, and then trims them back the specified distance. If the two objects do intersect, the portions of the objects that extend beyond the chamfer or fillet normally are deleted. If you prefer, however, you can retain these portions.

# Chamfering Objects

The CHAMFER command connects two nonparallel objects by extending or trimming them and then joining them with a line to create a beveled edge. You can chamfer lines, polylines, rays, and infinite lines.

When creating a chamfer, you can specify how far to trim back the objects from their intersection, or specify the length of the chamfer and the angle that it forms along the first object. The CHAMFER command is particularly powerful when chamfering polylines, because you can chamfer not only the intersection of two polyline segments, but also the entire polyline.

To start the CHAMFER command, do one of the following:

- On the Modify toolbar, click Chamfer.

- From the Modify menu, choose Chamfer.

- At the command line, type **CHAMFER** (or **CHA**) and then press ENTER.

When you start the command, AutoCAD displays the current chamfer mode and distances, as well as a list of available options:

```
(TRIM mode) Current chamfer Dist1 = 0.5000, Dist2 = 0.5000
Select first line or [Polyline/Distance/Angle/Trim/Method]:
```

- **Polyline** Lets you chamfer an entire 2-D polyline in a single step. AutoCAD prompts you to select a 2-D polyline. After you select a polyline, a chamfer segment is added between the existing straight segments of the polyline, arc segments separating two straight segments are removed, and segments too short to be chamfered are ignored.

- **Distance** Lets you set the distance to trim back the first line and the distance to trim back the second line. AutoCAD prompts for each chamfer distance. After you set the chamfer distance, you can select the objects to be chamfered.

- **Angle** Lets you control the chamfer by specifying a length and an angle for the resulting chamfer line. AutoCAD prompts for the length and chamfer angle. After you set these values, you can select the objects to be chamfered.

- ■ **Trim**   Lets you toggle between the Trim and No Trim modes.
- ■ **Method**   Lets you explicitly toggle between the Angle and Distance chamfer methods. (Note that when you select the Distance or Angle options, the command automatically switches to that method.)

To chamfer two objects by specifying two different chamfer distances (see Figure 10-6):

1. Start the CHAMFER command.
2. Type **D** and then press ENTER, or right-click and select Distance from the shortcut menu.
3. Specify the first chamfer distance and then press ENTER.
4. Specify the second chamfer distance and then press ENTER.
5. Select the first object.
6. Select the second object.

To chamfer two objects by specifying one distance and the angle (see Figure 10-7):

1. Start the CHAMFER command.
2. Type **A** and then press ENTER, or right-click and select Angle from the shortcut menu.
3. Specify the chamfer length on the first line and then press ENTER.
4. Specify the chamfer angle from the first line and then press ENTER.
5. Select the first object.
6. Select the second object.

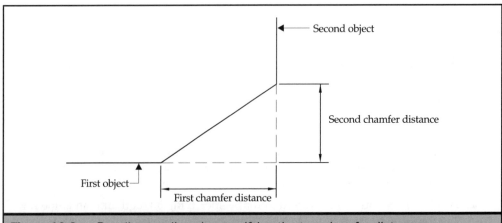

**Figure 10-6.**    *Beveling two lines by specifying the two chamfer distances*

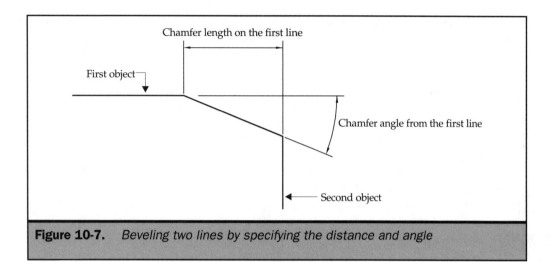

**Figure 10-7.**    *Beveling two lines by specifying the distance and angle*

MOVING BEYOND
THE BASICS

*Online*

**EXAMPLES**

### LEARN BY EXAMPLE

*Use the CHAMFER command to chamfer the four corners of a steel plate. Open the drawing Figure 10-8 on the companion web site.*

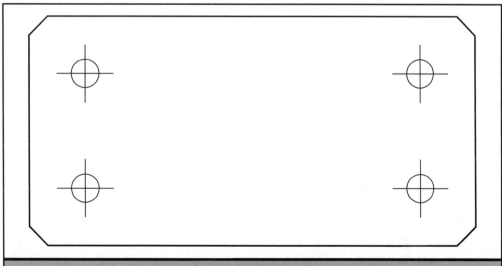

**Figure 10-8.**    *Beveling the corners of a steel plate by using the Polyline option of the CHAMFER command*

To chamfer the steel plate, consisting of a polyline previously drawn by using the RECTANGLE command, use the Polyline option of the CHAMFER command. Use the following command sequence and instructions:

```
Command: CHAMFER
(TRIM mode) Current chamfer Dist1 = 0.0000, Dist2 = 0.0000
Select first line or [Polyline/Distance/Angle/Trim/Method]: D
Specify first chamfer distance <0.0000>: .125
Specify second chamfer distance <0.1250>: ENTER
Select first line or [Polyline/Distance/Angle/Trim/Method]: P
Select 2D polyline: (click the rectangle)
4 lines were chamfered
```

*Remember that the RECTANGLE command has its own options for chamfer and fillet. The plate originally could have been drawn with chamfered corners, but this may have made it more difficult to locate the bolt holes.*

## Filleting Objects

The FILLET command connects two objects with an arc of a specified radius to create a rounded edge. You can fillet pairs of line segments, straight polyline segments, arcs, circles, rays, and infinite lines. You can also fillet lines, rays, and infinite lines that are parallel. The FILLET command is particularly powerful when filleting polylines, because you can fillet not only the intersection of two polyline segments, but also the entire polyline.

To start the FILLET command, do one of the following:

- On the Modify toolbar, click Fillet.
- From the Modify menu, choose Fillet.
- At the command line, type **FILLET** (or **F**) and then press ENTER.

When you start the command, AutoCAD displays the current fillet mode and radius, as well as a list of available options:

```
Current settings: Mode = TRIM, Radius = 0.5000
Select first object or [Polyline/Radius/Trim]:
```

- **Polyline**   Lets you fillet an entire 2-D polyline in a single step. AutoCAD prompts you to select a 2-D polyline.

- **Radius**   Lets you set the fillet radius. AutoCAD prompts for the radius. After you set the fillet radius, you can select the objects to be filleted.
- **Trim**   Lets you toggle between the Trim and No Trim modes.

**Tip**   *When you fillet a 2-D polyline, if the fillet radius isn't 0, a fillet arc of the current radius is added at each vertex where two line segments meet. If two line segments are separated by an arc segment, and the two line segments don't diverge as they approach the arc segment, the arc segment is removed and replaced with a fillet arc. If the fillet radius is 0, no fillet arcs are inserted. Any fillet arcs placed by a previous fillet command are removed, however, as are arc segments separating nondiverging line segments.*

To fillet two objects (see Figure 10-9):

1. Start the FILLET command.
2. Type **R** and then press ENTER, or right-click and select Radius from the shortcut menu.
3. Specify the fillet radius and then press ENTER.
4. Select the first object.
5. Select the second object.

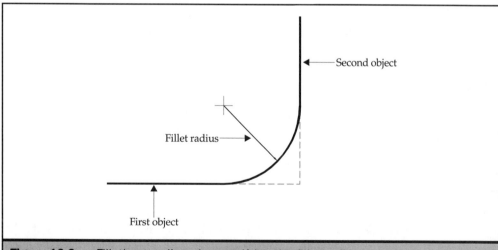

**Figure 10-9.**   *Filleting two lines by specifying the fillet radius*

**Figure 10-10.**    *Filleting the corners of a steel plate by using the Polyline option of the FILLET command*

*LEARN BY EXAMPLE*
*Use the FILLET command to fillet the four corners of the steel plate from the previous exercise. Open the drawing Figure 10-10 on the companion web site.*

To fillet the steel plate, use the following command sequence and instructions:

```
Command: FILLET
Current settings: Mode = TRIM, Radius = 0.0000
Select first object or [Polyline/Radius/Trim]: R
Specify fillet radius <0.0000>: .125
Select first object or [Polyline/Radius/Trim]: P
Select 2D polyline: (click the rectangle)
4 lines were filleted
```

When you fillet circles and arcs, more than one fillet can exist between the objects. As you can see in Figure 10-11, the point at which you select the objects determines the fillet.

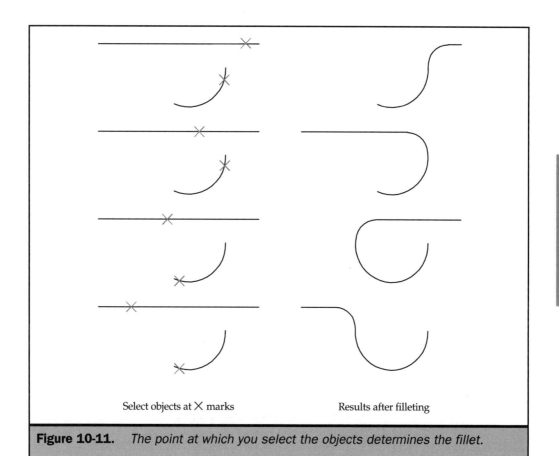

Select objects at ✕ marks                    Results after filleting

**Figure 10-11.**    *The point at which you select the objects determines the fillet.*

MOVING BEYOND
THE BASICS

You can fillet parallel lines, rays, and infinite lines. The first object must be a line or ray; the second object may be a line, ray, or infinite line. The diameter of the fillet arc is always equal to the distance between the parallel objects; the current fillet radius is ignored.

# Chapter 11

## Changing Properties

ometimes, rather than actually modifying the physical geometry of objects, you need to change their appearance. For example, you may discover that an object should be on a different layer or should be drawn with a different color or linetype. Rather than erasing and redrawing the object, you can simply change its properties.

Also, as you have learned in previous chapters, you can assign names to many drawing elements, such as layers, linetypes, views, and viewports. You can also assign names to blocks, dimension styles, text styles, and User Coordinate Systems (UCSs). You'll learn about these elements in later chapters. Once they are named, you can change the names of any of these elements. If you've created named items in your drawing that you subsequently do not use, you can remove them from the drawing in order to shorten the list of named elements.

This chapter explains how to do the following:

- Change an object's properties
- Match the properties of selected objects to those of another object
- Change an object's size or position
- Rename objects and elements
- Remove unused elements

## Changing Object Properties

AutoCAD provides the following two main tools for changing the properties of objects:

- **Object Properties toolbar**   Changes properties common to all objects, such as layer, color, linetype, lineweight, and plot style
- **Properties window**   Changes any properties of any objects

You already learned about using the controls on the Object Properties toolbar to set the current properties. When you select one or more existing objects when no command is active, however, you can use these same controls to view or change the properties of those objects.

**Note**   *To use the Object Properties toolbar to change the properties of existing objects, the Noun/Verb selection mode must be turned on. You learned about noun/verb selection and the PICKFIRST system variable in Chapter 7. If this mode is active, when you select an object, the object becomes highlighted and grips appear at key points on the object (as long as grips are also active). Both of these modes are active by default.*

Which information is displayed on the Object Properties toolbar depends on whether any objects are selected. When no objects are selected, the controls on the Object Properties toolbar show the current layer and layer properties as well as the current color, linetype, lineweight, and plot style. If a single object is selected, the controls show the properties associated with that object. When more than one object is selected, if all the objects

## Changing Properties from the Command Line

You can also change object properties by using two older commands: CHANGE and CHPROP. Both of these commands operate exclusively from the command line (rather than utilizing dialog boxes). The CHPROP command lets you change the color, layer, linetype, lineweight, or thickness of one or more objects, while the CHANGE command lets you change the elevation and linetype scale. The CHANGE command can also be used to alter text and attribute definitions (including the style, height, rotation angle, and text string), change the insertion point and rotation angle of blocks, and change the end points of lines and the radii of circles.

share the same property (such as the same color or linetype), that control on the toolbar shows the actual property associated with the objects. If the objects don't share the same property, however, that control is blank. For example, if you select three objects drawn on different layers, the Layer control will be blank.

**Note**    *This chapter deals with changing the layer, color, linetype, and linetype scale associated with one or more objects. You can also change the thickness of selected objects. When you work in plan view, an object's thickness is not visually apparent. You will learn about thickness in Chapter 19, during the discussion of drawing in three dimensions.*

## Changing Layers

Careful layer management is one of the most important skills that you can develop when working with AutoCAD. As you learned in Chapter 8, layers are one of the key tools used to organize drawing information. Despite your best efforts, however, you may find that you've drawn an object on the wrong layer, or, after careful consideration, you may decide that it makes more sense to draw an object on a layer other than the one on which it was initially created. While you can use the standards-checking tools you learned about in Chapter 8 to change layers, when you only need to make a small number of specific changes, you can use the Layer control on the Object Properties toolbar to easily change the layer of one or more objects to another existing layer.

To change the layer for one or more objects:

1. Select the object or objects whose layer you want to change.

2. On the Object Properties toolbar, choose the new layer from the Layer control.

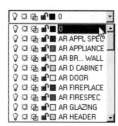

MOVING BEYOND
THE BASICS

*You can change objects to layers that are locked, frozen, or turned off, but you can't change them to an xref-dependent layer. Xref-dependent layers are displayed in the Layer control list as unavailable. You'll learn about xrefs in Chapter 15.*

Several powerful Express tools help you manage layer properties. You can learn more about the Change to Current Layer, Layer Delete, and Layer Merge tools on the companion web site.

## Changing Colors

Although you generally manage colors in your drawings by assigning colors by layer, as you learned in Chapters 2 and 8, you can also assign specific colors to objects, thus overriding the layer color. In some instances, it is useful to assign color on a per-object basis, such as to avoid having to create a new layer just to draw a few objects using a different color. After you draw objects, you can change their color on a per-object basis by using the Color control on the Object Properties toolbar.

To change the color of one or more objects:

1. Select the object or objects whose color you want to change.

2. On the Object Properties toolbar, choose the new color from the Color control.

3. If you don't see the color you want, choose Other, select the color you want in the Select Color dialog box, and then click OK.

## Changing Linetypes

You generally assign linetypes on a per-layer basis, similar to the way that you control color. But, as with colors, linetypes can be assigned on a per-object basis and can be changed by using the Linetype control on the Object Properties toolbar.

To change the linetype of one or more objects:

1. Select the object or objects whose linetype you want to change.

2. On the Object Properties toolbar, choose the new linetype from the Linetype control.

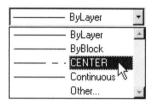

**Note** *You can change the linetype only to another linetype that has already been loaded into the current drawing. To use a linetype other than one that appears in the list, choose Other and then use the Linetype Manager to load the desired linetype from a linetype library file, as described in Chapter 8.*

## Changing Lineweights

As with linetypes, you generally assign lineweight on a per-layer basis. Lineweight can also be assigned on a per-object basis and can be changed by using the Lineweight control on the Object Properties toolbar.

To change the lineweight of one or more objects:

1. Select the object or objects whose lineweight you want to change.

2. On the Object Properties toolbar, choose the new lineweight from the Lineweight control.

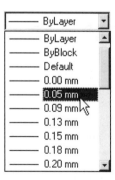

## Changing Plot Styles

AutoCAD's Plot Style object property is incredibly useful when you need to plot the same drawing in different ways—such as plotting an architectural floor plan and a lighting plan from the same drawing. When you plot your drawing, plot styles can

MOVING BEYOND THE BASICS

override an object's color, linetype, and lineweight. For example, on the lighting plan, you might plot the walls using a 50% screen.

The Plot Style object property works like the other object properties, such as linetype and color, but is only available when using named plot styles (not color-dependant plot styles). Although you generally assign a plot style on a per-layer basis, plot style can also be assigned on a per-object basis and can be changed by using the Plot Style control on the Object Properties toolbar.

To change the plot style of one or more objects:

1. Select the object or objects whose plot style you want to change.

2. On the Object Properties toolbar, choose the new plot style from the Plot Style control.

3. If you don't see the plot style you want, choose Other to display the Select Plot Style dialog box.

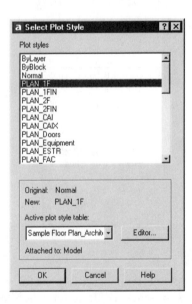

4. In the Select Plot Style dialog box, select the plot style you want from the current active plot style table; or, to select a plot style from a different plot style table, choose a different table from the Active Plot Style Table drop-down list. Then click OK.

# Using the Properties Window

Whereas the Object Properties toolbar is the most convenient way to change the properties common to one or more objects, AutoCAD's Properties window is the primary tool for changing any of the object-specific properties—such as the size and location of an object—as well as the layer, linetype, color, and so on.

To display the Properties window, do one of the following:

- On the Object Properties toolbar, click Properties.
- From either the Tools or Modify menu, choose Properties.
- At the command line, type **PROPERTIES** and then press ENTER, or press CTRL-1.

The Properties window, shown in Figure 11-1, is a very versatile tool. The window has two tabs: Alphabetic and Categorized. Both tabs contain the same information. The Alphabetic tab displays the object properties in alphabetical order, whereas the Categorized tab displays them by category. The Properties window is nonmodal, meaning

MOVING BEYOND
THE BASICS

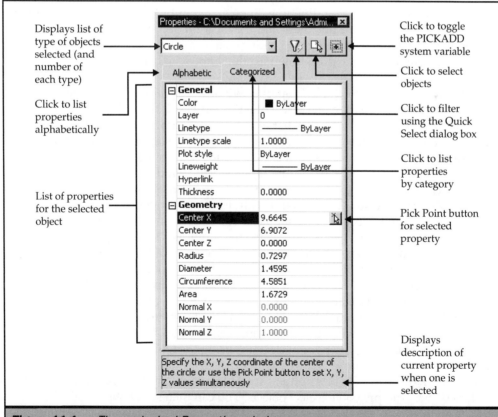

**Figure 11-1.**   *The undocked Properties window*

that it can be left open while you continue to work in your drawing. You can even dock the window to the right or left side of the drawing area.

To close the Properties window, do one of the following:

- On the Object Properties toolbar, click Properties.
- From the Tools menu, choose Properties.
- In the Properties window, click the Close button.
- Right-click in the Properties window and choose Hide from the shortcut menu.
- At the command line, type **PROPERTIESCLOSE** and then press ENTER, or press CTRL-1.

**Note**   *You can also use the controls in the shortcut menu displayed when you right-click within the Properties window to disable docking of the window, turn off the current property description, and undo the most recent changes you made to properties.*

When no objects are selected, the Properties window shows the current properties—such as the current layer, color, linetype, lineweight, and plot style—as well as other settings, such as the current thickness and the location of the UCS. You can use the controls in the window to change those settings. For example, you can change the current layer by selecting it from the drop-down list that appears when you select the Layer property within the window.

When you select an object, the Properties window displays the properties for that object. If you select more than one object, the Properties window displays the general properties and any other properties that they have in common. You can then use the controls within the window to change those settings. For example, you can change the color assigned to the selected objects by selecting it from the drop-down list that appears when you select the Color property within the window.

If you select multiple objects, you can use the drop-down list at the top of the Properties window to select a specific type of object. This list also indicates the number of objects of each type currently selected. For example, if you select a line, an arc, and a circle, you can select Arc (1) from this list to display just the properties for the arc. If you chose only one of a particular type of object, you can change any of the properties associated with that object. For example, if only one circle is selected, you can change any of the properties assigned to that circle, including its size and position. If you chose more than one object of a particular type, however, any changes you make will affect all the objects of that type. For example, if you selected two circles and then change the center point, both circles will be relocated to the same center point. However, you can click the Quick Select button to filter your selections using the Quick Select dialog box. You learned how to use this dialog box in Chapter 7.

If you click the Select Objects button, AutoCAD clears any objects that may already be selected and prompts you to select objects. You can then select objects using any selection method. The Toggle Value of PICKADD Sysvar button lets you control whether additional objects you select are added to the currently selected objects or replace the currently selected objects as a new selection set. The appearance of this button indicates the current status.

Selected objects are added
to the selection set

Selected objects replace the
previous selection set

Within the window, the method you use to modify a particular property depends on the type of property you wish to change. You can use any one of the following methods:

■ **Type a new value**   For properties that have a value—such as thickness, coordinate point, radius, area, and so on—you can change the value by typing a new value in the adjacent box.

■ **Select a value from a drop-down list**   For properties chosen from a list—such as layer, linetype, plot style, and so on—you can select a new value from the drop-down list that appears when you select that property.

■ **Change the property value using a dialog box**   For properties normally set or edited using a dialog box—such as a hyperlink, hatch pattern name, or the contents of a text string—you can change the value by using the appropriate object editing dialog box that appears when you click the ellipses button that appears when you select that property.

■ **Use the Pick Point button to change a coordinate value**   For positional properties—such as the position of objects—you can specify a new position by clicking the Pick Point button that appears when you select that property, and then clicking the new position within the drawing.

# Using Keyboard Shortcuts in the Properties Window

There are a number of standard Windows keyboard shortcuts that you can use when working with the Properties window. For example, you can use the TAB key to cycle through the various properties, and use the arrow keys and PAGE UP and PAGE DOWN keys to move vertically between the properties in this window. You can also use CTRL-Z to undo changes, and CTRL-X, CTRL-C, and CTRL-V to cut, copy, and paste properties, respectively.

In addition to those standard shortcuts, you can use the keyboard shortcuts shown in Table 11-1 to perform other actions related to the Properties window.

# Changing Properties

The Properties window makes it very easy to view and change the properties of one or more objects. You can select the objects and then open the Properties window, or open the window first and then select the objects whose properties you wish to change. The objects being changed are normally highlighted and grips appear at key points on those objects. You can use any of the methods you learned about in Chapter 7 to add or remove objects from the current selection set. The last portion of this section looks at using the Properties window to change several specific properties.

## Changing Linetype Scale

As you learned in Chapter 2, linetypes other than the continuous linetype consist of repeating patterns of dashes, dots, and open spaces, and can also include text and symbols. The appearance of these linetypes can also be adjusted by adjusting the linetype scale factor. The smaller the linetype scale, the more repetitions of the linetype pattern that are generated per drawing unit.

AutoCAD applies a global linetype scale factor to all objects drawn with a noncontinuous linetype. You can also control the linetype scale on a per-object basis by using the property modification tools.

| Keys | Action |
|---|---|
| CTRL-1 | Display or close the Properties window |
| HOME | Move to the first property in the list |
| END | Move to the last property in the list |
| CTRL-SHIFT-*alpha. character* | Move to the next property in the list beginning with the alphabetical character |
| ESC | Cancel a property change |
| ALT-DOWN ARROW | Open a settings drop-down list |
| ALT-UP ARROW | Close a settings drop-down list |

**Table 11-1.** *Keyboard Shortcuts in the Properties Window*

**EXAMPLES**

*LEARN BY EXAMPLE*
*Use the Properties window to change the linetype scale associated with an object. Open the drawing Figure 11-2 from the companion web site.*

To change the linetype scale of one of the lines shown in Figure 11-2, use the following instructions:

1. Start the PROPERTIES command.
2. Select the line whose linetype scale you want to change.
3. In the Properties window, select the Linetype Scale property, highlight the value (1.0000), type **2**, and then press ENTER. AutoCAD immediately updates the object.

*Remember that the per-object linetype scale factor is applied relative to the global linetype scale factor. Therefore, a line drawn with an object linetype scale factor of 0.5 in a drawing with a global linetype scale factor of 2 will appear the same as a line with an object linetype scale factor of 1 in a drawing with a global linetype scale factor of 1.*

## Changing Object Position and Size

Although you will usually modify the size and position of objects by using the object modification commands that you learned about in Chapters 7 and 10, you also can change the position and size of an object by using the Properties window. For example, if you select a circle, the Properties dialog box will include properties for the Center X, Center Y, and Center Z coordinates as well as the Radius and Diameter. To move the circle, you can either type new X, Y, and Z values or select one of these properties and then click the Pick Point button.

When you click the Pick Point button, AutoCAD prompts you to pick a point in the drawing. As you move the cursor, a rubber-band line extends from the current center point of the circle to the cursor position. As soon as you specify a new center point, AutoCAD moves the circle.

To change the size of the circle, select either the Radius or Diameter property and type a new value in the value box. As soon as you press ENTER (or select another property), AutoCAD immediately updates the size of the circle.

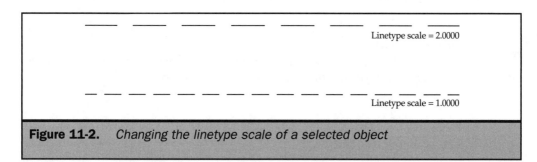

**Figure 11-2.**    *Changing the linetype scale of a selected object*

## Matching Object Properties

Another powerful way to change the properties associated with one or more objects is to match the properties with those of another object in the drawing, or in another open drawing. For example, suppose that you draw a line and then determine later that, although it is drawn on the proper layer, it should have the same linetype and color of another object. The Match Properties tool lets you apply all, or just some, of the properties from a source object to one or more destination objects.

To use the Match Properties tool, do one of the following:

- On the Standard toolbar, click Match Properties.
- From the Modify menu, choose Match Properties.
- At the command line, type **MATCHPROP** (or **MA**) and then press ENTER.

When you start the command, AutoCAD prompts you to select the source object—the object whose properties you want to copy to other objects. You must select this object by pointing; no other object selection methods will work. As soon as you select a source object, AutoCAD displays on the command line the types of properties that will be copied, and then prompts you to select the destination objects:

```
Current active settings: Color Layer Ltype Ltscale Lineweight Thickness
PlotStyle Text Dim Hatch
Select destination object(s) or [Settings]:
```

To copy all the properties from the source object to the destination object, you simply select the destination objects by using any selection method, and then press ENTER. If you want to copy only selected properties, type **S** and then press ENTER to display the Property Settings dialog box, similar to the one shown in Figure 11-3. You can then specify those properties that you want to copy, by selecting or clearing the appropriate check boxes. For example, to copy just the color of the source object to the destination objects, select the Color check box (so that a check mark appears) and clear the check marks from the other boxes.

*The Dimension, Text, and Hatch options have meaning only when applied to the appropriate type of object.*

*LEARN BY EXAMPLE*
*Use the Match Properties tool to copy selected properties from a source object to several destination objects. Open the drawing Figure 11-4 from the companion web site.*

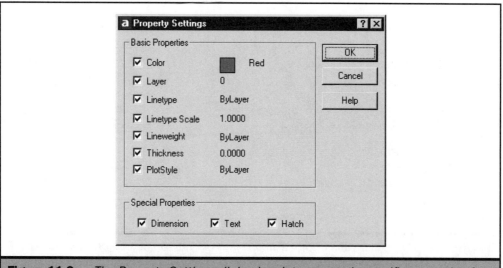

**Figure 11-3.**    *The Property Settings dialog box lets you apply specific properties from a source object to selected destination objects.*

To copy selected properties from the source object, shown in Figure 11-4, to the selected destination objects, use the following instructions:

1. Start the MATCHPROP command.

2. Select the object labeled Source Object in Figure 11-4.

3. Type **S** and then press ENTER.

4. In the Property Settings dialog box, make sure that Color is the only check box selected in the Basic Properties area, and then click OK.

5. Select the circles and identification numbers for details 1, 2, and 3, as well as the detail titles for details 2 and 3. Don't select the scale labels below the detail titles, however. After you select the six objects, press ENTER, or right-click and select Enter from the shortcut menu.

AutoCAD immediately applies the source object color to all six of the destination objects, but applies the text style only to the selected text objects. To change the color of the scale labels for details 1, 2, and 3 to match the rest of the detail identification, repeat this procedure, but this time make sure that the Text check box is not selected in the Property Settings dialog box.

The Layer Match Express tool matches the layer of selected objects to the layer of a selected destination object. You can learn more about this tool on the companion web site.

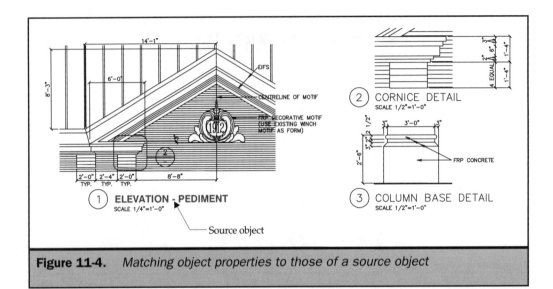

**Figure 11-4.** *Matching object properties to those of a source object*

# Renaming Objects and Elements

As you've learned, many AutoCAD objects—as well as elements such as layers, linetypes, views, and viewports—are assigned names when you create them. Other objects, such as arcs, circles, and lines, are unnamed. You often may discover that you need to change the name of a layer or other named element, either to make managing the element easier or because you made a typographical error when you originally named the element. Although you usually can change the name of any named object or element by using the same command that you used to create that element, using the general-purpose Rename dialog box, shown in Figure 11-5, is usually easier.

To display the Rename dialog box, do one of the following:

- From the Format menu, choose Rename.
- At the command line, type **RENAME** (or **REN**) and then press ENTER.

To rename any named object or element, select the element or object type in the Named Objects list and then select the name that you want to change in the Items list. When you select the item, its name appears in the Old Name edit box. To change the name, type a new name in the Rename To edit box, and then click the Rename To button to apply the change. You can then select another element to rename, or click OK to accept the changes that you made and complete the command.

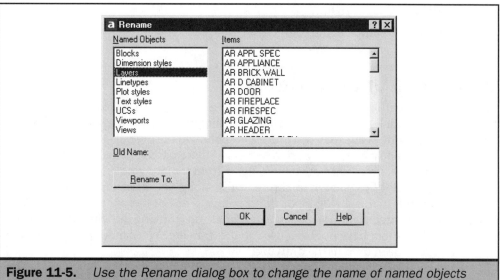

**Figure 11-5.**   *Use the Rename dialog box to change the name of named objects and elements.*

*If you make a mistake, you can abandon all the changes that you've made since starting the command by clicking the Cancel button or pressing ESC.*

You can't rename some standard elements, such as the layer 0 and the continuous linetype. You also can't use this tool to rename certain named objects, such as shapes and groups. The Rename dialog box includes some named elements that you haven't learned about yet. You'll learn about them in later chapters.

**LEARN BY EXAMPLE**

*Use the Rename dialog box to change the name of several layers in a drawing. Open the drawing Figure 11-6 on the companion web site.*

You can use wildcards in conjunction with the Rename dialog box to rename quickly a group of named objects with common characters. To rename to ARCH- all the layers that begin with the characters A-, use the following instructions:

1. Start the RENAME command.

2. In the Named Objects list, select Layer (if it isn't already selected).

3. In the Old Name edit box, type **A-\***, which is the wildcard combination that matches all layers that begin with the string A-.

4. In the Rename To edit box, type **ARCH-\***.

5. Click the Rename To button.

6. Click OK to complete the command.

# Removing Unused Items

There may be situations in which you create named items, such as layers or linetypes, that you don't actually use in your drawing. Although you can simply leave these items within your drawing, they do contribute slightly to the overall size of the drawing file. More importantly, they show up in listings of these named items, making the lists longer. If you wish, you can remove these items from the drawing.

The Purge dialog box, shown in Figure 11-6, enables you to remove unused items. Using the controls in this dialog box, you can remove one or more unused items. When you remove these items, they are permanently deleted from the drawing. AutoCAD only purges items that have not been used in the drawing. You cannot purge any item that has actually been used in the drawing.

To display the Purge dialog box, do one of the following:

■ From the File menu, choose Drawing Utilities | Purge.

■ At the command line, type **PURGE** and then press ENTER.

Normally, the Purge dialog box displays a hierarchical list of named items that you can purge from the drawing. To purge individual items, select them in the list and then click the Purge button. To purge all unused named items from the drawing, simply click the Purge All button.

**Note** *If you select the Purge Nested Items check box, AutoCAD removes all unused named items from the drawing even if they are contained within or referenced by other unused named items. But nested items are only purged if you select All Items or Blocks in the tree view or click the Purge All button.*

**Tip** *If you select the Confirm Each Item To Be Purged check box, after clicking the Purge or Purge All button, AutoCAD asks you to confirm each individual item before it is actually removed from the drawing. Instead of forcing yourself to make all of these additional confirmations, clear this check box and simply check the list in the dialog box to make sure you're not removing a named item you wish to keep prior to purging the items.*

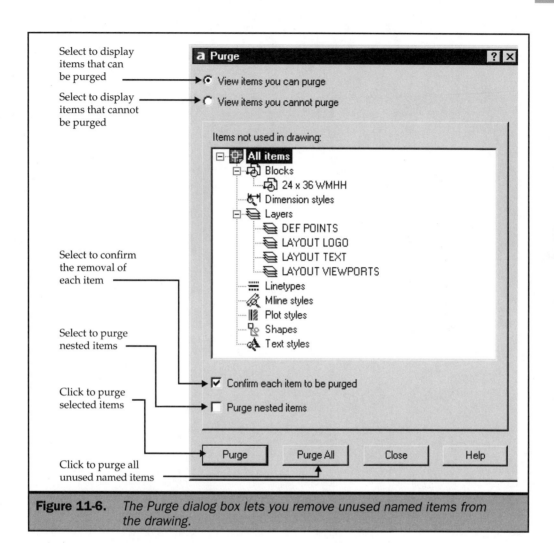

Select to display items that can be purged

Select to display items that cannot be purged

Select to confirm the removal of each item

Select to purge nested items

Click to purge selected items

Click to purge all unused named items

MOVING BEYOND THE BASICS

**Figure 11-6.**    *The Purge dialog box lets you remove unused named items from the drawing.*

# Chapter 12

## Adding Cross Hatching

A nother common way to convey information in a drawing is to fill areas with a solid color or repetitive pattern of lines. For example, you may need to represent the pattern of brick on the elevation of a building, or a type of soil or vegetation on a survey or map. Cross-sectional drawings of mechanical parts or structural components are also usually filled with a repetitive pattern of angled lines. These solid colors and repetitive patterns are called *hatch patterns.* Applying hatch patterns to areas of your drawing is a task easily accomplished by using AutoCAD's BHATCH (Boundary Hatch) and HATCH commands.

In AutoCAD, hatch patterns can be associative or nonassociative. When you apply an associative hatch pattern, if you later modify the shape of the boundary surrounding the hatch pattern, the pattern automatically updates so that it continues to fill the boundary. Nonassociative hatch patterns do not update automatically. Since associative hatch patterns provide much more flexibility, this chapter touches only briefly on nonassociative hatch patterns.

When you apply a hatch pattern, you choose the appearance of the hatch pattern by selecting either one of the predefined hatch patterns supplied in the AutoCAD hatch pattern library files or a hatch pattern from a custom library that you purchase from a third party or create yourself. You also determine the scale and angle at which the hatch pattern is applied. If you prefer, you can define a simple pattern of parallel lines on-the-fly, in which case you can control the spacing and angle at which the lines are applied. Hatch patterns are created on the current layer, so the color and linetype are controlled by the current color and linetype settings. After you add a hatch pattern to the drawing, you can modify any of its parameters.

This chapter explains how to do the following:

- Specify the appearance of hatch patterns
- Define hatch pattern boundaries
- Control hatch pattern styles
- Modify hatch objects
- Create your own hatch patterns

## Adding Hatch Objects

The hatch patterns that you apply to areas of a drawing are actually a special AutoCAD object called a *hatch object.* You add these hatch objects to the drawing by using the BHATCH command.

To add a new hatch object, do one of the following:

- On the Draw toolbar, click Hatch.
- From the Draw menu, choose Hatch.
- At the command line, type **BHATCH** (or **BH** or **H**) and then press ENTER.

When you start the command, AutoCAD displays the Boundary Hatch dialog box, shown in Figure 12-1. This dialog box provides controls for selecting the type of hatch pattern to apply and the scale and alignment of the pattern. You then select the boundary of the area to which the hatch pattern is to be applied. Because hatch objects can add a considerable number of lines to your drawing, once you select the boundary, you can click the Preview button to see what the hatch pattern will look like in your drawing, before you actually add it to the drawing. When you are satisfied with how the hatch pattern looks, you apply it. Other controls in the Boundary Hatch dialog box let you determine how AutoCAD determines the hatch boundaries, how the program deals with other objects that are enclosed within the hatch boundary, and whether the hatch pattern is applied as an associative or nonassociative hatch object.

MOVING BEYOND
THE BASICS

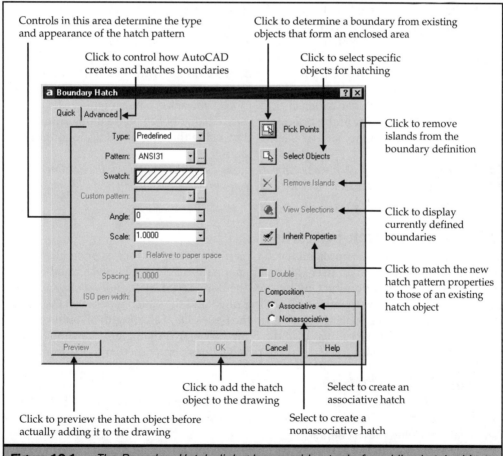

**Figure 12-1.** *The Boundary Hatch dialog box provides tools for adding hatch objects.*

# Selecting the Pattern Type

When you apply a hatch pattern, the first thing that you need to decide is the type of pattern that you want to apply. You can use one of AutoCAD's predefined patterns, a user-defined pattern, or a custom hatch pattern. You can also select the hatch pattern by matching the new hatch pattern to one that already exists in the drawing. Each of these options is described in the following sections.

## Predefined Hatch Patterns

AutoCAD supplies two hatch pattern libraries (ACAD.PAT and ACADISO.PAT), containing a total of 68 predefined hatch patterns. To select one of these predefined hatch patterns, choose Predefined from the Type drop-down list, and then either select the pattern name from the Pattern drop-down list or click the adjacent button or the Swatch box to display the Hatch Pattern Palette dialog box, shown in Figure 12-2.

The predefined hatch patterns are arranged on three separate tabs. The ANSI and ISO tabs contain all the ANSI and ISO standard hatch patterns shipped with AutoCAD, while the Other Predefined tab contains all the other hatch patterns supplied with the program. The Custom tab displays all the patterns defined in any custom hatch pattern files that you have added to the AutoCAD search path. To choose a pattern, either double-click the pattern or select it and then click OK. Once you select a hatch pattern, its image appears in the Swatch box in the Boundary Hatch dialog box. You then determine the size and angle at which the pattern is applied by using the Angle and Scale controls.

*When you select a predefined hatch pattern, its name also appears in the Pattern drop-down list in the Pattern Properties area. You can select the hatch pattern by name from this drop-down list.*

## User-Defined Hatch Patterns

The User-Defined selection applies a hatch pattern consisting of a series of parallel lines (see Figure 12-3). To apply a user-defined hatch pattern, choose User-Defined from the Type drop-down list on the Quick tab of the Boundary Hatch dialog box. You then determine the angle at which the pattern is applied and the spacing between the lines by using the Angle and Spacing controls on the Quick tab. Selecting the Double check box (located on the lower-right side of the Boundary Hatch dialog box, just above the Composition area) applies a second set of parallel lines perpendicular to the first.

## Custom Hatch Patterns

As mentioned previously, you can also use custom hatch patterns that you either created yourself or purchased from someone else. You'll learn about creating your own custom hatch patterns later in this chapter.

Custom hatch patterns generally are stored in individual hatch .PAT pattern files. To apply a custom hatch pattern, choose Custom from the Type drop-down list. Then, either type or select the name of the custom hatch pattern file in the Custom Pattern

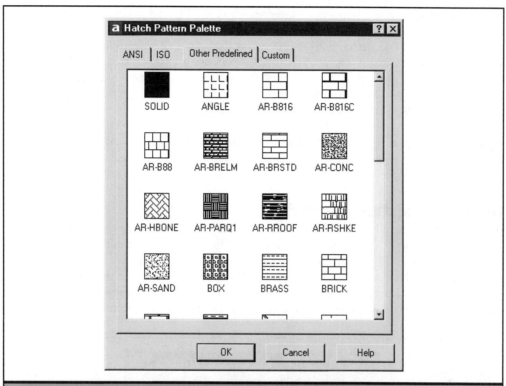

**Figure 12-2.** *Select predefined hatch patterns by using the Hatch Pattern Palette dialog box.*

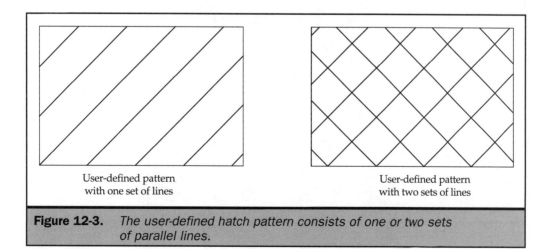

User-defined pattern
with one set of lines

User-defined pattern
with two sets of lines

**Figure 12-3.** *The user-defined hatch pattern consists of one or two sets of parallel lines.*

box, or click the adjacent button or the Swatch box to display the Custom tab of the Hatch Pattern Palette dialog box, shown in Figure 12-4. AutoCAD searches for .PAT files in its Support File Search Path and displays the names of custom hatch pattern files in the list on the left side of this tab. To display a preview image of a custom hatch pattern, select it in the list. After a brief pause, AutoCAD generates the custom hatch pattern image. After you select the custom hatch pattern, specify its size and angle by using the Angle and Scale controls in the Boundary Hatch dialog box.

*You can also add custom hatch patterns to one of AutoCAD's predefined hatch pattern library files. When you do so, the new hatch patterns appear at the end of the list of predefined patterns.*

## Matching Existing Hatch Patterns

Instead of specifying a hatch pattern, you can match one that already exists in the drawing. To do so, click the Inherit Properties button on the right side of the Boundary Hatch dialog box. AutoCAD prompts you to select an associative hatch object. As soon as you select a hatch object, AutoCAD displays its name, scale, and angle properties on the command line and prompts you to select an internal point. Click inside the area you want to hatch. If you wish to change any aspect of the acquired hatch pattern, press

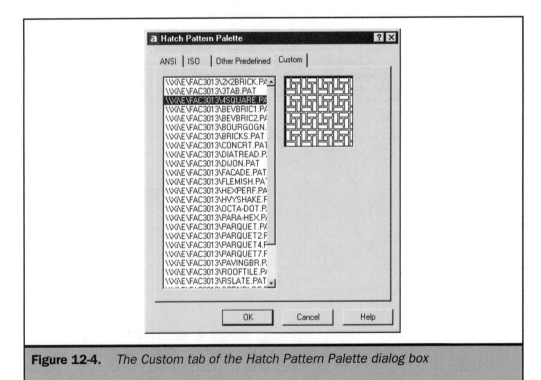

**Figure 12-4.** *The Custom tab of the Hatch Pattern Palette dialog box*

ENTER (or right-click) to return to the Boundary Hatch dialog box, either before or after specifying the area to be hatched. You can then make any necessary adjustments to the hatch pattern before you actually apply the new hatch.

# Controlling the Pattern Properties

After you select the type of hatch pattern to apply, you can control the appearance of the hatch pattern by changing its size or scale and the angle at which it is applied. These controls are located on the Quick tab of the Boundary Hatch dialog box.

## Pattern Scale and Size

Each hatch pattern definition contains information about the spacing of the lines that define the hatch pattern, much in the same way that linetype definitions contain information about the length and spacing between dashes and dots. Thus, changing the scale of a hatch pattern applies a scaling factor to the original hatch pattern definition. A value greater than 1 enlarges the hatch pattern; a value less than 1 reduces the hatch pattern from its original definition.

Some hatch patterns are designed to represent real materials. For example, AR-B816C represents 8×16-inch concrete blocks with mortar joints; applying this hatch pattern with a scale factor of 1 causes the blocks to be drawn using these dimensions. Other hatch patterns, such as EARTH and ZIGZAG, are simply representational. These patterns should be scaled, as appropriate, to produce the results that you want when you later plot your drawing.

When you apply a user-defined hatch pattern, the Scale edit box is disabled and the Spacing box becomes active. You then control the distance between the parallel lines of the user-defined pattern by typing a value into this edit box. The spacing between lines is measured in drawing units.

### Working with Metric Hatch Patterns

The hatch patterns whose names begin with the designation ISO are designed specifically for use in metric drawings. When you select one of these patterns, the ISO Pen Width drop-down list becomes active. You can then base the hatch pattern scale on a specific metric pen width. Like the metric linetypes, metric hatch patterns are designed for use in metric drawings. Although you can use them in any drawing, bear in mind that the original pattern definition is based on millimeters and will likely need to be scaled up considerably for use in nonmetric drawings.

**NOTE:** *The MEASUREINIT system variable (which is stored in the system registry) and the MEASUREMENT system variable (which is stored in the drawing and overrides the MEASUREINIT variable) normally determine whether hatch patterns are ANSI- or ISO-based.*

MOVING BEYOND THE BASICS

*If the scale or spacing value is too small, the entire area may be filled to such an extent that it appears you have used the Solid Fill pattern. Also, applying the hatch pattern at such a small scale may take a long time. If you use a value that is too large, the lines of the pattern may be spaced so far apart that none of the hatch pattern appears in the drawing.*

## Controlling the Angle

Each hatch pattern definition also contains information about the angle of the lines that comprise the hatch pattern. The image of the hatch pattern shows the alignment of the pattern when applied with an angle value of 0. To change the alignment at which the pattern is applied, type or select a new value in the Angle box. Remember that AutoCAD normally measures angles counterclockwise. To rotate the hatch pattern in a clockwise direction, you enter a negative value.

## Defining Hatch Boundaries

After you select the hatch pattern and make the necessary adjustments to its appearance, the next step is to define the boundaries of the area to be filled. This area must be completely enclosed by one or more objects, although, as shown in Figure 12-5, the objects themselves can overlap. You can define the boundaries either by picking points inside the enclosed area or by selecting the objects.

When a boundary is enclosed by a single closed object or by multiple objects that meet at their end points, you can select the boundary either by picking a point within that boundary or by selecting the individual objects. When a boundary is enclosed by multiple overlapping objects, however, you must select the boundary by picking a point within the boundary. Selecting the individual objects could produce very unexpected results.

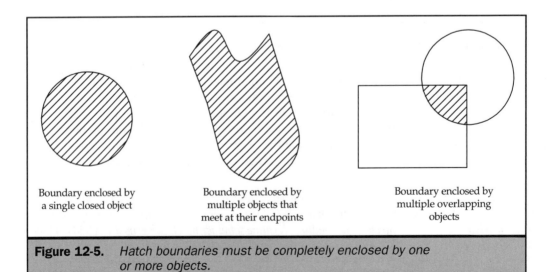

Boundary enclosed by a single closed object

Boundary enclosed by multiple objects that meet at their endpoints

Boundary enclosed by multiple overlapping objects

**Figure 12-5.**    *Hatch boundaries must be completely enclosed by one or more objects.*

## Picking Points

When you click the Pick Points button, AutoCAD determines a hatch boundary from existing objects that form an enclosed area. The Boundary Hatch dialog box disappears temporarily and AutoCAD prompts you to select an *internal point*, a point inside the area to be filled. As soon as you specify a point, AutoCAD analyzes the drawing and highlights the boundary objects. If any other enclosed objects or text objects are within the boundary, these *islands* are highlighted as well. The current settings on the Advanced tab, described later in this chapter, determine how AutoCAD deals with these islands. After you finish picking points, press ENTER to return to the Boundary Hatch dialog box.

*When you define boundaries by picking points, the Remove Islands button in the Boundary Hatch dialog box becomes active and can be used to remove objects that AutoCAD has defined as islands from the current boundary objects.*

## Selecting Objects

When you click the Select Objects button, AutoCAD determines the hatch boundary only from those objects that you select. The Boundary Hatch dialog box disappears temporarily and the program prompts you to select objects. Unlike the Pick Points option, the objects that you select must form completely enclosed areas. For AutoCAD to consider internal islands, you must explicitly select those objects as well. After you finish selecting objects, press ENTER to return to the Boundary Hatch dialog box.

*You can select multiple boundaries by using either or both methods, before actually applying the hatch pattern. Click the View Selections button to highlight the current boundaries.*

# Using the Boundary Options

As you've already learned, AutoCAD normally treats enclosed areas and text objects within hatch boundaries as islands. You can use the Advanced tab, shown in Figure 12-6, to determine how the program deals with these islands and to control other performance aspects of the BHATCH command.

The Advanced tab has four additional areas:

- **Island Detection Style**   These controls let you determine how AutoCAD deals with islands.

- **Object Type**   These settings specify whether to retain boundaries as objects, and the object type AutoCAD applies to those objects.

- **Boundary Set**   These controls let you limit the number of objects that AutoCAD analyzes when it creates a hatch boundary, thus improving the speed at which the BHATCH command operates.

- **Island Detection Method**   This setting specifies whether to include objects within the outermost boundary as boundary objects.

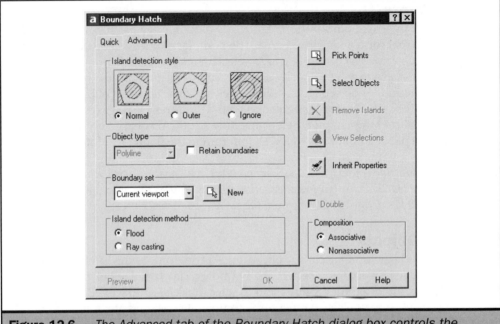

**Figure 12-6.**   *The Advanced tab of the Boundary Hatch dialog box controls the creation of hatch boundaries.*

## Dealing with Islands

The controls in the Island Detection Style area of the Advanced tab determine how AutoCAD deals with islands—and other objects that fall within the outermost hatch boundary. If you use the pointing method to determine the hatch boundary, AutoCAD automatically identifies these islands. If you select the boundary objects, you also must explicitly select the island objects.

Three boundary styles are available: Normal, Outer, and Ignore. You can choose one of these three styles by either selecting its radio button or simply clicking its icon. The individual icons show how AutoCAD treats islands, using each of these settings. If no islands are selected, the boundary style has no effect.

- **Normal**   Hatching begins at the outermost boundary. If no islands are found inside that boundary, everything within the boundary is hatched. If the boundary contains islands, AutoCAD hatches between the first and second boundaries, and if the second boundary contains an island, AutoCAD hatches between alternating boundaries. This is the default style.

- **Outer**   Hatching begins at the outermost boundary. If that boundary contains islands, hatching is applied only between the outer and the first inner boundary. This style is particularly useful when you need to hatch islands by using different hatch patterns.

■ **Ignore**  Hatching begins at the outermost boundary and ignores any islands. Everything within the outer boundary is hatched.

The two radio buttons in the Island Detection Method area determine whether AutoCAD includes objects within the outermost boundary as boundary objects. When you select the Flood radio button, AutoCAD includes islands as boundary objects. When you select the Ray Casting radio button, however, AutoCAD locates the boundary by running a line from the point you specify to the nearest object. It then traces the boundary in a counterclockwise direction, effectively excluding islands as boundary objects.

## Retaining Hatch Boundaries

Normally, a boundary object that you create by using the BHATCH command is discarded as soon as AutoCAD has added the new hatch object. If you select the Retain Boundaries check box, however, the boundary is added to the drawing as a new object, along with the new hatch object. In addition, when you select this check box, the Object Type drop-down list becomes active. You can then determine whether the new boundary object is created as a closed polyline or as a region.

## Defining the Boundary Set

When you pick a point to create a hatch boundary, AutoCAD normally analyzes all the objects on the screen. In very complex drawings, this can take several seconds. The controls in the Boundary Set area of the Advanced tab box let you limit the number of objects that AutoCAD analyzes.

By default, AutoCAD determines the boundary set from among all visible objects in the current viewport. If you click the New button, however, the dialog box temporarily disappears and AutoCAD prompts you to select the objects that you want it to consider when it creates the boundary set. When you finish selecting objects, press ENTER or right-click to return to the Boundary Hatch dialog box. The Boundary Set drop-down list now appears as Existing Set, thus limiting the boundary set selection to only those objects in the boundary set that you just created.

*LEARN BY EXAMPLE*
*Use the BHATCH command to add hatching to a drawing. Open the drawing Figure 12-7 on the companion web site.*

To add hatching to the cross-section view of the mechanical part shown in Figure 12-7, use the following instructions:

1. Start the BHATCH command.

2. In the Pattern drop-down list, choose STEEL, or click the adjacent button and select the STEEL hatch pattern from the Other Predefined tab of the Hatch Pattern Palette dialog box.

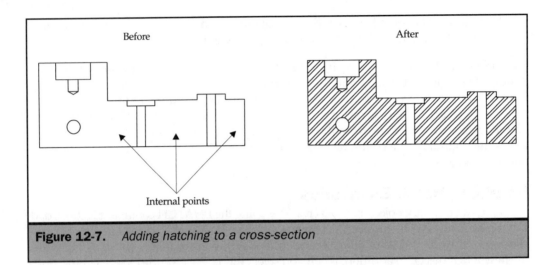

**Figure 12-7.**    *Adding hatching to a cross-section*

3. In the Scale drop-down list, choose 2.0000.

4. Click the Advanced tab and make sure that the Island Detection Style is set to Normal and the Island Detection Method is set to Flood.

5. Click the Pick Points button. The Boundary Hatch dialog box disappears and AutoCAD prompts you to select an internal point. Specify the points labeled in Figure 12-7. Notice that as you specify each point, AutoCAD highlights the boundaries. After you specify all three points, press ENTER or right-click and select Enter from the shortcut menu to return to the Boundary Hatch dialog box.

6. Click the Preview button if you want to see what the hatch pattern will look like, before you actually add it to the drawing. Next, press ENTER or right-click to return to the Boundary Hatch dialog box. You can then make any additional changes to the hatch pattern, such as choosing a different hatch pattern or changing the scale or angle, before you actually add the hatch object to the drawing.

7. Click the OK button to add the hatch object to the drawing and end the BHATCH command.

*If you apply a hatch pattern to an area that is located very far away from the drawing's origin point, you may get unexpected results, or AutoCAD's boundary detection may not work properly. When this occurs, move the User Coordinate System (UCS) origin closer to the boundary objects prior to using BHATCH, and then move it back.*

# Using the Direct Hatch Option of the HATCH Command

The HATCH command is an older AutoCAD command that creates nonassociative hatch objects. Because it provides much less flexibility than the BHATCH command, this command-line-driven command is not included in any of the toolbars or menus in AutoCAD 2002. To start the HATCH command, you must type **HATCH** (or **-H**) at the command line. In spite of its drawbacks, however, the HATCH command offers one capability that is not available when using BHATCH—the Direct Hatch or point acquisition option. With Direct Hatch, you can apply a hatch pattern without first creating the boundary objects.

When you start the HATCH command, AutoCAD prompts you to enter the name of the hatch pattern, the scale, and the angle at which to apply the pattern. AutoCAD then prompts:

```
Select objects to define hatch boundary or <direct hatch>,
Select objects:
```

At this point, if you select objects, the HATCH command behaves much like the BHATCH command, except that it creates nonassociative hatch objects. If you press ENTER, however, AutoCAD prompts you to specify points that define the hatch boundary, using options similar to those of the PLINE command. You can even instruct AutoCAD to retain the polyline boundary upon completion of the HATCH command.

*LEARN BY EXAMPLE*
*Use the HATCH command to apply a hatch pattern without first creating boundary objects. Open the drawing Figure 12-8 on the companion web site.*

To add the EARTH hatch pattern to the curb and gutter detail shown in Figure 12-8, use the following command sequence and instructions. Use the node object snap to select the 15 points indicated with X's.

```
Command: HATCH
Enter a pattern name or [?/Solid/User defined] <default>: EARTH
Specify a scale for the pattern <default>: 5
Specify an angle for the pattern <default>: 0
Select objects to define hatch boundary or <direct hatch>,
Select objects: ENTER
Retain polyline boundary? [Yes/No] <N>: ENTER
Specify start point: (select first point)
Specify next point or [Arc/Close/Length/Undo]: (select second point)
Specify next point or [Arc/Close/Length/Undo]:
```

Continue selecting each point in turn until you have selected all 15 points. As you select each point, the program adds a line segment between it and the previous point, thus indicating the hatch boundary. When you reach the last point, close the boundary and then complete the HATCH command using the following command sequence:

```
Specify next point or [Arc/Close/Length/Undo]: C
Specify start point for new boundary or <apply hatch>: ENTER
```

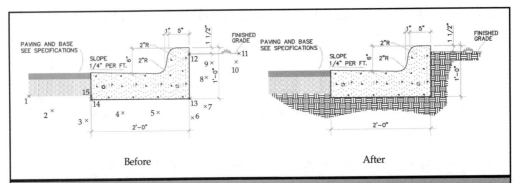

**Figure 12-8.** *Hatching added using the Direct Hatch option of the HATCH command*

**Note** *The Super Hatch Express tool enables you to use an image, block, xref, or wipeout object as a hatch pattern. You can learn more about this Express Tool on the companion web site.*

# Modifying Hatch Objects

Each hatch pattern is added to the drawing as a *hatch object*. As with other complex objects, AutoCAD provides a special command for modifying existing hatch objects. The HATCHEDIT command prompts you to select a hatch object. As soon as you do so, AutoCAD displays a Hatch Edit dialog box, similar to Figure 12-9, which is identical to the Boundary Hatch dialog box except that the controls for defining the hatch boundaries are disabled. Start the HATCHEDIT command by doing one of the following:

- Double-click the hatch object you want to modify.

- On the Modify II toolbar, click Edit Hatch.
- From the Modify menu, choose Object | Hatch.
- At the command line, type **HATCHEDIT** (or **HE**) and then press ENTER.

When the Hatch Edit dialog box appears, the values displayed are the current values of the hatch object that you selected. You can change the values to modify the scale and angle of the hatch object, click the Advanced tab to change the boundary style, or even specify a different pattern type. After you finish, click the OK button to update the hatch object, to reflect the new values.

**Tip** *You can also modify existing hatch objects by using the Properties window (displayed when you click the Properties button on the Object Properties toolbar). If you select a hatch object, AutoCAD displays the pattern within the Properties window. You can then change the pattern type, pattern name, angle, and scale. You learned about using the Properties window in Chapter 11.*

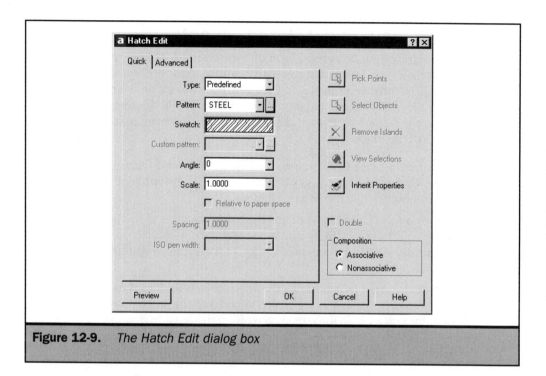

**Figure 12-9.** *The Hatch Edit dialog box*

## Modifying the Hatch Boundaries

As mentioned earlier in this chapter, one of the beauties of using associative hatches is that if you later modify the boundary surrounding the hatch pattern, the pattern itself automatically adjusts so that it continues to fill the boundary. If you modify any of the internal islands of an associative hatch object, by deleting an island for example, the hatch pattern also adjusts automatically.

*If you delete any of the boundary objects, the hatch object loses its associativity and thus loses its ability to adjust itself automatically. In addition, you should avoid moving an island outside the hatch boundary, because doing so can produce unpredictable results.*

## Exploding Hatch Objects

The EXPLODE command removes the hatch boundary associativity and converts hatch objects into individual lines. The individual lines remain on the layer on which the original hatch object was created, however, and retain the linetype and color originally assigned to the hatch object. Although you can modify the individual lines that make up the hatch pattern after you explode it, the loss of associativity, the increased effort of editing each line separately, and the increase in drawing size usually offset any advantage that you gain. In general, it's best to work with associative hatches. There is no way to convert nonassociative hatches into associative hatch objects.

# Controlling the Visibility of Hatch Objects

Hatch objects are drawn on the current layer and use the current linetype. If you create hatch objects on their own layer, you can easily control the visibility of those objects simply by freezing or turning off the layer. Creating hatch objects on their own layer also makes changing the hatch object's color or linetype easy. While creating hatch patterns on separate layers is a good practice and aids in efficient drawing management, you need to be aware of the following:

■ If you freeze the layer that contains the hatch objects and then modify the boundary objects, the hatch associativity is lost and the hatch object no longer adjusts automatically to boundary changes.

■ If you turn off the layer that contains the hatch objects and then modify the boundary objects, associativity is retained and the hatch objects continue to adjust automatically. You won't see the changes, however, until you turn on the hatch layer again.

■ If you lock the layer that contains the hatch objects and then modify the boundary objects, the associativity is retained, but the hatch objects do not adjust automatically. To force the hatch object to adjust to boundary changes, you must unlock the layer and then use the HATCHEDIT command. Select the hatch object and then click OK, without changing any of the settings.

You can also control the visibility of hatch objects by turning Fill on and off. When Fill is off, all hatch objects become invisible, regardless of the visibility of the layers on which they are drawn. The Fill setting also affects all other filled objects, however, such as wide polylines, filled multilines, and solids. You can turn this setting on and off by using any of the following methods:

■ Use the Apply Solid Fill check box on the Display tab of the Options dialog box, as described in Chapter 5.

■ Use the FILL command, as described in Chapter 5. At the command line, type **FILL**, press ENTER, and then type **ON** or **OFF**.

■ Change the FILLMODE value directly. At the command line, type **FILLMODE**, press ENTER, and then change the variable value (0 turns off FILLMODE, 1 turns on FILLMODE).

When you turn Fill on or off, you must regenerate the drawing before the change becomes visible. The current Fill setting has no effect on the hatch associativity.

*LEARN BY EXAMPLE*
*Change the scale of the hatch object and then use the STRETCH command to modify the boundaries of the associative hatch object that you created in the previous exercise. Open the drawing Figure 12-10 on the companion web site.*

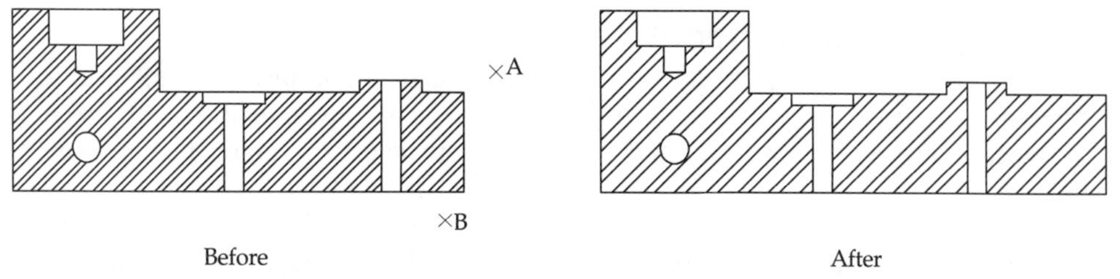

**Figure 12-10.** *Modifying the hatched cross-section*

To change the scale of the hatch object shown in Figure 12-10, use the following instructions:

1. Start the HATCHEDIT command.
2. When prompted to select a hatch object, click anywhere on the hatch object.
3. In the Hatch Edit dialog box, type **3** in the Scale edit box, and then click OK.

To stretch the boundary of the hatch object, use the following instructions:

1. Start the STRETCH command.
2. When you are prompted to select objects, use a crossing window to select the rightmost end of the cross-section by specifying the points shown in Figure 12-10, labeled A and B, and then press ENTER.

**Note** Be sure to select the points in the designated order (point A and then point B), to activate an implied crossing window.

3. When you are prompted to specify the base point or displacement, type **1,0** and then press ENTER.
4. When you are prompted to specify the second point of displacement, press ENTER.

**Note** Hatches created using AutoCAD Release 13 and earlier versions of AutoCAD aren't automatically converted into the new optimized format used in newer versions of AutoCAD. Instead, you should use the CONVERT command to change existing hatch objects into the new format. You learned about the CONVERT command in Chapter 4.

## Selecting Hatch Objects

Normally, when you select an associative hatch object, its boundaries are not selected. You can force AutoCAD to select the boundary objects as well as the hatch object by selecting the Associative Hatch check box on the Selection tab of the Options dialog box.

Selecting the hatch object's boundary without selecting the hatch object itself is a more difficult problem. The best way to accomplish this task is to either turn off the display of hatch patterns (using the Fill setting) or use object cycling (which was described in Chapter 7).

# Creating Custom Hatch Patterns

In addition to using AutoCAD's predefined hatch patterns or purchasing custom hatch patterns from a third party, you can create your own custom hatch patterns. You can add these new hatch patterns to one of AutoCAD's hatch pattern library files (ACAD.PAT or ACADISO.PAT), or save each new pattern to its own file. If you add the custom patterns to AutoCAD's own libraries, the new patterns appear in the list of predefined hatch patterns. If you create the custom hatch pattern in its own file, the file may contain only a single hatch pattern definition, the filename must be the same as the hatch pattern name, and the file must have the file extension .PAT. In addition, custom hatch pattern names can't be more than 32 characters in length.

Hatch patterns are defined in simple ASCII files, similar to linetype library files. As such, you can create new hatch patterns by using any ASCII text editor.

*LEARN BY EXAMPLE*
*One of the best ways to learn about creating hatch patterns is to look at the actual hatch pattern definitions in a hatch pattern library file. You can open this file by using a simple ASCII text editor, such as Windows Notepad. To ensure that you don't accidentally damage the hatch pattern library files that come with AutoCAD, download the file MYHATCH.PAT from the companion web site.*

To open MYHATCH.PAT by using Windows Notepad:

1. From the Start menu, choose Programs | Accessories | Notepad.

2. In Windows Notepad, choose File | Open.

3. In the Look-in drop-down list, select the folder to which you downloaded the hatch pattern from the companion web site.

4. In the File Name edit box of the Open dialog box, type **\*.pat**, select the file MYHATCH.PAT, and then click Open. A portion of the linetype file is shown in Figure 12-11.

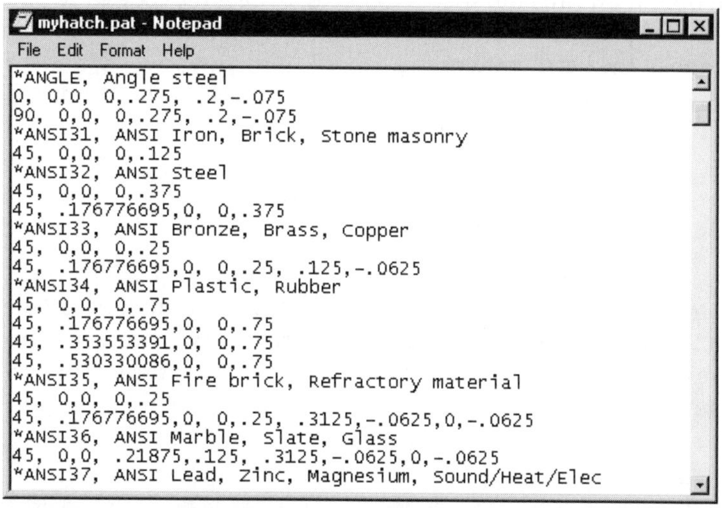

```
myhatch.pat - Notepad
File  Edit  Format  Help
*ANGLE, Angle steel
0,  0,0,  0,.275, .2,-.075
90, 0,0,  0,.275, .2,-.075
*ANSI31, ANSI Iron, Brick, Stone masonry
45, 0,0,  0,.125
*ANSI32, ANSI Steel
45, 0,0,  0,.375
45, .176776695,0,  0,.375
*ANSI33, ANSI Bronze, Brass, Copper
45, 0,0,  0,.25
45, .176776695,0,  0,.25, .125,-.0625
*ANSI34, ANSI Plastic, Rubber
45, 0,0,  0,.75
45, .176776695,0,  0,.75
45, .353553391,0,  0,.75
45, .530330086,0,  0,.75
*ANSI35, ANSI Fire brick, Refractory material
45, 0,0,  0,.25
45, .176776695,0,  0,.25, .3125,-.0625,0,-.0625
*ANSI36, ANSI Marble, Slate, Glass
45, 0,0,  .21875,.125, .3125,-.0625,0,-.0625
*ANSI37, ANSI Lead, Zinc, Magnesium, Sound/Heat/Elec
```

**Figure 12-11.** *A portion of the hatch pattern library file in Windows Notepad*

Leave this hatch pattern library file open in Windows Notepad, so that you can refer to it while you learn about the structure of a simple hatch pattern.

All hatch pattern definitions consist of a header line and one or more lines of descriptors. The header line begins with an asterisk, followed by the name of the hatch pattern, a comma, and an optional text description. This description usually consists of text that describes either the appearance or the typical use for the hatch pattern. For example, in Figure 12-11, notice that the ANGLE hatch pattern has the description "Angle steel."

The lines that follow the header contain the actual code for the hatch pattern. Each line defines one line in a family of hatch lines, and takes the following form:

```
angle, x-origin, y-origin, delta-x, delta-y [, dash-1, dash-2, ...]
```

Each line begins with an orientation angle and a pair of origin coordinates. The next data pair describes the distance that the line is to be offset in the X and Y direction for each repetition of the hatch pattern. If present, any remaining data describes pen up (positive numbers) and pen down (negative numbers) sequences, each separated by a comma, describing dashed-line patterns similar to linetype definitions. A dash length of 0 creates a dot. You can specify up to six dash lengths per line, and each line can be up to 80 characters in length.

The hatch pattern definition continues on subsequent lines, until AutoCAD reaches a new line that begins with an asterisk, indicating the start of the next hatch pattern. The program ignores blank lines and any text to the right of a semicolon. Although most hatch pattern definitions include spaces to make reading the code easier, the spaces are not needed and are ignored by AutoCAD. There must be a final carriage return, however, after the last line of the hatch pattern file.

Hatch patterns, such as ANSI31, which consist of just one line of descriptors, are composed of only one family of hatch lines:

```
*ANSI31, ANSI Iron, Brick, Stone masonry
45, 0,0 0,.125
```

The hatch pattern is drawn at a 45-degree angle and begins at the coordinate 0,0. The last pair of data, 0,.125, tells AutoCAD to repeat the pattern every .125 units in the Y direction.

More complex hatch patterns, such as ANSI38, illustrate the use of multiple hatch line families and dashed lines:

```
ANSI38, ANSI Aluminum
45, 0,0, 0,.125
135, 0,0, .25,.125, .3125,-.1875
```

The first line of the hatch pattern is drawn at a 45-degree angle and begins at the coordinate 0,0. Again, each copy of this line is offset .125 units in the Y direction. The second line is drawn at a 135-degree angle (perpendicular to the first line). This line also begins at the coordinate 0,0, but each copy is offset .25 units in the pattern's X direction and .125 units in the Y direction. In addition, this line consists of dashes .3125 units long with .1875 unit gaps between each dash.

As part of the pattern customization, you may want to view the resulting pattern in a drawing file. To update a hatch object after you make a change to the definition of the hatch pattern, use the HATCHEDIT command. Select the hatch object and then click Apply. The hatch updates to the new definition from the .PAT file.

**EXAMPLES**

*LEARN BY EXAMPLE*
*Create a new hatch pattern in its own .PAT file and then use it in an AutoCAD drawing.*

To create a new hatch pattern:

1. In Windows Notepad, choose File | New to create a new file.

2. On the first line, add the following hatch pattern name and description, and then press ENTER:

   ```
   *SQUARES, Box pattern
   ```

3. Add the following four lines of hatch pattern descriptor code, being sure to press ENTER at the end of each line, including the last line:

```
  0,  0,0,  2,2,  1,-1
 90,  0,0,  2,2,  1,-1
180,  1,1,  2,2,  1,-1
270,  1,1,  2,2,  1,-1
```

4. From the Windows Notepad File menu, choose Save. In the Save As dialog box, type the filename **SQUARES.PAT**. (Remember that when you save a custom hatch pattern to its own file, the filename must be the same as the hatch pattern name.) Make sure the Save As Type drop-down list is set to Text Documents, so that Notepad saves the hatch pattern definition as an ASCII file. You should also save custom hatch pattern files in either the AutoCAD Support folder or another directory included in AutoCAD's Support File Search Path (as defined on the Files tab of the Options dialog box). Then, click Save.

Now you can try out your new hatch pattern. Open the drawing file Figure 12-12 on the companion web site. To apply the custom hatch pattern, use the following instructions:

1. Start the BHATCH command.

2. On the Quick tab of the of the Boundary Hatch dialog box, select Custom from the Type drop-down list.

3. Click the button adjacent to the Custom Pattern drop-down list box. On the Custom tab of the Hatch Pattern Palette dialog box, select C:\Program Files\ AutoCAD 2002\support\squares.pat in the list of custom hatch patterns. An image of the pattern appears in the adjacent image area. Click OK.

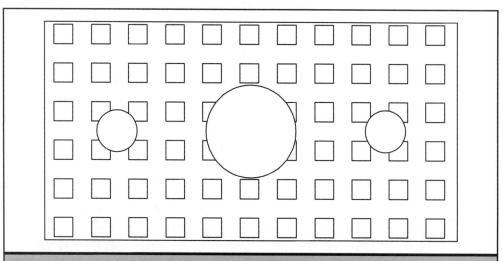

**Figure 12-12.** *The SQUARES custom hatch pattern*

4. Click the Pick Points button in the Boundary Hatch dialog box and then, referring to Figure 12-12, click anywhere inside the outer rectangular boundary and outside the circular islands.

5. Press ENTER to return to the Boundary Hatch dialog box, and then click OK.

 *When you provide a drawing that contains custom patterns to someone else, you have to also provide the pattern files that the drawing refers to. This is different from custom linetypes, in which the definitions are stored in the drawing.*

## Automating Hatch Pattern Creation

Although the creation of basic hatch patterns is relatively straightforward, it still requires some programming. Furthermore, creating more complex hatch patterns can become quite a tedious programming exercise. For that reason, several developers have created programs that automate the creation of custom hatch patterns. One such program, AUTOPAT.LSP, is provided on the companion web site, on the Chapter 12 page.

### Saving Hatch Pattern Images

In versions prior to AutoCAD 2000, you had to save images of custom hatch patterns as AutoCAD slide files and add those images to the ACAD.SLB slide library file in order for them to appear in the Boundary Hatch dialog box. This is no longer necessary. AutoCAD 2002 automatically generates images of custom hatch patterns on-the-fly regardless of whether the patterns are added to AutoCAD's standard hatch pattern library files or stored as individual .PAT files.

AutoCAD still uses the ACAD.SLB file, however, to speed up the display of hatch patterns in the Boundary Hatch dialog box. If you create a large number of custom patterns or wish to change the way the preview image appears within the Hatch Pattern Palette dialog box, you should continue to add slide images of your custom hatch patterns to the ACAD.SLB slide library. This slide library file is located in AutoCAD's SUPPORT directory. You'll learn more about slides and slide library files in Chapter 24.

You can use the SLIDELIB program supplied with AutoCAD to create slide library files. Unfortunately, this program lets you create only new slide library files. It does not provide a way to add new slides to an existing slide library file. To add slides of your custom hatch patterns to AutoCAD's existing slide library, therefore, you also need to create new slide images for all of AutoCAD's standard hatch patterns.

Although Autodesk's slide library program does not allow you to add new slides to, or delete existing slides from, an existing slide library file, several third-party programs are available that provide this capability. One such package, SlideManager, is provided on the companion web site. You'll find it on the page for Chapter 24.

AUTOPAT lets you draw a hatch pattern by using line and point objects. You must create the hatch pattern within an area measuring 1 unit square. After you draw the hatch pattern, run the AUTOPAT AutoLISP program to convert the lines and points into an actual hatch pattern definition. AUTOPAT either adds the new hatch pattern to AutoCAD's hatch pattern library file or creates a separate .PAT file. Complete instructions for using AUTOPAT are provided in a text file that accompanies the program.

MOVING BEYOND
THE BASICS

The
# Complete
# Reference

# Chapter 13

## Working with Text

A lthough the graphic elements are usually the most important parts of a drawing, drawings generally also contain lots of text. Text is used to identify the objects represented by the lines, arcs, and circles in your drawing, or to provide special notations, such as the type of material to be used to manufacture a part or the finish to be applied to a wall. Rather than being a tedious task, AutoCAD's text commands make it easy to add text to your drawings.

In addition to adding text, you can also control the text's appearance. For example, you can select the font used to draw the text, control the height and alignment (often called *justification*) of the text, and apply special formatting, such as bold, italic, and underline. When you add text to a drawing, you can add either individual lines of text or entire paragraphs. You can type the text as you're working on your drawing, or import the text from a word processor file. In addition, you can cut-and-paste or drag-and-drop text into your drawings.

This chapter explains how to do the following:

- Create line text
- Create multiline (paragraph) text
- Control text styles
- Format text
- Edit text
- Insert text from other sources
- Check for misspelled words

Dimensions and links to external databases can also add text to drawings. You'll learn about dimensions, including text with leaders, in Chapter 14. Using external databases to store and display textual data is discussed in Chapter 27.

## Creating Line Text

The simplest type of text that you can create is an individual line of text. A line of text can consist of a single letter, a word, or entire sentences. Although you can create multiple lines of text in this way, each line of text is created as a separate AutoCAD text object. You can use special control codes to underline individual letters or words, but the entire line text object uses a single text style, height, and color. You create line text objects by using the TEXT or DTEXT command.

**Note** *In AutoCAD R14 and earlier versions, the TEXT and DTEXT commands functioned differently. When you created text using the TEXT command, the text didn't appear in the drawing until you completed the command. Beginning with AutoCAD 2000, these two commands behave identically.*

As you create the text, you actually see a temporary representation of the text in your drawing as you create it, and you can easily create multiple lines of text. The text is generated using the current text style, unless you change the text style before you specify the starting point for the text. The starting point that you specify corresponds to the left end of the line of text—*left-justified*—unless you also change the justification before you specify the starting point. You'll learn about text styles and justification options later in this chapter. To simply add a line of left-justified text using the current text style:

1. Do one of the following:

   ■ On the Text toolbar, click Single Line Text.

   ■ From the Drawing menu, choose Text | Single Line Text.

   ■ At the command line, type **TEXT** (or **DTEXT** or **DT**) and then press ENTER.

2. Specify the starting point for the beginning of the line of text. (Note that you could instead choose the Style or Justify option at this point, to adjust those settings before you specify the starting point. These options are discussed later in this chapter.)

3. Specify the height of the text.

4. Specify the rotation angle.

5. Type the text, and then press ENTER at the end of each new line of text.

6. To complete the command, press ENTER again.

**Tip** *If you've already created some text and want to add a new line of text immediately below the previous text, start the TEXT command and, when prompted for the start point, press ENTER. The new text will keep the same height and rotation angle as the previous text.*

As you type the text, AutoCAD displays the text both in the drawing and on the command line. A text cursor appears on both the command line and in the drawing. Because the TEXT command shows the text in your drawing as you create it, it's easier to identify errors while typing the text. If you see an error on the line that you are typing, you can use the arrow cursor keys to move the text cursor back to the error to correct it on the command line. If the error is on a previous line, you can use the BACKSPACE key to go back and correct the error; however, this removes the letters that you BACKSPACE over. For this reason, it's usually easier to leave the error, finish entering the text, and then edit the text by using one of AutoCAD's text editing commands. You'll learn about these commands later in this chapter.

**Tip** *Although the text appears on the screen as you type, it may not appear in its final position until you complete the command. For example, text always initially appears left-justified, regardless of the justification option that you specify. You can select another point in the drawing at any time during the TEXT command. The text cursor immediately moves to that point in the drawing, and you can continue typing. AutoCAD creates a new text object at that location.*

MOVING BEYOND
THE BASICS

## Entering Special Characters

When you type text, you can enter special control codes to instruct AutoCAD to generate special symbols, such as the degree symbol (°), plus/minus symbol (±), or diameter symbol (Ø). You can also toggle overscore and underline modes on and off. These special characters are controlled by the code sequences shown in Table 13-1.

As you type the text, the control codes initially appear within the drawing as actual text. Once you complete the TEXT command, AutoCAD removes the control codes, replacing them with the actual symbols or formatting. Note that the overscore and underline modes can both be active at the same time. Both modes are turned off automatically when you press ENTER to end the line of text. If you need to continue these modes on a subsequent line, you need to turn them on again

## Setting Line Text Justification

If you've ever used a word processor, you probably know something about text justification. Although most text in a word processor is left-justified, you can also create text that is centered (the midpoint of each line of text aligns vertically), right-justified (the right end of all the lines of text align vertically), or fully justified (each line of text is stretched or compressed slightly so that both the left and right ends of each line of text aligns vertically). You can create similar types of alignments in AutoCAD, but since the TEXT command deals with individual lines of text, there are actually 14 other justification options in addition to the default left-justified option.

When you start the TEXT command, AutoCAD prompts:

```
Current text style:  "current" Text height:  current
Specify start point of text or [Justify/Style]:
```

| Control Code | Meaning |
|---|---|
| %%o | Toggles overscore mode on and off |
| %%u | Toggles underline mode on and off |
| %%d | Inserts a degree symbol (°) |
| %%p | Inserts a plus/minus symbol (±) |
| %%c | Inserts a diameter symbol (Ø) |

**Table 13-1.**   *Control Codes Available when Entering Single-Line Text*

If you type **J** (for Justify) and then press ENTER, AutoCAD prompts you to specify one
of the 14 options:

```
Enter an option [Align/Fit/Center/Middle/Right/TL/TC/TR/ML/MC/MR/BL/BC/BR]:
```

The Align and Fit options are special situations, because they determine not only
the location of the text, but also its rotation angle:

- **Align**    AutoCAD prompts you for two points and then aligns the text between
  these points. The angle of the line between these two points determines the text
  rotation angle. Because the text that you enter is sized to fit perfectly between
  these two points, AutoCAD doesn't prompt you to specify the text height. The
  height/width ratio of each letter remains the same.

- **Fit**    This is a variation on the Align option. AutoCAD prompts you for two points,
  which determine the rotation angle. But, unlike the Align option, AutoCAD still
  prompts you for the text height. AutoCAD then stretches or compresses the
  characters so that the line of text fits between these two points.

Each of the remaining options is described in the following list. Figure 13-1
illustrates these various justification options and indicates where the specification point
would be located in relation to the text. After you specify the justification, AutoCAD
prompts for the text height and rotation angle, and then prompts you to enter the text.

- **Center**    AutoCAD prompts you to specify the center point. The line of text is
  then aligned so that the midpoint of the text's baseline is placed at the point that
  you specify. AutoCAD then prompts for the text height and rotation angle.

- **Middle**    This option is similar to the Center option, but instead of placing the
  midpoint of the baseline at the selected point, AutoCAD centers the line of text
  both horizontally at the selected point, and vertically based on the top- and
  bottommost portions of the text, including the tops of capital letters and the
  descenders of letters such as *g*, *p*, and *y*.

- **Right**    As its name implies, this option is similar to left-justification, except
  that the point you specify determines the right end of the line of text. The length
  of the line of text extends to the left and is based entirely on the length of the
  line of text that you enter.

- **TL/TC/TR**    These three options stand for Top-Left, Top-Center, and Top-Right,
  respectively. AutoCAD aligns the text so that the topmost portion of the text aligns
  vertically with the point that you specify, and the line of text aligns horizontally,
  based on its left, center, or right end, depending on which of the three options
  that you select.

MOVING BEYOND
THE BASICS

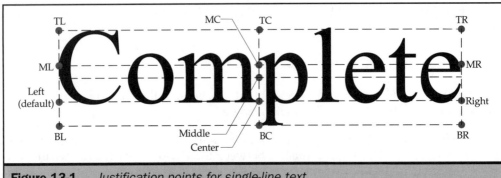

**Figure 13-1.** *Justification points for single-line text*

- **ML/MC/MR**   These three options stand for Middle-Left, Middle-Center, and Middle-Right, respectively, and are identical to the TL, TC, and TR options, except that the text aligns vertically at its midpoint rather than at its topmost point (the line halfway between the baseline and topmost part of the text). The length of the line of text then aligns horizontally, based on its left, center, or right end.

- **BL/BC/BR**   These three options stand for Bottom-Left, Bottom-Center, and Bottom-Right, respectively, and are also identical to the TL, TC, and TR options, except that the text aligns so that the bottommost portion of the line of text, including the descenders, aligns vertically with the point that you specify.

**Note**   *The point used to place the single-line text object is the object's insertion point, which can be selected by using the Insert object snap mode.*

## Setting Line Text Height

When you create line text, AutoCAD normally prompts you to specify the text height. As you learned in Chapter 2, when you specify the text height, you must take into account the scale at which your drawing will eventually be plotted. This is due to the fact that the text height is specified in drawing units, but those drawing units could represent millimeters, inches, miles, or any other unit that you decide to use. In addition, when you plot the drawing at a specific scale—such as 1/8"=1'-0"—you essentially instruct AutoCAD to reduce everything by a specific scale factor. In the case of a drawing plotted at 1/8"=1'-0", the scale factor is 1:96. Thus, anything drawn 1-foot high appears only 1/8-inch high in the printed drawing.

**Note**   *There are two instances in which AutoCAD doesn't ask for the text height: when you use the Align justification option, and when the text style uses a predefined text height. You'll learn about controlling the height as part of the text style definition later in this chapter.*

Table 2-2 in Chapter 2 shows some standard architectural and engineering scale ratios and the equivalent text heights that are required to create text that measures 1/8-inch high when you plot the drawing at the specified scale. Table 13-2 shows similar information as Table 2-2, expressed in a slightly different way. Table 13-2 shows the text height that you specify to produce text at various printed heights for several

| Drawing Scale When Plotted | Plotted Text Height | | | | | |
|---|---|---|---|---|---|---|
| | 3/32" | 1/8" | 3/16" | 1/4" | 3/8" | 1/2" |
| 1/16"=1'-0" | 18" | 24" | 36" | 48" | 72" | 96" |
| 1/8"=1'-0" | 9" | 12" | 18" | 24" | 36" | 48" |
| 3/16"=1'-0" | 6" | 8" | 12" | 16" | 24" | 32" |
| 1/4"=1'-0" | 4.5" | 6" | 9" | 12" | 18" | 24" |
| 3/8"=1'-0" | 3" | 4" | 6" | 8" | 12" | 16" |
| 1/2"=1'-0" | 2.25" | 3" | 45" | 6" | 9" | 12" |
| 3/4"=1'-0" | 1.5" | 2" | 3" | 4" | 6" | 8" |
| 1"=1'-0" | 1.125" | 1.5" | 2.25" | 3" | 4.5" | 6" |
| 1 1/2"=1'-0" | 0.75" | 1" | 1.5" | 2" | 3" | 4" |
| 3"=1'-0" | 0.375" | 0.5" | 0.75" | 1" | 1.5" | 2" |
| 1"=10' | 11.25" | 15" | 22.5" | 30" | 45" | 60" |
| 1"=20' | 22.5" | 30" | 45" | 60" | 90" | 120" |
| 1"=30' | 33.75" | 45" | 67.5" | 90" | 135" | 180" |
| 1"=40' | 45" | 60" | 90" | 120" | 180" | 240" |
| 1"=50' | 56.25" | 75" | 112.5" | 150" | 225" | 300" |
| 1"=60' | 67.5" | 90" | 135" | 180" | 270" | 360" |
| 1"=100' | 112.5" | 150" | 225" | 300" | 450" | 600" |

**Table 13-2.**   *Text Height Values to Use When Plotting at Various Drawing Scales*

often-used drawing scales. To determine the correct line text height, locate the row that corresponds to the printed scale of your drawing, and then select the column that corresponds to the desired printed text height.

You can calculate other text heights by multiplying the plotted text height by the scale factor for the printed scale (obtained from Table 2-2). Thus, to arrive at a plotted text height of 1/2-inch for drawings plotted at a scale of 1/8"=1'-0", the equation is $0.5 \times 96 = 48$.

### LEARN BY EXAMPLE
*Use the TEXT command to add line text to the title block of a drawing. Open the drawing Figure 13-2 on the companion web site.*

To add line text to the title block, as shown in Figure 13-2, use the following command sequence and instructions. Note that when you open the drawing, it is already zoomed in to the title block and the drawing is displayed in a paper space layout. You'll learn more about paper space and model space in Chapter 17.

```
Command: TEXT
Current text style:  "STANDARD" Text height:  0.2000
Specify start point of text or [Justify/Style]: C
Specify center point of text: 18.5,2.1
Specify height <0.2000>: .25
Specify rotation angle of text <0>: ENTER
Enter text: JACK ASSEMBLY
Enter text: ENTER
```

*Notice in the previous example that you can specify the justification point without first typing J to display the justification options. You can specify any of the justification options in this way, thus saving an extra step when you create single-line text.*

Identify your company name and address by adding two more lines of text to the title block, using the following command sequence, but substituting your company name and address, as appropriate:

```
Command: TEXT
Current text style:  "STANDARD" Text height:  0.2500
Specify start point of text or [Justify/Style]: C
Specify center point of text: 18.5,1.8
Specify height <0.2500>: .1875
Specify rotation angle of text <0>: ENTER
Enter text: Your Company Name
Enter text: Anytown, USA
Enter text: ENTER
```

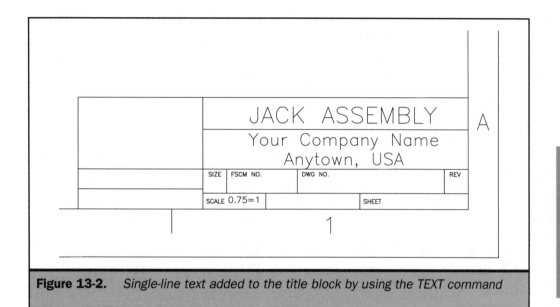

**Figure 13-2.**  *Single-line text added to the title block by using the TEXT command*

Notice that this step adds two separate line text objects. Next, add text to identify the drawing scale:

```
Command: TEXT
Current text style:  "STANDARD" Text height:  0.1875
Specify start point of text or [Justify/Style]: 16.65,.85
Specify height <0.1875>: .09375
Specify rotation angle of text <0>: ENTER
Enter text: 0.75=1
Enter text: ENTER:
```

## Creating Paragraph Text

Paragraph, or *multiline,* text consists of one or more lines of text or paragraphs that fit within a boundary width that you specify. Each multiline text (Mtext) object that you create is treated as a single object, regardless of the number of individual paragraphs or lines of text that it contains.

When you create multiline text, you first determine the paragraph's boundary width by specifying the opposite corners of a rectangle. Unlike line text, you don't need to press ENTER at the end of each line of text. Instead, the text automatically wraps so that it fits within the width of this rectangle. The first corner of the rectangle determines the default attachment point of the multiline text, although you can change the location of the attachment point in reference to the rectangle, and you can determine the direction

## Setting the Text Height Relative to Paper Space

Calculating the text height so that the resulting text later gets plotted at the desired height can often be problematic. This becomes even more true when you use paper space layouts to lay out your drawings exactly as you want them to be printed. You will learn about creating a layout to plot in Chapter 17.

AutoCAD 2002 provides a new command that automatically converts text height and other length values between model space and paper space, thus automating the text height calculation if, for example, you use AutoCAD's paper space feature to create layouts instead of plotting your drawings from model space. Although you can use it when no other command is active, the new SPACETRANS command is intended primarily to be invoked transparently when AutoCAD prompts for a text height or other length. To use the command, do one of the following:

- On the Text toolbar, click Convert Distance Between Spaces.
- At the command line, type '**SPACETRANS** and press ENTER.

For example, if you are currently working in a scaled model space viewport in a paper space layout and you use the TEXT command to create single-line text, when AutoCAD prompts you to specify the text height, you can use the SPACETRANS command transparently to specify the desired text height in paper space. AutoCAD automatically converts the text height you specify into model space units so that the resulting text is the correct height when you later plot your drawing. The following command sequence illustrates a typical use of this new command. In this instance, we wish to create text that will eventually be 1/8-inch high in a viewport whose contents will be plotted at a scale of 1/4"=1'-0".

```
Command: TEXT
Current text style: "Standard" Text height: 0'-0 3/16"
Specify start point of text or [Justify/Style]: (select start point)
Specify height <0'-0 3/16">: 'SPACETRANS
>>Specify paper space distance <1">: 1/8
Resuming DTEXT command.
Specify height <0'-0 3/16">: 6.0000
Specify rotation angle of text <0>:
```

Notice that when you use the SPACETRANS command in response to the text height prompt, AutoCAD prompts again, this time asking for the paper space distance. AutoCAD then converts the text height into the necessary model space dimension, and then continues the TEXT command.

*TIP: When you create paragraph text, you can't use the SPACETRANS command inside the Multiline Text Editor dialog box. Instead, you should set the Height option on the command line prior to displaying this dialog box. You'll learn how to do this later in this chapter.*

in which text flows within the rectangle. These options (described later in this chapter) are similar to the justification options for single-line text. You can also select the text style, text height, and rotation angle of the entire multiline text object. Multiline text provides more formatting options, however, than single-line text. For example, you can change the color, style, and height of individual letters or words within the paragraph.

You can create multiline text by using the MTEXT command. Unlike single-line text, multiline text never appears in your drawing until you complete the command. However, once you specify the paragraph's boundary width, AutoCAD displays a Multiline Text Editor dialog box, similar to Figure 13-3. This dialog box has four tabbed pages. You can edit text as you type it, change formatting characteristics such as text height and color, alter formatting properties such as text style and justification, control the spacing between lines of text, and even search for specified text strings and replace them with new text. You can also import text that is saved to an ASCII file or Rich Text Format (RTF) file, or paste text from the Windows Clipboard. This can save you a great deal of time—for example, you can add notes to your drawing that were previously entered into a word processor as part of a project manual.

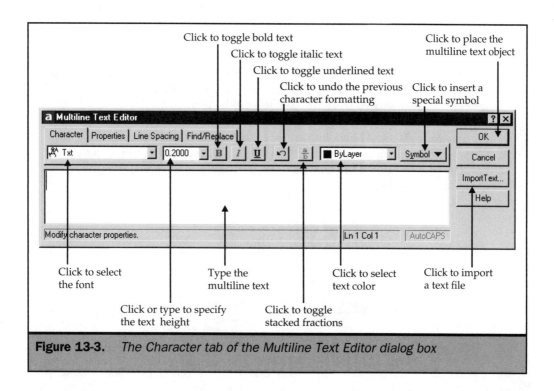

**Figure 13-3.** *The Character tab of the Multiline Text Editor dialog box*

To create multiline text:

1. Do one of the following:

- On the Draw or Text toolbar, click Multiline Text.
- From the Draw menu, choose Text | Multiline Text.
- At the command line, type **MTEXT** (or **MT** or **T**) and then press ENTER.

2. Specify the first corner of the text boundary rectangle.

3. Specify the width of the boundary rectangle by dragging the cursor to the left or right, or type a width value. As you drag the cursor to position the opposite corner of the rectangle, AutoCAD displays the rectangle on the screen, as well as an arrow indicating the direction of text flow. After you specify the boundary width, AutoCAD displays the Multiline Text Editor dialog box.

4. In the Multiline Text Editor dialog box, type the text that you want to appear in your drawing. Text wraps automatically when the cursor reaches the end of the boundary width. To create individual paragraphs within the multiline text object, press ENTER to end a paragraph, and then type the next paragraph.

5. Click OK to add the multiline text to your drawing and complete the command.

As you enter text into the dialog box, you can edit the text. Use your mouse to move the text cursor within the dialog box, or use the arrow keys on your keyboard to position the text cursor within the dialog box. You can also use the standard Windows control keys, shown in Table 13-3, to edit the text.

**Tip** *Pressing the right-mouse button when the cursor is within the Multiline Text Editor displays a shortcut menu that provides text editing options: Undo, Cut, Copy, Paste, Select All, Change Case (UPPERCASE or lowercase), Remove Formatting, and Combine Paragraphs.*

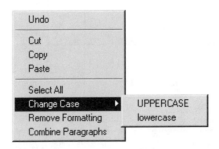

**Note** *You can change the text editor used when editing Mtext objects. By default, AutoCAD uses its internal mtext editor. To specify a different editor, change the value for the Text Editor Application under Text Editor, Dictionary, and Font File Names on the Files tab of the Options dialog box. You can also change the text editor by changing the MTEXTED system variable.*

| Key | Description |
|---|---|
| CTRL-A | Selects all text in the Multiline Text Editor |
| CTRL-B | Toggles bold formatting for selected text |
| CTRL-C | Copies selected text to the Windows Clipboard |
| CTRL-I | Toggles italic formatting for selected text |
| CTRL-SHIFT-L | Converts selected text to lowercase |
| CTRL-U | Toggles underline formatting for selected text |
| CTRL-SHIFT-U | Converts selected text to uppercase |
| CTRL-V | Pastes text from the Windows Clipboard into the text editor window |
| CTRL-X | Cuts selected text to the Windows Clipboard |
| CTRL-SPACEBAR | Removes character formatting for selected text |
| CTRL-SHIFT-SPACEBAR | Inserts a nonbreaking space |
| ENTER | Ends the current paragraph and starts a new line |

**Table 13-3.** *Windows Control Keys Used to Edit Text*

## Setting Options on the Command Line

When you start the MTEXT command, AutoCAD displays the current text style and text height, and then, as you just learned, prompts you to specify the corners of the boundary rectangle. The command displays several additional options, as well. The prompt appears on the command line as follows:

```
Current text style:  "STANDARD"  Text height:  0.2000
Specify first corner:
Specify opposite corner or [Height/Justify/Rotation/Style/Width]:
```

You can specify any of these options on the command line, before AutoCAD displays the Multiline Text Editor dialog box. To specify an option, type the first letter of the option and then press ENTER. The Width option is the only exception. You can specify the width without first selecting that option. After you specify the width, however, AutoCAD immediately displays the Multiline Text Editor dialog box. The following list explains the various available options:

■ **Height**   Lets you specify a new text height.

- **Justify**    Lets you specify the text justification and text flow in relation to the boundary rectangle. AutoCAD prompts you to enter the justification, and displays nine options: TL, TC, TR, ML, MC, MR, BL, BC, and BR. These options are similar to those for justifying line text, except that these options apply to the entire multiline text object. Text is left-, center-, or right-justified in relation to the width of the boundary rectangle, and is top-, middle-, or bottom-justified with respect to the total height of the resulting multiline text object. These relationships are shown in Figure 13-4. The dashed rectangle shows the original boundary; the dot shows the resulting text insertion point.

- **Rotation**    Lets you specify the rotation angle of the entire text boundary.

- **Style**    Lets you specify the text style to use for the multiline text.

- **Width**    Lets you specify the width of the text boundary by typing or picking a point in the drawing.

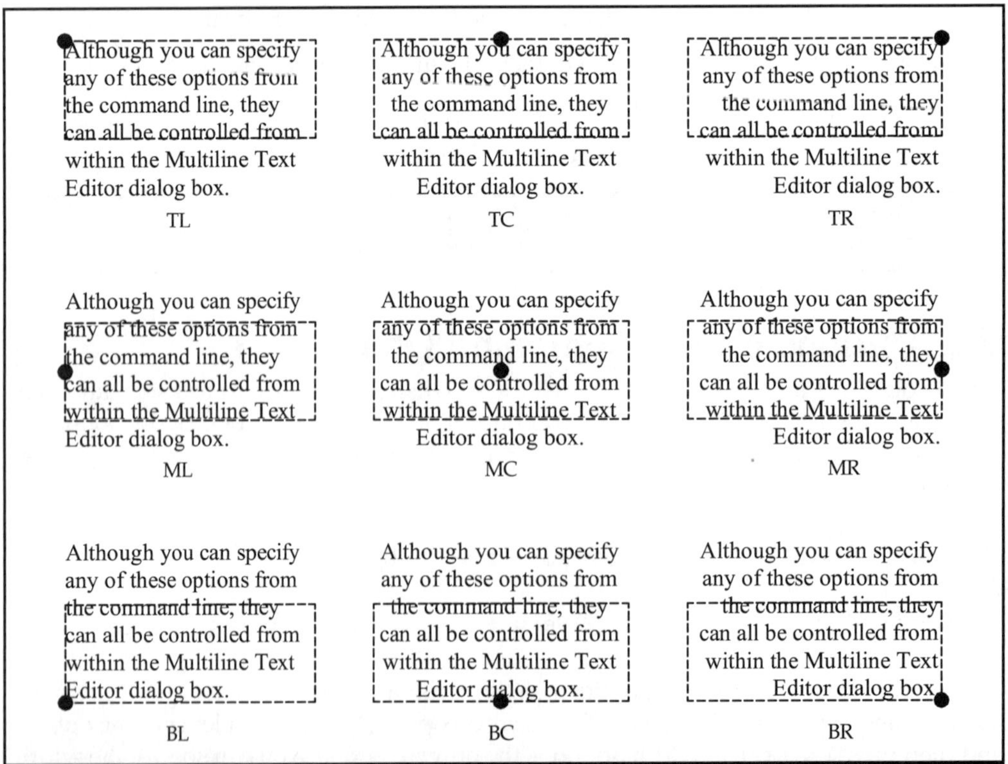

**Figure 13-4.**    *Multiline text offers nine justification options.*

Although you can specify any of these options from the command line, they can all be controlled from within the Multiline Text Editor dialog box. With the exception of setting the width of the boundary rectangle, it is often easier simply to accept the default values when you start the MTEXT command, and then make all the necessary changes from within the dialog box.

# Formatting Multiline Text

The various controls within the Multiline Text Editor dialog box make formatting text on-the-fly easy. The text appearing in the dialog box automatically assumes the current text style, but you can override these settings for selected characters, words, paragraphs, or the entire multiline text object. Characteristics such as the font, text height, underlining, and color can be applied to selected characters. Properties such as text style, justification, width, and rotation affect the entire multiline text object. These formatting options are controlled from the Character and Properties tabs, respectively.

## Formatting Individual Characters

As its name implies, the controls on the Character tab affect individual text characters. The formatting changes affect only the characters that you select; the current text style is not changed. Although you can change character formatting as you type, the easier way to change character formatting usually is to type the text first, select the characters or words that you want to change, and then select the formatting.

*To select one or more characters, click-and-drag the cursor over the characters. You can double-click to select a single word, or triple-click to select an entire paragraph.*

**Changing the Font**    The drop-down list for fonts (refer to Figure 13-3) shows all the TrueType fonts loaded on your system, arranged alphabetically by family name. AutoCAD compiled fonts are listed by the name of the file in which they are stored. You'll learn about AutoCAD fonts later in this chapter.

*Don't copy TrueType fonts (TTF fonts) into the AutoCAD Fonts folder. To be available in AutoCAD, TrueType fonts must be installed by the operating system so that they are listed in the system registry. Use the following steps to install TrueType fonts:*

*1. On the taskbar, select Start | Settings | Control Panel.*

*2. In the Control Panel, double-click Fonts.*

*3. In the Fonts dialog box, select File | Install New Font.*

*4. In the Add Fonts dialog box, locate the drive and folder containing the fonts you want to install, select them, and then click OK.*

MOVING BEYOND THE BASICS

*Alternately, you can use Windows Explorer to drag the appropriate files into the Windows/Fonts folder. If you move TrueType fonts from one location to another, you will have to repeat this procedure. You should never locate TrueType fonts in a drawing directory. Doing so could cause AutoCAD 2002 to crash.*

**Changing the Font Height**   The drop-down list for font height (refer to Figure 13-3) controls the character height in drawing units, which is the same unit of measurement that is used to specify the height of single-line text. To specify a height other than those in the list, simply type a new height value.

**Toggling Bold and Italic**   The Bold and Italic buttons on the Character tab toggle the bold and italic formatting of the selected text. These buttons are available only when the text characters that you select are drawn using a TrueType font that supports bold and italics.

**Toggling Underline**   The Underline button toggles underlining of the selected text and can be applied to both TrueType and AutoCAD compiled fonts.

**Creating Stacked Text**   The Stack button toggles the stacking of selected text, to create a diagonal fraction or a stacked fraction or tolerance values. To be stacked, the selected text must contain a caret (^), a forward slash (/), or a pound symbol (#). Text to the left of the symbol is stacked over the text to the right. Text containing a caret represents a tolerance value, and the caret is omitted when the text is stacked. Text containing a forward slash or pound symbol represents a fraction. A forward slash creates a vertical stack, and the forward slash is converted to a horizontal bar when the text is stacked. A pound sign creates a diagonal stack, and the pound sign is converted to a diagonal bar when the text is stacked. Samples of stacked text are shown in Figure 13-5. For example, when stacked, the text string 1+0.05^-0.02 appears similar to the example on the left. The text string 1'-0 3/16" appears similar to the center example, while the string 1'-0 3#16" appears similar to the example on the right.

You can also use the AutoStack feature to automatically stack fractions as you enter multiline text. When you use the MTEXT command and enter text consisting of numeric characters immediately before and after a slash, pound, or caret character, and then enter a nonnumeric character or press the SPACEBAR, AutoCAD displays the AutoStack Properties dialog box. You can use the controls in this dialog box to enable AutoStack, and

$$1 \; {}^{+0.05}_{-0.02} \qquad 1'\text{-}0\tfrac{3}{16}" \qquad 1'\text{-}0\;{}^{3}\!/_{16}"$$

**Figure 13-5.**   *Examples of stacked text*

then control how the AutoStack feature treats expressions containing these symbols. You can convert a slash to either a diagonal or horizontal fraction. The pound sign is always converted to a diagonal fraction, and the caret is always converted to a tolerance format.

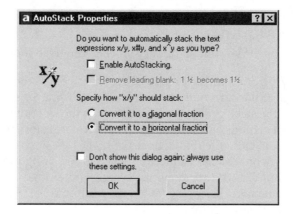

After enabling AutoStack, you can also instruct AutoCAD to remove blank spaces between a whole number and a fraction. After setting your AutoStack preferences, you can select the check box at the bottom of the AutoStack Properties dialog box to instruct AutoCAD to not display the AutoStack Properties dialog box in the future. If you later want to display this dialog box again, select any stacked text in the Multiline Text Editor (either while creating new text or editing some existing text), right-click and select Properties from the shortcut menu, and then click the AutoStack button in the Stack Properties dialog box.

When you display the Stack Properties dialog box, you can also change the upper and lower text independently, as well as change the appearance of the selected stacked text. The Style setting lets you change the stacked text to a Tolerance, Fraction (Horizontal), or Fraction (Diagonal). The Position setting controls the alignment of the stacked text.

MOVING BEYOND
THE BASICS

The actual alignment does not display in the Multiline Text Editor dialog box. Once you close the dialog box, however, AutoCAD adjusts the stacked text so that it uses the alignment you specified:

- ■ **Top**    Aligns the top of the stacked text with the top of the line of text.
- ■ **Center**    Centers the stacked text vertically with the center of the line of text. This is the default setting.
- ■ **Bottom**    Aligns the bottom of the stacked text with the bottom of the line of text.

The Text Size setting determines the size of the characters in the stacked text as a percentage of the size of the current text style. The default size for stacked text is 70%.

Once you change the stack properties, you can click the Defaults button and then either save the current settings as the new default values or restore the previous default values. Then, click OK to close the dialog boxes.

Note that AutoStack only stacks numeric characters occurring immediately before and after the slash, pound, and caret symbols. To stack nonnumeric characters or text that includes spaces, you must select the text in the Multiline Text Editor and then click the Stack button on the Character tab.

*Versions of AutoCAD prior to AutoCAD 2000 don't support diagonal stacked fractions. If you save your drawing in an earlier AutoCAD format, diagonal stacked fractions are converted to vertical fractions.*

**Changing Text Color**    Although text normally assumes the color of the layer on which it is created, you can override this setting by assigning colors to selected characters. To do this, select the text and then choose the color from the drop-down list (refer to Figure 13-3). If you choose Other, AutoCAD displays the standard Select Color dialog box that shows all 256 available colors.

**Inserting Special Characters**    From the Symbol drop-down list, you can insert special symbols for degrees, plus/minus, and diameter, or insert a nonbreaking space. If you choose Other, AutoCAD displays a standard Character Map dialog box that shows the entire character set for the current font, so that you can insert other special characters.

*You can also insert symbols by typing %%d (for degrees), %%p (for plus/minus), and %%c (for diameter), just as you can when entering single-line text. As previously noted, you can also add a nonbreaking space by pressing CTRL-SHIFT-SPACEBAR.*

**Reversing Changes**    Clicking the Undo button reverses the previous text-editing or character-formatting action. You can also press CTRL-Z to undo the most recent changes.

## Changing Multiline Text Properties

The Properties tab lets you change the properties affecting the entire multiline text object. This tab, shown in Figure 13-6, presents a different set of controls. You can use the controls on this tab to change the current text style, justification (refer to Figure 13-4), width, and rotation of the multiline text object.

*When you change the style assigned to multiline text, some of the character formatting that you applied may be lost. Any font, height, bold, or italic formatting previously applied is discarded. Stacking, underlining, and color formatting is retained. Styles having backward or upside-down effects aren't applied. Styles based on an AutoCAD compiled font that is assigned a vertical setting is displayed horizontally in the dialog box, but appears vertically in the drawing.*

## Changing Line Spacing

The Line Spacing tab, shown in Figure 13-7, lets you control the line spacing for new or existing multiline text objects. Like the text properties, the line spacing applies to the entire multiline text object, not to selected lines. The line spacing determines the distance between the baseline of one line of text and the baseline of the next line of text.

Two controls determine the line spacing. In the first drop-down list, select either At Least or Exactly. When you select At Least, AutoCAD uses the spacing factor and then automatically adjusts the line spacing to accommodate the height of the largest character in the line of text. This is the default setting, and it can result in lines of text that are not equally spaced. When you select Exact, the spacing between each line of

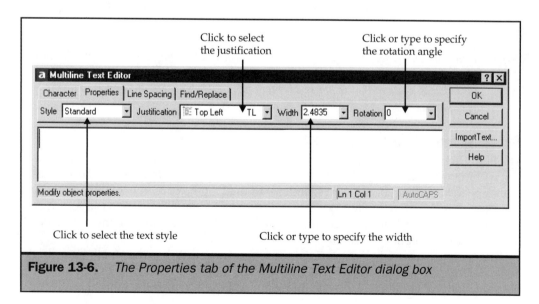

**Figure 13-6.**    *The Properties tab of the Multiline Text Editor dialog box*

Click or type to set line spacing

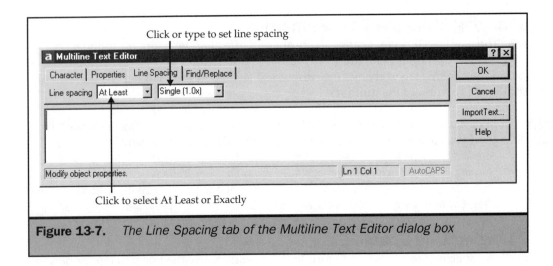

Click to select At Least or Exactly

**Figure 13-7.**   *The Line Spacing tab of the Multiline Text Editor dialog box*

text is the same regardless of any differences in text height caused by formatting that may be applied to individual characters.

The second control determines the actual spacing factor. AutoCAD provides three predefined options: Single (1.0x), 1.5 Lines (1.5x), and Double (2.0x). Single spacing is 1.66 times the height of the text characters. You can either select one of the options from the list, type a number followed by **x** to indicate a multiple of the standard line spacing, or type an absolute line spacing value (expressed in drawing units). For example, to specify a line spacing of exactly .75 units, select Exactly and then type **.75** in the spacing factor box.

 *While using an exact line spacing ensures that the line spacing is identical for all lines in a multiline text object, and can be quite useful when creating tables, if you reformat selected characters so that they use a larger font, those characters may overlap text on the adjacent lines.*

## Finding and Replacing Text

The Find/Replace tab, shown in Figure 13-8, provides controls to search for specified text within the Multiline Text Editor dialog box and replace it with new text. To use this tool, type the text that you want to search for in the Find box and the text that you want to replace it with in the Replace With box. Click the Find button to begin the search and then click the Replace button to replace the highlighted (found) text with the replacement text. Click the Find button again to continue the search.

The two check boxes help control the search parameters. When the Match Case check box is selected, AutoCAD finds text only if the case of all the text characters in the text object is identical to the case of the text in the Find box. If you select the Whole Word

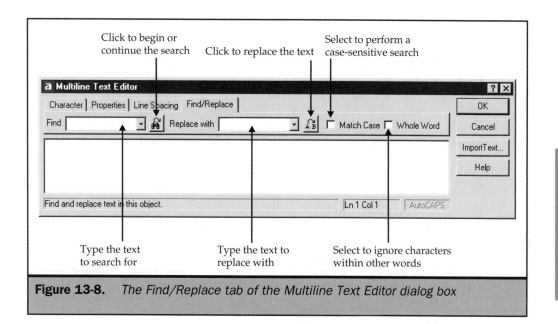

**Figure 13-8.** *The Find/Replace tab of the Multiline Text Editor dialog box*

check box, AutoCAD matches the text in the Find box only if it is a single word. If the text is part of another text string, it is ignored.

*The controls in the Find/Replace tab search for text only within the Multiline Text Editor dialog box. The FIND command finds and replaces text anywhere in the drawing. You'll learn about the FIND command later in this chapter.*

**LEARN BY EXAMPLE**
*Use the MTEXT command to add notes in the form of multiline text. Open the drawing Figure 13-9 on the companion web site.*

To add the notes as shown in Figure 13-9, use the following instructions:

1. Start the MTEXT command.
2. When prompted to specify the first corner, specify the coordinate 4,4 either by typing or pointing in the drawing.
3. When prompted to specify the opposite corner, specify the coordinate 8,3 either by typing or pointing in the drawing.
4. In the Multiline Text Editor dialog box, type **Notes:** and then press ENTER twice.

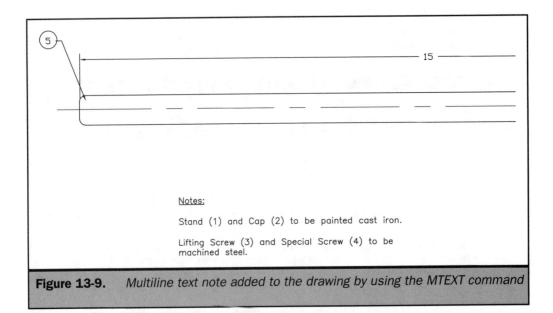

**Figure 13-9.** *Multiline text note added to the drawing by using the MTEXT command*

5. Type **Stand (1) and Cap (2) to be painted cast iron.** and then press ENTER twice.

6. Type **Lifting Screw (3) and Special Screw (4) to be machined steel.**

7. Click-and-drag the cursor to select the word Notes: and then click the Underline button.

8. Click OK.

Figure 13-9 shows a close-up of the drawing with the notes added.

*Displaying lots of text can really slow down AutoCAD's performance. To speed things up, you can enable Quick Text mode. You can control Quick Text mode from the Display tab of the Options dialog box or by changing the QTEXT system variable. When this mode is turned on and you regenerate the drawing, all text already placed in the drawing appears as rectangles that approximate the position and extents of each text object. Be forewarned, however, that when Quick Text mode is turned on, AutoCAD prints the rectangles instead of the actual text. Be sure to turn off Quick Text and regenerate the drawing, before you print it.*

# Inserting Text from Outside of AutoCAD

You can insert into your drawings any ASCII or RTF text files that were created using other word processors. AutoCAD provides three different methods for inserting such text: cut-and-paste, drag-and-drop, and importation into the Multiline Text Editor. All three methods produce multiline text objects. The Express Tools also include a powerful tool that lets you create a custom Rtext object linked to an external text file. This and the other Express text tools are described briefly at the end of the chapter. You can learn more about the Express Tools on the companion web site.

 *Text files that are inserted by using any of these methods are limited to a maximum file size of 16K.*

## Importing Text

You can import text into the Multiline Text Editor dialog box by clicking the Import button. AutoCAD displays a standard Open File dialog box. You can then choose any text file in ASCII or RTF format, and click Open.

Imported text retains its original character formatting and style properties, except that tabs in RTF files are converted to blank spaces and line spacing reverts to single-line. The imported text is inserted at the current cursor position in the dialog box. If any text in the dialog box was selected (highlighted) when you clicked the Import button, AutoCAD displays an Inserting Text dialog box.

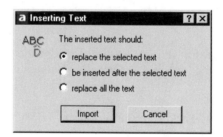

You can then choose to replace the selected text, insert the new text immediately after the selected text, or replace all the text currently displayed in the Multiline Text Editor dialog box with the new text.

## Dragging-and-Dropping Text

You can also use Windows' drag-and-drop capabilities to insert a text file into a drawing. If you drag a file with a .TXT file extension, AutoCAD inserts the text as a multiline text object, assigning it the current text style and text height. If you drag a text file with any other file extension, AutoCAD treats it as an OLE object. You already learned a bit about pasting OLE objects in Chapter 7. You'll learn more about OLE objects in Chapter 24.

 *Before you drag text into the drawing, position the Windows Explorer window and the AutoCAD window so that you can see both the file icon and the drawing in which you want to insert the file.*

The object's insertion point corresponds to the point where you drop it. The width of the resulting text object is based initially on the line breaks and carriage returns in the original file. After you insert a multiline text object into the drawing, you can use grips to move, rotate, or scale the object, or edit it to adjust its width and apply other formatting. You'll learn about editing text later in this chapter.

*If the text is inserted as an OLE object, you can use grips to move the object, but using grips to resize the object may cause the text to become unreadable. You can use UNDO to reverse any changes you have made. To edit the actual text, double-click the OLE object to open the text file in the program originally used to create it.*

## Cutting-and-Pasting Text

If the Windows Clipboard contains text, you can paste the text into the drawing by using the PASTECLIP command. To use this command, do one of the following:

- On the Standard toolbar, click Paste From Clipboard.
- From the Edit menu, choose Paste.
- At the command line, type **PASTECLIP** and then press ENTER.
- Press CTRL-V.

AutoCAD inserts the text in the upper-left corner of the graphics screen. After you insert the multiline text object into the drawing, you can use grips or text editing commands to modify it.

*You can also paste ASCII or RTF text directly into the Multiline Text Editor dialog box.*

# Working with Text Styles

As you've already learned, whenever you create text objects, either single-line text or multiline text, AutoCAD creates that text by using the current text style. The text style determines characteristics of the text, such as the font, size, angle, and its orientation.

When you create text, you can use any text style that has already been defined in the drawing. When you start the TEXT command, you can use the Style option to specify the desired text style. You can also use the Style option of the MTEXT command to specify the text style that you want to use, or you can change the text style later by using the Properties tab of the Multiline Text Editor dialog box.

Every drawing has a default text style, named Standard, which initially uses the TXT.SHX AutoCAD font. You can't delete or rename the Standard style, but you can modify its definition—for example, you can assign it a different font or change its angle and orientation. If you change the font or orientation of an existing text style, all existing text objects created using that style are automatically updated to reflect the new font or orientation. Changing any other characteristic has no effect on existing text. You can also create and use an unlimited number of additional text styles. A text style defines the characteristics of the text, as shown in Table 13-4.

Text styles can be defined and modified by using the Text Style dialog box, shown in Figure 13-10. This dialog box also lets you rename styles, other than Standard, and delete any styles that haven't been used yet in the drawing. To display this dialog box, do one of the following:

- On the Text toolbar, click Text Style.
- From the Format menu, choose Text Style.
- At the command line, type **STYLE** (or **ST**) and then press ENTER

| Characteristic | Default | Description |
|---|---|---|
| Style name | Standard | The name of the style, up to 31 characters. |
| Font file | TXT.SHX | The font file on which the style is based. |
| Font style | N/A | Font formatting applied to some TrueType fonts. |
| Big font file | None | A special text-shape definition file, used for non-ASCII character sets, such as Kanji. |
| Height | 0 | The character height. A value of 0 lets you control the text height at the time of text creation. |
| Width factor | 1 | The horizontal expansion or compression of the text. Values less than 1 compress the text; values greater than 1 expand the text. |
| Oblique angle | 0 | The slant of the text, in degrees. Negative values slant the text to the left; positive values slant it to the right. |
| Upside down | No | Determines whether text appears right-side up or upside down. |
| Backwards | No | Determines whether text appears backwards. |
| Vertical | No | Determines whether text has a horizontal or vertical orientation. |

**Table 13-4.**   *Text Style Characteristics*

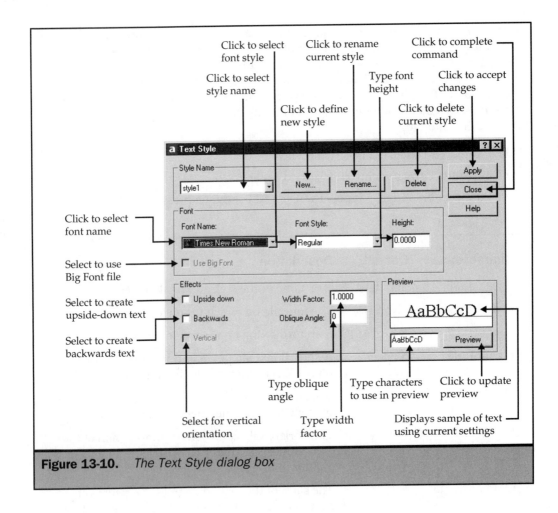

**Figure 13-10.** *The Text Style dialog box*

To define a new text style, click the New button. AutoCAD displays a New Text Style dialog box and a default style name.

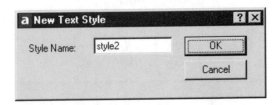

You can either accept the default name or enter a new style name. Click OK to create the new style.

When you initially create a new style, AutoCAD assigns it the settings and characteristics of the current style. You can then modify these settings as desired.

**Note** *You can use the PURGE command to remove unused text styles from a drawing.*

## Choosing the Font and Font Style

The first thing that you should decide when creating a new text style is the font file on which to base the style. Font files contain information about the shape and outline of the individual text characters. AutoCAD can use two different types of font files: TrueType fonts and AutoCAD compiled fonts. You are probably already familiar with TrueType fonts, because they are the standard fonts used by most Windows programs. Windows comes with many TrueType fonts, and you may have obtained others when you loaded other applications onto your computer. AutoCAD also comes with a collection of TrueType fonts, as well as 30 compiled AutoCAD fonts. TrueType fonts allow you to adjust style features, such as bold and italic, but take longer to display and plot. AutoCAD's compiled fonts (which have the file extension .SHX) are more efficient font files—often displaying and plotting much faster—but are more limited in appearance and don't offer bold or italic style control.

**Note** *AutoCAD Release 13 and earlier versions were also able to use PostScript fonts, and those versions of AutoCAD came with several PostScript fonts. Beginning with Release 14, AutoCAD no longer directly supported PostScript fonts; however, you can compile PostScript fonts into AutoCAD fonts for use with AutoCAD 2002 by using the COMPILE command. PostScript fonts used in drawings that were created in earlier versions of AutoCAD are automatically mapped to equivalent TrueType fonts. You'll learn more about font mapping and substitution later in this chapter.*

The Font Name drop-down list in the Text Style dialog box contains the names of all the AutoCAD compiled fonts, as well as all the TrueType fonts loaded on your system. The name of the current font is shown as the default. To change the font used by the current style, choose another font name from the list.

If you select a TrueType font, the Font Style drop-down list becomes active. If the selected font includes different styles—such as bold and italic—you can select them from this list. If the current font is an AutoCAD font, this list is unavailable.

**Note** *To accommodate foreign languages that use many more text characters than English, AutoCAD supports the Unicode character-encoding standard, which allows up to 65,535 characters. In addition, to support alphabets such as Kanji, AutoCAD fonts can use a special type of definition called Big Fonts. To set an SHX font to use a Big Font file, select the Use Big Font check box. The Font Style drop-down list box in the Text Style dialog box changes to Big Font, and the filename BIGFONT.SHX appears in the box.*

## TrueType Fonts and AutoCAD

Filled TrueType fonts always appear filled onscreen, but you can use the TEXTFILL system variable to determine whether the fonts are plotted as filled or as outlines when you print your drawing. The HIDE command also causes these fonts to appear as outlined fonts in the drawing, and if you remove hidden lines during plotting, the fonts appear as outlined fonts in printed drawings. In addition, you can adjust the smoothness of TrueType text characters at plot time by changing the Text Resolution value. This value is stored in the TEXTQLTY system variable.

If you change the effects applied to a TrueType font when defining a text style, the font may appear slightly different onscreen, due to the way AutoCAD displays TrueType fonts in the drawing. For example, the font may appear slightly bolder. This difference only affects the display, however; text created using that style will plot correctly.

In addition to affecting the way that AutoCAD plots text, AutoCAD uses TrueType fonts to display the text in the Multiline Text Editor dialog box. AutoCAD comes with TrueType font equivalents for all the AutoCAD compiled fonts supplied with the program. If you specify an AutoCAD font that you created or obtained from a third party, the font's appearance in the dialog box may be substantially different from its appearance in the drawing.

## Setting the Text Height

As you already learned, the text height controls the height of letters, measured in drawing units. If you accept the default text height of 0, you can control the text height at the time that you actually create the text. If you specify a text height other than 0, AutoCAD does not prompt you for the text height when you create the text. You can change the height of fixed-height text, however, after it has been inserted in the drawing.

 *Changing the height of an existing text style does not affect the appearance of existing text objects already created using that style.*

## Controlling Text Effects

The Effects area of the Text Style dialog box provides control of text characteristics, such as its width factor, obliquing angle, and the way that the text is displayed. The Upside Down, Backwards, and Vertical check boxes toggle these settings on and off. For example, if you select the Backwards check box, the text created with the current style appears backwards in the drawing. This might be useful if you need to plot onto the back side of transparent paper, so that the text is right-reading when viewed from the front. Figure 13-11 illustrates how these settings affect the appearance of text.

The Upside Down and Backwards selections are available for all TrueType and AutoCAD fonts. TrueType fonts can't be oriented vertically, however, and this selection is not available for some AutoCAD fonts. Changing the Upside Down or Backwards

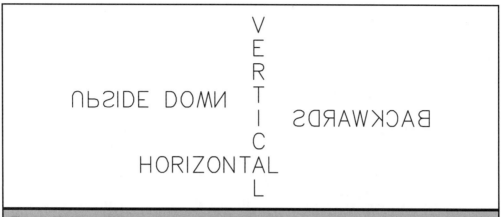

**Figure 13-11.** *Effects on text of the Backwards, Upside Down, and Vertical settings*

MOVING BEYOND
THE BASICS

setting of an existing style does not affect text objects already created using that style. Changing the Vertical setting, however, does cause any existing text to be updated.

> **Tip** *You don't need to create a special text style to draw text that is both upside down and backwards. You can simply rotate standard text 180 degrees when you create it.*

The Width Factor controls the horizontal scaling of the text characters. A value of 1 uses the width of the font as originally specified in the font file. A factor of 0.75 causes the text to be compressed 25 percent, whereas a factor of 1.50 results in the text being stretched 50 percent. Figure 13-12 illustrates how the width factor affects the appearance of text.

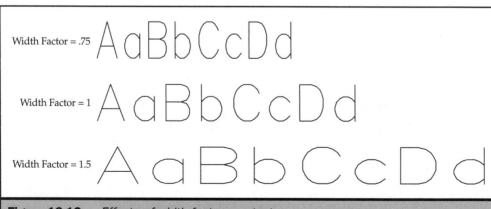

**Figure 13-12.** *Effects of width factors on text*

The Oblique Angle controls the angle or slant of the text characters. As shown in Figure 13-13, a positive value leans the letters the specified number of degrees to the right; a negative value makes them lean to the left.

**Tip**  *You can also use AutoCAD DesignCenter to copy text styles from another drawing. You'll learn about AutoCAD DesignCenter in Chapter 16.*

**Tip**  *Simply copying a missing SHX font to the Font folder, closing the drawing, and then reopening the drawing doesn't update text objects to the correct font. Exiting and then restarting AutoCAD 2002 does correct the problem. If you've opened multiple drawings, however, you may not want to exit from AutoCAD. Instead, use the following steps to resolve the problem:*

1. *Copy the SHX font file to the appropriate folder (such as the Fonts folder or another folder listed in AutoCAD's Support Search Path on the File tab of the Options dialog box).*

2. *At the command prompt, type **FONTALT** and press ENTER.*

3. *When prompted for the current alternate font, specify a different alternate font, ortype a period (.) and press ENTER (for none).*

4. *Regenerate the drawing. (Hint: Type **REGEN** and press ENTER.)*

*The text immediately updates to the correct font. Be sure to set the FONTALT system variable back to its original value. Note that this problem does not exist for TrueType fonts.*

**Figure 13-13.**    *Effects of oblique angles on text*

## Mapping and Substituting Fonts

Fonts, like linetypes and hatch patterns, are defined in files outside of your drawing. Unlike linetypes, however, once fonts are included in a drawing, they are still not stored within the drawing file. If you open a drawing file that references a font file that can't be located, AutoCAD might display a dialog box that prompts you to specify an alternative font file.

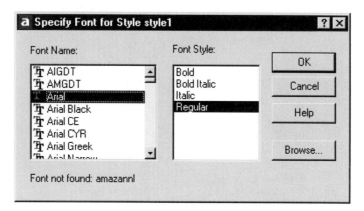

One way this can occur is if the drawing uses a nonstandard TrueType font or a custom AutoCAD font that is not loaded on your system. This often happens when you exchange drawings with consultants. Although you'll learn about a command in Chapter 26 that helps avoid this situation, AutoCAD does provide a way to substitute fonts automatically.

You can preassign an alternative font to be used in such situations by specifying the FONTALT system variable. This setting can also be controlled with the Alternate Font File setting under the Text Editor, Dictionary, and Font File Names selection on the Files tab of the Options dialog box. The default alternate font file is SIMPLEX.SHX.

Another way to specify alternative fonts is by using *font mapping,* which is particularly useful when you need to specify several alternative fonts, matching each with different font filenames. This is how AutoCAD replaces PostScript fonts used in pre-Release 14 drawings with equivalent TrueType fonts.

The FONTMAP system variable specifies the name of a font-mapping table file to be used. You can also control this setting from the Files tab of the Options dialog box, by changing the Font Mapping File setting under the Text Editor, Dictionary, and Font File Names selection. A font mapping file (which has the file extension .FMP) is a simple ASCII file. Each line of the file lists the name of the font being substituted for (without a file extension or path name), a semicolon (;), and the name of the substitute font. The substitute font name needs to include its file extension, such as .TTF for TrueType fonts. The following listing shows the default mapping file, ACAD.FMP, which contains substitutions for the PostScript fonts that are supplied with AutoCAD Release 13. Note that when the FONTMAP system variable is set, the fonts specified in the fontmap file are substituted, regardless of whether the original fonts can be located.

```
cibt;CITYB___.TTF
cobt;COUNB___.TTF
eur;EURR____.TTF
euro;EURRO___.TTF
par;PANROMAN.TTF
rom;ROMANTIC.TTF
romb;ROMAB___.TTF
romi;ROMAI___.TTF
sas;SANSS___.TTF
sasb;SANSSB__.TTF
sasbo;SANSSBO_.TTF
saso;SANSSO__.TTF
suf;SUPEF___.TTF
te;TECHNIC_.TTF
teb;TECHB___.TTF
tel;TECHL___.TTF
```

If AutoCAD can't find a substitute font in the fontmap file, it then tries to substitute the file specified by the FONTALT value. If it still can't find a valid substitution, it either substitutes a similar font (in the case of TrueType fonts) or prompts you for a new font.

# Editing and Changing Text

Text behaves similarly to other AutoCAD objects, and you can use most object-modification commands to copy, move, mirror, scale, and rotate text. Single-line and multiline text behave similarly when modified using most commands, but there are a few exceptions. For example, you can't explode single-line text by using the EXPLODE command, but you can explode multiline text, which reduces it to individual, single-line text objects. And although you can't stretch either type of text by using the STRETCH command, you can change the width of multiline text by using grips to stretch its boundary rectangle. You can also use grips to copy, mirror, rotate, and scale both single-line and multiline text.

*Although you can use grips to move multiline text, you must first switch to the Move mode. Otherwise, selecting a corner grip resizes the boundary box. A much easier way to move multiline text is to select the multiline text object and then click the object and hold down the pick button on your pointing device. The cursor changes to an arrow and a small box, and the multiline text boundary box becomes highlighted. You can then drag the multiline text to a new location, and then release the pick button on your pointing device.*

To edit the actual text, however, AutoCAD provides a specific text editing command—DDEDIT. In addition, the Property window (which you already learned about in Chapter 11) can be used both to change text properties and to edit the text's content.

The easiest way to start the DDEDIT command is to simply double-click the text object you want to edit. AutoCAD immediately displays a dialog box in which you can edit the text. You can also start the command by doing one of the following:

- On the Text toolbar, click Edit Text
- From the Modify menu, choose Object | Text | Edit.
- At the command line, type **DDEDIT** (or **ED**) and then press ENTER.

When you start the command in this way, AutoCAD first prompts you to select an annotation object. The dialog box displayed by the DDEDIT command varies depending on the type of text object that you select.

*Another type of text object that may appear in drawings is an attribute. Attributes are text entries saved as part of a block definition and can be exported to other programs, such as databases. If you select an attribute object, AutoCAD displays yet a third variation on the text editing dialog box. You'll learn about attributes in Chapter 15.*

## Editing Single-Line Text

If you select a single-line text object, AutoCAD displays an Edit Text dialog box, similar to Figure 13-14. Initially, the entire line of text is selected in the Text box and will be replaced by whatever you type. To edit the text that is already in the Text box, use the mouse or the arrow keys to position the text cursor within the edit field, or click-and-drag to select specific portions of the text. Then, edit the text as you would within any edit box.

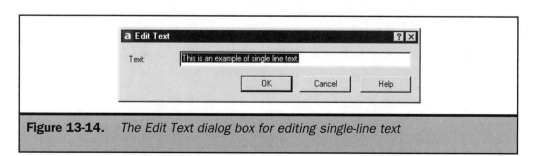

**Figure 13-14.** *The Edit Text dialog box for editing single-line text*

After you finish editing the text, click OK. AutoCAD updates the text in the drawing and then prompts you to select another annotation object. To complete the command, press ENTER.

## Editing Multiline Text

If you select a multiline text object, AutoCAD displays the same Multiline Text Editor dialog box that you used when you created the text (refer to Figure 13-3). You can then edit the text and change any of the character or property formatting. The controls on the Find/Replace tab (refer to Figure 13-8) prove particularly helpful at this point.

## Modifying Text and Text Properties

The Property window may prove to be more useful than the DDEDIT command, because you can also change other text properties, in addition to the text string itself. If you select a single-line text object, the Properties window appears similar to that shown on the left in Figure 13-15. In addition to editing the text in the Text box, you can change its color, layer, thickness, insertion point, height, rotation angle, width factor, obliquing angle, justification, style, and special effects, such as upside-down and backwards orientation.

If you select a multiline text object, the Properties window appears similar to the one on the right in Figure 13-15. Again, you can change the color, layer, insertion point, style, justification, direction, width, text height, and rotation angle all from within this dialog box. Although you can also edit the actual text, it's generally easier (except for minor changes) to click the adjacent button (which becomes visible when you select the contents field) and then edit the text using the same Multiline Text Editor dialog box that you used to create the text. When you view the text in the Properties window, the formatting codes are also visible in the Contents field.

 *Although you can also change the linetype and linetype scale, these settings have no visual effect on text objects.*

## Scaling Text

Although you could use the SCALE command to scale text, in order to ensure that each text object retains its original location, you would need to select each object individually. Doing so could be a very tedious process. AutoCAD 2002 adds a new command that changes the scale of one or more text objects (including single-line text, multiline text, and attributes) in a single action. The SCALETEXT command lets you specify a relative scale factor or an absolute text height, or you can scale selected text to match the height of existing text. Each text object is scaled using the same scale factor, and each maintains its current location.

To start the SCALETEXT command, do one of the following:

- On the Text toolbar, click Scale Text.
- From the Modify menu, choose Object | Text | Scale.
- At the command line, type **SCALETEXT** and then press ENTER.

MOVING BEYOND
THE BASICS

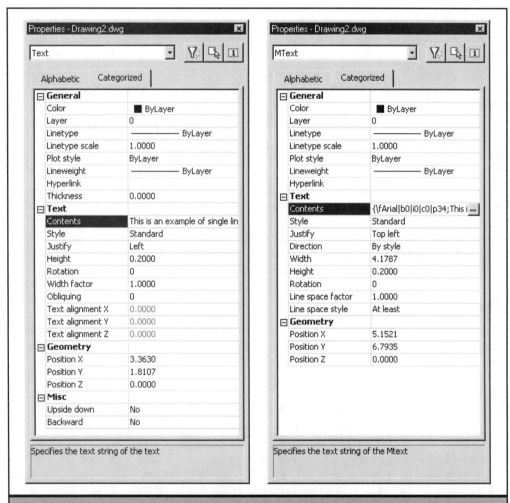

**Figure 13-15.** *The Properties dialog box, for modifying the properties of single-line text (left) and multiline text (right)*

When you start the command, AutoCAD prompts you to select objects. You can use any selection method you wish to select one or more text objects to be scaled. Non-text objects are ignored. When you finish selecting objects, AutoCAD prompts:

```
Enter a base point option for scaling
[Existing/Left/Center/Middle/Right/TL/TC/TR/ML/MC/MR/BL/BC/BR]
<current>:
```

Specify the location to serve as the base point for scaling the text by selecting one of the 14 options. The base point option you specify is applied individually to each selected text object, and is used only to establish the location for scaling the text objects. For example, if you choose the Middle option, each text object will be scaled individually based on the middle of each text object. The justification of the text is not changed. To scale each text object based on its original insertion point, choose the Existing option.

Once you choose the base point option, AutoCAD prompts:

```
Specify new height or [Match object/Scale factor] <current>:
```

Specify the new text height to be applied to all the selected text objects, or choose one of the other options. If you choose the Match Object option, AutoCAD prompts you to select a text object with the desired height. As soon as you select a text object whose height you want to match, the text height of all of the text objects you initially selected will change to match the target text height. If you choose the Scale Factor option, each of the selected text objects will be scaled by the same scale factor.

## Justifying Text

AutoCAD 2002 also provides a new command that enables you to change the justification of one or more text objects without changing their locations. To start the JUSTIFYTEXT command, do one of the following:

- On the Text toolbar, click Justify Text.
- From the Modify menu, choose Object | Text | Justify.
- At the command line, type **JUSTIFYTEXT** and then press ENTER.

When you start the command, AutoCAD prompts you to select objects. You can use any selection method you wish to select one or more text objects to be scaled. Non-text objects are ignored. When you finish selecting objects, AutoCAD prompts:

```
Enter a justification option
[Left/Align/Fit/Center/Middle/Right/TL/TC/TR/ML/MC/MR/BL/BC/BR] <current>:
```

Specify the location to serve as the new justification point by selecting one of the 15 options. The justification you specify is applied individually to each selected text object. The positions of the text objects do not change; only their justification (and therefore their insertion points) is altered.

# Finding and Replacing Any Kind of Text

While the Multiline Text Editor dialog box provides the ability to find and replace text in multiline text objects, the FIND command lets you find and replace single-line text, multiline text, block attribute values, text in dimensions, and text in hyperlink descriptions and hyperlinks. You can also use the FIND command to select and zoom in to text. The command locates text in both model space and in any layout defined in the current drawing. You can also narrow the search to a specific selection set. If you are working in a partially opened drawing, the command only considers the portion of the drawing that is currently open.

To start the FIND command, do one of the following:

- On the Standard or Text toolbar, click Find And Replace.
- From the Edit menu, choose Find.
- At the command line, type **FIND** and then press ENTER.

When you start the FIND command, AutoCAD displays the Find and Replace dialog box, shown in Figure 13-16. Type in the Find Text String field the text that you want to search for, and then type in the Replace With field the text that you want to replace it with. You can also choose one of the six most recently used strings from these drop-down lists.

From the Search In drop-down list, you can specify whether to search for the specified text in the entire drawing or only within the current selection set. Click the Select Objects button to define a new selection set.

You can also limit the types of objects included in the search and make the search case-sensitive. If you want to limit the types of objects included in the search, or make the search case-sensitive, click the Options button to display the Find and Replace Options dialog box, shown in Figure 13-17.

When you are ready to begin the search, click Find in the Find and Replace dialog box. In the Context area, AutoCAD displays and highlights the first instance of the found text, as well as the surrounding text. Click the Replace button to replace that occurrence of the text, or click Replace All to replace all instances of the text. If you simply want to find the text within the drawing, click the Zoom To button. AutoCAD immediately zooms in to display the found text.

**Note**   *Although AutoCAD searches in both model space and all layouts, it can only zoom in to display found text in the current model or layout tab.*

MOVING BEYOND THE BASICS

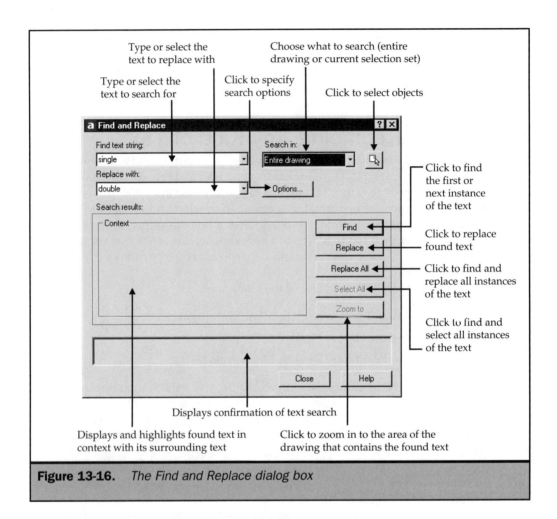

Type or select the text to replace with

Choose what to search (entire drawing or current selection set)

Type or select the text to search for

Click to specify search options

Click to select objects

Click to find the first or next instance of the text

Click to replace found text

Click to find and replace all instances of the text

Click to find and select all instances of the text

Displays confirmation of text search

Displays and highlights found text in context with its surrounding text

Click to zoom in to the area of the drawing that contains the found text

**Figure 13-16.**   *The Find and Replace dialog box*

# Checking Spelling

After you place text objects in your drawing, you can use AutoCAD's spelling checker to check the spelling of any of the text in the drawing, including text included in blocks. To start the spelling checker, do one of the following:

■ From the Tools menu, choose Spelling.

■ At the command line, type **SPELL** (or **SP**) and then press ENTER.

When you start the spelling checker, AutoCAD prompts you to select objects. After you finish selecting objects, press ENTER to continue. AutoCAD compares the words in the selected text objects against those in its main dictionary and a custom dictionary.

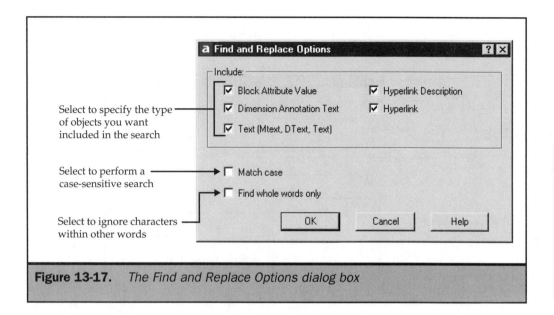

**Figure 13-17.** *The Find and Replace Options dialog box*

Select to specify the type of objects you want included in the search

Select to perform a case-sensitive search

Select to ignore characters within other words

If the program finds any words that may be spelled improperly, it displays the Check Spelling dialog box, shown in Figure 13-18. This dialog box displays the first potentially misspelled word, suggests several possible alternative spellings, and provides buttons for changing or ignoring the identified word. You can also add the word to the custom dictionary and change the dictionaries used to check the text. If no spelling errors are found, AutoCAD displays a message box telling you that the spelling check is complete.

*When you select the objects for the spelling checker, you can use any object selection method—such as All to select everything in the drawing—because AutoCAD ignores objects other than text objects.*

## Choosing a Dictionary

When you installed AutoCAD, the Setup program installed a main dictionary file. For most readers, the dictionary installed is probably the American English dictionary. The English language version of AutoCAD also includes four other dictionaries. Non-English language versions of AutoCAD are supplied with dictionaries that are appropriate to the respective versions. Whenever AutoCAD displays the Check Spelling dialog box, you can switch to a different dictionary by clicking the Change Dictionaries button. You can also specify the main dictionary setting under the Text Editor, Dictionary, and Font File Names selection on the Files tab of the Options dialog box.

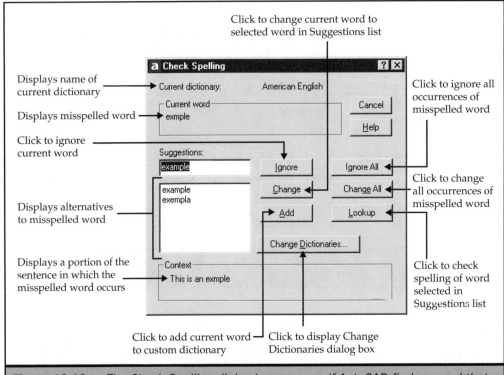

**Figure 13-18.**   *The Check Spelling dialog box appears if AutoCAD finds a word that isn't in its dictionaries.*

The main dictionaries supplied with AutoCAD contain thousands of common words. AutoCAD drawings often contain words that are not included in the main dictionaries, such as abbreviations, words particular to specific professions, or place names. When the spelling checker identifies a word that is not included in the main dictionary, you can click the Add button to add the word to a custom dictionary, so that AutoCAD skips this word in the future. This custom dictionary is named SAMPLE.CUS, by default, and is created in AutoCAD's Support directory. This file is a simple ASCII text file containing the additional words that you add, one word per line. Although you can't modify the main dictionary, you can freely edit the custom dictionary. You can also specify a different custom dictionary, either by clicking the Change Dictionaries button or by using the Options dialog box.

You occasionally may need to modify the custom dictionary. For example, if you accidentally add a misspelled word to the custom dictionary, you will want to remove that word. Although you could edit the custom dictionary file by using a text editor, such as Windows Notepad, an easier way does exist. When AutoCAD displays the Check Spelling dialog box, click the Change Dictionaries button to display the Change Dictionaries dialog box, shown in Figure 13-19. You can then use the controls in the bottom area of the dialog box to add new words to, or delete words from, the custom dictionary.

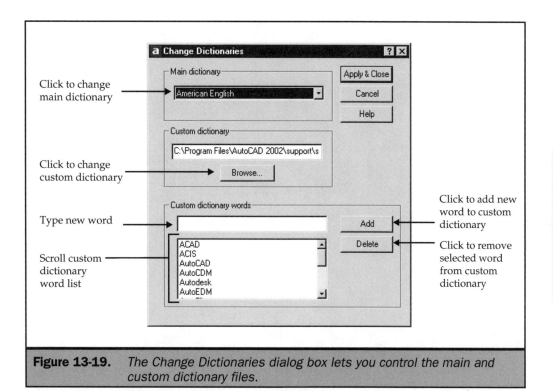

**Figure 13-19.** *The Change Dictionaries dialog box lets you control the main and custom dictionary files.*

# Using the Express Text Tools

The Express Tools include a collection of tools that you can use to enhance further your use of text in your drawings. These tools let you change the appearance of text, apply special effects, and edit or otherwise modify text already placed in your drawing:

- **RTEXT** Creates multiline text objects from either an ASCII text file or the value of a DIESEL expression, a macro language used to customize AutoCAD's status line.
- **TEXTFIT** Stretches or shrinks the width of single-line text objects so they fit between selected points.
- **TEXTMASK** Masks objects behind selected text objects, creating a clear area so that the text is more readable.
- **TXTEXP** Explodes text into a sequence of lines that can then be assigned thickness and elevation.
- **TXT2MTXT** Converts one or more lines of single-line text into a multiline text object.
- **ARCTEXT** Lets you create new text and place that text along an existing arc.

All of these tools are described in detail on the companion web site. Although they are all part of the Express Tools, which is now sold as extensions to AutoCAD, several of the individual tools can be downloaded free from the Autodesk web site.

# Chapter 14

## Dimensioning
## Your Drawing

Dimensions identify the size of the objects you draw, making accurate, legible dimensions one of the most important features of those drawings. AutoCAD's dimensioning tools let you add dimensional annotations to your drawing. You can quickly add dimensions simply by pointing to objects. You can also add tolerance symbols, notes, and leaders to your drawing by using the program's dimensioning tools.

AutoCAD's many dimensioning variables let you control the appearance of the dimensions. With dimension styles, you can save dimension variable settings, so that you can reuse them without having to re-create them. By creating *associative dimensions*, dimensions that are *tied* to the object that they annotate, if you subsequently modify the size of the object, the dimension updates automatically to reflect the changes, thus saving a lot of time and effort.

This chapter explains how to do the following:

- Create linear, angular, diameter, radius, and ordinate dimensions
- Dimension multiple objects
- Edit dimensions
- Create leaders and annotations
- Use and manage dimension styles and variables
- Add geometric tolerances

# Understanding Dimensioning Concepts

AutoCAD provides the tools needed to create five basic types of dimensions: linear, angular, radial, diameter, and ordinate. Figure 14-1 illustrates some of these basic dimensions. You can create dimensions for existing objects by selecting the objects to be dimensioned, or you can create dimensions by selecting points within the drawing. For example, you can create a linear dimension either by selecting the object to be dimensioned or by specifying the first and second extension line origins.

When you create a dimension, AutoCAD draws it on the current layer, using the current dimension style. Each dimension has an associated dimension style that controls the appearance of the dimension, such as the types of arrowheads, text style, and colors of the various components. You can modify existing dimension styles by changing the dimension variable settings and then update the dimension to reflect the new settings.

Each dimension you create consists of several parts. A *dimension line* shows where a dimension begins and ends. When you create an angular dimension, the dimension line is a dimension line arc that subtends the measured angle.

*Extension lines,* also called *projection lines,* are lines that extend away from the object for which you are creating a dimension, so that you can place the dimension line away from the object. Arrowheads form the termination at each end of the dimension line.

*Dimension text* contains the measured dimension and can also include prefixes, suffixes, tolerances, and other optional text. As you create dimensions, you can control the dimension text and specify its position and orientation. Figure 14-2 shows the various components of typical dimensions.

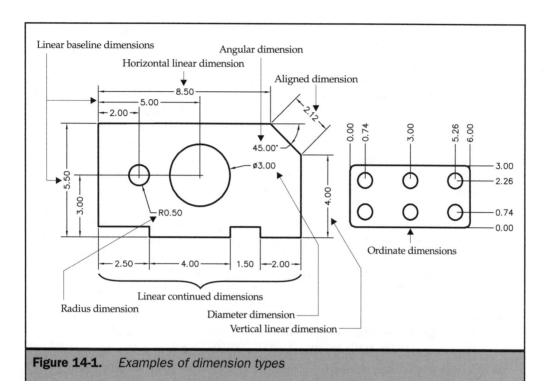

**Figure 14-1.** *Examples of dimension types*

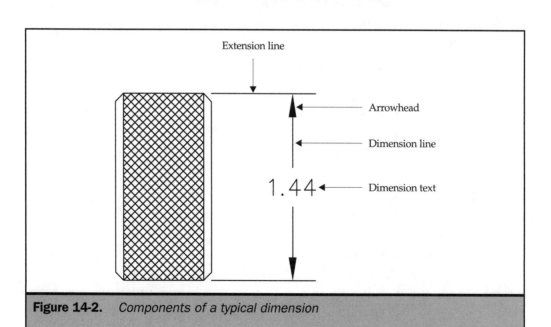

**Figure 14-2.** *Components of a typical dimension*

Dimensions can also contain other optional components. A *leader* is a line or curve leading from a feature of the drawing to an annotation. You can use leaders to place dimensions away from the dimension line or to add notes. Leaders generally begin with an arrowhead. AutoCAD automatically creates leaders for radial, diameter, and linear dimensions when the dimension text won't fit between the extension lines. Leaders and their annotations normally are associated, so that if you modify the annotation, the leader updates appropriately. When you create a radial or diameter dimension, you can include a center mark or center lines with the dimension. A *center mark* consists of a small cross that marks the center of a circle or arc. *Center lines* are crossing lines that extend out from the center of a circle or arc. These components are shown in Figure 14-3.

By default, AutoCAD creates *associative dimensions,* meaning that all the components— dimension lines, extension lines, arrowheads, and dimension text—are created as a single object, which is then tied to the object being measured. If you change the size of the object, the dimension updates automatically to reflect the changes. For example, if you change the length of an object from three inches to five inches, the dimension changes accordingly. AutoCAD actually lets you create three different types of dimensions, determined by the current value of the DIMASSOC system variable:

- **Associative dimensions**  Dimensions that automatically adjust their locations, orientations, and measurements when the geometric objects they are associated with are modified (DIMASSOC = 2).

- **Nonassociative dimensions**  Dimensions that do not automatically change when the geometric objects they measure are modified, but all dimension components are created as a single object (DIMASSOC=1).

- **Exploded dimensions**  Dimensions comprised of individual objects (DIMASSOC=0).

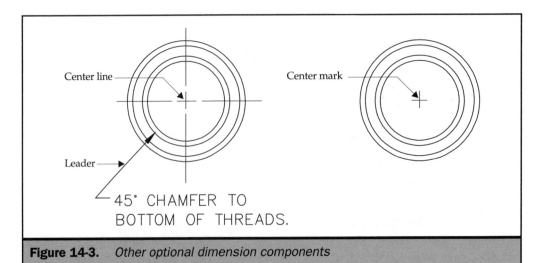

**Figure 14-3.**  *Other optional dimension components*

**Note** *The EXPLODE command converts associative and nonassociative dimensions into exploded dimensions. Once exploded, you cannot reassociate the objects into a dimension object except by undoing the action of the EXPLODE command.*

You can control the type of dimensions created by changing the value of the DIMASSOC system variable. Its value can also be controlled from the User Preferences tab of the Options dialog box. You can determine whether a dimension is associative or nonassociative by either selecting the dimension and looking at its properties in the Properties window or using the LIST command to display the properties of the dimension.

In releases of AutoCAD prior to AutoCAD 2002, the definitions of associative and nonassociative dimensions were different and were controlled by the DIMASO system variable. If you open a drawing file created in a previous version of AutoCAD, the DIMASSOC system variable value is based on the drawing's DIMASO value.

When you modify objects that have been dimensioned, you may inadvertently disassociate a dimension from its geometry. AutoCAD 2002 provides a new command for reassociating dimensions with the objects they annotate. You can also use this command to associate dimensions in drawings created in previous versions of AutoCAD. Associative dimensions do not support all objects, however. For example, dimensions annotating 2-D solids, 3DFaces, and multilines cannot be associative. In addition, associativity is not maintained between a dimension and a block reference if the block is redefined. You'll learn more about changing dimensioned geometry and changing dimension associativity later in this chapter.

**Note** *When you create any type of dimension, AutoCAD automatically creates a new layer named DEFPOINTS. This layer contains definition points, the points that control how and where a nonassociative dimension is tied to the object that it annotates. Objects drawn on the DEFPOINTS layer never get plotted, regardless of whether the layer is visible.*

## Creating Dimensions

For convenience, all of AutoCAD's dimensioning commands are located on the Dimension toolbar and Dimension menu. As with the program's other commands, you can start any dimensioning command by clicking the appropriate button on the toolbar, by choosing the command from the pull-down menu, or by typing the command at the command line.

When you start a dimensioning command, you can create dimensions in either of two ways:

- Select the object to dimension and specify the dimension line location
- Specify the extension line origins and the dimension line location

When you create dimensions by selecting an object, AutoCAD automatically places the extension line origins at the appropriate definition points, based on the type of object that you select. For example, the definition points are located at the end points of arcs, lines, and polyline segments. When you create dimensions by specifying the extension line origins, the points that you specify determine the definition points. To establish these points precisely, use object snaps.

# Creating Linear Dimensions

*Linear* dimensions annotate linear distances or lengths. They can be oriented horizontally, vertically, at a specified angle, or aligned parallel to either an existing object or the selected extension line origin points. After you create a linear dimension, you can subsequently add baseline or continued dimensions to it. A linear *baseline* dimension inserts an additional dimension from a common first extension line origin of a previous linear dimension. A linear *continued* dimension continues a linear dimension from the second extension line of a previous linear dimension, thus creating a chain of dimensions.

## Horizontal and Vertical Dimensions

The DIMLINEAR command is AutoCAD's primary linear dimensioning command. When you start the command, AutoCAD prompts:

```
Specify first extension line origin or <select object>:
```

You can either select the extension line origin points manually or let AutoCAD automatically determine the origin points of the first and second extension lines. Manual selection is the default. If you select the first extension point, AutoCAD prompts you to select the second, and the dimension is calculated as the distance between these two points. If you press ENTER when prompted to select the first extension line origin, AutoCAD prompts you to select the object to dimension. Figure 14-4 illustrates some typical linear dimensions.

To create a horizontal or vertical dimension:

1. Do one of the following:

- On the Dimension toolbar, click Linear Dimension.
- From the Dimension menu, choose Linear.
- At the command line, type **DIMLINEAR** (or **DLI** or **DIMLIN**) and then press ENTER.

2. Specify the first and second extension line origin points, or press ENTER and then select the object to dimension.

3. Specify the dimension line location either by moving the mouse to move the dimension visually into place and then clicking, or by typing a coordinate at the command line.

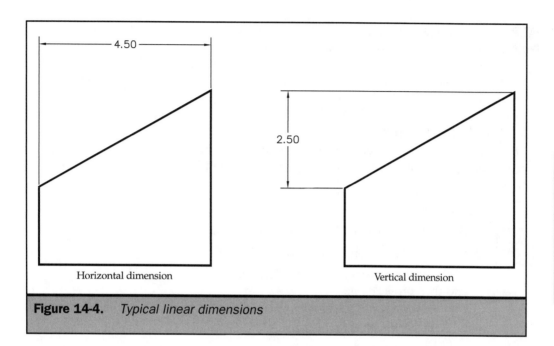

**Figure 14-4.**    *Typical linear dimensions*

When you create a linear dimension by using the DIMLINEAR command, AutoCAD automatically determines whether the dimension is a horizontal or vertical dimension, based on the position of the dimension line in relation to the extension line origin points. Before you specify the dimension line location, however, you can force the dimension to be horizontal or vertical. You can also override the dimension text or the dimension text angle. AutoCAD prompts:

```
Specify dimension line location or
[Mtext/Text/Angle/Horizontal/Vertical/Rotated]:
```

- To override the default dimension orientation, type **H** (for Horizontal) or **V** (for Vertical).

- To align the dimension to a specific angle, type **R** (for Rotated). AutoCAD then prompts you for the dimension line angle.

- To edit the dimension text on the command line, type **T** (for Text).

- To edit the dimension text by using the Multiline Text Editor dialog box (discussed in the previous chapter), type **M** (for MText).

- To change the dimension text orientation angle, type **A** (for Angle).

*When you edit dimension text by using either the Multiline Text Editor dialog box or the command line, the actual measured dimension is represented by a pair of angle brackets (<>). To include additional text as a prefix or suffix, type it before or after the angle brackets. If you omit the angle brackets, AutoCAD uses the text that you enter in lieu of the measured dimension.*

## Aligned Dimensions

*Aligned* dimensions are linear dimensions that align parallel to either a selected object or the selected extension line origin points, as shown in Figure 14-5. To create an aligned dimension:

1. Do one of the following:

   ■ On the Dimension toolbar, click Aligned Dimension.

   ■ From the Dimension menu, choose Aligned.

   ■ At the command line, type **DIMALIGNED** (or **DAL** or **DIMALI**) and then press ENTER.

2. Specify the first and second extension line origin points, or press ENTER and then select the object to dimension.

3. Specify the dimension line location.

Before you specify the dimension line location, you can edit the dimension text or change its orientation angle. Because the dimension line always aligns with either the object or extension line origin points that you specify, however, no options are given for changing the dimension line orientation.

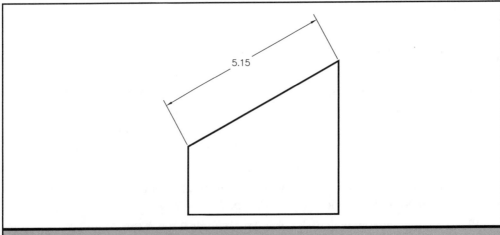

**Figure 14-5.**   *A typical aligned dimension*

## Linear Baseline Dimensions

A *baseline* dimension creates a second dimension from the same first extension line origin, or *baseline*, as the previous dimension, or another dimension that you select. This method of placing dimensions often is used in mechanical drafting to reference dimensions from a common point, or in architectural drafting to indicate the total length of a series of individual dimensions. Figure 14-6 shows a typical series of linear baseline dimensions. You can also create angular and ordinate baseline dimensions, as you will learn later in this chapter.

When you start the DIMBASELINE command, the prompts vary, depending on the type of dimension last created. If the previous dimension was a linear dimension, AutoCAD prompts:

```
Specify a second extension line origin or [Undo/Select] <Select>:
```

You can immediately select the next extension line origin or press ENTER to select a different base dimension. If you select a point, AutoCAD uses the first extension line of the previous dimension as the baseline of the dimension that you're creating, and repeats the prompt so that you can place multiple dimensions from the same baseline. To select a different existing dimension as a reference for the new baseline, press ENTER. AutoCAD then prompts you to select the base dimension. AutoCAD also displays this prompt when you start the command, if you haven't yet created a dimension during the current editing session. To create a baseline dimension:

1. Do one of the following:

- On the Dimension toolbar, click Baseline Dimension.
- From the Dimension menu, choose Baseline.
- At the command line, type **DIMBASELINE** (or **DBA** or **DIMBASE**) and then press ENTER.

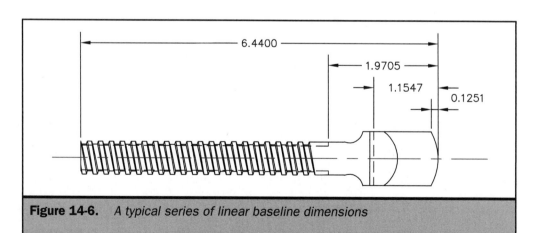

**Figure 14-6.** *A typical series of linear baseline dimensions*

2. Specify the second extension line origin, or press ENTER and select a different base dimension.

3. Continue specifying additional second extension line origins, press ENTER to select a different base dimension, or press ENTER twice to complete the command.

**Note** *AutoCAD automatically places the new baseline dimension above or below the previous dimension line. The distance between the two dimension lines is determined by the DIMDLI system variable. You can also control this distance by changing the Baseline spacing on the Lines and Arrows tab of the Dimension Style Manager dialog box, used to control dimension styles. You'll learn more about dimension styles later in this chapter.*

## Linear Continued Dimensions

A *continued* dimension creates a new dimension that continues from the second extension line origin of the previous dimension, or from another dimension that you select. This method of placing dimensions often is used in architectural drafting to create a string or chain of dimensions, such as those that locate a series of walls in a building. Figure 14-7 shows a typical string of continued dimensions. You can also create angular and ordinate continued dimensions, as you will learn later in this chapter.

**Tip** *It is often faster to create baseline or continued dimensions by using the Quick Dimension command, described a bit later in this chapter.*

When you start the DIMCONTINUE command, the prompts vary, depending on the type of dimension last created, similar to when you create baseline dimensions. If the previous dimension was a linear dimension, you can simply specify the second

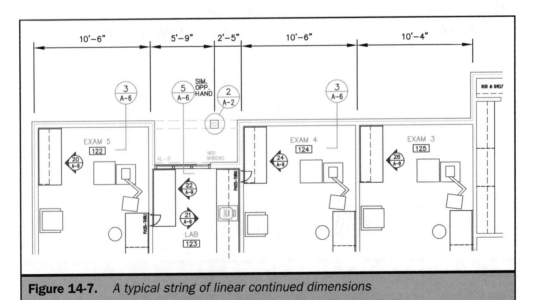

**Figure 14-7.** *A typical string of linear continued dimensions*

extension line origin to extend the new dimension line from its end, or press ENTER to select a different base dimension. To create a continued dimension:

1. Do one of the following:

   ■ On the Dimension toolbar, click Continue Dimension.

   ■ From the Dimension menu, choose Continue.

   ■ At the command line, type **DIMCONTINUE** (or **DCO** or **DIMCONT**) and then press ENTER.

2. Specify the second extension line origin, or press ENTER and select a different base dimension.

3. Continue specifying additional second extension line origins, press ENTER to select a different base dimension, or press ENTER twice to complete the command.

### LEARN BY EXAMPLE

*Add linear dimensions to the floor plan of a house. Open the drawing Figure 14-8 on the companion web site. (The floor plan was created using multilines, as described on the companion website. Refer to the pages for chapters 4 and 10.)*

To add the dimensions shown in Figure 14-8, use the following command sequence and instructions. Use the endpoint object snap to select the extension line origins. Generally, you will also want to create a new layer on which to draw dimensions; this has already been done for you in the drawing on the web site.

```
Command: DIMLINEAR
Specify first extension line origin or <select object>: (select point A)
Specify second extension line origin: (select point B)
Specify dimension line location or
[Mtext/Text/Angle/Horizontal/Vertical/Rotated]: (select point C)
Dimension text = 12' - 4
Command: DIMCONTINUE
Specify a second extension line origin or [Undo/Select] <Select>: (select point D)
Dimension text = 11' - 0
Specify a second extension line origin or [Undo/Select] <Select>: (select point E)
Dimension text = 12' - 4
Specify a second extension line origin or [Undo/Select] <Select>: (select point F)
Dimension text = 14' - 0
Specify a second extension line origin or [Undo/Select] <Select>: ENTER
Select continued dimension: ENTER
Command: DIMBASELINE
Specify a second extension line origin or [Undo/Select] <Select>: ENTER
Select base dimension: (select the first dimension near point C)
Specify a second extension line origin or [Undo/Select] <Select>: (select point F)
Dimension text = 49' - 8
```

```
Specify a second extension line origin or [Undo/Select] <Select>: ENTER
Select base dimension: ENTER
```

When you finish these steps, your drawing should look similar to Figure 14-8. On your own, use the same methods to add the necessary dimensions around the other three sides of the house.

# Creating Angular Dimensions

*Angular* dimensions annotate the angle measured between two nonparallel lines, the angle subtended by an arc, or the angle around a portion of a circle. You can also dimension the angle by selecting three points: the vertex and two end points. As you create the dimension, you can modify the dimension text and its orientation. After you create an angular dimension, you can add subsequent baseline or continued dimensions to it. When you start the angular dimension command, AutoCAD prompts:

```
Select arc, circle, line, or <specify vertex>:
```

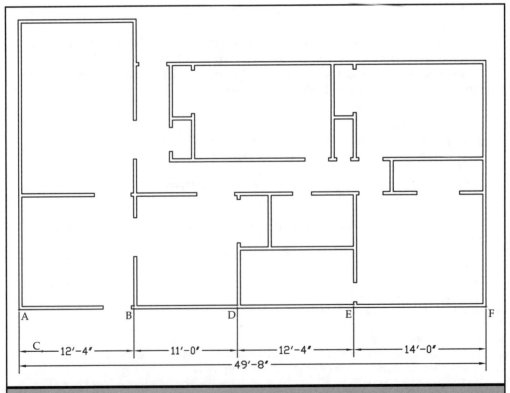

**Figure 14-8.** *Linear dimensions added to a plan for a small house*

The remaining prompts depend on the type of object that you select. For example, if you select an arc, AutoCAD measures the angle subtended by the arc, using the center of the arc as the vertex of the angle. The program prompts you for the location of the dimension arc line and draws extension lines from the end points of the arc. If you select a line or other linear object, AutoCAD prompts you to select a second line, and then it determines the angle between them. The program then prompts you for the location of the dimension arc line and determines which of the four possible angles to annotate, based on the location that you specify. Figure 14-9 illustrates some typical angular dimensions.

To dimension the angle subtended by an arc:

1. Do one of the following:

   ■ On the Dimension toolbar, click Angular Dimension.

   ■ From the Dimension menu, choose Angular.

   ■ At the command line, type **DIMANGULAR** (or **DAN** or **DIMANG**) and then press ENTER.

2. Select the arc.

3. Specify the dimension arc line location.

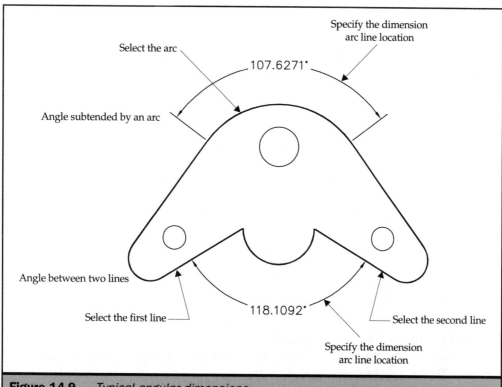

**Figure 14-9.**    *Typical angular dimensions*

To dimension the angle between two lines:

1. Do one of the following:
   - On the Dimension toolbar, click Angular Dimension.
   - From the Dimension menu, choose Angular.
   - At the command line, type **DIMANGULAR** (or **DAN** or **DIMANG**) and then press ENTER.

2. Select one line.

3. Select the other line.

4. Specify the dimension arc line location.

## Angular Baseline Dimensions

An *angular baseline* dimension creates a second angular dimension from the first extension line of a previous angular dimension.

*LEARN BY EXAMPLE*
*Create an angular dimension and a series of angular baseline dimensions. Open the drawing Figure 14-10 on the companion web site.*

To add the angular dimensions shown in Figure 14-10, use the following command sequence and instructions. Use object snaps to select points C and D.

```
Command: DIMANGULAR
Select arc, circle, line, or <specify vertex>: (select point A)
Specify dimension arc line location or [Mtext/Text/Angle]: (select point B)
Dimension text = 21.71
Command: DIMBASELINE
Specify a second extension line origin or [Undo/Select] <<Select>: ENTER
Select base dimension: (select the first dimension near point B)
Specify a second extension line origin or [Undo/Select] <Select>): (select point C)
Dimension text = 40.77
Specify a second extension line origin or [Undo/Select] <Select>): (select point D)
Dimension text = 64.75
Specify a second extension line origin or [Undo/Select] <Select>): ENTER
Select base dimension: ENTER
```

## Angular Continued Dimensions

An *angular continued* dimension creates a second angular dimension from the second extension line of a previous angular dimension.

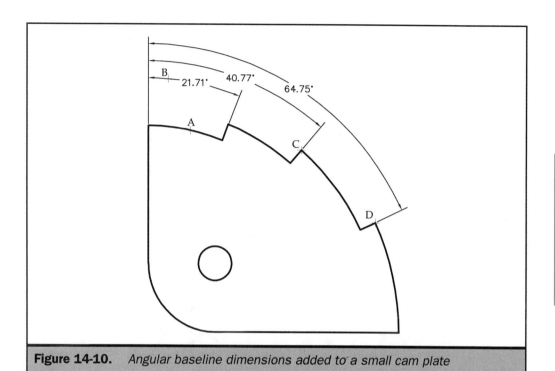

**Figure 14-10.**    *Angular baseline dimensions added to a small cam plate*

### LEARN BY EXAMPLE
*Create an angular dimension and a series of angular continued dimensions. Open the drawing Figure 14-11 on the companion web site.*

To add the angular dimensions shown in Figure 14-11, use the following command sequence and instructions. Use the center object snap to select the centers of the circles labeled A, B, C, D, E, F, and G.

```
Command: DIMANGULAR
Select arc, circle, line, or <specify vertex>: ENTER
Specify angle vertex: (select point A)
Specify first angle endpoint: (select point B)
Specify second angle endpoint: (select point C)
Specify dimension arc line location or [Mtext/Text/Angle]: (select point D)
Dimension text = 72.00
Command: DIMCONTINUE
```

```
Specify a second extension line origin or [Undo/Select] <Select>: (select point E)
Dimension text = 72.00
Specify a second extension line origin or [Undo/Select] <Select>: (select point F)
Dimension text = 72.00
Specify a second extension line origin or [Undo/Select] <Select>: (select point G)
Dimension text = 72.00
Specify a second extension line origin or [Undo/Select] <Select>: (select point B)
Dimension text = 72.00
Specify a second extension line origin or [Undo/Select] <Select>: ENTER
Select continued dimension: ENTER
```

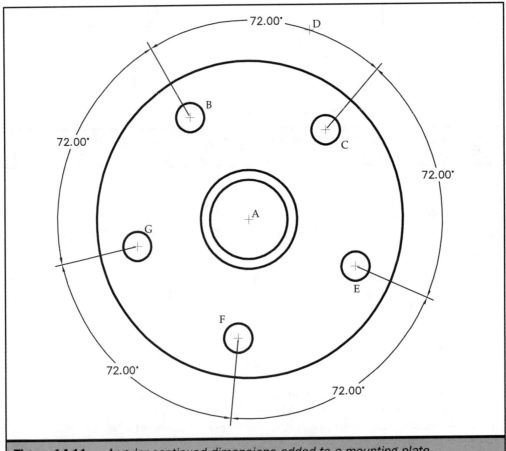

**Figure 14-11.**   *Angular continued dimensions added to a mounting plate*

# Creating Diameter and Radius Dimensions

*Diameter* and *radius* dimensions annotate the diameters and radii of arcs and circles, and can optionally include center lines or center marks. The DIMDIAMETER command creates diameter dimensions, and the DIMRADIUS command creates radial dimensions. When you start either of these commands, AutoCAD prompts you to select an arc or circle, which you must do by pointing to the object. After you select the arc or circle, AutoCAD prompts you to specify the dimension line location. Before you select the location, you can also modify the dimension text and its orientation.

To create a diameter dimension:

1. Do one of the following:

- On the Dimension toolbar, click Diameter Dimension.

- From the Dimension menu, choose Diameter.

- At the command line, type **DIMDIAMETER (or DDI or DIMDIA)** and then press ENTER.

2. Select the arc or circle.

3. Specify the dimension line location.

To create a radius dimension:

1. Do one of the following:

- On the Dimension toolbar, click Radius Dimension.

- From the Dimension menu, choose Radius.

- At the command line, type **DIMRADIUS (or DRA or DIMRAD)** and then press ENTER.

2. Select the arc or circle.

3. Specify the dimension line location.

When you create diameter and radius dimensions, the appearance of the dimension depends on the size of the circle or arc, the location that you specify for the dimension line, and the settings of various system variables. For example, the dimension line can be placed inside or outside the curve, the dimension can be placed inside or outside the curve, and the dimension text can be aligned with the dimension line. These variations are shown in Figure 14-12. Although you can control these system variables directly at the command line, it is much easier to establish the proper settings by using the Dimension Style Manager dialog box. You'll learn about dimension styles later in this chapter.

MOVING BEYOND THE BASICS

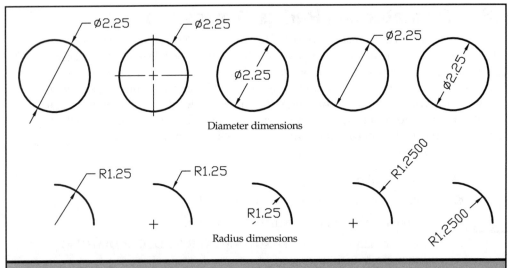

**Figure 14-12.** *The appearance of diameter and radius dimensions can vary, based on different system variable values.*

## Creating Ordinate Dimensions

An *ordinate* dimension annotates the perpendicular distance from an origin or base point (the origin of the current UCS). Ordinate dimensions consist of an X or Y coordinate and a leader. An X-ordinate dimension measures distances along the X axis; a Y-ordinate dimension measures distances along the Y axis. Ordinate dimensions are often used in machine design drawings. They prevent errors by dimensioning all features from the same reference datum.

When you begin to place an ordinate dimension, AutoCAD prompts you to select the feature to be dimensioned. You select the feature by specifying a point, usually by using an object snap. Once you select the feature, AutoCAD prompts:

```
Specify leader endpoint or [Xdatum/Ydatum/Mtext/Text/Angle]:
```

If you specify the leader end point, AutoCAD automatically determines whether the point is an X or Y ordinate, based on the direction of the end point in relation to the feature. Before you specify the leader end point, you can explicitly specify whether the datum represents an X or Y ordinate, or modify the dimension text by using either the Multiline Text Editor dialog box or the command line. Ordinate dimension text is normally aligned with the ordinate leader lines, but you can use the Angle option to override the text angle. Figure 14-13 shows some typical ordinate dimensions.

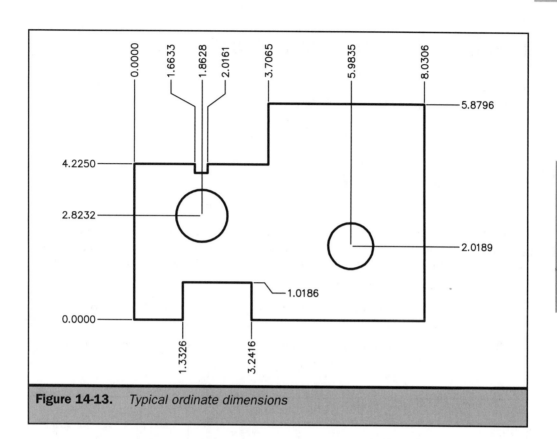

**Figure 14-13.** *Typical ordinate dimensions*

To create an ordinate dimension:

1. Do one of the following:

■ On the Dimension toolbar, click Ordinate Dimension.

■ From the Dimension menu, choose Ordinate.

■ At the command line, type **DIMORDINATE** (or **DOR** or **DIMORD**) and then press ENTER.

2. Specify the feature point.

3. Specify the ordinate leader end point.

# Dimensioning Multiple Objects

The QDIM command enables you to quickly create multiple baseline, continued, staggered, and ordinate dimensions in a single step. You can also dimension multiple circles and arcs. Figure 14-14 shows the various dimension types that can be created using the QDIM command.

To start the command, do one of the following:

- On the Dimension toolbar, click Quick Dimension.
- From the Dimension menu, choose QDIM.
- At the command line, type **QDIM** and then press ENTER.

When you start the command, AutoCAD prompts you to select the geometry you want to dimension. You can select multiple objects using any object selection method. After you select the objects, you can specify the type of dimension you want to create. The command displays the following options:

- **Continuous**   Creates a string or chain of linear dimensions (similar to continued dimensions)

- **Staggered**   Creates a series of staggered linear dimensions annotating each nearest pair of objects

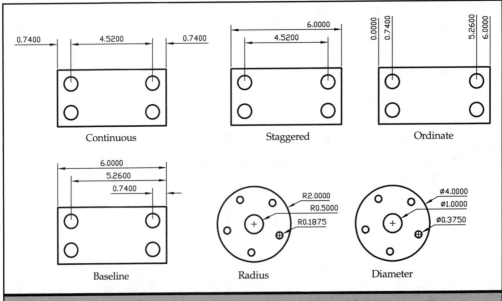

**Figure 14-14.**   *The QDIM command dimensions multiple objects at one time.*

- **Baseline**   Creates a series of linear dimensions from a common base point
- **Ordinate**   Creates a series of ordinate dimensions
- **Radius**   Creates radius dimensions of all selected arcs and circles
- **Diameter**   Creates diameter dimensions of all selected arcs and circles
- **Datum Point**   Sets a new datum point for baseline and ordinate dimensions
- **Edit**   Edits a series of dimensions by letting you add or remove dimension points

*LEARN BY EXAMPLE*
*Use the QDIM command to quickly add a series of dimensions. Open the drawing Figure 14-15 on the companion web site.*

To add the string of continuous dimensions shown in Figure 14-15, use the following command sequence and instructions:

```
Command: QDIM
Select geometry to dimension: (select point A)
Specify other corner: (select point B)
Select geometry to dimension: ENTER
Specify dimension line position, or
[Continuous/Staggered/Baseline/Ordinate/Radius/Diameter/datumPoint/Edit]
<Continuous>: E
```

Notice that when you select the Edit option, AutoCAD places a small X at each potential edit point. The command prompts:

```
Indicate dimension point to remove, or [Add/eXit] <eXit>:
```

Click each of the four points labeled C, D, E, and F to remove them from consideration. After you select each point, AutoCAD repeats the previous prompt. After selecting all four points, proceed with the following command sequence:

```
Indicate dimension point to remove, or [Add/eXit] <eXit>: ENTER
Specify dimension line position, or
[Continuous/Staggered/Baseline/Ordinate/Radius/Diameter/datumPoint/Edit]
<Continuous>: (select point G)
```

Next, add the radius dimensions, by using the following command sequence:

```
Command: QDIM
Select geometry to dimension: (select points H and I)
```

MOVING BEYOND
THE BASICS

```
Select geometry to dimension: ENTER
Specify dimension line position, or
[Continuous/Staggered/Baseline/Ordinate/Radius/Diameter/datumPoint/Edit]
<Continuous>: R
Specify dimension line position, or
[Continuous/Staggered/Baseline/Ordinate/Radius/Diameter/datumPoint/Edit]
<Radius>: (select point J)
```

**Note**    *After creating a series of dimensions using the QDIM command, you can use the command again to edit them. When editing existing dimensions, you can convert them to a different type of dimension—for example, changing them from continuous to staggered—or use the Edit option to add or remove edit points. Dimensions created with the QDIM command are not associative but they may be associated individually by using the DIMREASSOCIATE command. You'll learn about this command later in this chapter.*

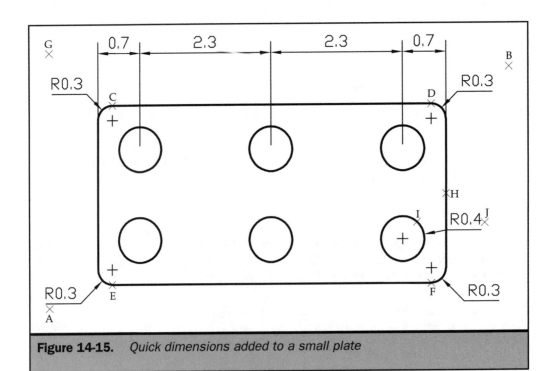

**Figure 14-15.**    *Quick dimensions added to a small plate*

# Editing Dimensions

There are many different ways to edit dimensions and the objects that they measure. You can modify the object, and thus its dimension. You can also edit the dimension text or the appearance of the dimension itself.

## Changing Dimensioned Geometry

The easiest way to edit objects and their dimensions is by using grips, but you can also use most of AutoCAD's other object modification commands, such as STRETCH, ROTATE, SCALE, TRIM, and EXTEND. The methods you use to modify the geometry and its dimensions varies, however, depending on whether the dimensions are associative, nonassociative, or exploded.

### Changing Associative Dimensions

Associative dimensions retain their associativity to the geometry they annotate as you modify that geometry as long as both the geometry and the dimensions are selected and operated on using a single command. For example, if you trim or extend an object that has an associative dimension, the dimension immediately changes to reflect the new length. If you move, copy, or array the geometry and its associated dimensions, each dimension retains associativity with its respective geometry.

In the following circumstances, dimensions are disassociated:

- If you erase the associated geometry
- If you use a Boolean operation such as UNION or SUBTRACT to alter the associated geometry
- If you use grip editing to stretch a dimension parallel to its dimension line
- If you created the association to the geometry using an Apparent Intersection object snap and subsequently move the geometry so that the apparent intersection no longer exists

In certain circumstances, an associative dimension may also become partially disassociated. For example, if a linear dimension is associated with the end points of two objects and you erase one of those objects, the dimension becomes partially disassociated. The remaining association is preserved, however, and you can reassociate the disassociated end to another object by using the DIMREASSOCIATE command. AutoCAD displays a message on the command line if a dimension becomes disassociated.

MOVING BEYOND
THE BASICS

## Changing Nonassociative Dimensions

When you edit an object dimensioned using nonassociative dimensions, you must select both the object and the object's definition points, which AutoCAD placed on the DEFPOINTS layer. The location of these definition points depends on the type of dimension. For example, definition points are placed at the extension line origins of linear dimensions, as well as at the intersection of the first extension line and the dimension line. The definition points for radius dimensions are placed at the center point and the point that is used to select the curve.

To use the STRETCH command to change the size of the object and its nonassociative dimension, as shown in Figure 14-16:

1. Start the STRETCH command.

2. Use a crossing window to select both the object and the dimension, and then press ENTER.

3. Specify the base point or displacement and the second point of displacement.

To accomplish the same thing by using grips, you must select both the object and the dimension and then stretch the hot grips corresponding to both the object and the dimension. Since there are multiple grips to select, in this case, using the STRETCH command is easier.

 *When you stretch an object and its dimension, the original orientation of the dimension is retained.*

## Changing Exploded Dimensions

Since exploded dimensions are simply a collection of individual objects (lines, 2-D solids, and text), you can edit those objects as you would any other AutoCAD objects. There is no way to convert exploded dimensions into associative or nonassociative dimensions.

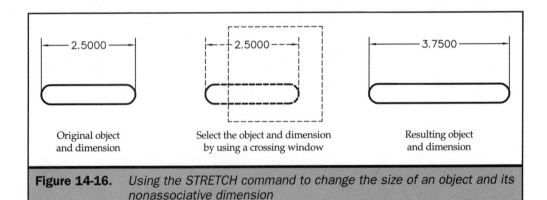

**Figure 14-16.** *Using the STRETCH command to change the size of an object and its nonassociative dimension*

# Changing Dimension Associativity

AutoCAD 2002 provides a new DIMREASSOCIATE command, enabling you to reassociate dimensions that have become disassociated or that were originally created as nonassociative dimensions. For example, if you use the PEDIT command to change a previously dimensioned polyline segment into a splined polyline, its dimensions become disassociated. Use the DIMREASSOCIATE command to reassociate the dimension to the correct segment of the splined polyline.

You can also use the command to redefine the associativity of dimensions. To start the command, do one of the following:

- From the Dimension menu, choose Reassociate Dimensions.
- At the command line, type **DIMREASSOCIATE** (or **DRE**) and then press ENTER.

When you start the DIMREASSOCIATE command, AutoCAD first prompts you to select the dimensions you want to reassociate. After you select one or more dimensions, press ENTER to end object selection. AutoCAD then highlights each selected dimension in turn and prompts you to specify the association points for the highlighted dimension. The prompts vary depending on the type of dimension currently highlighted. A special marker appears at the association point for each association prompt. The marker indicates whether the definition point of the dimension is associative or nonassociative. A square with an X in it indicates that the point is associated with a location on an object. An X without a square indicates that the point is not associated with an object. You can do one of the following:

- Use an object snap to specify a new association for the extension line origin point.
- Type **S** and press ENTER to select a geometric object to associate with the dimension.
- Press ENTER to skip to the next extension line origin point.

If you specify a new association point or skip to the next association point, the marker moves to the next definition point and AutoCAD prompts you to specify its association point. If you type **S** and press ENTER, AutoCAD prompts you to select an object. As soon as you select the object, the dimension becomes associated to that object.

This process repeats until you have stepped through all the selected dimensions. At any time during this process, you can press ESC to end the command. Any associations you've made up to that point are retained.

*If you pan or zoom with a wheel mouse while the DIMREASSOCIATE command is active, the current marker disappears.*

AutoCAD also provides a DIMDISASSOCIATE command, enabling you to change selected associative dimensions into nonassociative dimensions. You may want to remove associativity from dimensions in drawings that will be opened by someone using a pre-AutoCAD 2002 version of AutoCAD who doesn't want any *proxy* objects in the drawing. A proxy object is a custom object created by an application that is not present

MOVING BEYOND THE BASICS

when the drawing is opened. AutoCAD 2002 associative dimensions are treated as proxy objects when opened in earlier versions of AutoCAD.

To disassociate one or more dimensions:

1. At the command line, type **DIMDISASSOCIATE** (or **DDA**) and press ENTER.

2. Select the dimensions you want to disassociate, using any selection method, and then press ENTER.

AutoCAD immediately disassociates the selected dimensions, and displays a message indicating the number of dimensions that were disassociated.

You can also use the DIMREGEN command to manually update the locations of all associative dimensions. This is necessary in the following three situations:

- To update associative dimensions created in paper space after panning or zooming with a wheel mouse in a layout with model space active

- To update associative dimensions that have been modified in a drawing that was edited using an earlier version of AutoCAD

- To update associative dimensions created in the current drawing but associated to externally referenced geometry that has been modified

To manually update the locations of associative dimensions, at the command line, type **DIMREGEN** and then press ENTER.

 *With trans-spatial objects (objects drawn in model space but dimensioned in a layout), the dimension value is sometimes not displayed after you use the DIMREASSOCIATE command. When this occurs, you can select Dim | Update from the Dimension menu and then select the dimension to correct this problem.*

## Making Dimensions Oblique

Linear dimension extension lines normally are created perpendicular to the dimension line. You can change the angle of the extension lines, however, so that they tilt relative to the dimension line, as shown in Figure 14-17. To make oblique extension lines:

1. Do one of the following:

- On the Dimension toolbar, click Dimension Edit.

- From the Dimension menu, choose Oblique.

- At the command line, type **DIMEDIT** (or **DED** or **DIMED**) and then press ENTER.

If you start the command by using the toolbar button or by typing the command, you must then type **O** (for Oblique) and press ENTER. Starting the command from the menu automatically selects the Oblique option.

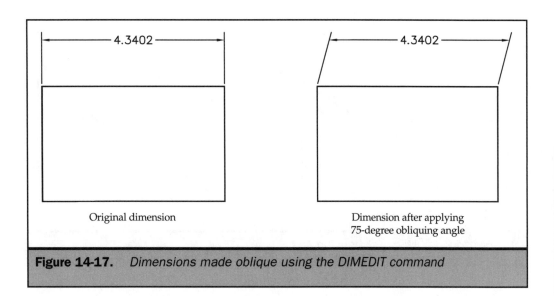

**Figure 14-17.** *Dimensions made oblique using the DIMEDIT command*

2. Select the linear dimension and then press ENTER.

3. Specify the obliquing angle and then press ENTER.

## Modifying Dimension Text

In the previous section, you learned how to use the DIMEDIT (Dimension Edit) command to make dimensions oblique. You can also use this command to rotate dimension text, replace selected dimension text with new text, or restore dimension text to its original position as defined by the current dimension style.

When you use the DIMEDIT command, you specify the change first and then select one or more dimensions to which to apply the change. All the selected dimensions are updated simultaneously.

You can also use the DIMTEDIT (Dimension Text Edit) command to modify a selected dimension text object, by changing its position or rotation angle. When you use this command, you first select a single dimension and then apply the change. The DIMTEDIT command can also restore the selected dimension text to its original *home* position.

To rotate dimension text (see Figure 14-18):

1. Do one of the following:

   ■ On the Dimension toolbar, click Dimension Edit.

   ■ At the command line, type **DIMEDIT** (or **DED** or **DIMED**) and then press ENTER.

2. Type **R** (for Rotate) and then press ENTER.

MOVING BEYOND
THE BASICS

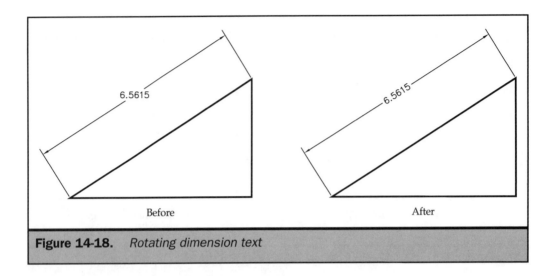

**Figure 14-18.** *Rotating dimension text*

   3. Specify the new dimension text angle and then press ENTER.

   4. Select the dimensions to be rotated and then press ENTER.

To move dimension text (see Figure 14-19):

   1. Do one of the following:

   ■ On the Dimension toolbar, click Dimension Text Edit.

   ■ At the command line, type **DIMTEDIT** and then press ENTER.

   2. Select the dimension that you want to move.

   3. Drag the dimension text to its new position and then click.

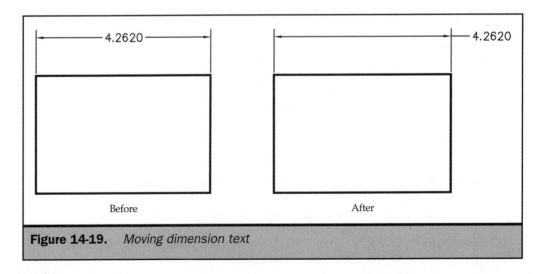

**Figure 14-19.** *Moving dimension text*

**Note** *You can also select the Left or Right options to place the dimension at the left or right end of the dimension line. If you start the command from the Dimension menu by choosing Align Text, you can place the dimension at the left, right, or center of the dimension line.*

You can also use the shortcut menu to move dimension text or change certain dimension properties (such as the dimension precision and dimension style). For example, the shortcut menu makes it easy to move the dimension text without altering the dimension line, or to move the dimension text and connect it back to the dimension line by using a leader.

To move the dimension text without altering the dimension line:

1. Select the dimension you want to move (so that its grips become visible).

2. Right-click to display the shortcut menu.

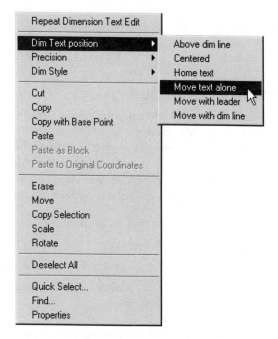

3. From the shortcut menu, choose Dim Text Position | Move Text Alone.

4. Move the dimension text to its new position and then click.

**Tip** *You can also move dimension text using grips. Select the dimension so that its grips are displayed. Click the grip located at the dimension text, making it the hot grip. The Stretch grip editing mode becomes active. Move the dimension text to a new location, and then press ESC to release and clear the grips. If you move the dimension text so that it no longer requires the dimension line to be broken, the dimension line heals itself automatically.*

To replace existing dimension text with new text:

1. Do one of the following:
   - On the Dimension toolbar, click Dimension Edit.
   - At the command line, type **DIMEDIT** (or **DED** or **DIMED**) and then press ENTER.

2. Type **N** (for New) and then press ENTER.

3. In the Multiline Text Editor dialog box, enter the new text and any desired formatting, and then click OK.

4. Select the dimensions to be replaced and then press ENTER.

 *You can also use the DDEDIT command to edit a single-dimension text object.*

To restore a single-dimension text object to its home position, use the Home option of the DIMTEDIT command. To move several dimension text objects back to their home position simultaneously, use the Home option of the DIMEDIT command.

## Changing Dimension Properties

You can use the Properties window to edit any dimension property, including the size and appearance of dimension arrows, the dimension text, and the dimension units. These properties are all determined by the dimension style in effect when the dimension was created. You'll learn more about dimension styles later in this chapter.

After altering any of the dimension properties, you can use the shortcut menu to save the modified dimension properties to a new style by doing the following:

1. Select the modified dimension (so that its grips become visible).

2. Right-click to display the shortcut menu.

3. From the shortcut menu, choose Dim Style | Save As New Style.

4. In the Save As New Dimension Style dialog box, enter the name of the new dimension style, and then click OK.

 *If you save your changes to an existing dimension style name, the changes may alter the appearance of any existing dimensions using that style.*

# Creating Leaders and Annotations

*Leaders* consist of a line, a series of lines, or a spline that connects an annotation to a feature in the drawing. Generally, the leader includes an arrowhead at the starting point of the line, with the arrow pointing toward the feature. An *annotation,* consisting of text, a feature control frame, a block reference, or a copy of an existing annotation, is placed immediately adjacent to the last point of the leader line. If the leader includes text, the text can consist of one or more lines created as a multiline text object, and you can enter the text at the command line or by using the Multiline Text Editor dialog box. A small line, called a *hook line,* is added to the leader if the last segment of the leader line is drawn at an angle greater than 15 degrees from horizontal. Figure 14-20 shows some typical leaders and annotations.

Leaders and their annotations are *associative*—if you modify the annotation, the leader updates as necessary. For example, if you move the annotation, the leader updates

MOVING BEYOND THE BASICS

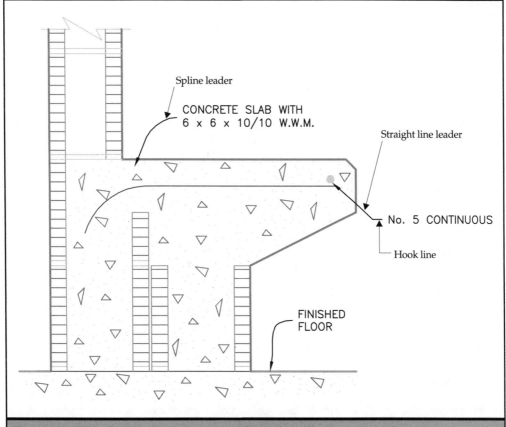

**Figure 14-20.**    *Typical leaders and annotations*

so that the end point of the last leader line segment remains adjacent to the annotation. When associative dimensioning is turned on and object snaps are used to locate the leader arrowhead, the leader is associated with the object to which the arrowhead is attached. If you move the object, the annotation remains in place and the leader updates so that the arrowhead remains attached to the object.

The entire leader object is drawn on the current layer, but the color is determined by the current dimension line color. The type and size of the arrowhead, as well as the overall scale of the leader, are also determined by the current dimension style.

AutoCAD actually provides two different commands for creating leaders. The QLEADER command uses a dialog box for controlling leader settings, while all LEADER command options must be specified on the command line. For that reason, most users find the QLEADER command to be much more powerful. To start the QLEADER command, do one of the following:

- On the Dimension toolbar, click Quick Leader.
- From the Dimension menu, choose Leader.
- At the command line, type **QLEADER** and then press ENTER.

When you start the QLEADER command, AutoCAD prompts:

```
Specify first leader point, or [Settings] <Settings>:
```

To begin creating a leader using the current settings, simply specify the first point of the leader line. To view or change the current leader settings, select the Settings option to display the Leader Settings dialog box. This dialog box has three separate tabs. The Annotation tab (shown in Figure 14-21) enables you to define the type of annotation to be attached to the leader, by selecting one of the following:

- **Mtext**   After creating the leader, AutoCAD prompts you to create an Mtext annotation.
- **Copy an Object**   After creating the leader, AutoCAD prompts you to select an existing text, Mtext, tolerance, or block reference object to use as the annotation.
- **Tolerance**   After creating the leader, AutoCAD displays the Tolerance dialog box for creating a feature control frame. You'll learn more about feature control frames later in this chapter.
- **Block Reference**   After creating the leader, AutoCAD prompts you to insert a block.
- **None**   The command ends after creating the leader.

When creating an Mtext annotation, you can also specify whether the command prompts you for the width of the multiline text object. You can also force all Mtext annotations to be left-justified, and create multiline text annotations surrounded by a rectangular frame.

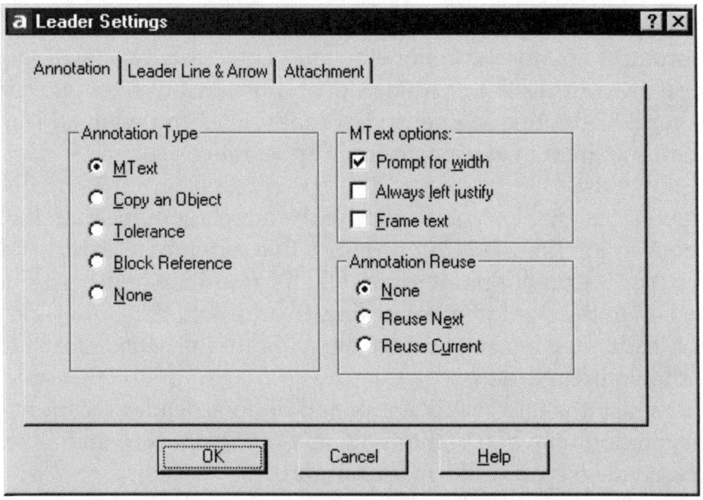

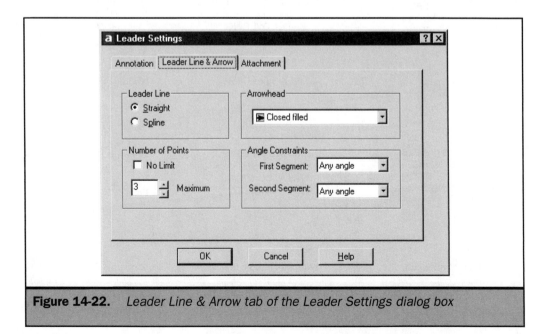

**Figure 14-21.**    *Annotation tab of the Leader Settings dialog box*

The Leader Line & Arrow tab (shown in Figure 14-22) determines the leader line and arrowhead format. You can create straight line or spline leaders and select the type of arrowhead. The controls in the Number Of Points area predetermine the number of

**Figure 14-22.**    *Leader Line & Arrow tab of the Leader Settings dialog box*

MOVING BEYOND
THE BASICS

points that the QLEADER command prompts you to specify before prompting you for the annotation. For example, if you set the number of points to 3, the command automatically prompts you for the annotation after you create two leader line segments. You can also use the controls in the Angle Constraints area to set angle constraints for the first and second leader line segments. For example, if you want all leader lines to start at a 45-degree angle (to have a consistent appearance), select 45° from the First Segment drop-down list.

The Attachment tab enables you to control the attachment location for leader lines and multiline text annotation. This tab provides five different attachment locations, and you can specify the attachment point separately for multiline text positioned to the left of a leader line and to the right of a leader line. In addition, the Underline Bottom Line option creates a leader line segment under the width of the Mtext annotation. These variations are shown in Figure 14-23.

After you establish the QLEADER command options, leader creation goes much faster, because you don't have to bother with any of the options, and all leaders and annotations you create have a uniform appearance.

To draw a simple leader with text annotation:

1. Start the QLEADER command.

2. Specify the starting point of the leader (the position of the arrow).

3. Specify the end point of the leader line segment, and then press ENTER.

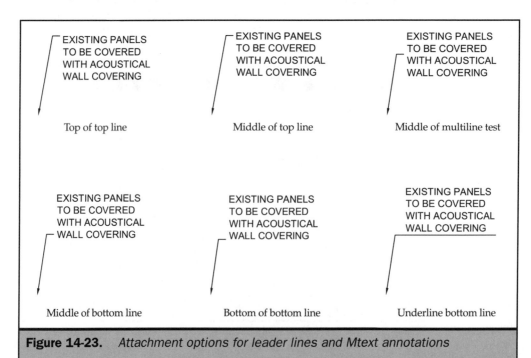

**Figure 14-23.** *Attachment options for leader lines and Mtext annotations*

4. Enter the first line of text. To type multiple lines of text, press ENTER at the end of each line.

5. Press ENTER again to end the command.

To draw a spline leader with text annotation entered by using the Multiline Text Editor dialog box:

1. Start the QLEADER command.

2. Press ENTER to display the Leader Settings dialog box.

3. On the Leader Line & Arrow tab, under Leader Line, select Spline, and then click OK.

4. Specify the starting point of the leader.

5. Specify the end point of the first leader line segment.

6. Specify additional leader line end points. You can see the appearance of the resulting spline leader as you specify the points. After you specify the last end point, press ENTER.

7. Specify the text width.

8. Press ENTER to open the Multiline Text Editor dialog box.

9. Enter the annotation text and then format it as you want it to appear. When finished, click OK.

**Note** *When prompted to specify the text width, if you specify a width of 0, the Width setting in the Multiline Text Editor dialog box is set to No Wrap. You can still create a multiline text annotation that fits within a specific width by specifying the Width value in the dialog box. Settings from the QLEADER command are applied as overrides to the current dimension style.*

## Modifying Leaders

As you've already learned, moving an annotation causes the last leader line segment to move so that it remains adjacent to the annotation. If you move the annotation past the midpoint of the last leader line segment, the justification of the annotation changes. For example, if the leader originally extended to the right, the resulting annotation text is left-justified. Moving the annotation to the left, beyond the midpoint of the leader line, causes the annotation to reformat so that it is right-justified.

Although leaders are associated with annotations, the reverse is not true. If you move the leader, the annotation does not move with it. Rather, the leader is moved independently. The associativity remains, however, so that if you then move the annotation, the leader adjusts so that the annotation remains at the same position, relative to the modified leader location, unless you move the annotation past the midpoint of the last leader line segment. In that case, the leader automatically reattaches itself to its original default position in relation to the annotation.

If you copy both the leader and its annotation, the copy retains its associativity. Removing either the leader or the annotation from the drawing—by erasing or exploding the objects or including them in a block—permanently breaks their associativity. When the associativity is broken, AutoCAD removes the hook line from the leader.

Other than the associativity between the annotation and its leader, they exist in the drawing as separate objects. Editing either—such as changing the color or linetype of the leader—has no effect on the other. You can change the appearance of leaders, however, by changing dimension style values affecting them and then updating the leader to reflect those changes. You can also modify leaders by using standard object modification commands, text editing commands, or the Properties window.

*Any modification to the annotation that changes its position or attachment point affects the position of the end point of the leader. For example, rotating the annotation causes the leader hook line to rotate.*

*LEARN BY EXAMPLE*
*Create multiple leaders from the same annotation. Open the drawing Figure 14-24 on the companion web site.*

You can use grip editing to create multiple leaders from the same annotation by using the Copy option. Use the following instructions:

1. Click the leader so that the grips become visible.

2. Click the grip at the arrowhead to make it the hot grip. AutoCAD prompts:

```
** STRETCH **
Specify stretch point or [Base point/Copy/Undo/eXit]:
```

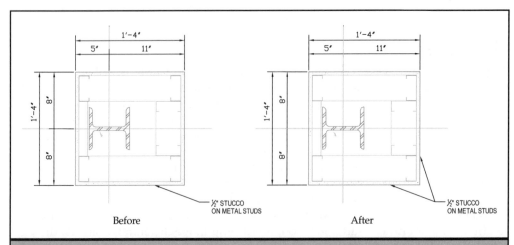

**Figure 14-24.** *Using grip editing to create multiple leaders from the same annotation*

3. Type **C** and press ENTER to select the Copy option.

4. Click to position the arrowhead of the new leader.

5. Press ENTER to exit from grip editing, and then press ESC to clear the grips.

# Understanding Dimension Styles and Variables

All dimensions that you place in the drawing are created using the current dimension style. Dimension styles provide a way for you to change various settings that control the appearance of dimensions. You can save those settings as a named dimension style and then restore them for reuse. If you don't define a dimension style before you create dimensions, AutoCAD uses the STANDARD dimension style, which stores the default dimension variable settings.

**Note**    *The STANDARD dimension style is based on the American National Standards Institute (ANSI) dimensioning standards. It doesn't completely conform to these standards, however. If you start a new drawing and select metric units, AutoCAD uses the ISO-25 (International Standards Organization) dimension style. The DIN (German) and JIS (Japanese Industrial Standards) styles are provided in the AutoCAD DIN and JIS drawing templates.*

You can control dimension system variables either by entering the variable name directly at the command line or by using the Dimension Style Manager dialog box, shown in Figure 14-25. Generally, using the dialog box is much easier than having to remember the names of AutoCAD's 70 dimension system variables. To display the Dimension Styles dialog box, do one of the following:

- On the Dimension toolbar, click Dimension Style.

- From the Dimension menu, choose Style.

- At the command line, type **DIMSTYLE** (or **D**, **DST**, or **DIMSTY**) and then press ENTER.

Each dimension style consists of a saved set of dimension properties that enables you to control the layout and appearance of dimensions by assigning the values of relevant dimension system variables. When you create a new dimension style, you can base the properties of the new style on those of an existing style. This helps you to create groups of related dimension styles. You can also create a substyle of an existing style, which specifies differences within a dimension style that are to be used only for a particular type of dimension. For example, within an existing style, you might specify a different type of arrowhead for angular dimensions, or a different dimension text color for ordinate dimensions.

MOVING BEYOND
THE BASICS

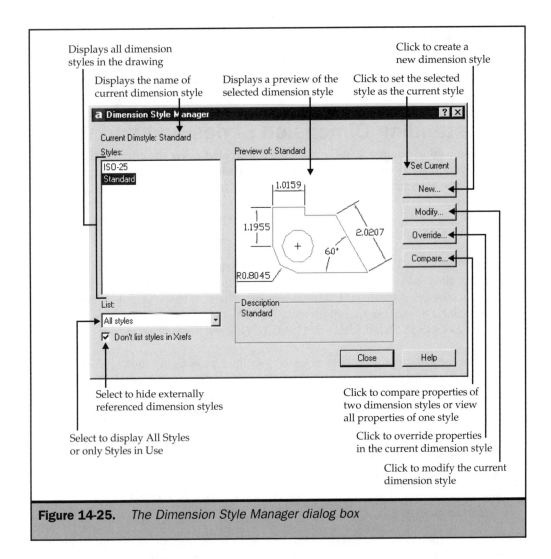

Displays all dimension styles in the drawing

Displays the name of current dimension style

Displays a preview of the selected dimension style

Click to create a new dimension style

Click to set the selected style as the current style

Select to hide externally referenced dimension styles

Click to compare properties of two dimension styles or view all properties of one style

Select to display All Styles or only Styles in Use

Click to override properties in the current dimension style

Click to modify the current dimension style

**Figure 14-25.** *The Dimension Style Manager dialog box*

When you create a new dimension style for a particular type of dimension, it appears in the Styles list in the Dimension Style Manager dialog box under the dimension style to which it belongs. For example, suppose you defined a new dimension style for angular dimensions based on the STANDARD style, in which you specified that arrowheads should appear as small dots. When you use the STANDARD style for angular dimensions, each arrowhead will appear as a small dot, but for all other dimensions, the arrowhead will remain as defined for the main STANDARD style (a closed filled arrow).

To create a dimension style:

1. Start the DIMSTYLE command.

2. In the Dimension Style Manager dialog box, click New. AutoCAD displays the Create New Dimension Style dialog box, shown in Figure 14-26.

3. In the Create New Dimension Style dialog box, enter the new style name.

4. Select the style you want to use as the basis for the new style.

5. Specify the dimension type for the new style (or use All Dimensions to define a dimension style for all the dimension types).

**Note**    *If you specify a dimension type, the New Style Name box becomes unavailable, because you are defining a substyle of an existing style.*

6. Click Continue. AutoCAD displays the New Dimension Style dialog box, in which you specify the various properties of the dimension. These properties are organized on six tabs, each controlling a different aspect of the dimension.

7. Choose any of the following tabs to control the dimension settings for the new style:

   ■ **Lines and Arrows**   Controls the appearance and properties for dimension lines, extension lines, arrowheads, and center marks

   ■ **Text**   Controls the format, placement, and alignment of dimension text

   ■ **Fit**   Controls the placement of dimension text, arrowheads, leader lines, and dimension lines

   ■ **Primary Units**   Controls the format and precision of primary dimension units and the prefixes and suffixes for dimension text

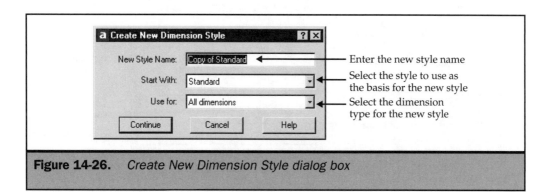

**Figure 14-26.**   *Create New Dimension Style dialog box*

- **Alternate Units**   Controls the format, precision, scale, and placement of alternate dimension units
- **Tolerances**   Controls the format and display of tolerances

You'll learn about each of these settings in the following sections.

8. After making changes on the various tabs of the New Dimension Style dialog box, click OK to return to the Dimension Style Manager dialog box.
9. Click Close to close the Dimension Style Manager dialog box.

## Controlling Dimension Lines and Arrows

The Lines and Arrows tab, shown in Figure 14-27, controls the appearance and properties for dimension lines, extension lines, arrowheads, and center marks. As you change any of the settings, the preview area immediately updates so that you can see your changes.

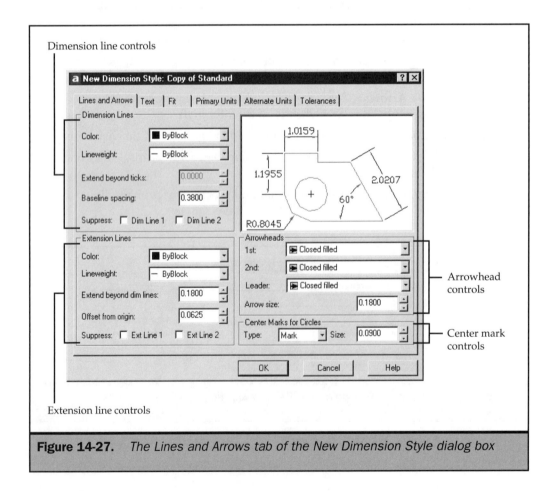

**Figure 14-27.**   *The Lines and Arrows tab of the New Dimension Style dialog box*

*Although its controls remain the same, the title of this dialog box varies depending on whether you are creating a new style, modifying an existing style, or overriding the current dimension style.*

## Dimension Line Controls

The Color and Lineweight drop-down lists determine the dimension line color and lineweight, either of which you can also specify explicitly. Extend Beyond Ticks determines how far the dimension line extends beyond the extension lines when you use oblique arrowheads. The Baseline Spacing value controls the distance between successive dimension lines when you create baseline or staggered dimensions. The two Suppress check boxes determine whether dimension lines are created. For example, when the Dim Line 1 check box is selected, the first dimension line is not created. AutoCAD determines the first and second dimension lines based on the order of selection of the extension line origins. All of these options are illustrated in Figure 14-28.

## Extension Line Controls

The controls in the Extension Lines area of the Lines and Arrows tab, also illustrated in Figure 14-28, are similar to those in the Dimension Lines area. The Color and Lineweight drop-down lists determine the extension line color and lineweight, respectively. Extend Beyond Dim Lines determines how far the extension line extends beyond the dimension line, while Offset From Origin determines how far the extension line is offset from its origin point. The two Suppress check boxes determine whether AutoCAD creates the first and second extension lines.

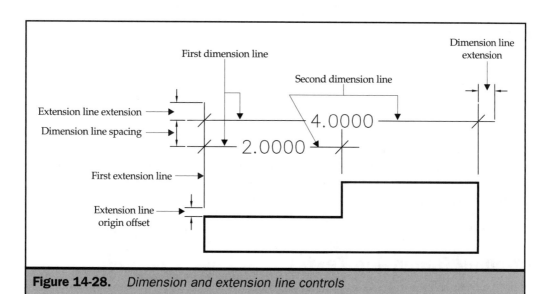

**Figure 14-28.** *Dimension and extension line controls*

MOVING BEYOND
THE BASICS

## Arrowhead Controls

The controls in the Arrowheads area of the Lines and Arrows tab determine the appearance of the arrowheads that are drawn when you create dimensions. AutoCAD provides 19 different arrowheads; you can also create your own or specify that no arrowhead be drawn.

To select one of AutoCAD's predefined arrowheads, choose it from the drop-down list. When you select the first arrowhead, AutoCAD automatically selects the same arrowhead as the second arrowhead. To choose different first and second arrowheads, select the first arrowhead and then the second one. Note that although the first arrowhead is shown in the preview area as the left-hand or bottom arrow, the actual determination of first and second arrowheads is based on the order in which you select the extension line origins.

You can also specify a user-defined arrowhead by choosing User Arrow from the drop-down list. AutoCAD then displays the Select Custom Arrow Block dialog box. From the drop-down list, you can select any block already defined in the current drawing, and then click OK. The block you select is then used as the specified arrowhead.

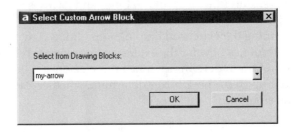

*When you create your own arrowheads, the overall size of the arrowhead block should be one drawing unit, so that the resulting arrowhead scales properly when it is used in a drawing. Create the right-hand arrowhead. AutoCAD automatically reverses it when it occurs at the left end of a dimension line. To ensure that your user-defined arrowheads are always available when you create a new drawing, save the arrowhead block as part of your template drawing.*

## Center Mark Controls

The controls in the Center Marks For Circles area of the Lines and Arrows tab determine the appearance of center marks and centerlines for any diameter and radius dimensions. From the Type drop-down list, choose Mark to create a center mark, Line to create a centerline, or None to suppress the creation of center marks. The value in the Size box controls the size of the center mark or centerline, as shown in Figure 14-29.

# Controlling Dimension Text

The Text tab, shown in Figure 14-30, controls the format, placement, and alignment of dimension text. As you change any of the settings, the preview area immediately updates so that you can see your changes.

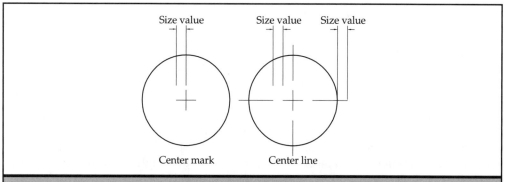

**Figure 14-29.**    *Center mark and centerline controls*

MOVING BEYOND
THE BASICS

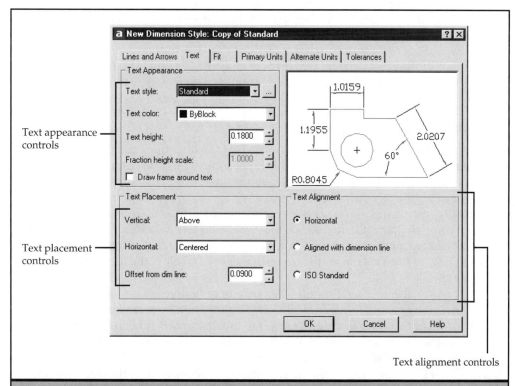

**Figure 14-30.**    *The Text tab of the New Dimension Style dialog box*

## Text Appearance Controls

The controls in the Text Appearance area of the Text tab determine the appearance properties of dimension text. Many of these controls are similar to the controls that you already learned about for creating text objects. From the Text Style drop-down list, you can choose any text style that is already defined in the drawing. The Text Color drop-down list lets you explicitly specify the dimension text color. To define a new text style, click the button adjacent to the Text Style drop-down list to display the Text Style dialog box. The Text Height value determines the height of dimension text. If the primary dimension units are either architectural or fractional, you can also specify the Fractional Height Scale, which determines the scale of fractions relative to the height of the dimension text.

**Caution**  *If the text style you select has a fixed height (that is, the height of the text is explicitly specified within the text style definition as a value greater than 0), the explicit text height overrides the text height specified as part of the dimension style. In that case, dimension text will default to the text height defined in the text style, and the size of dimension text objects won't update when you change the overall dimension scale (controlled by the DIMSCALE system variable). To avoid this problem, you should assign dimensions a text style that has a text height value of 0.*

If you select the Draw Frame Around Text check box, AutoCAD draws a rectangle around the dimension text. The size of this rectangle is determined by the DIMGAP system variable, which is controlled by the Offset From Dim Line value under Text Placement.

## Text Placement Controls

The controls in the Text Placement area of the Text tab determine the placement of dimension text. You can control the vertical and horizontal position of the text separately in relation to the dimension line. The Vertical controls let you position the text centered, above, or outside the dimension line, or you can choose to have it conform to the Japanese Industrial Standards (JIS). When centered, the dimension text splits the dimension line. The Above option places the text above the dimension line, whereas the Outside option places the dimension text on the side of the dimension line that is farthest from the extension line origin. This places the text either above or below the dimension line, as appropriate.

The Horizontal controls determine the position of dimension text along the length of the dimension line and in relation to the extension lines. Text can be centered along the dimension line, placed at the end closest to the first or second extension line, or placed over the first or second extension lines. You can choose any of the five options from the drop-down list.

The Offset From Dim Line value determines the space between the dimension text and the dimension line when the text splits the dimension line. This value is also used to determine whether the dimension text fits in the space between the extension lines, and to control the size of the rectangle drawn around the dimension text when you select the Draw Frame Around Text check box under Text Appearance.

## Text Alignment Controls

The controls in the Text Alignment area of the Text tab determine the orientation of the dimension text based on its position in relation to the extension lines and orientation of the dimension line. By selecting one of the three buttons, you can specify that the text is forced to remain horizontal, aligned at the same angle as the dimension line, or set to conform to ISO standards (text is aligned with the dimension line when placed inside the extension lines, but aligned horizontally when placed outside the extension lines).

 *Many of the dimension text orientation and placement options are interrelated. For example, although the Above option places dimension text above the dimension line, when the dimension line isn't horizontal, and the dimension text inside the extension lines is forced to be horizontal, the dimension text splits the dimension line.*

# Controlling the Fit of Dimension Text and Arrowheads

The Fit tab, shown in Figure 14-31, controls the placement of dimension text, arrowheads, leader lines, and dimension lines. As you change any of these settings, the preview area immediately updates so that you can see the effects of your changes.

## Dimension Fitting Options

The controls in the Fit Options area of the Fit tab determine the placement of dimension text and arrowheads inside or outside the extension lines, based on the space available between those extension lines. When there is enough space between the extension lines for both the text and arrowheads, AutoCAD always places both between the extension lines. When there isn't enough room, however, the controls in this area determine which components of the dimension are placed between the extension lines. You can choose one of the following settings:

- **Either the Text or the Arrows, Whichever Fits Best**   As its name implies, when there isn't enough room for both the text and arrowheads between the extension lines, AutoCAD determines which of these components fits best. If there isn't room for either the text or the arrowheads, both are placed outside the extension lines.

- **Arrows**   When there isn't enough room for both the text and arrowheads, AutoCAD places the arrowheads outside the extension lines. If there isn't enough room for the arrowheads, both the text and arrowheads are placed outside the extension lines.

- **Text**   When there isn't enough room for both the text and arrowheads, AutoCAD places the text outside the extension lines. If there isn't enough room for the text, both the text and arrowheads are placed outside the extension lines.

- **Both Text and Arrows**   When there isn't enough room for both the text and arrowheads between the extension lines, both are placed outside the extension lines.

MOVING BEYOND THE BASICS

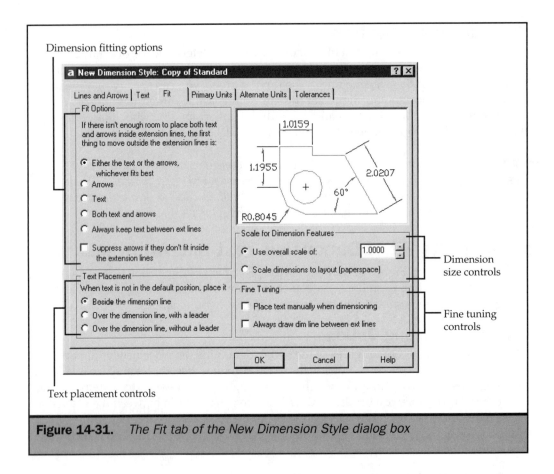

**Figure 14-31.** *The Fit tab of the New Dimension Style dialog box*

■ **Always Keep Text Between Ext Lines**   This setting forces AutoCAD to always place the text between the extension lines, even if it doesn't fit.

In addition to these five buttons, you can also select the check box at the bottom of the Fit Options area to suppress the arrowheads if there isn't enough space for them between the extension lines. Figure 14-32 shows the effects of these various settings.

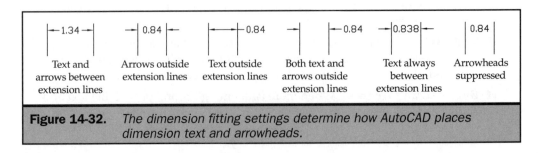

**Figure 14-32.** *The dimension fitting settings determine how AutoCAD places dimension text and arrowheads.*

## Text Placement Controls

The controls in the Text Placement area of the Fit tab determine the placement of dimension text when it is moved from its default position as defined by the dimension style. You can choose one of the following settings:

- **Beside the Dimension Line**   If the dimension text is moved away from the dimension line, it is placed alongside the dimension line.

- **Over the Dimension Line, With a Leader**   If the dimension text is moved away from the dimension line, AutoCAD creates a leader connecting the text to the dimension line.

- **Over the Dimension Line, Without a Leader**   If the dimension text is moved away from the dimension line, the text is not connected to the dimension line.

You can move dimension text from its default position by using grip editing. Figure 14-33 shows the effects of these various settings.

## Scale Controls

The controls in the Scale For Dimension Features area of the Fit tab determine the overall size of dimension geometry relative to the objects in the drawing. The dimension scale determines the size, distance, or spacing of the dimension components, such as the dimension text height, arrowhead size, and offset distances. The scale value has no effect on the actual measured dimension.

The option you choose and the scale value you use depends on how you lay out your drawing and create dimensions. There are three different methods you can use:

- **Dimension in model space and plot in model space**   If you create dimensions in model space and plot your drawing in model space, you should select the Use Overall Scale Of radio button and then specify a Scale value appropriate to the scale at which you will eventually plot your drawing. Generally, the Scale value should be the inverse of the intended plot scale. For example, if you plan to plot your drawing at 1/8 of its actual size, you should set the Scale value to 8. For a drawing to be plotted at 1/8"=1'-0", set the Scale value to 96. Refer to Table 2-2 for other appropriate Scale value settings.

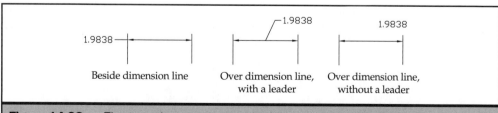

**Figure 14-33.**   *The text placement controls determine how AutoCAD places dimension text when it is moved from its default position.*

■ **Dimension in model space and plot in paper space** If you create dimensions in model space and then create a paper space layout for plotting, you should select the Scale Dimensions To Layout (Paperspace) radio button. This causes AutoCAD to determine a scale factor based on the scaling between the current model space viewport and paper space. This was the preferred method for complex, multiple-view drawings prior to AutoCAD 2002, and is still the method you should use when the dimensions in a drawing need to be referenced by other drawings or when dimensioning 3-D isometric views. When you use this method, in order to prevent dimensions in one layout viewport from being displayed in other layout viewports, you must create a separate dimensioning layer for each layout viewport that is frozen in all of the other layout viewports. You'll learn more about this and about paper space in general in Chapter 17.

■ **Dimension in layouts** In AutoCAD 2002, when using associative dimensions, you can simply create dimensions in paper space by selecting model space objects or specifying object snap points on model space objects. The associativity between paper space dimensions and model space objects is maintained. You don't need to change any of the dimension scale controls. AutoCAD automatically adjusts the dimension values based on the scale of each viewport.

## Fine Tuning

The two check boxes in the Fine Tuning area of the Fit tab let you determine additional fit options. When you select the Place Text Manually When Dimensioning check box, AutoCAD ignores any horizontal justification settings and places the text at the position you specify when you create the dimension. When the Always Draw Dim Line Between Ext Lines check box is selected, AutoCAD always draws the dimension line between the extension lines, even when the arrowheads are placed outside the extension lines.

# Controlling the Primary Dimension Units Format

The Primary Units tab, shown in Figure 14-34, controls the format and precision of primary dimension units and the prefixes and suffixes for dimension text. As you change any of these settings, the preview area immediately updates so that you can see the effects of your changes.

## Linear Dimension Controls

The controls in the Linear Dimensions area of the Primary Units tab determine the format and precision for linear dimensions. You can select the type of units—Scientific, Decimal,

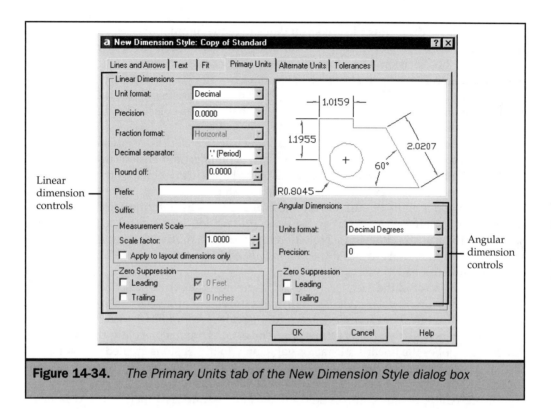

**Figure 14-34.** *The Primary Units tab of the New Dimension Style dialog box*

Engineering, Architectural, Fractional, or Windows Desktop (which uses the settings on the Windows Regional Settings Control Panel)—from the Unit Format drop-down list.

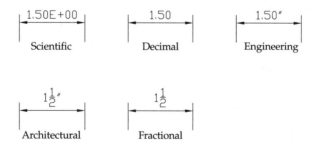

MOVING BEYOND
THE BASICS

The Precision drop-down list lets you determine the number of decimal places displayed in the dimension text. If you are using either architectural or fractional dimensions, the Precision drop-down list lets you select the fractional denominator, and the Fraction Format control becomes available so that you can specify the format for fractions: Diagonal, Horizontal, or Not Stacked. If you are using decimal units, the Decimal Separator control becomes available so that you can select a Period, Comma, or Space as the separator for decimal formats.

The Round Off value determines the rounding rules for all dimension measurements except angular dimensions. For example, you can round linear dimensions to the nearest 0.25 units. Note that the number of digits displayed after the decimal point depends on the Precision value, however, not the Round Off value.

You can add a prefix or suffix to all linear dimensions by entering the desired text in the Prefix or Suffix boxes. You can use special characters and control codes to enter special prefix or suffix characters. Note that if you enter a prefix, it replaces the radius or diameter symbol that AutoCAD automatically adds when you create radius or diameter dimensions.

The Scale Factor in the Measurement Scale area sets the actual scale factor for linear dimension measurements. AutoCAD multiplies the actual dimension measurement by the value in this box. For example, if you enter a scale factor of 5, a 1-inch dimension displays as 5 inches. If you select the Apply To Layout Dimensions Only check box, AutoCAD applies the linear scale factor value only to dimensions created in layouts, so that the length scale factor reflects the zoom scale factor for objects in a model space viewport. You'll learn more about model space viewports in layouts in Chapter 17.

The Zero Suppression controls determine whether AutoCAD includes zero values in the dimension text. The Leading and Trailing check boxes suppress zeros in scientific, decimal, engineering, and fractional dimensions, while the Feet and Inches check boxes control whether AutoCAD includes a zero when the feet or inches value is zero when the unit format is engineering or architectural. For example, by suppressing trailing zeros in decimal dimensions, a dimension that normally appears as 2.5000 is drawn as 2.5. Similarly, by suppressing zero inches, an architectural dimension that normally appears as 14'-0" simply appears as 14'.

## Angular Dimension Controls

The controls in the Angular Dimensions area of the Primary Units tab determine the angle format for angular dimensions. You can select the type of units—Decimal Degrees, Degrees Minutes Seconds, Gradians, or Radians—from the Units Format drop-down list. The Precision selections let you determine the number of decimal places displayed for angular dimensions.

The two check boxes in the Zero Suppression area let you suppress leading and trailing zeros for angular dimensions. For example, by suppressing trailing zeros in a decimal angular dimension, a dimension that normally appears as 60.00° is drawn as 60°.

# Controlling Alternate Dimension Units

The Alternate Units tab, shown in Figure 14-35, controls the format, precision, scale, and placement of alternate dimension units. As you change any of these settings, the preview area immediately updates so that you can see the effects of your changes.

If you select the Display Alternate Units check box, AutoCAD creates both alternate units and primary units. You use alternate units when you need to dimension objects by using two systems of measurement, such as English and metric units. Alternate units appear inside square brackets.

## Alternate Units Controls

The controls in the Alternate Units area are similar to many of those in the Linear Dimensions area of the Primary Units tab. For example, you can select the type of units—Scientific, Decimal, Engineering, Architectural Stacked, Fractional Stacked, Architectural, Fractional, or Windows Desktop—from the Unit Format drop-down list. Similarly, the Precision selections let you determine the number of decimal places or the denominator for alternate units.

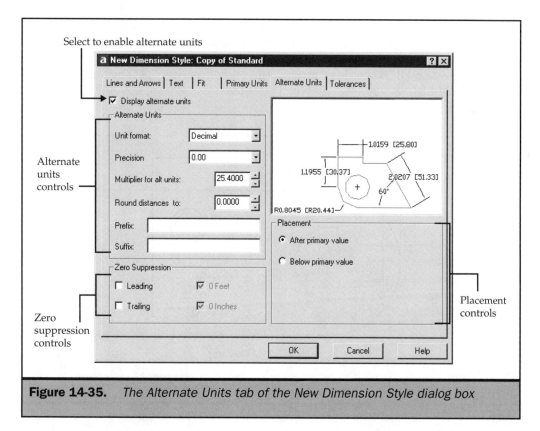

**Figure 14-35.** *The Alternate Units tab of the New Dimension Style dialog box*

MOVING BEYOND
THE BASICS

One control that is different, however, is the Multiplier For Alt Units. This value specifies the multiplier for conversion between the primary and alternate units. This value is initially set to 25.4 units, which is the number of millimeters per inch. AutoCAD calculates alternate units by multiplying the measured distance by this scale factor. The Round Distances To value determines the rounding rules for all alternate dimensions. You can add a prefix or suffix to all alternate dimensions by entering the desired text in the Prefix or Suffix boxes. Figure 14-36 illustrates some typical dimensions with alternate units. The mm suffix is included by specifying it in the Suffix box; the inch symbol is included by selecting the engineering format for the primary units.

*Including the special formatting character \X in a dimension prefix or suffix causes any text preceding that character to be aligned above the dimension line, and causes any text following the character to be aligned below the dimension line. To include additional dimension text below the dimension line, include the special formatting character \P after the text.*

## Zero Suppression

The controls in the Zero Suppression area of the Alternate Units tab determine whether AutoCAD includes zero values in the alternate dimensions. The Leading and Trailing check boxes suppress zeros in scientific, decimal, engineering, and fractional dimensions, while the 0 Feet and 0 Inches check boxes control whether AutoCAD includes a zero when the feet or inches value is zero when the alternate unit format is engineering or architectural.

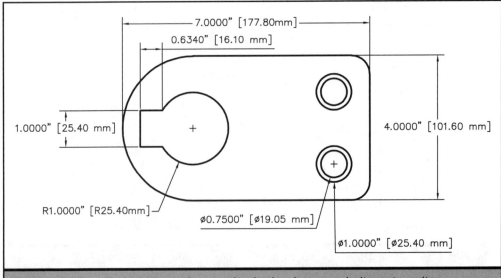

**Figure 14-36.** *Typical dimensions, using both primary and alternate units*

## Placement

The controls in the Placement area of the Alternate Units tab control the placement of the alternate units. You can specify that the alternate units be placed after the primary units or below the primary units.

# Controlling Dimension Tolerances

The Tolerances tab, shown in Figure 14-37, controls the format and display of tolerances. As you change any of these settings, the preview area immediately updates so that you can see the effects of your changes.

## Tolerance Format Controls

The controls in the Tolerance Format area of the Tolerances tab control the creation and format of dimension tolerances for primary dimensions. You can choose one of the following five tolerance methods from the Method drop-down list (see Figure 14-38):

- **None**   No tolerance values are added to the dimension.
- **Symmetrical**   A plus-or-minus sign (±) is appended to the dimension, along with a single tolerance value determined by the Upper Value.

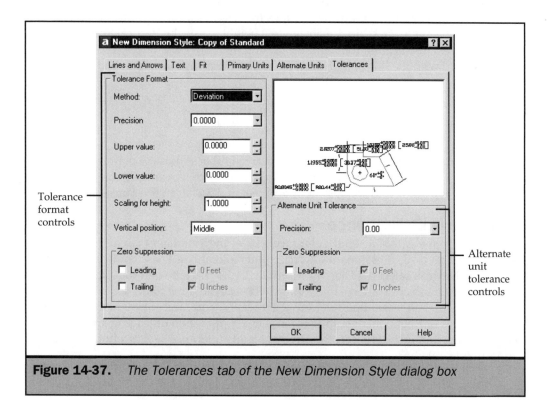

**Figure 14-37.**   *The Tolerances tab of the New Dimension Style dialog box*

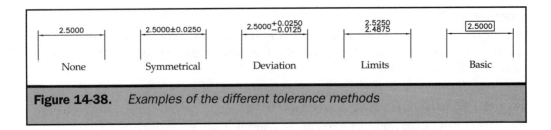

**Figure 14-38.** *Examples of the different tolerance methods*

- **Deviation** Separate upper and lower tolerance values, determined by the Upper and Lower Values, respectively, are appended to the dimension, along with plus (+) and minus (−) signs.
- **Limits** Separate upper and lower limit dimensions, determined by adding the Upper and Lower Values, are created instead of the primary dimension.
- **Basic** No tolerance values are added, but the dimension is created with a box around it, to indicate that it is a basic dimension.

The Precision selection lets you determine the number of decimal places or fractional denominator, depending on the type of primary units specified on the Primary Units tab. The Scaling For Height value determines the height for tolerance text as a ratio of the tolerance height to the main dimension text height, while the Vertical Position setting controls the text justification for symmetrical and deviation tolerances. The tolerance can be aligned with the Bottom, Middle, or Top of the main dimension text.

The controls in the Zero Suppression area are similar to those you already learned about on the Primary Units and Alternate Units tabs. In this case, they control the suppression of leading and trailing zeros and feet and inches in tolerances.

## Alternate Unit Tolerance

The controls in the Alternate Unit Tolerance area of the Tolerances tab determine the precision and zero suppression for alternate units tolerances. These controls are only available when you select the Display Alternate Units check box on the Alternate Units tab. When alternate units are created, AutoCAD uses the same tolerance method for both primary and alternate unit tolerances. However, you can control the number of decimal places or fractional denominator and the suppression of leading and trailing zeros and feet and inches within the alternate units tolerance separately from the primary units tolerance.

### Exporting and Importing Dimension Styles

The dimension styles that you create are saved only within the current drawing. Ideally, once you define dimension styles for specific types of drawings, such as architectural floor plans or mechanical part drawings, you save them in a template drawing. That way, when you need to create another, similar type of drawing, you can begin by using that template drawing.

Sometimes, however, you may create a dimension style in a particular drawing and want to reuse that style in another drawing. The DIMEX Express tool lets you export named dimension styles from a drawing to a separate file. You can then use the DIMIM Express tool to import dimension styles into another drawing. You can also simply drag and drop dimension styles from any drawing by using AutoCAD DesignCenter. The Express tools are described in detail on the companion web site. You'll learn about AutoCAD DesignCenter in Chapter 16.

# Managing Dimension Styles

When you create a new dimension, it is created using the current dimension style. If you have several dimension styles defined in a drawing, you should select the desired current dimension style before creating a new dimension. With multiple dimension styles, it is not uncommon to create a dimension using a different style than the one you wanted, and you therefore need a way to change the dimension style assigned to an existing dimension. You sometimes may also need to change the appearance of a few particular dimensions, but decide that the changes don't really warrant the creation of a new dimension style. AutoCAD enables you to override specific dimension variables to create the unique dimension. In yet a third situation, you may decide to modify some of the style variables for all the dimensions drawn using that style. You can then update all the dimensions that were already created with that named style.

To set the current dimension style, do one of the following:

- On the Dimension toolbar, select the dimension style from the Dim Style Control drop-down list.

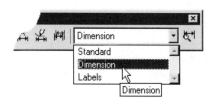

■ Start the DIMSTYLE command, select the dimension style in the Styles list, and click Set Current. You can also simply double-click the dimension style in the Styles list.

*You can use the command-line version of the DIMSTYLE command to set the current style to that of an existing dimension:*

1. *At the command line, type* **-DIMSTYLE** *and press* ENTER.

2. *Type* **R** *(for Restore), and press* ENTER *twice.*

3. *When AutoCAD prompts you to select a dimension, click the dimension whose style you want to make current.*

*Note that this method even works for dimension style overrides.*

To rename a dimension style:

1. Start the DIMSTYLE command.

2. In the Dimension Style Manager dialog box, right-click the dimension style name in the Style list, and then choose Rename in the shortcut menu.

3. Type the new dimension style name and press ENTER.

You can also use the shortcut menu to delete a dimension style, or simply select the dimension style name in the Style list, and press DELETE. Note, however, that you can't delete a dimension style if it is the current style, if dimensions in the drawing use that style, or if the style has substyles associated with it.

## Changing the Dimension Style Assigned to a Dimension

The shortcut menus make it easy to change the dimension style assigned to an existing dimension.

To change the dimension style assigned to a dimension:

1. Select the dimension whose style you want to change.

2. Right-click and choose Dim Style in the shortcut menu.

3. Choose the dimension style you want to assign from the shortcut submenu.

*The shortcut submenu displays only six dimension styles. If the drawing has more than six dimension styles, choose Other to display the Apply Dimension Style dialog box, which lists the names of all the dimension styles in the drawing.*

To update a dimension to the current style:

1. Do one of the following:

- On the Dimension toolbar, click Dimension Update.
- From the Dimension menu, choose Update.
- At the command line, type **-DIMSTYLE**, press ENTER, type **A** (for Apply), and then press ENTER again.

2. Select the dimensions to be updated, and then press ENTER.

# Overriding Dimension Variables

The most direct way to override a dimension variable is to type the variable name and then change its value as you create the dimension. Another way is to use the DIMOVERRIDE command. Unfortunately, both methods require that you know which of AutoCAD's 70 dimension system variables need to be altered. If you do want to use this method, you'll find a complete list of system variables, along with their default values and meanings, in Appendix B.

A more practical way to override dimension variables is to use the Dimension Style Manager dialog box. To create dimension style overrides:

1. In the Dimension Style Manager, select the dimension style that you want to use, and then click Override.

2. On the appropriate tabs, change the values that you want to override, and then click OK. AutoCAD displays <style overrides> below the dimension style name in the Styles list.

3. Click Close to close the Dimension Style Manager dialog box.

To create a new dimension by using the dimension overrides, simply create the dimension as you normally would. The overrides are automatically applied to the new dimensions. To update an existing dimension to use the override settings, update the dimension as described in the previous section.

You can't set the overrides as the current dimension style, however, because they are already current. If you select another style as the current dimension style, the overrides are discarded. You can save the overrides by either renaming the overrides to a new style or saving the settings to the current style.

To save overrides to the current style:

1. In the Dimension Styles Manager dialog box, right-click <style overrides> in the Styles list.

2. From the shortcut menu, choose Save To Current Style.

MOVING BEYOND THE BASICS

To rename the overrides to a new style:

1. In the Dimension Styles Manager dialog box, click New.

2. In the Create New Dimension Style dialog box, enter a name for the new dimension style. Notice that <style overrides> appears in the Start With box. Then, click Continue.

3. In the New Dimension Style dialog box, click OK.

# Modifying Dimension Styles

If you make changes to dimension style settings as overrides or save your changes as a new dimension style, those changes appear only in new dimensions or in those that you explicitly update. When you modify an existing dimension style, those changes are also normally reflected only in subsequent dimensions created using that style. If you modify the current dimension style, however, any changes you make are immediately reflected in any existing dimensions that were drawn using that style. For example, if you change the dimension text color to red for the current dimension style, all the dimensions in your drawing created using that style will update so that the dimension text color is red.

To modify a dimension style:

1. In the Dimension Styles Manager dialog box, select the style you want to modify in the Styles list, and click Modify.

2. In the Modify Dimension Style dialog box, change the dimension style settings.

3. When you are finished making changes, click OK, and then click Close to close the Dimension Styles Manager dialog box.

*You can also use the MATCHPROP command to match object dimension properties.*

# Comparing Dimension Styles

You may find that when you have multiple dimension styles defined in a drawing, you forget what a particular style is supposed to be used for. In such situations, it is often helpful to be able to review the dimension variable settings for a particular dimension style, or to compare the differences between two styles. You can use the Compare function in the Dimension Style Manager to compare two dimension styles or list all the properties for a single dimension style.

When you compare dimension styles, AutoCAD displays the Compare Dimension Styles dialog box, shown in Figure 14-39. In the two drop-down lists, select the dimension styles you want to compare. If you select the same style in both lists, AutoCAD displays all the properties for that style. If you select two different styles, the program lists only the differences between the two styles. After you compare the dimension styles, you can click the Copy To Clipboard button to copy the information displayed in the dialog box to the Windows Clipboard. You can then paste that information into another program.

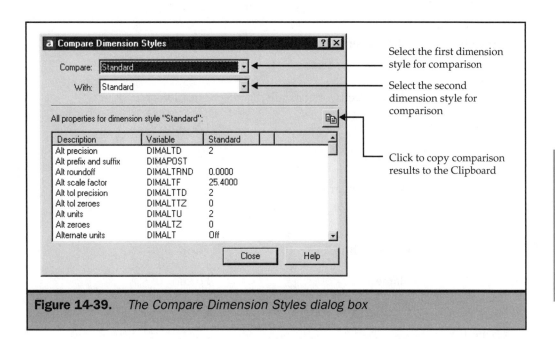

**Figure 14-39.**   *The Compare Dimension Styles dialog box*

Select the first dimension style for comparison

Select the second dimension style for comparison

Click to copy comparison results to the Clipboard

MOVING BEYOND
THE BASICS

# Adding Geometric Tolerances

When you create drawings for mechanical parts, you need to realize that no part can be manufactured to exact dimensions. Rather, the designer needs to consider the amount of variation that is acceptable in a given part. Geometric tolerances (sometimes called *feature-based* tolerances) define a tolerance zone that is specific to a particular feature and that establishes the degree to which the feature can vary and still be considered acceptable. This drafting convention uses a specific set of tolerance symbols for controlling the geometric tolerance characteristics for position, concentricity, symmetry, parallelism, perpendicularity, angularity, cylindricity, flatness, roundness, straightness, and so on.

   *This book doesn't attempt to explain how to use geometric tolerances properly, but instead simply describes how to add feature control frames to your drawings. To understand fully how to apply geometric tolerances, you should refer to a textbook on engineering drafting.*

   Geometric tolerances use a feature control frame consisting of a rectangle that is divided into compartments. Each feature control frame consists of at least two compartments. The first compartment contains a symbol that indicates the type of geometry being controlled, such as location, orientation, or form. The geometric tolerance symbols and their characteristics are shown in Table 14-1.

| Symbol | Characteristic | Type |
|--------|----------------|------|
| ⊕ | Position | Location |
| ◎ | Concentricity or coaxiality | Location |
| ⩵ | Symmetry | Location |
| // | Parallelism | Orientation |
| ⊥ | Perpendicularity | Orientation |
| ∠ | Angularity | Orientation |
| ⌭ | Cylindricity | Form |
| ▱ | Flatness | Form |
| ○ | Circularity or roundness | Form |
| — | Straightness | Form |
| ⌒ | Profile of a surface | Profile |
| ⌒ | Profile of a line | Profile |
| ↗ | Circular runout | Runout |
| ↗↗ | Total runout | Runout |

**Table 14-1.** *Geometric Tolerance Symbols*

**Note** *These symbols are obtained from the GDT.SHX font support file.*

The second compartment contains the tolerance value. When appropriate, this value is preceded by a diameter symbol and followed by a material condition symbol. Material conditions apply to features that can vary in size. Table 14-2 shows the material condition symbols and their meanings.

These two compartments can be followed by up to three datum references, each of which can optionally include material conditions at each datum. Datum references generally are used as reference tolerances to one of up to three perpendicular planes from which a measurement is to be made, although datum reference letters can also indicate an exact point or axis. Figure 14-40 shows a typical feature control frame for a hole.

| Symbol | Meaning |
|--------|---------|
| Ⓜ | At maximum material condition (MMC), a feature contains the maximum amount of material stated in the limits. |
| Ⓛ | At least material condition (LMC), a feature contains the minimum amount of material stated in the limits. |
| Ⓢ | Regardless of feature size (RFS) indicates that the feature can be any size within the stated limits. |

**Table 14-2.**   *Material Condition Symbols*

MOVING BEYOND
THE BASICS

*The color and text style settings used by the TOLERANCE command are set in the Lines and Arrows tab and Text tab of the Dimension Style Manager dialog box, respectively. Be sure to use a text style with a height value of 0 rather than a text style with a fixed text height, so that the height can be determined when you create the actual tolerance.*

In this example, the first symbol indicates that the geometric characteristic being controlled is the position of the hole. The next symbol specifies the size of the tolerance zone, which, in this case, is a diameter of 0.025 around the perfect position of the hole. The M within the circle following this value means that the tolerance is measured at the maximum material condition. In other words, when the hole is at its smallest, the tolerance must apply, but when the hole is larger, a greater tolerance zone may be calculated, based on the actual size of the hole. The remaining box, the datum reference, indicates that the position of the feature is measured from datum surfaces A, B, and C.

**Note**   *Not every feature control frame requires a datum. For example, flatness can simply refer to a feature, without referring to another surface. But features such as perpendicularity, which by their nature refer to another surface, must include a datum to identify the other surface.*

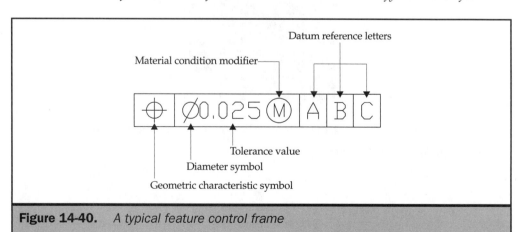

**Figure 14-40.**   *A typical feature control frame*

As you already learned, you can place a feature control frame at the end of a leader that has been created by using the LEADER command. In many instances, however, the feature control frame is simply placed below or adjacent to another dimension, to indicate the allowable tolerances for the dimensioned feature. For example, when you dimension the diameter of a hole in a machined part, you first dimension the diameter of the hole and then place a feature control frame below the diameter, to show the allowable dimensional variation of the hole.

**Note** *When you use geometric tolerances to define the allowable positional variation, you create the dimension that locates the position of the hole as a basic dimension (which, as you learned previously, has a box around it) and then use a leader to attach a feature control frame to the hole, defining the tolerance zone for locating the hole.*

When two tolerances apply to the same geometry, you can also add a *composite tolerance,* consisting of a primary tolerance value followed by a secondary tolerance value. To make a tolerance even more specific, it can also contain a *projected tolerance,* consisting of a height value followed by a projected tolerance symbol. For example, you can use a projected tolerance to indicate the perpendicularity of an embedded part.

To add a feature control frame:

1. Do one of the following:

   ■ On the Dimension toolbar, click Tolerance.

   ■ From the Dimension menu, choose Tolerance.

   ■ At the command line, type **TOLERANCE** (or **TOL**) and then press ENTER.

   AutoCAD displays the Geometric Tolerance dialog box, shown in Figure 14-41.

2. To include a geometric tolerance symbol, click inside the Sym box to display the Symbol dialog box, and then click to select the appropriate symbol.

3. Under Tolerance 1, click the diameter box to add a diameter symbol.

4. In the value box, type the first tolerance value.

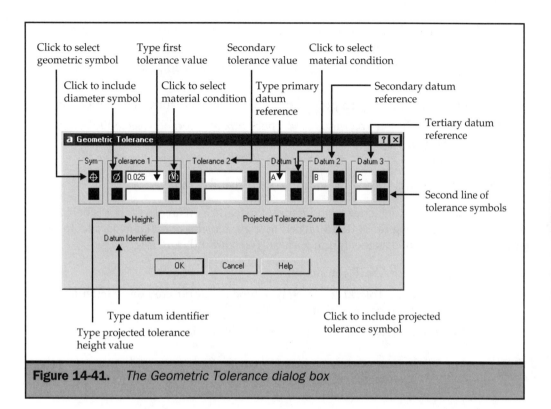

**Figure 14-41.** *The Geometric Tolerance dialog box*

5. To include a material condition symbol, click the material condition box to display the Material Condition dialog box, and then click to select the appropriate material condition symbol.

6. Under Tolerance 2, repeat Steps 3 through 5 to add a secondary tolerance value, if appropriate.

7. Under Datum 1, type the primary datum reference letter in the Datum box.

8. To include a material condition for the datum, click the material condition box to display the Material Conditions dialog box, and then click to select the appropriate material condition symbol.

9. Repeat Steps 7 and 8 to add secondary and tertiary datum under Datum 2 and Datum 3, if appropriate.

10. On the second line, repeat Steps 2 through 9 to add composite tolerances, if appropriate.

11. In the Height box, type a projected tolerance zone height value, if appropriate.

12. Click the Projected Tolerance Zone check box to insert a projected tolerance zone symbol, if appropriate.

13. Click OK.

14. In the drawing, specify the location of the feature frame.

*Datum flags* are boxed letters that identify the feature on an object that is being used as a datum. Datum flags are often attached to the extension line that is parallel to the surface being identified as the datum surface. To create a datum flag:

1. Start the TOLERANCE command.

2. In the Geometric Tolerance dialog box, type the datum reference letter in the Datum Identifier box and then click OK.

3. In the drawing, specify the location of the datum flag.

**Tip**  *You can modify tolerance values by double-clicking the geometric tolerance or by using the DDEDIT command or the Properties window.*

# Chapter 15

## Working with Blocks, Attributes, and External References

B locks, attributes, and external references provide you with mechanisms to manage objects in your drawings and include additional information with standard drawing objects. Using blocks, you can combine numerous objects into a single object and then reuse it, by inserting multiple copies. With attributes, you can associate text, such as part numbers or prices, as part of a block and then extract that attribute information to a separate file, such as a database, for further analysis. External references enable you to link separate reference drawing files to the current drawing, to combine information without actually adding the contents of the reference drawings to the current drawing. If you need to make changes, you can edit the external reference from within the drawing that references it. And if you make changes to the referenced file, all the references are updated automatically.

This chapter explains how to do the following:

■ Create, insert, and redefine blocks

■ Create, edit, and insert attributes

■ Extract attribute data to a separate file

■ Attach and work with external references

■ Edit external references in place

# Working with Blocks

Blocks generally consist of several objects combined into a single object that you can insert into a drawing and manipulate as single object. A block can consist of visible objects, such as lines, arcs, and circles, as well as visible or invisible data, called *attributes*. Blocks are stored as part of the drawing file.

Blocks can help you better organize your work, quickly create and revise drawings, and reduce your drawing file size. Using blocks, you can create a library of frequently used symbols. After you create these symbols, you can insert them as blocks, instead of redrawing a particular symbol from scratch each time that you want to use it.

When you insert a block into a drawing, AutoCAD actually inserts a reference to the original block definition, rather than a complete copy of the block itself. In this way, overall drawing size is reduced. When you insert the block reference, you can specify the insertion point, scale, and rotation angle. Exploding a block reference reduces that instance of the block into copies of the original objects that comprise the block, but does not alter the block definition. If you redefine the block, however, all references of that block are immediately updated to reflect your changes.

When you create a block, the layers, colors, linetypes, and lineweights on which the original objects were created play a special role in the appearance of the block. For example, if you insert a block that contains objects originally drawn on layer 0 and assigned the color, linetype, and lineweight BYLAYER, the block is placed on the current layer and assumes the color, linetype, and lineweight of that layer. If you insert a block that contains objects originally drawn on other layers or with explicitly specified colors, linetypes, or lineweights, the block retains those original settings.

If you insert a block that contains objects originally assigned color, linetype, and lineweight BYBLOCK, and the block itself has the color, linetype, and lineweight BYLAYER, those objects adopt the color, linetype, and lineweight of the layer on which they are inserted. If the block is assigned an explicit color, linetype, or lineweight, such as red or dashed, those objects adopt those qualities.

A *nested* block occurs when you include other blocks in a new block that you are creating. Nesting is useful when you want to combine and include small components, such as nuts and bolts, into a larger assembly and need to insert multiple instances of that assembly into an even larger drawing. The same rules regarding layer, color, linetype, and lineweight apply to nested blocks. Blocks nested in another block can themselves contain nested blocks. The only limitation is that you can't create blocks that reference themselves.

| Tip |
|-----|

*Controlling layers, colors, linetypes, and lineweights can become confusing if you don't plan ahead when creating blocks. To minimize potential problems, follow these simple guidelines:*

- *To control the layer on which blocks are inserted, and assign all the objects in the block to the color, linetype, and lineweight of the layer on which the block is inserted, create all block components on layer 0, with the color, linetype, and lineweight set to BYLAYER.*

- *To make specific block objects always use the same color, linetype, or lineweight, assign those properties explicitly to the objects, before combining the objects into a block.*

- *To control explicitly the color, linetype, and lineweight of a block instance at the time of insertion, create the original block objects with the color, linetype, and lineweight set to BYBLOCK.*

## Creating Blocks

You can create blocks in two different ways:

- By saving a block within the current drawing
- By saving a block as a separate drawing file that you can insert into other drawings

| Note |
|------|

*With AutoCAD DesignCenter, you can easily insert blocks stored within a drawing into any other drawing. You'll learn about AutoCAD DesignCenter in Chapter 16.*

When you create a block, you specify its name, insertion point, and the objects that comprise it. The *insertion point* is the base point for the block and serves as the reference point when you later insert an instance of the block into the drawing. Before you can create the block, you must draw the objects that will comprise the block.

## Saving Blocks Within the Current Drawing

The BLOCK command saves blocks for use within the current drawing. When you start the command, AutoCAD displays the Block Definition dialog box, shown in Figure 15-1. To create a block for use within the current drawing by using the BLOCK command:

1. Do one of the following:

   - On the Draw toolbar, click Make Block.
   - From the Draw menu, choose Block | Make.
   - At the command line, type **BLOCK** (or **B**) and then press ENTER.

2. In the Block Definition dialog box, type the name of the block in the Name field.

> **Note** *Unlike versions prior to AutoCAD 2000, block names can now include spaces and can be up to 255 characters in length if the EXTNAMES system variable is set to 1.*

3. Under Base Point, specify the base or insertion point by doing one of the following:

   - Click the Pick Point button and then select the base point within the drawing.
   - Type the X, Y, and Z coordinates in their respective boxes.

4. Under Objects, click the Select Objects button, select the objects to include in the block by using any object selection method, and then press ENTER.

> **Tip** *You can also click the Quick Select button to create a filter for your selection set, or select objects before starting the BLOCK command.*

5. Under Objects, specify which of the following actions you want applied to the objects selected for inclusion in the block:

   - **Retain**   Keeps the selected objects in the drawing as individual objects at their current locations
   - **Convert to Block**   Converts the selected objects into an instance of the new block
   - **Delete**   Removes the selected objects from the drawing after you create the block

6. Under Preview Icon, specify whether you want to include a preview image of the block in the block definition.

> **Note** *Preview images are used with AutoCAD DesignCenter. You'll learn about DesignCenter in Chapter 16.*

7. In the Insert Units drop-down list, select the units to which the block should be scaled when dragged from AutoCAD DesignCenter.

8. In the Description box, type a description to help you identify the block.

**Note**    *Block descriptions can be viewed by using AutoCAD DesignCenter and can be used when searching for blocks.*

9. Click OK.

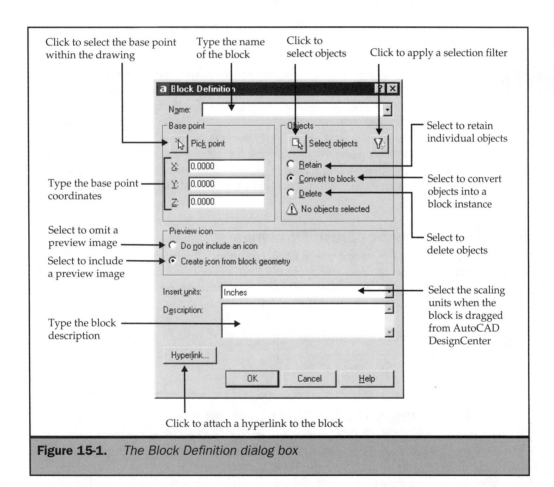

Click to select the base point within the drawing

Type the name of the block

Click to select objects

Click to apply a selection filter

Type the base point coordinates

Select to retain individual objects

Select to convert objects into a block instance

Select to omit a preview image

Select to include a preview image

Select to delete objects

Type the block description

Select the scaling units when the block is dragged from AutoCAD DesignCenter

Click to attach a hyperlink to the block

**Figure 15-1.**    *The Block Definition dialog box*

You can also create a block for use within the current drawing by using the BLOCK command from the command line:

1. Type **-BLOCK** (or **-B**) and then press ENTER.

2. Type the block name and then press ENTER.

3. Specify the insertion point for the block.

4. Select the objects that you want included in the block and then press ENTER.

    The block is immediately created. The objects that you select are removed from the drawing and are now part of the block.

5. To restore the original object to the drawing while retaining the new block, type **OOPS** and then press ENTER.

| Note | *Although the Block Definition dialog box makes the creation of blocks more understandable and offers the capability of including a preview image and descriptive text, many AutoCAD users still prefer to create blocks from the command line. For example, the dialog box uses the current drawing base point (usually 0,0,0) as the insertion point, unless you explicitly change it, whereas the command line always prompts you to specify the insertion point. The command line version always removes from the drawing the original objects that you include in the block, necessitating the use of the OOPS command to restore them, whereas the dialog box lets you retain the original objects or immediately convert them into a block insertion.* |

## Saving Blocks as Separate Drawings

When you create a block by using the BLOCK command, the block can be used only within the current drawing. You will probably find lots of instances in which you want to use those blocks in other drawings. After all, part of the power of AutoCAD is its capability to draw something once and then reuse it many times.

| Note | *You can use AutoCAD DesignCenter to drag blocks from one drawing into any open drawing.* |

### Creating Block Icons for Older Drawings

You can use the BLOCKICON command to create preview icons for blocks created using AutoCAD Release 14 or older versions. Simply open the drawing containing blocks that you want to update, type **BLOCKICON**, and press ENTER. AutoCAD prompts you to enter the names of the blocks that you want updated. You can specify individual block names, or simply select the default to update all the blocks in the current drawing.

The WBLOCK (or Write Block) command writes all or a portion of the drawing to a new drawing file. And, as you soon learn, you can insert a drawing file into a drawing as easily as you can insert a block that has already been defined within that drawing.

When you start the WBLOCK command, AutoCAD displays the Write Block dialog box, shown in Figure 15-2. You must specify the source objects and the destination. Under Source, select one of the following buttons:

- ■ **Block**   Specifies the name of a block that has already been defined in the current drawing, and saves it to a file. Select the block name in the adjacent drop-down list.

- ■ **Entire Drawing**   Selects the entire drawing as a block and saves it to a file. This is similar to saving the drawing by using the SAVEAS command, except that only objects included in the current space (model space or paper space) are included, and symbols that have been defined but not yet used, such as block definitions, linetypes, and text styles, are not included. Named views and User Coordinate Systems are not included in the new file, either.

- ■ **Objects**   Lets you select the individual objects you want saved to the new file, similar to the Block Definition dialog box. You must also specify the base point.

In the Destination area, you must specify the filename and the folder in which it is to be saved. You can also select the units to be used when the new file is inserted as a block when dragged from AutoCAD DesignCenter.

To save a block as a separate drawing file:

1. Type **WBLOCK** (or **W**) and then press ENTER.

2. In the Write Block dialog box, specify the source object as Block, Entire Drawing, or Objects, and supply the necessary information as follows:

   - ■ If you select Block, choose the block name in the adjacent drop-down list.

   - ■ If you select Entire Drawing, proceed to Step 3.

   - ■ If you select Objects, supply the information required under Base Point and Objects.

3. Under Destination, type the filename and specify the folder in which it is to be saved.

**Note**   *If you select a block as the source object, AutoCAD automatically uses the block's name for the new filename, but you can provide a different name if necessary.*

4. In the Insert Units drop-down list, select the units to which the block should be scaled when dragged from AutoCAD DesignCenter.

5. Click OK.

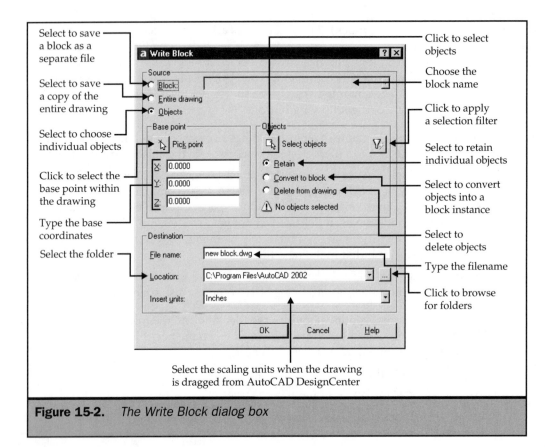

Select to save a block as a separate file

Select to save a copy of the entire drawing

Select to choose individual objects

Click to select the base point within the drawing

Type the base coordinates

Select the folder

Click to select objects

Choose the block name

Click to apply a selection filter

Select to retain individual objects

Select to convert objects into a block instance

Select to delete objects

Type the filename

Click to browse for folders

Select the scaling units when the drawing is dragged from AutoCAD DesignCenter

**Figure 15-2.** *The Write Block dialog box*

*LEARN BY EXAMPLE*
*Create a simple block to represent a door in plan view. Start a new drawing from scratch in which to draw the door.*

To draw the Door block shown in Figure 15-3, use the following command sequence and instructions. You'll draw the block one unit in size so that later, when you insert the block into a floor plan, you'll be able to specify the actual width of the door in inches.

```
Command: LINE
Specify first point: 0,0
Specify next point or [Undo]: 0,1
Specify next point or [Undo]: ENTER
Command: ARC
```

```
Specify start point of arc or [CEnter]: CE
Specify center point of arc: 0,0
Specify start point of arc: 1,0
Specify end point of arc or [Angle/chord Length]: 0,1
```

Next, combine the line and arc into a block:

```
Command: -BLOCK
Enter block name or [?]: DOOR
Specify insertion base point: 0,0
Select objects: (select the line and the arc)
Select objects: ENTER
```

Finally, use the WBLOCK command to save the door block as a separate drawing file, so that you can insert it into other drawings:

1. Type **WBLOCK** (or **W**) and then press ENTER.
2. In the Write Block dialog box, under Source, select Block.
3. In the adjacent drop-down list, choose Door. Notice that AutoCAD automatically fills in the filename.
4. Specify the folder in which you want to save the new file.
5. Click OK.

You can discard the current drawing, because you saved the door block to its own drawing and thus don't need the original door objects any longer.

**Note**   *In the previous exercise, you could have eliminated the separate step of combining the line and arc into a block prior to using the WBLOCK command simply by selecting Objects as the source in the Write Block dialog box. You would then have had to specify the base point and select the two objects.*

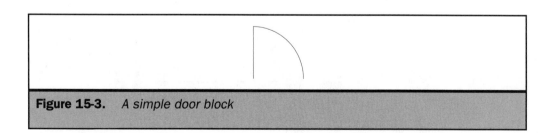

**Figure 15-3.**   *A simple door block*

# Inserting Blocks

You can insert blocks or other drawings into the current drawing. When you insert a block, it is treated as a single object. When you insert a drawing, it is added to the current drawing as a block and treated as a block. You can then insert multiple instances of that block, without reloading the original drawing file. The inserted drawing file actually becomes part of the current drawing, however. If you change the original drawing file, those changes have no effect on the current drawing unless you redefine the block by reinserting the changed drawing. You'll learn more about this later in this chapter.

When you insert a block or drawing, you must specify the insertion point, scale, and rotation angle. The block's insertion point corresponds to the base point that you specified when you first created the block. When you insert a drawing file as a block, AutoCAD uses the base point of that drawing (generally the 0,0,0 coordinate) as its insertion point, although you can change that base point by first opening the original drawing and redefining its base point, using the BASE command.

The INSERT command inserts block instances and, like the BLOCK command, provides both a dialog box and a command line method for doing so. The Insert dialog box, shown in Figure 15-4, lets you insert either a block or a file, which you can specify by choosing either from the Name drop-down list or from a File dialog box, accessed by clicking the Browse button. You can then either specify the insertion point, scale, and rotation angle by using the appropriate controls in the dialog box, or wait and specify them onscreen or at the command line when you actually place the block in the drawing.

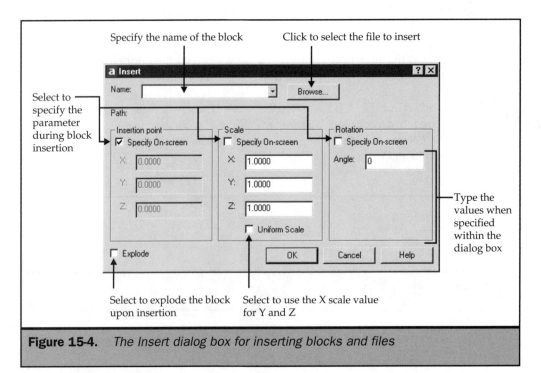

**Figure 15-4.** *The Insert dialog box for inserting blocks and files*

When you use the command line method, you must respond to all the prompts either by typing or by using the shortcut menus. Generally, the dialog box is easier to understand and simpler to use.

**Note**   *You can also copy-and-paste blocks between drawings. When you do so, however, you simply make a copy of the block insertion, complete with any existing attribute values. This is similar to using the COPY command to make copies of an existing block insertion in the current drawing, except that you also add the original block definition to the drawing to which it is pasted. You can then use the INSERT command actually to insert new instances of the block. You can also drag-and-drop blocks between open drawings or insert blocks by dragging them into the current drawing from AutoCAD DesignCenter.*

To insert a block by using the Insert dialog box:

1. Do one of the following:

   ■ On the Draw toolbar, click Insert Block.

   ■ From the Insert menu, choose Block.

   ■ At the command line, type **INSERT** (or **I**) and then press ENTER.

2. In the Insert dialog box, specify the block or file to be inserted.

3. If necessary, change the values under Insertion Point, Scale, and Rotation.

**Tip**   *Generally, it's easier to specify the insertion point when you actually insert the block. Depending on the nature of the block, it's often easier to preset the scale and rotation angle within the dialog box.*

4. Click OK.

5. Specify the insertion point for the block.

**Note**   *When you insert an external file, its WCS is aligned parallel to the XY plane of the current UCS in the current drawing. Therefore, you can align a block reference from an external file anywhere in 3-D space by first orienting the UCS before inserting the file.*

The command line version of the INSERT command functions similarly to the dialog box, except that you specify everything at the command line. AutoCAD prompts you for the name of the block to be inserted. If you've already inserted a block, its name appears as the default. Thus, to insert another instance of the same block, you can just press ENTER. If you specify a block name that isn't defined in the current drawing, AutoCAD searches its library path for a file of the same name. If it finds the file, it loads that drawing into the current drawing and then inserts an instance of it as a block.

*The command line version of the INSERT command also provides some features not found in the dialog box. The Pscale, PX, PY, PZ, and Protate options let you preview the block insertion at different scales and rotation angles before actually inserting the block. When you use these options, the preview image of the block, which is attached to the cursor, is adjusted to reflect the preview values that you specify. After specifying the insertion point, you are prompted for the actual scale and rotation angle.*

To insert a block from the command line:

1. Type **-INSERT** (or **-I**) and then press ENTER.

2. Type the name of the block, or the complete path and filename of the drawing to be inserted, and then press ENTER.

3. Specify the insertion point for the block.

4. Specify the scaling factors and rotation angle, or press ENTER to accept the default values.

*When AutoCAD prompts you for the name of the block, you can type a question mark to display a list of blocks already defined in the drawing. To insert a file by using a dialog box, type a tilde (~) and then press ENTER to display the Select Drawing File dialog box.*

When inserting a block using either the dialog box or command line, rather than inserting a block instance, you can insert the individual objects that comprise the block. When you use the dialog box, you can accomplish this by selecting the Explode check box. To accomplish the same thing from the command line, prefix the block name or filename with an asterisk (*). When you explode a block upon insertion, the objects are scaled only by the X scale factor. You'll learn more about exploding blocks later in this chapter.

### LEARN BY EXAMPLE

*Insert instances of the Door block (saved earlier as a separate drawing) into the floor plan of a house. Open the drawing Figure 15-5 on the companion web site. (The floor plan was created using multilines, as described on the companion web site. Refer to the pages for Chapters 4 and 10.)*

To add the doors shown in Figure 15-5, use the following command sequence and instructions:

1. Start the INSERT command.

2. In the Insert dialog box, click the Browse button and then select the DOOR.DWG file that you created previously. Make sure that all three Specify On-screen check boxes are selected, and then click OK.

3. Use the endpoint object snap to select the door jamb at point A.

4. When AutoCAD prompts you to specify the X scale factor, type **36** and then press ENTER.

5. Press ENTER to accept the same Y scale factor.

6. Make sure Ortho mode is on, drag the door until it is oriented properly, and then click to complete the block insertion.

*You can also use AutoCAD DesignCenter to quickly locate the DOOR.DWG and insert it as a block. You'll learn about AutoCAD DesignCenter in Chapter 16.*

Next, add the small closet door at point B by using the command line version of the INSERT command. Use the following command sequence and instructions:

```
Command: -INSERT
Enter block name or [?] <DOOR>: ENTER
Specify insertion point or [Scale/X/Y/Z/Rotate/PScale/PX/PY/PZ/
PRotate]: (select point B)
Enter X scale factor, specify opposite corner, or [Corner/XYZ] <1>: 30
Enter Y scale factor <use X scale factor>: ENTER
Specify rotation angle <0>: 90
```

## Inserting Arrays of Blocks

The MINSERT command inserts multiple instances of a block or file in a rectangular pattern—basically performing the combined functions of the INSERT and ARRAY commands. The command first prompts you for the block name, its insertion point, and the scale factors and rotation angle. The command then asks for the number of rows and columns, and the distance between the rows and columns. The rotation angle controls the orientation of the resulting array. Once created, the entire array is treated as a single object.

While the MINSERT command is quite powerful, it has several limitations. First, it creates rectangular arrays only. Second, because the entire array is treated as a single object, the resulting array can't be modified or exploded, although you can use the Properties window to change the number of rows and columns or the spacing between them. If you need to change the position of one of the block instances, you have to erase the entire array and re-create it. If you redefine the original block, however, all the instances of it are automatically updated.

The big advantage to using the MINSERT command is its effect on file size. Using MINSERT to create an array of blocks, rather than inserting the block and then using the ARRAY command, results in a much smaller drawing file. Consider using the MINSERT command in instances where you need to create a large rectangular array of blocks that you know you won't later need to modify.

Finally, add the back door at point C. This time, you need to insert the door with a negative X scale factor, to reverse the swing direction of the door. Use the following command sequence and instructions:

```
Command: -INSERT
Enter block name or [?] <DOOR>: ENTER
Specify insertion point or [Scale/X/Y/Z/Rotate/PScale/PX/PY/PZ/
PRotate]: (select point C)
Enter X scale factor, specify opposite corner,
or [Corner/XYZ] <1>: -36
Enter Y scale factor <use X scale factor>: 36
Specify rotation angle <0>: 0
```

Your drawing should now look similar to Figure 15-5. On your own, use the same methods to add the remaining doors.

## Exploding Blocks

A *block* is simply a collection of objects treated as a single object. If you need to modify a block, you can explode any instance of an inserted block back into its original component objects. The exploded objects retain the scale factor that was applied when the block was inserted. When you explode a block, only that single instance of the block is affected.

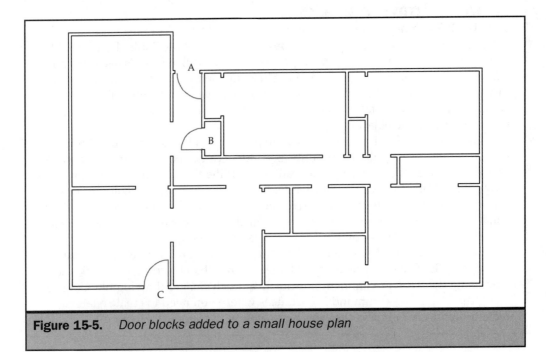

**Figure 15-5.**    *Door blocks added to a small house plan*

The original block definition remains in the drawing, and you can still insert additional copies of the original block.

 *If you explode a block that contains attributes (which you will learn about in the next section), the attribute values are lost, but the original attribute definitions remain.*

To explode a block:

1. Do one of the following:

- On the Modify toolbar, click Explode.
- From the Modify menu, choose Explode.
- At the command line, type **EXPLODE** (or **X**) and then press ENTER.

2. Select the block and then press ENTER.

Exploding a block reduces that block into its individual components. If that block contains nested blocks, they are unaffected, although you can repeat the EXPLODE command to explode nested blocks or reduce component objects, such as polylines, to their next simpler level of complexity.

 *You can't explode nonuniformly scaled blocks (blocks inserted using different X, Y, or Z scale factors) unless the value of the EXPLMODE system variable is 1.*

## Editing Block Descriptions

Blocks can include descriptive text. You can also use the BLOCK command to edit a block description previously saved with a block definition.

To edit a block description:

1. Start the BLOCK command.

2. In the Block Definition dialog box, under Name, select the block whose description you want to modify.

3. In the Description box, type the desired changes to the block description.

4. Click OK.

 *Block descriptions can be viewed using AutoCAD DesignCenter. In addition, you can use DesignCenter to search for blocks based on their descriptions. You'll learn more about AutoCAD DesignCenter in Chapter 16.*

## Redefining Blocks

You can redefine all instances of a block within the current drawing, to update the appearance of all occurrences of that block. To redefine a block that was created in the current drawing, you create a new block by using the same name as the block that you want to redefine. If the block was inserted from a separate drawing file that subsequently has been updated, you must reinsert that block to update all instances of the block in the current drawing.

To redefine a block in the current drawing by using the BLOCK command:

1. Start the BLOCK command.

2. In the Block Definition dialog box, under Name, select the block you want to redefine.

3. Under Base Point, specify the base insertion point.

4. Under Objects, click the Select Objects button, select the objects to be included in the redefined block, and then press ENTER.

5. Redefine the Insert Units and Description, as needed.

6. Click OK. AutoCAD displays a dialog box asking if you really want to redefine the block.

7. Click Yes. AutoCAD immediately updates all instances of the redefined block in the current drawing.

You can use the INSERT command to update all instances of a block that was inserted from a separate drawing file that has been modified. Open the drawing containing the block. To update all block instances based on a separate drawing file:

1. Start the INSERT command.

2. In the Insert dialog box, click the Browse button and specify the separate drawing file.

3. Click OK. AutoCAD displays a dialog box asking if you really want to redefine the block.

4. Click Yes. AutoCAD immediately updates all instances of the redefined block in the current drawing.

**Note** *If any of the Specify On-screen check boxes were selected, the program prompts you for the required information, as if you were inserting a new instance of the block. At that point, you can simply press ESC. If none of the Specify On-screen check boxes were selected, however, the command ends after updating all the existing instances of the redefined block.*

# Working with Attributes

An *attribute* is a special type of object that you can save as part of a block definition. Attributes consist of text-based data. You can use attributes to track such things as part numbers and prices. Attributes have either fixed or variable values. When you insert a block containing attributes, AutoCAD adds the fixed values to the drawing along with the block, but prompts you to supply any variable values.

After you insert blocks that contain attributes, you can use that information in a spreadsheet or database—to produce a parts list or bill of materials—by extracting the attribute information to a separate file.

In addition to having fixed or variable values, the attributes themselves can be visible or hidden. As their name implies, *hidden* attributes are not displayed or plotted, whereas *visible* attributes are visible. In either case, the values are still stored in the drawing and written to a file when you extract them.

 *You can also link nongraphical data from an external database to objects in your drawing. You'll learn more about connecting a database to your AutoCAD drawings in Chapter 27.*

## Creating Attributes

You add an attribute to a drawing by first defining the attribute and then saving it as part of a block definition. To define an attribute, you specify the characteristics of the attribute, including its name, prompt, and default value; the location and text formatting; and the optional modes (hidden, fixed, validate, and predefined).

As with the commands for creating and inserting blocks, you can create attributes either by using a dialog box or by working entirely at the command line. In the case of attributes, however, the dialog box–based version of the command is much easier to use. The ATTDEF command displays the Attribute Definition dialog box, shown in Figure 15-6, which provides all the attribute creation options in one easy-to-use interface.

To create an attribute:

1. Do one of the following:
   - From the Draw menu, choose Block | Define Attributes.
   - At the command line, type **ATTDEF** (or **AT**) and then press ENTER.

2. In the Mode area, select the attribute modes that you want to use. If none are selected, the resulting attribute will be visible and variable.

   A *verified* attribute causes AutoCAD to ask you to verify the value. When you create a *preset* attribute, AutoCAD does not prompt you for the value, but unlike a constant attribute, you can edit its value after insertion.

3. In the Attribute area, type the tag, prompt, and default value.

   The *tag* is the name that AutoCAD uses to identify the attribute, and cannot be left blank. The *prompt* field can contain any prompt that you want AutoCAD to display when the block is inserted. If left blank, AutoCAD uses the tag entry as the prompt. The *value* field contains the attribute's default value.

4. In the Insertion Point area, type the X-, Y-, and Z-coordinate values or click Pick Point to specify a point in the drawing.

5. In the Text Options area, use the various controls to determine the justification, text style, text height, and rotation angle that should be used to display the attribute when it is visible.

6. Click OK to complete the creation of the attribute.

After you define an attribute, if you use the ATTDEF command again, you can select the Align Below Previous Attribute Definition check box. This places the next attribute below the previous attribute and assigns to it the same text options as the previous attribute. In that case, the Insertion Point and Text Options areas of the dialog box become unavailable.

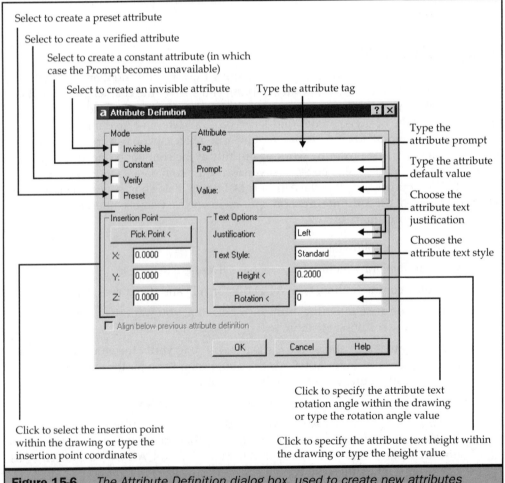

**Figure 15-6.** *The Attribute Definition dialog box, used to create new attributes*

*When you first define attributes, the attribute tags all appear in the drawing, including those defined as invisible. Once you associate them with a block and insert the block into the drawing, however, only those attribute values that are specified as being visible actually appear in the drawing.*

## Editing Attribute Definitions

Don't worry if you make a mistake while creating attribute definitions. You can edit an attribute definition before you associate it with a block and before it is saved as part of a block definition.

To edit the attribute tag, prompt, or default value, use the DDEDIT command, which you learned about in Chapter 13. AutoCAD prompts you to select an annotation object. You can also simply double-click the attribute definition. When you select an attribute definition, the program displays the Edit Attribute Definition dialog box.

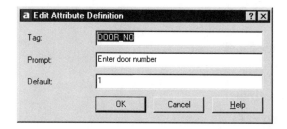

You can also edit the attribute tag, prompt, default value, or any of the other attribute definition characteristics—such as the attribute text style, justification, and insertion point—by using the Properties window.

 *You can also edit attribute definitions by using the CHANGE command.*

## Attaching Attributes to Blocks

After you create one or more attributes, they are treated just like any other AutoCAD object until you associate them with a block. You can associate one or more attributes with a block when you define or redefine the block, by selecting the attributes among the objects to be included in the block. Once the attributes are incorporated into the block, AutoCAD prompts you each time that you insert the block, so that you can specify different values for the attributes.

 *When you create the block, the order in which you select the attributes determines the order in which AutoCAD prompts you for attribute values when you insert the block. You can use the Block Attribute Manager, however, to change the order. You'll learn about this new tool later in this chapter.*

MOVING BEYOND
THE BASICS

Attributes can be used to keep track of information within your drawing. They can also be used to create standard drafting symbols, such as detail callouts, wherein the symbol's appearance remains the same but the values change.

*LEARN BY EXAMPLE*
*Use attributes to create a standard callout symbol. Open the drawing Figure 15-7 on the companion web site.*

To create the callout symbol block shown in Figure 15-7, use the following command sequence and instructions:

1. Start the ATTDEF command.

2. In the Attribute Definition dialog box (refer to Figure 15-6), specify the following values:

    **Tag:** detail

    **Prompt:** Enter detail number

    **Value:** 0

    **Justification:** Center

    **Text Style:** STYLE1

    **Height:** 1/8"

    **Rotation:** 0

3. Click the Pick Point button and select a convenient point near the center of the drawing.

4. Click OK.

    The word DETAIL should appear, centered at the point that you specified.

Attribute definition   Inserted block

**Figure 15-7.**   *An architectural callout symbol created by using attributes*

5. Start the ATTDEF command again.

> **Tip** *You can press* ENTER *or the right-mouse button to restart the previous command.*

6. In the Attribute Definition dialog box, specify the following values:

   **Tag:** sheet

   **Prompt:** Enter sheet number

   **Value:** A1

7. Select the Align Below Previous Attribute Definition check box.

8. Click OK.

   The word SHEET should now appear directly below the word DETAIL. Now, draw the remainder of the callout symbol.

9. Start the CIRCLE command.

10. Specify the center point of the circle by picking a point below the letter T in the word DETAIL, so that the center point is centered approximately between the two attributes.

11. Type **.25** to specify the radius of the circle and then press ENTER.

12. Start the LINE command.

13. Use the quadrant object snap to specify the starting point of the line at the left side of the circle.

14. Use the quadrant object snap to specify the end point of the line at the right side of the circle.

15. Press ENTER to end the command.

Your drawing should now look similar to the symbol on the left side of Figure 15-7. The attribute text flows beyond the circle. Don't worry about that for now. When you insert the callout symbol block, the values that you enter will be much shorter than the tags themselves. Now, create the block by using the following command sequence. When prompted for the insertion point, use the midpoint object snap to select the horizontal line. When prompted to select the objects to include in the block, be sure to select the attributes in the order in which you created them, so that AutoCAD later prompts for the detail number and then the sheet number.

```
Command: -BLOCK
Enter block name or [?]: DETAIL
Specify insertion base point: (select the midpoint of the horizontal line)
```

```
Select objects: (select the circle and the line, and then select the DETAIL and SHEET attributes,
in that order)
Select objects: ENTER
```

Now, you can insert the block by using the following command sequence:

```
Command: -INSERT
Enter block name or [?]: DETAIL
Specify insertion point or [Scale/X/Y/Z/Rotate/PScale/PX/PY/PZ/
PRotate]: (select any point in the drawing)
Enter X scale factor, specify opposite corner, or [Corner/XYZ] <1>: ENTER
Enter Y scale factor <use X scale factor>: ENTER
Specify rotation angle <0>: ENTER
Enter attribute values
Enter detail number <0>: 1
Enter sheet number <A1>: ENTER
```

Notice that when you insert the block, after specifying the insertion point, scale factor, and rotation angle, AutoCAD displays the prompts that you originally entered when you first created the attributes. You entered a value of 1 for the detail number, but accepted the default value of A1 for the sheet number. Your drawing should look similar to the symbol on the right side of Figure 15-7.

## Attribute Prompts Using Dialog Boxes

In the previous example, when AutoCAD prompted for the attribute values, the prompts appeared at the command line. You can easily reconfigure AutoCAD so that the attribute prompts appear in a dialog box, instead. This is controlled by the ATTDIA system variable. When its value is 0, attribute prompts appear at the command line. If you change its value to 1, however, AutoCAD displays a dialog box when it prompts for attribute information.

*LEARN BY EXAMPLE*
*Observe how the ATTDIA value affects the procedure for inserting a block containing attributes. Continue working with the drawing from the previous exercise.*

To see how the ATTDIA value changes the procedure for inserting a block with attributes, use the following command sequence and instructions to change its value, and then insert the Detail block again:

```
Command: ATTDIA
Enter new value for ATTDIA <0>: 1
```

1. Start the INSERT command.

2. In the Insert dialog box, under Name, select the Detail block.

3. Under Insertion Point, make sure that the Specify On-screen check box is selected. Under the Scale and Rotation areas, make sure that the Specify On-screen check boxes are not selected. The X, Y, and Z scale factors should all be 1 and the rotation angle should be 0.

4. Click OK.

5. When prompted to specify the insertion point, select any point in the drawing.

At this point, instead of prompting you for the detail number and sheet number, AutoCAD displays an Enter Attributes dialog box, similar to the one shown in Figure 15-8. To complete the insertion of the callout block, use the following instructions:

1. In the Enter Detail Number box, type **2**.

2. Click OK.

When the ATTDIA value equals 1, AutoCAD uses an Enter Attributes dialog box instead of prompting at the command line. While a dialog box didn't save any steps in this example, it can be quite useful when working with blocks containing many attributes,

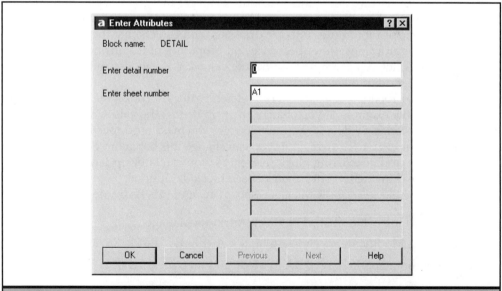

**Figure 15-8.**    *Attribute prompts displayed in the Enter Attributes dialog box*

MOVING BEYOND
THE BASICS

because you can see all the attributes at a glance instead of having to step through each attribute in sequence. The dialog box also makes it easier to go back and change a value before you click the OK button. If the block contains more than eight attributes, the Next and Previous buttons become available, so that you can page through multiple screens of attributes.

*When you use a dialog box, you must complete the block insertion by clicking the OK button. If you click the Cancel button, the entire block insertion is canceled.*

# Editing Attributes

You'll probably encounter situations in which you enter the wrong attribute information while inserting a block. You can edit the attribute values of a block that has been inserted into the drawing. In fact, AutoCAD provides several different ways to edit attributes. You can edit one or more of the attribute values attached to a single block, change the appearance of individual attributes, or globally change several attribute values at once. There may also be times in which you need to change the attributes in a block definition after that block has been inserted into the drawing. AutoCAD provides tools enabling you to do this as well.

## Changing Attributes in a Block Definition

AutoCAD 2002 provides a new Block Attribute Manager that enables you to change attributes in a block definition, even if the block containing those attributes has already been inserted one or more times into the drawing. You can change the attribute tag, prompt, and default values; the appearance and properties of the attribute text; and the order in which AutoCAD prompts for attribute values. You can also remove attributes from the block definition.

Changing the properties of existing block references does not affect the values assigned to those blocks. For example, if you change an attribute tag, the values assigned to that attribute do not change. By default, attribute changes you make are applied to all instances of the block in the current drawing, but you can change this behavior. If you decide not to automatically update existing block references, you can later update all the block references in the drawing to match their revised block definitions.

To display the Block Attribute Manager (shown in Figure 15-9), do one of the following:

- On the Modify II toolbar, click Block Attribute Manager.
- From the Modify menu, choose Object | Attribute | Block Attribute Manager.
- At the command line, type **BATTMAN** and then press ENTER.

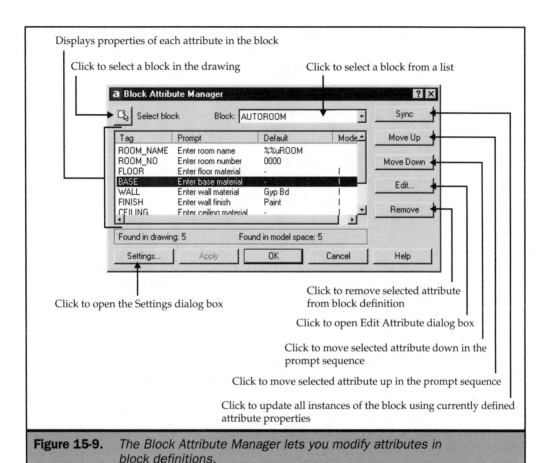

Displays properties of each attribute in the block

Click to select a block in the drawing

Click to select a block from a list

Click to open the Settings dialog box

Click to remove selected attribute
from block definition

Click to open Edit Attribute dialog box

Click to move selected attribute down in the
prompt sequence

Click to move selected attribute up in the prompt sequence

Click to update all instances of the block using currently defined
attribute properties

**Figure 15-9.** *The Block Attribute Manager lets you modify attributes in block definitions.*

MOVING BEYOND
THE BASICS

Initially, the Block Attribute Manager displays the definition of the first block defined in the drawing. You can select a different block definition by either clicking the Select Block button and then picking a block in the current drawing, or by selecting the block from the Block drop-down list.

The dialog box contains a list of all the attributes defined in the current block definition. By default, this list shows the tag, prompt, default value, and modes for each attribute. To control which attribute properties are displayed in the list, click the Settings button to display the Settings dialog box. Only the properties selected in the upper portion of the dialog box are displayed in the list.

The check boxes in the lower portion of the Settings dialog box determine whether AutoCAD highlights duplicate attribute tags in the Block Attribute Manager and whether changes you make apply to existing instances of the block. If you don't apply changes to existing blocks, the changes you make using the Block Attribute Manager apply only to new instances of the block.

 *Since duplicate tag names can lead to unpredictable results, you should always let AutoCAD highlight duplicate tag names, and then change those names to eliminate the duplication.*

The attribute tag list indicates the order in which AutoCAD prompts for attribute values whenever you insert an instance of the block. To change the order, select an attribute tag in the list and then click the Move Up button to move the selected attribute earlier in the prompt sequence or click the Move Down button to move it later in the prompt sequence.

 *The Move Up and Move Down buttons are unavailable for attributes with constant values (Mode=C).*

To modify any aspect of an attribute, select the attribute in the list and then click Edit to display the Edit Attribute dialog box, shown in Figure 15-10. This dialog box has three tabs that enable you to modify the following:

- **Attribute**   Provides controls for changing the attribute mode, or the tag, prompt, or default value

- **Text Options**   Provides controls for changing the appearance of the attribute text

- **Properties**   Provides controls for changing the layer, linetype, color, lineweight, and plot style values

When the Auto Preview Changes check box is selected, any changes you make are immediately reflected in the drawing. When this check box is cleared, the changes don't appear in the drawing until you click Apply or OK in the Block Attribute Manager dialog box.

To remove an attribute from a block definition and all block references, select the attribute in the list and click the Remove button. Although the attribute is immediately removed from all instances of the block, the attributes don't actually disappear from the drawing until you regenerate the drawing using the REGEN command.

AutoCAD 2002 also provides a new command to update block references in the current drawing with any changes you may have made to the block definition. For example, you may have used the Block Attribute Manager to change attribute properties but chose not to automatically update existing block references when you made those changes. To apply those changes to all instances of the block in the current drawing:

1. Do one of the following:

   - On the Modify II toolbar, click Synchronize Attributes.

   - At the command line, type **ATTSYNC** and then press ENTER.

   AutoCAD prompts:

   ```
   Enter an option [?/Name/Select] <Select>:
   ```

2. Do one of the following:

   - Type **N**, press ENTER, and then enter the name of the block whose references you want to update.

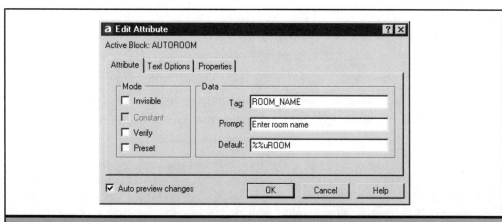

**Figure 15-10.**   *The Attribute tab of the Edit Attributes dialog box*

- Type **?** and press ENTER to view the names of all the blocks in the current drawing. You can then use the Name option to specify the name of the block you want to update.

- Press ENTER and then click on an instance of the block you want to update.

  If you specify a block name, AutoCAD immediately updates all instances of the block. If you select a block in the drawing, AutoCAD first displays the block name and asks whether you want o update the block.

*Although changing attribute properties does not normally cause the existing attribute values to be altered, this is not the case if you change an attribute tag and don't apply the changes to existing references at the time you make the changes. If you later use the Sync button in the Block Attribute Manager or the ATTSYNC command, the attribute value will change to the default value in all instances of the block.*

## Changing Attributes Attached to a Single Block

The most common and easiest change that you're likely to make to attribute values is to alter one or more attribute values attached to a single instance of a block. AutoCAD 2002 provides a new tool, making it easier than ever to accomplish this. In addition to changing attribute values, you can modify the appearance and properties of the attribute text. To edit attributes attached to a block, do one of the following:

- On the Modify II toolbar, click Edit Attribute.

- From the Modify menu, choose Object | Attribute | Single.

- At the command line, type **EATTEDIT** and then press ENTER.

AutoCAD prompts you to select a block. Select the instance of the block whose attributes you want to change. As soon as you select the block, AutoCAD displays the Enhanced Attribute Editor (shown in Figure 15-11). The dialog box contains the following three tabs:

- **Attribute**   Provides controls for changing the attribute value

- **Text Options**   Provides controls for changing the appearance of the attribute text

- **Properties**   Provides controls for changing the layer, linetype, color, lineweight, and plot style values

You can change one or more of the values, text options, or properties, and then click OK to complete the command. If you want to change the attributes in another block instance, click the Apply button to apply the changes and then click the Select Block button. AutoCAD prompts you to select a block. Click in the drawing to select the next block instance.

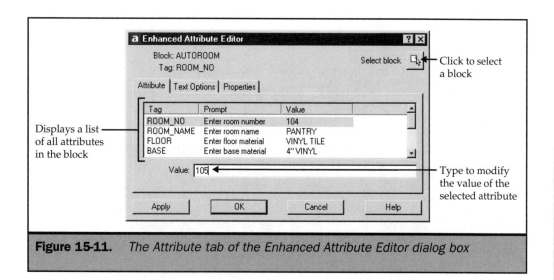

**Figure 15-11.**     *The Attribute tab of the Enhanced Attribute Editor dialog box*

## Changing Multiple Attribute Values

Another commonly encountered situation is to discover that you need to change a particular attribute value that has been specified in several instances of a block. For example, suppose that you specify that a particular set of details—indicated by using the callout block that you created in the previous example—can be found on sheet A-3, but then you renumber the sheets so that the details now appear on sheet A-4. Rather than individually change each instance of the callout block that references sheet A-3, you can change them all at once.

The command line version of the ATTEDIT command offers numerous options. For example, you can edit just those attributes that are visible on the screen or all the attributes in the drawing whose value matches the string that you specify. You can also specify the block name, attribute tag, and attribute value, to further limit your selection. These options are useful when the attribute value string that you want to replace may occur in other blocks or may be assigned to other tags besides the ones that you specifically want to change. In addition, if you specify that you want to edit only attributes that are visible on the screen, AutoCAD prompts you to select the attributes to be changed, letting you further limit the blocks selected. When editing all the attributes in the drawing, AutoCAD doesn't prompt you to select attributes.

To globally edit an attribute value:

1. Do one of the following:

   ■ From the Modify menu, choose Object | Attribute | Global.

   ■ At the command line, type **-ATTEDIT** (or **-ATE** or **ATTE**) and then press ENTER.

AutoCAD prompts:

```
Edit attributes one at a time? [Yes/No] <Y>:
```

2. Type **N** (for No) and then press ENTER. AutoCAD prompts:

```
Performing global editing of attribute values.
Edit only attributes visible on screen? [Yes/No] <Y>:
```

3. Press ENTER to edit only those attributes that are visible on the screen. Optionally, you can type **N** (for No) and then press ENTER to edit all the attributes in the drawing. Next, AutoCAD prompts:

```
Enter block name specification <*>:
```

4. Press ENTER to edit all the blocks in the drawing. Optionally, you can type a specific block name to limit your selection. AutoCAD then prompts:

```
Enter attribute tag specification <*>:
```

5. Again, press ENTER to edit all attributes, or type a specific attribute tag name to limit your selection. AutoCAD prompts:

```
Enter attribute value specification <*>:
```

6. Press ENTER to edit all attributes, or type a specific attribute value to limit your selection. If you previously specified that you were editing only attributes visible onscreen, AutoCAD now prompts you to select attributes.

7. Select the attributes that you want to change, using any selection method, and then press ENTER. AutoCAD prompts:

```
Enter string to change:
```

8. Type the string that you want to change (in this case, the old sheet number, A-3) and press ENTER. AutoCAD then prompts:

```
Enter new string:
```

9. Type the new string (in this case, the new sheet number, A-4) and press ENTER.

*The Global Attribute Edit Express tool, described on the companion web site, makes it even easier to change attribute values for some or all insertions of a specific block.*

## Changing the Visibility of Attributes

Although you may have used invisible attributes for data such as part numbers or room finish information that should not be visible or plotted, you sometimes may want to see that data. Or, you may want to hide temporarily all attributes information, including those attributes that normally are visible. The ATTDISP command changes the display of attributes and provides three options:

- **Normal**   The visibility of attributes is determined by the attribute mode
- **On**   All attributes are visible
- **Off**   All attributes are invisible

To display all attributes, do one of the following:

- From the View menu, choose View | Display | Attribute Display | On.
- At the command line, type **ATTDISP**, press ENTER, type **ON**, and then press ENTER.

## Redefining Blocks Containing Attributes

If you redefine a block that contains attributes, any previous instances of the block are updated visually to reflect the changes that you make. Although any new instances of the redefined block will contain any new attributes that you have added, the existing instances of the block are not updated to incorporate the new attributes. AutoCAD provides a special command that eliminates this problem, letting you update existing instances of blocks to include the new attributes.

*TIP:   Before you can redefine the block, you must first create the objects and attributes that you want to combine into the new block definition. If you didn't save the original objects, you can insert an instance of the block and then explode it back into its individual components.*

To redefine a block that contains attributes and update attributes in existing instances of that block:

1. Type **ATTREDEF** and then press ENTER.
2. Type the name of the block that you want to redefine and then press ENTER.
3. Select the objects that you want to combine into the new block definition and then press ENTER.
4. Specify the insertion base point of the new block.

As soon as you specify the insertion base point, AutoCAD begins updating existing instances of the redefined block. This may take a few minutes, depending on the number of times the block was inserted in the drawing. If you included new attributes in the redefined block, their default values will appear in all the updated instances. You can then edit individual instances, as necessary. If you deleted any attributes from the redefined block, those attributes will be deleted from all the updated block instances.

*There are numerous Express Tools that help you work with blocks and attributes. The BLOCK? Express tool lists the objects in a selected block definition. You can use BCOUNT to count the number of instances of each block in a selection set or in the entire drawing. BURST explodes block instances, converting attribute values into text objects. All of these Express Tools are described on the companion web site.*

# Extracting Attribute Information

After you create attributes, attach them to block definitions, and insert those blocks into your drawing, you'll want to extract all that data to use with a database program. Extracting attribute information doesn't affect the drawing itself—it simply creates a separate file that contains the data.

AutoCAD 2002 introduces a new Attribute Extraction wizard that steps you through the process of extracting attribute information. With it, you can specify which block data to extract, and then save the data to an external file in comma-delimited ASCII text format (CVS), tab-delimited ASCII text format (TXT), or in either Microsoft Excel or Access file format (if either of these programs is installed on your system).

 *Previous versions of AutoCAD used a different method for attribute extraction that saved attribute data in either a comma-delimited format (CDF), space-delimited format (SDF), or drawing data exchange format (DXF). To save the data in CDF or SDF format, you were first required to manually create an attribute extraction template. You then used the ATTEXT command to extract the attribute data. While that method remains available, the new Attribute Extraction wizard is much easier to use and more flexible in its capabilities.*

To start the new Attribute Extraction wizard to extract attribute information, do one of the following:

- On the Modify II toolbar, click Attribute Extract.
- From the Tools menu, choose Attribute Extraction.
- At the command line, type **EATTEXT** and then press ENTER.

AutoCAD displays the first step of the Attribute Extraction wizard, shown in Figure 15-12. Choose one of the following options:

- **Select Objects**    Extract attributes attached to a set of blocks within the current drawing that you select. Click the adjacent button to temporarily close the dialog box, select the desired blocks using any selection method, and then press ENTER to return to the wizard.

- **Current Drawing**    Extract attributes attached to all the blocks in the current drawing.

## Exporting to a Microsoft Access Database File

In order to be able to export to a Microsoft Access Database (MDB) file, you must first install the Microsoft Data Access Objects (DAO). To do so, insert the AutoCAD 2002 CD in your CD-ROM drive, and use Windows Explorer to browse to the Support folder. Double-click the archive file DA35.ZIP and extract the contents of the file to a temporary folder on your hard drive (such as C:\TEMP). Then, in Windows Explorer, browse to the folder where you extracted the files and double-click SETUP.EXE to run the DAO setup program to install Data Access Objects version 3.5.

■ **Select Drawings** Extract attributes attached to all the blocks in one or more drawing files that are not currently open. Click the adjacent button to display a Select Files dialog box, specify the drawing files, and then click Open to return to the wizard.

**Caution** *You should not use the Select Drawings option to determine the total number of blocks in multiple drawings, because the results may be incorrect. To get accurate results, extract the blocks in each drawing and then combine the extracted data in a spreadsheet.*

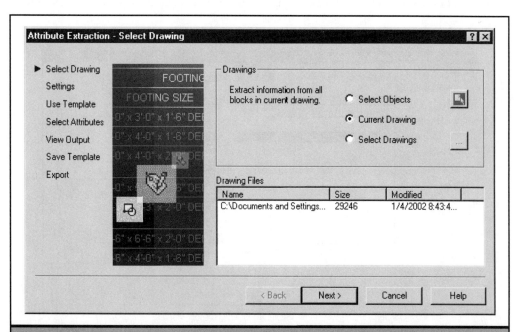

**Figure 15-12.** *In the first step of the Attribute Extraction wizard, you specify the blocks from which you want to extract attribute information.*

MOVING BEYOND
THE BASICS

After you select the drawing or specify the blocks, click Next. AutoCAD displays the Settings page, which offers two check boxes, both of which are selected by default:

- **Include Xrefs**    Extracts attribute information from blocks in any externally referenced files included in the drawings you selected

- **Include Nested Blocks**    Extracts attribute information from blocks nested within other blocks in the drawings you selected

After you select these settings, click Next. AutoCAD displays the Use Template page of the Attribute Extraction wizard. In order to extract attribute data, AutoCAD needs a template file. The template tells AutoCAD what data you want to extract and how that data should be structured. When you use the Attribute Extraction wizard, AutoCAD actually creates a temporary template on-the-fly. In a later step, you can save that template as a separate file (a special file with a .BLK file extension) so that you can reuse the same settings again later. If you've previously saved an attribute extraction template that you want to reuse, select the Use Template radio button, click the Use Template button, and then locate the template file you want to use. Otherwise, select the No Template radio button, and then click Next to proceed to the next step.

The Select Attributes page (shown in Figure 15-13) displays two lists. The Blocks list on the left includes all the blocks from which you can extract attribute data, as well as the number of instances of each block. The Attributes For Block list on the right includes

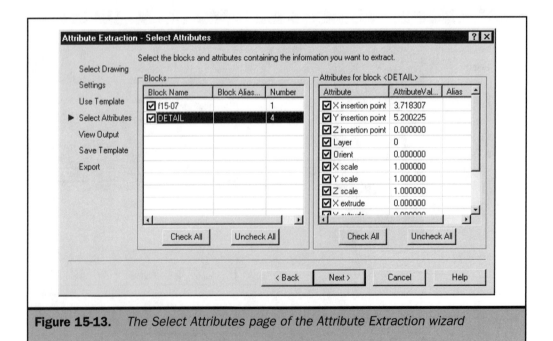

**Figure 15-13.**    *The Select Attributes page of the Attribute Extraction wizard*

all the attributes corresponding to the block selected in the Blocks list. Notice that in addition to the actual attribute tags, AutoCAD includes 11 additional fields, which contain elements such as the X, Y, and Z coordinates of the block's insertion point, the layer on which it is inserted, and so on.

Each line in both lists includes a check box. In the Blocks list, select the check boxes for blocks whose attributes you want to extract. Clear the check box for any block whose data you don't want to include. Then, for each block selected in the Blocks list, select those attributes in the adjacent list that you want included in the resulting file and clear the check boxes for any attributes you don't want to include. You can use the Check All or Uncheck All buttons below the lists to quickly select or clear all of the check boxes.

**Caution**    *The attributes selected in the Attributes For Block list apply only to the block selected in the Block list. You need to repeat this process for each block included in the Block list.*

Notice that the Blocks list also includes a Block Alias column. You can type a name in this column for a selected block to use that name in the resulting attribute extraction file rather than the actual block name. The Attributes For Block list includes a similar column so that you can assign an alias to the block attribute.

After you select the blocks, and the attributes to be extracted for each block, click Next to display the View Output page. This page displays a table showing a preview of the attributes to be extracted. Use this table to ensure that the blocks and attributes shown in the list are the ones you want to extract. The page provides two different views of the information, which you can toggle between by clicking the Alternate View button. One view shows the attributes for each block by block name, with each attribute for each block shown on its own line. The other view shows the values for each attribute by block name, with one line for each instance of the block.

**Note**    *When extracting attribute data from multiple drawings or xrefs, if AutoCAD encounters blocks with the same name but different attributes, the instances of the block in the first drawing will appear in the attribute extract file using its actual block name. Subsequent instances of the block in the other drawings will appear in the attribute extract file using the block name followed by a tilde character (~) and the full path and filename in which the block was found.*

If necessary, click the Back button to return to the previous page so you can adjust your selections. When you are satisfied with the data to be written to the attribute extract file, click Next.

**Tip**    *You can also click the Copy To Clipboard button to copy all or selected portions of the table to the Clipboard. You can then paste that data into another application. If you simply want to include the attribute data as text within your drawing, you can copy the table to the Clipboard, cancel the Attribute Extraction wizard, and then simply paste the data into your drawing as Mtext.*

The Save Template page enables you to save the attribute extraction settings you have made to a template file (with a .BLK file extension) so that you can reuse those same settings again in the future. If you don't want to save the settings, simply click Next. To save the template file, click the Save Template button. AutoCAD displays a Save As dialog box. Specify the location and filename, and then click Save. You are then returned to the Attribute Extraction wizard. Click Next to continue.

The Export page (shown in Figure 15-14) is where you actually create the attribute extraction file. Specify the type of data file to be created by selecting it from the drop-down list, assign it a filename, and then click Finish.

***LEARN BY EXAMPLE***
*Practice using the Attribute Extraction wizard by extracting attributes from blocks in a typical drawing. You'll find the drawing, along with step-by-step instructions, on the Chapter 15 page on the companion web site.*

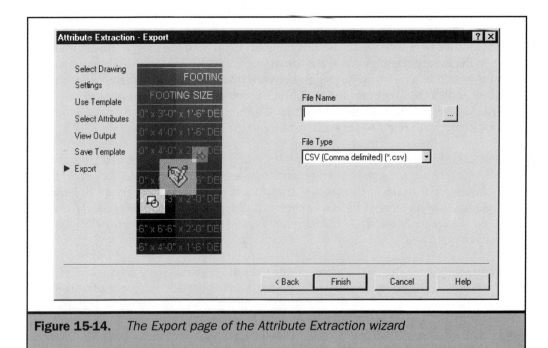

**Figure 15-14.** *The Export page of the Attribute Extraction wizard*

# Working with External References

When you insert a block that is saved as a separate drawing, you actually add the contents of that drawing to your current drawing and then place an instance of the block at the specified location. That technique is fine when the inserted drawings are relatively small. But, when you insert large drawings, the size of your current drawing can grow quickly. In addition, as you have already seen, any changes that you make to the original inserted drawing are not automatically reflected in the drawing in which they're inserted. You must manually update the block by reinserting it from the file.

In response to these concerns, AutoCAD provides another way to add information from one drawing into another—external references, or *xrefs*. An xref links another drawing to the current drawing. Unlike inserting a drawing as a block, xrefs attach a pointer to the external file. The objects in the xref appear in the current drawing, but the objects themselves are not added to the drawing. Thus, attaching an xref does not significantly increase the size of the current drawing file.

External references provide additional capabilities that are not available when you insert a drawing as a block. Unlike blocks inserted from other drawings, when you attach an xref, any changes that you make to the original drawing file are reflected in the drawings that reference it. These changes appear automatically each time that you open the drawing containing the xref. In addition, if you know that the original drawing was modified, you can reload the xref at any time that you're working on a drawing that references it.

External references are particularly useful for assembling master drawings from component drawings. You can also use xrefs to coordinate your work with others in a group. For example, an architect may draw the basic floor plan of a building. The electrical and mechanical engineers could then reference that drawing as the base drawing for their electrical, lighting, heating, and plumbing plans, instead of having to redraw the floor plan. If the architect makes changes to the floor plan, those changes show up immediately in the engineer's drawings.

In addition, you can define a clipping boundary around an xref so that only a portion of it displays in the current drawing. This enables you to use an xref of a portion of a large drawing as the basis for a larger-scale detail drawing.

Using xrefs has one potential downside, however. Unlike blocks, which become part of the current drawing, xrefs remain separate. Thus, if you do need to exchange drawings with others, you must be sure to send them both the master drawing and all the drawings that it references. Otherwise, when they attempt to load the master drawing, AutoCAD will display an error message, reporting that it cannot find the externally referenced drawings. Thankfully, AutoCAD's ETRANSMIT command creates a transmittal set of an AutoCAD drawing and all its related files, including xrefs. You'll learn about this tool in Chapter 26.

MOVING BEYOND
THE BASICS

*When you complete all work on the master drawing, you can also bind the xrefs into the drawing. Binding the xrefs makes them a permanent part of the drawing, which is basically the same as inserting the drawings as blocks. Binding the xrefs adds all the objects to the master drawing, thus increasing its size, but eliminates the possibility of breaking the links to their external references.*

Because they are separate, external references help reduce a drawing's file size and ensure that you are always working with the most recent version of a drawing.

## Attaching External References

Attaching a separate drawing to the current one creates an xref. The xref appears in the drawing as a special type of block definition, but the drawing objects are linked rather than added to the current drawing. If you modify the linked drawing, those changes are reflected in the current drawing whenever you open it.

When you attach an xref, its layers, linetypes, text styles, and other elements are not added to the current drawing. Rather, these elements are also loaded from the referenced file each time that you open the host drawing or reload the xref. Attached xrefs can themselves contain other, nested xrefs. When you attach an xref, any nested xrefs contained in the file also appear in the current drawing.

You can attach as many copies of an external reference file as you want. Each copy can have a different position, scale, and rotation angle.

To attach an xref:

1. Do one of the following:

   - On the Reference toolbar, click External Reference Attach.
   - From the Insert menu, choose External Reference.
   - At the command line, type **XATTACH** (or **XA**) and then press ENTER.

   AutoCAD displays a Select Reference File dialog box, so that you can select a file to attach as an external reference.

2. In the Select File to Attach dialog box, select the file to attach as an external reference, and then click Open. AutoCAD displays the External Reference dialog box, shown in Figure 15-15. Note that, by default, the Reference Type is set as Attachment and the Retain Path check box is selected. You'll learn about the Overlay xrefs later in this chapter.

3. Under Insertion Point, either specify the insertion point or select the Specify On-screen check box so that AutoCAD prompts you for this information when you actually insert the external reference. Do the same for the values under Scale and Rotation.

4. Click OK.

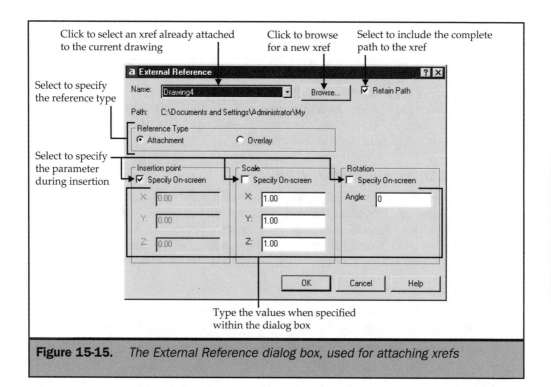

**Figure 15-15.**    *The External Reference dialog box, used for attaching xrefs*

*When you include the path while attaching the xref, you create an explicit link to the xref. AutoCAD always looks for the xref in the specified directory. If you don't include the path, AutoCAD searches for the xref based on its Support file search parameters. You'll learn more about the reasons for including or not including the path later in this chapter.*

Depending on the information specified under Insertion Point, Scale, and Rotation, AutoCAD may prompt you at the command line for additional information to complete the xref insertion. For example, by default, the scale and rotation angle are predetermined in the dialog box, but AutoCAD prompts you for the insertion point.

**Tip**   *You can also attach an xref by clicking the Attach button in the Xref Manager dialog box, which is described later in this chapter.*

## Overlaying External References

As you may have noticed in the External Reference dialog box, you actually have two different ways available to reference an xref: as an attachment or as an overlay. When you *attach* an xref, all nested xrefs are also attached. When you *overlay* an xref, however, any nested xrefs are ignored.

You should use overlays rather than attachments when information that you are referencing in your current drawing isn't needed by anyone else who might reference your drawing. For example, an electrical engineer might need to reference a furniture plan, to properly place electrical receptacles, but that furniture plan likely isn't needed by any of the other consultants who will be referencing the engineer's electrical plan. Therefore, the electrical engineer should reference the furniture plan as an overlay rather than as an attachment. That way, anyone who later references the electrical plan does not also see the furniture plan as part of the electrical plan.

# Managing External References

The Xref Manager dialog box provides a unified interface for managing all the xrefs in the current drawing. This dialog box provides two different views of your xrefs: the List View and the Tree View. Initially, when you start the command, AutoCAD displays the List View, shown in Figure 15-16. This view shows a list of all the xrefs attached to the current drawing. Information about the xrefs is arranged into the following six columns:

- **Reference Name**   Lists the names of attached xrefs.
- **Status**   Indicates whether the xref is loaded, unloaded, unreferenced, not found, unresolved, orphaned, or marked for unloading or reloading.
- **Size**   Shows the size of the referenced drawing file (in K).
- **Type**   Indicates whether the xref is an attachment or an overlay.
- **Date**   Shows the date that the xref was last modified. (This field is blank if the xref is unloaded, not found, or unresolved.)
- **Saved Path**   Shows the saved path of the xref (which is not necessarily where the xref was found).

*By default, this list is sorted alphabetically by xref name. You can sort the list on any other parameter by clicking the appropriate column heading. You can also change the width of the columns by clicking the line between the headings, holding down the left-mouse button, and then dragging the line left or right. In addition, you can toggle xrefs between Attach and Overlay by double-clicking either word in the Type field.*

To display the External References dialog box, do one of the following:

- On the Reference toolbar, click External Reference.
- From the Insert menu, choose Xref Manager.
- Select an xref, right-click in the drawing area, and then choose Xref Manager from the shortcut menu.
- At the command line, type **XREF** (or **XR**) and then press ENTER.

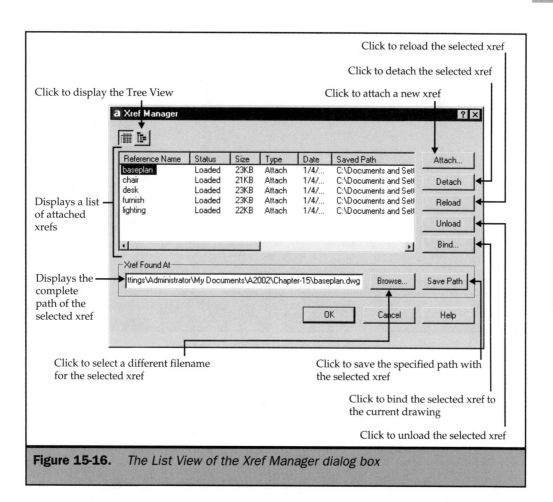

**Figure 15-16.**  *The List View of the Xref Manager dialog box*

The Tree View, shown in Figure 15-17, also shows the xrefs that are attached to the current drawing, but the Tree View displays them in a tree-structured view that shows the nesting level of the xrefs. External references attached directly to the current drawing appear at the top level of the tree, while xrefs nested in other xrefs are shown at subsequently lower levels. When multiple levels are shown, you can expand or compress the tree by clicking the plus or minus sign. To display the Tree View, click the Tree View button near the top of the dialog box, or press F4. To switch back to the List View, click the List View button, or press F3.

In addition to the List View and Tree View of the attached xrefs, the Xref Manager dialog box provides access to most of the other functions that you need to manage xrefs. You already learned how to attach xrefs. Clicking the Attach button displays either the External Reference dialog box previously shown in Figure 15-15 (if an xref is selected in

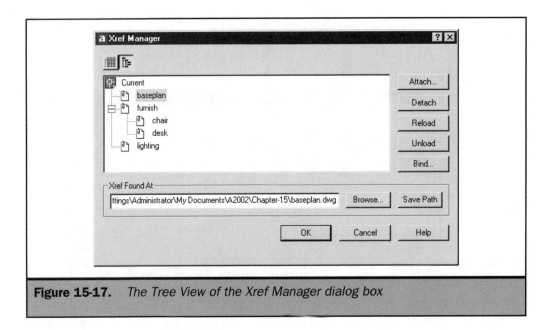

**Figure 15-17.** *The Tree View of the Xref Manager dialog box*

the list or tree) or the Select Reference File dialog box (if no xref is selected). In addition to attaching xrefs, you can do the following to xrefs:

- **Detach** Removes the xref from the current drawing and removes all copies of the xref from the drawing. Only xrefs attached or overlaid directly to the current drawing (those appearing in the Tree View at the top level) can be detached. Nested xrefs can't be detached.

- **Reload** Reads the most recently saved version of the xref into the current drawing.

- **Unload** Unloads the selected xrefs from the current drawing so that they don't display, thus improving AutoCAD's performance. The xrefs remain attached, however, and you can redisplay them by reloading them. You should unload rather than detach xrefs when you no longer need them in the current drawing but may need them again later, such as when you need to plot another copy of the drawing.

- **Bind** Makes the xref a permanent part of the current drawing. The xref is added to the drawing as a block. If the externally referenced drawing is subsequently updated, those changes are not reflected in the current drawing. When you bind an xref by using this option, you can additionally determine whether the xref is bound to the current drawing, by using the either the Bind or Insert option. The option that you choose determines how symbol names, such as layers, linetypes, and text styles, are added to the drawing. When you use the Bind option, symbol names that are added from the xref remain unique, whereas when you use the Insert option, the values for any symbol names in the current drawing that match those in the xref are retained and take precedence over those defined in the xref. For example, if both the current drawing and the xref have a layer named GRID, the color and linetype assigned to the GRID layer in the current drawing determine the color and linetype of objects inserted from the xref drawn on the similarly named layer.

## Using Demand Loading to Improve Xref Performance

Demand loading can increase the performance of AutoCAD when you work with external references. When demand loading is enabled, if layer or spatial indexes have been saved in the reference drawing, AutoCAD loads only the data from the reference drawing that is needed to regenerate the current drawing. For example, AutoCAD loads only objects on layers that are thawed when the xref has layer indexes, and loads only objects within clipping boundaries when the xref has spatial indexes. You'll learn more about clipping xrefs later in this chapter.

Demand loading is controlled by the XLOADCTL system variable, which, in turn, can be controlled from the Open and Save tab of the Options dialog box. The Demand Load Xrefs drop-down list has three options: Disabled, Enabled, and Enabled With Copy.

When demand loading is enabled (XLOADCTL equals 1), AutoCAD opens the xref whenever the drawing that references the xref is open, so that the program can read in any geometry that it needs *on demand*. Unfortunately, when a drawing is opened in this way, other users working on the network can't save changes to the referenced drawing, because they can open it only as a read-only drawing.

If other users need to be able to modify the referenced drawing, you can instruct AutoCAD to demand load a copy of the drawing, by selecting Enabled With Copy in the Options dialog box (XLOADCTL equals 2). That way, you can still benefit from demand loading and allow others to edit the referenced drawing. For example, User A opens a drawing containing an xref (with XLOADCTL equal to 2). A copy of the xref is made locally. User B then edits the xref. In order for User A to see the changes, User B must save the changes and then User A must manually reload the xref.

When demand loading is disabled, AutoCAD reads in the entire reference drawing, regardless of its layer visibility or the use of clipping boundaries.

*TIP:* *When you use the Enabled With Copy option to demand load xrefs across a network, you can use the XLOADPATH system variable to instruct AutoCAD to create the copies of the reference files on your local machine. On the other hand, to minimize the number of temporary files created when several users reference the same drawing, you can use the XLOADPATH variable (or the Temporary External Reference File Location setting on the Files tab of the Options dialog box) to point to a common directory on the network, so that all users share the same temporary copies of the reference drawings.*

To benefit from the performance improvement of demand loading, you must also have saved layer and spatial indexes along with the referenced drawings. The creation of layer and spatial indexes is controlled by yet another system variable, INDEXCTL. Rather than having to remember this variable name, you can control the creation of these indexes by selecting Options from the Tools drop-down list in the Save Drawing As dialog box. This displays the Saveas Options dialog box, which has four possible settings, listed in the table that follows the illustration.

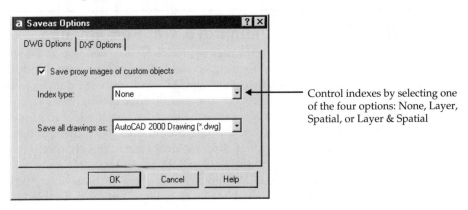

Control indexes by selecting one of the four options: None, Layer, Spatial, or Layer & Spatial

| Setting | INDEXCTL Value | Meaning |
| --- | --- | --- |
| None | 0 | No indexes created |
| Layer | 1 | Layer index created |
| Spatial | 2 | Spatial index created |
| Layer & Spatial | 3 | Both layer and spatial indexes created |

As already described, layer indexes ensure that AutoCAD doesn't need to load objects from the referenced drawing that are drawn on frozen layers. Spatial indexes organize objects based on their location within the referenced drawing in three-dimensional space, ensuring that AutoCAD doesn't load objects that are outside a clipping boundary. You can also define front and back clipping planes by using the XCLIP command's Clipdepth option. You'll learn more about the XCLIP command later in this chapter.

*TIP: Layer and spatial indexes are useful only when a drawing will be used as an xref. If you include these indexes when saving drawings that won't be referenced, you unnecessarily increase the size of drawing files. For that reason, AutoCAD sets the default INDEXCTL value to 0. You should change this setting only when you save drawings that will later be referenced.*

*Sometimes, you may need to know some information about the objects contained in an xref, such as their layer, color, or type of object. The XLIST Express tool, described on the companion web site, displays a dialog box that lists nested xrefs and block objects, displaying their object type, block name, layer, color, and linetype properties.*

## Detaching External References

Although you can erase individual instances of xrefs, if you no longer need the xref, you can completely remove it and all instances of it by detaching it from the current drawing. Detaching an xref also removes any dependent symbols, such as layer names and linetypes, that were added to the current drawing.

 *If you erase all references of an xref in the current drawing, AutoCAD detaches it from the drawing the next time that you open the drawing.*

To detach an xref:

1. Start the XREF command.
2. In the Xref Manager dialog box, select the xref to detach.
3. Click the Detach button.
4. Click OK.

**Note** *Remember that you can't detach nested xrefs.*

## Unloading and Reloading External References

As you learned earlier, when you unload a referenced file that is no longer needed, you can improve AutoCAD's performance, because the program no longer has to read or draw the objects from the xref or any nested xrefs that it contains. The referenced file remains attached, however, so that it can be reloaded if necessary.

To unload an xref:

1. Start the XREF command.
2. In the Xref Manager dialog box, select the xref to unload.
3. Click the Unload button.
4. Click OK.

Although AutoCAD automatically reloads all xrefs each time that you open a drawing containing externally referenced drawings, as well as each time that you print or plot a drawing containing xrefs, there are two instances in which you may need to reload an xref explicitly. Obviously, you must reload an xref that has been unloaded and is subsequently needed in the current drawing. You may also need to reload an xref when you're working on a network and other users may have opened and modified the reference drawing.

MOVING BEYOND
THE BASICS

To reload an xref:

1. Start the XREF command.

2. In the Xref Manager dialog box, select the unloaded xref to reload.

3. Click the Reload button.

4. Click OK.

 **Note** *Remember that other users can't save changes to reference drawings that have been opened by using demand loading, unless you instruct AutoCAD to open a copy of the demand-loaded drawing.*

## Binding External References to Drawings

As you already learned, xrefs are not part of the drawing. Rather, they are links to an externally referenced file. Sometimes, it is useful to bind the xrefs to the current drawing, particularly when you need to send drawings to someone else or are finished working on the drawing and need to archive it. Rather than send or archive the master drawing and all of its xrefs, you can bind those xrefs to the drawing, making them a permanent part of the drawing.

To bind an xref:

1. Start the XREF command.

2. In the Xref Manager dialog box, select the xref to bind.

3. Click the Bind button. AutoCAD displays the Bind Xrefs dialog box so that you can determine whether to bind the xref and all of its dependent symbols into the drawing, or detach the xref and then insert it into the drawing as a block.

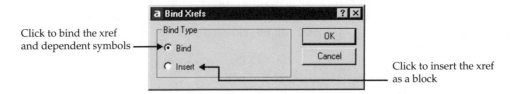

4. Click either the Bind or Insert radio button and then click OK.

5. Click OK to complete the command.

Binding an xref to the drawing inserts all of its objects and dependent symbols. Sometimes, however, you may simply want to bind a dependent symbol, such as a text style, into the current drawing, so that you can use it even if the xref is no longer attached. The XBIND command lets you bind any dependent symbol—blocks, dimension styles, layers, linetypes, and text styles—without having to bind the entire referenced drawing.

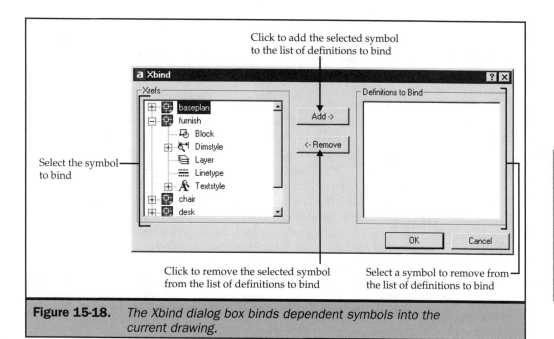

Click to add the selected symbol
to the list of definitions to bind

Select the symbol
to bind

Click to remove the selected symbol
from the list of definitions to bind

Select a symbol to remove from
the list of definitions to bind

**Figure 15-18.** *The Xbind dialog box binds dependent symbols into the current drawing.*

MOVING BEYOND
THE BASICS

The XBIND command displays the Xbind dialog box, shown in Figure 15-18, which displays a tree-structured list of all the xrefs in the current drawing. Click the plus sign (+) adjacent to an xref to expand that branch of the tree, to see the xref's dependent symbols. To bind a dependent symbol to the current drawing, select the symbol that you want to bind and then click the Add button to add it to the Definitions To Bind list. If you add an object to this list by mistake, you can select it in the right-hand list and then click the Remove button. When you're ready to bind the symbols into the current drawing, click OK.

To start the XBIND command, do one of the following:

- On the Reference toolbar, click External Reference Bind.
- From the Modify menu, choose Object | External Reference | Bind.
- At the command line, type **XBIND** (or **XB**) and then press ENTER.

## Controlling Dependent Symbols

When you attach an xref to the drawing, AutoCAD also adds the xref's dependent symbols, such as layers, linetypes, and text styles. To ensure that AutoCAD doesn't add a duplicate symbol name, it creates a new listing in the symbol table, consisting of the xref name and the symbol name, separated by the pipe symbol ( | ).

For example, if you attach an xref named FURNISH, containing a layer named DIMENSION, AutoCAD adds a new layer that appears in the layer listing as FURNISH | DIMENSION, so that no confusion occurs if another layer in the current drawing is named DIMENSION.

Although you can neither create new text by using a text style that is referenced in the current drawing as a dependent symbol nor make a dependent layer the current layer, you can control the visibility, color, and linetype associated with dependent layers. If you do make changes to dependent layers, however, you may have problems the next time that you load the xref, if your visibility, color, or linetype settings have reverted back to the settings established in the referenced drawing. The VISRETAIN system variable lets you control dependent layer properties. When the VISRETAIN value is set to 1, any changes that you make to the dependent layers in the current drawing take precedence over those established in the referenced drawing. When VISRETAIN equals 0, however, any changes that you make to dependent layers remain in effect only until the xref is reloaded. The next time that you open the host drawing, those settings revert back to the settings most recently saved in the referenced drawing. You can also control the VISRETAIN setting by using the Retain Changes To Xref Layers check box on the Open and Save tab of the Options dialog box.

Dependent symbols also come into play when you bind symbols or the entire xref into the current drawing. If you bind the xref or any of its dependent symbols, AutoCAD removes the pipe symbol and constructs a new name for the dependent symbol, consisting of the xref name and the symbol name, separated by a dollar sign ($), a number, and a second dollar sign. Thus, if you bind the FURNISH | DIMENSION layer to the current drawing, AutoCAD renames it FURNISH$0$DIMENSION. If a layer with that name already exists in the current drawing, AutoCAD increases the number between the dollar signs until no matching layer name exists (in other words, FURNISH$1$DIMENSION).

After a symbol name is bound to the current drawing, you can use standard AutoCAD commands to rename the symbol and then use it as you would any other symbol. For example, after the FURNISH$0$DIMENSION layer is bound to the current drawing, you can rename the layer FURN-DIM, make it the current layer, and draw new objects on that layer.

When you bind an xref to the drawing by inserting it as a block, the situation is a bit different. AutoCAD disregards the xref portion of the symbol name and simply adds the symbol name to the current drawing. This can cause problems. For example, if you already have a DIMENSION layer in the current drawing, the DIMENSION layer that is inserted from the FURNISH drawing takes on the color and linetype of the identically named layer in the current drawing. Of greater concern is the situation that arises when you insert an xref that itself contains a block with the same name as a block in the current drawing. When that happens, AutoCAD disregards the block definition from the xref, and all instances of the block that are inserted from the bound xref are redefined by using the identically named block in the current drawing. Unless you are sure that the identically named blocks are, in fact, identical, you should use the bind option rather than the insert option when you bind xrefs.

## Clipping Blocks and External References

After you attach an xref or insert a block, you can define a clipping boundary to isolate a portion of a particular block or xref instance. When the clipping boundary is applied, only the portion of the block or xref that is within the boundary remains visible. The boundary can consist of planar straight-line segments or an existing polyline. The boundary can also consist of front and back clipping planes, in which case objects aren't displayed that are outside the volume defined by the boundary and the specified depth. If the clipping boundary is turned off, the entire block or xref is displayed.

After you clip an instance of a block or xref, you can edit the instance just like any other unclipped block or xref. If you move or copy the clipped block or xref, the boundary moves with it. Xrefs that contained nested clipped xrefs appear clipped in the drawing, and if you clip an xref that contains nested xrefs, the nested xrefs are also clipped.

You apply a clipping boundary by using the XCLIP command. When you start the command, AutoCAD prompts you to select the objects that you want to clip. You can use any object selection method; AutoCAD ignores all objects other than blocks and xrefs. When you finish selecting objects and press ENTER, the XCLIP command offers six options:

```
Enter clipping option
[ON/OFF/Clipdepth/Delete/generate Polyline/New boundary] <New>:
```

When you first use the command, you generally will want to specify the clipping boundary. After the clipping boundary is applied, you can use the other options to turn the boundary on and off, set front and back clipping planes, or generate a new polyline object that matches the clipping boundary.

When you define a new boundary, AutoCAD prompts:

```
Specify clipping boundary:
[Select polyline/Polygonal/Rectangular] <Rectangular>:
```

Press ENTER to select the Rectangular option, in which case AutoCAD prompts you to specify the opposite corners of a rectangular boundary. If you select the Polygonal option, the program prompts you to specify the points that define the polygonal boundary, whereas if you choose the Select Polyline option, the program prompts you to select an existing polyline for use as the boundary.

MOVING BEYOND
THE BASICS

*If you use a polyline as the clipping boundary, and the polyline has arc segments or has been fit-curved, the clipping boundary is applied as though the polyline were composed of straight segments. If the polyline has been spline-curved, however, the spline is used as the clipping boundary.*

In addition to the New Boundary option, the XCLIP command offers five other options:

- **ON**  Turns on the clipping boundary if you previously turned it off.
- **OFF**  Turns off the clipping boundary, causing the entire block or xref to be displayed.
- **Clipdepth**  Lets you set the front and back clipping planes, thus defining a three-dimensional volume. You can specify the distance to the front and back clipping planes either by entering a value in reference to the clipping boundary or by selecting the front and back clip points. The clipping planes have the same shape as the clipping boundary and are parallel to it. If you previously defined front or back clipping planes, you can use the Remove suboption to eliminate one or both of them.
- **Delete**  Removes a previously defined clipping boundary from the selected block or xref.
- **Generate Polyline**  Creates a new polyline object that corresponds to the clipping boundary. The polyline is drawn on the current layer, using the current linetype and color settings.

To clip an xref or block:

1. Do one of the following:

   - On the Reference toolbar, click External Reference Clip.
   - From the Modify menu, choose Clip | Xref.
   - At the command line, type **XCLIP** (or **XC**) and then press ENTER.

2. Select the objects to clip and then press ENTER.

 *You can also select the xref you want to clip, right-click in the drawing area, and then choose Xref Clip from the shortcut menu.*

3. At the prompt, press ENTER to select the New Boundary option.

4. Select the polyline, polygonal, or rectangular option and then either select the existing polyline to use for the clipping boundary or define the polygonal or rectangular boundary.

*LEARN BY EXAMPLE*
*Apply a rectangular clipping boundary to a block or xref to isolate a portion of that block or xref as a large-scale detail. Open the drawing Figure 15-19 on the companion web site.*

To apply the rectangular clipping boundary, as shown in Figure 15-19, use the following command sequence and instructions:

```
Command: XCLIP
Select objects: (select the xref)
Select objects: ENTER
Enter clipping option
[ON/OFF/Clipdepth/Delete/generate Polyline/New boundary] <New>: ENTER
Specify clipping boundary:
[Select polyline/Polygonal/Rectangular] <Rectangular>: ENTER
Specify first corner: (select point A)
Specify opposite corner: (select point B)
```

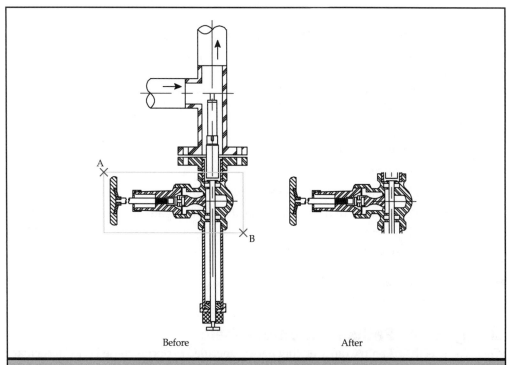

Before                              After

**Figure 15-19.**    *A rectangular clipping boundary, created by using the XCLIP command*

*If you attempt to define a second clipping boundary on an existing xref or block instance, AutoCAD asks you whether you want to delete the old boundary and apply a new boundary. If you answer yes, the old boundary is removed. If you answer no, the XCLIP command simply terminates. The only way that you can create the appearance of adding a second clipping boundary to the same xref is to make a copy of the xref and apply a new boundary to the copy.*

After you create a clipping boundary, that boundary normally is not visible. You can make the boundary visible, however, by changing the XCLIPFRAME system variable. A value of 0 makes all clipping boundaries invisible; a value of 1 makes them all visible. To change the value, do one of the following:

- On the Reference toolbar, click External Reference Clip Frame.
- From the Modify menu, choose Object | External Reference | Frame.
- At the command line, type **XCLIPFRAME** and then press ENTER.

*You can also use the CLIPIT Express tool, described on the companion web site, to clip a block, image, or xref by using a circle, polyline, or arc as the clipping boundary.*

# Changing the Path for External References

If you open a drawing that has an attached xref that has been moved to a different directory or renamed from what it was originally named when it was attached to the drawing, AutoCAD won't be able to load the xref and will report its status as *not found.* You can reestablish the link to the file by changing the path for the xref.

To change the xref path:

1. Open the Xref Manager dialog box.

2. Select the xref whose path you want to change.

3. In the Xref Found At area, click the Browse button.

4. In the Select New Path dialog box, specify the new path and then click Open. The new path is displayed in the Save Path field and the Xref Found At edit box.

5. Click OK to complete the command.

*If you change the path to a nested xref, you can save that change only if the VISRETAIN system variable is set to 1.*

## Changing the Paths for Multiple Xrefs

When you specify the path to an xref, that path is specified explicitly, based on a particular hard drive. This setup is fine, as long as you're the only one trying to load the drawing—but what if you load a drawing that contains xrefs located across a network or, worse yet, move the drawing and its xrefs to a different directory? AutoCAD may no longer be able to resolve those xrefs, because the paths are different. You can solve this dilemma by changing the paths. But, if there are several xrefs, this procedure can become time-consuming, because you must change the paths one at a time when using the Xref Manager dialog box.

Although you can't change multiple xref paths simultaneously from within the Xref Manager dialog box, you can do so from the command line. To use the XREF command at the command line, type **-XREF** (or **-XR**) and then press ENTER. AutoCAD prompts:

```
Enter an option [?/Bind/Detach/Path/Unload/Reload/Overlay/Attach] <Attach>:
```

When you select the Path option, AutoCAD prompts:

```
Edit xref name(s) to edit path:
```

MOVING BEYOND
THE BASICS

You can then enter a single xref, type several xrefs separated by commas, or use wildcard combinations. For example, to change the paths for all the xrefs in the current drawing, type an asterisk (*) and then press ENTER.

Next, AutoCAD displays the xref name and the old path (the path saved when the xref was originally attached) and prompts you for the name of the new path. You can then enter the new path. If you specify more than one xref to change, AutoCAD repeats these prompts for each xref.

When you enter the new path, you must enter the actual filename, but you don't need to type the complete path. Instead, you can specify the path as an *implicit path,* which contains just the path information needed for AutoCAD to locate the file. For example, suppose that the complete path to the host drawing is C:\DRAWINGS\JOBS\98-01, and that the xrefs are located at C:\DRAWINGS\JOBS\98-01\XREFS. You can specify the implicit path as **XREFS\\*drawing*,** in which *drawing* is the actual xref drawing name.

## Specifying Xref Search Paths

When you open a drawing that contains xrefs, AutoCAD first tries to load the xref from the path originally specified when the reference was attached. If the xref is not found, AutoCAD next searches for the file in the current directory, and then in the directory in which the current drawing is located. If AutoCAD still can't find the xref, it usually searches all the directories in the Support File path. AutoCAD includes an additional feature—the *project path*—which specifies another path for the program to search before searching the support paths.

The project path lets you manage xrefs on a per-project basis. If the PROJECTNAME system variable is set and a corresponding project name has been established in the Options dialog box, AutoCAD searches for the xref along the project search path.

To use the project name feature, you must first add a project name and then specify the project search paths in the Options dialog box. Because this information is stored as an entry in the Windows System Registry, no other method of specifying the project path is allowed.

To add a project name:

1. Do one of the following:

   ■ From the Tools menu, choose Options.

   ■ At the command line, type **OPTIONS** (or **OP** or **PR**) and then press ENTER.

2. On the Files tab of the Options dialog box, expand the Project Files Search Path entry and then click Add. AutoCAD creates a new project folder, assigning it the name Project*x*, wherein *x* is the next available project number.

3. Either enter a new, more descriptive project name and then press ENTER or just press ENTER to accept the default project name.

4. Click OK to complete the command, or click Apply to complete the creation of the project name and remain in the Options dialog box.

**Note**   *After you create a project name, you can rename it by selecting it (or pressing F2) and simply typing a new name. To delete a project name, select it and then click Remove. Project names can't be more than 31 characters in length.*

After you create a project name, you can specify the search path for that project. You can create multiple project names, and each name can have its own unique search paths. To add a search path to a project name:

1. Start the Options command.

2. On the Files tab of the Options dialog box, select the project name and then click Add.

3. Type the search path below the project name, or click Browse and select a path by using the Browse for Folder dialog box.

4. Click OK to complete the command, or click Apply to complete the addition of the search path and remain in the Options dialog box. You can then add additional search paths to the project name.

When you have more than one search path assigned to a project name, the order in which the paths appear below the project name determines the order in which AutoCAD searches those paths. You can change the order by selecting the path and then clicking the Move Up or Move Down button.

**Note**   *After you create a search path, you can change it by selecting it (or pressing F2) and then simply typing a new path or clicking Browse. To delete a path, select it and then click Remove.*

After you establish a project name and search paths, you can use the Options dialog box to specify that the project name be the current active project. Once that has been done,

AutoCAD searches the paths associated with that project for xrefs that were not found in their specified paths.

To set the current project:

1. Start the Options command.

2. On the Files tab of the Options dialog box, expand the Project Files Search Path entry.

3. Select the project name and then click Set Current.

4. Click OK or Apply.

*You can also set the current project name by typing **PROJECTNAME** at the command line and then entering the name of the project, thus changing the value of the PROJECTNAME system variable directly.*

# Editing References in Place

Although external references can help you combine drawings and manage complex sets of drawings, they would pose a problem if you needed to make changes to objects in a drawing that were actually part of an external reference. AutoCAD solves this potential problem with a function called *in-place editing*—the ability to edit external references from within a drawing to which they are attached. You can also use in-place editing to edit block definitions from within the current drawing.

When working with external references, in-place editing provides the advantage of being able to modify the xref within the context of the current drawing, instead of having to open the xref drawing and go back and forth between the drawing files. Using in-place editing to modify a block definition enables you to make the changes directly to an instance of the block rather than to the original objects you used to define the block (which may have been deleted when the block was first created).

*In-place editing is meant primarily for making minor changes to external references. If you need to make major revisions, you should open the xref file and make the changes directly in that file. Making major changes to an xref using in-place editing can temporarily increase the size of the current drawing file. Also note that you can't use in-place editing when a drawing has been opened using the Partial Open feature.*

## Editing Blocks and External References

You use the REFEDIT command to initiate in-place editing. AutoCAD prompts you to select the reference you want to edit. Click to select the block or external reference you want to edit. You must select the reference by pointing. As soon as you select a block or xref, AutoCAD displays the Reference Edit dialog box, similar to Figure 15-20, which

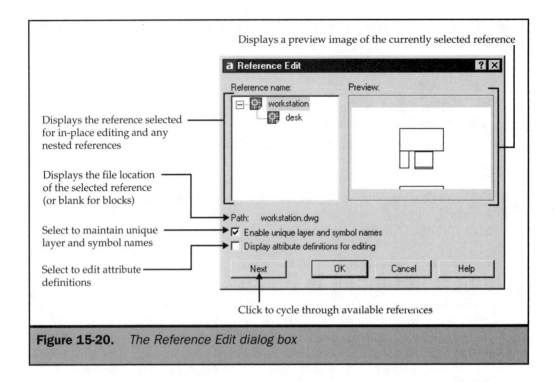

**Figure 15-20.**    *The Reference Edit dialog box*

displays the name of the reference and any other references nested within the selected reference. You can then select the reference you want to edit.

To start the REFEDIT command, do one of the following:

- Double-click the block or external reference.

- On the Refedit toolbar, click Edit Block or Xref.
- From the Modify menu, choose In-place Xref And Block Edit | Edit Reference.
- At the command line, type **REFEDIT** and then press ENTER.

**Note**    *You can't edit a block reference that was inserted using the MINSERT command.*

Nested references are displayed only if the object you selected is part of a nested reference. For example, if you select an xref named DESK, which is nested within an xref named WORKSTATION that also contains a second nested xref called CHAIR, you will see both the WORKSTATION and DESK xrefs in the Reference Edit dialog box, but not CHAIR. You can then modify either the WORKSTATION or DESK xref. Since the CHAIR xref is nested within WORKSTATION, however, it can also be selected for modification within the context of WORKSTATION. In other words, you can move the CHAIR within WORKSTATION, but you can't modify the CHAIR xref itself.

**Note**    *If you are editing an xref that contains OLE objects, the OLE objects are displayed but can't be edited. You'll learn more about OLE objects in Chapter 24.*

If you select a block reference, you can display attribute definitions within the reference and make them available for editing, by selecting the Display Attribute Definitions For Editing check box. The attribute values become invisible, and the attribute definitions become available for editing along with the block geometry. If you make changes to the attribute definitions, those changes only affect subsequent insertions of the block. The attributes in existing instances of the block are not affected.

The Enable Unique Layer And Symbol Names check box controls whether layer and symbol names of objects extracted from the reference remain unique or are altered. You'll learn more about this later in this chapter.

After selecting the reference you want to edit, click OK. AutoCAD then prompts you to select the nested objects. Select those objects that you want to edit, and then press ENTER. The objects you select become part of the *working set*, the group of objects temporarily available for modification. AutoCAD visually indicates those objects available for in-place editing by displaying them normally while all the other objects in the drawing are faded. By default, all the other objects in the drawing are faded by 50 percent. You can control the degree of fading by changing the XFADECTL system variable. If you change XFADECTL from its default value of 50 to 75, all the objects in the drawing that are not included in the working set are displayed at 25 percent of their normal intensity.

**Tip**    *The XEDIT system variable determines whether the current drawing can be edited in-place when referenced by another drawing. You can also control this variable by changing the Allow Other Users To Refedit Current Drawing setting on the Open and Save tab of the Options dialog box.*

## Adding or Removing Objects from the Working Set

After you select the objects to be included in the working set, AutoCAD displays the Refedit toolbar, which shows the name of the selected reference and provides access to the other in-place editing commands. This toolbar is displayed automatically when you begin in-place editing and is closed automatically when you save changes back to the reference or discard those changes.

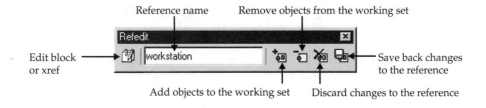

Reference name    Remove objects from the working set

Edit block or xref    Save back changes to the reference

Add objects to the working set    Discard changes to the reference

While performing in-place editing, you can add objects to or remove objects from the working set, by clicking the respective buttons on the Refedit toolbar. These functions are also available from the Modify menu. In most situations, if you create new objects, they are automatically added to the working set. Any objects not added to or removed from the working set display as faded within the drawing. If you remove an object from the working set, it is removed from the reference and added to the current drawing itself when you save the in-place editing changes back to the reference. If you add an existing object to the working set, it is removed from the current drawing and added to the reference itself when you save the changes back to the reference. You can tell whether an object is in the working set or the host drawing by the way it is displayed.

To add objects to the working set:

1. Do one of the following:

- On the Refedit toolbar, click Add Objects To Working Set.
- From the Modify menu, choose In-place Xref And Block Edit | Add To Working Set.
- At the command line, type **REFSET** and then press ENTER twice.

2. Select the objects you want to add, and then press ENTER.

To remove objects from the working set:

1. Do one of the following:

- On the Refedit toolbar, click Remove Objects From Working Set.
- From the Modify menu, choose In-place Xref And Block Edit | Remove From Working Set.
- At the command line, type **REFSET**, press ENTER, type **R**, and then press ENTER again.

2. Select the objects you want to remove, and then press ENTER.

*If Noun/Verb selection mode is enabled, you can also select the objects to be added to or removed from the working set before starting the REFSET command.*

Although an object in the working set may have been drawn on a locked layer in the xref, you can unlock its layer and make changes to the object when performing in-place editing. Any changes you make are saved when you save back your changes to the reference, but the original layer state is preserved within the reference drawing.

*You can also erase objects from either the working set or the host drawing. Objects that are erased from either the host drawing or the working set are not restored if you later discard the changes you made to the reference. You can use the UNDO command, however, to restore the drawing to its original state.*

# Saving Back Changes to References

After you complete your in-place editing, you can either save back or discard the changes you made to the reference. When you save back your changes, AutoCAD regenerates the drawing. All instances of the modified reference are regenerated to reflect those changes (except for changes you may have made to attribute definitions). If you discard your changes, the reference is restored to its original state.

To save back changes to the reference, do one of the following:

- On the Refedit toolbar, click Save Back Changes To Reference.
- From the Modify menu, choose In-place Xref And Block Edit | Save Reference Edits.
- Right-click in the drawing area and choose Close REFEDIT Session | Save Reference Edits from the shortcut menu.
- At the command line, type **REFCLOSE** and then press ENTER twice.

AutoCAD displays an alert dialog box, warning you that all reference edits will be saved. To save reference changes, click OK. To cancel the command and continue doing in-place editing, click Cancel.

To discard all changes made to the working set, do one of the following:

- On the Refedit toolbar, click Discard Changes To Reference.
- From the Modify menu, choose In-place Xref And Block Edit | Discard Reference Edits.
- Right-click in the drawing area and choose Close REFEDIT Session | Discard Reference Edits from the shortcut menu.
- At the command line, type **REFCLOSE**, press ENTER, type **D**, and then press ENTER again.

AutoCAD displays an alert dialog box, warning you that all reference edits will be discarded. To discard reference changes, click OK. To cancel the command and continue doing in-place editing, click Cancel.

*When you edit and save an xref using in-place editing, its original drawing preview image is no longer displayed in any of the dialog boxes. To restore the preview image, you must open and save the referenced drawing.*

When you save back your changes to the reference, objects added to the reference that inherited properties not originally defined in the reference retain those new properties. For example, if you added an object to the reference that was originally drawn in the host drawing on a layer that did not exist in the reference, that layer is added to the reference.

You can also control whether layer and symbol names of objects extracted from the reference and added to the host drawing are unique or altered. If you selected the Enable Unique Layer And Symbol Names check box in the Reference Edit dialog box, layer and symbol names added to the host drawing from the xref are assigned the prefix $#$, where # is a sequential number, similar to the way they are handled when you bind xrefs. If that check box was not selected, layer and symbol names are unchanged when added to the host drawing, similar to the names of inserted objects.

### *LEARN BY EXAMPLE*

*Use in-place editing to modify an external reference from within its host drawing. You'll find the drawing, along with step-by-step instructions, on the Chapter 15 page on the companion web site.*

# The Complete Reference

# Chapter 16

## Managing Content with AutoCAD DesignCenter

Auto CAD DesignCenter makes it easy to reuse both existing drawings and the individual objects and settings within those drawings. With DesignCenter, you can quickly find and open drawings, insert an entire drawing into the current drawing as a block, view object definitions (such as blocks, layers, linetypes, and text styles) within drawings and copy them into the current drawing, find and attach external references, and view and attach raster images to the current drawing. You can use this powerful tool to manage drawings and other files on your local computer, on other computers on your local area network, or even across the Internet. And, because DesignCenter works much like a web browser, you can save to a *Favorites* folder pointers to frequently used drawings so that you can quickly access them again in the future.

This chapter explains how to do the following:

- Understand the DesignCenter interface
- Open drawings using DesignCenter
- Add content to drawings using DesignCenter
- Save shortcuts to a Favorites folder

# Understanding the DesignCenter Interface

AutoCAD DesignCenter is designed to make it easy for you to find and organize drawing files and the objects contained within those files, and to reuse that information in new drawings. It can remain open while you use other AutoCAD commands. When you first start DesignCenter, it appears docked along the left side of the drawing area. This is its default position. When it is docked, you can change the width of DesignCenter by dragging the vertical split bar between DesignCenter and the drawing area or adjacent toolbars. You can also dock DesignCenter on the right side of the drawing area, or undock it so that it becomes a floating window, as shown in Figure 16-1.

To display AutoCAD DesignCenter, do one of the following:

- On the Standard toolbar, click AutoCAD DesignCenter.
- From the Tools menu, choose AutoCAD DesignCenter.
- At the command line, type **ADCENTER** (or **ADC**) and press ENTER, or press CTRL-2.

To close DesignCenter, do one of the following:

- On the Standard toolbar, click AutoCAD DesignCenter.
- From the Tools menu, choose AutoCAD DesignCenter.
- At the command line, type **ADCLOSE** and press ENTER, or press CTRL-2.

*If AutoCAD DesignCenter is currently displayed in a floating window, you can also close it by clicking its Close button.*

To dock DesignCenter to the right side of the drawing area, click any of its edges and then drag DesignCenter to the right side of the drawing area. To undock DesignCenter, click any of its edges, drag it away from the docking region at the right or left side of the AutoCAD window, and drop it anywhere on your screen. You can then change the height or width of the floating DesignCenter window. To dock it again, drag the DesignCenter window back over one of the docking areas.

**Tip** *You can also dock DesignCenter back to its previous docked position by double-clicking the title bar of the floating DesignCenter window.*

The AutoCAD DesignCenter window consists of a navigation pane, or *tree view*, on the left, and a content pane, or *palette*, on the right, separated by a vertical split bar. Across the top of the window is a toolbar. The tree view provides a hierarchical listing of the current content source (such as a drive on your computer), whereas the palette displays the items within the current source object selected in the tree view. For example, in Figure 16-1, a folder has been selected in the tree view, and the palette shows the content of that folder, including other folders and several drawing files.

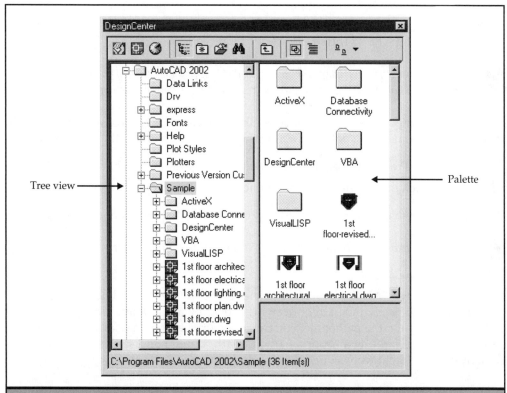

**Figure 16-1.**    *AutoCAD DesignCenter displayed as a floating window*

MOVING BEYOND
THE BASICS

# Using the Tree View

The tree view works similarly to the Windows Explorer directory tree. You can expand or compress the tree by clicking the plus or minus sign, respectively, in the small box adjacent to an item in the tree. The DesignCenter tree view shows a lot more information, however, than Windows Explorer. Notice that even drawing files have a plus sign adjacent to them. Clicking this plus sign expands the tree to display content within that drawing. When you select an item in the tree view, its content is displayed in the palette.

You can change the source of the content displayed in the tree view by selecting one of the buttons on the left side of the DesignCenter toolbar. These buttons enable you to select one of the following content sources:

■ **Desktop**   Lists local and network drives

■ **Open Drawings**   Lists all drawings currently open in AutoCAD

■ **History**   Lists the last 20 locations accessed using DesignCenter

You can also select the content source from the shortcut menu displayed when you right-click the palette background.

 *If you currently have a custom application loaded that provides custom content to AutoCAD (such as an Object ARX application), the toolbar also displays a Custom Content button.*

After you locate and select the content source, you can hide the tree view, if you wish, to make more room for the palette. To hide the tree view, do one of the following:

■ Click the Tree View Toggle button on the DesignCenter toolbar.

■ Right-click the palette background and choose Tree from the shortcut menu.

When the tree view is hidden, you can restore it by repeating either of these steps.

 *The Tree View Toggle button isn't active when you select the History content source.*

You can also type **ADCNAVIGATE** at the AutoCAD command line and then type the pathname to specify the path or drawing file you want to load. When AutoCAD prompts you to enter the pathname, you can enter any valid path, including a directory path, a pathname and filename, or a network path. When you use this command, DesignCenter automatically switches to the desktop view.

## Using the Palette

The palette displays the content of the current source, and you can select that source either by navigating the tree view or by using the Load DesignCenter Palette dialog box, a dialog box nearly identical to the Select File dialog box used to open existing drawings (see Figure 1-15). When you select a folder containing drawings in the tree view, the palette displays either a list of those drawings or preview images of the drawing files, as well as an indication of the other types of files in the selected folder. If you select a drawing file in the tree view, the palette displays either a list of named objects or icons representing the types of named objects within the drawing, such as blocks, layers, linetypes, and text styles. If you expand the tree view and select one of these named objects within the drawing—such as blocks—the palette displays either a list of all the blocks or preview images of all the blocks contained within the drawing. You can then use DesignCenter to insert any of those blocks into the current drawing.

To load the palette by using the Load DesignCenter Palette dialog box:

1. Do one of the following:

- On the DesignCenter toolbar, click Load.

- Right-click the palette background and choose Load from the shortcut menu.

2. Select the item whose content you want to load, and click Open.

The controls in the Load DesignCenter Palette dialog box enable you to quickly locate files in the History, My Documents, or Favorites folder, on your Desktop, or anywhere else on your computer, a local area network, an FTP site, or any other previously saved location. You can use the controls in the Tools drop-down menu to find files or add selected locations to your Favorites folder, or use the Search The Web function to locate content at a remote location.

You can also load the palette by using Windows Explorer. After using Windows Explorer to navigate to the item that you want to load into the palette, simply drag-and-drop that item into the palette.

The AutoCAD Today window provides yet another way to load the DesignCenter palette. In the Today window, when you select the Symbol Libraries tab in the My Drawings area, AutoCAD displays a list of DesignCenter symbol libraries. Click one of the symbol libraries to immediately load it into the DesignCenter palette.

Instead of using the tree view, you can move up or down through the hierarchy of drives, folders, files, and drawing content within the palette. To display the contents of any item within the palette, double-click that item. To move up a level in the hierarchy, click the Up button on the DesignCenter toolbar.

**Caution**     *When you reach the lowest level of the hierarchy (such as a list of all the blocks within a drawing), double-clicking an item at that level inserts it into the current drawing. You'll learn about using DesignCenter to insert drawing content later in this chapter. Also note that all blocks, including nested blocks, are shown at a single level.*

MOVING BEYOND THE BASICS

## Changing the Palette Display

You can display information in the DesignCenter palette in one of four different ways:

- **Large Icons**   Displays previews of drawings and blocks, and displays other items and file types as large icons, each with its name shown below it.

- **Small Icons**   Displays a single-column list of icons identifying the type of item (folder, drawing, block, and so on), each with its name adjacent to it on the same line.

- **List**   Displays a multicolumn list whose content is otherwise similar to the one displayed by the Small Icons view.

- **Details**   Displays a list similar to the Small Icons view, with additional columns showing file size and file type (when the palette displays the hierarchy at the folder level). You can sort the list by clicking the heading of any of the columns. For example, to sort the list by file size, click the File Size heading.

To select the way in which you want the palette displayed, click the View button on the DesignCenter toolbar and then choose the desired view option. You can also select the view option by right-clicking the palette background, choosing View, and then choosing the desired view option from the shortcut menu.

**Note**   *When you browse for drawings using the Large Icons view, DesignCenter creates thumbnail views for every drawing in the directory in which you are browsing, and stores those thumbnails in a file with the file extension .CDC. If a directory contains many drawings, the first time you use the Large Icons view within that directory, the display of those icons may be slower while this CDC file is created. Subsequent viewing of that directory will be much faster, however, because DesignCenter reads the CDC file to display the large icons, rather than having to re-create them.*

## Displaying Descriptions and Previews

Below the palette, you can also display a preview and a description of a selected drawing or block. The drawing description consists of the information stored using the Drawing Properties dialog box, which you learned about in Chapter 2. The block description is the one specified in the Description area of the Block Definition dialog box when the block was originally defined.

To display the preview area, click the Preview button on the DesignCenter toolbar, or right-click the palette background and choose Preview from the shortcut menu. Similarly, to display the description area, click the Description button on the DesignCenter toolbar, or right-click the palette background and choose Description from the shortcut menu. Figure 16-2 shows a typical preview image and description of a block contained within a drawing.

The preview image is scaled to fill the preview area. The larger the preview area, the larger the preview image. You can change the height of both the preview and description areas by dragging the bars separating them from the rest of the palette, and you can change the width by changing the width of the palette or the entire DesignCenter window.

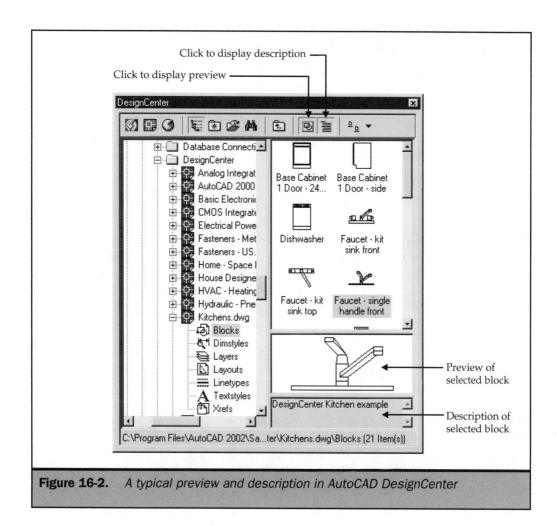

**Figure 16-2.**  *A typical preview and description in AutoCAD DesignCenter*

MOVING BEYOND
THE BASICS

You can copy the description to the Windows Clipboard. You can't edit the description, however, from within AutoCAD DesignCenter. To edit the drawing description, you must open the drawing and use the Drawing Properties command, as described in Chapter 2. To edit the block description, you must edit the block within the drawing, as described in Chapter 15.

**Note** *If you load the palette with other file formats recognized by AutoCAD, such as a raster file, you can also display a preview image and description of the selected file. If you select an AutoCAD hatch pattern file (such as ACAD.PAT) in the tree view, its hatch patterns are displayed in the palette, and a preview and description of a selected pattern in the preview and description areas.*

# Using AutoCAD DesignCenter

AutoCAD DesignCenter provides a powerful, visual tool to do any of the following tasks:

- Open drawings
- Find drawings and other files
- Find specific content—blocks, layer definitions, dimension styles, and so on— in existing drawings and reuse it in the current drawing
- Insert blocks and attach xrefs and images
- Drag-and-drop hatch patterns

## Opening Drawings

You can use AutoCAD DesignCenter to open any AutoCAD drawing. The drawing can be located anywhere on your computer, on a local area network, or on the Web.

To open a drawing using DesignCenter, navigate to the location of the drawing, so that the drawing's icon is visible in the DesignCenter palette, and then do one of the following:

- Drag the icon of the drawing into the background of AutoCAD's drawing area.
- Right-click the drawing icon in the palette and choose Open In Window from the shortcut menu.

**Caution** *When opening a drawing by dragging, be sure to drag the drawing file icon into the background of AutoCAD's drawing area (the gray area outside of individual drawing windows), not into an open drawing. Dragging a drawing into an open drawing window inserts the drawing being dragged into the open drawing as a block.*

# Finding Content

Although you can use the DesignCenter tree view and palette to visually locate content, AutoCAD DesignCenter provides additional tools to help you search for drawings and other content, such as blocks, text styles, and layer definitions. DesignCenter's Find dialog box, shown in Figure 16-3, enables you to search for drawings and the content within drawings.

To display the Find dialog box, do one of the following:

■ On the DesignCenter toolbar, click Find.

■ Right-click in a blank area anywhere in DesignCenter and choose Find from the shortcut menu.

MOVING BEYOND
THE BASICS

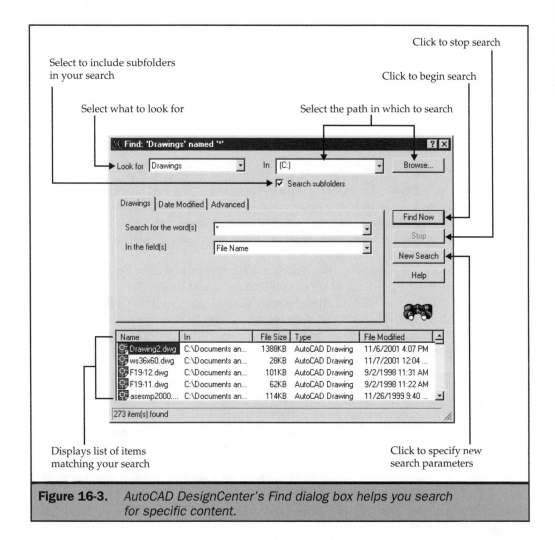

**Figure 16-3.**  AutoCAD DesignCenter's Find dialog box helps you search for specific content.

When you use the Find dialog box, the first step is to specify the type of content you want to search for and the path in which you want to search. In the Look For drop-down list, select the type of content. You can search for any of the following content:

- **Blocks**   Searches drawings for a specified block
- **Dimension styles**   Searches drawings for a specified dimension style
- **Drawings**   Searches for a drawing matching the search parameters
- **Drawings and Blocks**   Searches for a drawing or a block within a drawing
- **Hatch Pattern Files**   Searches for hatch pattern files matching the search parameters
- **Hatch Patterns**   Searches hatch pattern files for a specified hatch pattern name
- **Layers**   Searches drawings for a specified layer name
- **Layouts**   Searches drawings for a specified layout
- **Linetypes**   Searches drawings for a specified linetype
- **Text styles**   Searches drawings for a specified text style
- **Xrefs**   Searches drawings for a specified external reference

**Note**   *You can also search for customized content if a custom application is loaded.*

Depending on the type of content you specify, the Find dialog box provides additional tools to help you in your search. For example, if you search for a block, the dialog box displays a Blocks tab in which you can specify the name of the block you wish to find. Similarly, if you search for a text style, the dialog box displays a Textstyles tab in which you can specify the name of the text style.

When searching for a drawing, the dialog box displays three tabs. On the Drawing tab, you can specify the filename or other text that you want to search for, and the field in which you want to search. In addition to searching for an actual filename, you can search for words in the title or subject, a keyword, or the name of the author of the drawing. To locate a drawing based on its title, subject, author, or keywords, this information must have been previously saved as part of the drawing's properties.

**Caution**   *At present, DesignCenter's search capabilities are quite limited. For example, when performing a keyword search, you can only enter one keyword at a time. In addition, you need to include an asterisk at the beginning and end of the word for which you are searching. For example, to search for the keyword Transportation, you must type* ***\*Transportation\**** *in the Search For The Word(s) box.*

You can also search for a drawing file based on the date it was created or modified. The controls on the Date Modified tab let you specify a specific date or date range. On the Advanced tab, you can specify additional search parameters, such as specified text found within the drawing file as a block definition name, attribute value, or block or drawing description. You can also search based on a maximum or minimum file size.

> **Note** *To use the Date Modified or Advanced tab, you must also specify the appropriate search parameters on the Drawings tab. For example, to search among all drawings for a file containing the words "site plan" in a block or drawing description, you must also specify on the Drawings tab that you're searching among all the drawings. To do so, in the Search For The Words box, type an asterisk (\*), and select File Name in the In The Field(s) drop-down list.*

After you specify your search parameters, click Find Now. DesignCenter begins searching along the path you specified for the content matching your search parameters. When it locates items matching those parameters, it displays them at the bottom of the Find dialog box. As soon as you locate the item you're searching for, you can click the Stop button to end the search. You can then load the found items into the DesignCenter palette by doing one of the following:

- Drag the item from the search results list into the palette.
- Right-click the item in the search results list and choose Load Into Palette from the shortcut menu.

Depending on the type of content you were searching for, the shortcut menu also displays commands enabling you to add that content into the current drawing, as you will learn in the next section.

> **Tip** *You can make drawings and blocks easier to find by adding descriptions and other information to them when you create them. For example, when you create a block, you should save a description as part of the block definition so that you can later search for the block based on its description. You should also save a preview image of the block when you create it. You learned about defining blocks in Chapter 15. Similarly, when you save a drawing, if you also include drawing properties, such as title, subject, author, keywords, and comments, you can later search for the drawing based on any of those properties. You learned about saving drawing properties in Chapter 2.*

## Adding Content to Drawings

After you locate content using AutoCAD DesignCenter, you can add that content to an open drawing by doing any of the following:

- Drag the content from the palette or Find dialog box into an open drawing.
- Copy the content from the palette or Find dialog box to the Windows Clipboard and then paste it into an open drawing.

Depending on the type of content you are adding, there are additional methods you can use. The method you choose depends on the type of content you insert.

*You can't add any content from DesignCenter into a drawing while another command is active. If you attempt to do so, DesignCenter displays a message stating that it can't place content when the drawing is busy. You must first end the active command.*

## Inserting Blocks

You can use DesignCenter to insert block definitions into a drawing. AutoCAD DesignCenter provides two different methods for inserting blocks into a drawing. If you simply drag or copy the content from DesignCenter into a drawing, AutoCAD compares the units of the drawing (as specified by the Insert Units value in the Units dialog box) with those of the block (as specified by the Insert Units value in the Block Definition dialog box) and automatically scales the block to match the scale of the drawing. The block is then inserted using the default rotation at the point you specify.

You can also specify the coordinates, scale, and rotation when you insert the block. To do so, do one of the following:

- In the palette, double-click the block.
- In the palette or Find dialog box, right-click the block, drag it into the drawing, and select Insert Block from the shortcut menu.
- In the palette or Find dialog box, right-click the block and select Insert Block from the shortcut menu.

When you use any of these methods, AutoCAD displays the Insert dialog box. You can then specify the insertion point, scale, and rotation angle either within the dialog box or onscreen, as you learned in Chapter 15.

## Attaching Raster Images

You can use DesignCenter to attach a raster image to an open drawing. When you attach a raster image to a drawing, you specify its insertion point, scale factor, and rotation angle, much the same as you would when inserting a block or attaching an external reference. You can specify this information either on the command line or within a dialog box, depending on the method you use to attach the image file. If you simply drag or copy the raster image file from DesignCenter into a drawing, AutoCAD prompts you on the command line for the insertion point, scale factor, and rotation angle.

To attach a raster image by using a dialog box, do one of the following:

- In the palette, double-click the icon of the raster image.
- In the palette, right-click the icon of the raster image, drag it into the drawing, and select Attach Image from the shortcut menu.
- In the palette, right-click the icon of the raster image and choose Attach Image from the shortcut menu.

You'll learn more about attaching raster images in Chapter 23.

## Attaching External Reference Files

As you learned in the previous chapter, attaching a drawing as an external reference is much the same as inserting it as a block, except that instead of adding the contents of the drawing to the current drawing, AutoCAD simply includes a pointer or link to the external file. You can use DesignCenter to attach an external reference to an open drawing. When you attach an external reference to a drawing, you specify its insertion point, scale factor, and rotation angle. You can specify this information either on the command line or within a dialog box, depending on the method you use to attach the xref. If you simply drag or copy the external reference from DesignCenter into a drawing, AutoCAD prompts you on the command line for the insertion point, scale factor, and rotation angle.

To attach an external reference by using a dialog box, do one of the following:

- In the palette, double-click the icon of the external reference.

- In the palette, right-click the icon of the external reference, drag it into the drawing, and select Attach Xref from the shortcut menu.

- In the palette, right-click the icon of the external reference and choose Attach Xref from the shortcut menu.

When you use any of these methods, AutoCAD displays the External Reference dialog box. You can then specify whether you want to attach the xref as an attachment or as an overlay, and specify the insertion point, scale, and rotation angle either within the dialog box or onscreen, as you learned in Chapter 15.

## Adding Cross Hatching

You can use DesignCenter to add cross hatching to an open drawing. As previously mentioned, when you select an AutoCAD hatch pattern file in the tree view, the individual hatch patterns defined in that file appear in the palette. If you simply drag-and-drop a selected hatch pattern onto a bounded area in the drawing, AutoCAD immediately applies the hatch pattern in that area using its default settings. You can also right-click and drag the selected hatch pattern onto a bounded area in the drawing. AutoCAD then displays a shortcut menu, enabling you to do one of the following:

```
Apply Hatch
Apply and Edit Hatch...
Apply Hatch to Multiple Objects
BHATCH...

Cancel
```

- Select Apply Hatch to apply the hatch pattern within the boundary using its default settings.

- Select Apply And Edit Hatch to apply the hatch pattern and then display the Hatch Edit dialog box so you can modify any of the default settings.

- Select Apply Hatch To Multiple Objects to apply the hatch pattern within multiple boundaries. The hatch pattern is applied using its default settings. After applying the hatch pattern within one boundary, AutoCAD prompts you to specify another insertion point. The command repeats until you press ESC.

- Select BHATCH to open the Boundary Hatch dialog box using the selected hatch pattern. You can then change any of the settings and apply the hatch pattern as you normally would when using the BHATCH command.

**Tip** *You can also display the Boundary Hatch dialog box with a hatch pattern preselected by double-clicking the hatch pattern in the palette or right-clicking the hatch pattern and choosing BHATCH from the shortcut menu. You can also select Copy from the shortcut menu to copy the hatch pattern to the Windows Clipboard. You can then use the PASTE command to paste the hatch pattern within bounded areas in the drawing.*

## Copying Other Content Between Drawings

In addition to blocks, raster images, and external references, you can copy any of the other content that is displayed in AutoCAD DesignCenter. For example, you can copy dimension styles, layers, layouts, linetypes, and text styles. Since AutoCAD doesn't have to prompt you for the insertion point, scale, or rotation angle of this type of content, the content is simply added to the database of the drawing. You can therefore use any of the following methods to copy the content:

- In either the palette or Find dialog box, drag the content into the open drawing.

- In either the palette or Find dialog box, copy the content to the Windows Clipboard and then paste it into the open drawing.

- In either the palette or Find dialog box, right-click the content and select the Add option from the shortcut menu.

- In the palette, double-click the icon representing the content.

**Tip** *When copying any type of content other than blocks, raster images, or external references, you can use the SHIFT or CTRL key when selecting objects to copy more than one object at a time.*

## Retrieving Frequently Used Content

AutoCAD DesignCenter maintains a Favorites folder, which works much like the Favorites folder in your web browser. In fact, this folder—called Autodesk—is actually added to your Favorites list in Microsoft Internet Explorer and saved within the Autodesk folder in your Windows system Favorites folder.

## Copying Content with Duplicate Names

When you add content from AutoCAD DesignCenter into an open drawing, AutoCAD first checks whether an object with the same name already exists in the open drawing. For example, if you attempt to copy a layer named WALLS into a drawing that already contains a WALLS layer, AutoCAD detects this duplication and displays a warning on the command line that a duplication exists. What happens next depends on the type of content you are adding to the drawing.

When you insert a block or attach an external reference that has the same name as an existing block or xref, AutoCAD warns you of the conflict and asks whether you want to redefine the existing block or xref with the one that is being inserted.

If you copy any other type of content from DesignCenter into an open drawing, AutoCAD simply ignores any duplicate definitions. For example, when you attempt to copy the WALLS layer from DesignCenter into the current drawing, AutoCAD ignores the operation. The WALLS layer in the current drawing retains all of its original settings.

When you attach a raster image, however, AutoCAD simply adds another instance of the image to the drawing.

As you work with AutoCAD DesignCenter, you can add drawings and folders to the Autodesk Favorites list. When you add a drawing or folder to your Favorites list, DesignCenter adds a shortcut to the Autodesk Favorites folder. This folder can contain shortcuts to drawings and folders on your computer, on other machines on a local area network, or anywhere on the Internet.

To add items to your Favorites list:

- In the DesignCenter tree view or palette, right-click the item and choose Add To Favorites from the shortcut menu.

- While using the Load DesignCenter Palette dialog box, select the file or folder and then choose Add To Favorites from the Tools drop-down list.

You can also add items to your Favorites list from within the Select Hyperlink window or from within Microsoft Internet Explorer by using the Add To Favorites function to add the current Internet location to the Autodesk folder.

 *When using DesignCenter, you can also add all the items currently loaded in the palette to your Favorites list. To do so, right-click the palette background and choose Add To Favorites from the shortcut menu.*

To display the contents of your Favorites folder in AutoCAD DesignCenter, simply click the Favorites button. You can also use the tree view to navigate to the Autodesk folder in your Windows system Favorites folder on your computer.

Since the contents of your Favorites folder are simply shortcuts stored within the Autodesk folder in your Windows system Favorites folder, you can use Windows Explorer to move, copy, or delete those shortcuts. You can quickly access that folder from within DesignCenter so that you can organize and manage its contents. To organize your Favorites folder, right-click the palette background and choose Organize Favorites from the shortcut menu. Windows Explorer immediately displays the contents of the Autodesk Favorites folder.

| Note | *You can also use Microsoft Internet Explorer to manage the items in your Autodesk Favorites folder.* |

The Complete Reference

AutoCAD 2002

# Chapter 17

## Creating a Layout to Plot

Most uses of AutoCAD drawings eventually involve the creation of a paper drawing. To use your drawing to actually produce the building, mechanical part, or other objects that you have created, you need to print a hard copy. Your goal when producing paper drawings is to make a print that looks exactly the way that you want it to look.

To print a copy of your drawing, you can simply print the drawing exactly as you created it. But often, drawing sheets need to include additional information, such as a border and title block identifying your company, the name of the project, and information such as the date, revision dates, and who created the drawing. Actual printed drawings often also contain more than one drawing, at different scales, or need to include only a portion of what you have drawn. For example, suppose that you have drawn the floor plan of a large building, but you need to create a bordered sheet that shows only a portion of that plan at a large scale. Although you could accomplish this task by using the AutoCAD commands that you've already learned, you would probably have to delete objects or use layers to isolate just the portion of the drawing that you want to print. There is a better way.

AutoCAD actually provides a second working environment, called *paper space,* in which to lay out the drawing exactly as you want it to be printed. You can switch to paper space and create areas, called *floating viewports,* which can then display various views of the drawing at different scales. Paper space lets you control which portion of your drawing is printed, and at what scale it is printed. In fact, you can create multiple layouts, each of which simulates an individual sheet of paper on which you will print your drawing. Each layout can contain different views of the drawing, utilize different plot scales, and even simulate different paper sizes.

This chapter explains how to do the following:

- Switch between paper space and model space
- Create and use layouts
- Add title blocks and borders
- Reuse layout templates
- Control layout settings
- Create, arrange, and control floating viewports

In the following chapter, you'll learn how to print or plot your drawings.

# Using Paper Space and Model Space

When you first start a drawing, you begin working in model space. *Model space* is AutoCAD's name for its drawing environment in which you create two-dimensional and three-dimensional objects, based on either the World Coordinate System (WCS) or a User Coordinate System (UCS). Your view of this area initially is a single viewport that completely fills the current drawing window. As you have learned, you can divide this viewport into multiple *tiled* viewports. Each of these viewports can display the same or different 2-D or 3-D views, all of which are also tiled with their edges touching. You can work in only one of these viewports at a time, and can print only the current viewport.

AutoCAD's paper space environment, as its name implies, represents the paper layout of your drawing. In this working environment, you can create and arrange different views of your drawing in a way that is similar to the way in which you arrange detail drawings or orthogonal views on a sheet of paper. You can also add annotations, create a border, and create a title block.

The paper space views that you create are *floating* rather than tiled. In paper space, you can place the viewports anywhere on the screen; their edges can either touch or not touch; and you can print them all simultaneously. Think of paper space as a literal representation of a sheet of paper. You do not need to use paper space to print your drawing, but as Figure 17-1 shows, it offers several advantages.

This paper space or layout environment is much more visual than in pre-AutoCAD 2000 versions, and enables you to create multiple layouts. In addition, each layout can have its own unique page or plot settings. Figure 17-2 shows a typical layout. A shadowed rectangular outline indicates the paper size. Within this sheet of paper, a dashed rectangle indicates the margins of the printable area of the paper. The paper size and margin are in turn based on whichever plotter is currently selected.

**Note**     *You can control whether the margins, paper background, and shadow appear in layouts, by changing the Layout Elements settings on the Display tab of the Options dialog box, shown in Figure 17-3. You can also control the background color of layouts and the plot preview from the Color Options dialog box, displayed when you click the Colors button.*

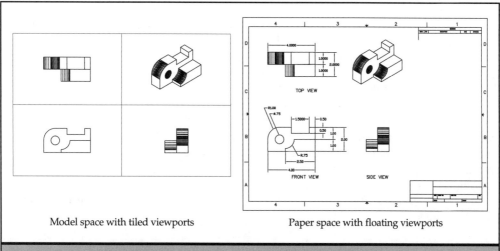

Model space with tiled viewports                    Paper space with floating viewports

**Figure 17-1.**    *The difference between tiled and floating viewports*

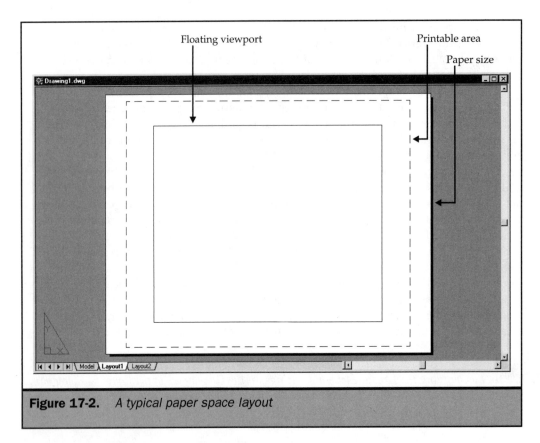

**Figure 17-2.** *A typical paper space layout*

At the bottom of the drawing window are tabs that let you switch between model space and your various paper space layouts. When you first start a drawing, the Model tab is active, meaning that you are working in model space. After working in a layout, when you switch back to the Model tab, it means that you have switched back to model space. As you will learn shortly, you can also work in model space within any of the floating viewports that you create in a layout.

There are also two layout tabs (Layout1 and Layout2) visible at the bottom of the drawing window. Although you don't have to use multiple layouts—and those existing layouts aren't active until you use them—AutoCAD displays two layout tabs just to remind you that you can create more than one paper space layout. When you are ready to set up your drawing for plotting, you can switch to one of the layout tabs and begin to create a paper space layout.

When you create and ultimately print a drawing, you generally proceed through the following steps:

1. Create the drawing in model space.

2. Activate a paper space layout.

3. Specify the page settings, such as the plotting device, paper size and orientation, and plot scale.

4. Add a title block and border.

5. Create floating viewports and position them in the layout.

6. Set the scale of each floating viewport.

7. Add any other necessary objects and annotations in the layout.

8. Plot the layout.

Select to display paper background

Select to display paper margins

Select to display tabs at bottom of drawing window

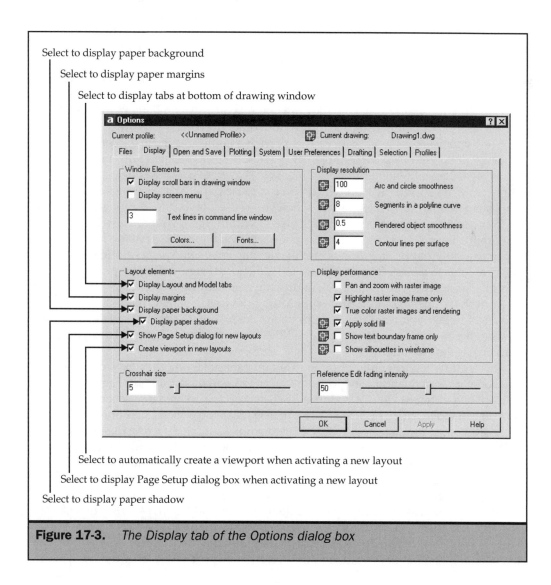

Select to automatically create a viewport when activating a new layout

Select to display Page Setup dialog box when activating a new layout

Select to display paper shadow

**Figure 17-3.** *The Display tab of the Options dialog box*

> **Note** *Before you create a layout, you must have at least one plotter or printer already configured for use with AutoCAD. Available plotting devices include Windows system printers and nonsystem printers and plotters supported by the Autodesk Heidi plotter drivers. If you have access to more than one plotter, you can choose the plotter you want to use as part of the page settings you assign to the layout. You'll learn more about adding and configuring plotters in Chapter 18. You'll also find information about configuring plotters in Appendix C.*

Each layout tab provides a separate paper space environment in which you can create viewports and add a border and title block. The settings within a layout that control the type of plotter, the paper size and orientation, and the plot scale are collectively called a *page setting*. When you first select a layout tab, AutoCAD displays a Page Setup dialog box so that you can specify these settings for that layout. You can also give the page settings a unique name and save the setup. Page setups are nothing more than a way of saving the plot settings so that they can be reused in other layouts or in other drawings. Each named page setup is saved in the drawing. As you will learn later in this chapter, you can also import and reuse page setups from other drawings.

When you create a layout, you can also give it a name and reuse it in other drawings along with its page settings. The page settings associated with the layout are reused regardless of whether those settings were saved as a named page setup. This simply provides another way in which you can reuse paper space settings you already established in other drawings. In fact, when you reuse a layout by copying or inserting it into another drawing, the paper space objects within that layout are also included.

Within a layout, you can change the page settings to plot the drawing in a different way or to a different printer. For example, you may create one page setting for plotting your drawings to a large format plotter, and a different page setting for printing test plots to a laser printer. By simply changing the page setting assigned to a layout, you can print that same layout on different plotting devices. You also have the option of choosing whether you want the changes you make to a plot setting saved as part of the drawing. If you don't want to save the settings, clear the Save Changes To Layout check box in the Plot dialog box.

After you switch from model space to a paper space layout, the next step is to create one or more floating viewports and arrange them on the layout. Each viewport can display a different view of your drawing, and each is treated as a separate object that you can move, copy, or delete. You can also create objects, such as borders, title blocks, and other annotations, directly in the layout, without affecting the model itself.

While in paper space, you can't modify the individual objects within the viewports, and zooming or panning the drawing affects the entire paper space layout. To edit the objects in the drawing rather than the paper space layout, you can switch back to model space. Each viewport then functions as a window into your model space drawing. You can click within any of the floating viewports to make it the current viewport, and then add or modify the objects in that viewport. The changes that you make in one viewport can be restricted in such a way that they appear only within that viewport, and zooming or panning in the current viewport affects only that viewport.

## Saving to Earlier Versions

When you save a drawing to a version prior to AutoCAD 2000, if that drawing contains multiple layouts, the program creates a single paper space environment based on the last layout tab that you used. When you open that drawing in Release 13 or 14 and switch to paper space, AutoCAD displays that layout. The additional layouts are still stored within the drawing file, however, to ensure round-trip compatibility. When you open the drawing again in AutoCAD 2000 or 2002, all of the layouts will still be available.

**Note** *You can also set and lock the scale of individual floating viewports, so that the scale of the viewport doesn't change in relation to the layout. You'll learn more about this feature later in this chapter.*

# Switching to Paper Space

As soon as you select a Layout tab, AutoCAD switches to paper space. There are actually several different ways to switch to paper space. You can switch to paper space using any of the following methods:

- Select a Layout tab.
- On the status bar, click MODEL.
- At the command line, type **TILEMODE**, press ENTER, type **0**, and press ENTER again.

When you switch to paper space by using any of these methods, AutoCAD activates a layout. If that layout is activated for the first time, the program normally displays the Page Setup dialog box, which contains two tabs. When you first activate a layout, the Layout Settings tab is displayed. This tab determines the paper size, drawing orientation, plot area, plot scale, and other plotting options. You'll learn more about this dialog box later in this chapter.

**Note** *The Page Setup dialog box normally displays whenever you select a layout tab for the first time. Some users don't want to deal with this dialog box every time they activate a new layout, preferring instead to change the layout settings later. To disable the automatic display of the Page Setup dialog box, clear the Show Page Setup Dialog For New Layouts check box under Layout Elements on the Display tab of the Options dialog box, shown earlier in Figure 17-3.*

You can also use the new Create Layout wizard to switch to paper space and set up a new layout, complete with all the page settings. You'll learn all about this powerful wizard in the next section.

**Note** *The ability to work in paper space is controlled by the TILEMODE system variable. When you work in tiled viewports, this system variable is turned on (TILEMODE=1). When TILEMODE is off, you can switch between paper space and model space. When you switch to paper space, the program turns this variable off. You can then work in paper space or in any of the floating model space viewports. When you switch back to the Model tab, the program turns the TILEMODE variable on again. When TILEMODE is on, you can only work in model space.*

When you are working in paper space, AutoCAD displays the paper space icon in the lower-left corner, instead of the UCS icon. In addition, the crosshairs move across the entire screen, and the MODEL option on the status bar changes to PAPER.

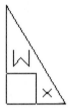

**Tip** *In early versions of AutoCAD, when you switched to paper space for the first time, your drawing seemed to disappear. Although this was normal, many users found it very disconcerting. In AutoCAD 2000 and 2002, when you activate a layout for the first time, AutoCAD automatically adds a floating viewport in the new layout. You can disable this default by clearing the Create Viewport In New Layouts check box on the Display tab of the Options dialog box, shown earlier in Figure 17-3.*

## Switching to Model Space

When in paper space, you can create or modify only paper space objects. To edit the model itself, you need to switch back to model space. You can easily switch back and forth between model space and paper space, as necessary. You can either switch back to the Model tab to work on your model, or modify it directly within a floating viewport.

To switch back to the Model tab, do one of the following:

- Select the Model tab.
- At the command line, type **MODEL** and press ENTER.
- Change the TILEMODE value to 1.

To switch to model space within one of the floating viewports, simply double-click within that viewport. When working in model space in a floating viewport, AutoCAD displays a model space UCS icon in the lower-left corner of each of the floating viewports. In addition, the PAPER option on the status bar changes back to MODEL. The current viewport is indicated by a bold border, and the crosshairs move only within that viewport.

You can also switch to model space in a floating viewport by doing one of the following:

■ On the status bar, click PAPER.

■ At the command line, type **MSPACE** (or **MS**) and then press ENTER.

When you use either of these methods to switch to model space, AutoCAD automatically selects the last active viewport. You can then change viewports by simply clicking inside the viewport.

To switch back to paper space, do one of the following:

■ Double-click over any area of the layout outside of the floating viewports.

■ On the status bar, click MODEL.

■ At the command line, type **PSPACE** (or **PS**) and then press ENTER.

## Creating Layouts

To work in the paper space environment, you need to create a layout. As you just learned, when you switch to paper space by activating a layout for the first time, AutoCAD normally displays the Page Setup dialog box so that you can specify all of the page setup parameters. Although this dialog box gives you complete control over both the appearance of the layout and the way it is ultimately plotted, this dialog box can be somewhat intimidating when first encountered in this way. You may also not want to deal with the layout settings right at that moment. In addition, until you become comfortable with using paper space, this dialog box is not the easiest method for creating layouts.

### Using the Create Layout Wizard

AutoCAD's Create Layout wizard guides you through the process of creating a layout and establishing its page settings in a logical, step-by-step approach. As you proceed through the eight steps, the wizard prompts you to do the following:

■ Name the layout

■ Select the printer from those available on your system

■ Select the paper size (based on those available for the selected printer)

■ Select the orientation of the drawing on the paper

■ Insert a title block

- Create floating viewports
- Locate those viewports on the paper
- Complete the layout

*In reality, these steps represent the same decisions you need to make when you establish the settings yourself using the Page Setup dialog box. For that reason, you should use the Create Layout wizard a few times to familiarize yourself with these settings before creating a layout using the Page Setup dialog box.*

To start the Create Layout wizard, do one of the following:

- From the Insert menu, choose Layout | Layout Wizard.
- From the Tools menu, choose Wizards | Create Layout.
- At the command line, type **LAYOUTWIZARD** and press ENTER.

When you start the Create Layout wizard, AutoCAD displays the Begin page. The first step is to specify the name for the layout you are creating. Since the drawing already has two layout tabs named Layout1 and Layout2, the wizard presents Layout3 as the default layout name. You can specify any name except that of an existing layout. Generally, you should give layouts descriptive names, such as Architectural Plan or 1st Floor Lighting. After you enter the layout name, click Next.

*As you proceed through the following steps, if you make a mistake or change your mind, you can click the Back button to return to any previous step, make the necessary changes, and then continue on to the next step.*

On the Printer page, you can select the plotting device you will ultimately use to plot your drawing when using this layout. The wizard displays a list of those plotting devices currently available on your system, including any Windows system printers. Simply select the plotter you want to use, and click Next.

*You should make sure that the plotter you want to use is already configured and available on your system. If the plotter you want to use doesn't appear in the list, you can cancel the wizard, and then add or configure a new plotter. To add a Windows system printer, use the Add Printer wizard in the Printers folder in the Windows Control Panel. To add a nonsystem printer supported by an Autodesk Heidi plotter driver, you can use the Add-A-Plotter wizard, which is accessible from the Plotting tab of the Options dialog box, by selecting Wizards | Add Plotter from the Tools menu or Plotter Manager from the File menu, or by typing **PLOTTERMANAGER** at the command line.*

On the Paper Size page (see Figure 17-4), you can select the paper size on which you want to print the drawing. The drop-down list displays all the paper sizes available

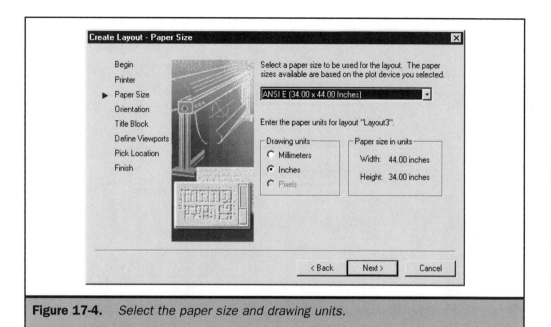

**Figure 17-4.**    *Select the paper size and drawing units.*

for the plotter you selected in the previous step. To specify the paper size, simply choose it from the list. You can also specify the type of units in which you want the paper space units displayed (inches or millimeters). Note that if the plotter you select is configured for raster output, you must specify the output size in pixels. When you have specified the paper size, click Next.

**Caution**    *If you plan to insert a title block, you should select the paper units that match the type of title block that you will use. Title blocks are drawn full-size and inserted at a 1:1 scale factor based on the paper units you select. For example, an ANSI D-size title block measures 32.5 units by 21 units. If you were to select millimeters instead of inches, the paper would measure 863.60mm by 558.80mm, but if you then insert an ANSI D-size title block drawing, it will occupy just a small corner of the sheet, because inserted at a 1:1 scale, it will be only 32.5mm by 21mm in size.*

On the Orientation page, specify the orientation of the drawing on the paper by selecting either Portrait or Landscape, and then click Next.

On the Title Block page (see Figure 17-5), you can specify the title block you want to use for the layout. The wizard displays a list of predefined title blocks. To include a title block, simply choose it from the list. As a title block is highlighted in the list, its image is displayed in the adjacent Preview box. If you don't want to include a title block, choose None.

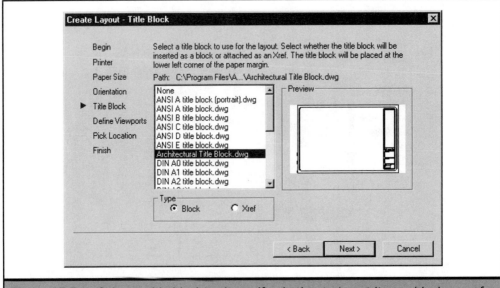

**Figure 17-5.** *Select a title block and specify whether to insert it as a block or xref.*

**Note** *You should select a title block that matches the paper size. Otherwise, the title block will not fit the selected paper size. The printable area for some plotters may actually be smaller than the standard title block size. For example, a standard ANSI D-size title block may not actually fit within the printable area of the plotter's D-size sheet of paper. You may need to add a custom paper size to accommodate the standard title block, or create custom title blocks to fit within the printable area of your plotter's standard paper size. You'll find information about adding custom paper sizes in Appendix C.*

You can also specify whether the title block is inserted as a block or attached as an external reference. In either case, the title block is created in paper space. When you have completed this step, click Next.

**Note** *The title blocks displayed on the Title Block page are simply AutoCAD drawings stored in AutoCAD's template directory. Any custom title block drawings you create and save to this directory also appear in the list.*

On the Define Viewports page (see Figure 17-6), you can add viewports to the layout, specify the setup type, assign a scale to the viewports, and specify the spacing between them. (Note that you can also add model space viewports later by using the

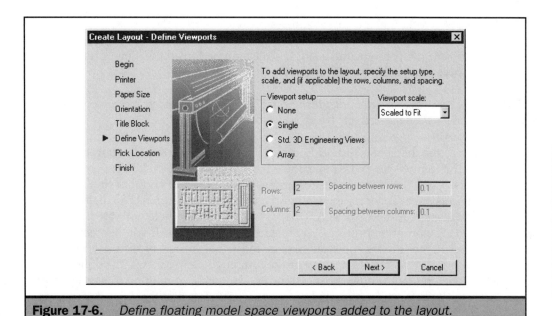

**Figure 17-6.** *Define floating model space viewports added to the layout.*

MVIEW or VPORTS commands.) There are several different series of choices to be made on this page, depending on the viewport setup that you choose:

■ If you don't want to add any viewports at this time, select None and then click Next.

*If you select None and then click Next, you actually move to the final step of the wizard, and there is no way to go back to a previous step to change your mind. If you made a mistake, you'll need to cancel the wizard and start again.*

■ To add a single viewport, select Single. You can then specify the scale of the viewport from the Viewport Scale drop-down list, or select Scaled To Fit to simply scale the model space extents to fit within the resulting viewport. Then, click Next to position the viewport.

■ To add four standard engineering views consisting of the top, front, right-side, and isometric views of a three-dimensional object, select Std 3D Engineering Views. You can then specify the single scale assigned to all four viewports and the spacing between them. Then, click Next to position the viewports.

■ To add a rectangular array of equal-sized viewports, select Array. You can then specify the number of rows and columns of viewports, the spacing between each row and column, and a single scale assigned to all the viewports. Then, click Next to position the viewports.

**Note** *If you select a scale factor other than Scaled To Fit, the view in each viewport will be centered within the viewport based on the model space extents.*

If you selected the Single, Std 3D Engineering Views, or Array viewport setup options on the Define Viewports page, the wizard displays the Pick Location page. This page lets you specify the opposite corners of the viewport configuration chosen on the previous page. To locate the viewports, click Select Location. The dialog box temporarily disappears and AutoCAD displays a layout complete with the title block you selected. The program prompts you to specify two corners. If you're creating a single viewport, the corners you select will be the actual corners of the viewport. If you're creating either a viewport array or the four engineering views, the points you specify will determine the outermost extents of a rectangle containing those viewports. The actual viewports are then created within that rectangle and their size depends on the number of viewports and the spacing between them. If you skip this step and simply click the Next button on the Pick Location page, the viewports you specified will be sized to fit within the printable area of the paper.

**Caution** *If you need to change any of the parameters specified on a previous page, click the Back button now. Once you begin to locate the viewports, there is no way to go back to a previous step to make changes. If you make a mistake or change your mind, you'll need to cancel the wizard and start again.*

After you specify the location of the viewport, AutoCAD automatically proceeds to the final page of the Create Layout wizard. Simply click Finish to complete the layout. AutoCAD quickly creates the new layout, complete with the viewports you specified. When you save the drawing, AutoCAD saves the layout along with its page setup.

After you've created the layout, you can modify the layout by changing the viewports or adding additional objects to the layout. You can also use the PAGESETUP command to change the page setup.

## Using the Page Setup Dialog Box

As you've already learned, when you activate a layout for the first time, AutoCAD displays the Page Setup dialog box (shown in Figures 17-7 and 17-8). This dialog box enables you to select the plotting device assigned to the layout, and specify the layout settings, such as paper size, drawing orientation, and plot scale. The information you

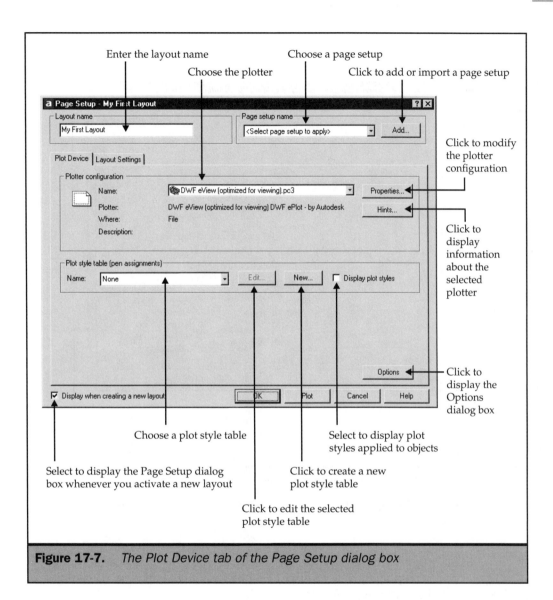

Enter the layout name

Choose the plotter

Choose a page setup

Click to add or import a page setup

Click to modify the plotter configuration

Click to display information about the selected plotter

Click to display the Options dialog box

Select to display plot styles applied to objects

Click to create a new plot style table

Click to edit the selected plot style table

Choose a plot style table

Select to display the Page Setup dialog box whenever you activate a new layout

**MOVING BEYOND THE BASICS**

**Figure 17-7.**   *The Plot Device tab of the Page Setup dialog box*

need to provide within this dialog box to configure the layout is virtually the same as the information you specify when using the Create Layout wizard. It's just organized differently and provides additional options so that you can control all aspects of the page setup.

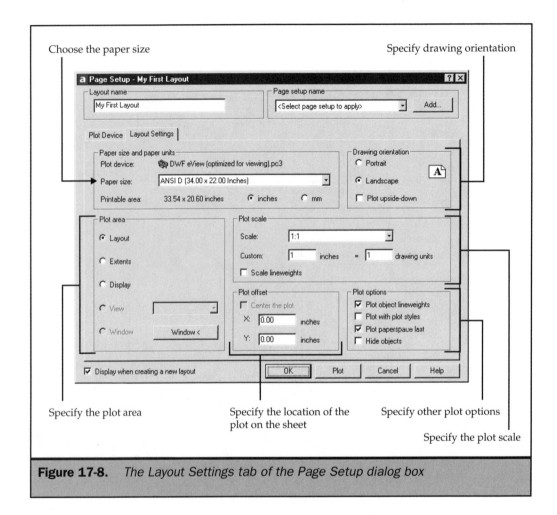

**Figure 17-8.** *The Layout Settings tab of the Page Setup dialog box*

In addition to displaying this dialog box whenever you activate a layout for the first time, you can display the Page Setup dialog box any time you want to modify the current page layout settings or create a new layout.

To display the Page Setup dialog box, do one of the following:

- On the Layouts toolbar, click Page Setup.
- From the File menu, choose Page Setup.
- At the command line, type **PAGESETUP** and then press ENTER.

*You can also display the Page Setup dialog box by right-clicking the layout tab whose settings you want to change, and then choosing Page Setup from the shortcut menu.*

The top of the dialog box lets you specify the layout name and the page setup name. To specify a more descriptive name for the layout, simply type the name in the Layout Name box. Notice that the layout name also appears on the title bar of the dialog box.

**Tip**    *To simply rename an existing layout, right-click its layout tab and choose Rename from the shortcut menu.*

The Page Setup Name box initially displays a message prompting you to select a page setup to apply to the layout. Since this is probably your first layout, you can ignore this box for the time being. Later, you can use this drop-down list and its adjacent button to save and reuse named page setups. Remember that a page setup name is simply the page setup information you specified for another layout and then saved as a named page setup.

The rest of the dialog box is logically divided into two tabs. On the Plot Device tab, shown in Figure 17-7, you select the plotter from a drop-down list of available devices, similar to the way in which you selected a plotter on the Printer page of the Create Layout wizard. When you select a plotter, the Plotter Configuration area displays additional information, including the name of the plotter to which the plot will be sent (such as a local or network port or a file).

If necessary, you can click the Properties button to display the Plotter Configuration Editor, a powerful dialog box in which you can modify all aspects of the plotter's configuration. The Hints button simply displays a Help topic about the selected plotter. In the lower portion of the Plot Device tab, you can select a plot style table, which is the way in which AutoCAD controls pen assignments. The adjacent Edit and New buttons let you modify or create plot style tables. When selected, the Display Plot Styles check box displays plot styles applied to objects in the drawing. You'll learn about configuring plotters and using plot style tables in the next chapter.

## Controlling the Paper Size

The Layout Settings tab, shown in Figure 17-8, is where you select the paper size, drawing orientation, plot scale, and most of the other settings you selected when you created a layout by using the Create Layout wizard. In the Paper Size And Paper Units area, you select the paper size to be used for the layout and specify the units in which it is measured. Notice that the Plot Device field displays the name of the plotter you selected on the Plot Device tab. The Paper Size drop-down list contains just those sheet sizes available for the plotter you selected. You can also specify whether the paper is to be measured in inches or millimeters. As you already learned, if you plan to insert a predefined title block, you should select the units that match the units used to define

MOVING BEYOND THE BASICS

that title block. The Printable Area field displays the size of the printable area for the paper size you select.

| Caution |

*After establishing the paper size, if you change plotters to one that doesn't support the existing paper size, AutoCAD displays a dialog box warning you that the paper size is not supported. The program automatically changes the layout's paper size to the default paper size specified for the plotter you selected. You can disable this warning dialog box from within the dialog box itself or from the Plotting tab of the Options dialog box, both of which change the value of the PAPERUPDATE system variable. When this variable's value is 1, AutoCAD displays this dialog box and automatically switches to the plotter's default paper size when you select a different plotter. When PAPERUPDATE equals 0, AutoCAD uses the paper size specified on the Layout Settings tab, if possible. However, if the paper size is not supported by the plotter you select, the program still switches to that plotter's default paper size, but no longer displays a dialog box warning you that it has done so. Because of a bug in AutoCAD, however, once you disable this warning dialog box, there is no way to enable it again, other than to select the Show All Warning Messages check box on the System tab of the Options dialog box (which reactivates all of the other warning dialog boxes that you might have previously disabled). In addition, when loading a PC3 file, the paper size from the PC3 file is always used, even when PAPERUPDATE equals 0.*

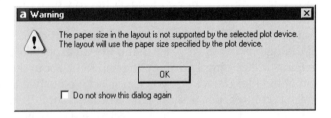

## Setting the Drawing Orientation

In the Drawing Orientation area, you select the orientation of the drawing on the paper. If you select Portrait, the image of the sheet of paper within the layout changes so that its long dimension is vertical, while the orientation of your AutoCAD objects remain unchanged. If you select Landscape, the sheet of paper appears with its long dimension horizontal. You can also select the Plot Upside-Down check box to plot the drawing upside down on the sheet of paper. However, this control works a bit differently than in R14 and earlier versions of AutoCAD. In early versions of AutoCAD, when you rotated a plot, the plot actually appeared rotated when you previewed the plot. Because each layout in AutoCAD 2002 provides a preview of the way the sheet appears when plotted, and you are likely to continue editing your drawing directly in a layout, it doesn't make sense to display the drawing rotated or upside down within the layout. Instead, when you select the Plot Upside-Down check box, the shadow image of the sheet of paper changes to indicate that the sheet of paper has been rotated 180 degrees in relation to the objects in the drawing.

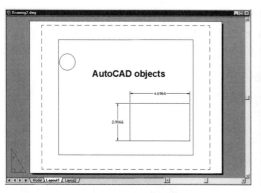

Landscape orientation

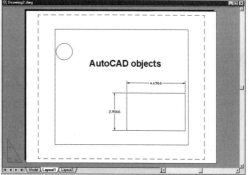

Upside down landscape orientation

## Setting the Plot Area

Under Plot Area, you select the area of the drawing that you want to plot. Generally, when plotting a paper space layout, you will plot the entire sheet, and AutoCAD therefore selects Layout as the default option. This option plots all the drawing objects within the margins of the specified paper size. The *plot origin,* the lower-left corner of the drawing, is placed at the 0,0 position, which is the lower-left corner of the printable area.

**Note** *When plotting the Model tab, this default changes to Display, which plots everything currently displayed in the AutoCAD window. You'll learn more about this in the next chapter.*

When plotting a layout, you can also plot a selected area of the layout by choosing one of the other options:

- **Extents**   Plots the paper space extents, including any drawing objects that extend beyond the paper size. This is the same as zooming to the extents and then plotting the display.
- **Display**   Plots only the objects currently displayed in the drawing window.
- **View**   Plots a previously saved view. This option is only available if you've already saved one or more named views, as described in Chapter 5. Choose the view from the adjacent drop-down list.
- **Window**   Plots a rectangular portion of the layout. Before you can select this option, you must first click the Window button and specify the window area to be plotted. After you specify the opposite corners of the window, the Window option is automatically selected.

## Setting the Plot Scale

In the Plot Scale area, you specify the scale at which you want the layout to be plotted. You can either specify an actual scale or let AutoCAD scale the drawing to fit onto the paper. Generally, when you plot a layout to the same size sheet of paper as what is displayed in the layout view, you will set the scale to 1:1—that is, 1 plotted unit equals 1 drawing unit—and that is the default when plotting a layout.

*If you want to make a quick test print onto a smaller size sheet of paper, you can change the paper size, select Extents under Plot Area to plot the entire drawing, and then choose Scaled To Fit from the Scale drop-down list to let AutoCAD automatically scale the drawing to fit onto the smaller paper.*

The Scale Lineweights check box determines whether lineweights are scaled proportionately in a layout with the plot scale. Generally, you want lineweights to plot at their actual width regardless of the plot scale. For example, when you draw an object with a 0.25mm lineweight, you want that object actually plotted with that exact width. A possible exception to this rule is when plotting a scaled-down version of your drawing, such as when plotting a half-size test plot. In such instances, select the Scale Lineweights check box to plot the lineweights at a reduced scale in proportion to the plot scale. Changing this setting changes the value of the LWSCALE system variable.

*The Scale Lineweights option is not available when plotting from the Model tab.*

## Changing the Plot Offset

Under Plot Offset, you can adjust the location of the plot within the sheet of paper. If you're plotting the entire layout, AutoCAD normally locates the lower-left corner of the drawing—the plot origin—at the lower-left corner of the printable area, the 0,0 position. You can move the drawing in relation to this point, however, by specifying negative values to move the drawing down or to the left, or positive values to move it up or to the right. If you specify the plot area as anything other than the Layout, you can also center the plot on the sheet.

## Controlling Other Options

The controls under Plot Options let you control other settings for lineweights, plot styles, and the current plot style table. This area provides the following four options:

- **Plot Object Lineweights** When selected, specifies that lineweights are plotted. This option is always selected (and unavailable for change) when you plot with plot styles.

- **Plot with Plot Styles** When selected, AutoCAD plots using the plot styles that are applied to objects and defined in the plot style table. Since plot styles can determine lineweights, selecting this option automatically selects the Plot Object Lineweights option as well. You'll learn about plot styles and plot style tables in the next chapter.

- **Plot Paperspace Last** When selected, AutoCAD plots all of the model space geometry before plotting the paper space geometry. In versions of AutoCAD prior to 2000, the paper space geometry was always plotted first.

*If you plot from the command line (by typing -PLOT), the Plot paper space last? [Yes/No] <No>: prompt is reversed. If you respond No, AutoCAD will actually plot the paper space objects last. If you answer Yes to the prompt, AutoCAD plots the paper space objects first.*

- **Hide Objects**  When selected, AutoCAD performs hidden-line removal on the objects drawn within the layout (or within model space when plotting the Model tab). This option has no effect on the removal of hidden lines within the individual floating viewports. Hidden-line removal of model space objects within floating viewports is controlled by the object properties for that viewport. You'll learn about controlling hidden-line removal within individual viewports later in this chapter.

## Working with Layouts

After you finish setting up the layout, AutoCAD displays the layout tab using the page settings you've established. The name you assigned to the layout is displayed on its layout tab. The layout tabs initially display the names of the layouts in the order in which they were created. Once you've created a layout, you can delete, rename, move, or copy the layout by right-clicking its layout tab and then choosing an option from the shortcut menu.

The Delete and Rename options, as their names imply, let you delete or rename the selected layout. The Move and Copy options deserve a bit more explanation. When you choose the Move or Copy option, AutoCAD displays the Move or Copy dialog box. To change the order of the layout tabs—for example, to move the selected layout so that it appears before or after one of the other layouts at the bottom of the drawing window—choose the layout you want the current layout to appear before, and then click OK. To move the layout to the right end of the layout tabs, choose Move To End, and then click OK. To create a new layout as a copy of the selected layout and place it at the chosen location, select Create A Copy, and then click OK.

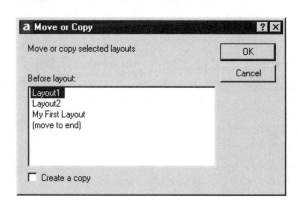

| Tip | *The Create A Copy option provides a handy way to create a duplicate layout using the same settings you've already established. For example, suppose you have drawn both the architectural and electrical lighting plans in a single AutoCAD drawing and then created a layout for the architectural plan, complete with its plot and layout settings, a title block, and a floating viewport containing the floor plan. In the architectural plan layout, you've selected the layers you want to plot. Now you're ready to create the electrical lighting plan. Everything about that plan is identical to the architectural plan except for the layers that are visible. Instead of repeating the steps you used to create another layout, you could create a copy of the architectural plan layout and then simply change the visibility of the layers within the floating viewport. You'll learn how to control the visibility of layers individually in each viewport later in this chapter.* |
|---|---|

## Naming and Saving a Page Setup

After you create a layout page setup, you can assign that setup a name. Named page setups are saved as part of the drawing. Once you name a setup, you can apply that setup to another layout or apply other named setups to the current layout. You can also import named page setups from other drawings.

For example, you can create a setup for plotting a drawing at a 1:1 scale on a D-size sheet of paper, and create a second setup for plotting the same layout to fit on an A-size sheet of paper. By naming and saving both of these page setups, you can easily apply either of them to any other layout before plotting.

### Modifying a Named Page Setup

Because the Page Setup dialog box doesn't have a Save button, many users wonder how to modify an existing named page setup. You can modify existing setups by using the following steps:

1. Select the existing named page setup you want to change.

2. Make the required changes.

3. Choose the Add button.

4. Double-click the named page setup you selected in Step 1, and then click OK.

5. AutoCAD displays an alert dialog box, telling you that Page setup *xxx* already exists and asking if you want to redefine it. Click Yes.

The existing page setup is immediately saved with the new settings you defined in the dialog box.

To name and save a page setup:

1. Display the Page Setup dialog box.

2. Use the controls in the Page Setup dialog box to establish your page settings.

3. In the Page Setup Name area, click Add. AutoCAD displays the User Defined Page Setups dialog box, shown in Figure 17-9.

4. In the New Page Setup Name box, enter a name for the page setup.

5. Click OK.

After you save the named page setup, its name appears as the current name under Page Setup Name in the Page Setup dialog box. You can switch to a different page setup by choosing it from the Page Setup Name drop-down list.

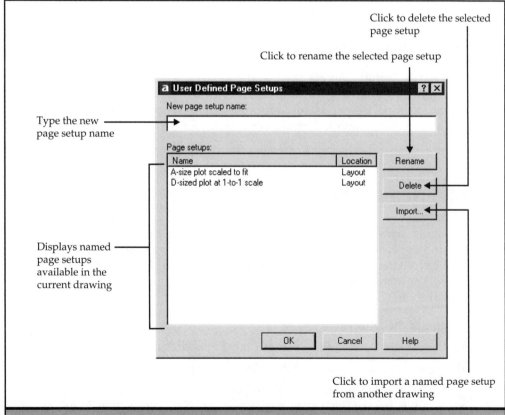

**Figure 17-9.** *The User Defined Page Setups dialog box*

## Utilities for Working with Page Setup

The Layout Utility commands—downloadable from the Autodesk Product Support web site and also included on the companion web site on the Chapter 17 Utilities page—provide several tools that are useful when you have a page setup in one drawing that you want to export and then import into several other drawings.

VIEWTOLAYOUT copies objects in a paper space view (along with current plot settings) and creates a new paper space layout in the same drawing. The utility converts drawings that contain multiple drawing sheets in a single paper space into multiple layouts, each with a single drawing sheet.

The PAGEOUT utility stores the plot settings in a named page setup and enables you to import the plot settings into other drawings. This utility creates a named page setup from the plot settings of the current model space or layout. The newly created named page setup contains the plot settings necessary to plot another drawing similar to the current model or layout. PAGEOUT creates a script that can then be used to import the named page setup into another drawing and make it current. After you run PAGEOUT, use PAGEIN (or the SCRIPT command) to import the plot settings into another drawing. The script created by PAGEOUT calls the -PSETUPIN command to import the page setup name, and calls the -PLOT command to make the page setup name current.

The PAGEIN utility imports page setups created by the PAGEOUT command into any drawing by reading the script file created using PAGEOUT. The utility then saves the imported page setup information to the current model or layout. After you run PAGEIN, your drawing is ready to plot.

# Importing a Named Page Setup

After you have saved named page setups in a drawing, you can import those page setups into other drawings. Once you import a named page setup, you can select that setup and apply it to a layout. The setup controls all the settings on the Plot Device and Layout Settings tabs of the Page Setup dialog box but does not affect the objects contained in the drawing.

To import a named page setup:

1. Do one of the following:

   ■ In the User Defined Page Setups dialog box, click Import.

   ■ At the command line, type **PSETUPIN** and press ENTER.

2. AutoCAD displays a Select File dialog box. Select the drawing file from which you want to import the named page setup, and then click Open. AutoCAD displays the Import User Defined Page Setup(s) dialog box. The list displays

the page setup names in the drawing you selected, and indicates whether each is a model space or layout page setup.

3. Select one or more named page setups that you want to import, and then click OK.

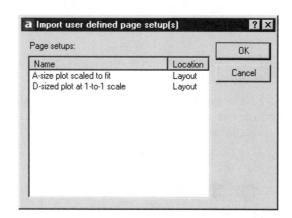

After you have imported a named page setup, its name appears in the Page Setup Name drop-down list in the Page Setup dialog box and can be assigned to any layout in the current drawing.

# Using Layout Templates

As you learned in Chapter 2, when you start a new drawing, you can base that drawing on a template that contains standard settings that you generally use. A template is simply a normal AutoCAD drawing that has been saved as a drawing template file (a file with a .DWT file extension). AutoCAD comes with numerous templates representing different standard bordered drawing sheets. In addition to being able to use one of these templates when you start a new drawing, you can insert a layout from one of these or any other AutoCAD template files, or from any drawing file.

## Inserting a Layout from a Template

You can insert any AutoCAD template or drawing file as a layout. When you create a new layout from a template or drawing, all the page settings in the existing template or drawing are used in the new layout. Any paper space geometry, including viewports, is also included and displayed in the new layout. No model space geometry is imported, however. This is a handy way of importing just the paper space objects from one drawing file to another.

**Note** *Whereas the PSETUPIN command imports named page setups, not layouts, when you insert a layout from a template, you insert the layout, not the page setups.*

To insert a layout from a template or drawing:

1. Do one of the following:

   ■ On the Layouts toolbar, click Layout From Template.

   ■ From the Insert menu, choose Layout | Layout From Template.

   ■ Right-click any layout tab and then choose From Template from the shortcut menu.

   ■ At the command line, type **LAYOUT** and press ENTER, and then type **T** and press ENTER (or right-click and select Template from the shortcut menu).

2. AutoCAD displays a Select File dialog box. Select the template file or drawing file you want to insert as a new layout, and then click Open. AutoCAD displays the Insert Layout(s) dialog box.

3. Select one or more of the layout names and then click OK.

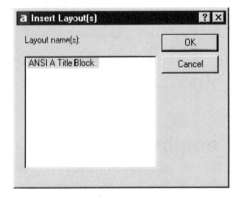

**Note**    *You can also insert a layout by clicking the Layout From Template button on the Layout toolbar. Because this macro uses a command line macro version of the LAYOUT command, however, AutoCAD doesn't display the Insert Layouts dialog box, as it does when you start the command from the command line or Insert menu. Instead, after selecting the template file, you are prompted to enter the name of the layout on the command line.*

AutoCAD immediately creates a new layout using the layout template you selected. The new layout is automatically assigned the next layout number in sequence (based on the appropriate naming convention), along with the name of the imported layout.

When you create a new layout, the name assigned to the layout tab depends on whether the drawing, template, or DXF file from which the layout was inserted uses the default layout tab names or custom names.

AutoCAD Release 14 and previous versions did not have multiple layouts, so layouts based on files created in earlier versions use the default naming convention, such as Layout1, Layout2, and so on. Any layouts based on AutoCAD drawing, template, or DXF files that use custom or user-defined layout names will produce new layouts based on those custom or user-defined names.

**Note**    *When you create a new layout from a drawing or template, any unreferenced symbol table information (such as unused layers) contained in that drawing or template is also copied into the new layout. After you create the layout, you can use the PURGE command to remove any unnecessary information.*

## Inserting a Layout Using AutoCAD DesignCenter

AutoCAD DesignCenter provides yet another way to reuse layouts you created in one drawing to create new layouts in the current drawing. To use DesignCenter to insert a layout from an existing drawing, you must first use the tree view or palette to navigate to the drawing from which you want to insert the layout, and then display the layouts in the palette. You can then use any of the following methods to insert a layout into the current drawing using AutoCAD DesignCenter:

- Select one or more layouts from the palette and drag-and-drop them into the current drawing.

- Select the one or more layouts from the palette, right-click, and choose Add Layout(s) from the shortcut menu.

- Select one or more layouts from the palette, right-click, and choose Copy from the shortcut menu. Then, in AutoCAD, switch to the drawing in which you want to insert the layouts, and use the PASTECLIP command to paste the layouts into the drawing.

- Double-click a layout in the DesignCenter palette.

When you use AutoCAD DesignCenter to insert a layout, AutoCAD ignores any unused symbol table information, thus eliminating the need to purge this information, as you do when you insert layouts directly from drawings by using the LAYOUT command.

## Saving a Layout Template

As you've already learned, any drawing can be saved as a template. When you use the SAVE command to save a drawing as a template, all the geometry and layout settings are also saved to the DWT file.

You can save a drawing to a template either by using the SAVEAS command and selecting AutoCAD Drawing Template File (*.dwt) as the Save As Type, or by using the

MOVING BEYOND THE BASICS

Saveas option of the LAYOUT command. When you save a template by using the SAVEAS command, AutoCAD includes all the layouts in the current drawing as part of the template file. But AutoCAD also includes all unused symbol table definitions, such as unused layer names (layers that don't contain any objects). If you create a new layout using a template created in this way, the new layout will include all the unused definitions, as well. When you save a template by using the LAYOUT command, however, AutoCAD only saves the layout you specify, and doesn't include any unused symbol table information.

To save a layout template using the LAYOUT command:

1. At the command line, type **LAYOUT** and press ENTER. AutoCAD prompts:

   ```
   Enter layout option [Copy/Delete/New/Template/Rename/SAveas/Set/?] <set>:
   ```

2. Type **SA** and press ENTER, or right-click and select SAveas from the shortcut menu. AutoCAD prompts:

   ```
   Enter layout to save to template <current>:
   ```

3. Press ENTER to save the current layout to a template, or type the name of the layout you want to save, and then press ENTER.

4. AutoCAD displays a Create Drawing File dialog box. Enter a name for the drawing template file you are saving. By default, the template will be saved in the template file directory, as specified under Drawing Template File Location on the Files tab of the Options dialog box, but you can save the template file to any folder you wish.

5. Click Save.

## Using PCP and PC2 Settings in a Layout Template

You can import layout and plot settings contained in both PCP and PC2 files created in an earlier version of AutoCAD and apply them to the current model or layout. A PCP file is a partial plot configuration file created in AutoCAD Release 12, 13, or 14, which contains device-independent plot configuration information, such as the plot area, rotation, paper size, plot scale, plot origin, and plot offset. A PC2 file is a complete plot configuration file created in AutoCAD Release 14, which can contain both device-independent and device-dependent plot configuration information, such as pen assignments and plotter calibration changes.

Although PCP and PC2 files have been largely supplanted in AutoCAD by plot style tables (PC3 files, which you will learn about in the next chapter), AutoCAD comes with a wizard that enables you to import plotting device and pen setting information from these files, so that you can reuse settings that you've already established.

## Understanding PCP, PC2, PC3, STB, and CTB Files

You can import PC2 file settings using the Add Plotter wizard (available in the Plotter Manager), which creates a PC3 file. You can also import the same PC2 file using the Add a Plot Style Table wizard (available in the Plot Style Manager), which creates either an STB or a CTB file. To import all the information stored in a PC2 file, you must use both wizards, to get all the PC2 file settings into the new plotter configuration file format and plot style table format.

PC3 files contain the following information from the PC2 file:

- Device driver name
- Model name
- Media type
- Paper size and orientation
- Resolution
- Plot destination (port or network share name)

The Plot Style file contains the following information obtained from the PC2 file:

- Pen assignments
- Scale information

It's a good idea to import the PC2 file to a CTB file instead of to an STB file, because the CTB file more closely emulates the way pen assignments are handled in AutoCAD Releases 13 and 14. PC2 files store what to plot, how to plot, and pen tables. Putting all three in one file caused a lot of problems, so in newer versions of AutoCAD, these settings have been split into three different files:

- Pen tables moved to CTB or STB files (also referred to as plot style tables)
- PC3 files store information about how to plot (settings such as paper size and loading orientation, paper source, destination, resolution, color depth, and so on), the make and model of the plotter, and how the plotter is connected.
- Layouts hold all the rest of the information about what to plot, including settings such as the plot scale, offset, rotation, and plot style table name. These settings are stored in the drawing because they are drawing-dependent.

The best way to make permanent changes in layout settings is by using the **Page Setup** dialog box for the layout you want to modify. Any changes you make will always be saved. The only overlap between PC3 files and layouts is that both store a paper size, and the layout remembers the name of the PC3 file used for the layout. A drawing allows you to save several sets of named page setups. You can import page setups from one drawing into another, using the PSETUPIN command. You can reuse both PC3 and CTB files with many different drawings.

**Note** *The paper size, plot area, plot scale, plot origin, and plot offset are immediately applied. Any pen assignment information can subsequently be saved to a plot style table by using the Plot Style Table wizard, which you will learn about in the next chapter. Optimization levels and plotter connection information can be saved to a PC3 file using the Add Plotter wizard during plotter configuration. Information about adding and configuring plotters is provided in Appendix C.*

To import settings from PCP or PC2 into the current layout:

1. Do one of the following:

   ■ From the Tools menu, choose Wizards ∣ Import Plot Settings.

   ■ At the command line, type **PCINWIZARD** and then press ENTER.

2. AutoCAD displays the Introduction page of the Import PCP or PC2 Plot Settings wizard. After reading the introductory information, click Next to display the Browse File Name page.

3. Click the Browse button, locate the PCP or PC2 file you want to import, and then click Next.

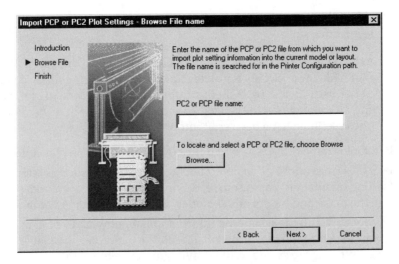

4. The wizard immediately updates the current layout using the information imported from the PCP or PC2 file. If you want to view or modify the new settings, click the Page Setup button to display the Page Setup dialog box. When you are finished, click OK to close the Page Setup dialog box.

5. Click Finish to complete the Import PCP or PC2 Plot Settings wizard.

# Creating Floating Viewports

When you first activate a layout, AutoCAD normally creates a single floating viewport within that layout. As you've also learned, when you use the Create Layout wizard, AutoCAD can automatically create one or more floating viewports. Once you become more comfortable with layouts, however, you will probably find it more convenient to create floating viewports manually.

You should think of floating viewports as objects that provide you with a viewport into model space. You can separately control the view, scale, and contents of each floating viewport. Floating viewports can overlap or be separate from one another. In addition, you can specify that individual layers are frozen in specific viewports. You'll learn more about this shortly.

## Placing Floating Viewports in a Layout

When you create floating viewports, you control the number of viewports created and the arrangement of the viewports. As with tiled viewports, AutoCAD provides several standard viewport configurations. You can create a single floating viewport or divide the drawing area into two, three, or four viewports, as shown in Figure 17-10. Although these viewport arrangements look nearly identical to the tiled viewports you created in Chapter 5, these are floating viewports. Once you create the floating viewports, you can move or change them in most of the same ways as you would any other drawing object.

If you create a single viewport, you can make it fill the entire printable area, or determine its size by specifying the opposite corners. When you create several viewports at one time, you specify the opposite corners of a rectangular area and then AutoCAD automatically arranges the viewports within that area. You can restore several types of standard or named viewport configurations and control the spacing between the viewports. Although all of these options create rectangular viewports, you can also create nonrectangular viewports, as you will learn later in this chapter.

| Tip | *You can download the ACLYUTIL.ARX utility from the Autodesk Product Support web site or the companion web site. This utility includes a VIEWTOLAYOUT command, which enables you to use a named paper space view and all the geometry that lies within the boundary of a specified view to create a new layout.* |
| --- | --- |

AutoCAD offers several different methods for creating floating viewports, the primary one being the VPORTS command. This is the same command you learned about in Chapter 5. When you use this command in a layout, however, the Viewports dialog box looks slightly different, as shown in Figure 17-11. The left side of the dialog box contains a list of standard viewport configurations, while the Preview area on the right side shows how the display will be divided based on the selections you make.

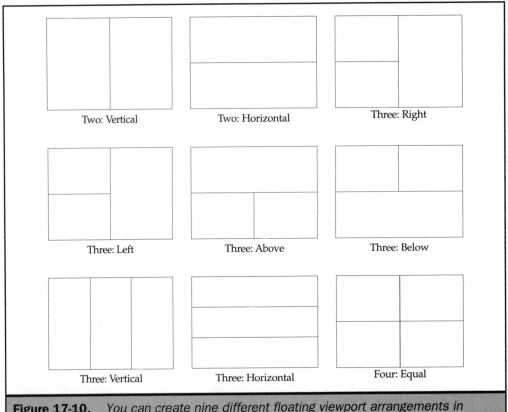

**Figure 17-10.** *You can create nine different floating viewport arrangements in addition to creating a single floating viewport.*

The current viewport is indicated by a double rectangle. If you compare this version of the dialog box to the one displayed when working in the Model tab, however, you'll notice a few differences: the list of standard viewports is slightly different, there's no box for you to save a named viewport configuration, and you can specify the spacing between viewports (since these viewports are floating rather than tiled).

If you select 2D in the Setup drop-down list, the floating viewports are all created with the current 2-D view in all the viewports. You can then change those views as needed. If you select 3D, however, a set of standard orthogonal views (such as top, front, side, and SE isometric) are applied to the viewports.

To create floating viewports in a layout:

1. Do one of the following:

- On the Layouts or Viewports toolbar, click Display Viewports Dialog.
- From the View menu, choose Viewports | New Viewports.
- At the command line, type **VPORTS** and press ENTER.

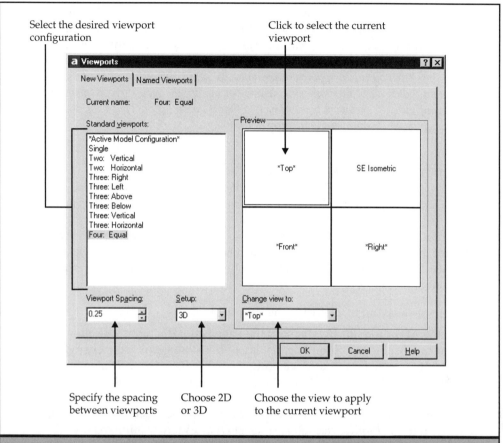

**Figure 17-11.** *The New Viewports tab of the Viewports dialog box*

2. In the Viewports dialog box, under Standard Viewports, choose the desired viewport configuration (such as Four: Equal).

3. In the Viewport Spacing box, specify the spacing you want between all the resulting viewports. (Remember that since the viewports are floating, you can move or resize them after they are created.)

4. Under Setup, choose either 2D or 3D.

5. Click OK. AutoCAD prompts:

```
Specify first corner or [Fit] <Fit>:
```

6. Do one of the following:

■ To arrange the viewports to fill the printable area of the layout, press ENTER (to select the Fit option).

■ Specify the corners of a rectangle in which to fit the viewports.

*When you create a floating viewport, AutoCAD creates the viewport border on the current layer. To make the viewport borders invisible, create a new layer before you create floating viewports, and then turn off that layer after you create the viewports. If you forget, you can always create a new layer, change the border objects to that layer, and then turn off the viewport border layer. To select the viewport borders, you must turn on that layer before you can rearrange or modify the viewports. Remember that you can configure the viewport border layer so that it doesn't plot. That way, it can remain visible but won't appear in your printed drawings.*

### LEARN BY EXAMPLE
*Switch to a new layout tab and create a bordered sheet complete with a title block and standard engineering views. Open the drawing Figure 17-12 on the companion web site.*

The AutoCAD drawing you just opened is a three-dimensional drawing of a nondescript mechanical part or widget. You currently see the part as it was created in model space. To begin creating the bordered sheet, use the following steps to activate a new layout and select a paper size:

1. Activate a layout by choosing the Layout1 tab.

2. AutoCAD automatically displays the Page Setup dialog box. (Remember that you can disable the automatic display of this dialog box whenever you activate a layout for the first time by clearing the Show Page Setup Dialog For New Layouts check box in the Layout Elements area on the Display tab of the Options dialog box. If you've turned this feature off, display the Page Setup dialog box now by right-clicking the layout tab and selecting Page Setup.) Switch to the Plot Device tab and select DWF eView (optimized for viewing).pc3 from the Name drop-down list.

*This example assumes that you've already installed and configured one or more plotting devices. In particular, it uses the DWF eView (optimized for viewing).pc3 plotter configuration. If you don't yet have this device installed on your system, you can install it now by clicking the Options button to display the Plotting tab of the Options dialog box. Click the Add Or Configure Plotters button and then start the Add-A-Plotter wizard. On the Begin page, select My Computer, and then click Next. On the Plotter Model page, under Manufacturers, select Autodesk ePlot (DWF); then, under Models, select DWF eView (optimized for viewing); and then click Next. On the remaining pages, simply click Next. When the wizard is finished, click Apply & Close to close the Options dialog box and return to the Page Setup dialog box. You should now see the DWF eView (optimized for viewing).pc3 selection in the drop-down list of available plotters.*

3. In the Layout Settings tab of the Page Setup dialog box, select ANSI expand D (34.00 × 22.00 Inches).

4. Make sure that the Plot Area is set to Layout, the Drawing Orientation is set to Landscape, and the Plot Scale is set to 1:1.

5. Click OK.

By default, AutoCAD automatically creates a floating viewport in the layout, containing the same 3-D view of the widget that you saw on the Model tab. Since you won't be using this viewport, erase it as you would any other drawing object.

> **Note**  *By default, AutoCAD creates a new floating viewport whenever you activate a layout for the first time. You can turn off this behavior by clearing the Create Viewport In New Layouts check box under Layout Elements on the Display tab of the Options dialog box. Remember that you can also disable the automatic display of the Page Setup dialog box by clearing the Show Page Setup Dialog For New Layouts check box.*

Next, you'll insert a standard border and title block. You've already learned several different ways to do this. In this example, you'll insert one of AutoCAD's predefined borders as a block.

1. From the Insert menu, choose Block.

2. In the Insert dialog box, click Browse, navigate to the AutoCAD 2002/Template folder, select ANSI D TITLE BLOCK.DWG, and click Open.

3. In the Insert dialog box, clear all three Specify On-Screen check boxes and make sure that the insertion point is set to X=1,Y=0, and Z=0; the scale is set to X=1, Y=1, and Z=1; and the rotation angle is set to 0. Then click OK.

Your drawing now contains a title block and border. Next, you'll create four floating viewports containing standard engineering views:

1. From the View menu, choose Viewports | New Viewports.

2. In the Viewports dialog box, in the Standard Viewports list, select Four: Equal. Under Setup, choose 3D. Notice that the Preview area shows the four standard engineering views: Top, Front, Right, and SE Isometric.

3. Under Viewport Spacing, specify a spacing value of 1, and then click OK. AutoCAD prompts:

```
Specify first corner or [Fit] <Fit>:
```

4. Make sure object snap is turned off, and then select a point slightly above and to the right of the lower-left corner of the border as the first corner of the bounding area. AutoCAD prompts:

```
Specify opposite corner:
```

5. Select a point slightly below and to the left of the revision area near the upper-right corner of the border.

## Using External References

When you are assembling a sheet in paper space, you can use xrefs to create a drawing that contains views of several other drawings. For example, suppose that you have drawn the elevations of a building in separate AutoCAD drawing files. To plot both the north and south elevations on the same sheet, you could create a new drawing and attach the two elevation drawings as external references. Create the sheet border in paper space (or attach it as an external reference) and then attach the two elevations as xrefs, each in its own floating viewport.

To improve AutoCAD's performance, each drawing that you create should contain only the objects comprising one printed sheet. You should also minimize the number of floating viewports to only those necessary for the sheet that you are assembling, and use the XCLIP command to display only the portion of the xref that needs to be included in the composite sheet. This reduces the amount of time that AutoCAD takes to load and regenerate the drawing, results in smaller files, and improves your overall productivity.

AutoCAD quickly creates the four viewports. Your drawing should look similar to Figure 17-12. Save your drawing. You'll use this drawing later in this chapter when you learn how to align the contents of floating viewports and control the scale of the objects within the viewports.

# Placing Saved Viewport Configurations in a Layout

In Chapter 5, you learned how to save and restore named viewport configurations in model space. If you previously created and saved a tiled viewport configuration using the VPORTS command in model space, you can also restore that named viewport configuration as floating viewports in the current layout.

To restore a named model space viewport configuration into a layout:

1. Start the VPORTS command.

2. In the Viewports dialog box, select the Named Viewports tab.

3. In the Named Viewports list, select the viewport configuration you want to restore. The viewport arrangement is displayed in the Preview area.

4. Click OK.

5. When AutoCAD prompts you to specify the location for the viewport in the layout, either specify the opposite corners of a rectangle in which to fit the viewports or choose the Fit option to fit the viewport configuration into the printable area of the layout.

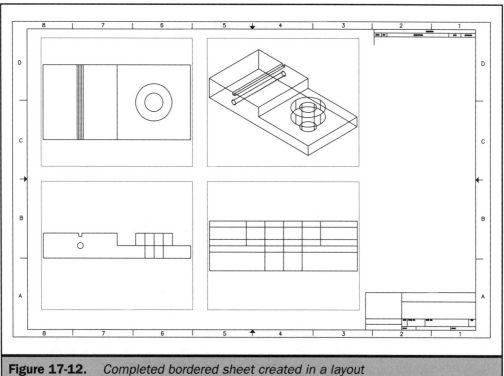

**Figure 17-12.** *Completed bordered sheet created in a layout*

**Note**

*You can also use the MVIEW command to create floating viewports and restore named model space viewport configurations. The operation of this command is very similar to the VPORTS command, but since it does not utilize a dialog box, the VPORTS command is generally much easier to use when creating floating viewports. The Viewports dialog box also lets you specify the spacing between the floating viewports, an option that is not available when using the MVIEW command. When it comes time to control the individual viewports, however, you will use the MVIEW command. You'll learn about using the MVIEW command to control floating viewports later in this chapter.*

## Changing the Properties of a Viewport

As mentioned earlier, after you create floating viewports, you can modify them as needed, by using standard AutoCAD commands. You can snap to the viewport borders by using object snaps. You can also use grip editing. To rearrange or otherwise modify a floating viewport, you must be in paper space and the viewport borders must be visible. To select a viewport, click its border.

MOVING BEYOND
THE BASICS

## Using Named Views in Paper Space

In Chapter 5, you also learned how to save and restore named views, which makes redisplaying a particular portion of the drawing faster and easier. You can use those same named views in paper space viewports. When you save a view in model space, you save the contents of the current model space viewport. If you restore a model space view in paper space, it displays as the contents of one of the floating viewports. You can also save and restore paper space views. Views saved in paper space simply restore the zoom and pan position of the entire layout.

The procedure for saving and restoring views in paper space is exactly the same procedure as for model space, with one exception: In addition to the list of saved views, the View dialog box indicates whether the view was saved in model space or in a layout.

View saved in a layout

View saved in model space

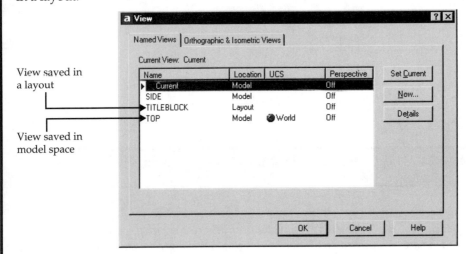

To restore a named view:

1. Do one of the following:

- On the View or Standard toolbar, click Named Views.

- From the View menu, choose Named Views.

- At the command line, type **VIEW** (or **V**) and then press ENTER.

2. In the View dialog box, select the named view that you want to restore.

3. Click Set Current, and then click OK to complete the command.

If you selected a model space view, AutoCAD prompts you to select the viewport in which to restore the view. As soon as you click the border of the floating viewport, the view is restored in that viewport, replacing its previous content.

*NOTE:  You can't restore a model view in a viewport that has been locked. You'll learn about locking viewports later in this chapter.*

You can copy, move, scale, stretch, and even erase viewports. When you resize a viewport, you change only the size of its borders, not the scale of its components. Changing the size of a viewport may make more of its contents visible or crop off some of what is displayed. Any changes that you make to the viewport while in paper space, including erasing the viewport itself, affect only the viewport, not the objects within the viewport.

### LEARN BY EXAMPLE

*Make the viewport borders invisible by placing them on a separate layer and turning off that layer.*

Using the same widget drawing that you worked on in the previous example, create a new layer and then change the viewport borders to that layer by following these instructions:

1. Use the LAYER command to display the Layer Properties Manager dialog box.

2. To create a new layer, click New, type **BORDER**, and then press ENTER.

3. Click the light bulb icon adjacent to the new BORDER layer to turn off that layer.

4. Click OK to close the Layer Properties Manager dialog box.

5. On the Standard toolbar, click Properties to display the Properties window.

6. Click to select the borders of the four floating viewports (so that you can see the grips at each corner).

7. In the Properties window, click Layer to make it active, and then choose BORDER from the Layer drop-down list. AutoCAD displays a warning dialog box, informing you that the objects you selected will be changed to a frozen layer or one that has been turned off.

8. Click OK.

9. Close the Properties window.

Your drawing should now be displayed without the borders surrounding the viewports. Remember, if you do need to modify the viewports themselves—for example, to move them—you first need to make them visible again by turning on the BORDER layer.

Of course, you can also use the Properties window to change other viewport properties, such as its color, linetype, linetype scale, plot style, and lineweight. Linetype and lineweight are displayed on nonrectangular viewports and ignored on rectangular viewports.

**Tip**

*To change the contents of the viewport, including its scale, you must switch to model space. You'll learn more about modifying the contents of a floating viewport later in this chapter.*

MOVING BEYOND THE BASICS

# Controlling Visibility in Floating Viewports

In addition to AutoCAD's other capabilities, when you create floating paper space viewports, you can control the visibility of layers individually in each viewport by freezing and thawing layers on a per-viewport basis. This is an important capability, considering what happens to dimensions and notes that you add to your drawings within model space views.

When a viewport contains 3-D objects, you may also want to remove hidden lines so that the image of your model is easier to understand. Removing hidden lines takes time, however, and may not be necessary in all viewports. You can therefore control this on a per-viewport basis, as well. You can also turn individual viewports on or off. When a viewport is turned off, its contents are no longer visible in the layout, nor present when you plot the layout (although the viewport borders will be visible and may plot unless you also turn off those layers).

**Note**  *You can also use plot style tables to accomplish other things when plotting a layout. For example, you can assign a screening value to an object so that it appears dimmer in the resulting plot. You can also use plot styles to control the color, lineweight, linetype, line end style, line join style, and line fill style of plotted objects. You'll learn about plot styles in the next chapter.*

## Controlling Layer Visibility in Floating Viewports

Normally, anything that you draw in one viewport is also visible in all the other viewports, assuming those viewports display the same portion of the drawing. While this is fine for the model itself, it can pose a serious problem when you add dimensions and notes. Consider the orthographic views in the previous example. If you add dimensions to the top view, those dimensions also appear in the other viewports, although they might not be readable. For example, the dimensions that you add to the top view are seen on-edge in the front view. This certainly isn't what you want. Since frozen layers are invisible, being able to freeze layers in specific viewports eliminates this problem.

**Note**  *In previous versions of AutoCAD, dimensioning in model space and then plotting from paper space was the preferred method for dimensioning complex, multiple-view drawings. While this is still the method you should use when the dimensions in a drawing need to be referenced by other drawings or when dimensioning 3-D isometric views, AutoCAD 2002's associative dimensioning (which you learned about in Chapter 14) enables you to create dimensions in paper space by selecting model space objects or specifying object snap points on model space objects, eliminating the need to create all the extra layers solely to make dimensions invisible in specific viewports. Since there may still be instances in which you will need to use the older method, the process of freezing layers in specific viewports is still explained here in detail.*

AutoCAD provides two different methods for freezing layers in specific floating viewports: using the Layer Properties Manager dialog box and using the VPLAYER command. Both methods have their advantages and disadvantages. When you use the Layer Properties Manager dialog box, you can simply click the Active VP Freeze icon to toggle the visibility of a particular layer on and off. Unfortunately, that changes the visibility only within the current viewport. Generally, you will want a particular layer (the one on which the dimensions are drawn) to be thawed in only one viewport and frozen in all the others. To do this using the dialog box, you need to repeat the process for each viewport.

The VPLAYER command lets you freeze and thaw layers in multiple viewports at one time. The drawback to using this command, however, is that you must type the layer names at the command line.

*The global freeze/thaw setting overrides the settings for individual viewports. For example, if a layer is thawed in a specific viewport but frozen globally, it will be frozen in all viewports.*

When you freeze specific layers in the current floating viewport, you must first switch to model space and then use the LAYER command to freeze the specific layers.

### *LEARN BY EXAMPLE*
*Freeze several layers in the current viewport. Open the drawing Figure 17-13 on the companion web site.*

In Figure 17-13, several layers were added to the drawing from the previous example, one each for the dimensions in the top (upper-left), front (lower-left), and side (lower-right) views. As you can see, these dimensions also appear in the isometric (upper-right) view. To freeze these layers in the viewport containing the isometric view:

1. Double-click inside the isometric viewport to switch to model space and make it the active viewport.

2. Start the LAYER command.

3. In the layer list, choose the DIM-FRONT, DIM-SIDE, and DIM-TOP layers. (Remember that you can select several layers at once by pressing the CTRL or SHIFT key while selecting them.)

4. Click the Active VP Freeze icon to freeze those layers.

5. Click OK to close the Layer Properties Manager dialog box.

The dimension layers are no longer visible in the isometric viewport. Those layers are still visible in the other three viewports, however, whereas you really want them visible only in their specific viewports. You could use the VPLAYER command to freeze the dimensioning layers in all the viewports, and then thaw just the appropriate

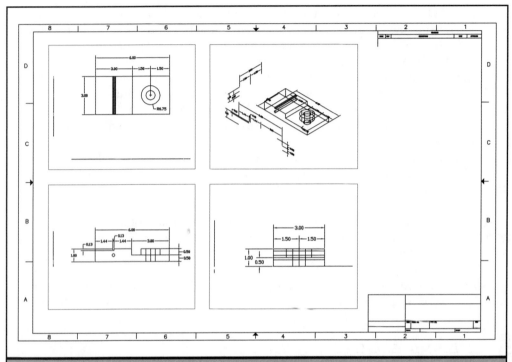

**Figure 17-13.** *Dimensions added to one view appear in all the views until you freeze the appropriate dimension layers in specific viewports.*

layer in the top, front, and side views. To do this, use the following command sequence and instructions:

```
Command: VPLAYER
Enter an option [?/Freeze/Thaw/Reset/Newfrz/Vpvisdflt]: F
Enter layer name(s) to freeze: DIM*
Enter an option [All/Select/Current] <Current>: A
Enter an option [?/Freeze/Thaw/Reset/Newfrz/Vpvisdflt]: T
Enter layer name(s) to thaw: DIM-TOP
Enter an option [All/Select/Current] <Current>: S
Switching to Paper space.
Select objects: (select the top view viewport border)
Select objects: ENTER
Switching to Model space.
Enter an option [?/Freeze/Thaw/Reset/Newfrz/Vpvisdflt]: (repeat for each layer or press
ENTER to end the command)
```

You could continue this process to thaw the appropriate layer in each of the other two viewports, but since you must type the layer name each time, it is probably easier to use the VPLAYER command to freeze the layers in all the viewports, and then use the LAYER command to thaw the appropriate layer in the individual viewport.

When you create a layer that should be visible only in the current viewport, you can specify that it be frozen in any new viewports that you create, so that you don't have to repeat the previous process. You can do this by using either the Layer Properties Manager dialog box or the VPLAYER command's Newfrz option, but using the dialog box is easiest.

To freeze or thaw layers in all new viewports:

1. Start the LAYER command.

2. In the layer list, choose the layers to freeze or thaw.

3. Click the New VP Freeze icon to freeze or thaw those layers.

4. Click OK to close the Layer Properties Manager dialog box.

## Removing Hidden Lines When Plotting Viewports

If your drawing contains three-dimensional objects—such as 3DFaces, meshes, extruded objects, surfaces, or solids—you can instruct AutoCAD to perform hidden-line removal on the objects within selected viewports when you plot the layout. Each floating viewport has its own Hideplot property that affects only the plotted output of that viewport. The Hideplot setting doesn't change the screen display. This property is also separate from the more general Hide Objects control on the Plot Settings tab of the Page Settings dialog box. When you select that check box, AutoCAD only hides objects created in paper space when plotting a layout, or in model space when plotting the Model tab.

You can control the Hideplot setting for individual viewports using any of the following methods:

■ Change the Hide Plot setting in the Properties window

■ Use the MVIEW command

■ Use a shortcut menu

To change the Hideplot setting using the Properties window, do the following:

1. Display the Properties window.

2. Click to select the border of the floating viewport (so that you can see the grips at each corner).

3. In the Properties window, click Hide Plot to make it active, and then choose Yes (to instruct AutoCAD to perform hidden-line removal within that viewport) or No (to turn off hidden-line removal).

4. Close the Properties window.

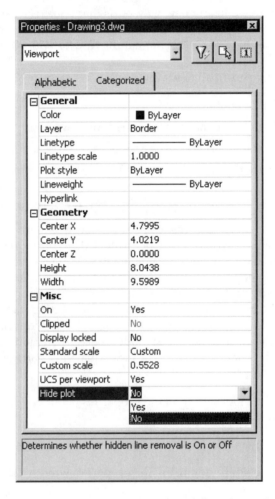

To change the Hideplot setting using a shortcut menu, select the border of the floating viewport, right-click in the drawing area to display the shortcut menu, choose Hide Plot, and then choose either Yes or No.

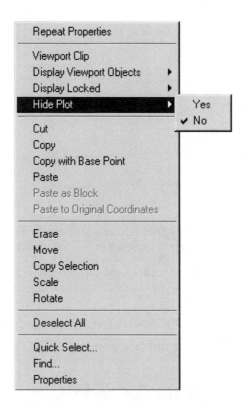

To change the Hideplot setting using the MVIEW command, at the command line, type **MVIEW** (or **MV**) and press ENTER. AutoCAD prompts:

```
Specify corner of viewport or
[ON/OFF/Fit/Hideplot/Lock/Object/Polygonal/Restore/2/3/4] <Fit>:
```

Type **H** (or right-click and select Hideplot from the shortcut menu). AutoCAD then prompts:

```
Hidden line removal for plotting [ON/OFF]:
```

Type either **ON** or **OFF** (or right-click and select the option from the shortcut menu). AutoCAD then prompts:

```
Select object:
```

MOVING BEYOND THE BASICS

Select the viewports for which you want to turn the Hideplot setting on or off, and then press ENTER.

*When using the MVIEW command, it's obvious that you can change the Hideplot setting for more than one viewport at a time. You can also select multiple viewports when changing the Hideplot setting using the Properties window, but not when using the shortcut menu.*

*LEARN BY EXAMPLE*
*Turn on hidden-line removal for a viewport.*

Using the same widget drawing that you worked on in the previous example, turn on hidden-line removal in the isometric viewport by following these instructions:

1. Switch back to paper space. (*Hint:* You can double-click anywhere in the layout outside of a viewport or click MODEL on the status bar to switch it back to PAPER.)

2. Select the upper-right viewport (so that the grips are visible at its corners).

3. Right-click anywhere in the drawing area, and select Hide Plot | Yes from the shortcut menu.

## Turning Floating Viewports On and Off

Displaying a large number of viewports can slow down the performance of your computer when regenerating the content of those viewports. New viewports are always turned on by default. You can turn viewports off when you don't need to see their contents, thus improving performance. When a viewport is off, it remains blank; AutoCAD won't regenerate or plot its contents until you turn it back on. You can still modify the viewport object (by moving, copying, or resizing its border) even when the viewport is turned off, as long as its borders are still visible.

You can control the visibility of viewports by using the same methods you just learned about to control the Hideplot setting.

To change the visibility of one or more viewports using the Properties window, do the following:

1. Display the Properties window.

2. Select the border of the floating viewport.

3. In the Properties window, click On to make it active, and then choose either No (to turn the viewport off) or Yes (to turn it on).

To change the visibility using a shortcut menu, select the border, right-click in the drawing area to display the shortcut menu, choose Display Viewport Objects, and then choose either Yes or No. You can also turn viewports on or off using the ON or OFF option of the MVIEW command.

# Editing in Floating Viewports

After you create floating viewports, you can modify the objects in the viewport by switching back to model space, by double-clicking inside a viewport. Once back in model space, you can still see all the viewports, but now one of the viewports becomes the current active viewport. You can then switch between viewports simply by clicking anywhere within the viewport border. Then, you can modify the objects in the viewport by using any AutoCAD command. With the exception of changes made to objects on layers that aren't visible in particular viewports, any changes that you make in one viewport are immediately reflected in all the other viewports. Changes that you make to the grid, snap settings, zoom magnification, or viewport orientation, however, affect only the current active viewport.

## Scaling the Viewport Contents

As you've already learned, you normally draw objects full size when working in model space. And when you plot a layout, which represents an actual sheet of paper, you typically plot that layout at 1:1 scale—in other words, full size. When you create floating paper space viewports within your layout, therefore, you need to scale the contents of those viewports so that the objects you drew in model space will fit within the confines of your paper. This is similar to drawing to a particular scale, such as 1/4-inch to the foot, when you draw on an actual sheet of paper.

If you change the size of the viewport object itself, by scaling or stretching its border, you only change the size of the viewport, not the scale of the objects within that viewport. Rather, to change the scale of the objects within the viewport, you must scale the contents of that viewport relative to paper space. You do this by applying a zoom scale factor relative to paper space units.

The scale factor for each viewport is really a ratio between the actual size of the model displayed in the viewport and the scale of the layout. If you plot the layout at 1:1 scale, this ratio is determined simply by dividing the paper space units by the scale factor you want displayed within the viewport. For example, if you've drawn the floor plan of a house and want it to be plotted at 1/4"=1'-0" on the layout, and the layout will be plotted at 1:1, you would set the scale of the viewport to 1/48 (1 layout unit divided by 1/48th of a foot).

In earlier versions of AutoCAD, you accomplished this by using the special *XP* option of the ZOOM command, which instructed AutoCAD to set the zoom scale factor relative to paper space. In versions beginning with AutoCAD 2000, however, there is an easier way. You can change the plot scale of the viewport by using either the Properties window or the Viewports toolbar.

To change the viewport scale using the Properties window:

1. Display the Properties window.

2. Select the viewport whose scale you want to change.

3. In the Properties window, select Standard Scale to make it active, and then choose the scale from the drop-down list.

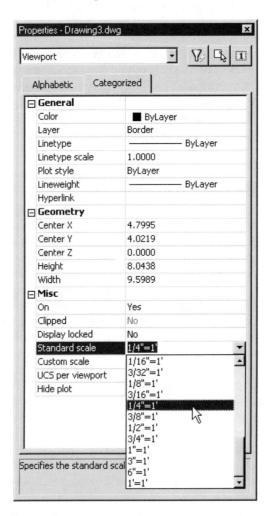

The scale you select is immediately applied to the viewport.

To change the viewport scale using the Viewports toolbar:

1. Display the Viewports toolbar.

2. Select the viewport whose scale you want to change.

3. Choose the scale from the drop-down list.

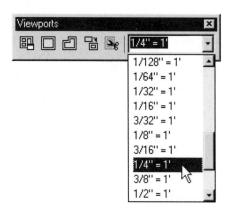

*LEARN BY EXAMPLE*
*Set the viewport scale for the four viewports in the previous example.*

Using the same widget drawing that you worked on in the previous example, set the scale factor for all four viewports to 1:1 by doing the following:

1. Display the Viewports toolbar. (*Hint*: If the toolbar isn't already displayed, right-click any toolbar button or title bar and choose Viewports in the shortcut menu, so that a check mark appears adjacent to its name.)

2. Select all four viewports (so that grips appear at each corner).

3. Choose 1:1 from the drop-down list in the Viewports toolbar.

## Locking the Viewport Scale

Although AutoCAD now makes it much easier to establish the viewport scale, in reality, the scale is still set by zooming the viewport by a specific scale factor relative to paper space. If you use the ZOOM command within the viewport (in other words, while working in model space within the floating viewport), you immediately change the scale factor for that viewport. If you hadn't meant to do this, you had to either undo the ZOOM command (if you caught your mistake quickly) or reestablish the viewport scaling. Obviously, this was a real problem in previous versions of AutoCAD.

To prevent this situation from happening, once you've established the viewport scale, you can lock the scale so that any changes you make no longer affect the viewport scale. You can turn on viewport scale locking on a per-viewport basis by using either the Properties window or a shortcut menu. After you lock the viewport scale, most of the commands that affect the way you would change the view within that viewport—such as DVIEW, PLAN, VIEW, VPOINT, and 3DORBIT—no longer can be used within that viewport. In addition, when you attempt to use the PAN or ZOOM commands within that viewport, AutoCAD automatically switches to paper space for the duration of the command. This ensures that you can't accidentally change the scale of the viewport. If you attempt to zoom in to view something within the viewport at a greater magnification, AutoCAD temporarily switches to paper space. Your magnification change is applied to the entire layout. As soon as the ZOOM command is complete, AutoCAD switches back to model space. You thus accomplish the change in magnification without changing the viewport scale.

To lock the viewport scale using the Properties window:

1. Display the Properties window.

2. Select the viewport whose scale you want to lock.

3. In the Properties window, select Display Locked to make it active, and then choose Yes from the drop-down list.

*If you select a nonrectangular viewport, the viewport is not initially displayed in the Properties window, because nonrectangular viewports consist of both a viewport and an AutoCAD object. To change the display locking for nonrectangular viewports, after displaying the Properties window and selecting the viewport, choose Viewport (1) from the drop-down list at the top of the Properties window. The Display Locked option is then displayed in the Properties window. You'll learn more about nonrectangular viewports later in this chapter.*

To lock the viewport scale using the shortcut menu, select the border of the viewport, right-click in the drawing area to display the shortcut menu, and choose Display Locked | Yes.

### LEARN BY EXAMPLE

*Lock the viewports so you can see the way AutoCAD behaves when you attempt to zoom within a locked viewport.*

Using the same widget drawing that you worked on in the previous example, lock several of the viewports by doing the following:

1. Select the viewports containing the top, front, and side views (so that grips appear at their corners).

2. Right-click anywhere in the drawing area to display the shortcut menu, and then choose Properties.

3. In the Properties window, select Display Locked to make it active, and then choose Yes from the drop-down list.

4. Close the Properties window, and press ESC to remove the grips.

Now, try to use the ZOOM command to zoom within one of the locked viewports. Double-click within the viewport in the upper-left corner to make it the current active viewport. Then, type **ZOOM** and press ENTER. Notice that AutoCAD immediately switches to paper space. Any zoom you attempt to perform in the locked viewport actually affects the entire layout.

Cancel the ZOOM command, double-click in the viewport in the upper-right corner to make the unlocked isometric viewport the current viewport, and try the ZOOM command again. Notice that this time, AutoCAD lets you zoom within the viewport, because that viewport wasn't locked. But, any change in magnification within the viewport actually affects the scale of the viewport relative to the rest of the layout.

## Scaling Linetype Patterns in Paper Space

The scale of the floating viewport also affects the scaling of noncontinuous linetypes, and the linetype scale can be based on the drawing units of the space in which the object was created. When you create a layout that contains several viewports whose contents are displayed at different scales, the same linetype will display differently in each viewport, based on the scale of the viewport. This is probably not what you want.

AutoCAD provides a system variable that lets you control whether linetypes are scaled based on the space in which they were created or based exclusively on paper space drawing units. When PSLTSCALE equals 0, linetypes are scaled based on the space in which they were created. When this value equals 1, however, the linetypes are scaled based only on the paper space units, allowing viewports to have different magnifications, while linetypes in each viewport scale identically. Although you can control the PSLTSCALE system variable directly by typing it at the command line, you can also simply turn on paper space linetype scaling using the Linetype Manager dialog box, by selecting the Use Paper Space Units For Scaling check box in the Details area of the Linetype Manager dialog box, as shown in Figure 17-14.

## Aligning and Rotating Views in Floating Viewports

When you create standard engineering drawings, you normally align the views. You can accomplish the same thing in AutoCAD by aligning the view in one floating viewport with the view in another viewport. You can align the views horizontally, vertically, or at an angle by establishing the base point in one viewport and then aligning a point in the other viewport with that base point. When you align viewports, you are actually panning the contents of one viewport relative to the other.

You can also rotate an entire view within a floating viewport by a specified angle. This is different from the ROTATE command, which rotates individual objects. When you rotate a view within a viewport, you rotate the view itself, not the viewport borders.

Both of these functions are accomplished by using the MVSETUP command. This command has many other functions, most of which can be performed more easily using other commands. When it comes to aligning and rotating views within floating viewports, however, the MVSETUP command is still the best method.

MOVING BEYOND THE BASICS

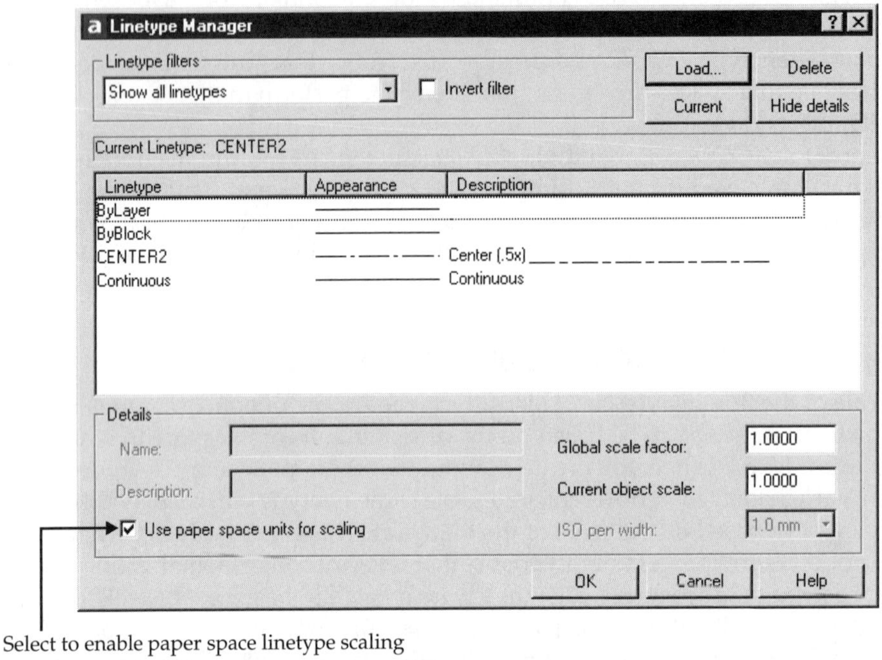

Select to enable paper space linetype scaling

**Figure 17-14.**    *You can control the scaling of linetype patterns in paper space by using the Linetype Manager dialog box.*

When you start the command, AutoCAD prompts:

```
Initializing...
Enter an option [Align/Create/Scale viewports/Options/Title block/Undo]:
```

In this case, the only option we're interested in is the Align option. When you select this option, AutoCAD prompts:

```
Enter an option [Angled/Horizontal/Vertical alignment/Rotate view/Undo]:
```

You can then choose one of the options:

■ **Angled**  Aligns a point in one viewport at a specified distance and angle from a base point in another viewport.

■ **Horizontal**  Aligns a point in one viewport horizontally with a base point in another viewport.

- **Vertical**   Aligns a point in one viewport vertically with a base point in another viewport.
- **Rotate**   Rotates the contents of a viewport around a base point by a specified angle.

**Note**   *To align the view in one floating viewport with the view in another viewport, or to rotate the contents of a viewport, the viewport being moved can't be locked.*

**LEARN BY EXAMPLE**
*Align the contents of a viewport with another viewport.*

Using the same widget drawing that you worked on earlier, align the top, front, and side views by doing the following:

1. Turn off the viewport locking for the three viewports that you locked in the earlier example.

2. Double-click in the lower-left viewport to make it the current active viewport.

3. Use the PAN command to center the front view within this viewport without changing its current scale. (*Hint*: On the Standard toolbar, click Pan Realtime, and then move the view until all of its dimensions are visible within the viewport.)

4. Use the MVSETUP command to align the top and side views with the front view, by using the following command sequence and instructions:

```
Command: MVSETUP
Enter an option [Align/Create/Scale viewports/Options/Title block
/Undo]: A
Enter an option [Angled/Horizontal/Vertical alignment/Rotate view
/Undo]: H
Specify basepoint: (use the endpoint object snap to select the lower-left corner of
the widget in the front view)
Specify point in view to be panned: (use the endpoint object snap to select
the lower-left corner of the widget in the side view)
Enter an option [Angled/Horizontal/Vertical alignment/Rotate view
/Undo]: V
Specify basepoint: (use the endpoint object snap to select the lower-left corner of
the widget in the front view again)
Specify point in view to be panned: (use the endpoint object snap to select
the lower-left corner of the widget in the top view)
Enter an option [Angled/Horizontal/Vertical alignment/Rotate view
/Undo]: ENTER
Enter an option [Align/Create/Scale viewports/Options/Title block
/Undo]: ENTER
```

Once the objects in the viewports are aligned, you should turn viewport locking on again so that you don't accidentally change the scale of these viewports. Remember to save your work.

# Creating Nonrectangular Viewports

Thus far, you've created rectangular viewports in paper space to display different views of your model. You can also create nonrectangular boundaries. AutoCAD actually provides two different ways to create nonrectangular boundaries: you can define polygonal boundaries when you create a new viewport, or you can associate a closed object—a circle, ellipse, polyline, region, or spline—with a rectangular viewport and use it to clip the geometry in the viewport to the edge of that boundary object. You can even use a region with a hole in it; the viewport is visible within the region boundary but hidden by the hole.

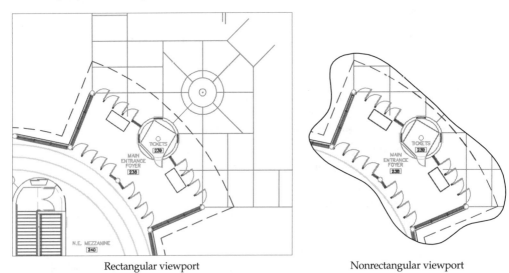

Rectangular viewport                     Nonrectangular viewport

AutoCAD provides several different commands for creating nonrectangular viewports. For example, when you use the MVIEW command (or type **-VPORTS** to use the VPORTS command at the command line), AutoCAD prompts:

```
Specify corner of viewport or
[ON/OFF/Fit/Hideplot/Lock/Object/Polygonal/Restore/2/3/4] <Fit>:
```

The Polygonal option lets you define the viewport boundary as you create the viewport. AutoCAD prompts you to select the starting point for the viewport polyline. Once you specify the first point, the program prompts:

```
Specify next point or [Arc/Close/Length/Undo]:
```

You can then define the polyline boundary exactly the same way that you create a polyline. As soon as you close the polyline, AutoCAD creates the new viewport and clips it to the polyline boundary.

**Note**   *When you define a polygonal viewport boundary, AutoCAD simply creates a new polyline and then immediately clips the new viewport to that polyline. When you create nonrectangular viewports in this way, you don't get the chance to smooth the polyline, and once associated with a viewport, you can't use the PEDIT command to smooth the polyline. Thus, the resulting viewport boundary will always have sharp corners at the vertices unless you create arc polyline segments. To create a viewport with a smooth flowing boundary using fit curve or splined polylines, you must create the boundary first and then use it to clip the underlying viewport. It's also generally easier to create a rectangular viewport first and then create a boundary to use for clipping that viewport, since you can better visualize the relationship of the boundary and the viewport.*

Notice that the MVIEW command and the command line version of the VPORT command also provide an Object option. When you select this option, AutoCAD prompts you to select the viewport object you want to use to clip the viewport. As soon as you select the object, AutoCAD creates a new viewport clipped to those objects. In this case, the viewport boundary must exist before you actually create the viewport. The scale of the resulting viewport is calculated by determining the extents of the viewport boundary object and the underlying viewport. Again, because it's more difficult to visualize the relationship between the underlying rectangular viewport and the viewport boundary object until after you create both objects, you typically create the viewport first, adjust its scale, and then create the viewport boundary and use it to clip the viewport.

To clip an existing viewport to a nonrectangular viewport object:

1. Create a floating viewport.

2. Create the object you want to use to clip that viewport (a circle, ellipse, closed polyline, region, or closed spline).

3. Start the VPCLIP command by doing one of the following:

   ■ At the command line, type **VPCLIP** and press ENTER.

   ■ Select the viewport you want to clip, right-click to display the shortcut menu, and choose Viewport Clip.

   AutoCAD prompts:

   ```
   Select clipping object or [Polygonal] <Polygonal>:
   ```

4. Select the object you want to use to clip the viewport.

**Note**   *The VPCLIP command also provides a Polygonal option, which lets you define a polyline boundary on-the-fly after you've created a floating viewport.*

The nonrectangular viewport object is automatically associated with the viewport, and the two objects remain connected as long as the nonrectangular viewport boundary exists. While they're connected, if you erase the viewport boundary, you erase the entire viewport object. To remove the viewport boundary, thus restoring the underlying rectangular viewport, you can use the VPCLIP command to delete the boundary, as shown in the following command sequence:

```
Command: VPCLIP
Select viewport to clip: (select an existing nonrectangular viewport)
Select clipping object or [Polygonal/Delete] <Polygonal>: D
```

AutoCAD immediately deletes the viewport boundary.

**Tip**
*If you want to replace one nonrectangular viewport with a different nonrectangular boundary, instead of selecting the Delete option, simply select the new boundary. AutoCAD automatically deletes the old boundary and clips the underlying viewport to the new boundary.*

Since the nonrectangular boundary is itself an AutoCAD object, you can use standard object modification commands to change that boundary just like any other object. You can also use grips to edit the boundaries. To edit a boundary by using grips, select the boundary (so that its grips become visible) and then edit the boundary as you would any other object. For example, if the boundary object is a polyline, AutoCAD displays grips at each polyline vertex. To move a vertex, select a grip (to make it the hot grip) and then move the location of the grip. As you adjust the boundary, the program clips the underlying viewport contents in real time. Similarly, when you Pan or Zoom within a nonrectangular viewport, the model space geometry in the underlying viewport is clipped in real time.

**Caution**
*If you freeze the layer on which the viewport boundary object was drawn, the boundary is not displayed and the underlying viewport is no longer clipped. If the boundary layer is turned off but not frozen, the viewport remains clipped. You should also note that complex nonrectangular viewports result in slower plotting performance.*

# Chapter 18

## Plotting Your Drawing

After you establish exactly what you want to appear on your printed drawing, you are ready to print or plot the drawing. You can plot to any previously configured plotter, including any Windows system printer. You can also add new printers or plotters. If you prefer, you can create an electronic plot by plotting to a file saved in Autodesk's Drawing Web Format (DWF). DWF files can then be published to the Internet for others to see. You'll learn about DWF files in Chapters 25 and 26.

When you plot your drawing, AutoCAD lets you determine the portion of the drawing to plot. The printed drawing includes only what is visible in the drawing and within the print area that you specify. Objects on layers designated not to be plotted will not appear in the resulting print. You can plot either the entire drawing or a selected portion. You can plot the Model tab or one or more of the layout tabs you've created.

As you learned in the previous chapter, you can use plot styles to control the appearance of objects when they are plotted, and you can create and apply different plot styles. AutoCAD also uses plotter configurations to store information about the plotter and all of its settings, as well as information about the paper sizes accommodated by each plotter.

Since plotting can be a time-consuming process, and since most modern plotters and printers can run with little or no user intervention, AutoCAD comes with a batch plot utility that enables you to construct a list of drawings to be plotted. You can then print the drawings in that list at another time.

This chapter explains how to do the following:

- Print or plot your drawing
- Use plot styles
- Use the batch plotting utility

## Understanding Plotting

As you've learned, plotting has been significantly changed since pre-AutoCAD 2000 versions. If you're a long-time AutoCAD user, the changes may seem a bit confusing at first. But in reality, by separating the layout and actual plotting processes, you now have much greater control over your final output and can save and reuse settings to produce plots that look exactly the way you intend them to look.

AutoCAD 2000 introduced several new concepts:

- **Layouts**   Store the view of the drawing along with a page setup, plot style table, and plotter configuration
- **Plot styles**   Control how objects are plotted
- **Plot style tables**   Contain collections of plot styles that can be assigned to objects and layers
- **Plotter configurations**   Contain information about a plotter and its settings as well as the available paper sizes

When you create and plot a drawing, you should proceed through the following steps:

1. Create the drawing in model space.

2. Switch to a layout tab.

3. Select the plot device and specify page settings, such as paper size, orientation, and scale.

4. Add a title block and border and arrange viewports and views for plotting.

5. Optionally, create plot styles and assign them to objects or layers to control pen settings and special effects such as screening, lineweights, plot colors, and fill styles.

6. Plot your drawing.

# Understanding Layouts

In the previous chapter, you learned about layouts. Layouts simulate the sheet of paper on which you will plot your drawing. If you plot from paper space, each layout tab displays exactly what will be printed. Layouts also store settings for the page setup, including the plot device, plot style table, print area, rotation, plot offset, paper size, and scale. Each drawing can contain multiple layouts, and each layout can display different views of the drawing and have its own plot device and plot style table. As you will learn later in this chapter, if you wish, you can plot all of the layouts associated with the drawing using a single command.

You can also save the plot device, plot style table, and page setup settings as a named page setup. Then, when you're ready to plot, you can select a named page setup and use it to replace the current settings in the Page Setup dialog box. Although you don't need to use named page setups, doing so lets you save and reuse settings. By applying different named page setups to the same layout, you can use that layout to plot the same drawing to different plotters and on different paper sizes.

# Understanding Plot Styles

A plot style controls the appearance of objects in your drawing when they are plotted. For example, a plot style might specify that a solid object is actually plotted using a fill pattern, or that all of the objects in the drawing actually print as black lines even though they appear in the drawing in different colors. A plot style consists of a set of overrides that determine the color, dithering, grayscale, pen assignments, screening, linetype, end styles, join styles, and fill styles applied to objects or layers in your drawing.

The use of plot styles is optional. By default, AutoCAD doesn't assign a plot style to your drawing. As you learned in Chapter 8, when you create a new drawing, you can choose whether the drawing uses color-dependent or style-dependent plotting. Color-dependent plotting determines the appearance of objects when they are plotted based on the color in which they were drawn, similar to AutoCAD Release 14 and previous versions. Style-dependent plotting enables you to control the appearance of plotted objects on a per-object or per-layer basis. If you don't apply a plot style, either method prints objects the way they appear in the drawing.

Plot styles are stored in plot style tables, and each table can contain multiple plot styles. In color-dependent plotting, you can specify overrides to be applied on a per-color basis. For example, you can specify that objects that are red in the drawing actually print as wide lines with a crosshatch pattern. You can create new color-based plot style tables, each containing its own set of color-dependent overrides. Color-dependent plot style tables are stored in files with the extension .CTB.

> **Note** *In AutoCAD 2000 and later versions, you can specify the lineweight on a per-layer or per-object basis. You no longer need to use pen assignments to control lineweight.*

For style-dependent plotting, each plot style table contains a default plot style called NORMAL, which plots objects the way they appear in the drawing. You can create new style-dependent plot style tables, and each can contain an unlimited number of plot styles that specify different sets of overrides. Style-dependent plot style tables are stored in files with the extension .STB.

In order to use a plot style, you must select a plot style table and attach it to the Model tab or a layout. You do this by selecting the plot style table on the Plot Device tab of the Page Setup or Plot dialog box when the particular layout tab or the Model tab is active. If your drawing uses named plot styles, you can then attach a plot style from the selected plot style table to an object or layer in your drawing. If your drawing uses color-dependent plotting, the color assigned to the object or layer determines the appearance of the drawing when it is plotted.

You'll learn more about plot styles and plot style tables later in this chapter.

## Understanding Plotter Configurations

You can plot your drawings to any configured Windows system printer. You can also plot to printers other than Windows system printers by creating a plotter configuration for that printer. Plotter configurations contain information about a plotter and all of its settings, as well as its available paper sizes. Each plotter configuration is stored in a file with the extension .PC3.

Generally, to plot to most Windows system printers, particularly desktop printers, you don't need to create a plotter configuration file. All of the available Windows system printers appear in the plotter list on the Plot Device tab of the Page Setup or Plot dialog box. To plot to a nonsystem printer or to large-format plotters, however, you need to create a plotter configuration.

>  *Although optional, you can create a plotter configuration for a Windows system printer. This may be useful if you often print to the Windows system printer with settings other than the default settings.*

Plotter configuration files are created and managed from the Autodesk Plotter Manager, a window into AutoCAD's Plotters directory. You can display the Plotter Manager by doing any of the following:

- From the File menu, choose Plotter Manager.
- At the command line, type **PLOTTERMANAGER** and press ENTER.

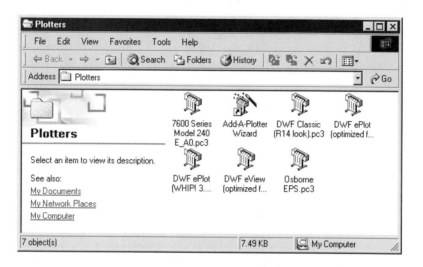

You can also display the Autodesk Plotter Manager from the Windows Control Panel or by clicking the Add Or Configure Plotters button on the Plotting tab of the Options dialog box. The Plotter Manager's Add-A-Plotter wizard greatly simplifies plotter configuration by stepping you through the configuration process. The wizard creates a PC3 file that you can then edit using the Plotter Configuration Editor.

**Note**    *When configuring a nonsystem printer, you are prompted to specify the manufacturer of your plotter and then choose from a list of supported models. The lists displayed are based on the available plotter drivers already loaded onto your system, in the form of Autodesk Heidi drivers. These drivers may be located in any of the directories specified in AutoCAD's Device Driver File Search Path. AutoCAD comes with a number of specialized Heidi drivers for printers other than Windows system printers. Your particular printer or plotter may have also been supplied with a Heidi driver, which you will need to copy into AutoCAD's driver directory. AutoCAD no longer uses Autodesk Device Interface (ADI) plotter drivers. The new Heidi drivers have an .HDI file extension. By default, these drivers are copied to the AutoCAD 2002/Drv directory during installation.*

# Plotting a Drawing

When you are ready to plot your drawing, you simply start the PLOT command. AutoCAD displays the Plot dialog box, shown in Figure 18-1. This dialog box looks very similar to the Page Setup dialog box that you learned about in Chapter 17. In fact, if you already used the Page Setup dialog box to specify the plot device and plot settings, those settings are already selected for you, and you may not need to change anything before clicking OK to send your plot to the specified plot device. If you're plotting from the Model tab, however, AutoCAD assigns the default plot device as specified on the Plotting tab of the Options dialog box. If you need to change this or any of the other settings, refer to the description of these controls in the previous chapter.

To display the Plot dialog box, do one of the following:

- On the Standard toolbar, click Plot.
- From the File menu, choose Plot.
- At the command line, type **PLOT** (or **PRINT**) and then press ENTER, or press CTRL-P.

**Note**    *You can also display the Plot dialog box by right-clicking one of the tabs along the bottom of the drawing area and selecting Plot from the shortcut menu.*

As with the Page Setup dialog box, the Plot dialog box contains an area across the top for the layout and page setup name, and two tabs for controlling the plot device and plot settings. If you make any changes to the layout, you can save those changes. The layout name is displayed on the left side of this area, and the Page Setup Name box controls enable you to add a new page setup or select one from the drop-down list.

On the Plot Device tab, shown in Figure 18-1, you select the plotter from a drop-down list of available devices. As with the Page Setup dialog box, the Plotter Configuration area also displays information about the selected plotter. The other controls in this area are also similar to those you learned about in the previous chapter.

In the Plot Style Table area, you can select the plot style table you want to use for the current tab. The adjacent Edit and New buttons let you modify or create plot style tables. You'll learn about plot styles later in this chapter.

**Note**    *The Display Plot Styles check box that appears in the Plot Style Table area in the Page Setup dialog box is not included in the Plot dialog box. When setting up your drawing to use named plot styles, you select this check box to see plot styles applied to objects in the drawing.*

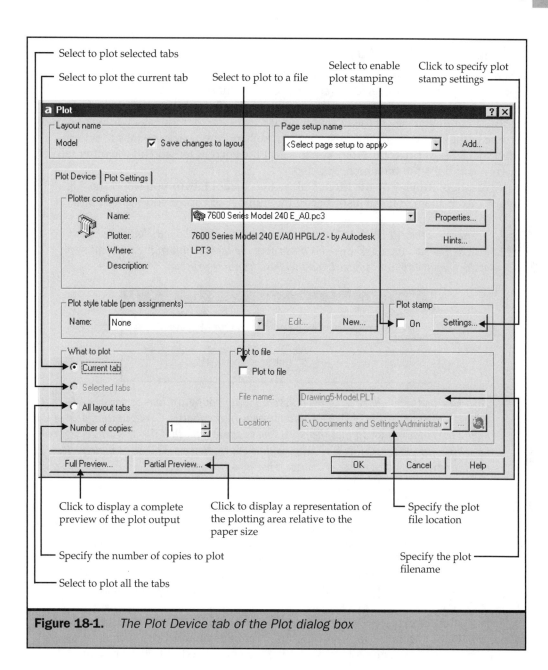

**Figure 18-1.** *The Plot Device tab of the Plot dialog box*

## Selecting What to Plot

When you plot your drawing, AutoCAD normally plots the current tab. You can also plot all the layout tabs or selected tabs. The controls under What To Plot on the Plot Device tab let you control which tabs are plotted. To plot selected tabs, before starting the PLOT command, select the tabs you want to plot (so that the tabs you want to plot become highlighted). You can press the CTRL key and then click the tabs to select individual tabs, or press the SHIFT key to select the current tab and all the other tabs between it and the tab on which you click.

If you are plotting only the Model tab as the current tab, or you select the All Layout Tabs button, you can also specify the number of copies you want to plot.

*If you select the All Layouts Tab button and not all of the layouts have previously been activated, AutoCAD displays a warning dialog box, informing you that some of the layouts are uninitialized and therefore won't be printed.*

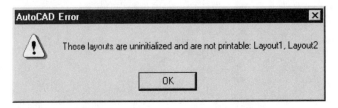

*Also note that if you have assigned different plotter configurations to various tabs that you select to plot, AutoCAD displays a message in the Plotter Configuration area of the dialog box, informing you that the plotter name varies and that the tabs are configured for a different plotter. If this isn't what you intended, you can choose a single plotter configuration for all the tabs, or change what you are plotting.*

## Stamping Each Plot

Users who plot copies of drawings while their work is still in progress sometimes find it difficult to later determine which plot is the most recent. To solve this problem, AutoCAD users have developed several methods for adding date and time stamps on their drawings. Autodesk has also included various plot stamping utilities in previous versions of AutoCAD. In AutoCAD 2002, this capability is provided as part of the PLOT command.

To turn plot stamping on, select the check box in the Plot Stamp area of the Plot Device tab. Once you turn plot stamping on, it remains on until you turn it off by clearing the check box. The plot stamp settings are not saved with the drawing.

To configure the plot stamp settings, click the Settings button to display the Plot Stamp dialog box. The dialog box contains controls for specifying the fields to be included in the plot stamp. You can also save plot stamp settings to a plot stamp parameter file (a file with a .PSS file extension) or load a previously saved PSS file.

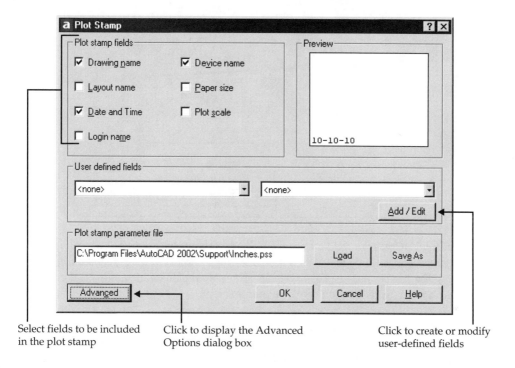

Select fields to be included in the plot stamp

Click to display the Advanced Options dialog box

Click to create or modify user-defined fields

**MOVING BEYOND THE BASICS**

**Caution**  *You may find that the options in the Plot Stamp dialog box are unavailable, or AutoCAD may report that it cannot save any changes you made to the plot stamp settings to the plot stamp parameter file (INCHES.PSS or MM.PSS) because it is tagged as a read-only file. To correct this problem, use Windows Explorer to locate these files, which are normally copied into the AutoCAD 2002/support folder. In Windows Explorer, right-click the PSS file and select Properties from the shortcut menu. In the File Properties dialog box, clear the Read-only check box so that the file is no longer marked as a read-only file.*

You can click the Advanced button to display the Advanced Options dialog box. This dialog box contains options for controlling the location, orientation, and appearance of the plot stamp. You can also save plot stamp information to a log file.

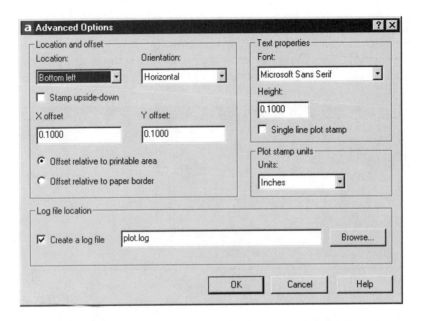

**Note** *The Preview image in the Plot Stamp dialog box shows the relative location and orientation of the plot stamp based on the settings selected in the Advanced Options dialog box. It does not display a preview of the actual plot stamp, however, nor is the plot stamp visible when you display a plot preview.*

You can also click the Add/Edit button to create user-defined fields and values, text that can optionally be included in the plot stamp. For example, you can add fields for job names or media types. Once you have configured the plot stamp, click OK to close the dialog box.

**Note** *You can also configure the plot stamp by using the PLOTSTAMP command. This command displays the same Plot Stamp dialog box displayed when you click the Settings button in the Plot Stamp area of the Plot dialog box. You cannot turn plot stamping on or off, however, by using this dialog box. That capability is only available from either the Plot dialog box or by using the command line version of the PLOTSTAMP command (by typing –PLOTSTAMP).*

## Plotting to a File

If you prefer to save your output to a file for subsequent printing rather than printing directly to the named plotter, select the Plot To File check box. If you're plotting an electronic version of your drawing (a DWF file) or have selected one of AutoCAD's raster format plotter drivers, this check box is selected automatically. When you plot to a file, you need to specify the name of the plot file you want to save as well as the location in which to save it. By default, AutoCAD provides the plot filename, using the current drawing name and appending either -Model (if plotting the Model tab) or

*-Layout* (if plotting a Layout tab), where *Layout* is the name of the layout tab. The file extension depends on the type of plot file being created. For DWF and raster formats, AutoCAD automatically appends the file extension associated with that type of file (such as .DWF for electronic plots, or .TIF for raster files saved in TIFF format). Plot files for configured plotters and system printers are given the file extension .PLT. You can change the name from the default assigned by AutoCAD simply by editing the proposed name in the File Name box prior to actually creating the plot file.

You also need to specify where you want to save the plot file. By default, AutoCAD saves the plot file in the same directory as the drawing file being plotted. You can use the controls on the Location line to specify a different location. For example, you can type a different path in the Location box or choose a previously used location from the drop-down list. Click the adjacent button (the one with three dots) to display a Browse for Folder dialog box, in which you can select a different folder on your computer or network in which to save the plot file. If you click the Browse The Web button (the rightmost button), AutoCAD displays the Browse the Web dialog box. You can then navigate to the location on the Internet to which you want to save the plot file.

**Caution**  *If you save a plot file with the same name as a previous plot file, AutoCAD displays a dialog box warning you that you are about to overwrite the existing file.*

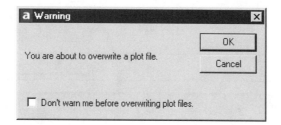

## Controlling Plot Settings

The controls on the Plot Settings tab, shown in Figure 18-2, are identical to those of the Page Setup dialog box. You use these controls to control the paper size, drawing orientation, plot area, plot scale, plot offset, and other plot options. You've already learned about these controls in the previous chapter.

## Previewing the Output

Since the layout tabs may not accurately show the way the drawing appears on the current paper—for example, lineweights may not be displayed and nonplotting layers remain visible—it's still a good idea to make sure that the drawing really will end up plotting the way that you want it to. AutoCAD provides a plot preview so that you can visually check the plot before actually outputting the drawing onto paper.

There are actually two different Preview modes: Partial Preview and Full Preview. To preview the drawing, click either of these buttons, located at the bottom of the Plot dialog box.

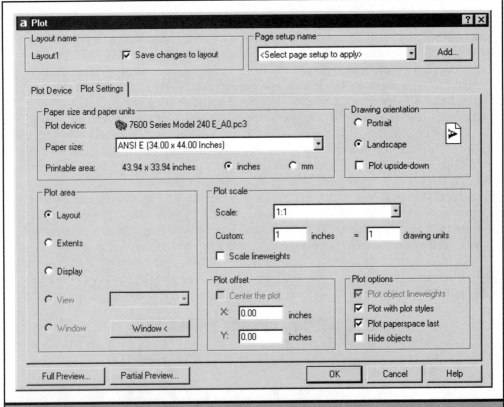

**Figure 18-2.**    *The Plot Settings tab of the Plot dialog box is identical to the Layout Settings tab of the Page Setup dialog box.*

## Viewing a Partial Preview

Clicking the Partial Preview button instructs AutoCAD to display a preview image that graphically shows the relationship of the plot area on the paper (see Figure 18-3). This display lets you quickly determine whether the plot will fit on the paper. The paper size is shown as a white rectangle, while the area containing the printed drawing is shown as a blue shaded area. A small red triangle icon indicates the drawing origin.

The Partial Plot Preview window also displays information about the paper size, printable area, and effective plotting area, as well as warnings of any problems related to the current plot. For example, if the plot won't fit on the paper, AutoCAD displays a message informing you that the plotting area exceeds the printable area. Similarly, if you move the origin, AutoCAD may warn you that the origin forced the effective area off the paper. You can then correct these problems before proceeding with your plot.

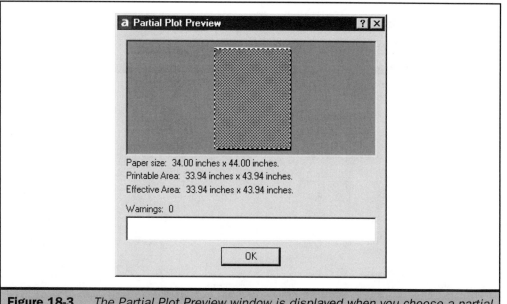

**Figure 18-3.** *The Partial Plot Preview window is displayed when you choose a partial preview.*

## Viewing a Full Preview

Clicking the Full Preview button instructs AutoCAD to display the drawing exactly as it will appear when plotted on the paper, including all lineweight and linetype settings. This preview requires a regeneration of the drawing, which can take a considerable amount of time if the drawing is very large, particularly if AutoCAD must remove hidden lines. Once the preview image has been generated, you see the preview display and its relationship to the paper, as shown in Figure 18-4.

The cursor changes to a magnifying glass with plus and minus signs, indicating that you can increase or decrease the displayed magnification of the preview image, similar to the real-time zoom. You can switch to Pan mode by right-clicking to display the shortcut menu. To exit from Preview mode, display this shortcut menu and then click Exit.

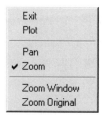

Panning and zooming the preview image has no effect on how the drawing will be plotted. These functions are provided only so that you can better observe the preview

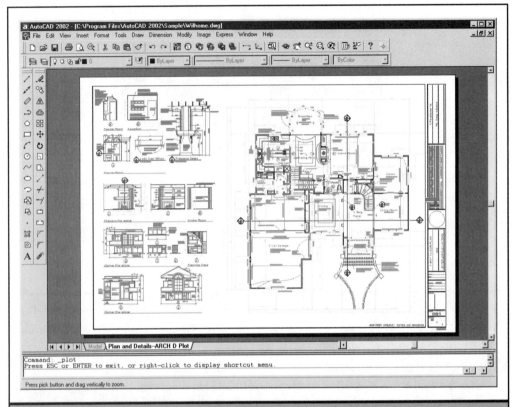

**Figure 18-4.**    *The Full Preview displays the drawing exactly as it will look on the paper when plotted.*

image. When plotting a layout, although the preview looks almost identical to the layout itself, it does enable you to make sure that AutoCAD will perform any desired hidden-line removal during plotting. You can also view the effect of any plot styles that aren't visible in the layout (because the Display Plot Styles check box on the Page Setup tab isn't selected).

*You can also display a Full Preview by doing either of the following:*

■ *From the File menu, choose Plot Preview.*

■ *At the command line, type* **PREVIEW** *and then press* ENTER.

# Using Plot Styles

As you've already learned, AutoCAD provides an object property called plot style that can change the appearance of an object when it is plotted. Every object and layer has a plot style property. By assigning a plot style to an object or layer, you can override its color, linetype, and lineweight. Plot styles also enable you to assign end, join, and fill styles, as well as control output effects such as dithering, grayscale, pen assignment, and screening.

Plot styles are defined and stored in plot style tables. By attaching different plot style tables to layouts, you can plot the same drawing in different ways. If you assign a particular plot style to a layout or object and then detach or delete the plot style table in which that style is defined, however, the plot style will have no effect on the appearance of the drawing.

## Understanding Plot Style Modes

AutoCAD actually has two different plot style modes—Color-Dependent and Named—and you should determine which style you want to use when you first create the drawing or first open a drawing in AutoCAD 2002 that was originally created in R14 or an earlier version of AutoCAD. You can control which plot style mode is used in new drawings from the Plotting tab of the Options dialog box (shown in Figure 18-5). Under Default Plot Style Behavior For New Drawings, select either Use Color Dependent Plot Styles or Use Named Plot Styles. This setting is stored in the PSTYLEPOLICY system variable.

If you select color-dependent plot style tables, you can specify the default plot style table to assign whenever you start a new drawing. Color-dependent plot style tables, as their name implies, assign plot properties based on color, similar to R14 and earlier versions of AutoCAD. A single color-dependent plot style table is assigned for the entire drawing. Because different named plot styles can be assigned to specific objects or layers, however, when using named plot styles, you can also specify the default plot style assigned for the default (0) layer and for objects. These settings are stored in the DEFLPLSTYLE and DEFPLSTYLE system variables, respectively. Clicking the Add Or Edit Plot Style Tables button displays the AutoCAD Plot Style Manager, so that you can modify existing plot styles or create new ones. You'll learn about the Plot Style Manager later in this chapter.

**Note** *Changing the default plot style behavior in the Options dialog box has no effect on the current drawing nor on any existing drawings created using AutoCAD 2000 or newer. This setting only affects new drawings and drawings created in pre-AutoCAD 2000 versions when they are first opened and then saved in AutoCAD 2000 format. Since named plot styles didn't exist prior to AutoCAD 2000, by default, drawings created using earlier versions use color-dependent plot styles. However, when you open a pre-AutoCAD 2000 drawing and have already selected Use Named Plot Styles in the Options dialog box, the drawing is automatically converted from color-dependent styles to named plot styles. You can also use the CONVERTPSTYLES command to convert a drawing that uses color-dependent plot styles into a drawing that uses named plot styles, and vice versa.*

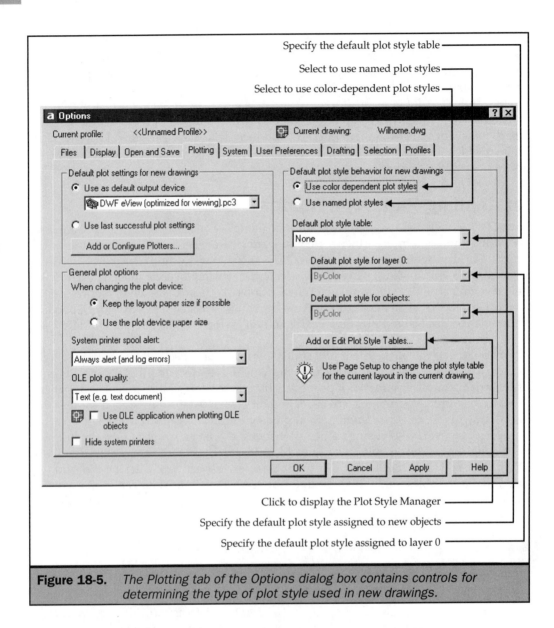

**Figure 18-5.** The Plotting tab of the Options dialog box contains controls for determining the type of plot style used in new drawings.

## Understanding Color-Dependent Plot Style Mode

Color-dependent plot styles, as their name implies, are based on the color of the object within the AutoCAD drawing. Since a layer or object can be any one of 255 possible colors, there are 255 color-dependent plot styles within a color-dependent plot style table, one for each possible color in the drawing. In a color-dependent plot style table, you can control the settings—such as color, grayscale, pen number, linetype, lineweight, and fill type—for each color individually, but you can't add, delete, or rename the color-dependent plot styles. These settings are controlled from the Plot Style Table Editor, shown in Figure 18-6. When using color-dependent plot styles, the only way to change the plot style used for a particular object or layer is to change the color assigned to that object or layer. AutoCAD 2002 comes with a number of predefined color-dependent plot style tables, and you can create and save your own. Color-dependent plot style tables are stored in files with the extension .CTB.

If you are familiar with using pen assignments in earlier versions of AutoCAD, you already know the basics of using color-dependent plot styles. In versions of AutoCAD prior to AutoCAD 2000, you used the color of an object to control its pen number, linetype, and lineweight. In AutoCAD 2002, you simply have more properties that you can control.

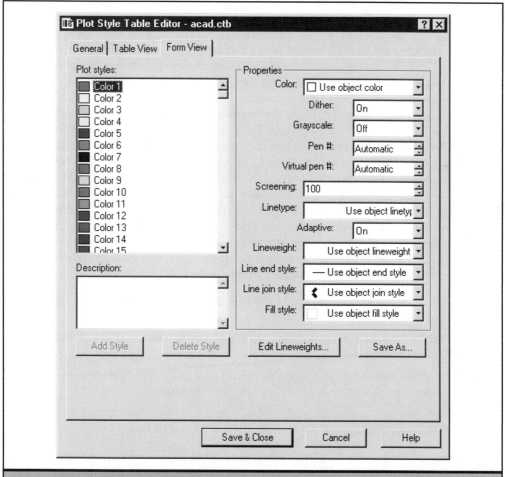

**Figure 18-6.**    *The Plot Style Table Editor for color-dependent plot styles*

Using color-dependent plot styles limits the ways in which you can use color in your drawings, because by associating color with a specific pen, you lose the ability to work with color independent of lineweight, linetype, and the other plot properties. However, because users are likely to be familiar with plotting from previous versions of AutoCAD, by default, AutoCAD 2002 uses color-dependent plot styles.

 *If you've saved pen settings to PCP or PC2 files in earlier versions of AutoCAD, you can use the Add-A-Plot Style Table wizard to import those settings. You'll learn more about this wizard later in this chapter.*

## Understanding Named Plot Style Mode

Named plot styles remove the limitation inherent when working with color-dependent plot styles. With named plot styles, you can assign plot styles to layers or individual objects independent of their color. With named plot styles, you can use an object's color in the same way that you use any other property. In a named plot style table, you can create an unlimited number of named plot styles and control the settings for each style individually. AutoCAD 2002 comes with several predefined named plot style tables, and you can create and save your own. Named plot style tables are stored in files with the extension .STB.

As with color-dependent plot styles, you control these settings from the Plot Style Table Editor. When you first display the Plot Style Table Editor for named plot styles, however, AutoCAD displays the Table View tab, as shown in Figure 18-7. This is simply a different way to view plot styles, and you can easily switch to this tab when working with color-dependent plot styles. Notice, however, that when working with named plot styles, the Add Style button is active, indicating that you can add new plot styles to the current plot style table. In addition, if you select a plot style, the Delete Style button also becomes active, allowing you to delete existing plot styles. These two functions are not available when working with color-dependent plot styles.

Regardless of whether you're using color-dependent or named plot styles, you can display the Plot Style Editor by doing any of the following:

- From the Plot Device tab of either the Page Setup or Plot dialog box, select the plot style table from the drop-down list, and then click Edit.

- From the Current Plot Style dialog box (displayed by the PLOTSTYLE command), select the plot style table from the drop-down list, and then click Editor.

- In the Plot Style Manager, double-click an STB or CTB file (or right-click the file and then choose Open from the shortcut menu).

- When creating a new plot style table using the Add-A-Plot Style Table wizard, on the Finish page, click the Plot Style Table Editor button.

You'll learn more about the Plot Style Editor and these other commands later in this chapter.

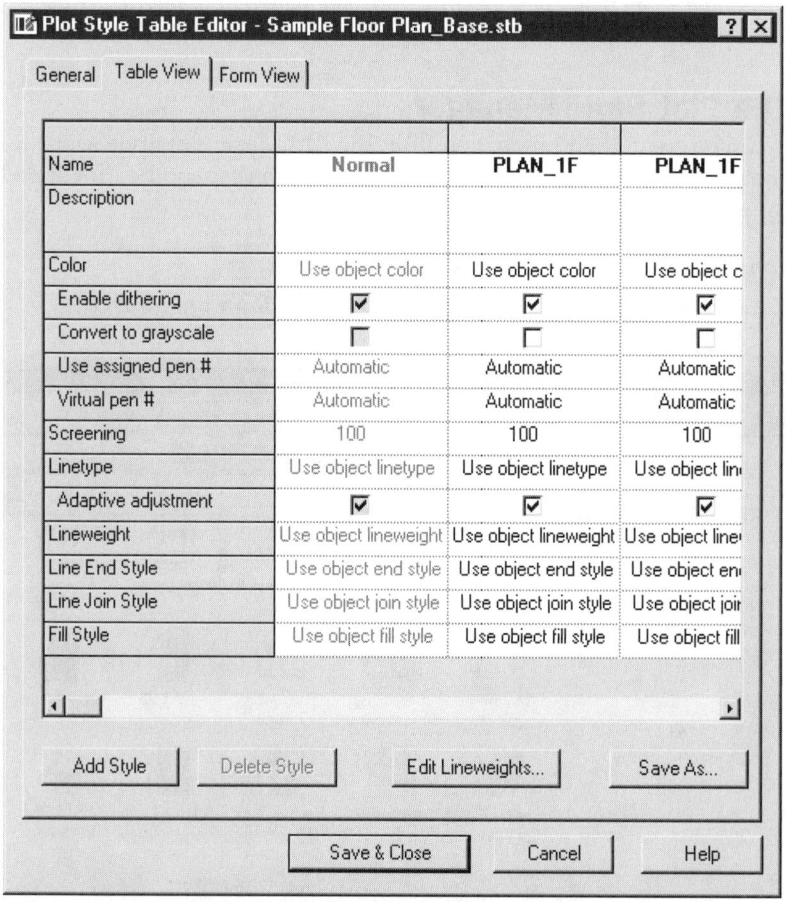

**Figure 18-7.** *AutoCAD displays the Table View tab of the Plot Style Editor when using named plot styles.*

## Displaying Plot Styles in a Drawing

As you learned in the previous chapter, when you use plot styles, you have the option of displaying objects in the drawing based on the plot styles applied to them. The application of plot styles to the drawing is determined by the Display Plot Styles check box on the Plot Device tab of the Page Setup dialog box. When this check box is selected, the effect of plot styles is visible in the selected layout of the drawing.

Displaying plot styles in the drawing can slow down AutoCAD's performance, however. For that reason, you may decide not to display plot styles in the drawing. You can still see the effect of plot styles prior to plotting by displaying a Full Preview of the current tab.

**Note** *When you display plot styles in a drawing, you may need to regenerate the drawing before the effect of the plot styles becomes visible in the drawing.*

## Using the Plot Style Manager

Plot styles can be created and managed from the Autodesk Plot Style Manager, a window into AutoCAD's Plot Styles directory. You can display the Plot Style Manager by doing any of the following:

- From the File menu, choose Plot Style Manager.
- At the command line, type **STYLESMANAGER** and press ENTER.

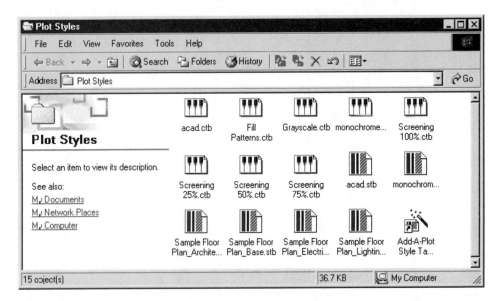

You can also display the Autodesk Plot Style Manager from the Windows Control Panel or by clicking the Add Or Edit Plot Style Tables button on the Plotting tab of the Options dialog box. The Plot Style Manager's Add-A-Plot Style Table wizard greatly simplifies the creation of new plot style tables by stepping you through the creation process. You'll learn about this wizard in the next section of this chapter.

## Creating a Plot Style Table

You create new plot style tables by using the Add-A-Plot Style Table wizard. This wizard guides you through the process of creating a new plot style table. You can use it to create both color-dependent and named plot style tables. When you create a color-dependent plot style table, the resulting table contains 255 plot styles, named Color 1 through Color 255. Because each plot style is tied directly to an AutoCAD Color Index (ACI) number, you can't add, delete, or rename these styles. When you create a named plot style table, however, the table is initially created with a single default style named Normal. You can then add additional plot styles to the table. Each plot style you add to a named plot style table is initially named Style 1, Style 2, and so on. After adding each plot style, however, you can change its name to something more descriptive, such as AS-BUILT, ELECTRICAL PLAN, and so on.

The Add-A-Plot Style Table wizard enables you to create a new plot style from scratch or use an existing plot style table as the basis for the new table. You can also import the pen table properties from an AutoCAD R14 configuration file or from an existing PCP or PC2 file.

To start the Add-A-Plot Style Table wizard, do one of the following:

- From the Tools menu, choose Wizards | Add Plot Style Table.

- In the Autodesk Plot Style Manager, double-click the Add-A-Plot Style Table Wizard icon.

The Add-A-Plot Style Table wizard runs as a separate application, in its own window. When you start the wizard, the first page provides an introduction that explains the wizard's basic operation. Click Next to display the Begin page, shown in Figure 18-8.

On the Begin page, you need to choose how you want to create the new plot style table. Choose one of the following options:

- **Start from Scratch**   Creates a new plot style table.

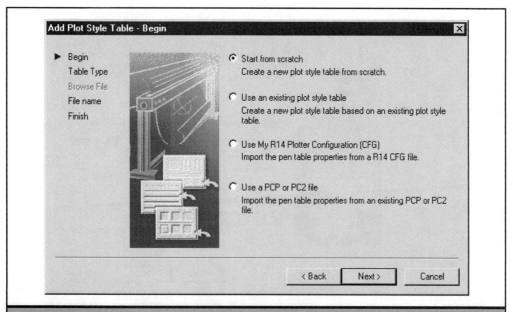

**Figure 18-8.**   *The Begin page of the Add-A-Plot Style Table wizard lets you choose how you want to create the new plot style table.*

- **Use an Existing Plot Style Table**   Creates a new plot style table by making a copy of an existing plot style table. The new plot style table contains all the plot styles from the original. The wizard skips the Table Type step, since the determination of whether the new table is color-dependent or uses named styles is made automatically based on the type of the existing plot style table.

- **Use My R14 Plotter Configuration (CFG)**   Creates a new plot style table using the pen assignments stored in the ACADR14.CFG file. Choose this option if you want to import settings but don't have a PCP or PC2 file.

- **Use a PCP or PC2 File**   Creates a new plot style table using the pen assignments stored in an existing PCP or PC2 file.

*You can also create a new plot style table based on an existing table by clicking the Save As button in the Plot Style Table Editor dialog box. You'll learn more about this dialog box later in this chapter.*

After you choose the method by which you will create the new plot style table, click Next. For all the choices except when using an existing plot style table, the wizard displays the Table Type page, which appears similar to Figure 18-9. This page varies somewhat based on the method you're using to create the new plot style table. Select the appropriate option to specify whether you're creating a color-dependent plot style table or a named plot style table, and then click Next.

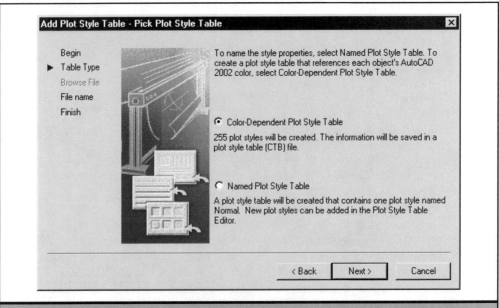

**Figure 18-9.**   *The Pick Plot Style Table page lets you choose the type of plot style table you are creating.*

If you're creating a new plot style table from scratch, the wizard skips the Browse File page. Otherwise, the Browse File Name page appears. This page varies depending on the method you're using to create the new plot style table. If you're using an existing plot style table, the wizard displays a dialog box with a drop-down list containing the names of all the plot style tables saved in AutoCAD's Plot Styles directory. To base the new plot style on one of these styles, simply select the style from the list. If the plot style on which you're basing the new style is saved in a different directory, click the Browse button and then use the Select File dialog box to locate the plot style table (either a CTB or an STB file). When you base a new plot style table on an existing PCP or PC2 file, the Browse File Name page is very similar.

If you're creating the new plot style from the R14 configuration file, the dialog box contains a field and an adjacent Browse button for specifying the CFG file. Because the CFG file can contain pen mapping information for more than one plotter, a drop-down list lets you choose the plotter for which you want to import the pen mapping information.

After specifying the file on which you're basing the new plot style, click Next. On the next page of the wizard, the File Name page, you enter the filename for the new plot style table you're creating. The new file is automatically assigned the file extension .CTB if you're creating a color-dependent plot style table, or .STB if the new style table uses named plot styles. Then, click Next.

On the Finish page of the Add-A-Plot Style Table wizard, shown in Figure 18-10, you can click the Plot Style Table Editor button to immediately display the Plot Style

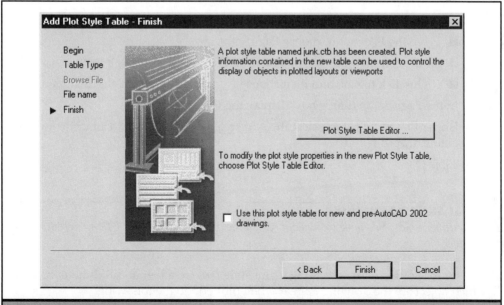

**Figure 18-10.** *The Finish page lets you access the Plot Style Table Editor and attach the new plot file to all new drawings.*

Table Editor, so that you can modify the existing plot styles in the table. In the case of named plot styles, you can also add, delete, or rename the styles. This final page also contains a check box. Unless you want the plot style table you just created to be attached to all new drawings you create and all pre-AutoCAD 2000 drawings you open in AutoCAD 2002, this check box should remain cleared.

Click Finish to save the new plot file table to AutoCAD's Plot Styles directory. The resulting CTB or STB file appears in the Plot Style Manager.

**Note** *You can also start a slightly modified version of the Add-A-Plot Style wizard either by clicking the New button on the Plot Device tab of the Plot dialog box or from the Wizards submenu of the Tools menu. If the current drawing uses color-dependent plot styles, the submenu selection says Add Color-Dependent Plot Style Table, whereas if the drawing uses named plot styles, the selection appears as Add Named Plot Style Table. The wizard preselects the type of style table you're creating based on the type of plot style table already used in the drawing.*

## Attaching Plot Style Tables to Layouts

You can attach any plot style table to the Model tab or to any layout tab by using the Page Setup dialog box. To attach a plot style table:

1. Choose the Model tab or the layout tab to which you want to attach the plot style table.

2. Display the Page Setup dialog box by doing any of the following:

   ■ On the Layouts toolbar, click Page Setup.

   ■ From the File menu, choose Page Setup.

   ■ At the command line, type **PAGESETUP** and then press ENTER.

   ■ Right-click the tab and then choose Page Setup from the shortcut menu.

3. In the Page Setup dialog box, display the Plot Device tab.

4. Under Plot Style Table, select the plot style table you want to attach from the Name drop-down list.

5. Click OK.

*LEARN BY EXAMPLE*
*Attach a plot style table to a layout. Open the drawing SAMPLE-01 on the companion web site.*

To see for yourself how attaching a plot style table to a layout can change the way in which objects in the drawing appear when plotted, do the following:

1. Select the Architectural-monochrome tab.

2. Right-click the Architectural-monochrome tab and then choose Page Setup from the shortcut menu.

3. On the Plot Device tab, under Plot Style Table, choose monochrome.ctb from the drop-down list to attach a color-dependent plot style table to produce monochrome plots.

4. Click to select the Display Plot Styles check box, so that the plot style table settings are visible in the Architectural Title Block tab.

5. Click OK.

On your own, select the Architectural-color tab. Notice that the plot style you applied to the Architectural-monochrome tab only affects that tab. You can attach a different plot style table to each tab in order to change the appearance of the drawing for different purposes. For example, in an architectural renovation project, you may attach one plot style to a layout so that new construction is plotted with a dark, solid line, while existing conditions are shown using a screened effect. On another layout of the same plan, you may attach a plot style that displays demolition work using solid lines, and displays new construction with a screened background.

**Note** *When you attach a plot style table to the Model tab, AutoCAD displays a dialog box asking if you want to assign the plot style table to all the layouts. Click Yes to apply the style table to all the layouts, or click No to apply the style table only to the Model tab. Remember that you can always change the style table assigned to any tab, including the Model tab.*

# Adding, Renaming, and Deleting
# Plot Styles in a Named Plot Style Table

For named plot style tables, you can add new plot styles to a plot style table and delete named styles from the plot style table. When you add a new plot style to a table, AutoCAD initially assigns it the name Style 1, Style 2, and so on. You can rename any named plot style to something more descriptive.

**Note** *You can't add, rename, or delete plot styles in color-dependent plot style tables. You can change the properties assigned to color-dependent plot styles, however, as you will learn later in this chapter.*

To add a plot style to a named plot style table:

1. Display the Plot Style Table Editor by doing one of the following:

   ■ From the Plot Device tab of either the Page Setup or Plot dialog box, select the plot style table from the drop-down list, and then click Edit.

   ■ From the Current Plot Style dialog box (displayed by the PLOTSTYLE command), select the plot style table from the drop-down list, and then click Editor.

■ In the Plot Style Manager, double-click an STB or CTB file (or right-click the file and then choose Open from the shortcut menu).

2. In the Plot Style Table Editor, choose either the Table View tab or Form View tab.

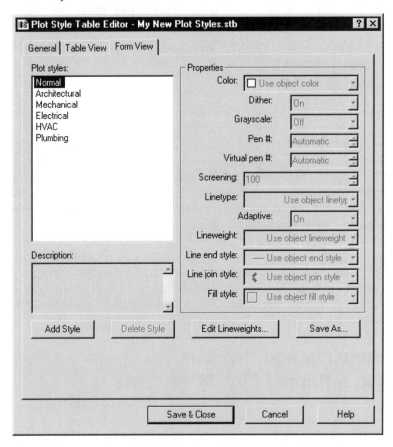

3. Click Add Style, or right-click in the dialog box and choose Add Style from the shortcut menu.

4. In the Add Plot Style dialog box, specify the name for the new plot style (or accept the default), and then click OK.

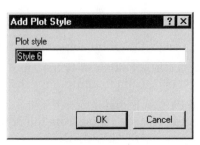

**Note**

*The Add Plot Style dialog box only appears if you click Add Style in Step 3 when viewing the Form View tab of the Plot Style Table Editor. If you click the Add Style button when viewing the Table View tab, a new style is automatically added and the name immediately highlighted so that you can type a different style name.*

5. Click OK to close the Plot Style Table Editor.

To delete a plot style from a named plot style table, select the plot style in either the Table View or Form View tab of the Plot Style Table Editor, and then click Delete or right-click and choose Delete from the shortcut menu.

**Tip**

*To select the plot style in the Table View tab, click the header cell above the plot style name. In the Form View tab, select the style name.*

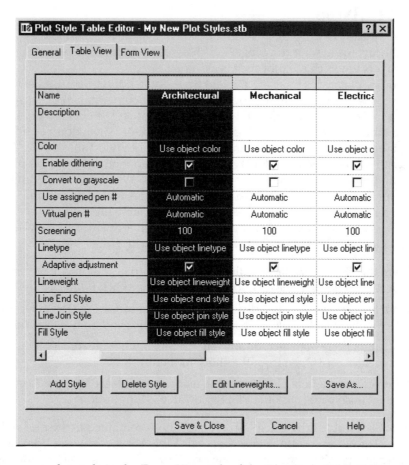

To rename a plot style in the Form View tab of the Plot Style Table Editor, select the plot style and then either right-click and choose Rename Style from the shortcut menu, or press F2. To rename a plot style in the Table View tab, click the style name and then edit the name as you would edit text in any text field in a dialog box. You can also rename named plot styles by using the RENAME command.

**Tip** *You can also hide plot styles in the Table View tab of the Plot Style Table Editor dialog box. To hide one or more plot styles, select the styles you want to hide. You can select multiple adjacent styles by pressing SHIFT while selecting the plot styles. Then, right-click and choose Hide from the shortcut menu. You can repeat this procedure to hide other plot styles. To unhide a hidden plot style, select the plot style adjacent to a hidden plot style, right-click, and choose Unhide from the shortcut menu. You can also drag the vertical divider in the column header above the plot style name to 0-resize the columns in the Table View tab, or drag the horizontal divider in the row header on the left to resize the rows.*

## Editing Plot Styles

You can modify any of the plot styles in the plot style table attached to the Model tab, or any of the layouts in your drawing. Any changes you make to a plot style affect the objects or layers to which the plot style is assigned.

You modify plot styles by using the Plot Style Table Editor. Because the Plot Style Table Editor is actually a separate program, you can open more than one instance of the editor at a time. For example, in the Autodesk Plot Style Manager, double-click a plot style table to display it in the Plot Style Table Editor. Then, double-click another plot style table to open it in another instance of the Plot Style Table Editor. When working with color-dependent plot style tables, you can modify any of the plot styles and copy-and-paste settings between plot style tables. When working with named plot styles, in addition to being able to modify plot styles within a table and copy plot styles between tables, you can also add, delete, and rename plot styles.

Regardless of whether you're using color-dependent or named plot styles, you can display the Plot Style Editor by doing any of the following:

■ From the Plot Device tab of either the Page Setup or Plot dialog box, select the plot style table from the drop-down list, and then click Edit.

■ From the Current Plot Style dialog box (displayed by the PLOTSTYLE command), select the plot style table from the drop-down list, and then click Editor.

■ In the Plot Style Manager, double-click a STB or CTB file (or right-click the file and then choose Open from the shortcut menu).

■ When creating a new plot style table using the Add-A-Plot Style Table wizard, on the Finish page, click the Plot Style Table Editor button.

The Plot Style Table Editor displays the plot styles contained in the plot style table you opened. The editor has three tabs: General, Table View, and Form View. You've already learned a bit about the Table View and Form View tabs. The General tab, shown in Figure 18-11, displays the plot style table name, an optional description, the

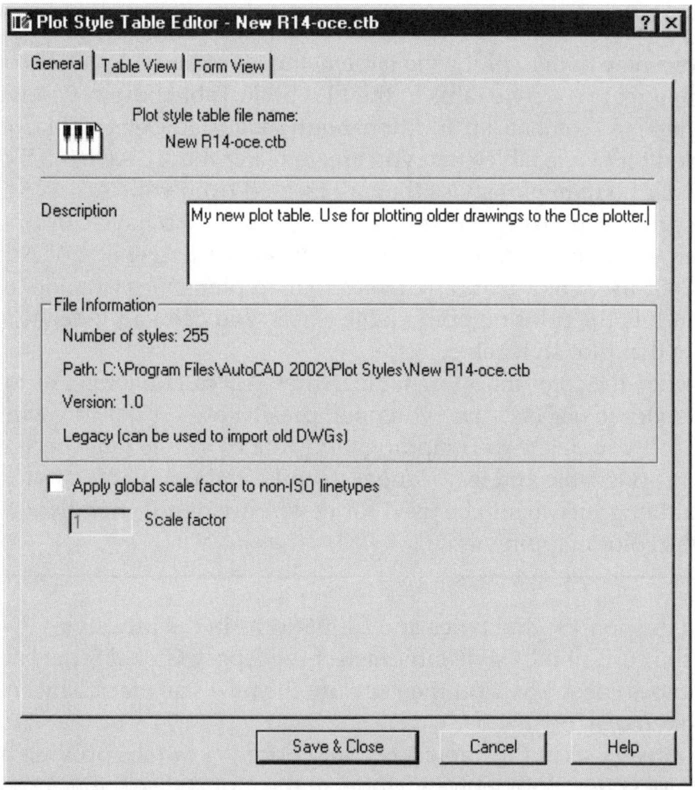

**Figure 18-11**    *The General tab of a typical plot style table*

number of styles contained in the table, the complete path for the CTB or STB file, a version number, and an indication of whether the style table is a regular (named) or legacy (color-dependent) style table as well as whether it can be used to import drawings created using earlier versions of AutoCAD. You can also apply scaling to non-ISO lines and fill patterns.

To create or change the description of the plot style table:

1. Open the plot style table whose description you want to change.

2. In the Plot Style Table Editor, select the General tab.

3. Type the new description in the Description field.

4. Click Save & Close to save the changes to the plot style table.

> ## Deleting the AutoCAD Release 14 Color Mapping Table
> If you create a new named plot style table using an older R14 plotter configuration and then open that plot style table in the Plot Style Table Editor, you will notice that the General tab contains an additional button labeled Delete R14 Color Mapping Table. Named plot style tables that you create using the ACADR14.CFG file or a PCP or PC2 file contain plot styles that are created from your AutoCAD Release 14 pen mappings. Color-dependent plot style tables also have color mapping tables. When you open a drawing created with an earlier version of AutoCAD, AutoCAD 2002 uses color mapping tables to map plot styles to colors or objects of each color. While the color mapping table exists, you can't add, delete, or rename plot styles in that plot style table.
>
> If you delete the color mapping table, AutoCAD can no longer automatically assign plot styles to objects when you open pre-AutoCAD 2000 drawing files for the first time. If you delete the mapping table, the plot style table becomes an ordinary plot style table and is no longer useful for applying plot styles to older drawings, although it can still be used for new drawings. Generally, you should not delete the color mapping table.

To scale all the non-ISO linetypes and fill patterns in the plot styles of objects controlled by the current plot style table, select the Apply Global Scale Factor To Non-ISO Linetypes check box, and then specify the new scale factor by entering it in the Scale Factor box.

As you've already seen, the Table View and Form View tabs provide two different ways to view plot styles. Both views include all the plot styles in the current plot style table. The Table View tab is best for working with plot style tables containing a relatively small number of styles, while the Form View tab is more convenient for working with plot style tables containing a large number of plot styles, such as color-dependent plot style tables. In the Table View tab, each plot style appears in its own column, with style properties arranged in rows. The Form View tab displays a list containing all the plot styles in the table. The adjacent Properties area shows the properties assigned to the style selected in the Plot Styles list.

## Copying a Plot Style or Plot Style Setting to Another Plot Style Table

You can use standard Windows cut-and-paste functions to copy a plot style or plot style setting from one plot style table to another. To copy a style or setting from one table to another:

1. From the File menu, choose Plot Style Manager.

2. In the Autodesk Plot Style Manager, open the plot style table from which you want to copy the style or setting by double-clicking the CTB or STB file.

3. Open the plot style table to which you want to copy the style or setting by double-clicking the CTB or STB file.

4. In the plot style table you're copying from, select the Table View tab.

5. Select the plot style setting or settings you want to copy. (*Note*: You can copy multiple adjacent settings by selecting the first setting in the group, pressing the SHIFT key, and then selecting the last setting in the group.) To select the entire plot style, click the header cell above the plot style name.

6. Right-click and choose Copy from the shortcut menu.

7. Switch to the plot style table you're copying to, select the Table View tab, and then select the plot setting, group of plot settings, or the column corresponding to the plot style you want to replace. (*Note*: To create a new plot style using the styles copied from the other table, first add a new plot style and then select the column corresponding to the new plot style.)

8. Right-click and choose Paste to paste the settings or style to the table. To paste one or more settings to all the styles in the table, right-click and choose Paste To All Styles.

9. Click Save & Close to save the changes to the target plot style table. You can click Cancel to close the other plot style table.

| Tip | *You can also use drag-and-drop to copy a plot style from one table to another.* |

## Changing a Plot Style Description

Each plot style in both color-dependent and named plot style tables can have a description. You can change the plot style description from either the Table View or Form View tab of the Plot Style Table Editor dialog box. To change the plot style description, display the Plot Style Table Editor for the plot style whose descriptions you want to change, and then do either of the following:

■ In the Table View tab, click in the Description field for the style whose description you want to change, and then type the new description.

■ In the Form View tab, select the plot style whose description you want to change, click in the Description box, and then type the new description.

## Assigning Plot Style Color

By assigning a plot style color to a plot style, you can override the color assigned to an object or layer when you plot the drawing. Within the Plot Style Table Editor, you can change the color assigned to both color-dependent and named plot styles. With the exception of AutoCAD's monochrome plot style tables, the default color setting is to plot objects using the color in which they appear in the drawing.

To change the color assigned to a plot style, do either of the following:

- In the Table View tab, click in the Color field for the style whose color you want to change, and then select the new color from the drop-down list.
- In the Form View tab, select the plot style whose color you want to change in the Plot Styles list, and then, in the Properties area, select the new color from the Color drop-down list.

## Enabling Dithering

*Dithering* is the combining of colored dots to give the impression of displaying more colors than are actually available. Some plotters use dithering to plot drawings with more colors than it would otherwise be able to print based on the ink colors available in the plotter. By default, dithering is enabled for all plot styles. If your plotter doesn't support dithering, this setting is ignored.

In some instances, the use of dithering results in some plotted lines appearing to have been plotted with a dotted linetype. If this occurs, you can turn dithering off to eliminate this condition. You may also turn dithering off to make dim colors more visible. When you turn dithering off, AutoCAD maps colors to the nearest color, thus reducing the number of colors the plotter is capable of printing.

To disable dithering, do either of the following:

- In the Table View tab, clear the Enable Dithering check box for the style you want to change.
- In the Form View tab, select the plot style whose setting you want to change, and then select Off from the Dither drop-down list.

## Converting to Grayscale

The Grayscale setting enables you to convert the specified color to grayscale, if the plotter supports grayscale. With the exception of AutoCAD's Grayscale plot style table, this setting is normally off.

To enable grayscale, do either of the following:

- In the Table View tab, select the Convert To Grayscale check box for the style you want to change.
- In the Form View tab, select the plot style whose setting you want to change, and then select On from the Grayscale drop-down list.

## Assigning Pen Numbers

The Pen Number setting enables you to assign a pen to be used for each plot style in the plot style table by selecting from a range of pen numbers from 0 to 32. If you specify a value of 0, AutoCAD updates the Pen Number setting to read Automatic and then uses the information provided under Physical Pen Characteristics in the Plotter Configuration Editor to select the pen closest to the color of the object you are plotting.

When you add or edit a plotter configuration (a PC3 file) for a pen plotter, you can specify information about the color, speed, and width of each pen. This information is provided under Physical Pen Characteristics on the Device and Document Settings tab of the Plotter Configuration Editor. You can then assign a pen number to a plot style to instruct AutoCAD to use that pen when plotting.

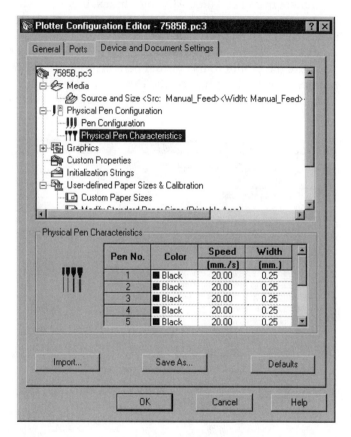

To assign a pen number to a plot style, do either of the following:

- In the Table View tab, select the Use Assigned Pen # field for the style you want to change, and then either type a new pen number or click the arrows to select a new pen number.

- In the Form View tab, select the plot style whose pen number you want to change, and then either type a new pen number or click the arrows to select a new pen number.

## Using Virtual Pens

Although many plotters don't actually use pens, some nonpen plotters can simulate pen plotters by using virtual pens. For some plotters, the Virtual Pen setting enables you to control the pen's width, fill pattern, end style, join style, color, and screening from the front panel on the plotter. The Virtual Pen setting in the plot style enables you to assign a virtual pen number for each plot style in the plot style table by selecting from a range of pen numbers from 1 to 255. If you specify a value of 0, AutoCAD updates the Virtual Pen setting to read Automatic, which specifies that AutoCAD should make the virtual pen assignment from the AutoCAD Color Index. If your plotter is configured to use virtual pens, the Virtual Pen setting overrides the other settings for the plot style. Otherwise, both the virtual pen and physical pen number are ignored.

When you add or edit a plotter configuration (a PC3 file) for a nonpen plotter, you may be able to configure the plotter to use virtual pens. You'll find the controls for this setting under Graphics | Vector Graphics on the Device and Document Settings tab of the Plotter Configuration Editor. Under Color Depth, select 255 Virtual Pens from the drop-down list.

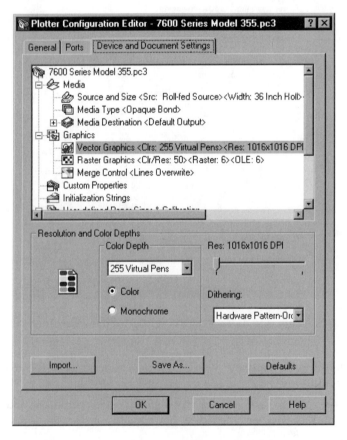

To specify a virtual pen number, do either of the following:

- In the Table View tab, select the Virtual Pen # field for the style you want to change, and then either type a new virtual pen number or click the arrows to select a new virtual pen number.

■ In the Form View tab, select the plot style whose virtual pen number you want to change, and then either type a new virtual pen number or click the arrows to select a new virtual pen number.

## Using Screening

Screening specifies a color intensity setting that determines the amount of ink AutoCAD places on the paper when plotting. You can select screening intensities in the range from 0 to 100 percent for each plot style in the plot style table. A Screening value of 0 reduces the color to white, meaning that no ink is placed on the paper. A value of 100 causes the color to print at its full intensity.

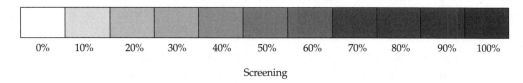

Screening

To specify a screening value, do either of the following:

■ In the Table View tab, select the Screening field for the style you want to change, and then either type a screening value or click the arrows to select a new screening value.

■ In the Form View tab, select the plot style whose screening value you want to change, and then either type a screening value or click the arrows to select a new screening value.

## Assigning Plot Style Linetypes

Although you normally assign a linetype to objects or layers in your drawing, you can use the plot style Linetype setting to override the linetype when you plot your drawing. You can assign a linetype for each plot style in the plot style table. The default setting is to use the linetype assigned in the drawing.

To specify a plot style linetype, do either of the following:

■ In the Table View tab, select the Linetype field for the style you want to change, and then select the desired linetype from the drop-down list.

■ In the Form View tab, select the plot style whose linetype you want to change, and then select the desired linetype from the drop-down list.

## Completing Linetype Patterns

The Adaptive Adjustment setting adjusts the scale of linetypes to complete the linetype pattern. When this setting is turned off, lines drawn using a linetype might end before the end of the pattern, but the scale of the linetype will be preserved. If this setting is turned on, the scale of the linetype is adjusted so that linetype patterns are complete for each line drawn with that linetype.

To use adaptive adjustment, do either of the following:

- In the Table View tab, select the Adaptive Adjustment check box for the style you want to change.
- In the Form View tab, select the plot style whose setting you want to change, and then select On from the Adaptive drop-down list.

## Assigning Plot Style Lineweight

Although you normally assign the lineweight to objects or layers in your drawing, you can use the plot style Lineweight setting to override the lineweight when you plot your drawing. You can assign a lineweight for each plot style in the plot style table. The default setting is to use the lineweight assigned in the drawing.

To specify a plot style lineweight, do either of the following:

- In the Table View tab, select the Lineweight field for the style you want to change, and then select the desired lineweight from the drop-down list.
- In the Form View tab, select the plot style whose lineweight you want to change, and then select the desired lineweight from the drop-down list.

## Assigning a Line End Style

You can assign a line end style to each plot style in the plot style table. The default setting is to use the object end style. The Line End Style drop-down list provides four additional selections: Butt, Square, Round, and Diamond. If you select one of these options, the setting overrides the object's line end style when you plot your drawing.

To specify a line end style, do either of the following:

- In the Table View tab, select the Line End Style field for the style you want to change, and then select the desired style from the drop-down list.
- In the Form View tab, select the plot style whose line end style you want to change, and then select the desired style from the drop-down list.

## Assigning a Line Join Style

You can also assign a line join style to each plot style in the plot style table. The default setting is to use the object join style. The Line Join Style drop-down list provides four additional selections: Miter, Bevel, Round, and Diamond. If you select one of these options, the setting overrides the object's line join style when you plot your drawing.

To specify a line join style, do either of the following:

- In the Table View tab, select the Line Join Style field for the style you want to change, and then select the desired style from the drop-down list.
- In the Form View tab, select the plot style whose line join style you want to change, and then select the desired style from the drop-down list.

## Assigning Fill Style

AutoCAD normally plots objects using the fill style specified in the drawing. You can apply a fill style setting to plot styles in the plot style table, however, to override the object's fill style when you plot your drawing. The fill style is applied to polylines, donuts, solids, 3-D faces, and objects hatched with a solid hatch pattern. AutoCAD provides nine additional fill styles, shown in Figure 18-12.

## Editing Lineweights

There are 28 lineweights available for use in the plot style table, including the Use Object Lineweight setting. You can modify these lineweights to create a value not currently available, but you can't add or delete them. If you change a lineweight value, any other plot styles that use the lineweight are also changed.

To edit a lineweight, click the Edit Lineweights button. AutoCAD displays the Edit Lineweights dialog box. The Lineweights list displays the 27 lineweight values and indicates which ones are already in use in the drawing. The Units For Listing controls determine whether the lineweights are displayed in millimeters or inches.

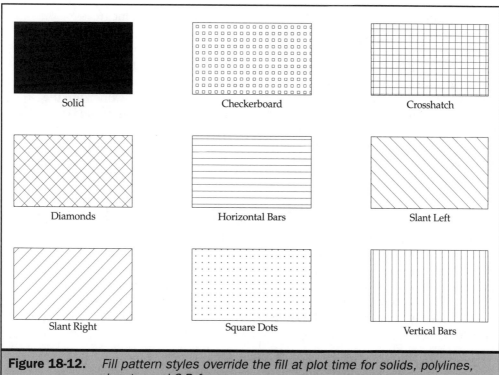

Solid     Checkerboard     Crosshatch

Diamonds     Horizontal Bars     Slant Left

Slant Right     Square Dots     Vertical Bars

**Figure 18-12.** *Fill pattern styles override the fill at plot time for solids, polylines, donuts, and 3-D faces.*

To edit a lineweight, select it in the list, and then do any of the following:

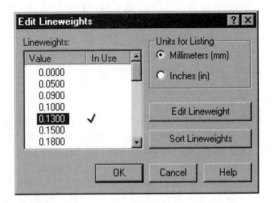

■ Click Edit Lineweight.

■ Press ENTER.

■ Press F2.

■ Right-click and choose Edit Lineweight from the shortcut menu.

■ Click the lineweight value a second time.

You can then type a new lineweight value. The value you enter is rounded and displayed with a precision of four decimal places. Lineweight values must be 0 or positive numbers. A lineweight of 0 causes AutoCAD to plot the line with the smallest width that the plotter can create. The maximum lineweight value is 200 millimeters (7.874").

After changing one or more lineweights, you can click the Sort Lineweights button to re-sort the list in size order. Click OK to close the Edit Lineweights dialog box and return to the Plot Style Table Editor.

# Changing the Plot Style Property for an Object or Layer

As you've already learned, every object in your drawing has a plot style property. For drawings that use color-dependent plot styles, the plot style property is always ByColor. In drawings that use named plot styles, however, you can assign a plot style to each layer and also assign a plot style to individual objects, similar to the way in which you can assign a color to a layer but can also assign colors on a per-object basis.

For drawings that use named plot styles, you can assign any of the following plot styles to objects:

■ **Normal**   Uses the object's default properties

■ **BYLAYER**   Uses the plot style assigned to the layer on which the object is drawn

■ **BYBLOCK**   Uses the plot style of the block that contains the object

■ **Named plot style**   Uses the plot style you specify for the object, where *named plot style* is the actual plot style name

You can change the plot style assigned to an object from the Properties window, the same way you would change any other property. You learned about changing properties in Chapter 11. The default plot style for objects is BYLAYER.

Each layer in a drawing also has a plot style property. For drawings that use color-dependent plot styles, the plot style property always matches the color of the layer. For example, if you create a layer and assign it the color red (color number 1), the layer's plot style will be Color_1. You can't change the plot style assigned to layers in drawings that use color-dependent plot styles. The style is always tied to the color assigned to the layer. For drawings that use named plot styles, however, you can also specify the plot style assigned to the layer. By default, AutoCAD assigns the plot style NORMAL. When the drawing is plotted, objects drawn on the layer plot using the other properties assigned to the object. You can change the plot style assigned to a layer from the Layer Properties Manager dialog box, which you learned about in Chapter 8.

For drawings that use named plot styles, you can also set the current plot style, which then overrides the layer plot style setting. New objects are then created using the current plot style. You can set the current plot style by selecting it from the Plot Style Control drop-down list on the Object Properties toolbar. You can also use the PLOTSTYLE command to set the current plot style. The command displays the Current Plot Style dialog box, shown in Figure 18-13. To change the current plot style, select it

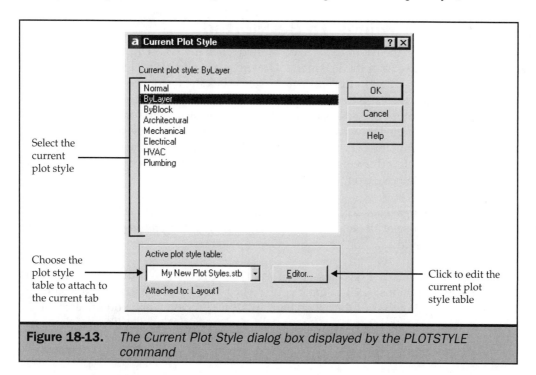

Select the current plot style

Choose the plot style table to attach to the current tab

Click to edit the current plot style table

**Figure 18-13.**   *The Current Plot Style dialog box displayed by the PLOTSTYLE command*

from the list and then click OK. You can also attach a different plot style table to the current tab by choosing it from the Active Plot Style Table drop-down list and display the Plot Style Table Editor to edit the current active plot style table. To start the command, type **PLOTSTYLE** and press ENTER.

*The PLOTSTYLE command is not available in drawings that use color-dependent plot styles.*

## Using the Batch Plot Utility

AutoCAD 2002 is supplied with a Batch Plot Utility that creates a list of drawing files to be plotted in a batch and then plots those drawings from that list without additional user intervention. You can specify the layouts for each drawing to be plotted. If desired, you can also override the page setup, plot device, plot settings, and layers plotted from those already established within a drawing file.

To load the Batch Plot Utility, click Start | Programs | AutoCAD 2002 | Batch Plot Utility.

When you start the Batch Plot Utility, it loads its own instance of AutoCAD and then displays the AutoCAD Batch Plot Utility window, shown in Figure 18-14. This window contains its own menus and toolbar, providing access to the various batch plot commands.

### Selecting the Drawings To Be Plotted

To add a drawing to the list of those to be plotted, click the Add Drawing button or choose Add Drawing from the File menu. Drawings are normally plotted in the order in which they were added to the list. You can also click the header of any of the columns to reorder the list based on any of the other properties, such as drawing filename, path, layout, page setup, or plot device. For example, to sort the drawings alphabetically, click the Drawing File header. There is no other provision for specifically reordering the list other than deleting drawings from the list and adding them again. To delete a drawing from the list, select it and then click the Remove button or choose Remove from the File menu.

**Tip**    *When plotting to more than one device, you may want to reorder the list based on the plot device, to send all the drawing to a specific device before plotting drawings destined to another device.*

You can add as many drawings as you want to the list. After adding them, you can control the other options, such as the layouts to be plotted and the various plot settings. Unless you are going to start the actual batch plotting process when you finish setting up all the necessary parameters, you must save the batch plot file list. To save the list, click the Save List button, choose Save List from the File menu, or press CTRL-S.

MOVING BEYOND
THE BASICS

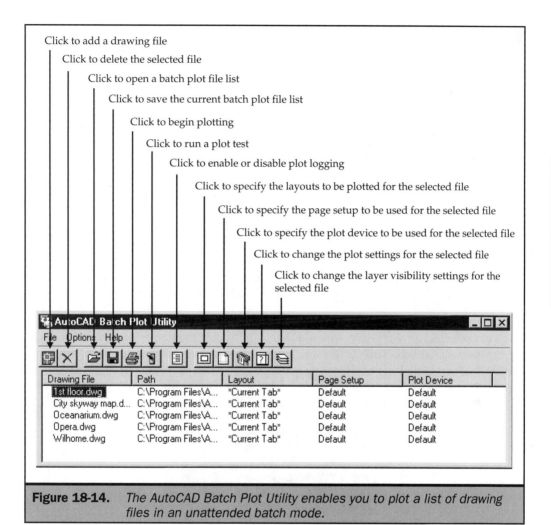

**Figure 18-14.**    *The AutoCAD Batch Plot Utility enables you to plot a list of drawing files in an unattended batch mode.*

AutoCAD saves batch plot file lists with the file extension .BP3. After saving a list, you can open it again later by clicking the Open List button, choosing Open List from the File menu, or pressing CTRL-O.

**Note**    *Since the Batch Plot utility doesn't have an option to specify the number of copies you want, to plot multiple copies of a drawing, you must add the drawing to the batch plot list multiple times, once for each copy.*

To clear the current list, choose New List from the File menu. You can also append a previously saved list to the current list by choosing File | Append List. The drawings in the saved list are appended below the drawings already displayed in the dialog box.

*If you close the Batch Plot Utility without saving a batch plot file list, all of your settings will be lost. If you haven't already saved the contents of the list of drawings, you will be reminded to do so.*

## Selecting Layouts To Be Plotted

For each drawing in the list, you can specify which layouts you want to plot. To do so, select the drawing and then either click the Layouts button, choose Options | Layouts, right-click and choose Layouts from the shortcut menu, or simply click anywhere in the Layout column below the header.

AutoCAD displays the Layouts dialog box. Initially, this dialog box contains three selections:

- ■ **\*Current Tab\***   Plots the tab that was current when the drawing was last saved
- ■ **Model Tab**   Plots the Model tab
- ■ **Last Active Layout Tab**   Plots the tab that was most recently active when the drawing was last saved, other than the Model tab

To plot a specific layout, click the Show All Layouts button. The Batch Plot Utility loads the selected drawing into AutoCAD and then adds the names of all the layouts to the Layouts dialog box. You can then select one or more layouts to be plotted. You can use the SHIFT or CTRL key to select multiple layouts. When you select multiple layouts, a separate instance of the drawing file is added to the batch plot file list for each layout so that you can also control the other options individually for each layout.

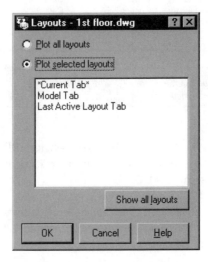

You can also select the Plot All Layouts option to plot all the layouts in the drawing. In that case, however, you can't override the other plot settings individually for each layout.

# Controlling Page Setups

Although the page setup for each tab is normally specified within the drawing file, you can change the page setup used to plot the specified layouts for each drawing in the batch file list. To do so, select the drawing and then either click the Page Setups button, choose Options | Page Setups, right-click and choose Page Setups from the shortcut menu, or simply click anywhere in the Page Setup column below the header.

AutoCAD displays the Page Setups dialog box. Initially, this dialog box contains the page setups in the selected drawing. To load a page setup from a different drawing or template, click the adjacent button and then select the drawing or template containing the named page setup that you want to use.

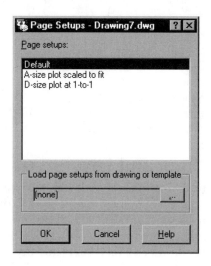

# Selecting the Plot Device

The plot device is also normally specified within the drawing file. You can specify that the drawing be plotted using a different plot device when batch plotted, however. To do so, select the drawing and then either click the Plot Devices button, choose Options | Plot Devices, right-click and choose Plot Devices from the shortcut menu, or simply click anywhere in the Plot Device column below the header.

AutoCAD displays the Plot Devices dialog box, which contains a list of all the currently configured plotters and available Windows system printers. Select the plot device you want to use, and then click OK. If the plotter you want to use does not appear in the list, you can click the Browse button and then locate its PC3 file.

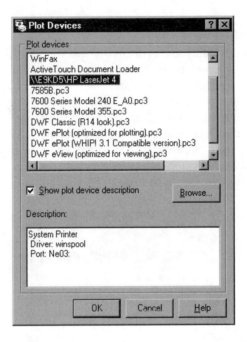

## Controlling Plot Settings and Layers To Be Plotted

You can use the controls in the Plot Settings dialog box to override the plot settings and layers specified by the settings saved in the drawing file. To display this dialog box, select the drawing and then click either the Plot Settings or Layers button. You can also display this dialog box from the Options menu or the shortcut menu. The button or menu selection you choose determines which tab is initially displayed.

The Plot Settings tab lets you specify the plot area and plot scale, and a filename and location if you're plotting the drawing to a file.

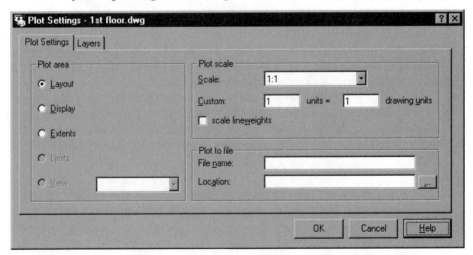

The Layers tab lists all the layers in the selected drawing and indicates whether the layer is to be plotted or not based on the current drawing settings. You can then turn the plotting of individual layers on and off by selecting them in the list and clicking the On or Off button.

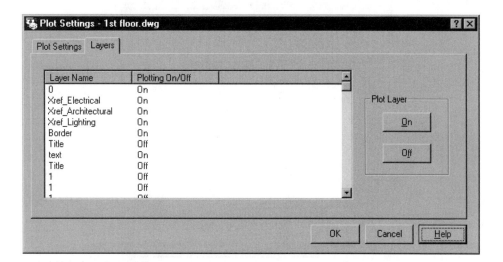

## Logging Batch Plot Operations

By default, the Batch Plot Utility generates a plot journal log that records information about batch plotting operations. The log file created is an ASCII file that can be opened and read using any text editor or word processor. It contains information about the date and time at which each drawing was plotted. In addition, the program generates an error log if any errors were encountered during batch plotting. Like the journal log, the error log is an ASCII file that contains information about the error.

You can control batch plot logging from the Logging dialog box, shown in Figure 18-15. To display this dialog box, click the Logging button or choose File | Logging.

## Plotting Using the Batch Plot Utility

After you add all the drawings to the batch plot list and configure all the parameters to your specifications, you are ready to actually begin the batch plotting process. To do so, simply click the Plot button. AutoCAD immediately loads each drawing in order and plots it according to the settings you've established. As each file is processed, the list is updated. A green check mark appears to the left of each drawing that was successfully plotted, while a red X is added to indicate drawings that weren't plotted.

*To cancel the batch plotting process, click inside the AutoCAD window and then click the Cancel button in the Plot Progress dialog box.*

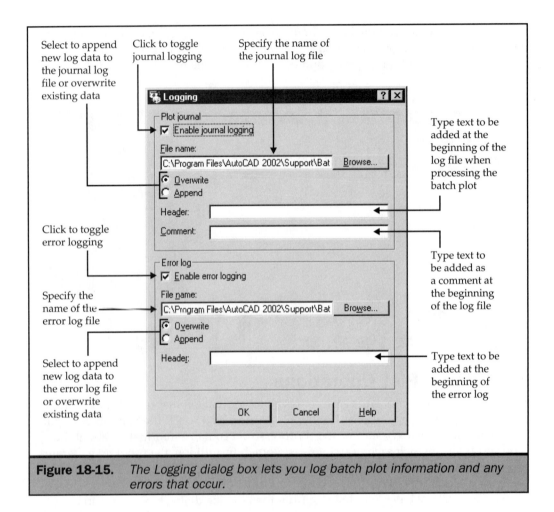

Select to append
new log data to
the journal log
file or overwrite
existing data

Click to toggle
journal logging

Specify the name of
the journal log file

Type text to be
added at the
beginning of the
log file when
processing the
batch plot

Click to toggle
error logging

Type text to
be added as
a comment at
the beginning
of the log file

Specify the
name of the
error log file

Select to append
new log data to
the error log file
or overwrite
existing data

Type text to be
added at the
beginning of
the error log

**Figure 18-15.**    *The Logging dialog box lets you log batch plot information and any errors that occur.*

If you prefer, you can test the batch plotting process before you actually output any of the drawing files, either by clicking the Plot Test button or by choosing Plot Test from the File menu. When the plot test has been completed, AutoCAD displays a Plot Test Results dialog box. You can then analyze the results to determine any potential problems, prior to starting the actual batch plot.

**Note**    *When you close the Batch Plot Utility, AutoCAD also closes. Because the utility opened and made changes to the drawing files that you plotted, AutoCAD displays an alert dialog box asking if you want to save the changes.*

## Plotting Using Script Files

You can also create script files to automate plotting. You create script files outside of AutoCAD by using a word processor or text editor, and then save the script as an ASCII file with the file extension .SCR. The script file must contain the exact sequence of commands needed to be sent to AutoCAD in order to set up and plot the drawing. You'll learn more about script files in Chapter 29.

When you plot using scripts, you must use the command line version of the PLOT command, specified by including a hyphen prefix: -PLOT. When plotting using a script file, you can specify the layout name, page setup name, output device name, and a filename (if plotting to a file).

You can also use the ScriptPro utility, supplied as part of the AutoCAD Migration Assistance package, to apply a single script file to a list of drawings. You'll learn more about the ScriptPro utility in Appendix D.

# The Complete Reference

# Part III

## Becoming an Expert

# Chapter 19

## Drawing in Three Dimensions

aper drawings typically represent two-dimensional views of three-dimensional objects. Although this representation of objects is typical in conventional drafting, it requires that the person who created the drawing and the people looking at the paper copy of it both understand how to interpret the information in the drawing. In addition, when the drawing includes 2-D views of 3-D objects in which each view is created independently, the possibility for errors greatly increases. In particular, if the object being represented is modified, you must change each view separately.

Although creating 3-D models can be more difficult than creating 2-D objects, 3-D models offer several advantages:

- You can view a 3-D model from any point in space.
- You can automatically create standard top, front, side, and auxiliary views.
- You can remove hidden lines and produce shaded images.

In addition, depending on how the objects are modeled, you can check for interference, do engineering analysis, and extract manufacturing data from 3-D models.

In Chapter 17, you learned how to use paper space layouts to create multiple viewports, each displaying a different view (top, front, side, and isometric) of a 3-D object. In this chapter, you'll learn how to use AutoCAD to create 3-D models. Then, in the two chapters that follow, you'll learn how to edit those models in 3-D space and how to modify 3-D solids.

This chapter explains how to do the following:

- View objects in three dimensions
- Create three-dimensional wireframe objects
- Create three-dimensional meshes
- Create three-dimensional solids

# Viewing Objects in Three Dimensions

In model space, you can view an AutoCAD drawing from any position in 3-D space. From any selected viewing position, you can add new objects, modify existing objects, and generate hidden-line, shaded, and rendered views.

When you create or view objects in 3-D space, you specify 3-D coordinates instead of 2-D coordinates. As you learned in Chapter 6, 3-D coordinates are practically identical to 2-D coordinates. The only difference is that you include a Z-coordinate value in addition to the X and Y values. Chapter 6 also includes important information about entering spherical and cylindrical coordinates, using point filters, and creating User Coordinate Systems. You may want to review this information before continuing with the remainder of this chapter.

When you are working in a layout, AutoCAD always displays the drawing sheet in plan view, although the individual floating viewports can each display your model from a different viewpoint.

## Understanding Standard Views and Projections

When working in 3-D, one of the most common drafting practices is the use of standard views and projections. When you create two-dimensional representations of three-dimensional objects, you normally create three standard views: top, front, and right. If the object is particularly complex, you might create additional views—such as bottom, left, or back.

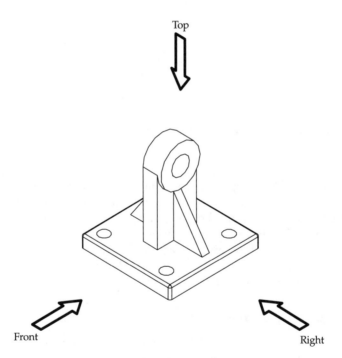

Each of these six standard views shows a 2-D view of the 3-D model and illustrates only two of the object's three possible measurements: length, width, and height. Whenever multiple standard views are arranged either onscreen or on paper, the views are arranged so that they align and the adjacent views share one of these measurements. For example, the top and front view share the same width; the front and right-side view share the same height.

BECOMING AN EXPERT

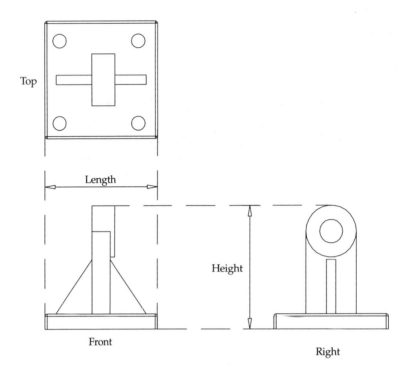

These views are said to "project," because you can visualize their creation by placing the 3-D object in the center of a cube and then projecting the object onto each face of the cube. If you then unfold the cube, you get the standard views or *projections*.

Of course, AutoCAD lets you view 3-D models from any viewpoint, and provides tools for setting the current viewpoint. In the sections that follow, you'll learn how to quickly reorient your model to any of the standard 2-D projections. You can also view your model from a standard isometric viewpoint.

## Setting a Viewing Direction

You view 3-D drawings by setting the *viewing direction*. The viewing direction establishes the *viewing position*, the Cartesian coordinate that corresponds to the viewpoint looking back at the origin point, the 0,0,0 coordinate.

AutoCAD provides several commands that you can use to change the viewing direction. You can use the Viewpoint Presets dialog box, shown in Figure 19-1, to set the current viewing angles in relation to the X axis and the XY plane, specify the coordinate of the viewpoint, or change the view by manipulating a compass and axis tripod. You can also switch to a standard viewing direction, such as a front or side view, or a particular isometric view, by selecting these views from AutoCAD's menus or the Viewpoint toolbar.

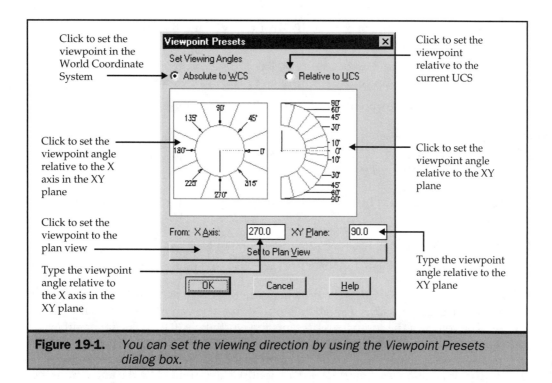

Click to set the viewpoint in the World Coordinate System

Click to set the viewpoint angle relative to the X axis in the XY plane

Click to set the viewpoint to the plan view

Type the viewpoint angle relative to the X axis in the XY plane

Click to set the viewpoint relative to the current UCS

Click to set the viewpoint angle relative to the XY plane

Type the viewpoint angle relative to the XY plane

**Figure 19-1.**   *You can set the viewing direction by using the Viewpoint Presets dialog box.*

BECOMING AN EXPERT

To display the Viewpoint Presets dialog box, do one of the following:

■ From the View menu, choose 3D Views | Viewpoint Presets.

■ At the command line, type **DDVPOINT** (or **VP**) and then press ENTER.

The current viewing angle is shown as solid lines within the left and right image tiles, and as values in the two edit boxes. The left-hand image tile lets you set the viewing angle in the XY plane. The right-hand image tile lets you set the viewing angle relative to the XY plane. You can set the new viewing angle to one of the 45-degree presets by clicking the dividers within the image tiles, by typing the angles explicitly into the edit boxes, or by setting the current viewing angle so that AutoCAD displays the plan view. The Absolute To WCS and Relative To UCS radio buttons determine whether the new viewing angle is based on the World Coordinate System (WCS) or the current User Coordinate System (UCS).

To set the viewing direction by specifying the coordinate of the viewpoint:

1. Do one of the following:

■ From the View menu, choose 3D Views | VPOINT.

■ At the command line, type **VPOINT** (or **-VP**) and then press ENTER. AutoCAD prompts

```
Current view direction: VIEWDIR=0.0000,0.0000,1.0000
Specify a view point or [Rotate] <display compass and tripod>:
```

2. Type the new viewpoint coordinate and then press ENTER.

**Note** *You can also use the Rotate option of the VPOINT command to specify the viewpoint as an angle in the XY plane and an angle from the plane. This is similar to using the Viewpoint Presets dialog box.*

To set the current view by using AutoCAD's compass and tripod, do one of the following:

■ From the View menu, choose 3D Views | VPOINT.

■ At the command line, type **VPOINT** (or **-VP**) and then press ENTER twice.

AutoCAD temporarily clears the drawing from the screen and displays the compass and tripod. The compass has a small cursor in it that you can move by moving the mouse. The center of the compass represents the North Pole (0,0,1). The inner circle is the equator (n, n, 0). The outer circle is the South Pole (0,0,-1). Thus, if the cursor is at any point between the center and the inner circle, the view of the 3-D object is from above. The orientation (east, north, west, or south) is determined by the angle from the center to the cursor. If the cursor is exactly on the inner circle, the view of the 3-D object is from ground level. If the cursor is between the inner and outer circles, the view of the object is from the underside, as if the ground were transparent. If the cursor is on the outer circle, the view is straight up at the object from directly below it.

As you move the compass cursor, the tripod also changes its orientation, to further assist you in setting the current viewpoint. Figure 19-2 shows several examples of the compass and tripod that are used to set the current viewpoint.

You can also switch to one of ten preset viewpoints—top, bottom, left, right, front, back, SW isometric, SE isometric, NE isometric, or NW isometric—by either clicking the appropriate button on the View toolbar, shown here, or choosing View | 3D Views, and then selecting the preset viewpoint from the submenu.

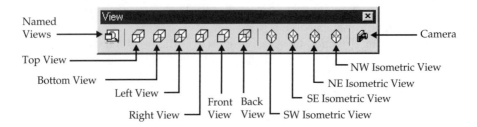

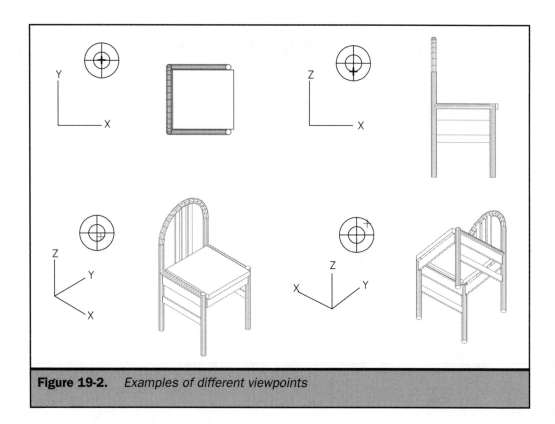

**Figure 19-2.**   *Examples of different viewpoints*

BECOMING AN EXPERT

Each view reorients the viewpoint so that you are looking at the model from the indicated orientation. For example, the right viewpoint changes the viewpoint position to 1,0,0, and the NE Isometric viewpoint changes the viewpoint position to 1,1,1.

You can also use the Orthographic & Isometric Views tab of the View dialog box, shown in Figure 19-3, to reorient the viewpoint to one of these ten preset viewpoints. To use this dialog box:

1. Do one of the following:

   ■ On the View toolbar, click Named Views.

   ■ From the View menu, choose Named Views.

   ■ At the command line, type **VIEW** and then press ENTER.

2. Choose the Orthographic & Isometric Views tab.

3. In the Current View list, select the name of the view you want to use, and then click the Set Current button. A small arrow appears adjacent to the view name in the list.

4. Click OK.

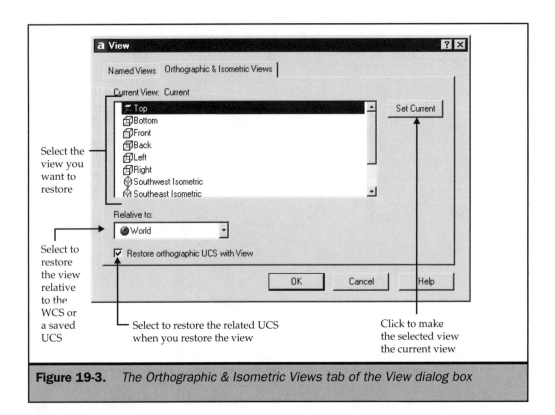

**Figure 19-3.** The Orthographic & Isometric Views tab of the View dialog box

> **Note** When you change the viewpoint, AutoCAD zooms to the extents of the drawing when viewed from the new viewpoint.

## Displaying a Plan View

When you display a *plan view* of your model, you are simply viewing the drawing from the default viewpoint (0,0,1). If you have set up a UCS, you can also set the current view to the plan view of the UCS. To display a plan view, do one of the following:

- From the View menu, choose 3D Views | Plan View.
- At the command line, type **PLAN** and then press ENTER.

In either case, you must then specify whether to display the plan view relative to the current UCS, the world UCS, or a named UCS.

# Manipulating 3-D Views Interactively

In addition to the various commands for establishing preset viewpoints, AutoCAD 2002 provides other commands that enable you to manipulate 3-D views interactively. The 3DORBIT command activates this interactive view in the current viewport. When this command is active, AutoCAD displays an arcball—a circle with four quadrants marked off by smaller circles. You can use your pointing device to manipulate the view of the model.

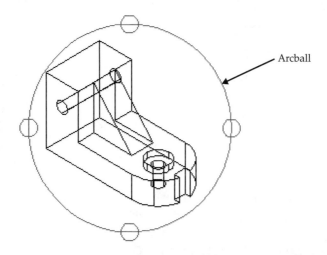

Arcball

The point at the center of the arcball, the point you are viewing, is the *target*. The point you are viewing from is the *camera*. While the 3DORBIT command is active, the target remains stationary. By moving your pointing device, you move the camera around the target, thus changing the viewpoint. When you end the 3DORBIT command, the drawing remains reoriented at the new viewpoint.

Sometimes, it is helpful to move the camera location prior to starting the 3DORBIT command. To change the camera and target points:

1. Do one of the following:

- On the View toolbar, click Camera.
- At the command line, type **CAMERA** and then press ENTER.

2. Specify the camera location by either selecting a point in the current viewport or typing its X,Y,Z coordinate.

3. Repeat Step 2 to specify the target location.

BECOMING AN EXPERT

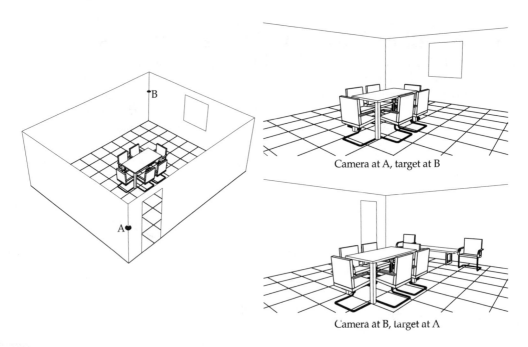

Camera at A, target at B

Camera at B, target at A

 *You can also change the camera and target locations by using the DVIEW command, which you'll learn about later in this chapter.*

# Using the 3DORBIT Command

While the 3DORBIT command is active, you can't enter any other command at the command line. If the 3DORBIT command is not active, starting any of the following commands also starts the 3DORBIT command and activates one of the command options:

- **3DCLIP**  Opens the Adjust Clipping Planes window
- **3DCORBIT**  Lets you set objects into continuous motion
- **3DDISTANCE**  Makes objects appear closer or farther away
- **3DPAN**  Lets you pan objects by dragging the cursor
- **3DSWIVEL**  Simulates the effect of turning the camera
- **3DZOOM**  Lets you zoom in and out by dragging the cursor

To start the 3DORBIT command, do one of the following:

- On the Standard toolbar or 3D Orbit toolbar, click 3D Orbit.
- From the View menu, choose 3D Orbit.
- At the command line, type **3DORBIT** (or **3DO** or **ORBIT**) and press ENTER.

As you've already learned, when the 3DORBIT command is active, an arcball is displayed in the active viewport. In addition, if the UCS icon is on, the icon appears shaded. The X axis is shown in red, the Y axis in green, and the Z axis in blue.

To manipulate the 3-D view, click-and-drag the cursor. Notice that as you move the cursor over different parts of the arcball, the cursor icon changes to indicate the following four different types of rotation that you can apply:

    When the cursor is inside the arcball, it appears as a small sphere surrounded by two ellipses. If you click-and-drag the cursor, you can manipulate the view freely, moving the viewpoint around the target.

    When the cursor is outside the arcball, it appears as a small sphere surrounded by a circular arrow. If you click-and-drag the cursor, you can rotate the view around an imaginary axis that extends through the center of the arcball perpendicular to the screen.

    When the cursor is directly over the circle at the left or right quadrant of the arcball, it appears as a small sphere surrounded by a horizontal ellipse. If you click-and-drag the cursor, you can rotate the view around an imaginary axis that extends vertically between the top and bottom quadrant points of the arcball.

    When the cursor is directly over the circle at the top or bottom quadrant of the arcball, it appears as a small sphere surrounded by a vertical ellipse. If you click-and-drag the cursor, you can rotate the view around an imaginary axis that extends horizontally between the left and right quadrant points of the arcball.

**Note**    *You can't edit any objects while the 3DORBIT command is active. To exit from the command, press ENTER or ESC, or right-click and choose Exit from the shortcut menu. Selecting most other AutoCAD commands from either a menu or by clicking a toolbar button also cancels the 3DORBIT command. The 3DORBIT command itself can be used transparently, however, when another command is active.*

When the 3DORBIT command is active, you can use the shortcut menu to select any of its command options. For example, you can change the view to one of the six standard orthographic views—top, bottom, front, back, left, or right—or one of the four standard isometric views, by choosing Preset Views from the shortcut menu and then choosing the desired view.

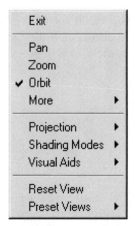

You can also use Pan and Zoom functions similar to AutoCAD's standard PAN and ZOOM commands. For example, to Pan in the 3-D orbit view:

1. Do one of the following:

- On the 3D Orbit toolbar, click 3D Pan.
- At the command line, type **3DPAN** and press ENTER.
- Start the 3DORBIT command, right-click, and then choose Pan from the shortcut menu.

The cursor changes into a hand, similar to the Realtime Pan feature.

2. Click-and-drag the cursor.

3. Release the pick button.

4. Right-click again to redisplay the shortcut menu, so you can select a different option.

To Zoom in the 3-D orbit view:

1. Do one of the following:

- On the 3D Orbit toolbar, click 3D Zoom.
- At the command line, type **3DZOOM** and press ENTER.
- Start the 3DORBIT command, right-click, and then choose Zoom from the shortcut menu.

The cursor changes into a magnifying glass, similar to the Realtime Zoom feature.

2. Click-and-drag the cursor.

3. Release the pick button.

4. Right-click again to redisplay the shortcut menu, so you can select a different option.

You can also choose More from the 3D Orbit shortcut menu, and then choose Zoom Window or Zoom Extents.

*To reset the 3-D orbit view to the view that was current when you first started the 3DORBIT command, choose Reset View from the 3D Orbit shortcut menu.*

While the Pan and Zoom options operate similarly to AutoCAD's regular commands, the Distance option works a bit differently. Instead of zooming the view, the Distance option moves the camera location along the line of sight between it and the target, as shown in Figure 19-4. To change the distance between the camera location and the target:

1. Do one of the following:

   - On the 3D Orbit toolbar, click 3D Adjust Distance.
   - At the command line, type **3DDISTANCE** and press ENTER.
   - Start the 3DORBIT command, right-click, and then choose More | Adjust Distance from the shortcut menu.

 The cursor changes into a horizontal line with a small arrow pointing up and a large arrow pointing down.

2. Click-and-drag the cursor toward the top of the screen to move the camera toward the target (thus making the objects appear larger) or toward the bottom of the screen to move the camera away from the target (thus making the objects appear smaller).

3. Release the pick button.

4. Right-click again to redisplay the shortcut menu, so that you can select a different option.

Instead of moving the camera around the target, you can also swivel the camera so that it is looking in a different direction, at a different target, as shown in Figure 19-5. This is similar to turning a camera on a tripod. To swivel the camera:

1. Do one of the following:

   - On the 3D Orbit toolbar, click 3D Swivel.
   - At the command line, type **3DSWIVEL** and press ENTER.
   - Start the 3DORBIT command, right-click, and then choose More | Swivel Camera from the shortcut menu.

 The cursor changes into a camera with an arched arrow around it.

2. Click-and-drag the cursor to swivel the camera. For example, drag the cursor to the right of the screen to move the target to the right. The objects appear to move toward the left.

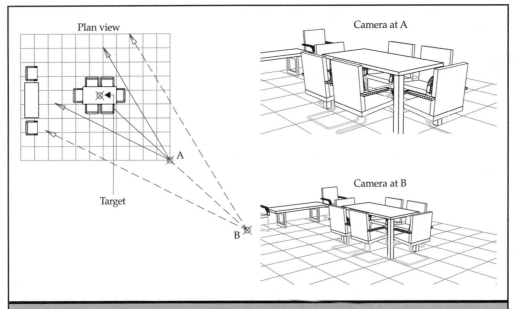

**Figure 19-4.** *The Distance option or 3DDISTANCE command moves the camera along the line of sight.*

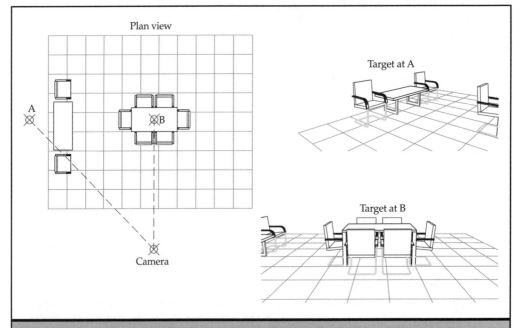

**Figure 19-5.** *The Swivel Camera option or 3DSWIVEL command changes the location of the target without changing the camera position.*

3. Release the pick button.

4. Right-click again to redisplay the shortcut menu, so that you can select a different option.

You can also click-and-drag the 3-D orbit view to set it into continuous orbiting motion. This sometimes can help you to visualize your model, and can also be used for demonstrations. To start a continuous orbit:

1. Do one of the following:

   - On the 3D Orbit toolbar, click 3D Continuous Orbit.
   - At the command line, type **3DCORBIT** and press ENTER.
   - Start the 3DORBIT command, right-click, and then choose More | Continuous Orbit from the shortcut menu.

   The cursor changes into a sphere with two continuous ellipses encircling it.

2. Click-and-drag the cursor in the direction that you want the camera to orbit.

3. Release the pick button. The camera continues orbiting around the target in the direction you indicated. The speed at which you moved the cursor determines the speed at which the camera moves.

4. To change the orbital direction, click-and-drag in a new direction, and then release the pick button.

**Note**   *While the continuous orbit is active, you can change the view by using the Reset View or Preset Views options from the shortcut menu. You can also change the projection mode, shading, and visual aids (which you will learn about in the following sections). If you use the Pan or Zoom options, however, the continuous orbit ends.*

## Setting Perspective and Parallel Projections

When you use the 3DORBIT command, AutoCAD lets you determine whether you want to view the current drawing in a perspective or parallel projection view. When you view objects in a perspective projection view, parallel lines appear to converge at one point as they recede into the distance. Objects that are closer to the camera appear larger and some objects may appear distorted if they are very close to the camera. A perspective projection view is analogous to what you see through the viewfinder of a camera.

In a parallel projection view, parallel lines remain parallel. Objects remain the same regardless of where they are positioned relative to the camera, and they never appear distorted. This is AutoCAD's default projection mode.

BECOMING AN EXPERT

Parallel projection

Perspective projection

To change the projection mode:

1. Start the 3DORBIT command.

2. Right-click and choose Projection from the shortcut menu.

3. In the submenu, choose either Parallel or Perspective.

**Caution** *The projection mode you choose remains in effect even after you exit from the 3DORBIT command. After a drawing is displayed in perspective view in a particular viewport, many commands—such as PAN and ZOOM—won't operate in that viewport until you switch back to a parallel projection. In addition, although you can create and modify objects by typing coordinates, you can't specify points by using your cursor within a perspective view.*

## Shading and Hiding Objects

Another set of options available from the 3D Orbit shortcut menu enables you to change the way objects are shaded. Using shading makes it much easier to visualize and work with three-dimensional objects.

By default, AutoCAD displays objects in wireframe. You can also select any of the following shading modes by choosing Shading Modes from the 3D Orbit shortcut menu:

- ■ **Wireframe**  Displays objects in 3-D views with all edges visible

- ■ **Hidden**  Displays objects in 3-D views with all edges that are hidden behind other surfaces removed from view

- ■ **Flat Shaded**  Displays objects in 3-D view with shaded polygonal faces

- **Gouraud Shaded**   Displays objects in 3-D view with smooth shaded surfaces
- **Flat Shaded, Edges On**   Displays objects in 3-D view with shaded polygonal faces and visible edges highlighted
- **Gouraud Shaded, Edges On**   Displays objects in 3-D view with smooth shaded surfaces and visible edges highlighted

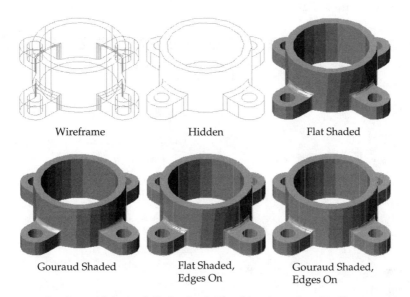

| Wireframe | Hidden | Flat Shaded |

| Gouraud Shaded | Flat Shaded, Edges On | Gouraud Shaded, Edges On |

When you shade an object while in the 3-D orbit view, the shading remains even after you exit from the 3DORBIT command. Unlike earlier versions of AutoCAD, you can select and snap to objects even though they are shaded.

**Note**   *Use the SHADEMODE command to change the shading when 3DORBIT is not active. If the UCS icon is visible, it appears shaded whenever you turn on one of the shading modes. To restore the icon to its nonshaded appearance, change the SHADEMODE setting to 2D Wireframe. Also note that you can still use the HIDE command and the pre-AutoCAD 2000 version of the SHADE command to create hidden lines and shaded images that can be saved as bitmaps or AutoCAD slide files. You'll learn more about hiding and shading objects using these commands in Chapter 22.*

## Improving 3DORBIT Performance

When you use the 3DORBIT command to manipulate a large model, the model may change from a shaded image to a wireframe image, or even to a simple box representation. This is called *adaptive degradation* and occurs whenever AutoCAD

BECOMING AN EXPERT

can't update the shaded image quickly enough. You can adjust several settings to eliminate this situation by doing the following:

1. Display the Options dialog box and select the System tab.

2. Click the Properties button to display the 3D Graphics System Configuration dialog box.

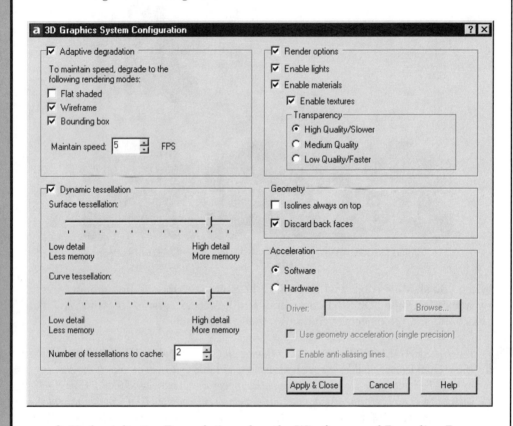

3. Under Adaptive Degradation, clear the Wireframe and Bounding Box check boxes.

4. In the Maintain Speed box, reduce the value to 1 FPS (frame per second).

5. Click Apply & Close.

6. Click OK to close the Options dialog box.

*NOTE:  AutoCAD never reduces models to a wireframe or bounding box representation when using the 3DCORBIT command, regardless of the settings in the 3D Graphics System Configuration dialog box, because that command never uses adaptive degradation.*

The following steps will also speed up AutoCAD's performance when using the 3DORBIT command:

■ Enable 3-D hardware acceleration. If your computer uses a video card with hardware support for OpenGL, you can instruct AutoCAD to let the graphics board perform the 3-D manipulations on the board. To enable hardware support, do the following:

1. Display the Options dialog box and select the System tab.

2. Click the Properties button to display the 3D Graphics System Configuration dialog box.

3. Under Acceleration, select Hardware. AutoCAD immediately displays a Select File dialog box.

4. In the Select File dialog box, choose WOPENGL7.HDI and then click Select.

5. Click Apply & Close.

6. Click OK to close the Options dialog box.

*NOTE: If your graphics board was supplied with its own HDI driver, you should use that driver instead of the WOPENGL7.HDI file supplied with AutoCAD.*

■ Turn off the display of silhouettes by setting the DISPSILH system variable value to 0.

■ Turn off dynamic tessellation by clearing the Dynamic Tessellation check box in the 3D Graphics System Configuration dialog box. Note that drawings will still display static tessellation lines when you're not using the 3DORBIT command.

■ Prior to starting the 3DORBIT command, select only those objects you want to manipulate. If you preselect objects, the 3DORBIT command manipulates only those objects you select. If you don't select any objects, however, 3DORBIT manipulates all the objects in the drawing.

*CAUTION: Visual problems sometimes occur with objects clipped by using the XCLIP command when SHADEMODE is set to a value other than 2D Wireframe or when using the 3DORBIT command. These problems occur when a polygonal xclip boundary has been applied to external reference objects. Symptoms may include missing objects (particularly 3-D solids), or 2-D objects that bleed beyond the xclip boundary. If you notice this, use a rectangular xclip boundary instead of a polygonal boundary.*

BECOMING AN EXPERT

# Using Clipping Planes

You can position front and back *clipping planes* and use them to help you visualize objects in your three-dimensional views. Clipping planes are planar boundaries positioned perpendicular to the line of sight between the camera and target points. Objects that are behind the back clipping plane or in front of the front clipping plane are hidden from view except when the clipping planes are turned off. Figure 19-6 shows the effect of establishing clipping planes.

To adjust the clipping planes in the 3-D orbit view, do one of the following:

- On the 3D Orbit toolbar, click 3D Adjust Clip Planes.
- At the command line, type **3DCLIP** and press ENTER.
- Start the 3DORBIT command, right-click, and then choose More | Adjust Clipping Planes from the shortcut menu.

AutoCAD displays the Adjust Clipping Planes window, shown in Figure 19-7. The window displays the objects in the 3-D orbit view, rotated at a 90-degree angle. A black horizontal line near the bottom of the window represents the front clipping plane. A similar green line toward the top of the window represents the back clipping plane. The cursor changes to an extruded cube with arrows pointing along either side.

As you adjust the clipping planes, the 3-D orbit view updates to reflect your changes. The five buttons across the top of the Adjust Clipping Planes window correspond to the following five options:

- **Adjust Front Clipping**   Moves the front clipping plane. Click-and-drag the black line representing the front clipping plane. The front clipping plane must be on in order to see the results in the main 3-D orbit view.

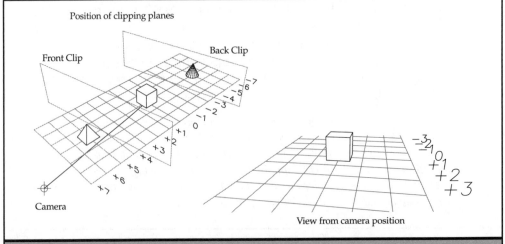

**Figure 19-6.**   *The front clipping plane hides objects between it and the camera. The back clipping plane hides objects beyond the plane.*

■ **Adjust Back Clipping**   Moves the back clipping plane. Click-and-drag the green line representing the back clipping plane. The back clipping plane must be on to see the results in the main 3-D orbit view.

 ■ **Slice**   Moves the front and back clipping planes simultaneously. Click-and-drag either clipping plane line. The planes remain spaced at their current distance. Both the front and back clipping planes must be on to see the results in the main 3-D orbit view.

 ■ **Front Clipping On/Off**   Toggles the front clipping plane on and off.

 ■ **Back Clipping On/Off**   Toggles the back clipping plane on and off.

 *You can also select these five options from the shortcut menu, displayed when you right-click in the Adjust Clipping Planes window. The Front Clip On/Off and Back Clip On/Off buttons also appear on the 3D Orbit toolbar.*

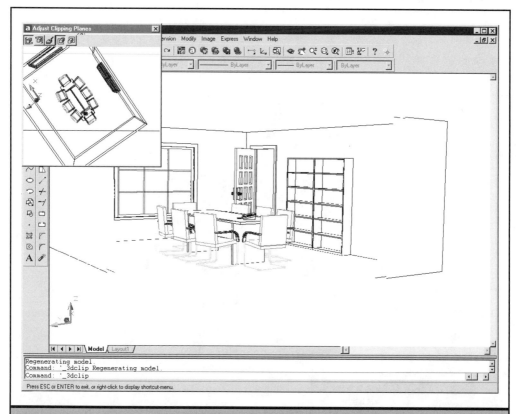

**Figure 19-7.**    *As you adjust the clipping planes in the Adjust Clipping Planes window, the 3-D orbit view updates to reflect your changes.*

BECOMING AN EXPERT

# Using the 3-D Orbit Visual Aids

In addition to the arcball, which is always displayed, the 3-D orbit view provides the following additional visual aids to help you visualize the orientation of objects when manipulating 3-D views:

- **Compass**   Three arcs that appear within the arcball, illustrating the orientation of the X, Y, and Z axes
- **Grid**   An array of lines on a plane corresponding to the current UCS and extending to the drawing limits

**Note**   *This grid is similar to the normal drawing grid you learned about in Chapter 2, except that AutoCAD divides the major grid line spacing (determined by the X and Y grid spacing) into ten subdivisions.*

- **UCS Icon**   The shaded version of the UCS icon

You can toggle these options on and off from the Visual Aids submenu of the 3D Orbit shortcut menu.

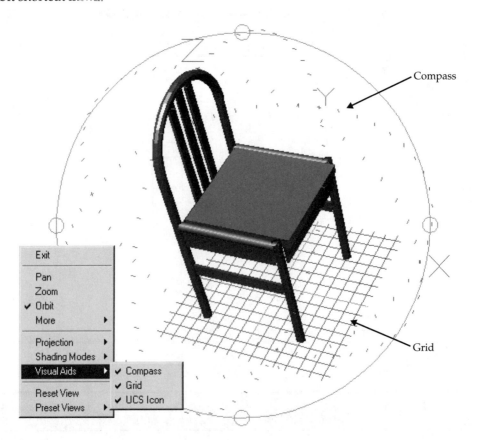

## Using the DVIEW Command

AutoCAD's older Dynamic View, or DVIEW, command is largely replaced by the 3DORBIT command. Like 3DORBIT, DVIEW lets you position the camera and target points, view the current drawing as a perspective or parallel projection, and control the clipping planes. Although the DVIEW command is much less interactive than 3DORBIT, it provides a few controls that are not available when using 3DORBIT.

To start the DVIEW command, at the command line, type **DVIEW** (or **DV**) and then press ENTER. Like 3DORBIT, the DVIEW command affects only the current viewport and can be used only in model space. When you start the command, AutoCAD prompts:

```
Select objects or <use DVIEWBLOCK>:
```

Selecting objects lets you limit the objects regenerated on the screen as you manipulate the view, something particularly useful when you are working with large drawings. If you simply press ENTER, AutoCAD displays an image of a small house with a gable roof. Sometimes, manipulating this image rather than your actual drawing is easier.

After you select the objects to be regenerated, press ENTER to continue. AutoCAD prompts:

```
Enter option
[CAmera/TArget/Distance/POints/PAn/Zoom/TWist/CLip/Hide/Off/Undo]:
```

The options are listed in the order in which you are likely to use them. For example, you'll generally want to change the line of sight first, by manipulating the position of the camera and target. After each change, this prompt is repeated. When you are satisfied with the way the view is displayed, press ENTER to exit from the DVIEW command. AutoCAD immediately regenerates the entire drawing using the new viewpoint and perspective setting.

To aid you in manipulating the objects onscreen, some of the DVIEW options display slider bars along the top of the current viewport. The current value is always indicated on the slider bar by a solid line. The position of the slider, indicated by a diamond, moves as you move the mouse. In addition, AutoCAD displays on the status line the value related to the current slider position. This value is constantly updated as you move the slider. As you move the mouse, the image on the screen also changes to reflect the changing values. You can also enter values at any time by typing them at the keyboard.

BECOMING AN EXPERT

You can select any of the following DVIEW options by typing the capital letters corresponding to the option:

- **Camera** Rotates the camera point around the target point, similar to dragging the cursor around the arcball. Objects remain in view at all times, because you're always looking back at the target point.

- **Target** Rotates the target point around the camera, similar to the 3DSWIVEL command. The prompts are the same as for the Camera option, except that the camera remains stationary and the target moves, as if you are standing in one place and turning your head.

- **Distance** Moves the camera along the line of sight toward or away from the target (similar to the 3DDISTANCE command). This option also turns on perspective viewing.

- **Points** Lets you specify both the camera and target points, similar to the CAMERA command.

- **Pan** Shifts the position of the image onscreen parallel to the current line of sight (similar to the 3DPAN command).

- **Zoom** Changes the perceived magnification of the image onscreen, but has two different effects, depending on whether the current image is displayed in a parallel or perspective projection. If perspective is off, the Zoom option prompts for a zoom scale factor and behaves similarly to the 3DZOOM command. If perspective viewing is on, however, the Zoom option lets you adjust the magnification and field of view, using a camera-lens analogy. AutoCAD prompts you for the focal length of the camera lens. Changing the focal length changes the field of view exactly as it would if you were using a camera. A shorter focal length lens lets you see more of the image at a particular camera-to-target distance, and a longer lens displays less of the image. The default lens length is 50mm and is analogous to a 50mm lens on a 35mm camera. Shorter focal length lenses are wide-angle lenses; longer lenses are telephoto lenses. Figure 19-8 shows the effect of changing the focal length of the lens.

*TIP:* *This option is not available when using the 3DORBIT command. Experiment with using the 3DORBIT command to switch to perspective view and establish the desired orientation of your 3-D objects, and then use the Zoom option of the DVIEW command to change the focal length of the lens to create more artistic compositions.*

- **Twist** Rotates the image about the line of sight.

- **Clip**   Lets you position front and back clipping planes perpendicular to the line of sight and then turn these clipping planes on and off.

- **Hide**   Removes hidden lines from the objects currently displayed by the DVIEW command. It has no effect on the display of objects when you exit from the command.

- **Off**   Turns off perspective projection and returns to parallel projection.

- **Undo**   Reverses the action of the last DVIEW option.

**Tip**   *You can use the VIEW command to save the views you create using the 3DORBIT or DVIEW commands (just as you would any other views of your drawing) so that they can be restored at a later time.*

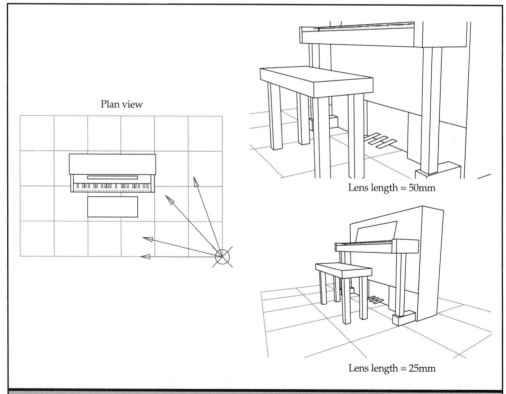

Plan view

Lens length = 50mm

Lens length = 25mm

**Figure 19-8.**   *The Zoom option changes the focal length of the camera lens when perspective viewing is on.*

BECOMING AN EXPERT

# Creating Three-Dimensional Objects

AutoCAD supports three different types of three-dimensional models:

- **Wireframe models**   Consist of lines and curves that define the edges of a 3-D object. You can create a wireframe model by drawing lines, arcs, polylines, and other 2-D objects anywhere in 3-D space. Wireframe models have no surfaces; they always appear as outlines. Because you must draw and position individually each object that makes up a wireframe model, creating them can be exacting and time-consuming.

- **Surface models**   Consist of both edges and the surfaces between those edges. You create surface models either by applying elevation and thickness to 2-D objects or by using specific 3-D object-creation commands. Surface models consist of individual planes forming a faceted, polygonal mesh.

- **Solid models**   Consist of both the surfaces and the volume inside those surfaces. You create solid models either by using specific solid-modeling commands to create basic shapes, such as boxes, cones, cylinders, spheres, wedges, and tori, or by sweeping or revolving 2-D objects. You can then combine these shapes by using Boolean operations, such as subtract, union, and intersect, to create more-complex solids. Solids are actually the easiest type of 3-D object to create.

## Creating Three-Dimensional Wireframe Objects

You can create wireframe models by drawing any two-dimensional object in three-dimensional space. You can position 2-D objects in 3-D space by typing 3-D coordinates, by setting a UCS on which to draw the object, or by moving an object to a different location in 3-D space after you create it. You can also draw splines and polylines that exist in 3-D space.

## Creating Three-Dimensional Meshes

A *mesh* represents the surface of an object with planar facets that consist of a special type of polyline comprising a 3-D polygon grid. The accuracy and density of this grid can be controlled separately in both directions—often referred to as the M and N directions—and the individual vertices can be edited. You'll learn about editing these 3-D meshes in Chapter 20. AutoCAD provides one command for creating individual 3-D faces (3DFACE) and six different commands for creating meshes: 3DMESH, PFACE, RULESURF, TABSURF, REVSURF, and EDGESURF.

*The mesh created by the RULESURF, TABSURF, REVSURF, and EDGESURF commands is created as a new object, based on defining objects that must already exist in the drawing. Those defining objects remain after the mesh is created, but you can erase them if you no longer need them. You can control the density of meshes that are created by the RULESURF and TABSURF commands by changing the value of the SURFTAB1 system variable, which controls the mesh density in the M direction. The density of meshes created by the REVSURF and EDGESURF commands can be controlled in both the M and N directions by changing the values of the SURFTAB1 and SURFTAB2 system variables, respectively. You need to change these values before you create the mesh.*

In addition to these commands, AutoCAD provides individual commands that create box, cone, dish, dome, pyramid, sphere, torus, and wedge shapes as meshes. In most instances, however, you create these types of objects as solids. Because the commands used to create these shapes as solids are very similar to their mesh counterparts, the mesh commands are not included here. You'll learn how to create these shapes as solid objects in the next section. You can also simulate a mesh by applying thickness and elevation to 2-D objects.

*Meshes work best when you don't need to model the level of detail provided by solids, such as the ability to determine the mass, volume, and moments of inertia of the solid. Meshes are also useful for modeling terrain topography.*

## Applying Elevation and Thickness

By default, AutoCAD creates new 2-D objects with a zero elevation and thickness. Perhaps the easiest way to create a 3-D object is to change the elevation or thickness of an existing 2-D object.

The *elevation* of an object is its Z-coordinate position in relation to the XY plane on which the object is drawn. An elevation of 0 indicates that the object is drawn on the XY plane of the current UCS. Positive elevations are above this plane; negative elevations are below it.

The *thickness* of an object is the distance that it is extruded above or below its elevation. A positive thickness extrudes the object upward in the positive Z direction; a negative thickness extrudes it downward in the negative Z direction. The thickness is applied uniformly to the entire object, and the Z direction is determined by the orientation of the UCS at the time that the object was created. Once you change the thickness of an object, you can see the results in any view other than a plan view. Thickness changes the appearance of objects, but does not change the type of object. For example, a circle with thickness appears as a cylinder, but it is still a circle. Similarly, a line appears as a 3-D plane, and a rectangle appears as a box, as shown in Figure 19-9.

BECOMING AN EXPERT

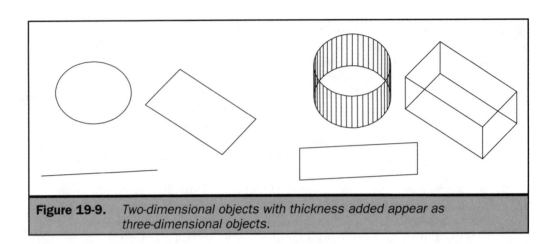

**Figure 19-9.**    *Two-dimensional objects with thickness added appear as three-dimensional objects.*

*Objects that have thickness can be hidden, shaded, and rendered. You can't apply thickness to text or attributes, dimensions, viewports, or other 3-D mesh or solid objects.*

You can change the default elevation and thickness settings in AutoCAD to create new 2-D objects with an elevation and thickness already applied to them. The current elevation and thickness remain in effect until you change them again, and are also applied as you change from one UCS to another.

To set the current elevation and thickness:

1. At the command line, type **ELEV** and then press ENTER.

2. Type the new elevation value and then press ENTER.

3. Type the new thickness value and then press ENTER.

*The current elevation and thickness are stored in the ELEVATION and THICKNESS systems variables, respectively. You can change the current value by changing these system variables.*

You can change the elevation and thickness of existing objects by using the Properties option of the CHANGE command. The CHPROP command lets you change the thickness of existing objects. You can also change both the elevation and thickness by using the Properties window.

*To change the elevation of an existing object, use the MOVE command to move it in the Z direction.*

## Creating 3DFaces

You can create 3DFace objects, which consist of a section of a plane in 3-D space, by specifying the X,Y,Z coordinates of three or more corners. After you specify the fourth point, AutoCAD continues to prompt you for additional faces by alternating prompts for the third and fourth point. This enables you to build a complex 3-D object. Each three- or four-sided plane is created as a separate 3DFace object.

To create a 3DFace object:

1. Do one of the following:

   - On the Surfaces toolbar, click 3D Face.
   - From the Draw menu, choose Surfaces | 3D Face.
   - At the command line, type **3DFACE** and then press ENTER.

2. Specify the first point of the 3DFace.

3. Specify the second, third, and fourth points.

4. Specify the third and fourth points to create additional 3DFaces.

5. To complete the command, press ENTER.

3DFaces are useful for creating planar surfaces, such as you might use to represent a wall of a room. Sometimes, you may want one or more of the edges of a 3DFace to be invisible. For example, if the wall of the room has a window, you need to create the wall by using several separate 3DFace objects to form the wall surfaces around the window opening. By making the edges shared by the individual 3DFaces invisible, you can hide the extraneous lines.

Any or all of the edges of a 3DFace can be invisible. To create an invisible edge, type **I** before you specify the starting point of the edge. Figure 19-10 shows a small house with openings in the walls, created by using 3DFaces with invisible edges.

*You can also use either the EDGE command or the Properties window to change the visibility of 3DFace edges after they have been created.*

## Creating Rectangular Meshes

You can create a 3-D rectangular mesh consisting of four-sided polygons. You determine the size of the mesh by first specifying the number of vertices along the primary (M-direction) and secondary (N-direction) mesh axes and then specifying the X,Y,Z coordinates for each vertex. Figure 19-11 shows a terrain model created as a 3DMesh object.

To create a rectangular mesh:

1. Do one of the following:

   - On the Surface toolbar, click 3D Mesh.
   - From the Draw menu, choose Surfaces | 3D Mesh.
   - At the command line, type **3DMESH** and then press ENTER.

2. Specify the number of vertices along the primary mesh axis.

3. Specify the number of vertices along the secondary mesh axis.

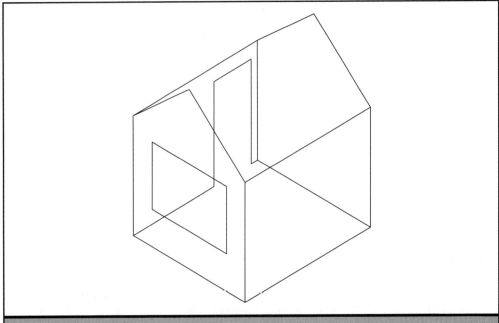

**Figure 19-10.** *An example of 3DFaces with invisible edges*

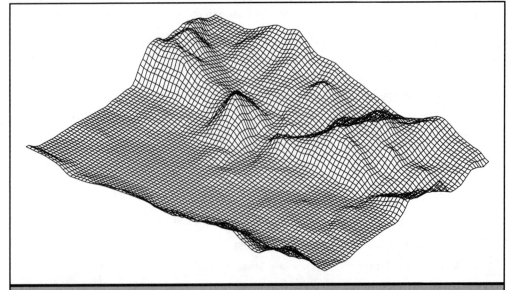

**Figure 19-11.** *A three-dimensional terrain model created as a 3DMesh object*

4. Specify the coordinates for each vertex. The command ends when you specify the coordinates for the last vertex.

 *Although creating rectangular meshes manually can be exacting, they are very useful for representing complex surfaces, such as 3-D terrain models. 3DMesh objects are most useful when used in combination with scripts, AutoLISP programs, or other programs that mathematically calculate the coordinates of the vertices.*

## Creating Polyface Meshes

You can create a polyface (polygon) mesh consisting of faces that connect three or more vertices. Creating a polyface mesh is similar to creating a rectangular mesh, except that you first specify the coordinates for each vertex and then define each face by entering the vertex numbers for all the vertices of that face. As you create each face, you can control the visibility and color of each edge and assign them to specific layers.

To create a polyface mesh:

1. Type **PFACE** and then press ENTER.

2. Specify the coordinates of each vertex. After each vertex that you specify, the next vertex number is displayed and you are prompted for the coordinates of each vertex. Specify the coordinates and then press ENTER. Continue to specify the coordinates for each numbered vertex.

3. To finish specifying vertex coordinates, press ENTER.

4. Specify the vertex numbers that define the first face. You specify the face by typing the vertex numbers that were previously defined when you specified coordinates in Step 2. Each face can be composed of three or more numbered vertices.

5. To finish defining the first face, press ENTER.

6. Specify the next face by entering its vertex numbers.

7. To complete the command, press ENTER.

*Tip* *To make an edge invisible, type the vertex number as a negative value.*

The PFACE command, like the 3DMESH command, is most useful when used in conjunction with a program that automates the creation of the vertices and faces. You'll find an example of a mesh created using the PFACE command on the Chapter 19 page of the companion web site.

## Creating Ruled Surface Meshes

The RULESURF command creates a *ruled* surface mesh, a 3-D polygon mesh that approximates the surface between two existing curves. You select the two objects that define the ruled surface, which can be arcs, circles, ellipses, elliptical arcs, lines, points,

BECOMING AN EXPERT

polylines, or splines. Both objects must be either open or closed, except that when one object is a point, the other object can be either open or closed. Figure 19-12 illustrates the creation of a simple ruled surface mesh.

To create a ruled surface mesh:

1. Do one of the following:

   - On the Surface toolbar, click Ruled Surface.
   - From the Draw menu, choose Surfaces | Ruled Surface.
   - At the command line, type **RULESURF** and then press ENTER.

2. Select the first defining object.

3. Select the second defining object.

*When you construct a ruled surface between two open curves, AutoCAD starts the surface from the end points of the defining curves nearest to the point that you used to select them. If you pick points near opposite ends, the resulting ruled surface may intersect itself. The point used to select the defining curves has no bearing on the resulting surface when the defining curves are closed, however. The direction in which the defining curve was generated can still cause the ruled surface to intersect itself. For example, circles are always generated counterclockwise, but the direction of generation for a closed polyline is determined from the direction of the last vertex. Thus, a ruled surface between a circle and a closed polyline may intersect itself, forming an hourglass figure.*

## Creating Tabulated Surface Meshes

The TABSURF command creates a *tabulated* surface mesh, a 3-D polygon mesh that approximates a surface generated by moving a path curve along a direction vector.

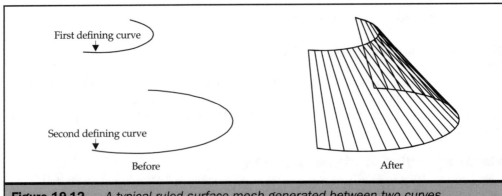

**Figure 19-12.** *A typical ruled surface mesh generated between two curves*

The length of the direction vector determines the distance that the path curve will be moved along the direction vector. You select the two objects that define the path curve and the direction vector. The path curve may be an arc, circle, ellipse, elliptical arc, line, polyline, or spline. The direction vector may be a line or polyline. If you choose a polyline, AutoCAD calculates the direction vector as the line from the first vertex to the last vertex, ignoring any intermediate vertices. The resulting mesh consists of a series of parallel polygons running parallel to the direction vector. Figure 19-13 illustrates the creation of a simple tabulated surface mesh.

To create a tabulated surface mesh:

1. Do one of the following:

   ■ On the Surface toolbar, click Tabulated Surface.

   ■ From the Draw menu, choose Surfaces | Tabulated Surface.

   ■ At the command line, type **TABSURF** and then press ENTER.

2. Select the path curve.

3. Select the object to be used as the direction vector.

## Creating Revolved Surface Meshes

The REVSURF command creates a *revolved* surface mesh, a 3-D polygon mesh that approximates the surface generated by rotating a 2-D profile around an axis. You select the two objects that define the profile and the axis. You also specify the starting angle

BECOMING AN EXPERT

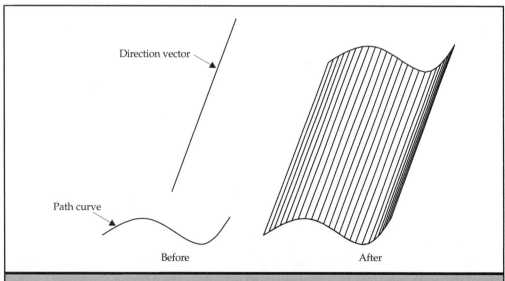

**Figure 19-13.**    *A typical tabulated surface generated by using the TABSURF command*

and the number of degrees to revolve the profile. The path curve may be an arc, circle, ellipse, elliptical arc, line, polyline, or spline. If you select a closed curve, AutoCAD closes the resulting mesh in the M direction. The axis can be a line or polyline. If you choose a polyline, AutoCAD calculates the axis as the line from the first vertex to the last vertex, ignoring any intermediate vertices. The axis of revolution determines the M direction of the mesh. Figure 19-14 illustrates the creation of a simple revolved surface mesh.

To create a revolved surface mesh:

1. Do one of the following:

   - On the Surface toolbar, click Revolved Surface.
   - From the Draw menu, choose Surfaces | Revolved Surface.
   - At the command line, type **REVSURF** and then press ENTER.

2. Select the path curve, the object to revolve.

3. Select the object to be used as the axis of revolution.

4. Specify the starting angle.

5. Specify the number of degrees to revolve the object.

## Creating Edge-Defined Coons Surface Patch Meshes

The EDGESURF command creates a surface called a *Coons* surface patch mesh, a 3-D polygon mesh defined as a bicubic surface (one curve in the M direction and another in the N direction) interpolated between four adjoining edges. You select the objects that

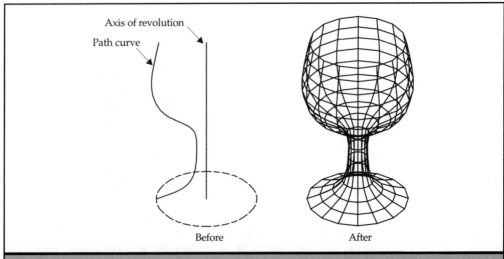

**Figure 19-14.** *A typical revolved surface mesh generated by using the REVSURF command*

define the edges. The four edge objects must form a closed loop and share end points, and may be arcs, elliptical arcs, lines, polylines, or splines. You can select the edges in any order. The first edge that you select determines the M direction of the mesh. Figure 19-15 illustrates the creation of a simple Coons surface patch mesh.

To create an edge-defined Coons surface patch mesh:

1. Do one of the following:

   - On the Surface toolbar, click Edge Surface.

   - From the Draw menu, choose Surfaces | Edge Surface.

   - At the command line, type **EDGESURF** and then press ENTER.

2. Select the first edge.

3. Select the second, third, and fourth edges.

## Creating Three-Dimensional Solids

Because a solid model represents both the surface of the object and the volume inside of those surfaces, it is the most complete and least ambiguous of the three methods for creating 3-D objects. This lack of ambiguity makes it very difficult to create a shape that can't actually exist in the real world. Constructing complex shapes by using solids rather than wireframes or meshes is also easier.

You create solids from one of the basic shapes—box, cone, cylinder, sphere, wedge, or torus—or by either extruding a 2-D object along a path (sometimes referred to as *lofting*) or revolving a 2-D object about an axis. AutoCAD provides eight different commands for creating solids in these ways, each of which is described in the following sections.

BECOMING AN EXPERT

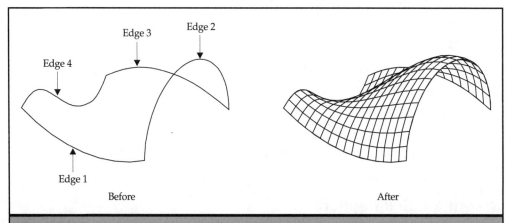

**Figure 19-15.** *A typical Coons surface patch mesh created by using the EDGESURF command. The resulting mesh is shown with hidden lines removed.*

After you create several solids, you can create more-complex shapes by combining solids through the use of Boolean operations: subtract, union, and intersect. Again, AutoCAD provides these functions as three separate commands, and each is described in this chapter. You can also further modify solids by filleting and chamfering their edges. You can also slice solids into two pieces or generate a 2-D cross-section of a solid. You'll learn how to edit solids in Chapter 21.

Solids are displayed in your drawing as wireframes until you hide, shade, or render the drawing to better visualize the objects. You'll learn about creating these types of 3-D images in Chapter 22. In addition to displaying solids, you can analyze them to determine their mass properties, such as volume, moments of inertia, and center of gravity. You can also export solids to other applications, either to generate the numerical control (NC) code needed to mill the actual object, or to perform finite element analysis (FEA) on the object to determine its structural characteristic. For example, you can export drawings as ACIS (.SAT) or Stereolithography (.STL) files for use by other programs. You'll learn more about exporting drawings in other file formats in Chapter 24.

*The ISOLINES system variable controls the number of tessellation lines used to visualize the curved portions of solids, and the DISPSILH variable controls the display of silhouette curves of solids when displayed as wireframes.*

## Creating a Solid Box

The BOX command creates a solid box, with the bottom of the box parallel to the XY plane of the current UCS. AutoCAD prompts you to locate either the corner or the center of the box, with the corner being the default method. You can then complete the box by defining a second corner and the height; by defining the box to be a cube and then providing the length; or by specifying the length, width, and height. Figure 19-16 illustrates the creation of a typical solid box.

To create a solid box:

1. Do one of the following:

   - On the Solids toolbar, click Box.
   - From the Draw menu, choose Solids | Box.
   - At the command line, type **BOX** and then press ENTER.

2. Specify the first corner of the bottom or base of the box.

3. Specify the opposite corner of the base.

4. Specify the height.

## Creating a Solid Sphere

The SPHERE command creates a solid sphere, with its latitudinal lines parallel to the XY plane and its central axis parallel with the Z axis of the current UCS. AutoCAD

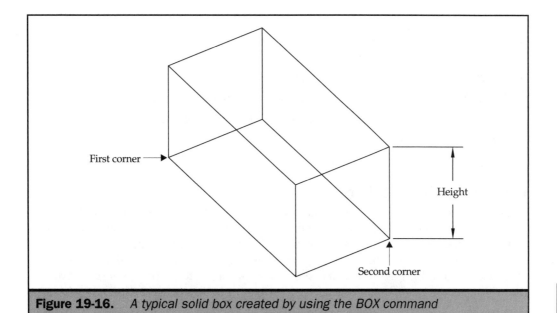

**Figure 19-16.**   *A typical solid box created by using the BOX command*

prompts you to locate the center of the sphere. You can then define the size of the sphere by specifying its diameter or radius. Figure 19-17 illustrates the creation of a typical solid sphere.

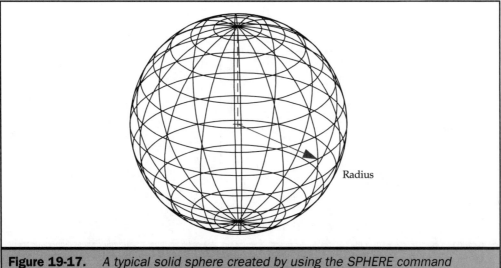

**Figure 19-17.**   *A typical solid sphere created by using the SPHERE command*

BECOMING AN EXPERT

To create a solid sphere:

1. Do one of the following:

- On the Solids toolbar, click Sphere.
- From the Draw menu, choose Solids | Sphere.
- At the command line, type **SPHERE** and then press ENTER.

2. Specify the center of the sphere.

3. Specify the diameter or radius of the sphere.

## Creating a Solid Cylinder

The CYLINDER command creates a solid cylinder with a circular or elliptical base. To create a circular cylinder, specify the center point of the circle forming the base. You can then specify either its radius or diameter. AutoCAD then prompts you to specify either the height of the cylinder as a positive or negative value along the Z axis (perpendicular to the base that you just defined) or the center of the other end, which then defines both the cylinder's height and orientation. Figure 19-18 illustrates the creation of a cylinder with a circular base.

To create a solid cylinder with a circular base:

1. Do one of the following:

- On the Solids toolbar, click Cylinder.
- From the Draw menu, choose Solids | Cylinder.
- At the command line, type **CYLINDER** and then press ENTER.

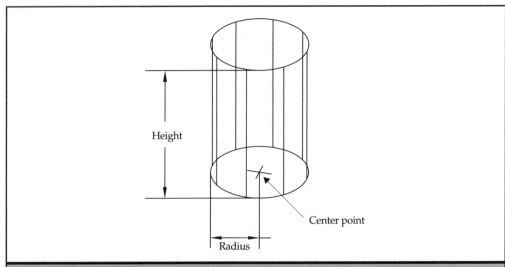

**Figure 19-18.** *A typical solid cylinder with a circular base created by using the CYLINDER command*

2. Specify the center of the base point.

3. Specify the radius or diameter of the base.

4. Specify the height or the center point of the other end.

**Tip**  *You can also create a solid cylinder by extruding a circle, using the EXTRUDE command. To create a cylinder with a feature such as a groove running along its length, first create the profile as a closed polyline, and then use the EXTRUDE command to extrude the profile. The EXTRUDE command is described later in this chapter.*

You can also use the Ellipse option of the CYLINDER command to create an elliptical cylinder. AutoCAD prompts you to define the dimensions of the ellipse, similar to the way in which you would create an ellipse by using the ELLIPSE command, and then prompts for the height of the cylinder or the center of the other end, the same as when creating a circular cylinder. Figure 19-19 illustrates the creation of a cylinder with an elliptical base.

To create a solid cylinder with an elliptical base:

1. Start the CYLINDER command.

2. Type **E** (for Elliptical) and then press ENTER.

3. Specify an axis end point.

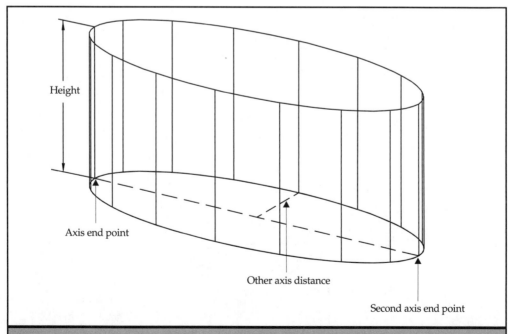

**Figure 19-19.**  *A typical solid cylinder with an elliptical base created by using the CYLINDER*

BECOMING AN EXPERT

4. Specify a second axis end point.

5. Specify the other axis distance.

6. Specify the height or the center point of the other end.

## Creating a Solid Cone

The CONE command creates a solid cone with a circular or elliptical base. The resulting cone then tapers symmetrically to a point perpendicular to its base. To create a circular cone, specify the center point of the circle forming the base, and then specify the size of the base by specifying either its radius or diameter. You can then specify either the height of the cone, as a positive or negative value along the Z axis (perpendicular to the base that you just defined), or the apex point, which then defines both the height and orientation of the Z axis and base. Figure 19-20 illustrates the creation of typical solid cones.

To create a solid cone with a circular base:

1. Do one of the following:

   - On the Solids toolbar, click Cone.

   - From the Draw menu, choose Solids | Cone.

   - At the command line, type **CONE** and then press ENTER.

2. Specify the center of the base point.

3. Specify the radius or diameter of the base.

4. Specify the height or the center point of the other end.

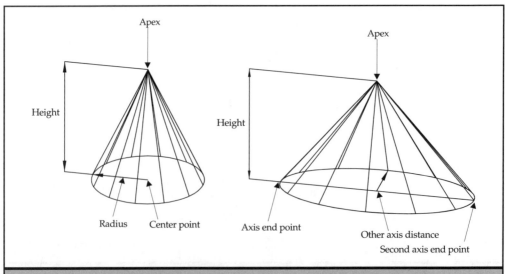

**Figure 19-20.** *The CONE command draws solid cones with either circular or elliptical bases.*

You can also create a truncated cone: draw a circle, use the EXTRUDE command to extrude the circle so that it tapers along its Z axis, and then use the SUBTRACT command to subtract a box from the tip of the cone. You'll learn about the EXTRUDE and SUBTRACT commands later in this chapter.

You can also use the Ellipse option of the CONE command to create an elliptical cone. The process is similar to creating a circular cone, except that AutoCAD prompts you to define the dimensions of the ellipse, similar to the way in which you would create an ellipse by using the ELLIPSE command, and then prompts you for the height of the cone or its apex.

## Creating a Solid Wedge

The WEDGE command creates a solid wedge, in which the bottom of the wedge is parallel to the XY plane of the current UCS and the wedge tapers along the X axis. The height of the wedge, which can be positive or negative, is parallel to the Z axis. AutoCAD prompts you to locate either the corner or center of the wedge, with the corner being the default method. You can then complete the wedge either by defining a second corner and the height; by defining the wedge based on a cube having a given length; or by specifying the length, width, and height. Figure 19-21 illustrates the creation of a typical solid wedge.

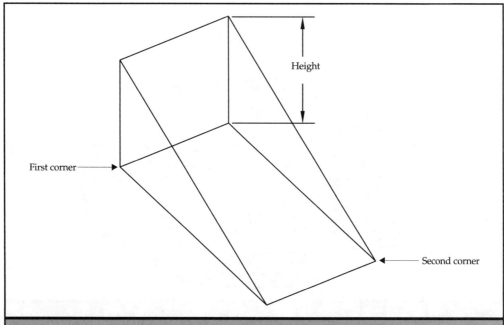

**Figure 19-21.**   *A typical solid wedge created by using the WEDGE command*

BECOMING AN EXPERT

To create a solid wedge:

1. Do one of the following:

- On the Solids toolbar, click Wedge.
- From the Draw menu, choose Solids | Wedge.
- At the command line, type **WEDGE** and then press ENTER.

2. Specify the first corner of the bottom or base of the wedge.

3. Specify the opposite corner of the base.

4. Specify the height.

## Creating a Solid Torus

The TORUS command creates a donut-shaped solid, defined by two radius values: one measured from the center of the torus to the center of the tube, and the other measured from the center of the tube to its outer surface. The resulting torus is created parallel to the XY plane of the current UCS, with its center axis parallel with the Z axis. Figure 19-22 illustrates the creation of a typical torus.

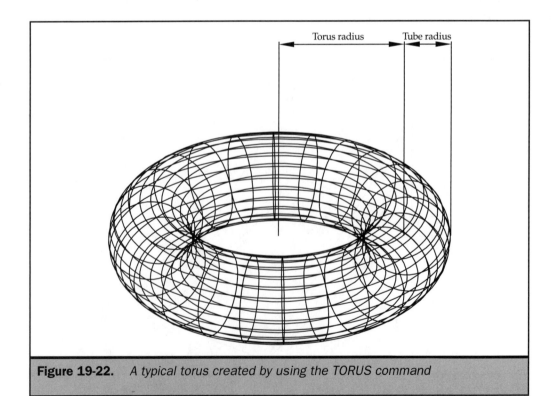

**Figure 19-22.** *A typical torus created by using the TORUS command*

To create a solid torus:

1. Do one of the following:

   ■ On the Solids toolbar, click Torus.

   ■ From the Draw menu, choose Solids | Torus.

   ■ At the command line, type **TORUS** (or **TOR**) and then press ENTER.

2. Specify the center of the torus.

3. Specify the radius or diameter of the torus.

4. Specify the radius or diameter of the tube.

**Note**   *You can create a self-intersecting torus, which has no center hole, by specifying a tube radius greater than the radius of the torus itself. If both radii are positive and the radius of the tube is greater than the radius of the torus, the result looks like a sphere with a depression at each pole. If the radius of the torus is negative and the radius of the tube is greater than the absolute value of the radius of the torus, the resulting torus looks like a sphere with pointed poles.*

BECOMING AN EXPERT

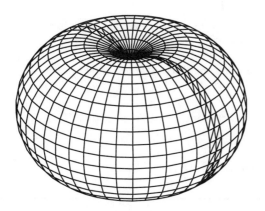

## Creating an Extruded Solid

The EXTRUDE command creates unique 3-D solids by extruding, or *lofting,* an existing 2-D object. You can extrude the object a specified distance along either the object's Z axis or a selected path. Closed 2-D objects, such as circles, ellipses, donuts, polygons, closed polylines, rectangles, regions, and closed splines, can be extruded.

**Caution**   *You can't extrude 3-D objects, objects within a block, polylines that have intersecting segments, or polylines that aren't closed.*

AutoCAD first prompts you to select the object that you want to extrude, and then asks you to specify either the height of the extrusion or a path along which to extrude the objects. If you specify the height, you can also specify an extrusion taper angle. Figure 19-23 illustrates the extrusion of an object by using a specified height and taper angle.

To create a solid by extruding an object to a specified height:

1. Do one of the following:

   ■ On the Solids toolbar, click Extrude.

   ■ From the Draw menu, choose Solids | Extrude.

   ■ At the command line, type **EXTRUDE** (or **EXT**) and then press ENTER.

2. Select the objects to extrude and then press ENTER.

3. Specify the extrusion height.

4. Specify the extrusion taper angle.

*If the taper angle that you specify would cause the object to taper to a point before reaching the specified height, AutoCAD warns*

```
Extrusion creation failed.
Modeling Operation Error:
     The Draft angle results in a self intersecting body.
Unable to extrude the selected object.
```

*and cancels the command. In that case, you should try again, using a smaller taper angle.*

To create a solid by extruding an object along a path (see Figure 19-24):

1. Start the EXTRUDE command.

2. Select the objects to extrude and then press ENTER.

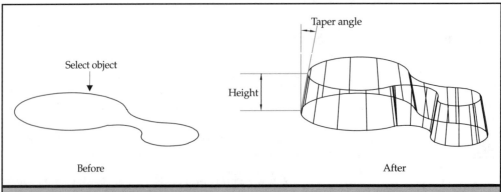

**Figure 19-23.**    *Creating a 3-D solid by extruding an object using a specified height and taper angle*

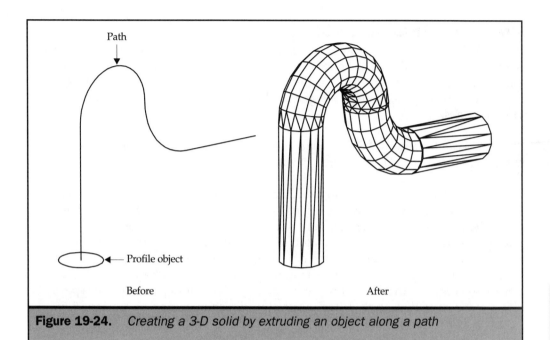

Path

Profile object

Before

After

**Figure 19-24.**   *Creating a 3-D solid by extruding an object along a path*

BECOMING AN EXPERT

3. Type **P** (for Path) and then press ENTER.

4. Select the object to use as the path.

**Tip**   *The EXTRUDE command is a handy way to create a 3-D solid quickly from a 2-D profile of an object, particularly when the profile is complex. For example, the EXTRUDE command quickly creates a 3-D gear from a 2-D profile of the gear.*

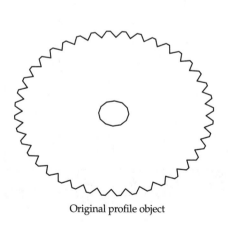

Original profile object

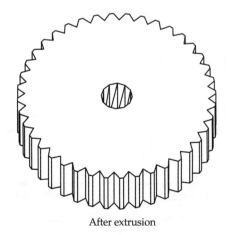

After extrusion

## Creating a Revolved Solid

The REVOLVE command creates a 3-D solid by revolving an existing closed 2-D object about an axis. The axis can be defined by the X or Y axis of the current UCS, by an existing object, or by the line between two specified points. You can revolve 2-D objects, such as circles, ellipses, donuts, polygons, closed polylines, rectangles, regions, and closed splines.

**Caution** *You can't revolve 3-D objects, objects within a block, polylines that have intersecting segments, or polylines that aren't closed.*

AutoCAD first prompts you to select the object that you want to revolve and then asks you to specify the axis of revolution. Again, that axis can be defined by a line or a polyline, the X or Y axis of the current UCS, or two points that you specify. Once you specify the axis of revolution, AutoCAD prompts you to specify the angle of revolution. This value determines the number of degrees that the object is revolved around the axis. A positive value revolves the object counterclockwise, based on the positive direction of the axis of revolution. The positive axis depends on the type of axis that you specified, and is either the positive direction of the X or Y axis or the direction away from either the first point specified or the point used to select the axis object. Figure 19-25 illustrates the creation of a revolved solid.

**Caution** *Although AutoCAD appears to allow you to select more than one object to revolve, only the first object selected is actually revolved. You should select the objects one at a time.*

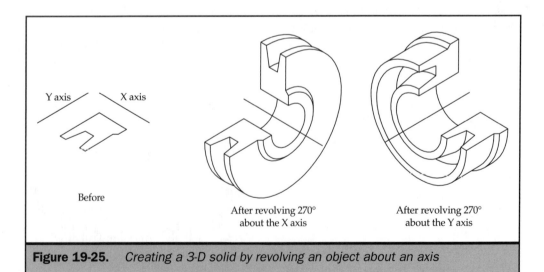

**Figure 19-25.** *Creating a 3-D solid by revolving an object about an axis*

To revolve an object about an axis:

1. Do one of the following:

   - On the Solids toolbar, click Revolve.
   - From the Draw menu, choose Solids | Revolve.
   - At the command line, type **REVOLVE** (or **REV**) and then press ENTER.

2. Select the object to revolve and then press ENTER.

3. Specify the axis of revolution.

4. Specify the angle of revolution.

*When you revolve or extrude a 2-D object to create a 3-D solid, AutoCAD normally deletes the original 2-D object. The DELOBJ system variable controls whether the original object is retained or replaced by the resulting solid. To retain the original object, change the DELOBJ value. If you don't retain the original object, it's generally a good idea to make a copy of the original objects and either put them on a different layer or move them off to the side of the drawing somewhere, in case you need to modify and reuse them to generate a new 3-D solid.*

## Creating a Composite Solid

After you create two or more solids, you can use the Boolean operations subtract, union, and intersect to create more complex, composite solids.

*You learned in Chapter 4 how to use the Boolean commands UNION, SUBTRACT, and INTERSECT to create composite regions. The use of these commands to create composite solids is very similar.*

The UNION command combines two or more solids to create a single, composite solid. AutoCAD prompts you to select the objects to be combined and then creates a new object consisting of the combined volume, as shown in Figure 19-26.

To create a composite solid by union:

1. Do one of the following:

   - On the Solids Editing toolbar, click Union.
   - From the Modify menu, choose Solids Editing | Union.
   - At the command line, type **UNION** (or **UNI**) and then press ENTER.

2. Select the solids to combine and then press ENTER.

The SUBTRACT command creates a composite solid by removing the common volume of one or more solids from other solids. AutoCAD first prompts you for the solids from which you want to subtract other solids, and then prompts for the solids

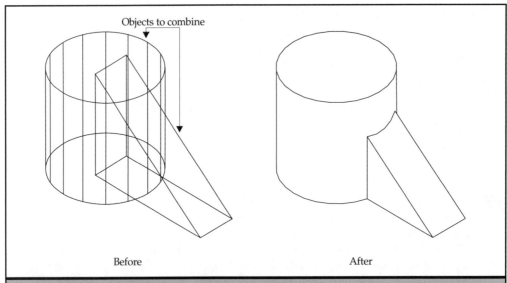

**Figure 19-26.** *The UNION command combines solids to create a composite solid.*

to be removed from the first set. The program then creates a new solid consisting of the remaining volume, as shown in Figure 19-27.

To create a composite solid by subtraction:

1. Do one of the following:

   ■ On the Solids Editing toolbar, click Subtract.
   ■ From the Modify menu, choose Solids Editing ∣ Subtract.
   ■ At the command line, type **SUBTRACT** (or **SU**) and then press ENTER.

2. Select the solid from which you want to subtract, and then press ENTER.

3. Select the solid to subtract.

The INTERSECT command creates a composite solid consisting of the common volume of two or more intersecting solids. AutoCAD prompts you to select the intersecting solids, and then creates a new object consisting of the volume common to those objects, as shown in Figure 19-28.

To create a composite solid by intersection:

1. Do one of the following:

   ■ On the Solids Editing toolbar, click Intersect.
   ■ From the Modify menu, choose Solids Editing ∣ Intersect.
   ■ At the command line, type **INTERSECT** (or **IN**) and then press ENTER.

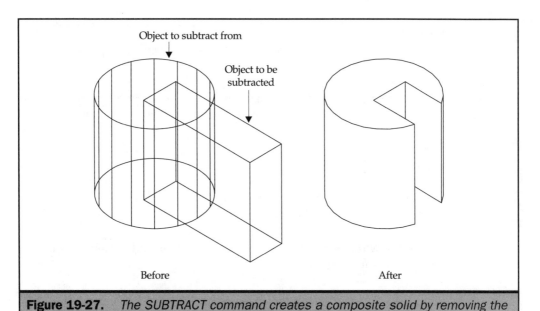

**Figure 19-27.** *The SUBTRACT command creates a composite solid by removing the volume of one set of solids from another.*

2. Select the objects to be intersected and then press ENTER.

*If you use the INTERSECT command on solids that do not actually overlap, AutoCAD deletes the solids and creates a null solid. Use the UNDO command to restore the original solids to the drawing.*

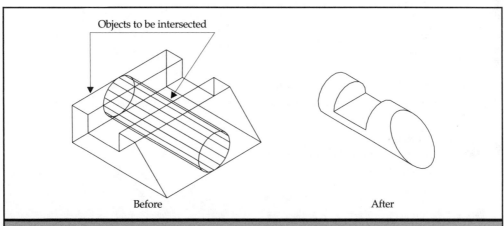

**Figure 19-28.** *The INTERSECT command creates a new composite solid consisting of the common volume of two or more solids*

BECOMING AN EXPERT

**EXAMPLES**

*LEARN BY EXAMPLE*

*Use the solid commands and Boolean operations to create a complex composite solid.*
*Open the drawing Figure 19-29 on the companion web site.*

To create the solid, begin by extruding the base profile (shown in Step 1 of Figure 19-29) to a height of 1 unit by using the following command sequence and instructions:

```
Command: EXTRUDE
Select objects: (select base object)
Select objects: ENTER
Specify height of extrusion or [Path]: 1
Specify angle of taper for extrusion <0>: ENTER
```

Your drawing should now resemble Step 2 in Figure 19-29. Next, add a solid box by using the following command sequence:

```
Command: BOX
Specify corner of box or [CEnter] <0,0,0>: 0,0,1
Specify corner or [Cube/Length]: 1.5,2.5,1
Specify height: 2
```

Your model should now match Step 3. Add the solid wedge by using the following command sequence:

```
Command: WEDGE
Specify first corner of wedge or [CEnter] <0,0,0>: 1.5,.75,1
Specify corner or [Cube/Length]: L
Specify length: 1.75
Specify width: 1
Specify height: 2
```

That completes the construction of the basic shape. Your model should now match Step 4. You can combine the three separate solids into one composite solid, as follows:

```
Command: UNION
Select objects: (select all three solids)
Select objects: ENTER
```

Next, create the cylinders that will be subtracted to form the holes:

```
Command: CYLINDER
Specify center point for base of cylinder or
  [Elliptical] <0,0,0>: 4,1.25,0
Specify radius for base of cylinder or [Diameter]: .25
Specify height of cylinder or [Center of other end]: .75
Command: CYLINDER
Specify center point for base of cylinder or
  [Elliptical] <0,0,0>: 4,1.25,.75
Specify radius for base of cylinder or [Diameter]: .5
Specify height of cylinder or [Center of other end]: .25
```

Your drawing should now look like Step 5. The next cylinder, the one drilled laterally through the block, poses a bit of a problem. Remember that AutoCAD normally creates cylinders such that their axis is parallel with the Z axis of the current UCS, but the CYLINDER command also offers the option of setting the axis by specifying the center of both ends of the cylinder. You could either create a new UCS or explicitly specify the center of both ends. In this case, it's just as easy to specify the two center points:

```
Command: CYLINDER
Specify center point for base of cylinder or
  [Elliptical] <0,0,0>: .75,0,2
Specify radius for base of cylinder or [Diameter]: .25
Specify height of cylinder or [Center of other end]: C
Specify center of other end of cylinder: .75,2.5,2
```

Finally, subtract the cylinders from the solid to create the finished composite solid:

```
Command: SUBTRACT
Select solids and regions to subtract from...
Select objects: (select the composite solid)
Select objects: ENTER
Select solids and regions to subtract...
Select objects: (select the three cylinders)
Select objects: ENTER
```

Use the HIDE command to remove hidden lines. Your drawing should now look similar to Step 6 in Figure 19-29.

BECOMING AN EXPERT

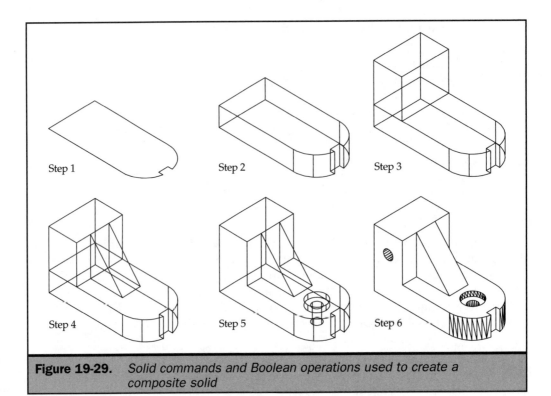

Step 1     Step 2     Step 3

Step 4     Step 5     Step 6

**Figure 19-29.**    *Solid commands and Boolean operations used to create a composite solid*

## Finding Mass Properties of Solids

After you create a solid, you can use the MASSPROP command to display its mass properties.

To find the mass properties of a solid object:

1. Do one of the following:

- On the Inquiry toolbar, click Mass Properties.
- From the Tools menu, choose Inquiry | Mass Properties.
- At the command line, type **MASSPROP** and then press ENTER.

2. Select the object and then press ENTER.

AutoCAD displays the mass properties—such as mass, volume, and moments of inertia—in the Text Window. You may need to press ENTER to view the entire properties listing. The program then asks whether you want to write the values to a file. If you respond yes, the program displays a standard file dialog box so that you can specify the name of the file. The file created is a simple ASCII text file.

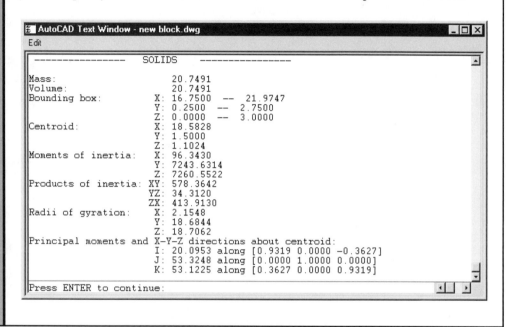

```
----------------    SOLIDS    ----------------

Mass:                   20.7491
Volume:                 20.7491
Bounding box:       X:  16.7500   --   21.9747
                    Y:  0.2500   --   2.7500
                    Z:  0.0000   --   3.0000
Centroid:           X:  18.5828
                    Y:  1.5000
                    Z:  1.1024
Moments of inertia: X:  96.3430
                    Y:  7243.6314
                    Z:  7260.5522
Products of inertia: XY: 578.3642
                    YZ:  34.3120
                    ZX:  413.9130
Radii of gyration:  X:  2.1548
                    Y:  18.6844
                    Z:  18.7062
Principal moments and X-Y-Z directions about centroid:
                    I:  20.0953 along [0.9319 0.0000 -0.3627]
                    J:  53.3248 along [0.0000 1.0000 0.0000]
                    K:  53.1225 along [0.3627 0.0000 0.9319]

Press ENTER to continue:
```

BECOMING AN EXPERT

# Chapter 20

## Editing in
## Three Dimensions

A s you've learned in previous chapters, creating objects is just part of the process of producing an AutoCAD drawing. You will probably spend as much as half of your time modifying the objects that you create, which actually is a good thing. A large proportion of the time-savings recognized from using AutoCAD effectively comes from reusing the objects that you have previously drawn.

When you edit three-dimensional objects, many of the commands that you use are the same as the commands that you use to modify two-dimensional objects. You can move, copy, scale, and delete 3-D objects just as easily as 2-D objects, using the same commands that you learned about in Chapter 7. For example, to copy an object so that the new object is located two units to the right and three units higher (in the Z direction) than the original object, you could specify the displacement as 2,0,3. You can also rotate and array these objects, although the commands that you've learned so far affect only the objects on a single 2-D plane, the current UCS. And some commands—such as BREAK, TRIM, and EXTEND—can't be used to modify 3-D meshes or solids.

The ability to modify objects in 3-D space is improved by the addition of several editing commands that are designed specifically for use in 3-D. For example, while the ROTATE command rotates objects in the current XY plane, the ROTATE3D command rotates objects about an axis defined anywhere in 3-D space. AutoCAD provides similar commands for arraying and mirroring objects in three dimensions.

In the previous chapter, you also learned that AutoCAD's 3-D mesh objects are actually a special form of polyline. You can edit these meshes by using the same polyline editing commands that you learned about in Chapter 10. This chapter explains how to do the following:

- Rotate objects in 3-D
- Array objects in 3-D
- Mirror objects in 3-D
- Align objects in 3-D
- Edit 3-D mesh objects

## Rotating in Three Dimensions

The ROTATE command that you learned about in Chapter 7 rotates objects about a specific 2-D point in the current UCS, and the rotation axis and direction of rotation is determined by the UCS. Although you can rotate objects in 3-D space by first reorienting the current UCS, the ROTATE3D command lets you eliminate that intermediate step.

The ROTATE3D command lets you rotate any object about an arbitrary 3-D axis. That axis can be based on an existing object, one of the axes of the current UCS, or two points that you specify anywhere in 3-D space. You first select the objects that you want to rotate, and then specify the axis of rotation, and, finally, specify the rotation angle, rotating the objects either relative to their current orientation or by referencing a base angle. When you rotate objects, their size remains the same.

After you select the objects to be rotated, AutoCAD prompts:

```
Specify first point on axis or define axis by
[Object/Last/View/Xaxis/Yaxis/Zaxis/2 points]:
```

The command provides seven options:

- **Object**  Aligns the axis of rotation with an existing object: a line, circle, arc, or 2-D polyline segment. If you select a line or polyline segment, the axis is aligned with the object (or passes through its end points). If you select an arc or circle, however, the axis is the line perpendicular to the object passing through its center point. The positive direction of the axis is based on the positive direction of the Z axis of the current UCS. If the axis lies in the XY plane of the current UCS, the positive direction is then based on the positive direction of the X and Y axes.

- **Last**  Lets you reuse the previous axis, if you've already used the ROTATE3D command.

- **View**  Aligns the axis of rotation with the viewing direction of the current viewport. AutoCAD prompts you for the point on the current view plane through which the axis passes. The positive direction of the axis points toward the front of the screen.

- **Xaxis/Yaxis/Zaxis**  Lets you align the axis of rotation parallel to one of the axes of the current UCS. AutoCAD prompts you for the point through which the axis passes. The positive axis direction matches the positive direction of the axis chosen.

- **2 points**  The default option, enabling you to define any axis passing through two 3-D points. The positive direction of the axis points from the first point specified to the second point.

After you select the two points, AutoCAD prompts you for the rotation angle:

```
Specify rotation angle or [Reference]:
```

To rotate the selected objects by a specified angle, simply type the angle of rotation. If you type **R** (for Reference) and then press ENTER, AutoCAD prompts you for the reference angle and the new angle. This is similar to the Reference option of the ROTATE command, which you learned about in Chapter 7. And, as you learned, it is sometimes easier to rotate objects in reference to another object, particularly when you want to rotate one object so that it aligns with another.

BECOMING AN EXPERT

*Remember that when a command displays a series of options, in addition to typing, you can select command options from the context-sensitive shortcut menu that is displayed when you right-click.*

To rotate an object about an arbitrary 3-D axis:

1. Do one of the following:

   ■ From the Modify menu, choose 3D Operation | Rotate 3D.

   ■ At the command line, type **ROTATE3D** and then press ENTER.

2. Select the objects to be rotated and then press ENTER.

3. Specify the first axis end point.

4. Specify the second axis end point.

5. Specify the rotation angle.

*LEARN BY EXAMPLE*
─────────────────────────────────────────────────
*Use the ROTATE3D command to rotate a 3-D object about an axis. Open the drawing Figure 20-1 on the companion web site.*

To rotate the lever arm 45 degrees about its pivot point, use the following command sequence and instructions. Use the center object snap to select the labeled point shown in Figure 20-1.

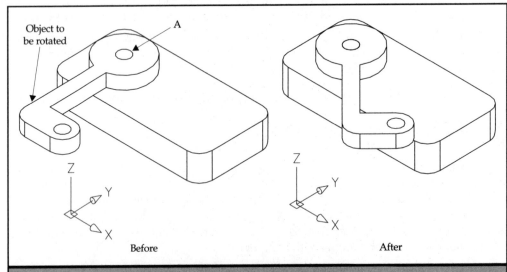

**Figure 20-1.** The ROTATE3D command lets you rotate objects about a specified axis located anywhere in 3-D space.

```
Command: ROTATE3D
Current positive angle:  ANGDIR=counterclockwise  ANGBASE=0
Select objects: (select the lever arm)
Select objects: ENTER
Specify first point on axis or define axis by
[Object/Last/View/Xaxis/Yaxis/Zaxis/2 points]: Z
Specify point on the Z axis <0,0,0>: CENTER
of (pick point A)
Specify rotation angle or [Reference]: 45
```

**Note** *The ANGDIR and ANGBASE values displayed on the command line when you first start the ROTATE3D command are actually system variable values that can also be set with the UNITS command. The ANGDIR value determines whether positive angles increase in a clockwise or counterclockwise direction, while the ANGBASE value determines the direction that AutoCAD uses as its base angle.*

## Arraying in Three Dimensions

Arraying objects in three dimensions is very similar to creating 2-D arrays. You can still create either rectangular or polar arrays. For a rectangular array, you control the number of copies in the array by specifying the number of rows and columns and the number of levels. You also specify the distance between each. Figure 20-2 illustrates how AutoCAD constructs a 3-D rectangular array.

For a polar array, you specify the axis about which to array the objects, the number of copies of the objects to create, and the angle subtended by the resulting array, as shown in Figure 20-3.

To create a three-dimensional rectangular array:

1. Do one of the following:

   ■ From the Modify menu, choose 3D Operation ∣ 3D Array.

   ■ At the command line, type **3DARRAY** and then press ENTER.

2. Select the objects to be arrayed and then press ENTER.

3. Type **R** (for Rectangular) and then press ENTER.

4. Type the number of rows in the array.

5. Type the number of columns in the array.

6. Type the number of levels.

7. Specify the distance (in the current Y direction) between rows.

8. Specify the distance (in the current X direction) between columns.

9. Specify the distance (in the current Z direction) between levels.

BECOMING AN EXPERT

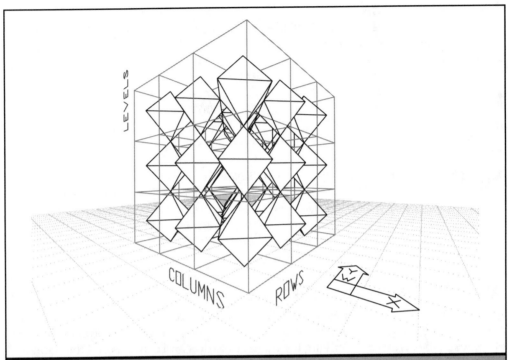

**Figure 20-2.** *When constructing 3-D rectangular arrays, you specify the number of rows, columns, and levels, and the distance between each.*

*LEARN BY EXAMPLE*
*Use the Rectangular option of the 3DARRAY command to create the first- and second-floor joists in a 3-D framing drawing. Open the drawing Figure 20-4 on the companion web site.*

To create both the first- and second-floor joists in a single step, use the following command sequence and instructions. Refer to Figure 20-4 when selecting the joist to be arrayed.

```
Command: 3DARRAY
Initializing...  3DARRAY loaded.
Select objects: (select the joist)
Select objects: ENTER
Enter the type of array [Rectangular/Polar] <R>: ENTER
Enter the number of rows (---) <1>: ENTER
Enter the number of columns (|||) <1>: 8
Enter the number of levels (...) <1>: 2
Specify the distance between columns (|||): 16
Specify the distance between levels (...): 8'10
```

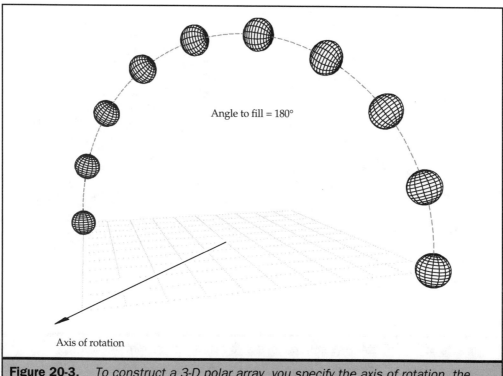

Angle to fill = 180°

Axis of rotation

**Figure 20-3.**   *To construct a 3-D polar array, you specify the axis of rotation, the number of copies, and the angle to fill.*

BECOMING AN EXPERT

To create a three-dimensional polar array:

1. Do one of the following:
   - From the Modify menu, choose 3D Operation | 3D Array.
   - At the command line, type **3DARRAY** and then press ENTER.
2. Select the objects to be arrayed and then press ENTER.
3. Type **P** (for Polar) and then press ENTER.
4. Type the number of items in the array.
5. Specify the angle to fill.
6. Specify whether the objects should be rotated as they are copied.
7. Specify the center point of the array.
8. Specify a second point along the central axis of the array.

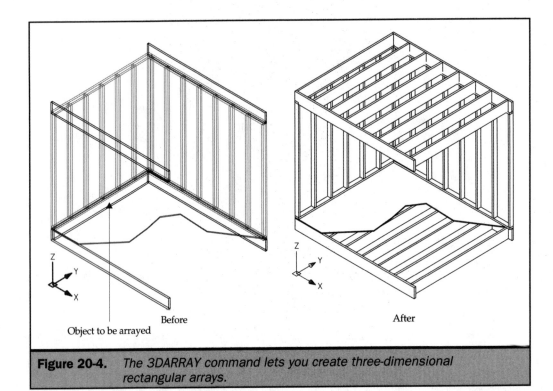

Before

Object to be arrayed

After

**Figure 20-4.** *The 3DARRAY command lets you create three-dimensional rectangular arrays.*

*LEARN BY EXAMPLE*
*Use the Polar option of the 3DARRAY command to create the five lug bolts. Open the drawing Figure 20-5 on the companion web site.*

The disk and one lug bolt have already been drawn. To create the remaining four bolts, use the following command sequence and instructions. Use the center object snap to select the labeled points shown in Figure 20-5.

```
Command: 3DARRAY
Initializing... 3DARRAY loaded.
Select objects: (select the lug bolt)
Select objects: ENTER
Enter the type of array [Rectangular/Polar] <R>: P
Enter the number of items in the array: 5
Specify the angle to fill (+=ccw, -=cw) <360>: ENTER
Rotate arrayed objects? [Yes/No] <Y>: ENTER
Specify center point of array: CENTER
of (select large cylinder at point A)
Specify second point on axis of rotation: CENTER
of (select small cylinder at point B)
```

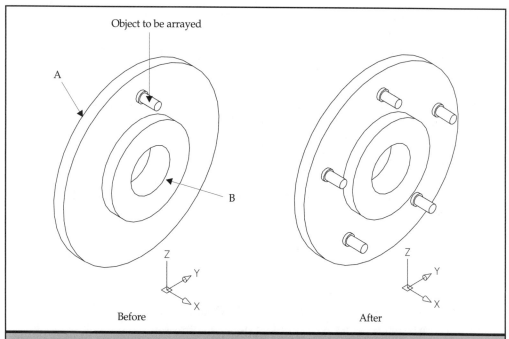

Object to be arrayed

A

B

Z

Y

X

Before

Z

Y

X

After

**Figure 20-5.** *The 3DARRAY command lets you create three-dimensional polar arrays.*

BECOMING AN EXPERT

# Mirroring in Three Dimensions

Mirroring objects in three dimensions is also very similar to mirroring objects in two dimensions. Instead of mirroring objects about a line defined in the current UCS, however, the MIRROR3D command lets you mirror objects about a mirroring plane specified anywhere in 3-D space. That plane can be based on an existing object, the current UCS, the current viewing direction, one of the three standard planes of the current UCS, or any three points that you specify in 3-D space. You first select the objects to be mirrored and then specify the mirror plane. Like the MIRROR command, the original objects can be retained or deleted.

After you select the objects to be mirrored, AutoCAD prompts:

```
Specify first point of mirror plane (3 points) or
[Object/Last/Zaxis/View/XY/YZ/ZX/3points] <3points>:
```

The command provides eight options:

- **Object** Lets you use the plane of an existing 2-D planar object—a circle, arc, or 2-D polyline—as the plane about which to mirror the objects.

- **Last** Lets you reuse the previous mirror plane, if you've already used the MIRROR3D command.

- **Zaxis** Defines the mirror plane by letting you select a point of the plane and a point on the Z axis of the plane. This option is similar to the Zaxis option of the UCS command, which you learned about in Chapter 6.

- **View** Aligns the mirror plane parallel to the viewing plane of the current viewport and passing through a selected point.

- **XY/YZ/ZX** Lets you align the mirror plane with one of the three standard planes of the current UCS. AutoCAD prompts you for the point through which the mirror plane passes.

- **3points** The default option, enabling you to define any plane passing through three 3-D points.

To mirror an object about an arbitrary plane:

1. Do one of the following:

   - From the Modify menu, choose 3D Operation | Mirror 3D.

   - At the command line, type **MIRROR3D** and then press ENTER.

2. Select the objects to be mirrored and then press ENTER.

3. Specify the first point on the mirror plane.

4. Specify the second point on the mirror plane.

5. Specify the third point on the mirror plane.

6. Press ENTER to retain the original objects, or type **Y** and then press ENTER to delete the old objects.

*LEARN BY EXAMPLE*
*Use the MIRROR3D command to create the other arm of a model of a drafting compass. Open the drawing Figure 20-6 on the companion web site.*

To mirror the arm of the drafting compass, use the following command sequence and instructions. To locate the mirror plane, use the center object snap to select the labeled point shown in Figure 20-6.

```
Command: MIRROR3D
Select objects: (select the arm of the compass)
Select objects: ENTER
Specify first point of mirror plane (3 points) or
[Object/Last/Zaxis/View/XY/YZ/ZX/3points] <3points>: ZX
Specify point on ZX plane <0,0,0>: CENTER
of (select point A)
Delete source objects? [Yes/No] <N>: ENTER
```

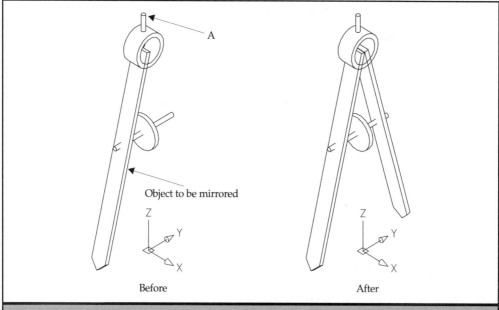

Object to be mirrored

Before

After

**Figure 20-6.**    *The MIRROR3D command lets you mirror objects about a specified plane located anywhere in 3-D space.*

## Aligning Objects in Three Dimensions

In Chapter 7, you learned how to use the ALIGN command to manipulate objects in two dimensions. As was noted in that chapter, the ALIGN command is used even more often to align objects in 3-D. When used in 3-D, the command is actually a combination of the MOVE and ROTATE3D commands, in a single step. The command first prompts you to select the objects that you want to move. It then prompts you for as many as three pairs of points. These points are referred to as *source points* and *destination points,* with the resulting realignment based on the correlation of these points.

When you align 3-D objects, you work with at least two pairs of points. The first pair defines the movement of the object. The second pair defines a 3-D rotation about one axis. As when using the ALIGN command to align 2-D objects, if you specify only two pairs of points, you can also use the ALIGN command to scale the objects being moved to match the alignment points. AutoCAD uses the distance between the first and second destination points to determine the scaling factor.

If you specify a third pair of points, you can also rotate the objects about a second axis. In that case, the objects are aligned so that the plane defined by the three source points aligns with the plane defined by the three destination points.

BECOMING AN EXPERT

To align an object in 3-D by specifying two pairs of points:

1. Do one of the following:

   ■ From the Modify menu, choose 3D Operation | Align.

   ■ At the command line, type **ALIGN** (or **AL**) and then press ENTER.

2. Select the objects to be aligned and then press ENTER.

3. Specify the first source point.

4. Specify the first destination point.

5. Specify the second source point.

6. Specify the second destination point.

7. Press ENTER to finish specifying points.

8. Specify whether to scale the objects based on the alignment points.

*LEARN BY EXAMPLE*
*Use the ALIGN command to realign a 3-D ring so that it fits around an axle post. Open the drawing Figure 20-7 on the companion web site.*

To realign the ring so that it fits around the axle post, use the following command sequence and instructions. Use the center object snap to select the labeled points shown in Figure 20-7.

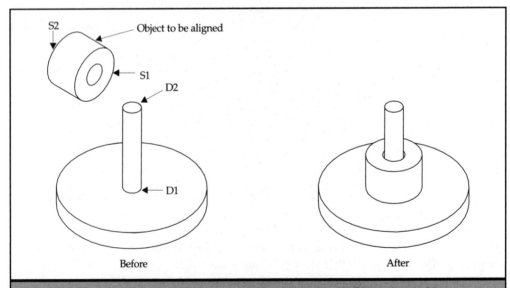

**Figure 20-7.**    *When used with two pairs of points, the ALIGN command lets you move and rotate 3-D objects, and can also scale them.*

```
Command: ALIGN
Select objects: (select the ring)
Select objects: ENTER
Specify first source point: CENTER
of (select point S1)
Specify first destination point: CENTER
of (select point D1)
Specify second source point: CENTER
of (select point S2)
Specify second destination point: CENTER
of (select point D2)
Specify third source point or <continue>: ENTER
Scale objects based on alignment points? [Yes/No] <No>: ENTER
```

To align an object in 3-D by specifying three pairs of points (see Figure 20-8):

1. Do one of the following:

   ■ From the Modify menu, choose 3D Operation I Align.

   ■ At the command line, type **ALIGN** (or **AL**) and then press ENTER.

2. Select the objects to be aligned and then press ENTER.

3. Specify the first source point.

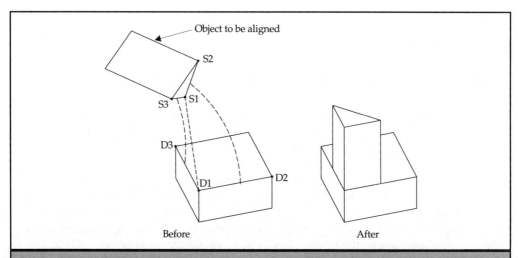

**Figure 20-8.**  *When used with three pairs of points, the ALIGN command lets you move and rotate 3-D objects so that the three source points align with the plane defined by the three destination points.*

BECOMING AN EXPERT

4. Specify the first destination point.

5. Specify the second source point.

6. Specify the second destination point.

7. Specify the third source point.

8. Specify the third destination point.

# Editing Three-Dimensional Mesh Objects

You can use the same PEDIT command that you learned about in Chapter 10 to modify any of the 3-D mesh objects created by using the 3DMESH, PFACE, RULESURF, TABSURF, REVSURF, and EDGESURF commands or the basic 3-D shapes created by using the 3-D command.

To edit a mesh, you must first start the PEDIT command and then select the mesh. To start the PEDIT command, do one of the following:

- On the Modify II toolbar, click Edit Polyline.

- From the Modify menu, choose Polyline.

- At the command line, type **PEDIT** (of **PE**) and then press ENTER.

When you are editing a mesh, the PEDIT command presents a smaller subset of command options:

```
Enter an option [Edit vertex/Smooth
surface/Desmooth/Mclose/Nclose/Undo]:
```

- **Edit vertex**   Presents a separate set of options for editing the individual mesh vertices.

- **Smooth surface**   Lets you fit a smooth surface to the mesh. The resulting surface is pulled toward the vertices, but passes through only the first and last vertices (in the case of open 3-D meshes). The original mesh is used as a frame. You can control the accuracy of the surface approximation by changing the SURFU and SURFV system variables. These variables control the density of the mesh in the M and N directions, respectively, similar to the SURFTAB1 and SURFTAB2 variables that control the density of the mesh when first created. The larger you set these values, the more precise the resulting mesh becomes, but, consequently, the mesh takes longer to generate and the overall drawing size increases.

A third system variable, SURFTYPE, controls the type of surface that is fitted to the mesh. A SURFTYPE value of 5 results in a quadratic B-spline, a value of 6 creates a cubic B-spline, and a value of 8 results in a curve approximating a Bezier surface.

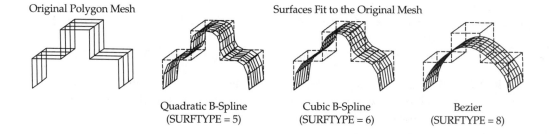

Original Polygon Mesh                    Surfaces Fit to the Original Mesh

Quadratic B-Spline          Cubic B-Spline          Bezier
(SURFTYPE = 5)              (SURFTYPE = 6)          (SURFTYPE = 8)

- **Desmooth**   Removes the smoothing applied by the Smooth Surface option.
- **Mclose/Nclose/Mopen/Nopen**   Open or close the mesh in the M or N direction. Either the open or close options appear, depending on whether the mesh is currently open or closed in each direction.

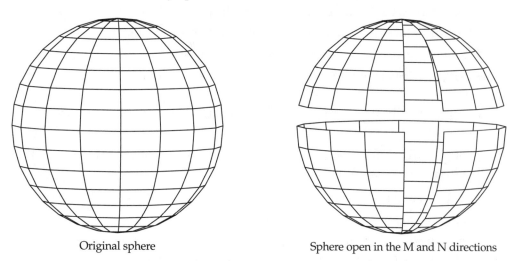

Original sphere                    Sphere open in the M and N directions

- **Undo**   Reverses the previous PEDIT option.
- **Exit**   Ends the PEDIT command.

## Editing 3-D Polyline Vertices

You can use the Edit Vertex option to modify individual mesh vertices. As you already learned, when you select this option, the PEDIT command switches to its special vertex editing mode and places an X marker on the first vertex. The X marker indicates the

BECOMING AN EXPERT

vertex that you are editing. When in vertex editing mode, you can edit the mesh one vertex at a time. The PEDIT command displays a slightly different set of prompts than those you learned about in Chapter 10:

```
Current vertex (m,n).
Enter an option [Next/Previous/Left/Right/Up/Down/Move/REgen/eXit] <N>:
```

where *m,n* are the current position of the X within the mesh in the M and N directions.

- **Next**  Moves the X marker to the next vertex in the N direction. When the X marker reaches the end of one row of vertices in the N direction, it moves to the next row in the positive M direction.

- **Previous**  Moves the X marker to the previous vertex.

- **Left**  Moves the X marker to the next vertex in the N direction (similar to the Next option).

- **Right**  Moves the X marker to the previous vertex in the N direction (similar to the Previous option).

- **Up**  Moves the X marker up one row in the positive M direction.

- **Down**  Moves the X marker down one row in the negative M direction.

- **Move**  Moves the current vertex to a different location.

- **REgen**  Regenerates the mesh and affects only the mesh, not the entire drawing.

- **Exit**  Ends the Edit Vertex option and returns to the main PEDIT options prompt.

Moving a mesh vertex is very similar to moving the vertex of a simple 2-D or 3-D polyline. To move a mesh vertex:

1. Start the PEDIT command.

2. Select the mesh.

3. Type **E** (for Edit Vertex) and then press ENTER. AutoCAD switches into vertex editing mode and displays an X marker on the first vertex.

4. Type **N** (for Next), **R** (for Right), or **U** (for Up) and then press ENTER to move the X marker until it reaches the vertex that you want to move.

5. Type **M** (for Move) and then press ENTER. AutoCAD prompts you to enter the new vertex location.

6. Specify the new location for the vertex. AutoCAD immediately moves the vertex, redraws the mesh, and repeats the vertex editing options.

7. Choose another option or type **X** (for Edit) to leave the vertex editing mode. When you leave vertex editing mode, AutoCAD repeats the main PEDIT options.

8. Choose another option or press ENTER to complete the command.

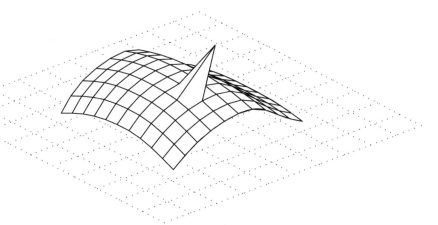

**Tip**  *You can move mesh vertices more easily by using grips. You need to use the PEDIT command, however, to perform other operations, such as mesh smoothing.*

BECOMING AN EXPERT

# Chapter 21

## Editing
## Three-Dimensional
## Solids

In the previous chapter, you learned some of the special commands for editing objects in three-dimensional space. You've also learned that many of the commands that you use to modify objects in 2-D work equally well for modifying objects in 3-D. In addition, in Chapter 19 you learned how to use the Boolean operations union, subtract, and intersect to create composite solids by using one solid to modify the shape of another solid.

When the time comes to modify the 3-D solids that you created in Chapter 19, however, AutoCAD's capabilities become a bit more limited. You can copy, move, scale, and delete these objects, but other operations, such as trim, extend, and stretch, cannot be performed on a 3-D solid. Instead, AutoCAD provides several other commands for cutting and sectioning solids, editing the faces and edges of solids, shelling solids to create thin walls, and separating composite solids back into their individual components. In addition, the FILLET and CHAMFER commands can be used to modify solids in ways that are both more complex than the capabilities available when working with 2-D objects and more appropriate to working with 3-D models. This chapter explains how to do the following:

- Chamfer solids
- Fillet solids
- Section solids
- Slice solids
- Edit faces
- Edit edges
- Separate solids
- Shell solids

**Note**   *AutoCAD provides just the basic solid-modeling capabilities. More advanced functions, such as the ability to alter the basic shapes (or features) originally combined into composite solids, is provided in more advanced programs, such as Autodesk's Mechanical Desktop.*

# Chamfering Solids

In Chapter 10, you learned how to connect two nonparallel objects with a line to create a beveled edge or chamfer. When you work with 3-D solids, you can also use the CHAMFER command to bevel one or more edges, creating a new surface between two nonparallel surfaces.

To chamfer a solid, you start the CHAMFER command and then select the solid by clicking an edge. After you select an edge, the program knows that you intend to chamfer a solid, and therefore presents a different series of prompts than would appear if you were modifying a 2-D object. AutoCAD highlights one of the surfaces adjacent to the edge that you selected, indicating that the highlighted surface is currently selected as the base surface. When chamfering a solid, you can define the chamfer based on distances only; there is no angle option. AutoCAD needs to differentiate between the two surfaces that are to be beveled, and thus designates one surface as the base surface. The first chamfer distance is measured perpendicular to the edge along this surface (the base surface distance); the second distance (the other surface distance) is measured perpendicular to the edge along the adjacent surface. If AutoCAD highlights the wrong surface, you can easily switch to the other surface.

Once you select the base surface, AutoCAD prompts you for the chamfer distances by asking for the base surface distance and the other surface distance. You can either accept the defaults or specify new distances. These distances determine the resulting chamfer. The final step is to specify the edges that you want to chamfer. If you don't want to chamfer all the edges of the base surface, you can select the edges individually. If you want to create a chamfer that bevels all the edges around the base surface, however, you can type **L** to switch to Loop mode. When in Loop mode, selecting any edge causes all the edges of the base surface to be selected. Figure 21-1 illustrates the difference between Edge mode and Loop mode.

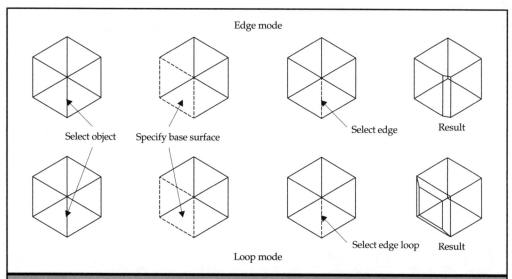

**Figure 21-1.**    *You can chamfer individual edges between adjacent surfaces or all the edges around a base surface of a solid.*

BECOMING AN EXPERT

To chamfer a solid:

1. Do one of the following:

   ■ On the Modify toolbar, click Chamfer.

   ■ From the Modify menu, choose Chamfer.

   ■ At the command line, type **CHAMFER** (or **CHA**) and then press ENTER.

2. Select the edge of the surface of the solid to be chamfered.

3. To select a different surface, type **N** (for Next) and then press ENTER, or right-click and select Next from the shortcut menu. To accept the highlighted surface as the base surface, press ENTER.

4. Specify the base surface chamfer distance.

5. Specify the other surface chamfer distance.

6. Do one of the following:

   ■ To chamfer all the edges around the base surface, type **L** (for Loop), press ENTER, click an edge of the base surface, and then press ENTER.

   ■ To chamfer individual edges, click those edges and then press ENTER.

---

*LEARN BY EXAMPLE*
*Use the CHAMFER command to bevel the mounting holes in a base plate. Open the drawing Figure 21-2 on the companion web site.*

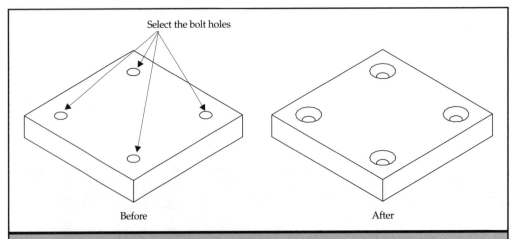

**Figure 21-2.**   *The CHAMFER command lets you quickly bevel the edges between two adjacent surfaces.*

To bevel the mounting holes in the base plate, use the following command sequence and instructions. Refer to Figure 21-2 when selecting the surfaces and edges to be chamfered. For clarity, the drawing on the right shows a hidden-line view.

```
Command: CHAMFER
(TRIM mode) Current chamfer Dist1 = 0.5000, Dist2 = 0.5000
Select first line or [Polyline/Distance/Angle/Trim/Method]: (select
one of the bolt holes)
Base surface selection...
Enter surface selection option [Next/OK (current)] <OK>: ENTER
Specify base surface chamfer distance <0.5000>: .25
Specify other surface chamfer distance <0.5000>: .25
Select an edge or [Loop]: (select one of the bolt holes)
Select an edge or [Loop]: (select a second bolt hole)
Select an edge or [Loop]: (select a third bolt hole)
Select an edge or [Loop]: (select the fourth bolt hole)
Select an edge or [Loop]: ENTER
```

## Filleting Solids

The FILLET command, another command that you learned about in Chapter 10, can also be used to modify solids, by rounding the edge that connects two adjacent surfaces. To fillet a solid, you start the FILLET command and then select the solid by clicking an edge. Once you select an edge, the program knows that you intend to fillet a solid, and therefore presents a different series of prompts than would appear if you were modifying a 2-D object.

AutoCAD highlights the edge that you selected and prompts you to specify the fillet radius. After you specify the radius or accept the default, you are prompted to select the edges that you want to fillet. To fillet just the edge that you initially selected, press ENTER. To fillet additional edges by using the same fillet radius, select each of them. You can also fillet a tangential sequence of edges by typing **C** to switch to Chain mode. When in Chain mode, selecting any edge causes AutoCAD to select all the other edges tangentially along the surface. Figure 21-3 illustrates the difference between Edge mode and Chain mode.

To fillet a solid:

1. Do one of the following:

   - On the Modify toolbar, click Fillet.
   - From the Modify menu, choose Fillet.
   - At the command line, type **FILLET** (or **F**) and then press ENTER.

2. Select the edge of the object to be filleted.

3. Specify the fillet radius.

4. Do one of the following:

- To fillet just the selected edge, press ENTER.
- To fillet all the tangential edges around a surface, type **C** (for Chain), press ENTER, click an edge to select the chain edge, and then press ENTER.
- To fillet individual edges, click those edges and then press ENTER.

*LEARN BY EXAMPLE*
*Use the FILLET command to round the top and corner edges of the base plate that you worked on in the previous example. Open the drawing Figure 21-4 on the companion web site.*

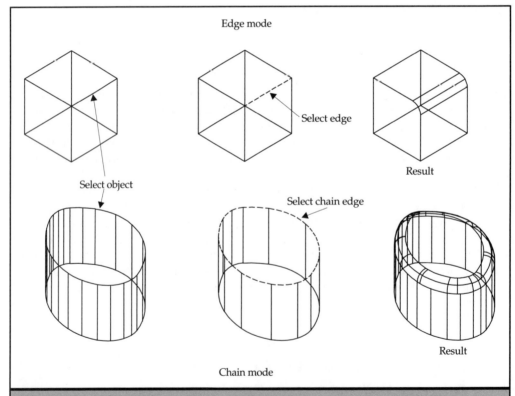

**Figure 21-3.** *You can fillet individual edges between adjacent surfaces or all the edges tangentially around a surface.*

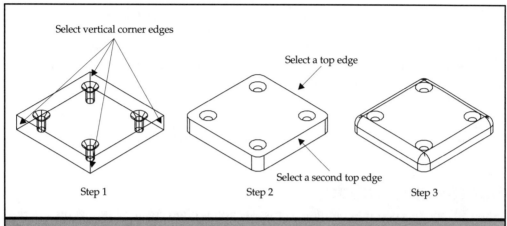

Select vertical corner edges

Select a top edge

Select a second top edge

Step 1                          Step 2                          Step 3

**Figure 21-4.**   *The FILLET command lets you quickly round the edges between
adjacent surfaces.*

To round the top and corner edges of the base plate, use the following command
sequence and instructions. Refer to Figure 21-4 when selecting the edges to be filleted.
Begin by filleting the four corner edges:

```
Command: FILLET
Current settings: Mode = TRIM, Radius = 0.5000
Select first object or [Polyline/Radius/Trim]: (select one of the vertical corner edges)
Enter fillet radius <0.5000>: ENTER
Select an edge or [Chain/Radius]: (select a second vertical corner edge)
Select an edge or [Chain/Radius]: (select a third vertical corner edge)
Select an edge or [Chain/Radius]: (select the fourth vertical corner edge)
Select an edge or [Chain/Radius]: ENTER
4 edge(s) selected for fillet.
```

When you finish filleting the four corner edges, your drawing should look like Step 2
in Figure 21-4. (For clarity, Steps 2 and 3 are shown with hidden lines removed.) Complete
the process by using the Chain mode to fillet the entire top edge in one step. Use the
following command sequence and instructions:

```
Command: FILLET
Current settings: Mode = TRIM, Radius = 0.5000
```

BECOMING AN EXPERT

```
Select first object or [Polyline/Radius/Trim]: (select a top edge)
Enter fillet radius <0.5000>: ENTER
Select an edge or [Chain/Radius]: C
Select an edge or [Chain/Radius]: (select a second top edge)
Select an edge or [Chain/Radius]: ENTER
8 edge(s) selected for fillet.
```

When you finish, your drawing should look like Step 3 in Figure 21-4.

# Sectioning Solids

When 3-D models become complex, a cross-section is often created to show the interior structure of the model along a particular cutting plane. The SECTION command can be used to create such a cross-section. The SECTION command creates a region based on the cross-section of a solid cut by a plane. Regions are created on the current layer and at the location of the cross-section. The region is created as a new object; the SECTION command does not alter the solid in any way. Once the region is created, you can easily relocate it, add cross-hatching, dimension it, or even use the new object as the basis for extruding a new solid.

You specify the location of the cutting plane based on an existing object, the current viewing direction, one of the three standard planes of the current UCS, or any three points that you specify in 3-D space. Figure 21-5 illustrates these different section-plane options. You first select the objects to be sectioned and then specify the section plane. After you select the objects to be sectioned, AutoCAD prompts:

```
Specify first point on Section plane by [Object/Zaxis/View/XY/YZ/ZX/3points] <3points>:
```

The command presents seven options:

- **Object** Lets you use the plane of an existing 2-D planar object—an arc, circle, ellipse, elliptical arc, spline, or 2-D polyline—as the section plane.

- **Zaxis** Defines the section plane by letting you select a point on the plane and a point on the Z axis of the plane (in other words, normal to the plane). This option is similar to the Zaxis option of the UCS command, which you learned about in Chapter 6.

- **View** Aligns the section plane parallel to the viewing plane of the current viewport and passing through a selected point.

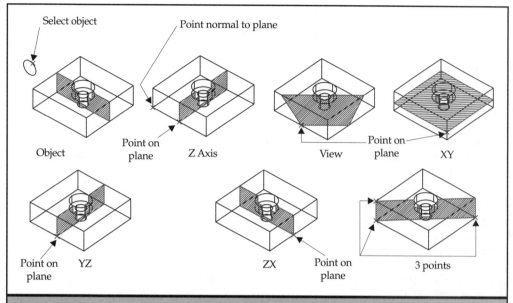

**Figure 21-5.** *You can specify the section plane by using any one of seven different options. Note that the cross-hatching was added separately after creating the region.*

- **XY/YZ/ZX** Let you align the section plane with one of the three standard planes of the current UCS. AutoCAD prompts you for the point through which the section plane passes.

- **3 points** The default option, enabling you to define any plane passing through three 3-D points.

*Although you can select the objects to be sectioned by using any object selection method, only solids can be sectioned. AutoCAD ignores any other objects that you select. If you select more than one solid, AutoCAD creates separate regions for each.*

To section a solid by using an arbitrary plane:

1. Do one of the following:

   - On the Solids toolbar, click Section.
   - From the Draw menu, choose Solids | Section.
   - At the command line, type **SECTION** (or **SEC**) and then press ENTER.

2. Select the objects to be sectioned and then press ENTER.

3. Specify the first point on the section plane.

BECOMING AN EXPERT

4. Specify the second point on the section plane.

5. Specify the third point on the section plane.

*LEARN BY EXAMPLE*

*Use the SECTION command to create a cross-section through a 3-D widget. Open the drawing Figure 21-6 on the companion web site.*

To create the cross-section, use the following command sequence and instructions. To locate the section plane, use the midpoint object snap to select the labeled point shown in Figure 21-6.

```
Command: SECTION
Select objects: (click anywhere on the widget)
Select objects: ENTER
Specify first point on Section plane by [Object/Zaxis/View/XY/YZ/ZX/
3points] <3 points>: ZX
Specify a point on the ZX-plane <0,0,0>: MIDPOINT
of (select point A)
```

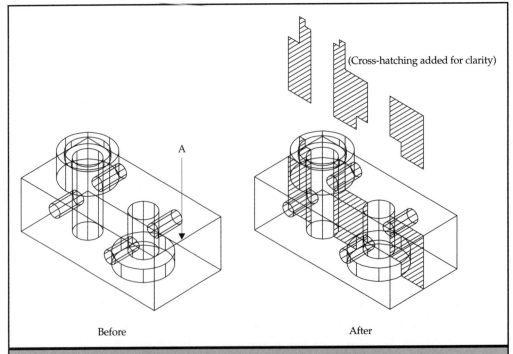

Before                                                          After

**Figure 21-6.**    *The SECTION command creates regions representing the cross-section through a solid at a specified section plane.*

After you create the cross-sectional region, it is positioned within the solid at the location of the section plane. Moving the region to a different location often is useful. Cross-hatching is also often added. In Figure 21-6, added cross-hatching makes the region more visible.

## Slicing Solids

While the SECTION command creates a cross-sectional region, sometimes you actually need to slice the solid open to illustrate its internal structure. The SLICE command does exactly that. Instead of creating a region, the command cuts through a solid along a specified plane. As when you use the SECTION command, you first select the objects to be sliced and then locate the plane. You specify the slicing plane exactly as you would locate a section plane. After you position the plane, however, you must indicate whether you simply want to separate the two sides of the solid (retaining both sides) or delete the solid on one side of the plane (retaining the solid only on the other side of the slicing plane).

When you slice a solid, AutoCAD creates new solids. The new sliced solids retain the layer and color properties of the original solids. To slice a solid using an arbitrary plane and retain only one side:

1. Do one of the following:

   ■ On the Solids toolbar, click Slice.

   ■ From the Draw menu, choose Solids | Slice.

   ■ At the command line, type **SLICE** (or **SL**) and then press ENTER.

2. Select the objects to be sliced and then press ENTER.

3. Specify the first point on the slicing plane.

4. Specify the second point on the slicing plane.

5. Specify the third point on the slicing plane.

6. Pick a point on the side of the slicing plane to indicate which portion of the solid object to retain.

*LEARN BY EXAMPLE*
*Use the SLICE command to cut the 3-D widget in half. Open the drawing Figure 21-7 on the companion web site.*

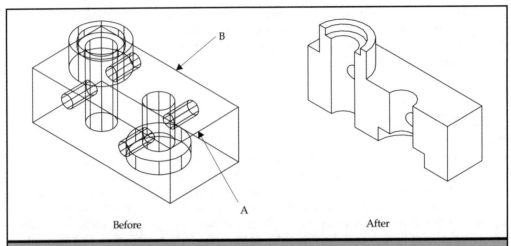

B

A

Before

After

**Figure 21-7.**   *The SLICE command uses a slicing plane to cut a solid into two pieces. You can then retain either both pieces or only the resulting solid on one side of the slicing plane.*

To slice the widget in half, use the following command sequence and instructions. To locate the slicing plane, use the midpoint object snap to select the labeled points shown in Figure 21-7.

```
Command: SLICE
Select objects: (click anywhere on the widget)
Select objects: ENTER
Specify first point on slicing plane by [Object/Zaxis/View/XY/YZ/ZX/
3points] <3 points>: ZX
Specify a point on the ZX-plane <0,0,0>: MIDPOINT
of (select point A)
Specify a point on desired side of the plane or [keep Both sides]:
(select point B)
```

# Editing Solids

AutoCAD provides several additional solids editing capabilities. You can edit solids by extruding, moving, rotating, offsetting, tapering, deleting, or copying individual faces or edges. You can also change the color of faces or edges within a solid. In addition, you can modify an entire solid by imprinting other geometry on the solid, separating a solid comprised of individual volumes into individual solid objects, shelling a solid to create a hollow, thin-walled solid, and cleaning a solid to remove shared edges or vertices. An additional function lets you verify that the resulting solid forms a valid solid object.

All of AutoCAD's face-editing capabilities are accomplished using the generalized SOLIDEDIT command. When you start this command, AutoCAD prompts:

```
Solids editing automatic checking:  SOLIDCHECK=1
Enter a solids editing option [Face/Edge/Body/Undo/eXit] <eXit>:
```

You can then select one of the available options:

- **Face**   Selects the face-editing mode
- **Edge**   Selects the edge-editing mode
- **Body**   Selects the body-editing mode
- **Undo**   Reverses all solid editing since starting the command
- **Exit**   Ends the command

Notice that the first thing the command asks you to do is determine whether you want to edit a face, an edge, or a body (the entire solid). Although you can certainly start the command from the command line and answer this initial prompt, it's much easier to start the command using a toolbar or menu. When you start the command using a toolbar or menu, AutoCAD skips this initial step by preselecting the Face, Edge, or Body option.

## Editing Faces of Solids

When you edit faces of a solid, AutoCAD prompts:

```
Enter a face editing option
[Extrude/Move/Rotate/Offset/Taper/Delete/Copy/coLor/Undo/eXit] <eXit>:
```

You can then select one of the available options:

- **Extrude**   Extrudes selected planar faces to a specified height or along a path
- **Move**   Moves selected faces to a specified height or distance
- **Rotate**   Rotates selected faces about a specified axis
- **Offset**   Offsets selected faces by a specified distance
- **Taper**   Tapers selected faces by a specified angle
- **Delete**   Removes selected faces
- **Copy**   Copies selected faces as a region or boundary
- **Color**   Changes the color of selected faces
- **Undo**   Reverses the previous action
- **Exit**   Exits the face-editing option and displays the main SOLIDEDIT prompt

BECOMING AN EXPERT

Before you can edit faces, you must select the faces you want to edit. You can select individual faces of the solid or select the faces you want to edit by creating a boundary set or by using a crossing polygon, crossing window, or fence. A *boundary set* is a set of faces defined by a closed boundary consisting of lines, arcs, circles, elliptical arcs, and spline curves. To define a boundary set on a solid, you first select a point within the solid. AutoCAD highlights the face as the boundary set. If you select the same point again, AutoCAD highlights the adjoining face. You can select additional faces, select all the faces, or remove selected faces. After you select the faces you want to edit, press ENTER to continue the particular face-editing command.

## Extruding Faces

After you select the faces you want to extrude, you can specify the height of the extrusion or extrude the faces along an existing path based on a line, arc, circle, ellipse, elliptical arc, polyline, or spline. Figure 21-8 illustrates several extruded faces.

Each face has a positive side, which is the side of the face in the same direction as the face's normal. When you specify an extrusion height, positive values extrude the face in its positive direction (normally outward from the solid), whereas negative values extrude it in the negative direction (normally inward). If you specify the extrusion height, you can also taper the face. A positive taper angle tapers the faces inward as you move in the extrusion direction, whereas a negative angle tapers the faces outward.

*If the taper angle you specify would cause the face to taper to a point before reaching the specified extrusion height, AutoCAD rejects the extrusion and repeats the face-editing option prompt.*

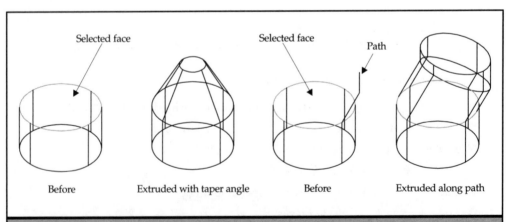

**Figure 21-8.** *You can extrude a face to a specified height or along a path.*

To extrude a face:

1. Do one of the following:

   - On the Solids Editing toolbar, click Extrude Faces.
   - From the Modify menu, choose Solids Editing | Extrude Faces.
   - Start the SOLIDEDIT command, select the Face option, and then select the Extrude option.

2. Select the faces you want to extrude, and then press ENTER to continue.

3. Specify the height of extrusion.

4. Specify a taper angle or press ENTER.

5. Press ENTER twice to end the command.

*LEARN BY EXAMPLE*
*Use the Extrude option to extrude a face along a path. Open the drawing Figure 21-9 on the companion web site.*

When you extrude a face along a path, all the profiles of the selected face extrude along the selected path. To extrude a face along a path, refer to Figure 21-9 and use the following command sequence and instructions:

```
Command: SOLIDEDIT
Solids editing automatic checking:  SOLIDCHECK=1
Enter a solids editing option [Face/Edge/Body/Undo/eXit] <eXit>: F
Enter a face editing option
[Extrude/Move/Rotate/Offset/Taper/Delete/Copy/coLor/Undo/eXit]
 <eXit>: E
Select faces or [Undo/Remove]: (select point A)
Select faces or [Undo/Remove/ALL]: ENTER
Specify height of extrusion or [Path]: P
Select extrusion path: (select path)
Solid validation started.
Solid validation completed.
Enter a face editing option
[Extrude/Move/Rotate/Offset/Taper/Delete/Copy/coLor/Undo/eXit]
 <eXit>: ENTER
Solids editing automatic checking:  SOLIDCHECK=1
Enter a solids editing option [Face/Edge/Body/Undo/eXit] <eXit>: ENTER
```

*When extruding along a path, make sure the path doesn't lie on the same plane as the selected face. Also, the path shouldn't have areas of high curvature.*

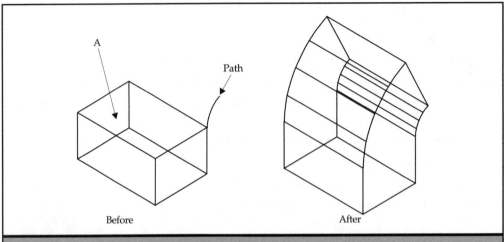

**Figure 21-9.** *Extruding a face along a path*

## Moving Faces

After you select the faces you want to move, you can specify a base point or displacement and a second point of displacement, similar to using the MOVE command. The program then moves the selected faces without changing their orientation.

To move a face:

1. Do one of the following:

   - On the Solids Editing toolbar, click Move Faces.
   - From the Modify menu, choose Solids Editing | Move Faces.
   - Start the SOLIDEDIT command, select the Face option, and then select the Move option.

2. Select the face you want to move, and then press ENTER to continue.

3. Specify the base point.

4. Specify the second point of displacement.

5. Press ENTER twice to end the command.

*LEARN BY EXAMPLE*

*Use the Move option to move a hole from one location to another in a 3-D solid. Open the drawing Figure 21-10 on the companion web site.*

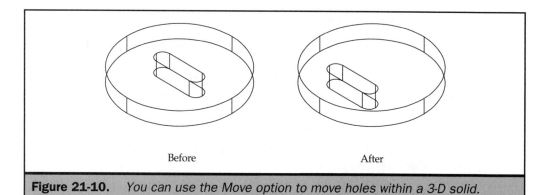

**Figure 21-10.** *You can use the Move option to move holes within a 3-D solid.*

To move the slotted hole from one location to another, as shown in Figure 21-10, use the following command sequence and instructions:

```
Command: SOLIDEDIT
Solids editing automatic checking:  SOLIDCHECK=1
Enter a solids editing option [Face/Edge/Body/Undo/eXit] <eXit>: F
Enter a face editing option
[Extrude/Move/Rotate/Offset/Taper/Delete/Copy/coLor/Undo/eXit]
 <eXit>: M
Select faces or [Undo/Remove]: (select a face)
Select faces or [Undo/Remove/ALL]: ALL
Select faces or [Undo/Remove/ALL]: R
Remove faces or [Undo/Add/ALL]: (select upper edge of circular plate)
Remove faces or [Undo/Add/ALL]: (select lower edge of circular plate)
Remove faces or [Undo/Add/ALL]: ENTER
Specify a base point or displacement: 0,-3
Specify a second point of displacement: ENTER
Solid validation started.
Solid validation completed.
Enter a face editing option
[Extrude/Move/Rotate/Offset/Taper/Delete/Copy/coLor/Undo/eXit]
 <eXit>: ENTER
Solids editing automatic checking:  SOLIDCHECK=1
Enter a solids editing option [Face/Edge/Body/Undo/eXit] <eXit>: ENTER
```

## Rotating Faces

The Face Rotate option lets you rotate faces about an axis by a specified angle, similar to AutoCAD's ROTATE3D command. After you select the faces you want to rotate, AutoCAD prompts:

```
Specify an axis point or [Axis by object/View/Xaxis/Yaxis/Zaxis] <2points>:
```

The command provides five options:

- **Axis by object** Aligns the axis of rotation with an existing object: a line, circle, arc, ellipse, polyline, or spline. If you select a line, polyline, or spline, the axis is aligned with the object (or passes through its end points). If you select an arc, circle, or ellipse, however, the axis is the line perpendicular to the object passing through its center point.

- **View** Aligns the axis of rotation with the viewing direction of the current viewport. AutoCAD prompts you for the point on the current view plane through which the axis passes.

- **Xaxis/Yaxis/Zaxis** Let you align the axis of rotation parallel to one of the axes of the current UCS. AutoCAD prompts you for the point through which the axis passes.

- **2points** The default option, enabling you to define any axis passing through two 3-D points.

After you select the axis of rotation, AutoCAD prompts you for the rotation angle:

```
Specify a rotation angle or [Reference]:
```

To rotate the selected faces by a specified angle, simply type the angle of rotation. If you type **R** (for Reference) and then press ENTER, AutoCAD prompts you for the reference angle and the new angle, similar to when using the ROTATE3D command.

To rotate selected faces about an axis:

1. Do one of the following:

   - On the Solids Editing toolbar, click Rotate Faces.
   - From the Modify menu, choose Solids Editing | Rotate Faces.
   - Start the SOLIDEDIT command, select the Face option, and then select the Rotate option.

2. Select the faces you want to rotate, and then press ENTER to continue.

3. Specify the method you want to use to define the axis of rotation, and then specify the axis.

4. Specify the rotation angle.

5. Press ENTER twice to end the command.

*LEARN BY EXAMPLE*
*Use the Rotate option to rotate a slotted hole in a solid plate. Open the drawing Figure 21-11 on the companion web site.*

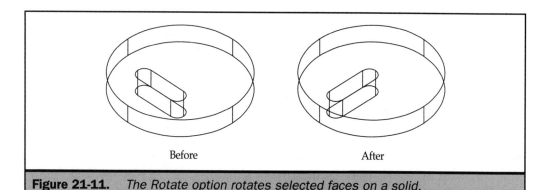

**Figure 21-11.**    *The Rotate option rotates selected faces on a solid.*

Before                    After

To rotate the slotted hole from the previous example about the Z axis passing through its center point, as shown in Figure 21-11, use the following command sequence and instructions:

```
Command: SOLIDEDIT
Solids editing automatic checking:  SOLIDCHECK=1
Enter a solids editing option [Face/Edge/Body/Undo/eXit] <eXit>: F
Enter a face editing option
[Extrude/Move/Rotate/Offset/Taper/Delete/Copy/coLor/Undo/eXit]
  <eXit>: R
Select faces or [Undo/Remove]: (select a face)
Select faces or [Undo/Remove/ALL]: ALL
Select faces or [Undo/Remove/ALL]: R
Remove faces or [Undo/Add/ALL]: (select upper edge of circular plate)
Remove faces or [Undo/Add/ALL]: (select lower edge of circular plate)
Remove faces or [Undo/Add/ALL]: ENTER
Specify an axis point or [Axis by object/View/Xaxis/Yaxis/Zaxis]
  <2points>: Z
Specify the origin of the rotation <0,0,0>: 0,-3,0
Specify a rotation angle or [Reference]: 90
Solid validation started.
Solid validation completed.
Enter a face editing option
[Extrude/Move/Rotate/Offset/Taper/Delete/Copy/coLor/Undo/eXit]
  <eXit>: ENTER
Solids editing automatic checking:  SOLIDCHECK=1
Enter a solids editing option [Face/Edge/Body/Undo/eXit] <eXit>: ENTER
```

## Offsetting Faces

The Face Offset option lets you offset selected faces by a specified distance. AutoCAD prompts you to specify the offset distance. New faces are then created by offsetting the

BECOMING AN EXPERT

selected faces at the specified distance from their original locations. Positive distances offset the faces in the direction normal to the positive side of the surface or face, increasing the volume of the solid. If you select a hole, however, a positive value decreases the size of the hole, whereas a negative value increases its size.

To offset a face on a solid:

1. Do one of the following:

   - On the Solids Editing toolbar, click Offset Faces.
   - From the Modify menu, choose Solids Editing | Offset Faces.
   - Start the SOLIDEDIT command, select the Face option, and then select the Offset option.

2. Select the faces you want to offset, and then press ENTER to continue.

3. Specify the offset distance.

4. Press ENTER twice to end the command.

*The direction of the offset depends on the faces you select. By carefully selecting the faces, you can change the size of the solid, change the size of a hole, or use the Offset option to extrude both the hole and solid.*

Before

Offset solid

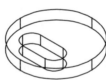
Offset hole

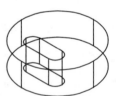
Offset solid and hole (similar to extrude)

*LEARN BY EXAMPLE*
*Use the Offset option to change the size of the holes in a solid plate. Open the drawing Figure 21-12 on the companion web site.*

To change the size of the holes in the solid plate, as shown in Figure 21-12, use the following command sequence and instructions:

```
Command: SOLIDEDIT
Solids editing automatic checking:   SOLIDCHECK=1
Enter a solids editing option [Face/Edge/Body/Undo/eXit] <eXit>: F
Enter a face editing option
```

```
[Extrude/Move/Rotate/Offset/Taper/Delete/Copy/coLor/Undo/eXit]
  <eXit>: O
Select faces or [Undo/Remove]: F
First fence point: (select point A)
Specify endpoint of line or [Undo]: (select point B)
Specify endpoint of line or [Undo]: (select point C)
Specify endpoint of line or [Undo]: (select point D)
Specify endpoint of line or [Undo]: ENTER
Select faces or [Undo/Remove/ALL]: ENTER
Specify the offset distance: -.5
Solid validation started.
Solid validation completed.
Enter a face editing option
[Extrude/Move/Rotate/Offset/Taper/Delete/Copy/coLor/Undo/eXit]
  <eXit>: ENTER
Solids editing automatic checking:  SOLIDCHECK=1
Enter a solids editing option [Face/Edge/Body/Undo/eXit] <eXit>: ENTER.
```

## Tapering Faces

The Face Taper option lets you taper faces by applying a draft angle along a direction vector. AutoCAD prompts you to specify the base point, another point along the axis of tapering, and the taper angle. When tapering a solid, positive angles taper the solid inward along the direction vector, whereas negative angles taper the solid outward.

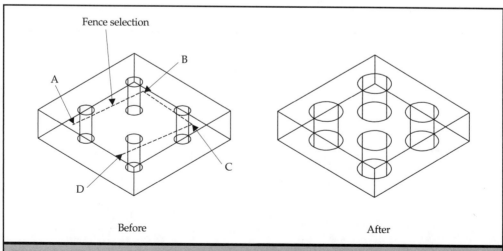

**Figure 21-12.**   *You can offset faces to change the size of holes in a 3-D solid.*

BECOMING AN EXPERT

If you are tapering a hole, however, positive angles taper the hole outward, while negative angles taper it inward.

**Note** *If the taper angle you specify would cause the face to taper to a point before reaching the specified extrusion height, AutoCAD rejects the extrusion and repeats the face-editing option prompt.*

To taper a face:

1. Do one of the following:

   - On the Solids Editing toolbar, click Taper Faces.
   - From the Modify menu, choose Solids Editing | Taper Faces.
   - Start the SOLIDEDIT command, select the Face option, and then select the Taper option.

2. Select the faces you want to taper, and then press ENTER to continue.

3. Specify the base point for the taper direction vector.

4. Specify another point along the taper direction vector.

5. Specify the taper angle.

6. Press ENTER twice to end the command.

**LEARN BY EXAMPLE**
*Use the Taper option to change the draft angle of the holes in a solid plate. Open the drawing Figure 21-13 on the companion web site.*

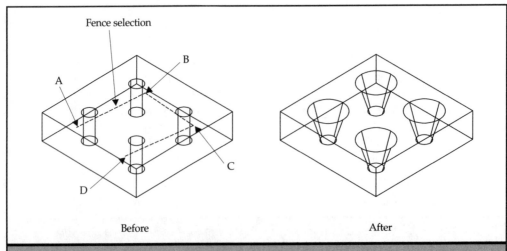

**Figure 21-13.** *The Taper option lets you taper faces along a direction vector.*

To change the draft angle of the holes in the solid plate, as shown in Figure 21-13, use the following command sequence and instructions:

```
Command: SOLIDEDIT
Solids editing automatic checking:  SOLIDCHECK=1
Enter a solids editing option [Face/Edge/Body/Undo/eXit] <eXit>: F
Enter a face editing option
[Extrude/Move/Rotate/Offset/Taper/Delete/Copy/coLor/Undo/eXit]
 <eXit>: T
Select faces or [Undo/Remove]: F
First fence point: (select point A)
Specify endpoint of line or [Undo]: (select point B)
Specify endpoint of line or [Undo]: (select point C)
Specify endpoint of line or [Undo]: (select point D)
Specify endpoint of line or [Undo]: ENTER
Select faces or [Undo/Remove/ALL]: ENTER
Specify the base point: CENTER
of (select the bottom of one of the holes)
Specify another point along the axis of tapering: CENTER
of (select the top of the same hole)
Specify the taper angle: 20
Solid validation started.
Solid validation completed.
Enter a face editing option
[Extrude/Move/Rotate/Offset/Taper/Delete/Copy/coLor/Undo/eXit]
 <eXit>: ENTER
Solids editing automatic checking:  SOLIDCHECK=1
Enter a solids editing option [Face/Edge/Body/Undo/eXit] <eXit>: ENTER
```

## Deleting Faces

The Face Delete option enables you to delete faces, fillets, and chamfers from an existing solid object. This option is quite powerful. It can even be used to remove holes from solids, as shown in Figure 21-14.

To delete a face:

1. Do one of the following:

   - On the Solids Editing toolbar, click Delete Faces.
   - From the Modify menu, choose Solids Editing | Delete Faces.
   - Start the SOLIDEDIT command, select the Face option, and then select the Delete option.

2. Select the faces you want to delete, and then press ENTER to continue.

3. Press ENTER twice to end the command.

BECOMING AN EXPERT

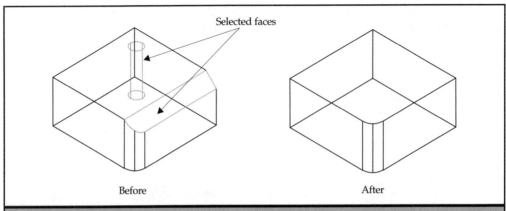

**Figure 21-14.** *The delete option lets you delete faces from a 3-D solid.*

## Copying Faces

The Face Copy option lets you copy one or more faces. After you select the faces you want to copy, you can specify a base point or displacement and a second point of displacement, similar to using the COPY command. The program then copies the selected faces without changing their orientation, as shown in Figure 21-15.

 *AutoCAD creates the copies as regions or bodies, however, not as solids. Although you can extrude a region to create a new solid, you can't modify bodies. Thus, you can't copy an existing hole to create another hole, because the copied faces become a body.*

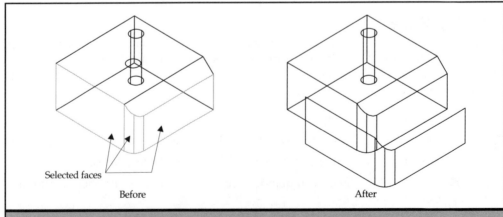

**Figure 21-15.** *The Copy option lets you copy one or more faces.*

To copy a face:

1. Do one of the following:

- On the Solids Editing toolbar, click Copy Faces.
- From the Modify menu, choose Solids Editing | Copy Faces.
- Start the SOLIDEDIT command, select the Face option, and then select the Copy option.

2. Select the faces you want to copy, and then press ENTER to continue.

3. Specify the base point.

4. Specify the second point of displacement.

5. Press ENTER twice to end the command.

## Coloring Faces

The Face Color option lets you change the color of one or more faces of a 3-D solid. After you select the faces you want to change, AutoCAD displays the Select Color dialog box. The color you specify overrides the color setting for the layer on which the solid was drawn.

To change the color of a face:

1. Do one of the following:

- On the Solids Editing toolbar, click Color Faces.
- From the Modify menu, choose Solids Editing | Color Faces.
- Start the SOLIDEDIT command, select the Face option, and then select the Color option.

2. Select the faces you want to change, and then press ENTER to continue.

3. In the Select Color dialog box, specify the new color and then click OK.

4. Press ENTER twice to end the command.

# Editing Edges of Solids

You can also edit individual edges of a 3-D solid. When you select the Edge option of the SOLIDEDIT command, AutoCAD prompts:

```
Enter an edge editing option [Copy/coLor/Undo/eXit] <eXit>:
```

You can then select one of the available options:

- **Copy**   Copies edges as lines, arcs, circles, ellipses, or splines
- **Color**   Changes the color of selected edges
- **Undo**   Reverses the previous action
- **Exit**   Exits the edge-editing option and displays the main SOLIDEDIT prompt

BECOMING AN EXPERT

Before you can edit edges, you must select the edges you want to change. You can select individual edges by pointing or by using the crossing, fence, or crossing polygon selection methods. You can select additional edges or remove selected edges from the selection set. After you select the edges you want to edit, press ENTER to continue the particular edge-editing command.

## Copying Edges

After you select the edges you want to copy, you can specify a base point or displacement and a second point of displacement, similar to using the COPY command. The program then copies the selected edges as lines, arcs, circles, ellipses, or splines, without changing their orientation, as shown in Figure 21-16.

To copy an edge:

1. Do one of the following:

   - On the Solids Editing toolbar, click Copy Edges.
   - From the Modify menu, choose Solids Editing | Copy Edges.
   - Start the SOLIDEDIT command, select the Edge option, and then select the Copy option.

2. Specify the edges you want to copy, and then press ENTER to continue.

3. Specify the base point.

4. Specify the second point of displacement.

5. Press ENTER twice to end the command.

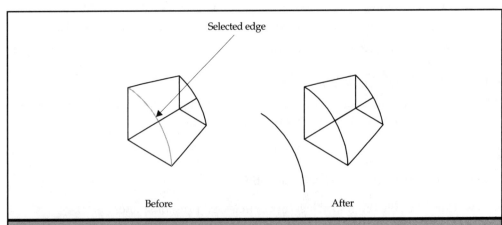

**Figure 21-16.** *The Edge Copy option lets you copy one or more edges.*

*You can use the Edge Copy option to create a copy of a hole in an existing 3-D solid. After you copy the edge representing the profile of the hole, use the EXTRUDE command to create a new solid representing the hole, and then use the SUBTRACT command to remove the new solid from the existing solid.*

## Coloring Edges

The Edge Color option lets you change the color of one or more edges of a 3-D solid. After you select the edges you want to change, AutoCAD displays the Select Color dialog box. The color you specify overrides the color setting for the layer on which the solid was drawn.

To change the color of an edge:

1. Do one of the following:

   - On the Solids Editing toolbar, click Color Edges.
   - From the Modify menu, choose Solids Editing | Color Edges.
   - Start the SOLIDEDIT command, select the Edge option, and then select the Color option.

2. Select the edges you want to change, and then press ENTER to continue.

3. In the Select Color dialog box, specify the new color and then click OK.

4. Press ENTER twice to end the command.

# Editing Entire Solids

When you edit the entire body of the solid, AutoCAD prompts:

```
Enter a body editing option
[Imprint/seParate solids/Shell/cLean/Check/Undo/eXit] <eXit>:
```

You can then select one of the available options:

- **Imprint**   Imprints an object on the selected solid
- **Separate solids**   Separates 3-D solids having independent volumes into independent solid objects
- **Shell**   Creates a hollow, thin wall of a specified thickness
- **Clean**   Removes shared edges or vertices having the same surface or curve definition
- **Check**   Validates the 3-D solid as a valid ACIS solid
- **Undo**   Reverses the previous action
- **Exit**   Exits the body-editing option and displays the main SOLIDEDIT prompt

BECOMING AN EXPERT

Before you can edit the solid, you must select it, either by pointing or by using the Last object selection method.

## Imprinting Solids

The Body Imprint option imprints a selected object onto the selected solid. AutoCAD prompts you to select a single object to imprint. You can imprint an arc, body, circle, ellipse, line, polyline, region, solid, or spline onto a solid. The object to be imprinted must intersect one or more faces on the selected solid. You can then specify whether to retain or delete the original imprint object. After you imprint the object onto the solid, you can select another object to imprint. Figure 21-17 shows an imprinted solid.

To imprint onto a 3-D solid:

1. Do one of the following:

   ■ On the Solids Editing toolbar, click Imprint.

   ■ From the Modify menu, choose Solids Editing | Imprint.

   ■ Start the SOLIDEDIT command, select the Body option, and then select the Imprint option.

2. Select the solid on which you want to imprint.

3. Select the object you want to imprint onto the solid.

4. Press ENTER to retain the original object, or type **Y** (for Yes) and press ENTER (or right-click and select Yes from the shortcut menu) to delete the original source object being imprinted.

5. Select another object to imprint or press ENTER.

6. Press ENTER twice to end the command.

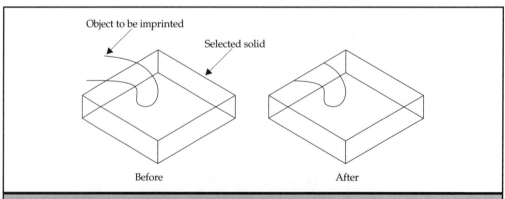

**Figure 21-17.**   *You can imprint objects onto a solid.*

## Separating Solids

As you've already learned, it is possible to create a single solid consisting of separate volumes. For example, if you select two solids that don't intersect when using the UNION command, AutoCAD creates a single solid consisting of the separate volumes. Similar solids can also be created using the SLICE command. The Body Separate option enables you to separate the volumes into individual solids, thus making each easier to modify.

To separate solids:

1. Do one of the following:

■ On the Solids Editing toolbar, click Separate.

■ From the Modify menu, choose Solids Editing | Separate.

■ Start the SOLIDEDIT command, select the Body option, and then select the Separate option.

2. Select the solid you want to separate.

3. Press ENTER twice to end the command.

## Shelling Solids

The Body Shell option converts a solid with a solid-filled volume into a shelled solid—a solid consisting of thin outer walls surrounding an internal empty volume. AutoCAD creates shelled solids by offsetting the existing outer surface of the solid. The program first prompts you to select the solid and then asks for the shell offset distance—the thickness of the shell. A positive offset distance offsets the inner shell face inward, whereas a negative value enlarges the solid by adding the shell thickness outward from the existing faces, as shown in Figure 21-18.

BECOMING AN EXPERT

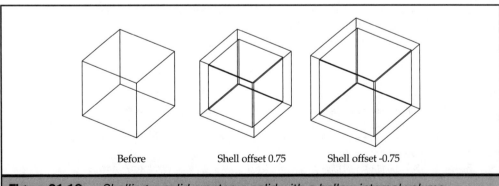

Before          Shell offset 0.75          Shell offset -0.75

**Figure 21-18.**   *Shelling a solid creates a solid with a hollow internal volume surrounded by a thin wall.*

**Caution** *You can only shell a solid one time. You can use the Face Extrude option, however, to change the shell thickness.*

To shell a solid:

1. Do one of the following:

■ On the Solids Editing toolbar, click Shell.

■ From the Modify menu, choose Solids Editing | Shell.

■ Start the SOLIDEDIT command, select the Body option, and then select the Shell option.

2. Select the solid you want to shell.

3. Select any faces to be excluded from shelling, and then press ENTER.

4. Specify the shell offset distance.

5. Press ENTER twice to end the command.

---

*LEARN BY EXAMPLE*
*Use the Body Shell and Face Extrude options to convert a solid box into a thin-walled box with an open top. Open the drawing Figure 21-19 on the companion web site.*

You can leverage the power of the Body Shell and Face Extrude options of the SOLIDEDIT command to convert a solid box into a thin-walled box with an open top, as shown in Figure 21-19. Use the following command sequence and instructions:

```
Command: SOLIDEDIT
Solids editing automatic checking:  SOLIDCHECK=1
Enter a solids editing option [Face/Edge/Body/Undo/eXit] <eXit>: B
Enter a body editing option
[Imprint/seParate solids/Shell/cLean/Check/Undo/eXit] <eXit>: S
Select a 3D solid: (select the solid)
Remove faces or [Undo/Add/ALL]: ENTER
Enter the shell offset distance: 1
Solid validation started.
Solid validation completed.
Enter a body editing option
[Imprint/seParate solids/Shell/cLean/Check/Undo/eXit] <eXit>: ENTER
Solids editing automatic checking:  SOLIDCHECK=1
Enter a solids editing option [Face/Edge/Body/Undo/eXit] <eXit>: F
Enter a face editing option
[Extrude/Move/Rotate/Offset/Taper/Delete/Copy/coLor/Undo/eXit]
 <eXit>: E
Select faces or [Undo/Remove]: (select point A, so that the top face
of the outer box is selected)
Select faces or [Undo/Remove/ALL]: (select point A again so that the
top face of the inner box is also selected)
```

```
Select faces or [Undo/Remove/ALL]: R
Remove faces or [Undo/Add/ALL]: (select the top face of the outer
box at point B to remove it from the set)
Remove faces or [Undo/Add/ALL]: ENTER
Specify height of extrusion or [Path]: -1
Specify angle of taper for extrusion <0>: ENTER
Solid validation started.
Solid validation completed.
Enter a face editing option
[Extrude/Move/Rotate/Offset/Taper/Delete/Copy/coLor/Undo/eXit]
 <eXit>: ENTER
Solids editing automatic checking:  SOLIDCHECK=1
Enter a solids editing option [Face/Edge/Body/Undo/eXit] <eXit>: ENTER
```

## Cleaning Solids

The Body Clean option removes edges or vertices that share the same surface or vertex definition on either side of an edge or vertex. This option checks the body faces or edges on the solid and merges adjacent faces that share the same surface, removing any redundant edges, including imprinted objects.

To clean a 3-D solid:

1. Do one of the following:

   - On the Solids Editing toolbar, click Clean.
   - From the Modify menu, choose Solids Editing | Clean.
   - Start the SOLIDEDIT command, select the Body option, and then select the Clean option.

2. Select the solid you want to clean.

3. Press ENTER twice to end the command.

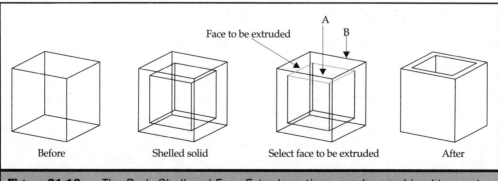

Before      Shelled solid      Select face to be extruded      After

**Figure 21-19.**  *The Body Shell and Face Extrude options can be combined to create a hollow box with an open top.*

BECOMING AN EXPERT

## Checking Solids

The Body Check option checks a solid to determine whether it is a valid 3-D solid object. Normally, AutoCAD doesn't create invalid solids. If you turn off the SOLIDCHECK system variable to improve performance, however, AutoCAD may create invalid solids. Valid solids can be modified without causing AutoCAD to display an ACIS failure error message, whereas invalid solids can't be modified. When you check a solid, AutoCAD simply reports whether the solid is valid or not.

To check a 3-D solid:

1. Do one of the following:

   - On the Solids Editing toolbar, click Check.
   - From the Modify menu, choose Solids Editing | Check.
   - Start the SOLIDEDIT command, select the Body option, and then select the Check option.

2. Select the solid object you want to validate. AutoCAD immediately displays a message informing you whether the object is a valid ACIS solid.

3. Press ENTER twice to end the command.

*When editing solids, the SOLIDCHECK system variable determines whether AutoCAD checks to ensure that the resulting solid is always a valid ACIS solid. When SOLIDCHECK equals 1, AutoCAD always performs solid validation. When the value equals 0, however, the program doesn't validate solids during editing operations, making it possible to create invalid solids.*

*When editing solids, you should try to orient the objects being edited as close to the origin as possible (within the range 0.0001 to 1000 units from the origin). Solids located a greater distance from the origin sometimes cause errors in AutoCAD 2002. After editing the solids, you can move them as needed.*

# Representing Solids Using SOLPROF, SOLVIEW, and SOLDRAW

Sometimes, the SECTION and SLICE commands you learned about earlier in this chapter just don't enable you to present visualizations of your models that clearly illustrate what they represent. Sometimes, to represent the solid object, you need to create orthographic or auxiliary views, the types of views that you would produce if you were using conventional drafting techniques. When that's the case, three special AutoCAD commands fill the bill. The key word here is *represent*. The SOLPROF, SOLDRAW, and SOLVIEW commands help you to do the work necessary to draw *representations* of solids that mimic conventional 2-D drafting.

The SOLVIEW and SOLDRAW commands are meant to be used in order. First, you use the SOLVIEW command to create the necessary floating viewports containing the various views of the 3-D solid, and then you use SOLDRAW to convert the contents of those viewports into 2-D representations of the solids.

# Setting Up Views

Begin the process by setting up the views. To start the SOLVIEW command, do one of the following:

- On the Solids toolbar, click Setup View.
- From the Draw menu, choose Solids | Setup | View.
- At the command line, type **SOLVIEW** and then press ENTER.

When you start the SOLVIEW command, AutoCAD turns off TILEMODE, switches to a paper space layout, and prompts you to select one of the main SOLVIEW options:

```
Enter an option [Ucs/Ortho/Auxiliary/Section]:
```

These options are meant to be used in the order shown. The first thing that you need to do is create a profile view relative to a User Coordinate System. Generally, you create the first view relative to the World Coordinate System, but you can select any named UCS or the current UCS. AutoCAD next prompts you for the scale of the view, which is also the paper space scale factor. After you establish the scale, you position the view by specifying its center, and then define the border of the viewport by specifying its clip corners. Generally, you will want to make the viewport large enough to contain the entire view of the solid, but you can use the borders to clip the view of the solid, to isolate just a portion of the object.

**Note**    *The SOLVIEW command automatically creates new layers that SOLDRAW uses to create its visible lines, hidden lines, and cross-sectional hatching for each view. SOLVIEW also creates layers that you use for adding dimensions. The command automatically defines these layers so that they are frozen in all but their specific viewports. The SOLVIEW command names these layers based on the view name that you specify when you name each viewport, by appending a hyphen and the additional three letters VIS, HID, DIM, or HAT for layers designated for visible lines, hidden lines, dimensions, and hatching, respectively. In addition, if the Hidden linetype is loaded, AutoCAD automatically assigns that linetype to the hidden line layers. The command also places all the viewport borders on a layer named VPORTS, so that you can later turn off the viewport borders when you plot the finished drawing. Since anything drawn on these layers is automatically modified when you use the SOLDRAW command, you should not draw anything else on these layers.*

BECOMING AN EXPERT

After you establish the first view, you will usually proceed to the Ortho option, which creates an orthographic view from an existing view. If the first view that you defined was a top view of the solid based on the WCS, the second view might be a front view. AutoCAD prompts you to pick the side of the viewport that you want to use for projecting the new view. Once you select an edge of the viewport, AutoCAD again asks for the view center and the clip corners, to define the next viewport.

Section and auxiliary views are defined in a similar fashion. When you specify a section view, the program first asks you to define the cutting plane and the side of the cutting plane from which you want to view, and then prompts you to specify the view center and clip corners. Auxiliary views, in which a view is projected onto an inclined plane, require you to define the plane by specifying two points and the side to view from. AutoCAD then asks for the view center and clip corners.

**Note**
*Remember that when AutoCAD first activates a layout, it normally creates a viewport within that layout. Since you will be creating those viewports using the SOLVIEW command, you should either disable the automatic creation of viewports in new layouts before starting the SOLVIEW command (by clearing the Create Viewport in New Layouts check box on the Display tab of the Options dialog box) or erase the viewport and then restart the SOLVIEW command.*

**LEARN BY EXAMPLE**
*Use the SOLVIEW command to set up top, front, side, and section views of the widget that you worked with in the previous exercises. Open the drawing Figure 21-20 on the companion web site.*

To set up the top, front, side, and section views, use the following command sequence and instructions:

```
Command: SOLVIEW
Regenerating layout.
Regenerating model.
Select an option [Ucs/Ortho/Auxiliary/Section]: U
Enter an option [Named/World/?/Current] <Current>: W
Enter view scale <1.0000>: .5
Specify view center: (locate view center near upper-left area of the screen)
Specify view center <specify viewport>: ENTER
Specify first corner of viewport: (specify lower-left corner of viewport)
Specify opposite corner of viewport: (specify upper-right corner of viewport)
Enter view name: TOP
UCSVIEW = 1  UCS will be saved with view
```

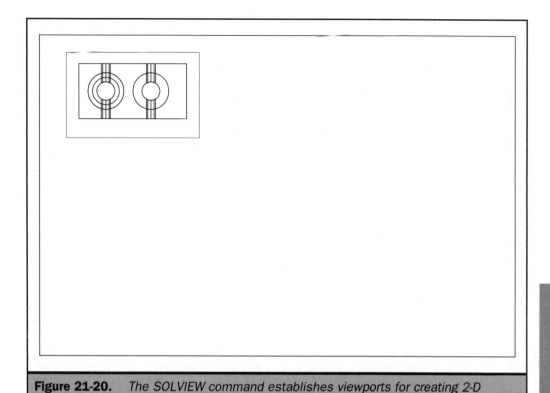

**Figure 21-20.**    *The SOLVIEW command establishes viewports for creating 2-D orthographic, section, and auxiliary profiles of solids. Here, the top view has been created.*

At this point, your drawing should look similar to Figure 21-20. Continue the SOLVIEW command to define the front view.

```
Enter an option [Ucs/Ortho/Auxiliary/Section]: O
Specify side of viewport to project: (click bottom edge of viewport border)
Specify view center: (specify a point below the top view)
Specify view center <specify viewport>: ENTER
Specify first corner of viewport: (specify lower-left corner of viewport)
Specify opposite corner of viewport: (specify upper-right corner of viewport)
Enter view name: FRO
UCSVIEW = 1   UCS will be saved with view
```

BECOMING AN EXPERT

Your drawing should now look similar to Figure 21-21. Continue the SOLVIEW command to define the side view.

```
Enter an option [Ucs/Ortho/Auxiliary/Section]: O
Specify side of viewport to project: (click right edge of front view viewport border)
Specify view center: (specify a point to the right of the front view)
Specify view center <specify viewport>: ENTER
Specify first corner of viewport: (specify lower-left corner of viewport)
Specify opposite corner of viewport: (specify upper-right corner of viewport)
Enter view name: SID
UCSVIEW = 1  UCS will be saved with view
```

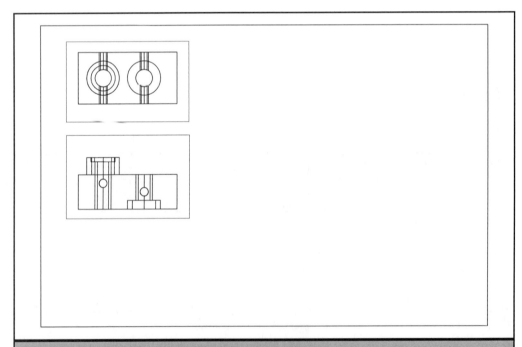

**Figure 21-21.** *In the second step, the SOLVIEW command has been used to set up a front view.*

Your drawing should now look similar to Figure 21-22. Continue the SOLVIEW command to define the section view. Referring to Figure 21-23, make the top view the current viewport before specifying the first and second points of the cutting plane.

```
Enter an option [Ucs/Ortho/Auxiliary/Section]: S
Specify first point of cutting plane: -.5,1.5
Specify second point of cutting plane: @7,0
Specify side to view from: (click below cutting plane in the top view)
Enter view scale <0.5000>: ENTER
Specify view center: (specify a point below the front view)
Specify view center <specify viewport>: ENTER
Specify first corner of viewport: (specify lower-left corner of viewport)
Specify opposite corner of viewport: (specify upper-right corner of viewport)
Enter view name: SEC
UCSVIEW = 1   UCS will be saved with view
Enter an option [Ucs/Ortho/Auxiliary/Section]: ENTER
```

BECOMING AN EXPERT

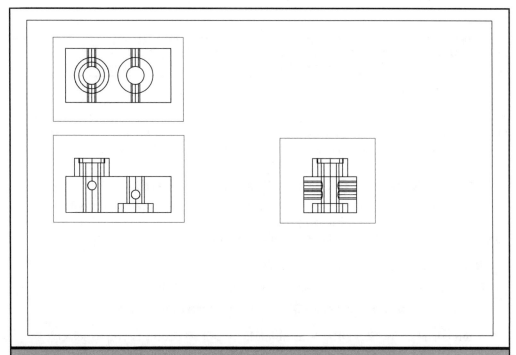

**Figure 21-22.**    *In the third step, SOLVIEW is used to establish a side view.*

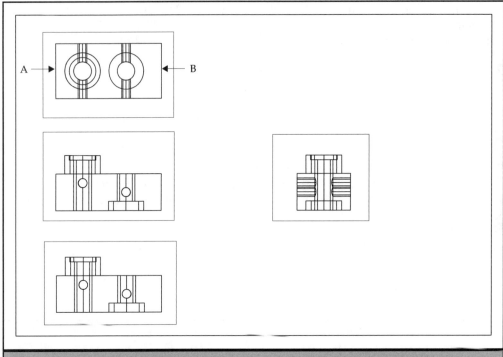

**Figure 21-23.** *In the fourth step, a section view is established by using the SOLVIEW command.*

This completes the creation of the views. Your drawing should now look similar to Figure 21-23. Now that you have established the views, you are ready to convert them into the proper representations of the orthographic views of the widget.

## Converting Views into Profiles and Sections

The next step in the process is to convert the views of the solid into representations of the proper 2-D orthographic projections. The SOLDRAW command does this automatically. To start the SOLDRAW command, do one of the following:

- On the Solids toolbar, click Setup Drawing.
- From the Draw menu, choose Solids | Setup | Drawing.
- At the command line, type **SOLDRAW** and then press ENTER.

When you start the SOLDRAW command, AutoCAD prompts you to select the viewports that you want to draw. After you select the viewports, AutoCAD immediately redraws them based on how they were defined using the SOLVIEW command.

### LEARN BY EXAMPLE
*Convert the views that you previously created into the actual profile and section drawings. Open the drawing Figure 21-24 on the companion web site.*

To convert the views created by using the SOLVIEW command into actual profile and section drawings, use the following command sequence and instructions:

```
Command: SOLDRAW
Select viewports to draw.
Select objects: (select all four viewports)
Select objects: ENTER
```

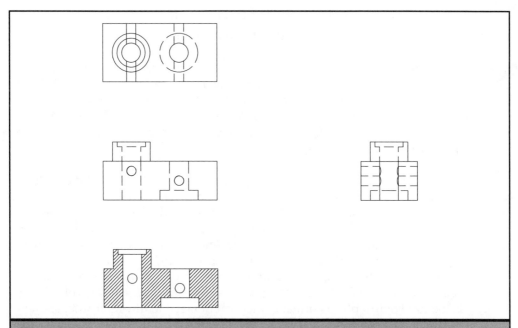

**Figure 21-24.**    *The SOLDRAW command converts the viewports into 2-D sections and profiles based on the settings established using the SOLVIEW command.*

BECOMING AN EXPERT

Use the LAYER command or the Layer Control drop-down list to freeze the VPORTS layer. Your drawing should then look like Figure 21-24.

# Creating Profiles

The SOLPROF command performs a similar function—it creates a profile image of selected 3-D solids. The image is produced based on the current viewpoint and can display hidden lines on a separate layer, so that you can assign them a dashed linetype. The resulting profile image displays only the edges and silhouettes of curved surfaces of the solids for the current view. When you use the SOLPROF command, however, you must first create the viewport and orient the solid the way that you want it, before you create the profile. The resulting profile is created as a block.

To start the SOLPROF command, do one of the following:

- On the Solids toolbar, click Setup Profile.
- From the Draw menu, choose Solids | Setup | Profile.
- At the command line, type **SOLPROF** and then press ENTER.

*The SOLPROF command is useful if you want to create a 2-D isometric view to augment the orthographic and section profiles that you created by using the SOLVIEW and SOLDRAW commands. Like SOLVIEW, SOLPROF creates new layers on which to place the profiles, and will create separate layers for visible and hidden lines. As with SOLVIEW, if the hidden linetype is already loaded, that linetype is automatically assigned to the hidden line layer.*

When you start the command, AutoCAD prompts you to select the solid objects for which you want to create profile images, and ignores any nonsolid objects that you might select. Next, the command asks whether you want to display hidden profile lines on a separate layer. If you respond yes, AutoCAD creates two separate blocks, one for the visible profile lines and the other for the hidden lines. Next, AutoCAD asks whether you want to project the profile lines onto a plane. Unlike the SOLVIEW and SOLDRAW command combination, SOLPROF can create either a 2-D or 3-D representation of the solid. If you project the profile lines onto a plane, AutoCAD creates a 2-D profile by projecting the solid onto a plane normal to the current viewing direction and passing through the origin of the current UCS. If you don't project the profile lines onto a plane, AutoCAD creates the profile lines in 3-D. In either case, you need to be sure that the solid is oriented properly, because the hidden lines are based on the current viewpoint.

Finally, AutoCAD asks whether you want to delete tangential edges. A *tangential edge* is the transition between two faces that meet at a tangent, such as the line formed at the end of fillet arcs when you fillet between two surfaces. Tangential edges generally aren't shown on most types of drawings, so the default response is to remove them. As soon as you respond to this final prompt, AutoCAD generates the profile drawing.

*LEARN BY EXAMPLE*

*Add an isometric profile drawing to the orthographic and section views that you created in the previous exercise. Open the drawing Figure 21-25 on the companion web site.*

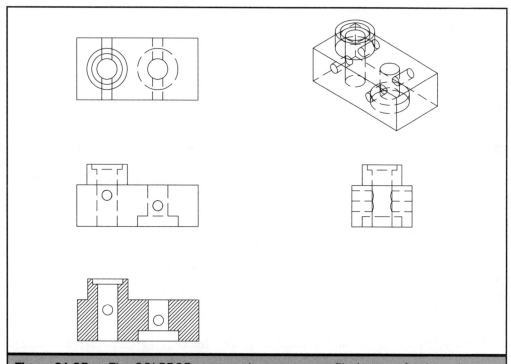

**Figure 21-25.**   *The SOLPROF command creates a profile image of selected 3-D solids.*

BECOMING AN EXPERT

To create a profile by using the SOLPROF command, use the following command sequence and instructions. First, you need to create a new floating viewport and set the desired viewpoint to properly orient the solid:

```
Command: MVIEW
Specify corner of viewport or [ON/OFF/Fit/Hideplot/Lock/Object/
Polygonal/Restore/2/3/4] <Fit>: (select first corner of viewport)
Specify opposite corner: (select opposite corner of viewport)
```

Next, switch to model space (remember, you can click PAPER on the status bar or double-click inside the viewport) and change the viewpoint to establish an isometric:

```
Command: VPOINT
Current view direction:  VIEWDIR=0.000,-1.0000,0.0000
Specify a viewpoint or [Rotate] <display compass and tripod>: 1,-1,1
```

**Tip** *You can also switch to this viewpoint either by clicking the SE Isometric View button on the View toolbar or by choosing 3D Views | SE Isometric from the View menu.*

Once the solid is oriented properly in the new viewport, use the SOLPROF command to create the profile by using the following command sequence and instructions:

```
Command: SOLPROF
Select objects: (select the solid object)
Select objects: ENTER
Display hidden profile lines on separate layer? [Yes/No] <Y>: ENTER
Project profile lines onto a plane? [Yes/No] <Y>: ENTER
Delete tangential edges? [Yes/No] <Y>: ENTER
```

In the viewport containing the isometric profile, freeze layer 0, and then turn off or freeze the VPORTS layer. Your drawing should now look similar to Figure 21-25. You could now add a title block, dimensions, and annotations.

# Chapter 22

## Creating Three-Dimensional Images

A s you've no doubt learned from working with the three-dimensional objects in the previous two chapters, three-dimensional objects are usually displayed in wireframe view, regardless of whether they were created as wireframe, surface, or solid objects. This makes it difficult to visualize your three-dimensional models. To better visualize the model in several of the exercises, you used the HIDE command to remove all the lines that were hidden behind other objects or surfaces when viewed from the current viewpoint.

Removing hidden lines is fine for simple models, but sometimes a more realistic image is needed to help you fully visualize the objects in your drawing. Shading goes a step further than hidden-line removal, by removing the hidden lines and then assigning flat colors to the visible surfaces, making them appear solid. Shaded images are useful when you want to quickly visualize your model as a solid object. In Chapter 19, you learned how to use the 3DORBIT command to create and manipulate shaded images.

Shaded images lack depth and realism, however. Rendering provides an even more realistic image of your model, and can be generated complete with multiple light sources, shadows, surface materials, and reflections, giving your model a photorealistic look.

When you remove hidden lines, shade, or render a model, AutoCAD treats the objects differently, depending on how you created them. Wireframe models always appear as transparent wireframes, because they have no surfaces. Because surface models and solid models have surfaces, they appear solid and can be hidden, shaded, and rendered. This chapter explains how to do the following:

- Create hidden-line images
- Create shaded images
- Create rendered images

## Creating Hidden-Line Images

Creating a hidden-line view of your drawing removes all the lines that are hidden behind other surfaces when you view it from the current viewpoint. In several of the exercises in the previous chapter, you used the HIDE command to remove hidden lines, and the HIDE command really doesn't offer much more than what you've already experienced.

The HIDE command affects only the current viewport. Because the drawing must be mathematically reconstructed to determine which lines are hidden behind visible surfaces, AutoCAD must regenerate the viewport whenever you use the HIDE command. The lines removed from view remain hidden until you regenerate the viewport. Figure 22-1 shows a typical hidden-line view.

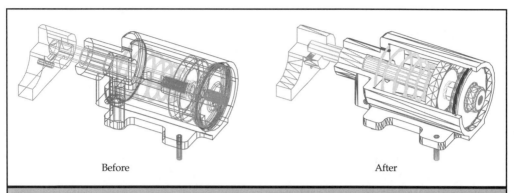

**Figure 22-1.** *The HIDE command removes lines that are hidden behind other surfaces.*

**Note**

*The behavior of the HIDE command just described is only true when SHADEMODE is set to 2D Wireframe. If you set SHADEMODE to any other value and then use the HIDE command, SHADEMODE gets set to Hidden, and regenerating the drawing will not change the display (in other words, the view remains hidden). You'll learn more about SHADEMODE later in this chapter. In addition, note that the HIDE command removes hidden lines only from the current view. To remove hidden lines from drawings when you plot, you need to specify the Hide-Lines option when plotting, or use the Hideplot option of the MVIEW command when using paper space.*

To create a hidden-line image when SHADEMODE is set to 2D Wireframe, do one of the following:

- On the Render toolbar, click Hide.
- From the View menu, choose Hide.
- At the command line, type **HIDE** (or **HI**) and then press ENTER.

**Note**

*If you use the HIDE command when SHADEMODE is set to anything other than 2D Wireframe, AutoCAD simply changes the SHADEMODE setting to Hidden.*

**Tip**

*The HIDE command still considers in its calculations objects on layers that are turned off. Thus, an object that is invisible because its layer is turned off may still obscure other objects in the drawing. To improve AutoCAD's performance and ensure that hidden lines are removed properly, freeze layers containing objects that don't need to be included in the hidden-line image.*

BECOMING AN EXPERT

## Hiding Solids

When hiding solids, AutoCAD first generates a mesh and then performs hidden-line removal on the mesh. If the DISPSILH system variable is on (its value is 1), AutoCAD draws only the silhouette of the solid, not the mesh.

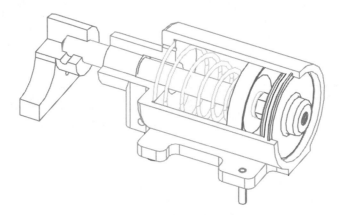

Several other system variables affect the number of facets that are used to describe the surfaces of 3-D solids when shading and rendering. The value set by the VIEWRES command controls the accuracy of curves. Higher values produce smoother-looking curves at the expense of longer regeneration time. The FACETRES system variable controls the smoothness of shaded and rendered solids. It is linked to the VIEWRES value and acts as a multiplier. For example, when FACETRES equals 1, there is a one-to-one correspondence between the display resolution of curves and the tessellation of solid objects. When FACETRES equals 2, AutoCAD displays twice the number of tessellation lines set by the VIEWRES value. The FACETRATIO system variable controls the aspect ratio of faceting for cylindrical and conic sections.

**Note**    *The HIDEPRECISION system variable controls the accuracy of hidden-line removal. AutoCAD normally uses single-precision math when calculating hidden lines. You can change the HIDEPRECISION value so that AutoCAD uses double-precision math, but doing so causes AutoCAD to use more memory and can significantly increase the time required to create hidden-line images, particularly when working with solids. Unless you detect errors in the resulting hidden-line images, you shouldn't change this value.*

**Tip**    *To hide text behind other geometry, you need to assign a thickness to the text, using the Properties window.*

# Creating Shaded Images

Shaded images increase the realism of your drawings one notch above that of hidden-line images. Creating a shaded image of your drawing removes hidden lines and then applies shading to the visible surfaces, based on their object color. You can determine whether the resulting image is flat or Gouraud shaded and whether the edges are visible, or you can simply create a hidden-line view of your model. For example, Figure 22-2 shows the same model from the previous illustration, but displayed as a Gouraud shaded image.

The shading method normally uses a single light source, which is located directly behind the viewpoint (in other words, behind your back as you look at the screen), and uses a fixed amount of ambient light. If you add light sources (as explained later in this chapter), however, those light sources are used in producing the shaded image. Similarly, the colors in shaded images are normally based on the AutoCAD color assigned to the objects in your model. If you apply materials (as explained later), however, the shaded image will include those materials.

While the addition of lights and materials makes shaded images more realistic than in previous versions, shaded images still do not include shadows or three-dimensional textures. For that level of realism, you will need to create rendered images. In spite of these limitations, however, shaded images are still often easier to visualize than hidden-line images, and are generated much faster than the renderings examined later in this chapter.

To create a shaded image, do one of the following:

- On the Shade toolbar, click one of the buttons corresponding to the desired shading option.

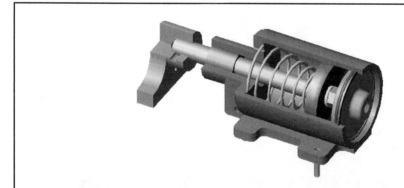

**Figure 22-2.**    *The SHADEMODE command creates shaded images in one of four modes, as well as hidden-line images, or restores models to their wireframe view.*

■ From the View menu, choose Shade and then choose one of the shading options from the submenu.

■ At the command line, type **SHADEMODE**, press ENTER, specify the desired option, and then press ENTER again.

The SHADEMODE command provides seven options:

■ **2D Wireframe**   Displays objects in 3-D views with all edges visible; raster and OLE objects, linetypes, and lineweights are also visible

■ **3D Wireframe**   Displays objects in 3-D views with all edges visible; raster and OLE objects, linetypes, and lineweights are not visible

■ **Hidden**   Displays objects in 3-D views; all edges hidden behind other surfaces are removed from view

■ **Flat Shaded**   Displays objects in 3-D view with shaded polygonal faces

■ **Gouraud Shaded**   Displays objects in 3-D view with smooth shaded surfaces

■ **Flat Shaded, Edges On**   Displays objects in 3-D view with shaded polygonal faces and visible edges highlighted

■ **Gouraud Shaded, Edges On**   Displays objects in 3-D view with smooth shaded surfaces and visible edges highlighted

Like the HIDE command, the SHADEMODE command affects only the current viewport. The drawing is regenerated when producing the shaded images, and that image remains in the viewport until you change the shading mode by using the SHADEMODE command again. Unlike the HIDE command when used in 2D Wireframe mode, however, regenerating the drawing has no affect on the shading. In addition, when you create a shaded image using the SHADEMODE command, you can edit the objects by selecting them as you normally would. When you select shaded objects, AutoCAD displays the wireframe image and grips on top of the shaded image. In addition, if you save and then close the drawing while it's displayed in a shaded mode, it will appear shaded when you open it again. This is considerably different from earlier versions of AutoCAD.

*Although AutoCAD doesn't offer any method for printing shaded images, you can capture and save a shaded image to an AutoCAD slide file for use in presentations. You'll learn about creating slide files in Chapter 24. In Chapter 29, you'll learn how to use scripts to produce self-running presentations using those slide files.*

## Using AutoCAD's Older Shading Methods

Versions of AutoCAD prior to AutoCAD 2000 used a simpler shading method that caused objects to have a continuous color across each surface, making them look flat. This method is still available by typing **-SHADE**. Images shaded using this method have only a single light source and cannot display materials. Shaded images only remain visible in the viewport until it is regenerated, and you can't select or snap to any object in the viewport until the viewport is regenerated. The appearance of the shaded image is further controlled by two system variables. SHADEDGE determines how faces are shaded and whether edges are visible. SHADEDIF controls the percentage of diffuse light reflected from each surface of your model.

*NOTE:* *Typing **SHADE** (without the dash) starts the SHADEMODE command and sets it to Flat Shaded, Edges On.*

 **Note**   *The 3-D graphics driver settings in the Options dialog box affect the performance of the SHADEMODE command. Autodesk provides a generic set of HDI files for most display cards. To obtain the best possible performance from your display card, you should obtain an HDI (Heidi Graphics System) driver from your video card manufacturer.*

# Creating Rendered Images

Creating rendered images of your drawings can increase their realism to near-photographic levels. With AutoCAD's rendering tools, you can add multiple light sources, assign materials to the objects in your drawing, and add people and landscaping to make your images look even more realistic.

Rendering is often the most time-consuming aspect of a project. Because of the subtleties involved, you can spend a lot of time adjusting the lighting and materials. You may spend considerably more time creating a rendering than you spend actually building the 3-D model. In addition, when you work with multiple light sources, each of which casts shadows, the computer calculations alone require a considerable amount of computer horsepower. For that reason, AutoCAD's rendering tools let you render just a portion of the image, to test your changes prior to creating a final, full-screen rendering.

The process of creating a computerized rendering involves four steps:

- Create the actual model
- Place lights
- Attach materials to objects in the drawing
- Render the image

BECOMING AN EXPERT

Other than creating the model, these procedures are conceptual rather than discrete sequential steps, and often are performed in an iterative process. For example, you may place some lights and create a test rendering. Then, after viewing the results of your test, you may change some of the lights and render the image again.

**Note** *Renderings can be displayed in True Color by selecting the True Color Raster Images And Rendering check box on the Display tab of the Options dialog box. When selected, AutoCAD displays raster images and renderings at their optimum quality, as determined by the operating system. The Display Properties Settings in Windows must also be configured to True Color in order for AutoCAD to display True Color images and renderings. Note that increasing the number of system display colors can decrease display performance, while clearing the True Color setting in AutoCAD or reducing the color palette in the system display setting optimizes performance.*

# Controlling Rendering

You control the actual rendering process by using the Render dialog box, shown in Figure 22-3. To display the Render dialog box, do one of the following:

- On the Render toolbar, click Render.
- From the View menu, choose Render | Render.
- At the command line, type **RENDER** (or **RR**) and then press ENTER.

**Note** *The first time that you use the RENDER command or any other rendering command, the AutoCAD Render application is automatically loaded into memory. The AutoCAD Render application is a separate ARX application that comes with AutoCAD. After you load it, you see the Render application on your Windows system tray. If necessary, you can unload AutoCAD Render from memory by using the Unload option of the ARX command.*

The Render dialog box contains numerous controls that enable you to determine the rendering type, options, procedures, destination, and so on. Each option is briefly reviewed now. Later in this chapter, you'll learn about some of the options that you'll use most often.

## Selecting the Rendering Type

AutoCAD actually produces one of three different types of renderings. In addition, independent developers can create other rendering processes that may be added to those provided by AutoCAD. You select the rendering type by choosing it from the Rendering Type drop-down list:

- **Render** Produces a flat, shaded image quickly (the basic AutoCAD rendering option). The results are similar to those produced by the SHADEMODE

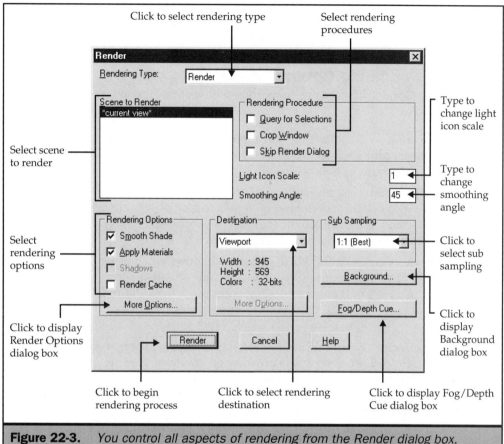

**Figure 22-3.** *You control all aspects of rendering from the Render dialog box.*

command except that surfaces are always shaded based on their AutoCAD color; materials are never visible.

- **Photo Real**  Produces photorealistic scanline images. This process is more time-consuming but can include bitmapped images and materials, and can simulate shadows cast by the various light sources.

- **Photo Raytrace**  Produces the most realistic images. Ray tracing calculates theoretical rays of light that extend from the current viewpoint into the model. AutoCAD's most time-consuming rendering process, ray tracing can generate realistic reflection, refraction, and more precise shadows.

## Choosing the Scene

The Scene To Render list contains the names of any previously saved scenes. When you create a rendering, you can save the viewpoint and a list of lights as a scene, which enables you to use a single drawing file to produce many different renderings. To render a specific scene, you simply select it from the list. You create scenes by using the SCENE command. You'll learn more about creating scenes later in this chapter.

## Selecting Rendering Options

The Rendering Options list contains numerous settings that let you fine-tune the rendering process:

- **Smooth Shade**   Smoothes out the rough edges of multifaceted surfaces. AutoCAD then blends colors across two or more adjacent faces. You will almost always want to select this option.

- **Apply Materials**   Applies materials attached to objects when producing the rendering. When this check box is not selected, all objects are assigned the color and material values defined for the *GLOBAL* material. This check box must be selected to see the materials that you attach to objects.

- **Shadows**   Causes AutoCAD to generate shadows when using the Photo Real or Photo Raytrace rendering types. Calculating shadows is probably the most time-consuming part of generating a rendering. When you're just trying out different lights and materials, you can speed things along by turning off this option.

- **Render Cache**   Causes AutoCAD to save rendering information to a cache file on the hard disk. As long as the drawing geometry or view remains unchanged, the cache file will be used for subsequent renderings, eliminating the need for AutoCAD to recalculate the tessellation lines. Selecting this option can be quite a time-saver, particularly when rendering solid objects.

## Selecting Rendering Procedures

The check boxes in the Rendering Procedures area of the Render dialog box enable you to control what happens when AutoCAD begins the actual rendering process:

- **Query for Selections**   If selected, when you start the actual rendering procedure, AutoCAD prompts you to select the objects to render. This lets you produce quick test renderings to see how a material looks on a particular object.

- **Crop Window**   If selected, when you start the actual rendering procedure, AutoCAD prompts you to specify the corners of an area of the screen to be rendered. This lets you produce quick test renderings on a small portion of the drawing.

- **Skip Render Dialog**   If selected, the next time that you start the RENDER command, AutoCAD immediately renders the current view without first displaying the Render dialog box.

## Controlling Faces

When you click the More Options button, AutoCAD displays a Render Options dialog box that is specific to the rendering type currently selected. The controls in this dialog box can be used to fine-tune the actions of the selected rendering type. Although you generally don't need to adjust these settings, depending on how your 3-D model was created, the Face Controls settings can save a considerable amount of time when you create renderings.

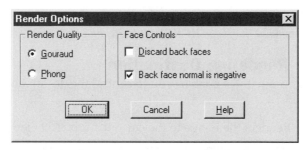

A large portion of AutoCAD's calculation involves removing surfaces hidden behind others. AutoCAD removes these surfaces before rendering, because rendering the hidden surfaces would be a waste of time. In addition, surfaces have both a front and back side, and AutoCAD normally determines which is the front side of a surface, based on the direction in which its vertices are drawn. Depending on how you construct your model, you may not need AutoCAD to consider the back sides of surfaces at all.

When you select the Discard Back Faces check box, AutoCAD ignores the back faces of all 3-D solid objects when it calculates a rendering, thus speeding up the entire rendering process. You should select this option only if you know for sure that none of the back faces need to be included in the rendering.

The Back Face Normal Is Negative check box is usually selected. This tells AutoCAD that the side of the face whose vertices are drawn counterclockwise is to be considered the front face. If you clear this check box, AutoCAD reverses the side of the face considered to be the front face. Because AutoCAD, and most add-on programs, creates objects such that their vertices are drawn counterclockwise, you should normally leave this check box alone.

*CAUTION:   You should always make sure to draw faces in a consistent manner. Mixing faces drawn counterclockwise and clockwise in the same model can produce displeasing renderings.*

BECOMING AN EXPERT

*The preceding option seems to prevent you from ever turning off the option again, because you no longer see the Render dialog box when you use the RENDER command. However, the Render Preferences dialog box contains all the same controls as the Render dialog box. When the Skip Render Dialog check box is selected, you can use the Render Preferences dialog box to turn off this option, so that the Render dialog box again is displayed when using the RENDER command. To display the Render Preferences dialog box, do one of the following:*

- *On the Render toolbar, click Render Preferences.*
- *From the View menu, choose Render | Preferences.*
- *At the command line, type **RPREF** (of **RPR**) and then press ENTER.*

## Selecting the Rendering Destination

The Destination drop-down list enables you to control the destination of the rendering. It offers three options:

- **Viewport**   Renders the image to the current AutoCAD viewport.
- **Render Window**   Renders the image to a separate window called the Render Window.
- **File**   Saves the rendering to a file. When you select this option, you can then click the More Options button to control the type of file, its resolution and color depth, and other settings.

*To create high-resolution images, save your rendering to a file. If you want to include an image of the rendering on a composite paper space drawing sheet, you first need to save the rendering to a file and then attach that image file to the drawing. You'll learn about attaching images in the next chapter.*

## Controlling Sub Sampling

The Sub Sampling drop-down list enables you to reduce rendering time at the expense of image quality, while retaining effects such as shadows. Sub sampling renders only a fraction of all the pixels. A sub sampling ratio of 1:1 renders all the pixels in the image, thus yielding the highest-quality results. A ratio of 8:1 yields the fastest rendering speed by rendering only 1 out of every 8 pixels.

## Controlling Other Rendering Options

The Render dialog box contains several other controls, some of which you can usually ignore, and two of which you'll learn more about later in the chapter. The Background and Fog/Depth Cue buttons display the Background and Fog/Depth Cue dialog boxes, respectively. These dialog boxes can also be displayed by using the BACKGROUND and FOG commands.

The Light Icon Scale setting simply determines the scale factor used to display the special light blocks that are inserted into the drawing when you add a light source. Although these blocks are visible in the drawing, they do not appear in renderings, and changing this value has no effect on the rendering at all.

The Smoothing Angle sets the angle at which AutoCAD interprets an edge. If two surfaces meet at an angle less than this value, AutoCAD smoothes the transition between those surfaces. If the surfaces meet at an angle greater than this value, AutoCAD considers the transition to be an edge. The default value is 45 degrees. Depending on the 3-D model being rendered, you may need to adjust this value.

*LEARN BY EXAMPLE*
*Create a quick study rendering. Open the drawing HOTEL-01 on the companion web site.*

Throughout this chapter, you'll be working with a 3-D building model created by using both AutoCAD's standard tools and an add-on program called Facade. To see the difference between AutoCAD's rendering types, use the RENDER command to produce your first rendering, by following these steps:

1. Display the Render dialog box.
2. Click the Render button.

AutoCAD immediately begins the rendering process, using the default Render type. AutoCAD displays information about this process on the command line as rendering progresses. When finished, your drawing should look similar to Figure 22-4. The image certainly isn't very realistic, because no lights are used in the model, nor are any materials assigned to the walls, windows, and doors. Even if you rendered the model again, using the Photo Real or Photo Raytrace rendering types, the image would look the same.

*If you want to cancel a rendering in progress for any reason, press the ESC key.*

# Working with Lights

Placing lights in your model begins to add a level of realism. The combination of carefully placed lighting and realistic materials quickly turns a flat, shaded image into a close approximation of reality. Controlling lights and materials is an iterative process, and you are likely to jump back and forth between the two, first adjusting the lighting, and then changing the way certain materials are attached. Lighting is discussed first.

You add and configure lights by using the Lights dialog box, shown in Figure 22-5. To display the Lights dialog box, do one of the following:

■ On the Render toolbar, click Lights.

BECOMING AN EXPERT

**Figure 22-4.** *The 3-D building model, rendered using the default rendering settings*

- From the View menu, choose Render I Light.
- At the command line, type **LIGHT** and then press ENTER.

This dialog box is divided into two sections. The left side of the dialog box lets you control individual lights. The right side of the dialog box controls the *ambient light,* which is the light that provides a constant illumination to every surface in the model. Ambient light does not come from any particular source and has no direction.

 *The first time that you add a light to the model, AutoCAD adds a new layer called ASHADE. AutoCAD stores all the lights on this layer as block objects.*

## Controlling Ambient Light

The controls on the right side of the dialog box determine the level of intensity and the color of the background light that provides constant illumination to all surfaces of the model. Ambient light alone can't produce a realistic image. Since all surfaces receive equal lighting, individual surfaces become indistinguishable from one another. You can use ambient light, however, to add just enough fill lighting to distinguish surfaces that are not directly illuminated by any of the individual light sources that you add to your model.

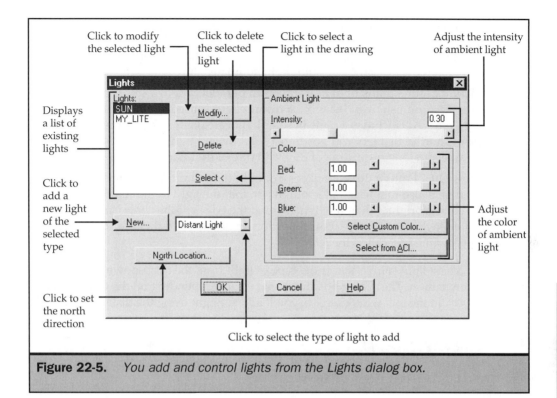

**Figure 22-5.** *You add and control lights from the Lights dialog box.*

BECOMING AN EXPERT

You can adjust the intensity of ambient light either by moving the slider or by typing a value in the adjacent Intensity edit box. Ambient light intensity ranges from 0 (no ambient light) to 1 (full brightness).

The sliders and adjacent edit boxes in the Color area let you adjust the red, green, and blue components of the ambient light. Click the Select Custom Color button to display a Colors dialog box, in which you can select a color or adjust the color by using the hue, luminance, saturation (HLS) or red, green, blue (RGB) components. The Select From ACI button displays the Select Color dialog box, in which you can select a color based on its AutoCAD Color Index (ACI).

*Keep ambient light levels as low as practical. Too high an intensity tends to give renderings a washed-out appearance.*

## Controlling Individual Lights

The Lights area of the dialog box contains a list of lights that are already placed within the drawing, along with controls for modifying existing lights and creating new ones. Initially, the list is empty, because the model contains no lights.

AutoCAD supports three different types of lights in addition to ambient light:

- **Point Light**   Radiates light in all directions, and the intensity of the light diminishes over distance. A point light simulates the light from a light bulb.

- **Distant Light**   Emits light that travels in parallel rays in a single direction and does not diminish over distance. A distant light is used to simulate sunlight, and AutoCAD includes a built-in solar light calculator (which you will learn about shortly) that lets you position a solar distant light source based on your latitude and longitude, time of day, and day of the year.

- **Spotlight**   Emits a directional cone of light that diminishes over distance. Spotlights also mimic actual spotlights in that the light in the central portion of the cone is brighter than the light around the edges of the cone. You can adjust the angle that defines the central portion, or *hotspot,* and the rate at which the light diminishes, called the *falloff.*

To add a new light source, select the type of light from the drop-down list and then click the New button. The dialog box displayed varies, depending on the type of light that you select. In the following exercise, you'll add sunlight using the distant light type.

*LEARN BY EXAMPLE*
*Add a solar light source to your model. If it isn't already open, open the drawing HOTEL-01 on the companion web site.*

When you add a new light source to your model, you must give it a name. So, you first need to add a distant light to represent the sun:

1. Display the Lights dialog box.

2. In the lights type drop-down list (to the right of the New button), choose Distant Light and then click New. AutoCAD displays the New Distant Light dialog box, shown in Figure 22-6.

**Note** *When you modify an existing distant light, AutoCAD displays a Modify Distant Light dialog box, which is identical to the New Distant Light dialog box, but already contains the settings for the selected light.*

3. In the Light Name edit box, type **SUN**.

Notice that some of the controls in this dialog box are similar to those in the Lights dialog box. For example, you can control the intensity and color of the light. The other controls, however, are very different. For example, because distant lights can cast shadows, you can turn shadows on and off. You can also control how shadows are created.

The controls on the right side of the dialog box let you control the position and direction of the light source. For example, you can adjust the azimuth and

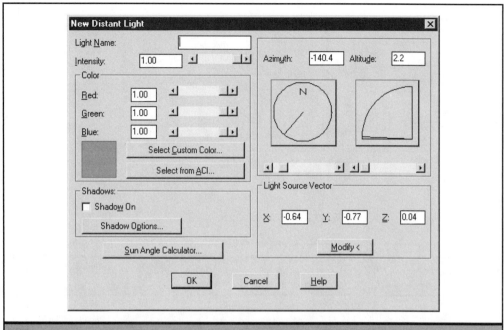

**Figure 22-6.**    *You add distant lights by using the New Distant Light dialog box.*

altitude of the light, or adjust its light source vector. Because the light that you are adding is meant to represent the sun, however, AutoCAD provides a much easier and intuitive method of placing the light source.

4. Click the Sun Angle Calculator button to display the Sun Angle Calculator dialog box, shown in Figure 22-7.

You can control both the placement of the distant solar light source, based on a particular date and time, and the location, based on latitude and longitude. These settings automatically control the Azimuth and Altitude settings. But what if you don't know the latitude and longitude? AutoCAD solves that problem with an additional dialog box.

5. Click the Geographic Location button to display the Geographic Location dialog box, shown in Figure 22-8.

Rather than specify the latitude and longitude, you can select the location of your model from a map or a list of major cities. The dialog box initially shows a map of North America, but you can display a different map by choosing the geographic area from the drop-down list located above the map. You can choose from North America, Canada, Europe, South America, Asia, Australia, the Asian subcontinent, or Africa. After you select the geographic area, you can

BECOMING AN EXPERT

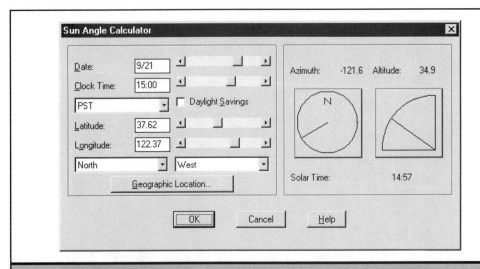

**Figure 22-7.** Use the Sun Angle Calculator dialog box to place a distant light, based on date, time, and location.

specify the exact location by clicking on the map or by selecting a major city from the City list. If the Nearest Big City check box is selected when you click on the map, the location automatically jumps to the nearest city in the adjacent list. The Latitude and Longitude values are automatically filled in for the location that you select.

6. From the City list, choose Seattle, WA, and then click OK.

7. In the Sun Angle Calculator dialog box, change the date to 10/1. You can leave the Clock Time set to 15:00 (3 P.M.). Then, click OK.

8. Click OK to close the New Distant Light dialog box, and then click OK to close the Lights dialog box.

You have just added your first light. Now, render the model to see how it looks:

1. Display the Render dialog box.

2. Select the Render rendering type.

3. Click Render.

When the rendering is displayed, you should see a bit of an improvement. The late afternoon sun illuminates the west wall of the building, and you can begin to see some of the more subtle details, but it still certainly isn't realistic at all. Notice too that there are no shadows.

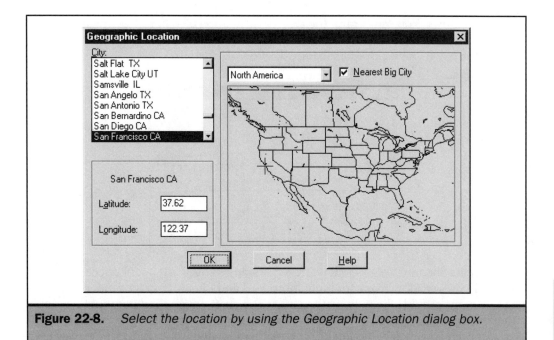

**Figure 22-8.**    *Select the location by using the Geographic Location dialog box.*

BECOMING AN EXPERT

---

**Tip**

*You can add cities or locations by editing the SITENAME.TXT file in the Support folder. If you do not know the latitude and longitude for the location you are adding, you can use the Geographic Location dialog box to establish a rough estimate, by clicking the location of the city or the location that you need. The Latitude and Longitude fields are immediately updated. Make a note of these values and then use them when editing SITENAME.TXT.*

## Adding Shadows

Remember that one of the controls in the New Distant Light dialog box lets you turn shadows on and off. Computing shadows adds a considerable amount of time to the rendering calculations. For that reason, you must explicitly turn on shadows for each individual light that you add to the drawing. In addition, when you generate the rendering, you can still turn shadows on and off. When enabled, AutoCAD calculates shadows only for those lights that also have their shadow selection enabled.

**Note**

*AutoCAD calculates shadows only when using the Photo Real or Photo Raytrace rendering options.*

In addition to enabling shadows, you can control the type of calculation that is used to compute those shadows on a light-by-light basis. AutoCAD supports three different types of shadows. You can control the type of calculation on a light-by-light basis by clicking the Shadow Options button in the New or Modify Light dialog box for the selected light, which displays the Shadow Options dialog box.

When the Shadow Volumes/Ray Traced Shadows check box is not selected, AutoCAD uses shadow maps for both Photo Real and Photo Raytrace rendering. Shadow maps provide the only method of creating soft-edged shadows, but they don't show the color that is cast by transparent or translucent objects. You can control the size of the shadow map generated for each individual light by selecting the size of the shadow map from the drop-down list. The larger the value, the greater the accuracy of the shadow map. You can also control the softness of the shadow's edge by adjusting the number of pixels along the edge that are blended into the underlying image. Values of 1 to 10 are possible, but the best results are usually obtained with a setting between 2 and 4.

When the Shadow Volumes/Ray Traced Shadows check box is selected, AutoCAD generates volumetric shadows rather than shadows based on shadow masks. The renderer calculates the volume of the shadow cast by an object. Volumetric shadows have hard edges, but their outlines are approximate. In addition, shadows cast by transparent or translucent objects are affected by the colors of those objects.

When you use the Photo Raytrace renderer, volumetric shadows are computed by using ray tracing, thus producing raytraced shadows. These shadows are generated by tracing the path of beams of light. They have hard edges, like other volumetric shadows, but their outlines are very accurate.

**Note**
*Shadows greatly increase rendering time. Volumetric shadows tend to be faster than raytraced shadows in simple models. In more complex models, however, raytraced shadows are quicker to calculate. Although faster, shadow maps are also quite time-consuming to calculate, and AutoCAD must calculate each shadow map individually. You can improve rendering time when using shadow maps, however, by hand-selecting just those objects that you want to cast shadows. To do this, click the Shadow Bounding Objects button in the Shadow Options dialog box, and then select just those objects in the drawing.*

*LEARN BY EXAMPLE*
*Add shadows to the rendering. If it isn't already open, open the drawing HOTEL-01 on the companion web site.*

When you added the sunlight source, you didn't select the Shadow On check box. To have the sunlight cast shadows, you must first turn on the shadows for this light source:

1. Display the Lights dialog box.
2. In the Lights list, select the SUN light and then click the Modify button.
3. In the Modify Distant Light dialog box, under Shadows, select the Shadow On check box.
4. Click the Shadow Options button.
5. In the Shadow Options dialog box, make sure the Shadow Volumes/Ray Traced Shadows check box is not selected.
6. From the Shadow Map Size drop-down list, choose 1024 and then click OK to close the Shadow Options dialog box.
7. Click OK to close the Modify Distant Light dialog box, and then click OK again to close the Lights dialog box.

Render the model again to see how it looks with shadows added. To see the shadows, you need to use either the Photo Real or Photo Raytrace renderer:

1. Display the Render dialog box.
2. From the Rendering Type drop-down list, choose Photo Real.
3. Under Rendering Options, make sure the Shadows check box is selected.
4. Click Render.

Notice that the rendering takes considerably longer. During the rendering process, AutoCAD displays a message stating that it is "Rendering shadow maps." When completed, your drawing should look similar to Figure 22-9. That's a considerable improvement, but still not very realistic. Now it's time to add some materials.

# Working with Materials

Attaching realistic materials to the objects in your model truly brings those objects to life. By attaching materials and adjusting the way that they appear on those objects, you make the flat surfaces appear to be made out of real brick and mortar. As with applying lights, attaching materials is an iterative process. You not only attach the materials, but you often must also adjust their scale and mapping. After each change, you will probably want to render the image. As the rendering times become more

BECOMING AN EXPERT

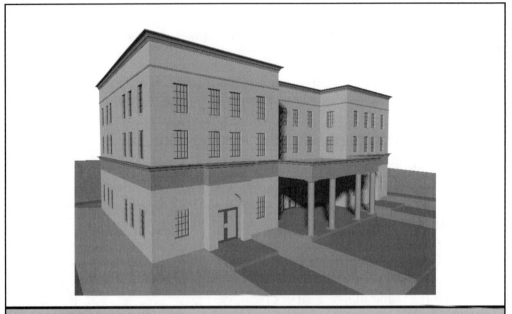

**Figure 22-9.** *The 3-D building model rendered with shadows*

lengthy, you'll see why AutoCAD lets you selectively render individual objects or small areas of the drawing prior to rendering the entire image.

You attach and manage materials by using the Materials dialog box, shown in Figure 22-10. To display the Materials dialog box, do one of the following:

- On the Render toolbar, click Materials.
- From the View menu, choose Render | Materials.
- At the command line, type **RMAT** and then press ENTER.

The Materials list contains the materials already available in the current drawing. The default material for objects with no other material attached is *GLOBAL*. Before you can attach materials to objects in the drawing, you must either load predefined materials or create your own.

AutoCAD comes with a materials library (RENDER.MLI), which is located in the Support directory. By default, the program uses this library file. You can also create

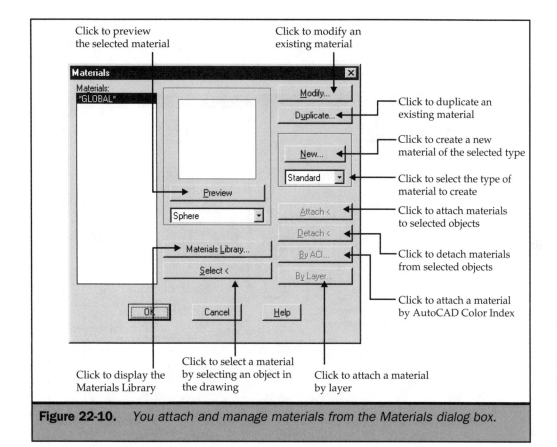

**Figure 22-10.** *You attach and manage materials from the Materials dialog box.*

and save your own materials library files and add custom materials to these files. To load predefined materials, click the Materials Library button to display the Materials Library dialog box, shown in Figure 22-11.

The Materials Library dialog box is split into two sections. The Current Drawing list on the left side contains the materials already loaded in the current drawing. The Current Library list on the right side contains the names of all the materials in the current library. To add one or more materials from the library to the current drawing, select them in the Current Library list and then click the Import button. Other controls in this dialog box let you save, remove, or delete materials from the various lists, save materials library files, or open different libraries.

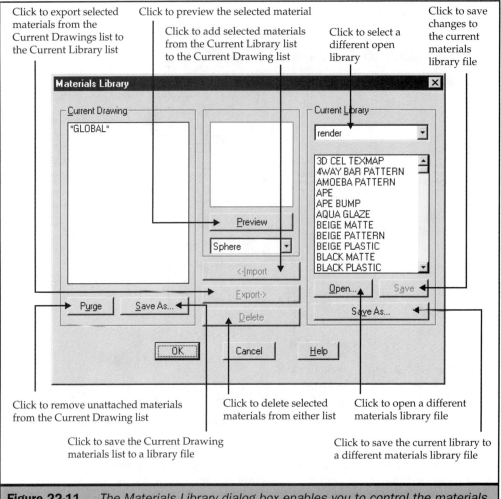

Click to export selected materials from the Current Drawings list to the Current Library list

Click to preview the selected material

Click to add selected materials from the Current Library list to the Current Drawing list

Click to select a different open library

Click to save changes to the current materials library file

Click to remove unattached materials from the Current Drawing list

Click to delete selected materials from either list

Click to open a different materials library file

Click to save the Current Drawing materials list to a library file

Click to save the current library to a different materials library file

**Figure 22-11.** *The Materials Library dialog box enables you to control the materials available in the current drawing.*

**Note**

*You can also display the Materials Library dialog box using the MATLIB command, by doing one of the following:*

- *On the Render toolbar, click Materials Library.*
- *From the View menu, choose Render | Materials Library.*
- *At the command line, type MATLIB and then press ENTER.*

*When the Materials Library dialog box is started in this way, you can add materials to the list of those available in the current drawing, but you can't attach those materials to objects in the drawing. You can attach materials only by using the Materials dialog box.*

The Preview area is particularly handy. Until you become more familiar with the many predefined materials, the only way you have of knowing what a particular material looks like is to preview it. This same Preview area occurs in all the other dialog boxes that deal with materials. To see what a material looks like, select it in either list and then click the Preview button. The material is mapped onto either a sphere or a cube, whichever you selected from the drop-down list.

*By default, AutoCAD stores its materials in the Textures directory. This is controlled by the Texture Maps Search Path on the Files tab of the Options dialog box. When you install AutoCAD, a typical installation doesn't install all the materials that come with AutoCAD. You must do a full or custom installation to install all the materials that come with the program.*

**LEARN BY EXAMPLE**
*Attach materials to the objects in your model. You will continue working with the drawing HOTEL-01 on the companion web site.*

AutoCAD provides three different methods for attaching materials to objects in your drawing:

■ By selecting the objects

■ By attaching materials to objects based on their AutoCAD Color Index (ACI) number

■ By attaching materials to objects based on the layer on which they are drawn

The method that you choose depends primarily on how you created your drawing. Obviously, in more complex drawings, attaching materials based on ACI number or layer is easier than selecting objects individually. The drawing that you are using in this exercise was carefully created so that materials can be assigned on a per-layer basis. Begin by attaching a brick material to the lower walls of the building:

1. Display the Materials dialog box.

2. In the Materials dialog box, click the Materials Library button.

3. In the Materials Library dialog box, under Current Library list, choose BROWN BRICK and then click Import. The BROWN BRICK material is added to the Current Drawing materials list.

4. From the Current Library list, choose SAND TEXTURE and then click Import.

*You can select more than one material at a time by pressing the* CTRL *key as you select the materials. Then, click Import to add the selected materials. You can also add consecutive groups of materials by holding down the* SHIFT *key while selecting the first and last material in the group.*

Although you will need to add other materials to the Current Drawing materials list, stop here momentarily and work with these two materials. So far, all you've done is add the materials to the Current Drawing materials list. The next step is to attach them to objects in your drawing.

5. Click OK to close the Materials Library dialog box.

6. In the Materials dialog box, under Materials, choose SAND TEXTURE and then click the By Layer button. AutoCAD displays an Attach by Layer dialog box.

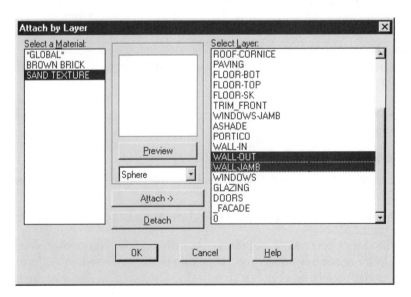

7. Press and hold down the CTRL key, choose WALL-OUT and WALL-JAMB in the Select Layer list, and then click the Attach button. The material is added to the layer list adjacent to these layers.

8. Repeat Step 7, this time selecting the BROWN BRICK material and attaching it to the WALL-LOW-OUT and WALL-LOW-JAMB layers.

9. Click OK to close the Attach by Layer dialog box, and then click OK to close the Materials dialog box.

Now, check the appearance of your rendering by following the steps listed next. Remember that materials are visible only when you use the Photo Real or Photo Raytrace rendering type option. Since you're only interested in the materials at this point, you can turn off the shadows to improve rendering time.

1. Display the Render dialog box.

2. Under Rendering Type, choose Photo Real.

3. Under Rendering Options, make sure Apply Materials is selected and Shadows is not selected.

4. Click Render.

When AutoCAD completes its calculations, the rendering is displayed. The materials are visible, but they don't look right. For one thing, the sand texture on the upper portion of the building looks blotchy.

BECOMING AN EXPERT

## More About Attaching Materials

If you attach materials on a per-object basis, you need to be aware of the type of object to which you attach the material. If the object is a block and that block is subsequently exploded, the material that was attached will no longer be attached to the individual objects. Materials that were attached to objects before being combined into a block, however, retain their respective materials within the block, and also retain them if the block is subsequently exploded.

When you copy an object that has a material attached, the material is copied as well. Unfortunately, materials are not applied when objects are attached as external references. You must bind the xref to regain its material properties. In addition, if you insert a block that has attached materials, you also need to import any custom materials from that drawing into the current drawing.

*You can use the SHOWMAT command to display the type of material attached to any object in an open drawing except objects in attached xrefs. The command also displays the method used to attach the material to that object. To start the command, type **SHOWMAT** and then press ENTER. AutoCAD prompts you to select an object. As soon as you select the object, AutoCAD displays the material name and attachment method.*

*If you select Tools | Options | System | Properties | Enable Materials, the colors of materials attached from the Materials Library display when SHADEMODE is set to anything other than 2D Wireframe mode, but not the actual material textures (bitmaps, bumpmaps, and the procedural granite, marble, and wood textures). Materials with textures that have dark colors might actually be more difficult to see onscreen, making this feature of limited use.*

## Adjusting Bitmaps

AutoCAD's materials fall into two distinct categories: bitmaps and procedural materials. Materials such as SAND TEXTURE are based on *bitmaps*, scanned raster images of actual materials. For the SAND TEXTURE, AutoCAD uses a file called SAND.TGA. The size at which those materials appear in your drawing is determined by the size of the bitmap image.

*Procedural materials*, such as granite, marble, and wood, are generated mathematically. You'll learn more about procedural materials later in this chapter.

You can adjust the bitmap, so that it is scaled more appropriately for your drawing, by selecting the material in the Materials dialog box and then clicking the Modify button. AutoCAD displays the Modify Standard Material dialog box for the selected material, as shown in Figure 22-12.

This dialog box provides controls that enable you to change the appearance of the material in many ways. The radio buttons in the Attributes area determine which aspect of the material is being changed. You can then adjust the values for the selected attribute by using the controls in the other areas of the dialog box. In the case of standard materials (those based on bitmaps), these attributes are as follows:

- **Color/Pattern**  Determines the color, bitmap blend, bitmap name, and bitmap scaling
- **Ambient**  Determines the shadow color of the material
- **Reflection**  Adjusts the reflective color of the material and lets you specify a reflection bitmap
- **Roughness**  Adjusts the roughness or shininess of the material
- **Transparency**  Adjusts the transparency of the material and lets you specify an opacity bitmap and a reflection map
- **Refraction**  Adjusts how refractive the material will be when rendering using the Photo Raytrace renderer

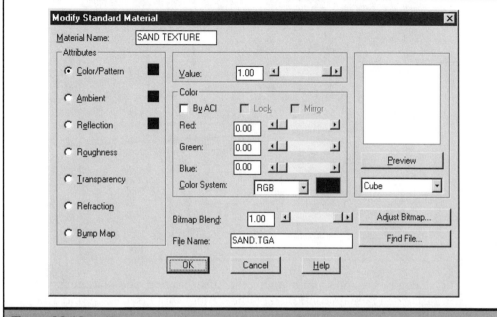

**Figure 22-12.** *You can modify the appearance of a material by using the Modify Standard Material dialog box.*

■ **Bump Map** Lets you specify the name of a bitmap file to be used as a bumpmap; *bumpmaps* make a material appear to have more three-dimensional depth

*LEARN BY EXAMPLE*
*Adjust the bitmaps for the materials attached to objects in your drawing. Continue working with the drawing HOTEL-01 on the companion web site.*

To make the sand texture appear at the correct scale, you need to change the scale of the bitmap:

1. Display the Materials dialog box.

2. In the Materials list, select SAND TEXTURE and then click the Modify button. AutoCAD displays the Modify Standard Material dialog box for the material named SAND TEXTURE.

3. Under Attributes, make sure the Color/Pattern radio button is selected, and then click the Adjust Bitmap button (in the lower-right corner of the dialog box). AutoCAD displays the Adjust Material Bitmap Placement dialog box.

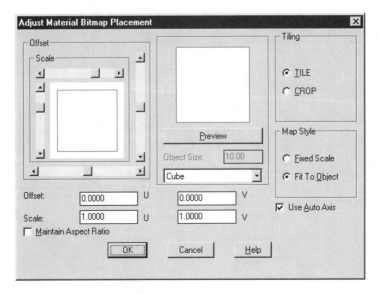

This dialog box enables you to adjust the offset, scaling, and placement values for the current bitmap, controlled by the settings for Offset, Scale, and Tiling, respectively. In this case, you want the bitmap to appear at a more appropriate scale. By default, AutoCAD maps the bitmap onto the surfaces by tiling, which means that it places multiple instances of the bitmap side by side on the surface, like a bunch of square tiles. The bitmap pattern is saved such that, when tiled in this fashion, the pattern matches at the edges of each tile, so that you can't discern the individual tiles. That's exactly what you want the sand texture pattern to do, so you can leave that setting alone. You also don't need to be concerned with the Offset value, because you want the sand texture pattern to cover each surface in its entirety. You do need to be concerned with the Scale value.

4. Under Scale, select the Maintain Aspect Ratio check box. This insures that the bitmap is scaled equally in both directions.

5. Change the Scale U value to 12, and then press the TAB key. Because you selected the Maintain Aspect Ratio, the Scale V value automatically changes to 12, as well.

6. Click OK to close the Adjust Material Bitmap Placement dialog box. Then, click OK to close the Modify Standard Material dialog box.

7. Click OK to close the Materials dialog box.

**Note** *When you modify the material, you change the parameters that affect the appearance of the material wherever it is attached in the drawing. You can also adjust bitmaps on an object-by-object basis. To revert back to the original material properties, you can reimport it by using the Materials Library dialog box.*

Check the appearance of your rendering again. The materials are now sized more appropriately.

 *Before you render the image this time, you may want to use the LIGHT command to adjust the ambient light value. Add a bit more general lighting so that you can see the sand material at the smaller scale. A value of 0.5 should yield good results without overly washing out the image.*

## Mapping Materials onto Objects

Sometimes, when you attach a material to an object, the material doesn't look right—instead of simply being the wrong scale, it appears as a horizontal or vertical streak of color. Such strange appearances are generally caused by the material not being properly aligned or projected onto the object to which it is attached.

The SETUV command lets you specify how materials are mapped onto objects. To start the command, do one of the following:

- On the Render toolbar, click Mapping.
- From the View menu, choose Render | Mapping.
- At the command line, type **SETUV** and press ENTER.

When you start the command, AutoCAD prompts you to select objects. As soon as you select the object whose material mapping you want to adjust, AutoCAD displays the Mapping dialog box.

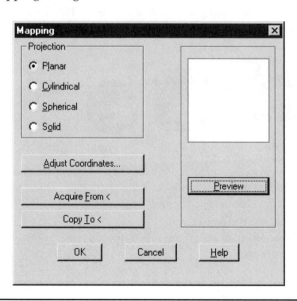

BECOMING AN EXPERT

The buttons in the Projection area let you specify the type of projection used to map the material onto the object. After selecting the projection, you can click the Adjust Coordinates button to display the Adjust Coordinates dialog box for the particular type of projection you selected, and adjust the way the material is mapped onto the object. You can also acquire material mapping from any existing object or apply the mapping you establish within the dialog box to any other objects in the drawing.

- **Planar**   Maps the material onto the object as though it were projected from a slide projector onto a flat plane. The material bitmap is scaled to fit the object, but it isn't distorted. You can select the reference plane, adjust the X and Y offset and rotation angle of the material, and adjust the bitmap placement.

- **Cylindrical**   Maps the material onto a cylindrical object so that the bitmap's horizontal edges are wrapped around the object. The height of the material bitmap is scaled along the cylinder's axis. You can select the parallel axis around which the material is wrapped, adjust the X and Y offset and rotation angle of the material, and adjust the bitmap placement.

- **Spherical**   Maps the material both horizontally and vertically. The left and right edges of the material bitmap are wrapped around the sphere. The top and bottom edges of the bitmap are compressed to points at the north and south poles of the sphere. You can select the parallel axis around which the material is wrapped, adjust the X and Y offset and rotation angle of the material, and adjust the bitmap placement.

- **Solid**   Applies materials based on three mapping coordinates—U, V, and W—enabling you to apply materials from any angle. You don't need to specify mapping coordinates for these materials. By changing the mapping coordinates for solids, however, you can skew the material pattern along any dimension of the solid.

Planar         Cylindrical    Spherical      Solid

*TIP:   To delete mapping coordinates assigned to an object, use the AutoLISP version of the SETUV command, as follows:*

```
Command: (c:setuv "D" (ssget))
Select objects: (select the objects whose mapping you want to delete)
Select objects: ENTER
```

## Adding Your Own Materials

In addition to using the materials that come with AutoCAD, you can create your own materials. While creating your own standard materials requires that you obtain bitmap images of materials (either by scanning in your own images or by obtaining images from a commercial library or some other source), creating your own procedural materials is relatively easy.

Procedural materials generate materials mathematically by using values that you control in a dialog box. The dialog box used to create a new material is similar to the one shown in Figure 22-12. Each type of material uses a slightly different set of controls. To begin creating a new material, select the type of material that you want to create from the drop-down list located below the New button in the Materials dialog box, and then click New.

*LEARN BY EXAMPLE*
*Create a new material for use in your rendering of the hotel.*

You will create a new granite material to attach to the cornice around the top of the building, above the first floor, and around the canopy and its supporting columns:

1. Display the Materials dialog box.

2. Select Granite from the drop-down list located below the New button and then click the New button. AutoCAD displays the New Granite Material dialog box.

3. In the Material Name edit box, type **MYGRANITE**.

   The Granite procedural material uses four different color settings, as well as reflection, roughness, sharpness, scale, and bumpmap settings. In the Granite Material dialog box, you will change the Second, Third, and Fourth color settings, as well as the scale. The other settings can remain at their default values.

4. Under Attributes, select Second Color, and then adjust the Value and Color settings as follows:

   Value: 0.50

   Red: 0.69

   Green: 0.50

   Blue: 0.40

BECOMING AN EXPERT

5. Make the following changes to the Third Color settings:

   Value: 0.50

   Red: 1.00

   Green: 0.98

   Blue: 0.61

6. Make the following changes to the Fourth Color settings:

   Value: 0.90

   Red: 0.88

   Green: 0.70

   Blue: 0.46

7. Under Attributes, select Scale and then change the Scale Value to 0.010.

8. Click OK to close the New Granite Material dialog box.

Now, you can attach the new material to objects in your drawing. Again, the easiest way is to attach the material to those objects drawn on particular layers.

1. In the Materials dialog box, under Materials, select the new material MYGRANITE.

2. Click the By Layer button to display the Attach By Layer dialog box.

3. In the Select Layer list, select the layers ROOF-CORNICE, TRIM_FRONT, and PORTICO, and then click the Attach button.

4. Click OK to close the Attach By Layer dialog box, and then click OK again to close the Materials dialog box.

Check the appearance of your rendering again. Try turning on the shadows again to see how your rendering is becoming more realistic with each successive step. Your drawing should look similar to Figure 22-13.

## Adding More Materials

Many objects in the drawing still do not have materials attached to them. Selecting the appropriate materials is a bit of a trial-and-error process. Most of the choices remaining for the hotel are pretty straightforward. You obviously want to assign a glass material to the windows. AutoCAD provides several standard glass materials. Here is where adjustments to the Reflection and Refraction settings can really make a difference. You also need to assign a material to represent the grass and the paving around the building.

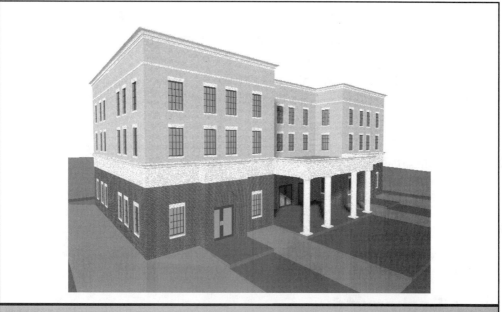

**Figure 22-13.** *The 3-D building with more appropriately scaled materials*

**EXAMPLES**

*LEARN BY EXAMPLE*
*Attach the remaining materials.*

On your own, add the materials shown in the following table, and then attach them to the layers indicated. In some cases, you also need to adjust the bitmap scaling, using the values shown in the table. An asterisk adjacent to the value in the table indicates that you can use the default scale or bitmap scale value.

| Material | Layers | Bitmap Scale |
| --- | --- | --- |
| BEIGE MATTE | WALL-LOW-IN, WALL-IN | 1.0* |
| BLACK MATTE | ROOF | 1.0* |
| BLUE GLASS | GLAZING | 1.0* |
| BRASS GIFMAP | HARDWARE | 1.0* |
| CLOVER PATTERN | FLOOR-1 | U=48, V=24 |

BECOMING AN EXPERT

| Material | Layers | Bitmap Scale |
|----------|--------|--------------|
| COPPER | DOORS-JAMB, WINDOWS-JAMB, DOORS | 1.0* |
| GRANITE PEBBLES | PAVING | 0.01 |
| PALM FROND | GRASS | 12.0 |
| WHITE MATTE | WINDOWS | 1.0* |

*You adjust the scale of the GRANITE PEBBLES procedural material in the Modify Materials dialog box by selecting the Scale radio button under Attributes and then changing its value. When you adjust the CLOVER PATTERN, make sure that the Maintain Aspect Ratio check box is not selected, so that you can apply the scaling factors individually to the U and V values.*

When you finish attaching these materials, render your model again. The image should now look pretty good, but it's still missing something.

## Adding a Background

Adding a realistic background is a wonderful and easy way to dress up a rendering. When you add a background, AutoCAD renders the model against the background. You add a background by using the Background dialog box, shown in Figure 22-14. To display the Background dialog box, do one of the following:

- On the Render toolbar, click Background.
- From the View menu, choose Render | Background.
- At the command line, type **BACKGROUND** and then press ENTER.

*You can also display the Background dialog box by clicking the Background button in the Render dialog box.*

You can add any one of four types of backgrounds. The four radio buttons across the top of the dialog box let you determine which of the background types to use:

- **Solid**  Uses a single background color, specified in the Colors area of the dialog box. This is the default option, and, by default, AutoCAD uses its graphic window background color as the background color for the rendering.
- **Gradient**  Uses a two- or three-color gradient. The controls in the Colors area enable you to define these gradient colors.
- **Image**  Uses a bitmap file as the background. You specify the bitmap file in the Image area. You can also specify a second bitmap as an environment map. This map is used to create reflections when you use the Photo Raytrace renderer.
- **Merge**  Uses the current AutoCAD image as the background.

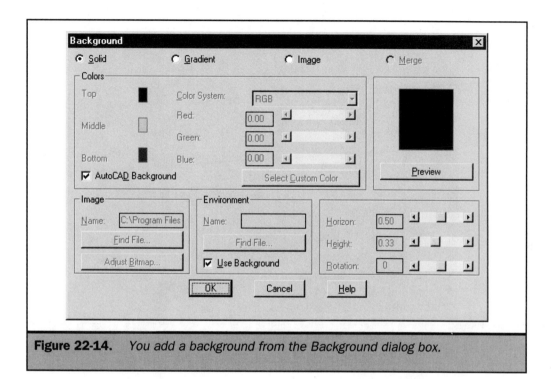

**Figure 22-14.** *You add a background from the Background dialog box.*

### LEARN BY EXAMPLE
*Add a background image to your model. Continue working with the drawing HOTEL-01 on the companion web site.*

1. Display the Background dialog box.
2. In the Background dialog box, select the Image radio button.
3. Under Image, click the Find File button.
4. In the Background Image dialog box, specify the CLOUD.TGA file located in AutoCAD's Textures directory, and click Open. (Note that you have to choose *.tga from the Files Of Type drop-down list before any of the files in the Textures directory are displayed.)
5. Click OK to close the Background dialog box.

Now, check the appearance of your rendering again. Your drawing should look similar to Figure 22-15. The rendering looks pretty good, but it still looks a bit empty.

BECOMING AN EXPERT

**Figure 22-15.** *The 3-D building with a background image*

## Adding Landscaping and People

The real world is a pretty cluttered place. It's filled with trees and bushes, and if you've ever tried to take a photograph of a building, you know that it's next to impossible not to include some people in the image. Rather than making the image look cluttered, however, landscaping makes it look more realistic, and including people in your rendering gives it a better sense of scale.

You add landscaping and people by using the Landscape New dialog box, shown in Figure 22-16. To display the Landscape New dialog box, do one of the following:

- On the Render toolbar, click Landscape New.
- From the View menu, choose Render | Landscape New.
- At the command line, type **LSNEW** and then press ENTER.

The left side of the dialog box contains a list of landscape objects contained in the current landscape library file. To add one of these objects to the drawing, select the object from this list and then click the Position button. When you click this button, the dialog box temporarily disappears and AutoCAD prompts you to choose the location of the base of the landscape object. You can specify the location of the object by pointing in the drawing (perhaps with the aid of an object snap) or by typing coordinates. Remember that the point you specify represents the base of the object.

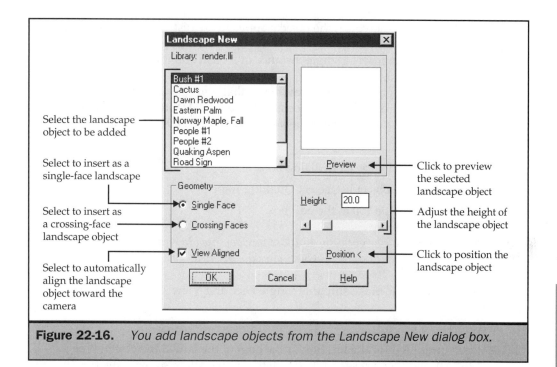

Select the landscape object to be added

Select to insert as a single-face landscape

Select to insert as a crossing-face landscape object

Select to automatically align the landscape object toward the camera

Click to preview the selected landscape object

Adjust the height of the landscape object

Click to position the landscape object

**Figure 22-16.**    *You add landscape objects from the Landscape New dialog box.*

When you insert a landscape object, you need to pay particular attention to the settings in the Height and Geometry areas. In the Height area, use either the edit box or the slider bar to adjust the height of the object, measured in drawing units. The controls in the Geometry area determine the geometry and alignment of the landscape object:

- **Single Face**   Maps the landscape object onto a single face. Single-face landscape objects render quickly, but may not look as realistic as those mapped onto crossing faces.

- **Crossing Faces**   Maps the landscape object onto two faces, which cross at a 90-degree angle. Crossing-face landscape objects often look more realistic, but take longer to render.

- **View Aligned**   Adjusts the faces onto which the landscape object is mapped so that one face is always aligned directly toward the camera.

*LEARN BY EXAMPLE*
*Add some landscape objects to your model. Continue working with the drawing*
*HOTEL-01 on the companion web site.*

Landscape objects are created as ARX objects. They appear in the AutoCAD drawing as triangles and are treated like any other AutoCAD object. When you use the RENDER command, however, the objects appear as realistic-looking trees, bushes, and people. Since AutoCAD treats these special objects as it does any other drawing object, however, you can insert a single instance of a bush and then copy or array that bush to quickly add foliage.

Before proceeding, regenerate the drawing so that you can see the wireframe image. Remember that SHADEMODE needs to be set to 2D Wireframe to display a wireframe image at this point. Create a new layer called **LANDSCAPE** and make it the current layer:

1. Display the Landscape New dialog box.

2. In the landscape object list, select Quaking Aspen.

3. Under Height, type **100** in the edit box.

**Note** *AutoCAD limits the height value to 100. Since you need to create a tree taller than 8 feet, you need to scale this object after it's inserted.*

4. Under Geometry, make sure that Single Face and View Aligned are both selected.

5. Click Position.

6. When prompted for the location of the base of the landscape object, type **-125,75,0**, and then press ENTER. (Note that you could also place the tree by pointing in the drawing. If you do this, however, you will find it much easier to first switch to a plan view and use .XY filtering, so that you can be sure that the landscape object is placed at the correct elevation in the drawing. You learned about .XY filtering in Chapter 6.)

7. Click OK to close the Landscape New dialog box.

Next, use the SCALE command to enlarge the tree to a more realistic size. Use the following sequence:

```
Command: SCALE
Select objects: LAST
Select objects: ENTER
Specify base point: -125,75,0
Specify scale factor or [Reference]: 3
```

Next, use the COPY command to place a copy of this tree. Use the following command sequence:

```
Command: COPY
Select objects: LAST
Select objects: ENTER
Specify base point or displacement, or [Multiple]: 75',-25'
Specify second point of displacement or <use first point as
displacement>: ENTER
```

Add some shrubs in front of the building, as follows:

1. Display the Landscape New dialog box.

2. In the landscape object list, select Wandering Yew.

3. Under Height, set the height to 48.

4. Under Geometry, select Crossing Faces.

5. Click Position.

6. When prompted for the location of the landscape object, type **22',-3'**, and then press ENTER.

7. Click OK to close the Landscape New dialog box.

Now, use the ARRAY command to create several copies of this bush. Create a rectangular array consisting of six rows and two columns. The row spacing should be **-3'** and the column spacing **3'**. When you finish this step, your drawing should look similar to Figure 22-17.

Now add some people. AutoCAD treats people the same as other landscape objects:

1. Display the Landscape New dialog box.

2. In the landscape object list, select People #1.

3. Under Height, set the height to 68.

4. Under Geometry, select Single Face.

5. Click Position, specify the location as **36',-16',-6**, and then press ENTER.

6. Click OK to close the Landscape New dialog box.

Render your drawing again. With each change that you make, your drawing becomes increasingly more realistic. On your own, continue adding landscape objects to fill in the empty spaces.

## Adding Landscape Objects

AutoCAD comes with a landscape library (RENDER.LLI), which is located in the Support directory. By default, the program uses this library file. You can also create

**Figure 22-17.** *Landscape objects appear as triangles when viewed in wireframe view.*

and save your own landscape libraries by using the Landscape Library dialog box, shown in Figure 22-18. To display this dialog box, do one of the following:

- On the Render toolbar, click Landscape Library.
- From the View menu, choose Render | Landscape Library.
- At the command line, type **LSLIB** and then press ENTER.

You can use the controls in this dialog box to modify existing landscape objects, create new ones, or delete objects from the landscape library file. You can also save and load custom landscape libraries. Landscape objects are created and modified by using a similar dialog box, shown in Figure 22-19.

This dialog box enables you to define the object's default geometry. The other parameters define the name of the object and specify two files that are used to control the appearance of the object. The Image and Opacity Map files are both bitmap image files. The Image file contains a picture of the landscape object. The Opacity Map file is a similar file, but rather than being an actual image of the landscape object, this file is a black-and-white image called a *mask*. The mask defines the areas of the image that you

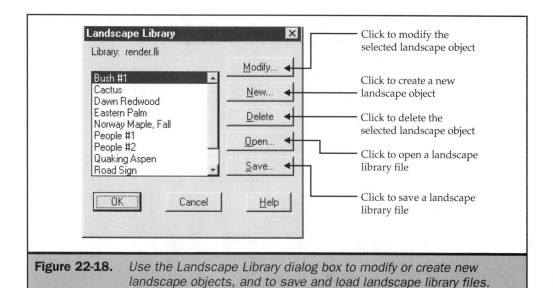

**Figure 22-18.**    Use the Landscape Library dialog box to modify or create new
landscape objects, and to save and load landscape library files.

BECOMING AN EXPERT

can see through. The black portion of the image represents the transparent areas, and
the white portion represents the opaque areas.

*LEARN BY EXAMPLE*
*Create a new landscape object for use in your rendering. You will use two Targa files,*
*GRAND-L.TGA and GRAND-O.TGA, supplied on the companion web site.*

You can take advantage of landscape objects to add a sign to the front of the hotel.
The two bitmaps provided on the companion web site were created using a simple

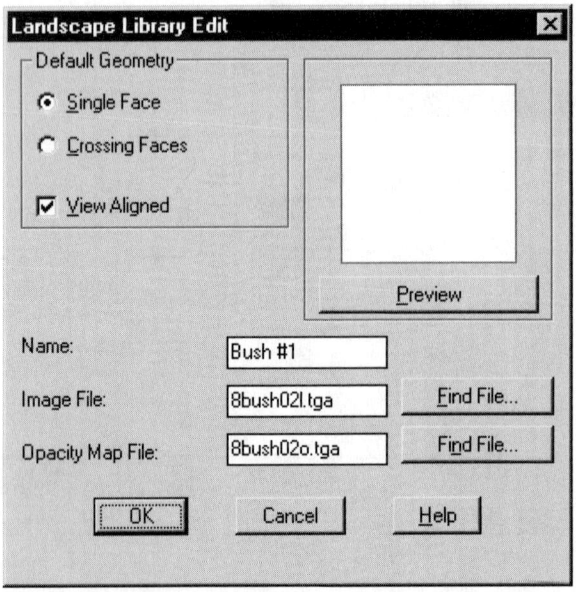

**Landscape Library Edit**  [×]

Default Geometry
- (●) Single Face
- ( ) Crossing Faces
- [✓] View Aligned

Preview

Name:         Bush #1

Image File:   8bush02l.tga        Find File...

Opacity Map File:  8bush02o.tga   Find File...

OK     Cancel     Help

**Figure 22-19.** *You edit existing landscape objects by using the Landscape Library Edit dialog box. A similar dialog box is used to create new landscape objects.*

paint program. One image is the actual image of the sign; the other is an opacity map, so that only the letters that spell out the words "Grand Hotel" will be opaque. Use the following steps to add the sign as a new landscape object:

*Copy the bitmaps used for the sign from the companion web site into AutoCAD's Textures directory on your hard disk, so that AutoCAD can find the bitmaps even when you're not connected to the Internet.*

1. Display the Landscape Library dialog box.

2. In the Landscape Library dialog box, click the New button.

3. In the Landscape Library New dialog box, under Default Geometry, make sure the Single Face radio button is selected and the View Aligned check box is not selected. You want the sign to always align with the face of the portico, rather than always face toward the camera.

**Note** *Don't despair if you make a mistake when you add a landscape object. You can always use the LSEDIT command to edit landscape objects that are already in the drawing. When you start this command, AutoCAD prompts you to select the landscape object that you want to edit. As soon as you select the object, AutoCAD displays the Landscape Edit dialog box, which is identical to the Landscape New dialog box. You can use the controls in the dialog box to modify the height, geometry, and position of the landscape object.*

4. In the Name edit box, type **Grand Hotel**.

5. Click the Find File button adjacent to the Image File edit box, select the file GRAND-L.TGA from the companion web site, and then click Open. (*Hint*: If you downloaded the file to your computer, use the copy you already downloaded. Otherwise, you can click the Search the Web button to connect to the companion web site.)

6. Click the Find File button adjacent to the Opacity Map File edit box, select the file GRAND-O.TGA from the companion web site, and then click Open.

7. Click OK to close the Landscape Library New dialog box.

8. Click OK to close the Landscape Library dialog box.

**Note** *AutoCAD displays a dialog box asking whether you want to save the new landscape object to the default landscape library file. Click Save Changes and then click Open. If you prefer, in the Landscape Library dialog box, you can click Save and then save the landscape objects to a new library file.*

Now that you've created the landscape object for the sign, you can add it to the front of the hotel's portico:

1. Display the Landscape New dialog box. Notice that the Grand Hotel object that you just added is now included in the list of landscape objects.

2. In the list of landscape objects, select the Grand Hotel landscape object.

3. Under Height, change the value to 36.

4. Click the Position button, specify the position as **45',-10'2,12'6**, and then press ENTER.

5. Click OK to close the Landscape New dialog box.

Render your drawing. The sign is now visible on the center of the portico, but the text is reversed. You can fix this quickly simply by rotating the sign object, as follows:

```
Command: ROTATE
Select objects: LAST
```

BECOMING AN EXPERT

```
Select objects: ENTER
Specify base point: 45',-10'2
Specify rotation angle or [Reference]: 180
```

Render the drawing again. The image should look similar to Figure 22-20. Your rendering has certainly come a long way.

## Using Scenes

So far, you've rendered the model from a single vantage point, using the single solar distant light source. Just as a single drawing can contain many named views, you can also define different scenes. In addition to saving the viewpoint, scenes also store a list of associated lights. This feature enables you to add additional lights to a drawing and use specific lights to render different scenes.

As you've noticed throughout this chapter, the Render dialog box contains a list of scenes, and you can select the scene that you want to render. Until now, you've always rendered the same scene, the current view. If you save additional scenes, when you use the RENDER command to render the drawing, you can select the scene that you want to render.

**Figure 22-20.**   *The rendering of the hotel with materials, landscaping, people, and signage*

You create scenes by using the Scenes dialog box, shown in Figure 22-21. To display the Scenes dialog box, do one of the following:

- On the Render toolbar, click Scenes.
- From the View menu, choose Render | Scene.
- At the command line, type **SCENE** and then press ENTER.

This dialog box contains a list of all the scenes defined in the current drawing. The buttons in this dialog box enable you to define new scenes, modify existing ones, and delete any scenes that you no longer need.

To create a new scene, click the New button to display the New Scene dialog box, shown in Figure 22-22. This dialog box contains three controls. The Scene Name edit box is where you specify the name of the scene. The Views list on the left side of the dialog box contains a list of all the named views defined in the current drawing. The Lights list on the right side contains a list of all the lights already created in the current drawing.

*When you modify an existing scene, AutoCAD displays a Modify Scene dialog box, which is identical to the New Scene dialog box.*

*LEARN BY EXAMPLE*
*Create a new scene and render it. Open the file HOTEL-02 on the companion web site.*

The version of the hotel model saved in HOTEL-02.DWG contains several additional lights. Point lights have been added on the inside of the building to illuminate the interior at each floor. Several lights have also been added on the outside of the building. A point light illuminates the area under the portico. Spotlights, used as

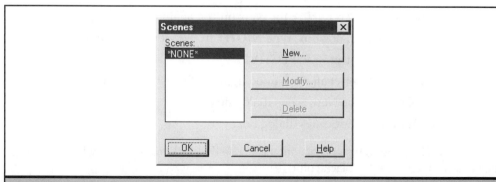

**Figure 22-21.**    *You control scenes by using the Scenes dialog box.*

BECOMING AN EXPERT

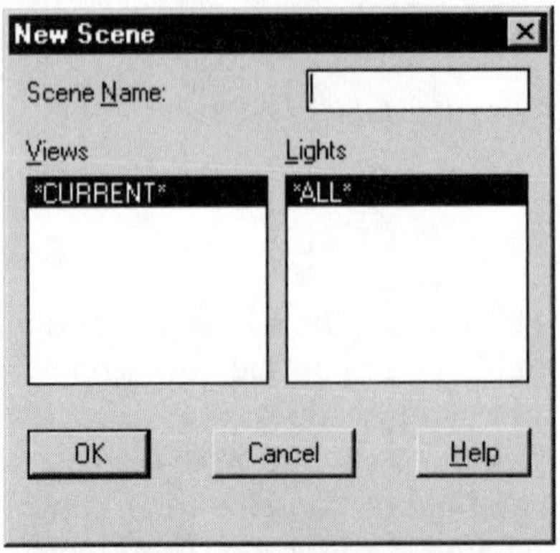

**Figure 22-22.**   *Create new scenes by using the New Scene dialog box.*

down-lights, illuminate the side entrance. And two relatively bright spotlights illuminate the front of the portico.

Use the SCENE command to create scenes for the daytime and nighttime views of the hotel:

1. Display the Scenes dialog box.

2. In the Scenes dialog box, click the New button.

3. In the New Scene dialog box, type **DAY** in the Scene Name edit box.

4. In the Views list, select PERSPECTIVE.

5. In the Lights list, select SUN.

6. Click OK to close the New Scene dialog box.

7. In the Scenes dialog box, click the New button again.

8. In the New Scene dialog box, type **NIGHT** in the Scene Name edit box.

9. In the Views list, select PERSPECTIVE.

10. In the Lights list, select all the named lights in the list except SUN.

11. Click OK to close the New Scene dialog box.

12. Click OK to close the Scenes dialog box.

Since this is going to be a nighttime view, use the Background dialog box to change the background to a solid-black color:

1. Display the Background dialog box.

2. Select the Solid radio button.

3. In the Colors area, make sure that the AutoCAD Background color check box is not selected. Set the Red, Blue, and Green values each to 0.00.

4. Click OK to close the Background dialog box.

Now render the drawing:

1. Display the Render dialog box.

2. In the Render dialog box, under Rendering Type, select Photo Raytrace.

3. In the Scene To Render list, select NIGHT.

4. In the Rendering Options area, make sure all the check boxes are selected.

5. Click Render.

Depending on the speed of your computer and the amount of memory available, this rendering may take a considerable amount of time. When the rendering is complete, your drawing should look similar to Figure 22-23.

You can also create scenes that produce renderings from different viewpoints. Create another view showing the front elevation of the hotel as seen by day:

1. Display the Scenes dialog box.

2. In the Scenes dialog box, click New.

BECOMING AN EXPERT

**Figure 22-23.**    *Rendering the NIGHT scene produces an image of the hotel by night.*

3. In the New Scene dialog box, type **EL-FRONT** in the Scene Name edit box.

4. In the Views list, select FRONT.

5. In the Lights list, select SUN.

6. Click OK to close the New Scene dialog box, and then click OK again to close the Scenes dialog box.

Now render the front elevation:

1. Display the Render dialog box.

2. In the Scene list, select EL-FRONT.

3. Click the Background button to display the Background dialog box.

4. In the Background dialog box, select the Image radio button. AutoCAD still remembers the file that you used previously as the background.

5. Click OK to close the Background dialog box.

6. Click Render.

Your drawing should look similar to Figure 22-24.

**Figure 22-24.** *Rendering the EL-FRONT scene produces a rendered elevation of the hotel by day.*

# Rendering to Other Destinations

As mentioned earlier, in addition to rendering directly to the current AutoCAD viewport, you can render to either of two other destinations: the Render Window or a file. The Render Window is actually a window created by AutoCAD's Render application, which, as you've already learned, is a separate application that is loaded into memory whenever you use any of AutoCAD's rendering commands. Rendering to a file enables you to save the rendering to a file rather than simply display it on the screen.

When you render to the Render Window, you can also subsequently save the image to a BMP file or print it to any configured Windows system printer. Rendering to the Render Window also has the advantage of enabling you to render an image to its own window, which in turn enables you to render several different views and see them all onscreen simultaneously, so that you can visually compare the differences (as shown in Figure 22-25). You can also copy the image to the Windows Clipboard and paste it into another program.

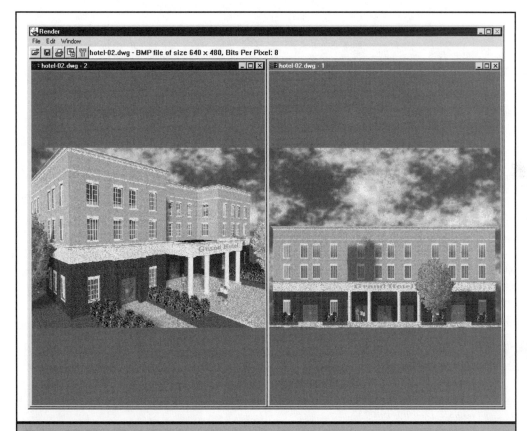

**Figure 22-25.**    *Each image rendered to the Render Window is displayed in its own window.*

When you render to a file, you can click the More Options button to display the File Output Configuration dialog box. You can then choose from any of five file types: BMP, PCX, PostScript, TGA, and TIFF. You can also specify the color depth (selecting anything from a monochrome image all the way up to 32-bit color) and resolution. This enables you to create renderings using much higher resolutions and greater color depth than that supported by your graphics card and monitor.

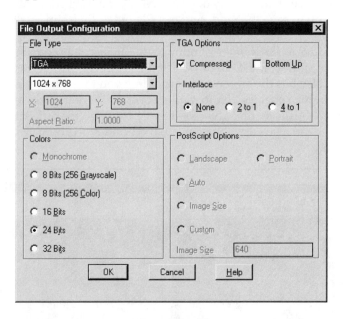

*If you plan to create photographic-quality prints of your renderings, you probably need to render to a file using a relatively high resolution, such as 2048×1566.*

To render to one of these destinations, select the destination from the drop-down list in the Render dialog box. Choose the other rendering options as you normally would. Then, click the Render button. If you selected the Render Window, AutoCAD automatically switches to the Render application window and, after completing its calculations, displays the rendering. If you render to a file, when you click the Render button, AutoCAD displays a File dialog box so that you can specify the name of the resulting file.

**Tip**

*You can use the SAVEIMG command to save a rendering from the current AutoCAD viewport to a BMP, TGA, or TIFF file, and use the REPLAY command to display any image saved in these formats in the current viewport. In addition to typing these commands on the command line, you can start them from the menu by choosing Tools | Display Image.*

# Chapter 23

## Using Raster Images

The ability to add raster images to your AutoCAD drawings opens up many capabilities. For example, if you want to include a rendered image of your model as part of the plotted drawing, you can save the rendering as a raster image (which you learned how to do in the previous chapter) and then attach that rendering to the drawing. You can also use raster images to include company logos, photographs, or scanned images in your drawings, so that they are part of the drawing when plotted.

Scanning an existing drawing opens up some interesting possibilities. For example, once a scanned image of an existing paper drawing is displayed in AutoCAD, you can use AutoCAD's drawing tools to trace the image, thus creating a new AutoCAD drawing from the old paper drawing.

This chapter explains how to do the following:

- Attach raster images to your drawing
- Manage, modify, and clip attached images

**Note** *In Chapter 22, you learned how to apply raster images as materials when creating rendered images of objects.*

# Attaching Raster Images

A raster image consists of a rectangular grid of dots, called *pixels,* and the size of a raster image is often referred to by the size of this grid. For example, a 640×480 raster image means that it's composed of a grid of dots consisting of 480 horizontal rows of dots with 640 dots in each row. As such, raster images are very different from the vector or line-based objects created by other AutoCAD commands. After raster objects are attached to a drawing, AutoCAD treats them just like other AutoCAD drawing objects, to the extent that they can be copied, moved, clipped, scaled, and so on. Many different raster formats exist, and AutoCAD supports all of the most common formats. Raster images and renderings can be displayed in AutoCAD in True Color by selecting the True Color Raster Images And Rendering check box on the Display tab of the Options dialog box. Note that the Display Properties Settings in Windows must also be configured to True Color in order to take advantage of this setting in AutoCAD.

When you attach a raster image, it is treated in much the same way that AutoCAD handles external references. Because raster image files can be quite large, AutoCAD does not actually add the image to the drawing, but instead attaches a link to the image file. That way, the image is displayed in the drawing, but does not greatly increase the size of the drawing file.

Since raster images are linked in this way, they present some of the same advantages and management issues that external references present. Any changes made to the raster image appear in the drawing automatically each time that you open the drawing containing the image. Because raster images are not part of the drawing, however, if you exchange drawings with others, you need to be sure to send them the raster images as well. And, as you learned when dealing with external references, you also must manage the paths that AutoCAD uses to locate the image files. If the path changes (for example, if you move the drawing or image file into a different directory), AutoCAD will not be able to find the attached image if you haven't updated the path.

Attaching a raster image to a drawing is so similar to attaching an external reference that even the dialog boxes look similar for these two procedures. Like external references, you can attach as many images and as many copies of each image as you want. Each copy can have a different position, scale, and rotation angle.

 *Unlike external references, you can't bind the raster image files into the AutoCAD drawing. Raster files always remain attached as external links.*

To attach a raster image:

1. Do one of the following:

   - On the Reference or Insert toolbar, click Image.
   - From the Insert menu, choose Image Manager.
   - At the command line, type **IMAGE** (or **IM**) and then press ENTER.

   AutoCAD displays the Image Manager dialog box, which you'll learn about in the next section. (Note that since the IMAGE command is stored in an external application file, its ability to load on demand is controlled by the DEMANDLOAD system variable. This variable can be controlled from the Open and Save tab of the Options dialog box.)

2. In the Image Manager dialog box, click Attach. AutoCAD displays the Select Image File dialog box.

 *When attaching an image, you can skip the Image Manager dialog box and go directly to the Select Image File dialog box by doing one of the following:*

- *On the Reference toolbar, click Image Attach.*
- *From the Insert menu, choose Raster Image.*
- *At the command line, type **IMAGEATTACH** (or **IAT**) and then press ENTER.*

3. In the Select Image File dialog box, select the raster image file that you want to attach, and then click Open. AutoCAD displays the Image dialog box, shown in Figure 23-1.

4. Under Insertion Point, either specify the insertion point or select the Specify On-Screen check box so that AutoCAD prompts you to specify the insertion point when you actually place the image in the drawing. Do the same for the scale factor and rotation angle.

5. Click OK.

BECOMING AN EXPERT

**Note** When you retain the path while attaching an image file, you create an explicit link to the image. AutoCAD always looks for the image in the specified directory. If you don't retain the path, AutoCAD searches for the image based on the Project Files Search Path stored in the PROJECTNAME system variable. If you haven't established this search path, AutoCAD bases its search on the Support File search parameters. This is identical to the way that AutoCAD resolves links to external references, and is described fully in Chapter 15.

Depending on the information specified under Insertion Point, Scale, and Rotation, AutoCAD may prompt you at the command line for additional information to complete the image insertion. For example, by default, the rotation angle is predetermined in the dialog box, but AutoCAD prompts you for the insertion point and scale factor.

**Tip** You can also attach raster images by using drag-and-drop, by dragging an image file from Windows Explorer into an open AutoCAD drawing. You can also attach images using AutoCAD DesignCenter.

**Note** If you click the Details button, the bottom of the Image dialog box expands to display the Image Information section. This section of the dialog box contains information about the size of the raster image, including the Current AutoCAD Unit. In Release 14, you could change this setting by selecting from the Current AutoCAD Unit drop-down list in the Attach Image dialog box. In Release 14, this setting was used only for raster images, whereas in AutoCAD 2002, the raster image drawing units variable has been combined with the AutoCAD DesignCenter Insert Units (INSUNITS) variable. For that reason, you can no longer change the Current AutoCAD Units from within the Details area when you use the IMAGEATTACH command. Instead, to set the current drawing units, use one of the following methods:

1. From the Format menu, select Units.

2. Under Drawing Units for DesignCenter Blocks, select the desired units from the drop-down list and then click OK.

or

1. At the command line, type **INSUNITS** and press ENTER.

2. Specify the new INSUNITS value and then press ENTER.

The default INSUNITS value is 0, which is unspecified (unitless). You can change the default value for this variable by modifying your default drawing template or the ACADDOC.LSP file. See Appendix B for a complete listing of the valid INSUNITS values.

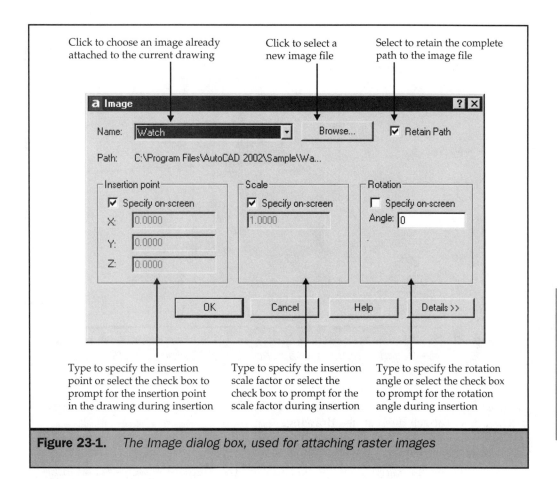

Click to choose an image already attached to the current drawing

Click to select a new image file

Select to retain the complete path to the image file

Type to specify the insertion point or select the check box to prompt for the insertion point in the drawing during insertion

Type to specify the insertion scale factor or select the check box to prompt for the scale factor during insertion

Type to specify the rotation angle or select the check box to prompt for the rotation angle during insertion

**Figure 23-1.**    *The Image dialog box, used for attaching raster images*

# Scaling an Image for Tracing

Raster images are particularly useful for bringing scanned paper drawings into AutoCAD so that they can be traced, thus re-creating portions of the existing drawing as an AutoCAD drawing. To trace the lines in the image as accurately as possible, however, you need to match the scale of the image to your AutoCAD drawing. This can usually be accomplished easily by using the Relative option of the SCALE command, which you learned about in Chapter 7.

*LEARN BY EXAMPLE*

*Insert a scanned drawing and scale it to full size so that it can be traced in AutoCAD. Open the drawing Figure 23-2 on the companion web site.*

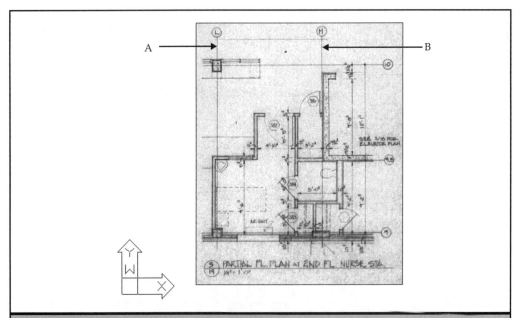

**Figure 23-2.** *You can scale a raster image by using a known relative distance, so that its size matches your current AutoCAD drawing units.*

Begin by attaching the scanned drawing. Use the following command sequence and instructions to attach the scanned drawing:

1. Start the IMAGEATTACH command.

2. In the Select Image File dialog box, look on the companion web site and locate the file NURSE.TIF. Select the file and then click Open.

3. In the Image dialog box, make sure that the Specify On-Screen check boxes under Insertion Point and Scale are selected and the one under Rotation is cleared, and then click OK.

4. When prompted for the insertion point, press ENTER to accept the default (0,0).

5. When prompted for the scale factor, drag the image so that it approximately fills the drawing.

You should now see the scanned image. Because you know that the distance between column grids L and M is exactly 13' 5", you can use this distance as a relative scale factor to accurately scale the scanned image. Use the following command sequence and instructions:

```
Command: SCALE
Select objects: (select the image by clicking its boundary)
Select objects: ENTER
Specify base point: 0,0
Specify scale factor or [Reference]: R
Specify reference length <1>: (select point A)
Specify second point: (select point B)
Specify new length: 13'5
```

Zoom to the drawing extents. The scanned image is now scaled so that distances on the scanned image match those in AutoCAD (see Figure 23-2). You can now draw over the scanned image by using standard AutoCAD commands. Of course, the accuracy of what you trace is dependent on the accuracy of the original drawing, so you may need to make some adjustments. Also, remember that you can use the DRAWORDER command to move the image behind any existing geometry in the drawing.

## Managing Raster Images

After you insert raster images into your drawing, you can view information about the images by using the Image Manager dialog box, which provides a unified interface for managing all the images in the current drawing. This dialog box provides two different views of your image files: the List View and the Tree View. Initially, when you start the command, AutoCAD displays the List View, shown in Figure 23-3. This view shows a list of all the images attached to the current drawing. Information about the images is arranged into the following six columns:

- **Image Name**   Lists the names of attached image files.
- **Status**   Indicates whether the image is loaded, unloaded, or not found. Images indicated as unloaded or not found are not displayed in the drawing.
- **Size**   Shows the size of the image file (in K).
- **Type**   Indicates the image file format.
- **Date**   Shows the date the image file was last modified. (This field is blank if the image is unloaded or not found.)
- **Saved Path**   Shows the saved path of the image file (which is not necessarily where the image file was found).

**Tip**   *By default, this list is sorted alphabetically by image filename. You can sort the list on any other parameter by clicking the appropriate column heading. You can also change the width of the columns by clicking-and-dragging the line left or right between the headings.*

BECOMING AN EXPERT

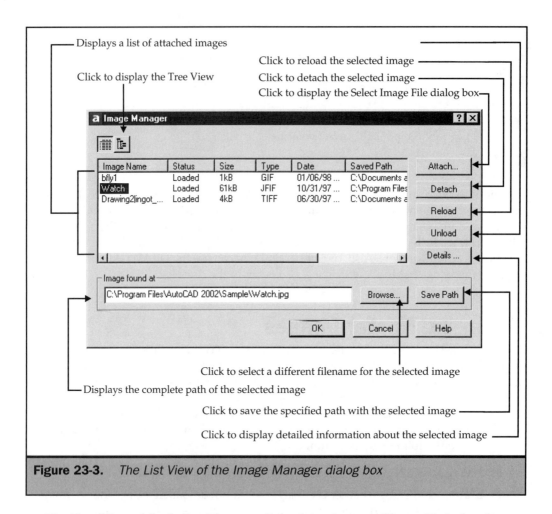

Displays a list of attached images

Click to reload the selected image

Click to display the Tree View

Click to detach the selected image

Click to display the Select Image File dialog box

Click to select a different filename for the selected image

Displays the complete path of the selected image

Click to save the specified path with the selected image

Click to display detailed information about the selected image

**Figure 23-3.** *The List View of the Image Manager dialog box*

The Tree View of the Image Manager dialog box, shown in Figure 23-4, also shows the images attached to the current drawing, but it displays them in a tree-structured view that shows the nesting level of the images within external reference files. Images attached directly to the current drawing appear at the top level of the tree, while images included as part of xrefs are shown at subsequently lower levels. When multiple levels are shown, you can expand or compress the tree by clicking the plus or minus sign. To display the Tree View, click the Tree View button near the top of the dialog box or press F4. To switch back to the List View, click the List View button or press F3.

In addition to the List and Tree Views of the attached images, the Image Manager dialog box provides access to most of the other functions that you need for managing images. You already learned how to attach images. Clicking the Attach button displays the Select Image File dialog box. In addition to attaching images, you can do the following:

■ **Detach** Removes the image and all copies of it from the current drawing. Only images attached directly to the current drawing (those appearing in the

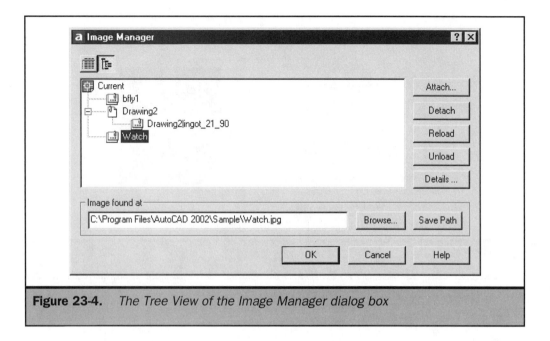

**Figure 23-4.**    *The Tree View of the Image Manager dialog box*

BECOMING AN EXPERT

Tree View at the top level) can be detached. Images included as part of an xref can't be detached.

- **Reload**    Reads the most recently saved version of the image into the current drawing.

- **Unload**    Unloads the selected images from the current drawing so that they don't display, thus improving AutoCAD's performance. The images remain attached, however, and you can redisplay them by reloading them. You should unload rather than detach images when you no longer need them in the current drawing but may need them again later, such as when you need to plot another copy of the drawing or perhaps trace another portion of a scanned paper drawing.

# Viewing Image File Details

The Details button displays the Image File Details dialog box, which contains information about the selected image, as shown in Figure 23-5. This dialog box displays the following information:

- Image name
- Saved path (path saved when the image was attached)
- Active path (where the image file is actually found)
- File creation date
- File size

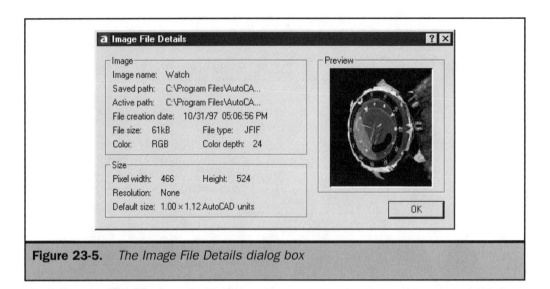

**Figure 23-5.** *The Image File Details dialog box*

- File type
- Color mode
- Color depth
- Image size, listed three different ways: the height and width measured in pixels, the resolution, and the default size measured in AutoCAD drawing units

The dialog box also displays a visual preview of the selected image.

**Note** *If the image resolution was not saved with the image file, the Resolution information appears as None.*

To view image file details:

1. Start the IMAGE command.
2. In the Image Manager dialog box, select the image to be viewed.
3. Click the Details button.

## Detaching Images

Although you can erase individual instances of images, if you no longer need the image in the drawing, you can completely remove the image and all instances of it by detaching it from the current drawing. To detach an image:

1. Start the IMAGE command.

2. In the Image Manager dialog box, select the image to be detached.

3. Click the Detach button.

4. Click OK.

*Detaching an image removes any links to the image, but does not affect the image file itself.*

## Unloading and Reloading Images

As you learned earlier, when you unload an image file that is no longer needed, you can improve AutoCAD's performance, because the program no longer has to read or display the image. The image file remains attached, however, so that it can be reloaded if necessary. To unload an image:

1. Start the IMAGE command.

2. In the Image Manager dialog box, select the image to be unloaded.

3. Click the Unload button.

4. Click OK.

Although AutoCAD automatically reloads all images each time you open a drawing that contains images, you may need to reload an image explicitly in two instances:

- If you subsequently need an image that was previously unloaded

- When the image has been modified while you are working on a drawing to which the image is linked

To reload an image:

1. Start the IMAGE command.

2. In the Image Manager dialog box, select the image to be reloaded.

3. Click the Reload button.

4. Click OK.

## Changing Image Filenames and Paths

If you open a drawing that has an attached image file that has been moved to a different directory or renamed from what it was originally named when it was attached to the drawing, AutoCAD may not be able to load the image. In that case, the program reports the image's status as not found. You can reestablish the link to the file by changing the path for the image.

BECOMING AN EXPERT

**Note**

*When AutoCAD searches for an attached image, it first looks for the image in the attached path, and then searches in the current directory, the directory in which the current drawing is located, and then in the directories specified in the Support File path. If a Project Files Search Path has been specified, AutoCAD also searches those directories to locate the attached image files. You learned how to establish the Project Files Search Path in Chapter 15.*

To change the image path:

1. Start the IMAGE command.
2. In the Image Manager dialog box, select the image whose path you want to change.
3. In the Image Found At area, click the Browse button.
4. In the Select Image File dialog box, specify the new path and then click Open. The new path is displayed in the Image Found At area.
5. Click the Save Path button. The new path is displayed in the Saved Path column.
6. Click OK to complete the command.

You can specify the image file path as either a complete path (one containing the complete path and filename) or an implicit or relative path (one containing just the path information needed for AutoCAD to locate the file). You learned how to specify implicit paths in Chapter 15.

**Note**

*You can use the REDIR Path Substitution Express tool to redefine hard-coded paths in external references, images, shapes, styles, and rtext objects. You'll find more detailed information about this Express tool on the companion web site.*

When you attach an image, AutoCAD uses the filename as the image name. The image name does not necessarily have to match the filename, however. You can change the image name without affecting the filename or path. To change the image name:

1. Start the IMAGE command.
2. In the Image Manager dialog box, click to select the image.
3. Click the image name again and then type the new image name.
4. Click OK to complete the command.

**Tip**

*You can also change the image name by pressing F2 and then typing the new image name.*

# Modifying Raster Images

All images have a rectangular boundary, which is used to select the image. When you attach an image, the image is inserted on the current layer and its boundary is created on that layer, as well, using the color and linetype settings currently in effect. You can modify the layer, color, or linetype as you would any other AutoCAD object. Similarly, if you turn off or freeze the layer on which the image was inserted, the image is no longer visible. You can change the image layer, boundary color, and boundary linetype by using the Properties window. You can also move, copy, rotate, scale, and otherwise modify the image by using any of AutoCAD's standard commands.

**Note** *If the attached image is bitonal (consists of only a foreground color and a background color), both the foreground pixels and the image boundary are drawn using the current color.*

In addition to changing the color, layer, or linetype, you can move an image or change its scale, rotation, size, or other parameters by using the Properties window. You can adjust the following properties in the Misc area of the Properties window (shown in Figure 23-6):

■ **Show Image** Displays both the image and its boundary (Yes) or only the boundary (No).

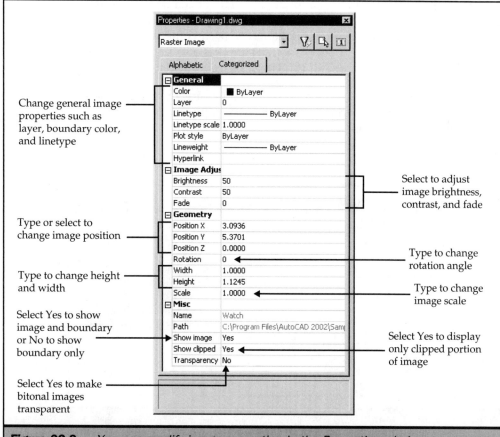

**Figure 23-6.** *You can modify image properties in the Properties window.*

BECOMING AN EXPERT

- **Show Clipped** Displays the entire image (No) or only the portion within a clipping boundary when such a clipping boundary has been applied (Yes). You'll learn about clipping images later in this chapter.

- **Transparency** Causes the background portion of bitonal images to be transparent (Yes) or opaque (No).

You can also adjust the Brightness, Contrast, and Fade values of the selected image.

 *When you change either the height or width of an image, AutoCAD always adjusts the other parameter to maintain the aspect ratio of the image. You can't scale or stretch images unequally.*

## Adjusting Images

You can adjust the brightness, contrast, and fade of an image by using the Image Adjust dialog box, shown in Figure 23-7. The adjustments that you make affect only the way the image displays and plots in AutoCAD, without affecting the original image. By adjusting the appearance of the images, you can improve the way that they display or plot. For example, adjusting the contrast may make an otherwise poor image easier to read. Adjusting the fade may make AutoCAD lines easier to see when displayed against a background image. The Image Adjust dialog box lets you adjust the following settings:

- **Brightness** Controls the brightness (and indirectly, the contrast) of the image in the range from 0 to 100. The higher the value, the brighter the image becomes (more pixels become white).

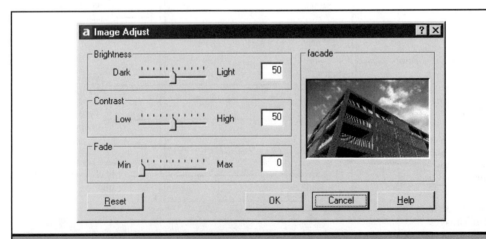

**Figure 23-7.** *You can adjust the image brightness, contrast, and fade by using the Image Adjust dialog box.*

- **Contrast** Controls the contrast (and indirectly, the fading effect) of the image in the range from 0 to 100. The higher the value, the more each pixel is forced to its primary or secondary color.

- **Fade** Controls the fading effect of the image in the range from 0 to 100. The higher the value, the more the image blends into the background. A value of 100 fades the image completely into the background.

- **Reset** Restores the original Brightness, Contrast, and Fade settings to 50, 50, and 0, respectively.

To adjust an image:

1. Do one of the following:

   - On the Reference toolbar, click Image Adjust.
   - From the Modify menu, choose Object | Image | Adjust.
   - At the command line, type **IMAGEADJUST** (or **IAD**) and then press ENTER.

*You can also select the image by using the Properties window and then clicking the button that appears in the Brightness, Contrast, or Fade box when you select any of these properties; or, select the image, right-click, and then choose Image | Adjust from the shortcut menu.*

2. Select the image to adjust.
3. In the Image Adjust dialog box, either move the appropriate slider bar or type a value in the adjacent edit box.
4. Click OK to complete the command.

*You can't adjust the brightness, contrast, or fade of bitonal images. You can make adjustments to the plotter configuration settings, however, to improve image plot quality. Start the PLOT command. In the Plot dialog box, select the Plot Device tab and then click the Properties button. In the Plotter Configuration Editor dialog box, select the Device and Document Settings tab and expand the Graphics node (by clicking the plus sign adjacent to the word Graphics). Select Raster Graphics to display the Raster Images area in the lower portion of the dialog box. This area contains three slider bars. (Note that the ability to change these settings depends on the device driver; not all drivers support this feature.) Move the Raster slider to Best. If the raster image is still not of a high enough quality, you can move the Trade-off bar toward Fewer colors. This will increase the sharpness of the image at the expense of reducing the number of colors, and is of more benefit when plotting to a monochrome plotting device.*

## Hiding Image Boundaries

In addition to using the Properties window to turn off the display of the image, you can hide the image boundaries. When image boundaries are hidden, you can't select the image by clicking it. This helps prevent the accidental modification of an image and also prevents the boundaries from being plotted. Turning image boundaries on and off affects all the images in the current drawing. To turn image boundaries on and off:

1. Do one of the following:

- On the Reference toolbar, click Image Frame.
- From the Modify menu, choose Object | Image | Frame.
- At the command line, type **IMAGEFRAME** and then press ENTER.

2. Type **ON** or **OFF** and then press ENTER.

 *When image boundaries are turned off, you can still select images by using selection methods other than pointing. For example, you can filter by object type or layer name. You learned how to apply filters in Chapter 7.*

## Controlling Image Transparency

If the image file format allows transparent pixels, you can control whether the background pixels in an image are transparent or opaque. When transparent, the other AutoCAD objects, such as lines, arcs, and circles, show through the transparent pixels. You can control transparency individually for each image.

 *Transparency also has an effect on the printed output. Bitonal images that have not been rotated or clipped and that have their transparency turned off always plot black. Turning on transparency corrects this problem. Another workaround is to create a polygonal clipping boundary for the image by using the IMAGECLIP command and then selecting the four corners of the image for the clipping boundary.*

To make background pixels transparent:

1. Do one of the following:

- On the Reference toolbar, click Image Transparency.
- From the Modify menu, choose Object | Image | Transparency.
- At the command line, type **TRANSPARENCY** and then press ENTER.

2. Select the image and then press ENTER.

3. Type **ON** (to make the background transparent) and then press ENTER.

 *You can also turn transparency on and off by using the Properties window, as described earlier; or, select the image, right-click, and choose Image | Transparency from the shortcut menu.*

## Controlling Image Quality

You can also control the image display quality. AutoCAD provides two different settings. When set to draft quality, images display faster but produce lower-quality images. At the high-quality setting, images appear less grainy, but at the expense of a slower display speed. Changing the image quality affects the display of all the images in the current drawing.

To change the image display quality:

1. Do one of the following:

- On the Reference toolbar, click Image Quality.
- From the Modify menu, choose Object | Image | Quality.
- At the command line, type **IMAGEQUALITY** and then press ENTER.

2. Type **D** (for Draft) or **H** (for High) and then press ENTER.

# Clipping Raster Images

BECOMING AN EXPERT

After you attach an image, you can define a clipping boundary to isolate a portion of the image, similar to the way that you can clip blocks and external references. When a clipping boundary is applied, only the portion of the image within the boundary remains visible and appears in plotted copies of the drawing. You define the boundary as either a rectangle or a polygon consisting of straight polyline segments. If the clipping boundary is turned off, the entire image is displayed.

The clipping boundary is applied only to the single instance of the image and doesn't affect the original image. If you move or copy the clipped image, the clipping boundary moves with it. To apply a clipping boundary, the image boundary must be visible.

You clip a raster image by using the IMAGECLIP command. When you start the command, AutoCAD prompts you to select the image that you want to clip. As soon as you select an image, AutoCAD displays four options:

```
Enter image clipping option [ON/OFF/Delete/New boundary] <New>:
```

When you first use the command, you will want to specify the clipping boundary. Once the boundary has been applied, you can use the other options to turn the boundary on and off, delete the boundary, or define a new boundary (which replaces the previous boundary):

- **ON**   Turns on the clipping boundary if you have previously turned it off
- **OFF**   Turns off the clipping boundary, causing the entire image to be displayed
- **Delete**   Deletes the clipping boundary
- **New Boundary**   Lets you define a new clipping boundary as either a rectangle or a polygon

To clip an image:

1. Do one of the following:

   - On the Reference toolbar, click Image Clip.
   - From the Modify menu, choose Object | Image Clip.
   - At the command line, type **IMAGECLIP** (or **ICL**) and then press ENTER.

*You can also select the image, right-click, and choose Image | Clip from the shortcut menu.*

2. Select the raster image to be clipped.

3. Type **N** (for New) and then press ENTER (or just press ENTER to accept the default option). AutoCAD prompts:

```
Enter clipping type [Polygonal/Rectangular] <Rectangular>:
```

4. Do one of the following:

   - To define a rectangular boundary, type **R** (for Rectangular), press ENTER (or just press ENTER to accept the default option), and then specify the two corners that define the rectangular clipping boundary.
   - To define a polygonal boundary, type **P** (for Polygonal), press ENTER, and then specify the vertices of the polygon. To complete the polygon, press ENTER.

*If you attempt to define a second clipping boundary on an existing image, AutoCAD asks whether you want to delete the old boundary and apply a new one. If you say yes, the old boundary is removed. The only way that you can create the appearance of adding a second clipping boundary to the same image is to make a copy of the image and apply a new boundary to the copy.*

After you create a clipping boundary, it takes the place of the overall image boundary. The visibility of clipping boundaries is then controlled by the same IMAGEFRAME command discussed earlier that controls the visibility of the other image boundaries.

*You can also use the CLIPIT Extended Clip Express tool to create a clipping boundary that approximates a circle or uses curves. You'll find more detailed information about this Express tool on the companion web site.*

Online

EXAMPLES

*LEARN BY EXAMPLE*
*Use the various image commands to call attention to a particular portion of an image.*
*Open the drawing Figure 23-8 on the companion web site.*

The image in Figure 23-8 shows a portion of downtown San Francisco with a three-block area highlighted. The image consists of two copies of the same image—one clipped and the other with a high fade value. This actually is a very simple trick to perform by using AutoCAD's image commands.

To insert the first instance of the image, use the following instructions:

1. Start the IMAGE command.

2. In the Image Manager dialog box, click Attach.

3. In the Select Image File dialog box, locate the file DOWNTOWN.JPG. (This file is located in the Chapter 23 page on the companion web site. You can either copy the file to your computer or click Search the Web and load the image from the web site.) After you select the file, click Open.

**Figure 23-8.** *This complex-looking image is actually two copies of the same image, combined with some simple AutoCAD image manipulation.*

BECOMING AN EXPERT

4. In the Image dialog box, clear all the Specify On-Screen check boxes. Set the Insertion Point to X=0, Y=0, and Z=0. Set the Scale to 3 and the Rotation Angle to 0.

5. Click OK to insert the image into the drawing.

6. Zoom to the drawing's extents.

Next, use the following command sequence and instructions to apply a clipping boundary around the desired portion of the image:

```
Command: IMAGECLIP
Select image to clip: (select the image by clicking on its boundary)
Enter image clipping option [ON/OFF/Delete/New boundary] <New>: ENTER
Enter clipping type [Polygonal/Rectangular] <Rectangular>: P
Specify first point: (specify the first polygon vertex)
Specify next point or [Undo]: (specify the second polygon vertex)
Specify next point or [Undo]: (specify the third polygon vertex)
Specify next point or [Close/Undo]: (specify the fourth polygon vertex)
Specify next point or [Close/Undo]: (continue specifying polygon vertices until the polygon is complete)
Specify next point or [Close/Undo]: ENTER
```

**Tip** *You can also use grips to adjust the image clipping boundaries.*

Your drawing should now show only the clipped portion of downtown San Francisco. Next, insert another copy of the raster image by following these steps:

1. Start the IMAGEATTACH command. Note that the DOWNTOWN.JPG image file is already selected in the Select Image File dialog box. Click Open.

2. In the Image dialog box, make sure the Insertion Point, Scale, and Rotation Angle settings are the same as for the first copy of the image, and then click OK.

The second copy of the image should fill your drawing area, hiding the previous clipped image. Next, use the IMAGEADJUST command to fade the second copy of the image:

1. Start the IMAGEADJUST command.

2. Select the second image that you attached by clicking its boundary, and then press ENTER.

3. In the Image Adjust dialog box, use the Fade slider or edit box to change the Fade value to 75, and then click OK.

## Changing the Display Order

The DRAWORDER command changes the order of drawing objects so that they are displayed or plotted in front of other objects. AutoCAD first prompts you to select the objects that you want to reorder and then lets you specify how you want to reorder those objects. You can move the objects above (in front of) or under (behind) selected objects, or simply move them to the front (top) or back (bottom) of the drawing.

To change the display order:

1. Do one of the following:

   - On the Modify II toolbar, click Draworder.

   - From the Tools menu, choose Display Order and then choose one of the menu suboptions.

   - At the command line, type **DRAWORDER** (or **DR**) and then press ENTER.

2. Select the objects to be reordered and then press ENTER.

3. Specify how to reorder the selected objects.

The second copy of the image now appears as you want it to, but the first copy (the one that you clipped) is still not visible, because, like most programs that display images, AutoCAD displays objects in the order in which they are created. In this case, because the second copy of the image, the faded copy, was inserted after the clipped copy, the faded copy appears in front of the clipped copy. You can change this, however, by using the DRAWORDER command.

To move the faded copy of the image so that it appears behind the clipped copy, use the following command sequence and instructions:

```
Command: DRAWORDER
Select objects: (select the faded image by clicking on its boundary)
Select objects: ENTER
Enter object ordering option [Above object/Under object/Front/Back] <Back>: ENTER
```

Complete the image by turning off the image boundaries and regenerating the drawing. Your drawing should now look similar to Figure 23-8.

**Note**
*You can use the Plot Paper Space Last option in the Plot Settings dialog box in conjunction with the DRAWORDER command to move a paper space object behind a model space object when plotting. This was not possible in versions prior to AutoCAD 2000. The DRAWORDER command does have a limitation, however, in that the draw order settings in external references are not honored. If you reference a drawing that has an attached image or other objects that have been edited with the DRAWORDER command, the draw order within the xref will not be in the correct draw order, even after regenerating the drawing. To correct this problem, use the following procedure:*

1. *Open the xref file.*
2. *Use the DRAWORDER command to set up the xref drawing.*
3. *Use the WBLOCK command to create a new drawing from the xref file. Type **ALL** when prompted to select objects.*
4. *Use the XATTACH command to attach the file created by using WBLOCK in Step 2.*

## Other Raster Image Controls

In R14 and earlier versions of AutoCAD, raster images were highlighted when they were selected. Beginning with AutoCAD 2000, images are no longer highlighted by default when selected. The IMAGEHLT system variable controls whether raster images are highlighted. To turn highlighting back on, change the value of this variable to 1.

The RTDISPLAY system variable controls whether raster images are displayed during real-time panning and zooming. If RTDISPLAY is set to its default value (1), the raster image is displayed only as an outline when performing real-time panning and zooming. If you want to see the raster image as you pan and zoom, change this value to 0. You can also control the RTDISPLAY variable from the Display tab of the Options dialog box, by selecting or clearing the Pan And Zoom With Raster Image check box. Note that if you enable this feature, AutoCAD's performance may be slower when panning and zooming drawings containing large raster images displayed in their entirety.

# Chapter 24

## Working with Other Applications

utoCAD is probably not the only program that you use. More likely, you also create documents by using a word processor or store data using a spreadsheet. Oftentimes, the information that you create and store by using these other programs relates to your drawings.

AutoCAD offers great flexibility in its ability to be used with other programs. You can include an AutoCAD drawing in a Microsoft Word document, for example, or insert a Microsoft Excel spreadsheet containing a parts list or door schedule into an AutoCAD drawing. To include AutoCAD drawings in other programs, and include documents from other programs in AutoCAD drawings, you either *link* or *embed* them. You can also save AutoCAD drawings in other file formats that can be used directly with other programs.

| Note | *You can also create DWF (AutoCAD Drawing Web Format) files that can be published to the Internet or included in files produced using other programs. In Chapter 25, you'll learn how to create DWF files. You'll learn about publishing those files in Chapter 26. You can also use external databases with AutoCAD, which you'll learn more about in Chapter 27.* |
|---|---|

This chapter explains how to do the following:

- Work with files saved in other file formats
- Insert markup data from Volo View
- Use object linking and embedding (OLE)
- Save and view slides

# Working with Other File Formats

When you work with other applications, you can save AutoCAD drawings in other formats by exporting part or all of a drawing to a format that is compatible with the other applications. You can also import files into your current AutoCAD drawing that are saved in various other formats.

## Exporting Drawings to Other Formats

You can export AutoCAD drawings in several different formats, for use with other programs. When you export to a different format, AutoCAD creates a new file in the designated format and, depending on the format, saves to that file either all the drawing objects or all the selected objects. The original drawing is not changed in any way.

You can export a drawing in any of the following formats:

- Windows Metafile (WMF)
- ACIS solid object file (SAT)
- Solid object stereo-lithography file (STL)

- Encapsulated PostScript file (EPS)
- Attribute extract file (DXX)
- Device-independent bitmap file (BMP)
- Autodesk 3D Studio file (3DS)
- AutoCAD Block (DWG)
- Drawing Exchange Format (DXF)
- Binary Drawing Exchange Format (DXB)
- DesignXML file (XML)
- Drawing Web Format (DWF)

With the exception of the DXF, DXB, XML, and DWF formats, you can export a drawing in any of these formats by using the EXPORT command. To start the EXPORT command, do one of the following:

- From the File menu, choose Export.
- At the command line, type **EXPORT** (or **EXP**) and then press ENTER.

AutoCAD displays the Export Data dialog box, shown in Figure 24-1. To export to a specific format, choose the format from the Files Of Type drop-down list, specify the filename and directory, click Save, and then follow the prompts. Exporting the drawing in DWG format duplicates the function of the WBLOCK command, which you learned about in Chapter 15.

**Note** *When exporting EPS files, you can control PostScript output options by selecting Options from the Tools pull-down menu in the Export Data dialog box.*

To save a drawing as a DXF file, you use the SAVEAS command and select one of the DXF formats from the Files Of Type drop-down list. You can save the file so that it is compatible with AutoCAD 2000 (which is the same as AutoCAD 2002 and 2000i), Release 14/LT98/LT97, Release 13/LT95, or Release 12/LT2.

**Note** *When saving a DXF file, you can control options such as the decimal precision and the format (ASCII or binary) from the Saveas Options dialog box, displayed by selecting Options from the Tools pull-down menu in the Save Drawing As dialog box.*

To save a drawing as a DXB file, you must first use the Add-A-Plotter wizard to add the AutoCAD DXB File plot configuration and then create the file by using the PLOT command with the DXB File selected as the current plot device. Similarly, to save a drawing as a DWF file, you select one of the DWF plot configurations when using the PLOT command. You'll learn more about creating and using DWF files in Chapters 25 and 26.

BECOMING AN EXPERT

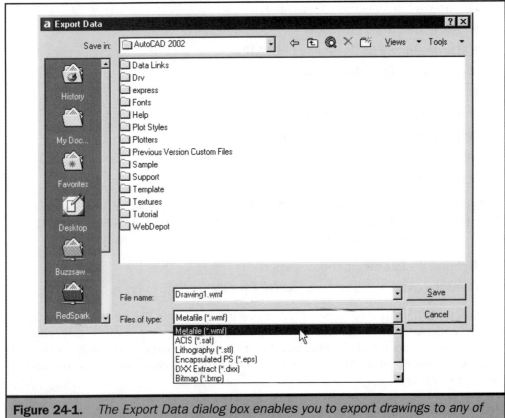

**Figure 24-1.**   The Export Data dialog box enables you to export drawings to any of eight different formats.

AutoCAD 2002 can also export and import DesignXML files. A DesignXML file is a special type of text file that is particularly well suited for Web-based applications. The file contains drawing information that can be read by both CAD and non-CAD applications. AutoCAD supports XML files that conform to the DesignXML and AcDbXML schema definitions.

You actually export DesignXML files using the WBLOCK command. When using the command, to create a DesignXML file instead of a DWG file, you simply specify the file extension as .XML instead of .DWG. You learned how to use the WBLOCK command in Chapter 15.

Table 24-1 lists all of the formats in which you can export files as well as the commands used to do so.

| Export File Type | Commands | Comments |
|---|---|---|
| 3DS | 3DSOUT<br>EXPORT | Saves objects having surface characteristics, as well as views, lights, and materials, to a 3DS file. AutoCAD prompts you to select objects. You can also specify how AutoCAD derives 3D Studio objects (by layer, color, or object type), whether blocks are split into components or converted into a single 3D Studio object, how smoothing is applied, and how vertices are welded. |
| BMP | BMPOUT<br>EXPORT | AutoCAD prompts you to select objects. Set the graphics background to white or the desired color using the Display tab of the Options dialog box before saving the BMP file. If you are exporting a hidden-line or shaded view, ensure that SHADEMODE is set to 2D Wireframe and then use the HIDE or –SHADE command prior to saving the BMP file. |
| DWF | PLOT | Saves plot to a DWF file. |
| DWG | WBLOCK | Saves all or a portion of the current drawing to a new drawing file, as described in Chapter 15. |
| DXB | PLOT | Saves plot to a DXB file. |
| DXF | DXFOUT<br>SAVEAS | In addition to specifying the AutoCAD version to be targeted, you can output all or selected objects, specify that the data be precise up to 16 decimal places, and determine whether the resulting file is a binary or ASCII DXF file. |
| DXX | ATTEXT<br>EXPORT | Creates a drawing exchange file, a subset of AutoCAD's standard DXF format, containing all the information about each block in the drawing, as described in Chapter 15. |

**Table 24-1.**   *Export File Types Supported by AutoCAD*

BECOMING AN EXPERT

| Export File Type | Commands | Comments |
|---|---|---|
| EPS | EXPORT PSOUT | You can specify what portion of the drawing is saved (window, extents, limits, and so on), include an optional preview image, set the scale and paper size, and include PostScript-coded procedure definitions, defined in a prolog section of the ACAD.PSF master support file. The PSPROLOG system variable determines what prolog portion is referenced when exporting the EPS file. |
| SAT | ACISOUT EXPORT | Stores regions and solids to an ASCII format ACIS file. AutoCAD prompts you to select objects. The ACISOUTVER system variable controls the ACIS version of the resulting file. |
| STL | EXPORT STLOUT | Stores a solid in a format compatible with stereo-lithography apparatus. AutoCAD prompts you to select objects. STLOUT enables you to create ASCII or binary files. The FACETRES system variable controls how AutoCAD triangulates the solid. |
| WMF | EXPORT WMFOUT | AutoCAD prompts you to select objects. |
| XML | WBLOCK | Saves selected objects. |

**Table 24-1.**    *Export File Types Supported by AutoCAD* (continued)

**Note**    *When you use the PSOUT command to create an EPS file, the lineweights are not exported. Instead, to apply line widths, you must edit the ACAD.PSF file to assign line widths to objects based on their color. Then, to assign different line widths, the objects in the drawing must be drawn using those colors. An easier alternative to this is to plot drawings to a file using the PostScript HDI driver.*

AutoCAD actually supports several other file formats. For example, when you use the RENDER command and save the rendering to a file, you can save the resulting rendering as a BMP, PCX, PostScript, TGA, or TIFF file. You can also use the SAVEIMG command to save both wireframe and rendered views to raster file formats such as BMP, TGA, and TIFF. If you plot by using the Raster File Formats HDI driver, you can create plot files in CALS, JPEG, BMP, PCX, PNG, TGA, or TIFF formats. The raster output HDI driver also enables you to control the background color of the resulting raster file.

*When you export a Windows metafile, the size of the resulting image file is based on the size of the current AutoCAD viewport. To control the size of the resulting image, either resize the drawing screen before exporting or use a paper space viewport to control the size of the image. In addition, by default the background color of the WMF file is the same as the background color of the AutoCAD drawing screen. You can use the WMFBKGND system variable, however, to control whether the background is included or transparent. A value of 1 saves the WMF file with the same background color as the current AutoCAD background color. In addition, when the background is transparent (WMFBKGND is equal to 0), you can use the WMFFOREGND variable to control the foreground color of AutoCAD objects. When WMFFOREGND is 0, AutoCAD swaps the foreground and background colors if necessary to ensure that the foreground color is darker than the background color, while a value of 1 swaps them so that the foreground color is lighter than the background color.*

# Importing Other Formats into the Current Drawing

AutoCAD can import files that were created by using other applications and that are saved in different formats, although AutoCAD imports fewer formats than it exports. Depending on the type of file being imported, AutoCAD converts the information in the file either into AutoCAD drawing objects or a single block object.

You can import files saved in the following formats:

- Windows Metafile (WMF)
- AutoCAD Drawing Exchange Format file (DXF)
- AutoCAD Drawing Exchange Binary file (DXB)
- ACIS solid object file (SAT)
- Autodesk 3D Studio file (3DS)
- DesignXML file (XML)

With the exception of the DXB, DXF, and XML formats, you can import any of these formats by using the IMPORT command. To start the IMPORT command, do one of the following:

- On the Insert toolbar, click Import.
- At the command line, type **IMPORT** (or **IMP**) and then press ENTER.

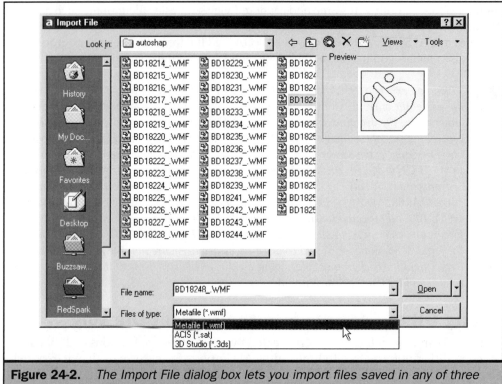

**Figure 24-2.** *The Import File dialog box lets you import files saved in any of three different formats.*

AutoCAD displays the Import File dialog box, shown in Figure 24-2. To import a specific type of file, choose the format from the Files Of Type drop-down list, select the file that you want to import, and then click Open.

You can import a DesignXML file by using the INSERT command, which you learned about in Chapter 15. Simply select DesignXML from the Files Of Type drop-down list. AutoCAD inserts the XML file as a block.

Table 24-2 lists all the formats in which you can import files, as well as the commands used to do so. With the exception of the DXFIN command, all the commands shown in the table can be started from the Insert menu or by typing the command name at the command line. The DXFIN command does not appear in the AutoCAD menu. You can use the OPEN command (available on the Standard toolbar or File menu), however, to import a DXF file.

*As you learned in Chapter 23, you can also use the IMAGE command to attach many types of raster image files to your drawings.*

| Import File Type | Commands | Comments |
|---|---|---|
| 3DS | 3DSIN<br>IMPORT | Imports files created with Autodesk 3D Studio, including meshes, materials, mapping, lights, and cameras. You can specify which objects within the file to import, how to assign those objects to layers, and how to handle objects consisting of multiple materials. |
| DXB | DXBIN | Imports a special binary-coded DXB file into the current drawing. |
| DXF | DXFIN<br>INSERT | You can only import a full DXF file into a new (blank) drawing. You can use the INSERT command, however, to import a partial DXF file into the current drawing. |
| SAT | ACISIN<br>IMPORT | Imports geometry objects from ASCII format ACIS files, converting them into body objects, solids, or regions. |
| WMF | IMPORT<br>WMFIN | Imports a Windows metafile into the current drawing as a block, issuing prompts similar to the INSERT command. The appearance of solid or wide lines is controlled by either the Options button in the dialog box or the WMFOPTS command. Select the Wire Frame (No Fill) check box if you don't want to include the background when importing a WMF file. |
| XML | INSERT | Inserts the contents of the XML file as a block. |

**Table 24-2.**    *Import File Types Supported by AutoCAD*

BECOMING AN EXPERT

# Inserting Markup Data from Volo View Express

When working in a design team, not everyone in the team may need the full power of AutoCAD. Some members may only need to review and comment on drawings. In the past, this often entailed plotting a check print, which the reviewer then marked up using a red pen. This check print would then be returned so that the changes could be manually incorporated into the AutoCAD drawing.

In order to eliminate the manual steps, Autodesk created two programs—Volo View and Volo View Express—that enable a non-AutoCAD user to open, view, and electronically mark up AutoCAD drawing files. A copy of Volo View Express is provided on the AutoCAD 2002 CD and can also be freely downloaded from Autodesk's web site. Volo View, a more powerful version of the program, can be purchased from Autodesk's online store.

The markups created using Volo View or Volo View Express can be saved to a special RedlineXML markup language (ML) file. These ML files can then be inserted back into your drawing files so that you can view the comments and revise your drawings accordingly.

Although users of Volo View Express can view and print both model and layout views of AutoCAD drawings and DXF files, the markup tools can only be used in layout views. Users of Volo View can also mark up model space views. When saving an ML file, the markups for the model space view and individual layouts are saved separately within the single ML file. When you later insert these markups into your AutoCAD drawing, you must have the proper model space or layout tab selected before inserting the markup.

To insert a markup into the current drawing:

1. Select the appropriate model space or layout tab.

2. Do one of the following:

   ■ From the Insert menu, choose Markup.

   ■ At the command line, type **RMLIN** and then press ENTER.

3. In the Insert Markup dialog box, select the ML file you want to insert, and then click Open.

AutoCAD immediately inserts the markup into the current model space or layout. If no markup data exists for the current layout, AutoCAD displays a dialog box informing you that no markup objects were found for the current layout. If the ML file contains markup data in more than one layout, you can switch to a different layout and then insert the ML file again to insert the additional markup data.

When you insert a markup, AutoCAD creates a special markup layer (called _Markup_) and then inserts the markup objects as AutoCAD objects. AutoCAD converts each Volo View markup object into the most appropriate AutoCAD object. For example, markup text is converted into AutoCAD multiline text. The inserted objects take on the properties of the markup layer (which by default is red and uses the continuous linetype). In addition, the markup layer is locked by default, preventing you from accidentally changing any of the notes or comments.

**Tip** *Although you can insert more than one ML file into a drawing, all the markups are always placed on the _Markup_ layer. In order to separate different markups, rename the existing _Markup_ layer before inserting a new ML file.*

# Working with Other Applications

Importing files that are created with other programs and exporting AutoCAD drawings to other formats are two useful methods of combining information that has been created with different programs. However, the data that you work with is *static*—after you create the import or export file, it can't be easily changed. Rather, if the information changes, you must repeat the import or export process.

By taking advantage of another Windows technology called OLE—Object Linking and Embedding—you can copy information from one program to another while retaining the ability to edit that information by using the same program that originally was used to create the information.

Although OLE may sound confusing, it actually is quite simple. AutoCAD supports OLE as both a source (or server application) and a destination (or client). Using it as a server, you can either embed AutoCAD drawings in or link them to documents that you create with other applications. When you *embed* a drawing in another document, you make a copy of that drawing. That copy is stored in the destination document, and none of the changes that you make to the copy affect the original drawing, nor do changes to the original affect the copy that you embed. Thus, embedding is similar in function to AutoCAD's INSERT command. When you *link* a drawing to another document, instead of placing a copy of that drawing in the document, you create a link, or reference, between the drawing and the document. If you change the source drawing, the changes are reflected in the document as soon as you update the links. Linking, therefore, is similar to using an external reference.

Similarly, you can embed another document, such as an Excel spreadsheet, in an AutoCAD drawing. A copy of the spreadsheet is stored in the AutoCAD drawing, and none of the changes that you make to the spreadsheet affect the original. However, if you link the spreadsheet to the AutoCAD drawing and later make changes to the spreadsheet in Excel, those changes appear in the AutoCAD drawing the next time that you update the links.

*Use embedding when you don't want the source document to be updated when you edit information in the compound document. Use linking when changes to the source document need to be reflected in the compound document. Linking is particularly useful when the same data is used in more than one compound document.*

## Using AutoCAD Data in Other Applications

AutoCAD provides several commands for linking and embedding drawings in other documents. The method that you choose depends on the capabilities of the other program and how you want to work with the AutoCAD drawing after you place it in the other

BECOMING AN EXPERT

document. AutoCAD provides the following commands for linking or embedding AutoCAD data into other applications:

- **COPYCLIP** Copies selected objects from an AutoCAD drawing so that they can be linked or embedded in a document in another Windows application.
- **COPYLINK** Copies the current view of a named drawing so it can be linked to a document in another Windows application.

**Note** *The CUTCLIP command copies selected objects from an AutoCAD drawing to the Windows Clipboard and erases them from the drawing. You can then paste the objects into another Windows application. The COPYBASE command (which you learned about in Chapter 7) copies selected objects from one AutoCAD drawing to another.*

## Linking AutoCAD Drawings to Other Documents

When you link an AutoCAD drawing to another document, the other document contains only a reference to the AutoCAD drawing file, rather than the actual drawing. As with external references, linking does not substantially increase the size of the compound document, but you must be sure to transfer the linked AutoCAD drawings when you exchange the compound document with others.

To link an AutoCAD drawing to a document in another application:

1. Save the AutoCAD drawing.
2. If the drawing contains more than one viewport, click inside the viewport that you want to link.
3. Do one of the following:
   - From the Edit menu, choose Copy Link.
   - At the command line, type **COPYLINK** and then press ENTER.
4. Start the destination application.
5. Paste the Windows Clipboard contents into the document, following the destination application's instructions for linking.

For example, to link an AutoCAD drawing into a Word document, you choose Paste Special from the Edit menu in Word, pick the AutoCAD Drawing Object from the list displayed in the Paste Special dialog box, select the Paste Link radio button, and then click OK.

**Note** *In any Microsoft Office-compatible programs, you can also choose Insert | Object. In the Object dialog box, click the Create from File tab, specify the name of the drawing file that you want to link, select the Link To File check box, and then click OK.*

The document in the other application now contains a linked copy of the AutoCAD drawing. All the changes that you make to the AutoCAD drawing are reflected in the linked copies.

## Editing Linked Drawings

You can edit linked drawings either inside the compound document or in AutoCAD. In both cases, AutoCAD must be installed on the system that contains the compound document that you're editing. To edit a linked drawing in a compound document:

1. Open the document containing the linked AutoCAD drawing.

2. Double-click the drawing within the document to start AutoCAD and open the linked drawing.

3. Make the changes to the drawing.

4. Save the changes to the drawing. The changes are automatically reflected in the compound document.

To edit a linked drawing in AutoCAD:

1. Start AutoCAD and open the linked drawing.

2. Make the changes to the drawing.

3. Save the changes to the drawing.

The next time that you open the compound document, either the link will update automatically or you will be asked if you want to update the link. The method used depends upon the application.

## Embedding AutoCAD Drawings in Other Documents

You can embed an AutoCAD drawing or selected AutoCAD objects into another document. When you embed an AutoCAD drawing into another document, it becomes part of the other program's document file. When you edit the drawing, you edit only the copy that is embedded in the other document. The advantage of embedding is that when you transfer the compound document to another computer, you don't have to worry about transferring the AutoCAD drawing as well; it is already embedded as part of the document. But, embedded objects increase the size of the compound document file, in much the same way that other drawings increase the size of the drawing file when they are inserted into the drawing file as blocks.

To embed selected AutoCAD objects into a document in another application:

1. Do one of the following:

   ■ On the Standard toolbar, click Copy To Clipboard.

   ■ From the Edit menu, choose Copy.

   ■ At the command line, either type **COPYCLIP** and press ENTER, or press CTRL-C.

2. Select the objects that you want to embed, and then press ENTER.

3. Start the destination application.

4. Paste the Windows Clipboard contents into the document, following the destination application's instructions for embedding.

BECOMING AN EXPERT

For example, to embed AutoCAD objects into a Word document, you could choose Paste Special from the Edit menu and then click OK, or just press CTRL-V.

**Note**

*To embed an entire AutoCAD drawing into any Microsoft Office-compatible program, you can choose Insert | Object, click the Create from File tab, click Browse, choose the file that you want to embed, click OK, make sure that Link To File is not selected, and then click OK.*

The document in the other application now contains a copy of the AutoCAD drawing. None of the changes that you make to that copy affect the original drawing.

**Tip**

*You can also use drag-and-drop to embed an entire AutoCAD drawing into another document. Since dragging-and-dropping does not use the Windows Clipboard, any data on the Clipboard is not affected. To drag an AutoCAD drawing into another document, position and resize the Windows Explorer window and the other application so that you can see both the file icon and the document in which you want to embed it. Within Windows Explorer, select the icon for the drawing file and drag it into the document.*

## Editing Embedded Drawings

You can edit embedded drawings inside the compound document as long as AutoCAD is installed on the system that contains the compound document that you're editing.

To edit an embedded drawing in a compound document:

1. Open the document containing the embedded AutoCAD drawing.
2. Double-click the drawing within the document to start AutoCAD and open the embedded drawing.
3. Make the changes to the drawing.
4. From the AutoCAD File menu, choose Update to save the changes to the embedded drawing.
5. Exit AutoCAD.

**Tip**

*When AutoCAD copies objects to the Windows Clipboard, it uses the extents of the drawing area or the current viewport to determine the clipped boundary size. Resize the AutoCAD graphics screen or viewport to include the desired amount of space surrounding the objects to be copied before copying the objects. You can also create a viewport in paper space to achieve the same results.*

# Using Data from Other Applications in AutoCAD

You can include data from other programs in AutoCAD drawings by using either embedding or linking. AutoCAD provides several commands for linking and embedding data in AutoCAD, just like it does when you include AutoCAD data in other applications.

The method that you choose depends on the type of object or file that you want to include, and what you want to do with it after it's been included. AutoCAD provides the following commands for linking or embedding data from other applications into AutoCAD drawings:

- **INSERTOBJ**   Inserts a linked or embedded object in the current AutoCAD drawing
- **OLELINKS**   Updates, changes, or cancels existing OLE links
- **PASTECLIP**   Inserts the Windows Clipboard contents into the current AutoCAD drawing
- **PASTESPEC**   Inserts the Windows Clipboard contents into the current AutoCAD drawing as either a linked or embedded object

*The PASTEORIG and PASTEBLOCK commands are used to paste data from other open drawings. You learned about these commands in Chapter 7.*

## Linking Objects into AutoCAD Drawings

If another program supports OLE, you can link its data to your AutoCAD drawings. Use linking when you want to include the same data in many drawings. When you update the data, all links to other files reflect the changes.

For example, if you create your company logo in an OLE-compatible drawing program, and you want to include it in the title block of every drawing that you create with AutoCAD, you can link the logo to each AutoCAD drawing. If you later change the original company logo in the drawing program, the AutoCAD drawings update automatically.

When you link data from another program, the AutoCAD drawing stores only a reference to the location of the file in which you created the data. You link the data from a saved file, so that AutoCAD can find the data and display it.

Because linking adds only a reference to a file, the data does not significantly increase the file size of the AutoCAD drawing. However, links require some maintenance. If you move any of the linked files, you need to update the links. In addition, if you want to transport linked data, you must include all the linked files.

You can update a linked object either automatically, every time that you open the drawing, or only when you specify. Any time that a link is updated, changes made to the object in its original file also appear in the AutoCAD drawing—and the changes also appear in the original file if they were made through AutoCAD.

To link data from another open application in an AutoCAD drawing by using the Windows Clipboard:

1. In the other application, select the data that you want to link and copy it to the Windows Clipboard.

2. Open the AutoCAD drawing.

BECOMING AN EXPERT

3. Do one of the following:

   ■ From the Edit menu, choose Paste Special.

   ■ At the command line, type **PASTESPEC** (or **PA**) and then press ENTER.

4. In the Paste Special dialog box, shown in Figure 24-3, make sure the proper data format is selected in the As list.

5. Select the Paste Link radio button to paste the contents of the Windows Clipboard into the current drawing and create a link to the original file.

6. Click OK.

To create an OLE link to a saved file that was created in another application:

1. Open the AutoCAD drawing.

2. Do one of the following:

   ■ On the Insert toolbar (or the flyout on the Draw toolbar), click OLE Object.

   ■ From the Insert menu, choose OLE Object.

   ■ At the command prompt, type **INSERTOBJ** (or **IO**) and then press ENTER.

3. In the Insert Object dialog box, shown in Figure 24-4, select the Create From File radio button.

4. Select the Link check box and then click Browse.

5. In the Browse dialog box, select the file that you want to link, and then click OK.

6. In the Insert Object dialog box, click OK.

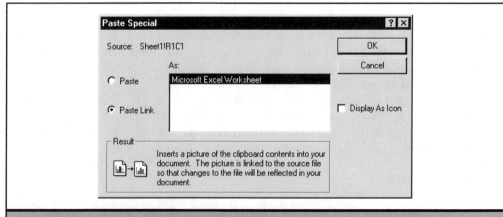

**Figure 24-3.**   *The Paste Special dialog box lets you link or embed objects from open applications into your drawing.*

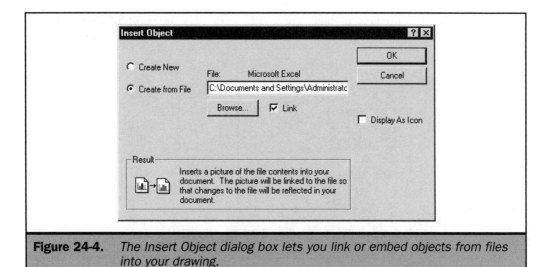

**Figure 24-4.**    *The Insert Object dialog box lets you link or embed objects from files into your drawing.*

    *When you link or embed into your drawing a file that contains multiple pages, only the first page (or the visible page, if using the Clipboard) is placed into the drawing. To link or embed multiple pages, you must use the Clipboard and explicitly copy-and-paste each page. You must then move and resize the pages to produce the desired appearance in your drawing.*

**Note**    *You can also use the INSERTOBJ command to create a new OLE object and embed it in the current drawing. To do so, select the Create New radio button in the Insert Object dialog box, select the application that you want to use in the Object Type list, and then click OK (see Figure 24-6, later in the chapter).*

## Editing Linked Objects

The AutoCAD drawing now contains a linked copy of the other data. All the changes that you make to the other document are reflected in the linked copies. You can edit the linked data by doing either of the following:

■ Double-click the linked object in AutoCAD to open the original application, make the changes, and then save the data.

■ Open the file containing the object in the application in which it was created, make the changes, save the data, and then update the linked data in AutoCAD.

BECOMING AN EXPERT

You can also click the linked object, right-click to display the shortcut menu, choose Linked Object (the name changes to reflect the type of linked object that you select), and then choose Edit.

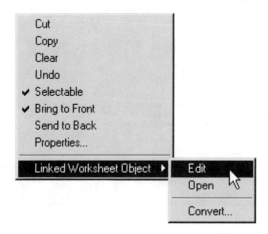

The application that created the object opens. You can then modify the object and choose Update from the server application's File menu to save the changes back to the object embedded in AutoCAD.

## Updating Links

AutoCAD normally updates links automatically whenever information in the original document changes. The OLELINKS command can be used to change this, however, so that you can update links manually.

To update a link manually:

1. Do one of the following:
   - From the Edit menu, choose OLE Links.
   - At the command line, type **OLELINKS** and then press ENTER.
2. In the Links dialog box, shown in Figure 24-5, select the link that you want to update.
3. Click Update Now.
4. Click Close.

*Use the Change Source function to reconnect links when the source file location changes. Use the Break Link function to break links to source files when you no longer want changes to be reflected in the AutoCAD drawing. Breaking a link does not remove the inserted objects, but rather converts the linked object into an embedded object.*

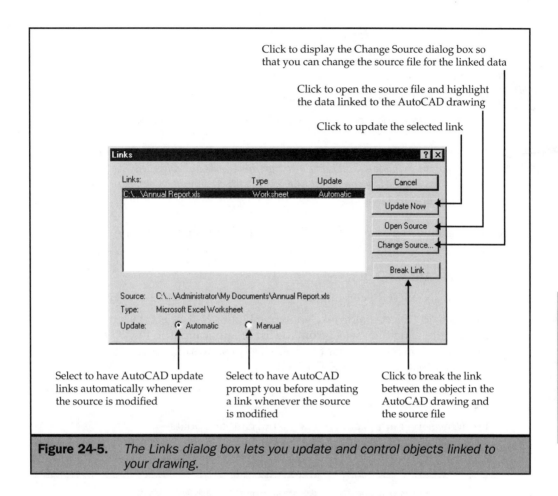

**Figure 24-5.** *The Links dialog box lets you update and control objects linked to your drawing.*

## Embedding Objects into AutoCAD Drawings

Embed an object into your AutoCAD drawing when you want to keep all the data that you work with in one file. Embedding also makes it easier to transfer files to another computer, because you don't also have to transfer the linked files. You can embed data from any other program that supports OLE.

When you embed data from another program, AutoCAD becomes the container for that data. The object embedded in the AutoCAD drawing becomes part of the drawing file. When you edit the data, you open from within the AutoCAD drawing the program in which the data was created. Any changes that you make to the embedded data exist only in the AutoCAD drawing.

To embed data from another open application into an AutoCAD drawing:

1. In the other application, select the data that you want to embed and copy it to the Windows Clipboard. (*Note:* Only data on the currently visible page is included.)

BECOMING AN EXPERT

2. Open the AutoCAD drawing.

3. Do one of the following:

- On the Standard toolbar, click Paste From Clipboard.
- From the Edit menu, choose Paste.
- At the command line, type **PASTECLIP** and then press ENTER.

*If the cursor is within the AutoCAD drawing, you can also press* CTRL-V *to paste the contents of the Windows Clipboard into the drawing as an embedded object. If the cursor is on AutoCAD's command line, however, text from the Clipboard is pasted at the current prompt.*

To create and embed data from another (server) application into an AutoCAD drawing:

1. Open the AutoCAD drawing.

2. Start the INSERTOBJ command.

3. In the Insert Object dialog box, shown in Figure 24-6, select the Create New radio button and then select from the list of applications in the Object Type list the application that you want to use. This list shows all the applications on your computer that support OLE. Click OK.

4. In the server application, create the data that you want to insert and save it in the server application.

5. From the server application's File menu, choose Update.

6. Close the server application.

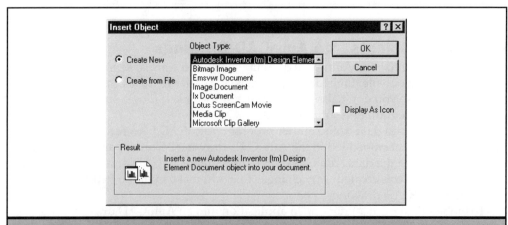

**Figure 24-6.** *The Insert Object dialog box lets you start server applications to create objects that are then linked or embedded into your drawing.*

**Note**

*You can also use the INSERTOBJ command to embed OLE objects from files previously created and saved in another application. To do so, select the Create From File radio button in the Insert Object dialog box and then select the file that you want to embed (refer to Figure 24-4).*

## Editing Embedded Objects

You can edit the embedded object within AutoCAD as long as the application used to create the object is loaded on your system.

To edit an embedded object in an AutoCAD drawing:

1. Open the AutoCAD drawing.

2. Do one of the following:

   ■ Double-click the object.

   ■ Select the object, right-click to display the shortcut menu, and then choose Object | Edit.

3. Modify the object.

4. From the server application's File menu, choose Update to save the changes to the embedded object.

5. Close the server application.

## Dragging Objects into AutoCAD

You can also drag selected objects and data from an active application into AutoCAD. When you drag objects from another application into AutoCAD, they are embedded, not linked. Both AutoCAD and the other application must be running and visible on your screen, and the other application must support drag-and-drop. You can also drag entire files from Windows Explorer into AutoCAD.

**Tip**

*Selecting and dragging objects from an open application into AutoCAD is the same as cutting and pasting the objects. The objects are cut (removed) from the other application and pasted into AutoCAD. To drag a copy of the objects into AutoCAD without removing them from the original application, press the* CTRL *key while you drag the objects.*

# Working with OLE Objects

Although most AutoCAD commands do not work with OLE objects, you can cut, paste, copy, and erase OLE objects that are present in your drawings and also move, stretch, and scale the objects. You can also move OLE objects so that they appear in front of or behind other objects, change the layer on which OLE objects are drawn, and control the display of OLE objects.

BECOMING AN EXPERT

When you perform any of these operations on an OLE object, you must first select the object, so that its object frame appears. The object frame has resize handles at its corners and midpoints.

To cut, copy, or clear an OLE object:

1. Right-click the OLE object to display the shortcut menu.

2. From the shortcut menu, choose one of the following options:

   - **Cut**   Permanently removes the object from the drawing and places it on the Clipboard

   - **Copy**   Places a copy of the object on the Clipboard

   - **Clear**   Permanently removes the object from the drawing, without placing it on the Clipboard

To move an OLE object:

1. Select the object.

2. With the cursor over the object so that the move cursor appears, drag the object to a new location.

To stretch or scale an OLE object:

1. Select the object.

2. Drag a resize handle to change the size of the object:

   - Drag a corner handle to scale the object, maintaining its height-to-width ratio.

   - Drag a midpoint handle to stretch the object in one direction.

AutoCAD also provides a special OLE Properties dialog box, shown in Figure 24-7, that enables you to change the size or scale of OLE objects. You can change the size of OLE objects based on drawing units or as a percentage of its current size. You can also scale text by assigning a point size to a font contained in the OLE object.

**Note**   *The OLE Properties dialog box displays automatically whenever you paste a new OLE object into a drawing. You can turn off this automatic display, however, by clearing the Display Dialog When Pasting New OLE Objects check box in the OLE Properties dialog box. You can also control the display of this dialog box by using the Display OLE Properties Dialog check box on the System tab of the Options dialog box.*

To display the OLE Properties dialog box, do one of the following:

- Select an OLE object, right-click to display the shortcut menu, and then choose Properties.

- Select an OLE object. At the command line, type **OLESCALE** and then press ENTER.

- Drag or paste an OLE object into your drawing (if the automatic display of the dialog box is enabled).

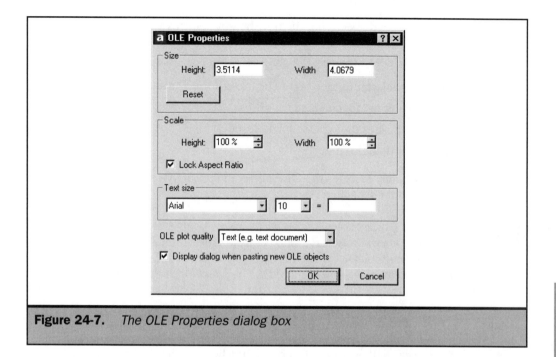

**Figure 24-7.** *The OLE Properties dialog box*

You can then change the size of the OLE object based on its height or width by doing either of the following:

- Under Size, enter a new height or width in drawing units.

- Under Scale, enter a new height or width as a percentage of the current height or width.

**Note** *When the Lock Aspect Ratio check box is selected, changing either the Height or Width value under Size or Scale automatically adjusts the other value to maintain the current aspect ratio between the two. If you want to change the height independent from the width, you must first clear this check box.*

You can use the OLE Properties dialog box to change the size of an OLE object by scaling the object based on its font, by doing the following:

1. Under Text Size, select a font from the drop-down list of fonts used in the selected OLE object.

2. Select a point size from the drop-down list of sizes used in the selected OLE object.

3. In the box adjacent to the equal sign, enter a text height value in drawing units. For example, if your drawing contains an OLE object that uses Arial 10-point text, you can specify that the text should appear .25 drawing unit in height.

BECOMING AN EXPERT

When you have finished specifying the new height, width, or text height, click OK to apply the changes.

To restore an OLE object to its original size:

1. Select the OLE object and then right-click to display the shortcut menu.

2. From the shortcut menu, choose Properties.

3. In the OLE Properties dialog box, click Reset.

4. Click OK to apply the changes.

AutoCAD immediately restores the OLE object to its original size when it was first dragged or pasted into the drawing.

**Note**    *If the original OLE object doesn't have a metafile containing size information, the Reset button will not appear in the OLE Properties dialog box. AutoCAD's UNDO command does not undo changes made in the OLE Properties dialog box. To undo a change made using this dialog box, right-click over the OLE object and then choose Undo from the shortcut menu.*

Although you've already learned how to use the DRAWORDER command to move AutoCAD objects in front of or behind other objects, you can't use the command to change the order of OLE objects. Rather, you must use the functions on the shortcut menu.

To move an OLE object to the back or front:

1. Right-click the OLE object to display the shortcut menu.

2. From the shortcut menu, choose Bring To Front or Send To Back, as appropriate.

**Note**    *Clicking an OLE object selects the object even when there is an AutoCAD object in front of or behind the OLE object. To select the AutoCAD object, you first need to suppress the selection of the OLE object. You can do this by right-clicking the OLE object and choosing Selectable, so that the check mark disappears. This acts as a toggle. To make the object selectable again, right-click it again, and choose Selectable.*

You may have instances in which you don't want OLE objects to be visible. For example, you may want them to appear and plot in paper space, but not be visible in model space. You can control the visibility of OLE objects by using the OLEHIDE system variable. Table 24-3 shows the possible OLEHIDE values and their associated settings.

To change the OLEHIDE value:

1. At the command line, type **OLEHIDE** and then press ENTER.

2. Type the desired value and then press ENTER.

The OLEHIDE value affects both the display and printing of OLE objects.

| Value | Setting |
|-------|---------|
| 0 | All OLE objects are visible in both model space and paper space |
| 1 | OLE objects are visible in paper space only |
| 2 | OLE objects are visible in model space only |
| 3 | OLE objects are not visible |

**Table 24-3.**  *Possible OLEHIDE Values and Their Meaning*

**Note** *When you insert a new OLE object into either model space or paper space, if the display of OLE objects was previously turned off, the OLEHIDE variable is automatically changed so that all OLE objects are displayed in that space.*

Including OLE objects in an AutoCAD drawing has certain limitations:

- OLE objects inserted in a drawing as part of a block or external reference aren't displayed or plotted.

- OLE objects are plotted only when you plot to a Windows System Printer or a raster-capable HDI plotter driver. HDI plotters denoted by NR are not raster-capable. The OLE object also doesn't rotate when you change the plot rotation. You should use the system printer's landscape mode instead.

- When you paste an Excel spreadsheet into an AutoCAD drawing, AutoCAD uses the metafile format. Thus, if the spreadsheet is too large, only a portion of it appears in the drawing.

**Tip** *When you drag or paste an OLE object into a drawing, the object is added to the current layer. You can't use AutoCAD's Properties window, however, to change the layer of the OLE object. Instead, to change the layer on which the OLE object is drawn, cut the object from the drawing, change the current layer, and then paste the object back into the drawing.*

## Saving and Viewing Slides

A *slide* is an image of the current AutoCAD graphics screen, captured to a special type of file called a *slide file*. Once captured, a slide can be quickly recalled to the screen for viewing. Slides capture images of the AutoCAD drawing exactly as it looks in the current viewport, including any hidden-line removal or shading. But a slide is only an image, not a drawing. Slides simply show what the screen looked like at the instant

BECOMING AN EXPERT

## Controlling OLE Plot Quality

The OLEQUALITY system variable controls the plotting quality level for newly inserted OLE objects. By default, this variable has a value of 1, which sets OLE objects so that they plot at a text quality level. The variable can be assigned one of the values listed in the following table:

| Value | Meaning |
|-------|---------|
| 0 | Line-art quality, such as an embedded spreadsheet |
| 1 | Text quality, such as an embedded Word document |
| 2 | Graphics quality, such as an embedded pie chart |
| 3 | Photograph quality |
| 4 | High-quality photograph quality |

Existing OLE objects in the drawing are not updated when you change this value. To change the quality level for individual OLE objects, select the object, right-click, and then choose Properties from the shortcut menu.

When plotting a drawing that contains OLE objects, you may find that the quality of the objects in the plotted output is poor, depending on the device used. You can control the quality of the plotted output of inserted OLE objects by using a combination of both the **OLE Properties** dialog box and the Plotter Configuration Editor. Select the OLE object, right-click, and then select Properties from the shortcut menu to display the OLE Properties dialog box. From the OLE Plot Quality drop-down list, select one of the following options:

- Line Art
- Text
- Graphics
- Photograph
- High Quality Photograph

Depending on the type of OLE objects in your drawing, you may need to experiment a bit to achieve the desired results. Start the PLOT command. In the Plot dialog box, select the Plot Device tab, and then click the Properties button. In the Plotter Configuration Editor dialog box, select the Device and Document Settings tab and expand the Graphics node (by clicking the plus sign adjacent to the word Graphics). Select Raster Graphics to display the Raster Images area in the lower portion of the dialog box. This area contains three slider bars. (Note that the ability to change these settings depends on the device driver; not all drivers support this feature.) Move the OLE slider to Best. If the OLE object is still not of a high enough quality, you can move the Trade-off bar toward Fewer colors. This will increase the sharpness of the image at the expense of reducing the number of colors, and is of more benefit when plotting to a monochrome plotting device.

the slide was created. They can't be edited in AutoCAD, and the slide file format is not compatible with many image editing programs. For that reason, the only practical way to modify a slide is to edit the drawing and then recapture it.

Although slides are limited in their capabilities, they do serve several purposes. They are the basis of icon menus and can also be used to display the appearance of hatch patterns in the various hatch pattern dialog boxes. You can also use slides to make presentations by showing slides of your drawings, and to reference a slide of a drawing while working on a different drawing. In Chapter 29, you'll learn how to automate slide presentations using scripts.

When you view a slide, it temporarily replaces the current drawing. When you refresh the display of the current drawing (for example, by redrawing, panning, or zooming the drawing), the slide image disappears and you are returned to the current drawing. You can't print a slide that is being displayed.

## Creating Slides

You create a slide by saving the current view as a slide. A slide does not include any objects on layers that are not currently visible. The content of the slide also depends on the current drawing space. In model space, the slide shows only the current viewport. In paper space, the slide contains all visible viewports.

To create a slide:

1. At the command line, type **MSLIDE** and then press ENTER.

2. In the Create Slide File dialog box, specify the name of the slide file that you want to create.

3. Click Save.

The current drawing remains onscreen and the slide is saved to the directory that you specify.

**Note**   *Before you create a slide, you must use AutoCAD's other commands to create the drawing and display it exactly the way that you want to capture it. If you want to capture a hidden-line or shaded view, set SHADEMODE to 2D Wireframe and then use the HIDE or -SHADE command before you create the slide. If you are creating the slide from model space with multiple viewports, you must make the lower-left viewport the current viewport and then create the slide. Using the MSLIDE command in any other viewport creates a blank slide file. Also note that the MSLIDE command will not capture a rendered image.*

# Viewing Slides

After you create a slide, you can view it within AutoCAD at any time by using the VSLIDE command, which displays any slide in the current viewport. To view a slide:

1. At the command line, type **VSLIDE** and then press ENTER.
2. In the Select Slide File dialog box, specify the name of the slide file that you want to view.
3. Click Open.

AutoCAD displays the slide in the current viewport.

**Note** *The slide image incorporates all the colors, the linetypes, and the screen resolution in effect at the time the slide is created. If you display a slide containing a shaded image at a higher resolution than what you used to create the slide, black lines may appear in the shaded image. To avoid this problem, maximize the AutoCAD screen and set your display to the highest resolution available when making slide files, particularly if those files will later be displayed on a different system.*

# Creating and Using Slide Libraries

Although AutoCAD uses slides as the basis for icon menus and hatch pattern previews, you won't find any slide files in AutoCAD's various directories, because the slides are actually stored in a *slide library* file. Slide libraries combine multiple individual slide files into a single file, which makes managing the individual slides easier.

AutoCAD provides a utility program (SLIDELIB.EXE) that combines several slide files into a slide library. To create a slide library, you must first create the individual slide files with the MSLIDE command and then run the SLIDELIB program, which reads a list of slide filenames from a list file. You can create this list file with any word processor that is capable of creating an ASCII text file, such as Windows Notepad or WordPad, or a word processor, such as Microsoft Word (by saving your document as a TXT file). Each individual slide file appears on a separate line, with or without its .SLD file extension. If the slide files are stored in different directories, however, you need to include the path. Because the resulting library contains the slides, the paths are not saved within the library. The following list illustrates a typical slide list file:

```
slide1
slide2
c:\sample\slide3
slide4.sld
```

Assuming that this list is saved as SAMPLE.TXT, you could create a slide library file named MYSLIDES.SLB by typing the following command sequence at the DOS prompt:

```
SLIDELIB MYSLIDES <SAMPLE.TXT
```

*Long filenames are not supported in slide libraries.*

Unfortunately, the SLIDELIB program makes no provision for updating slide library files. To add or delete a slide from a library file, you must edit the list and re-create the library file. And to do this, all the original slide files must be available, somewhat defeating the purpose of slide libraries. Luckily, several independent developers have created programs that circumvent this limitation. One such program, SlideManager, is provided on the companion web site.

*If you create slides and anticipate that you later will need to edit them, be sure to save the original drawing files from which they are generated.*

---

## Viewing Slides Saved in Library Files

When you use a select file dialog box with the VSLIDE command, to select the slide to be displayed, you can only select slide files. To view a slide stored in a slide library file, you must be able to enter the library name along with the slide filename. Although AutoCAD is able to do this as part of menus and AutoLISP and other add-on programs, to explicitly enter the library and slide name, you must run the VSLIDE command entirely at the command line. To do so, you must first disable the file dialog box that is normally displayed.

The FILEDIA system variable controls whether AutoCAD displays file dialog boxes. When set to its default value of 1, AutoCAD uses dialog boxes whenever it must prompt for a filename. If you change the FILEDIA value to 0, however, the dialog box is not displayed, enabling you to enter the filename at the command line, instead. After you change the FILEDIA value, you can display a slide stored in a slide library by using the VSLIDE command. Start the VSLIDE command and, when prompted for the slide filename, type the following:

*library(slidename)*

where *library* is the name of the library file (including any necessary path information), and *slidename* is the name of the individual slide in the library file. For example, if you want to view the slide TOPVIEW in the slide library MYSLIDES.SLB, stored in the SAMPLES directory on the D drive, you use the following command sequence:

```
Command: VSLIDE
Slide file <default>: D:\SAMPLES\MYSLIDE(TOPVIEW)
```

BECOMING AN EXPERT

# The
# Complete
# Reference

# Part IV

## Advanced Topics

# The Complete Reference

# Chapter 25

## Web-Enabling AutoCAD Drawings

The Internet and the World Wide Web, without a doubt, represent one of the most significant technological changes of the decade. In just a few years, tens of millions of people have become linked through this international computer network. While many value the Web for its entertainment and marketing potential, for CAD users, the Web represents a wonderful new way to share data and communicate in a collaborative design process.

Design has always been a collaborative process. For example, architects typically hire consultants from structural, mechanical, electrical, and other engineering disciplines to design specific building systems. When all the members of the design team moved from drafting boards to AutoCAD, they naturally exchanged information in the form of AutoCAD drawings. With the prevalence of the Web, they now often exchange those drawings via the Internet.

With its thorough integration with the Web, AutoCAD offers even more opportunity to leverage the Internet. Designers can now access drawings on the Web and use the Internet for tighter collaboration and review of contract documents.

AutoCAD 2002 is fully Web-enabled. With it, you can open or insert drawings from files located on the Internet the same as if they were located on your own computer or local area network (LAN). You can also attach external references, raster images, and other types of files from the Internet. Assuming that you have permission to do so, you can save drawings to files located on the Internet. You can attach a Uniform Resource Locator (URL) to objects or areas in an AutoCAD drawing and use those hyperlinks to open other documents. AutoCAD also enables you to plot drawings to a special Drawing Web Format (DWF) file for display on a web site.

In addition, Autodesk has developed several additional components and technologies that can be used to incorporate content from AutoCAD and other Autodesk programs on your web site. These technologies include:

- **WHIP!**   Enables viewing of DWF files embedded in HTML and other documents. WHIP! also supports drag-and-drop as a mechanism for opening DWG files or inserting DWG files into open drawings as blocks.

- **Volo View/Volo View Express**   Enables viewing of DWF, DWG, and DXF files embedded in HTML and other documents and also provides a stand-alone interface that adds the ability to mark up drawings and save those markups. You already learned about inserting markup data from Volo View in the previous chapter.

- **i-Drop**   Enables web content creators to add the ability to drag-and-drop design content from a web page directly into your designs.

**Note**   *AutoCAD's Internet capabilities are available only if you selected the Full install option or specifically selected them during installation, or if you have Microsoft Internet Explorer 5.0 or newer installed on your system. If necessary, you can add the missing components by rerunning AutoCAD's installation procedure, selecting the Custom install option, and selecting the Internet Tools option.*

The discussion of AutoCAD's Internet technologies is divided into two chapters. This chapter looks at the tools available for Web-enabling AutoCAD drawings themselves. In the following chapter, you'll learn how to use the Internet to transmit files, collaborate with others, and publish drawings to a web site.

This chapter explains how to do the following:

- Open and save drawing files across the Internet
- Add and manage hyperlinks in AutoCAD drawings
- Save drawings as Drawing Web Format (DWF) files

# Opening and Saving Files from the Internet

You can use AutoCAD to open and save files across the Internet. All of AutoCAD's commands—such as OPEN, SAVE, INSERT, and XREF—recognize any valid URL path to an AutoCAD file the same as if the file were located on your own computer or LAN.

When you open a drawing file across the Internet, the file is downloaded and then opened in AutoCAD. You can then edit the drawing and save it either locally or to any location on the Internet at which you have adequate permission to save files. If you know the URL of the file you want to open, you can type it in the File Name box in the Select File dialog box. If you've previously stored files on the Point A or Buzzsaw service, or at another location for which you've added an icon in the Places list in the Select File dialog box, you can select those locations directly and then select the file you want to open.

**Note**    *Although RedSpark still appears in the Places list of the Select File dialog box, that service is no longer available. Unfortunately, you cannot remove that icon from the Places list.*

To open a drawing from the Internet by using the Select File dialog box:

1. Start the OPEN command to display the Select File dialog box.
2. In the File Name box, type the complete URL to the file you want to open, and then click Open.

**Note**    *When opening a drawing file from the Internet, you must specify the file transfer protocol (such as HTTP:// or FTP://) and the extension (such as .DWG or .DWT) of the file you want to open.*

If you don't know the exact URL of the drawing you want to open, or to avoid having to enter a long URL every time you access various Internet locations, you can use the Browse the Web dialog box (shown in Figure 25-1) to open or save drawing files across the Internet. This dialog box, actually a miniversion of Microsoft Internet Explorer, enables you to navigate to a specific Internet location. You can display the

**Figure 25-1.** *The Browse the Web dialog box*

Browse the Web dialog box by clicking the Search The Web button in the File dialog box that is displayed by using commands such as OPEN, SAVEAS, IMPORT, EXPORT, XATTACH, or IMAGEATTACH.

 *You can also use your regular web browser to locate the URL of a drawing that you want to open, and then use cut-and-paste to copy the URL from your browser's address bar into the dialog box.*

When the Browse the Web dialog box is first loaded, it displays the default Internet location, as determined by the INETLOCATION system variable. You can use this dialog box as you would your regular web browser. For example, to navigate to a particular URL, you can type the URL in the Look In box, or select the URL from your

Favorites folder. To return to the default location, click the Home button. You can also use the Forward and Back buttons to navigate between pages after you have browsed through several web pages.

To open an AutoCAD drawing from the Internet by using the Browse the Web dialog box:

1. Start the OPEN command to display the Select File dialog box.

2. Click the Search The Web button to display the Browse the Web dialog box.

3. In the Browse the Web dialog box, do one of the following:

   ■ Click a hyperlink in the loaded HTML page.

   ■ In the Look In box, type a full or partial URL and then press ENTER.

4. Repeat Step 3 until you locate the file that you want to open.

5. Click the link to the file you want to open (so that the filename appears in the Name Or URL box).

6. Click Open.

You can use these same methods to open an AutoCAD drawing from an FTP site or a Web folder. You can also use these methods to save files to an Internet location, FTP site, or Web folder.

To save a drawing to an Internet location:

1. Start the SAVEAS command to display the Save Drawing As dialog box.

2. In the File Name box, type the complete URL of the file you want to save.

3. From the Save As Type drop-down list, select the type of file you want to save.

4. Click Save.

*When saving a drawing to the Internet, you must specify the file transfer protocol (which must be FTP://) and the file extension (such as .DWG or .DWT) of the file you want to save. You must also have the necessary permission to write to the specific Internet location.*

To save a drawing to an Internet location by using the Browse the Web dialog box:

1. Start the SAVEAS command to display the Save Drawing As dialog box.

2. Click the Search The Web button to display the Browse the Web dialog box.

3. In the Browse the Web dialog box, do one of the following:

   ■ Click a hyperlink in the loaded HTML page.

   ■ In the Look In box, type a full or partial URL and then press ENTER.

4. Repeat Step 3 until you navigate to the location at which you want to save the file.

ADVANCED TOPICS

5. In the Name Or URL File Name box, specify the name of the file (including the .DWG file extension and the FTP protocol; for example, ftp://mysite.com/ drawing.dwg), and then click Save.

If the Internet location to which you are saving the file requires authentication, AutoCAD displays a dialog box so that you can enter your username and password. As soon as you enter this information and click OK, AutoCAD copies the drawing file to the Internet location you specified. If you often access a particular FTP site, you can avoid having to always enter this information by adding the FTP site to a list of sites that can be directly accessed by AutoCAD (as explained in the next section).

## Setting Up FTP Accounts

Most FTP sites allow anyone to open or download files—known as an anonymous logon. Before you can save files to an FTP site, however, you will usually need to configure AutoCAD so that it can access the site. To do this, you must specify the name of the site, your username, and your password. You can add, modify, or remove FTP locations at any time using the Add/Modify FTP Locations dialog box, shown in Figure 25-2.

To add an FTP location that requires a username and password:

1. Display a file dialog box (for example, by using the OPEN or SAVEAS command).

2. In the file dialog box, choose Tools | Add/Modify FTP Locations.

3. In the Add/Modify FTP Locations dialog box, under Name Of FTP Site, enter the name of the FTP location (such as ftp:mycompany.com).

4. Under Log On As, select User.

5. In the User field, enter your username.

6. Under Password, enter your password. Note that the actual password will not appear, but rather will be replaced with asterisks.

7. Click Add and then click OK.

When you add an FTP site, it appears in the FTP folder of the Places list in all of AutoCAD's file dialog boxes. To open or save a drawing from an FTP site, select that folder and navigate to the desired location on the FTP server the same way as you would navigate to any folder on your own computer or LAN.

## Adding Folders to the Places List

When you add an FTP site, it appears in the FTP folder of the Places list in all of AutoCAD's file dialog boxes. For example, to open a drawing from an FTP site, you would display the Select File dialog box, select the FTP folder, double-click the FTP

Type to specify a username

Type to specify an FTP site

Select type of logon

Type to specify a password

Displays a list of configured FTP sites

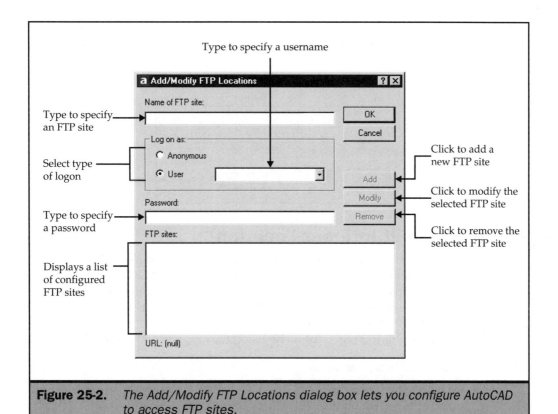

Click to add a new FTP site

Click to modify the selected FTP site

Click to remove the selected FTP site

**Figure 25-2.** *The Add/Modify FTP Locations dialog box lets you configure AutoCAD to access FTP sites.*

site in the Name list to connect to that site, navigate to the desired location on the FTP server the same way as you would navigate to any folder on your own computer or LAN, select the drawing file you want to open, and then click Open. It can be tedious, however, to constantly drill your way down through multiple folders on the FTP server every time you want to open or save a drawing. There is an easier way.

When you have located the folder on the FTP server in which the drawing files are located, you can add that folder to the Places list in AutoCAD's file dialog boxes (the list of named folders along the left side of the file dialog boxes). To add a folder, right-click the Places list and select Add Current Folder from the shortcut menu. AutoCAD immediately adds a new folder with the same name as the current folder on the FTP site.

When you close the file dialog box, AutoCAD asks whether you would like to save the changes you made to the Places list. Click Yes to save your changes. After saving the changes, you can go back into the file dialog box and change the name assigned to the folder to something more meaningful. To do so, right-click the folder in the Places

ADVANCED TOPICS

list and choose Properties from the shortcut menu. AutoCAD displays a Places Item Properties dialog box. In the Item Name field, specify the new name for the folder.

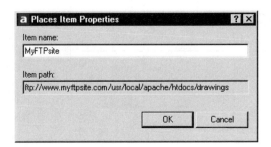

# Setting the Default Internet Location

As you learned previously, when you display the Browse the Web dialog box, it initially displays the default Internet location, as determined by the INETLOCATION system variable. By default, this location is the Autodesk web site (www.autodesk.com). You can change this location either by changing the INETLOCATION value or by specifying a different location in the Options dialog box.

To change the default Internet location by using the Options dialog box:

1. Start the OPTIONS command.

2. On the Files tab, expand the Menu, Help, And Miscellaneous File Names option.

3. Under this option, expand the Default Internet Location.

4. Select the current default Internet location and click Browse. AutoCAD displays the Select a File dialog box.

5. In the Select a File dialog box, do one of the following:

    ■ In the File Name box, type the URL to the Internet location you want to use as the default.

    ■ Click the Search The Web button to display the Browse the Web dialog box, navigate to the Internet location you want to use as the default, and then click OK.

6. Click OK to close the Options dialog box.

**Note** *If you upgraded from an earlier version of AutoCAD, the INETLOCATION value may be set to a location other than the Autodesk web site. AutoCAD 2000 was configured to use a file named HOME.HTM located in the main AutoCAD directory. This file is normally deleted, however, when you upgrade your software. If AutoCAD is unable to locate the URL specified in the INETLOCATION variable, you should change its value to a different URL.*

## Working with Other Commands Over the Internet

In addition to opening and saving drawing files across the Internet, you can insert drawings as blocks into an open drawing, import and export other file formats, and attach external references and raster image files across the Internet. For example, to attach an xref located on the Internet:

1. Start the XATTACH command.

2. In the Select Reference File dialog box, do one of the following:

   ■ In the File Name box, type the URL to the file you want to attach, including the transfer protocol (such as HTTP:// or FTP://).

   ■ Click the Search The Web button, and then use the Browse the Web dialog box to navigate to the file you want to attach.

3. Click Open.

4. In the External Reference dialog box, specify the pertinent information (such as reference type, scale, and so on) and then click OK.

5. If prompted, specify the insertion point, rotation, and so on to complete the xref attachment.

**Note** *Attaching external references stored on the Internet or an intranet provides a powerful way to maintain a set of drawings for use by a project team whose members are in many different locations. For example, an architect could maintain the master set of construction documents on a web site accessible by all the consulting engineers. By attaching external references of the underlying building plan, each member of the design team would be assured of always having the most up-to-date drawings. Such an arrangement requires a fast Internet connection, however, to assure that project team members don't have to wait a long time for large drawing files to download onto their systems.*

You use similar methods to attach raster images and insert blocks. In the next chapter, you'll learn how to insert blocks using drag-and-drop and Autodesk's new i-Drop technology.

## Working with Hyperlinks

A *hyperlink* is a pointer to another document. When you click a hyperlink, you jump to and display the linked document. Although you could attach hyperlinks in earlier versions of AutoCAD, those hyperlinks weren't really useable until you exported the drawing as a DWF file.

In AutoCAD 2002, you can attach a hyperlink to any object in your AutoCAD drawing. You can then use that hyperlink to jump to any associated file directly from within AutoCAD. For example, you could create a hyperlink to an Excel spreadsheet

ADVANCED TOPICS

that opens Excel and loads the spreadsheet file. You can also create hyperlinks that open other AutoCAD drawings. When you create the hyperlink, in addition to specifying the URL of the file you want to open, you can specify a named location to display within the file when you jump to that file. When linking to an HTML file, the named location can be a bookmark within that file. When you create a hyperlink to an AutoCAD drawing, however, you can specify a particular named view within that drawing. The ability to insert hyperlinks into your drawings provides a powerful way to associate different types of documents—such as other drawings, bills of materials, and project manuals—with an AutoCAD drawing.

When you create hyperlinks, those links can be *absolute* or *relative* links. An absolute link specifies the complete URL path to a file location (such as D:\MYDRAWINGS\ DRAWING1.DWG). A relative link, however, stores only a partial path to a file location, relative to the location of the current file or to the default base location (such as DRAWING1.DWG). Although absolute hyperlinks may work well, they present some significant limitations. Because the path to the linked file is absolutely defined, if you subsequently move the files referenced by the links, the links become broken and you will need to edit the links to reestablish the connections. If the drawing contains many links, this can be a time-consuming process. By creating relative hyperlinks, you can update the relative paths for all the hyperlinks in your drawing at one time.

**Note** *By default, AutoCAD resolves relative hyperlinks beginning from the location of the current drawing. For example, if you were to open a drawing in the MYDRAWINGS directory, AutoCAD would search for relative links in relation to that directory. You can also use the HYPERLINKBASE system variable to specify a default base directory from which AutoCAD resolves all relative hyperlinks. You can specify the hyperlink base from the Summary tab of the Drawing Properties dialog box (displayed by choosing Drawing Properties from the File menu or typing **DWGPROPS** at the command line) or by changing the HYPERLINKBASE system variable value. Although this variable doesn't exist in earlier versions of AutoCAD, the hyperlink base is maintained if you save your drawing in Release 14 format and subsequently open it again in a more recent version.*

You create and modify hyperlinks by using the Insert Hyperlink dialog box, shown in Figure 25-3. To display this dialog box, do one of the following:

- On the Standard toolbar, click Insert Hyperlink.
- From the Insert menu, choose Hyperlink.
- At the command line, type **HYPERLINK** and then press ENTER.
- Press the CTRL-K shortcut key combination.

When you start the command, AutoCAD prompts you to select objects. Select the object or objects to which you want to attach the hyperlink, and then press ENTER.

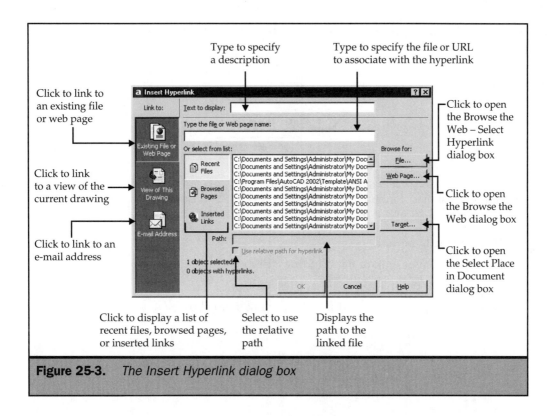

**Figure 25-3.** *The Insert Hyperlink dialog box*

AutoCAD then displays the Insert Hyperlink dialog box. The controls in the Link To area enable you to create a link to a file or web page, to a view of the current drawing, or to an e-mail address of a specified recipient. Type the filename or web page name, select from a list, or browse for a file or web page. Three controls let you display a list of recent files, web pages you recently browsed to, or recently inserted links. After you select the file or web page to which you want to link, you can click the Target button to display a dialog box in which you can select a named location within the linked file, such as a bookmark in an HTML file, or a named view in an AutoCAD drawing. In the Text To Display box, you can type a description for the hyperlink. This field is useful when the URL isn't particularly helpful in identifying the linked file.

Hyperlinks can point to files located on your computer, on another computer on your LAN, or anywhere on the Internet. After you create a hyperlink, AutoCAD displays a special hyperlink icon adjacent to the crosshairs whenever you move the cursor over an object that has an attached hyperlink. If you included a description, the description also appears in a ToolTip; otherwise, AutoCAD displays the actual URL in the ToolTip.

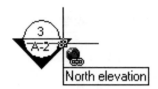

Whenever AutoCAD displays the hyperlink icon, you can click the object, right-click to display the shortcut menu, and choose Hyperlink. You can then use the Hyperlink shortcut menu to do any of the following:

- Open the hyperlink
- Copy the hyperlink to the Windows Clipboard
- Add the hyperlink to your Favorites folder
- Edit the hyperlink in the Insert Hyperlink dialog box

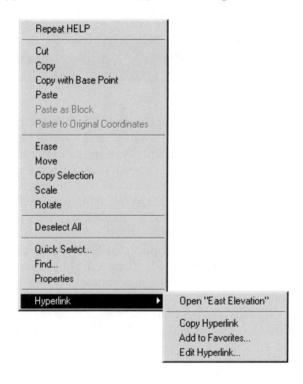

**Note** *The display of the hyperlink icon and ToolTip is controlled by the settings under Hyperlink on the User Preferences tab of the Options dialog box. In addition, the noun/verb selection mode must be enabled. This is controlled from the Selection tab of the Options dialog box or by changing the PICKFIRST system variable.*

When you create a hyperlink to another AutoCAD drawing, jumping to that hyperlink automatically opens the linked drawing in AutoCAD. If the link includes a named view, AutoCAD opens the drawing and displays the named view. Views that were created in model space are restored in the Model tab; views that were created in paper space are restored in the last active layout tab.

**Tip** *If you create a link to a drawing template (a DWT file) instead of an actual drawing file, when you jump to that link, AutoCAD creates a new drawing based on the template. This is a useful way to create a standard set of drawings based on a standardized drawing template.*

If you jump to a link to any other type of file, your computer will start the program file associated with the file type and then load the linked file. For example, if you jump to a linked Word document, your computer will launch Microsoft Word and open the specified file.

To create a hyperlink to another file:

1. Start the HYPERLINK command.

2. Select one or more objects to which you want to attach the hyperlink, and then press ENTER.

**Tip** *You can also select the objects first and then start the HYPERLINK command.*

3. In the Insert Hyperlink dialog box, do one of the following:

   ■ In the Type The File Or Web Page Name box, type the path and name of the file that you want to associate with the hyperlink.

   ■ Click File to display the Browse the Web – Select Hyperlink dialog box. Navigate to the file that you want to associate with the hyperlink, and then click Open.

4. If you are creating a link to an AutoCAD drawing, you can click Target to display the Select Place in Document dialog box in order to link to a named view in the drawing, if desired.

**Note** *To link to a bookmark or other named location in any other type of file, you must specify the named location according to the naming convention associated with the particular type of file. For example, if you're creating a link to a bookmark in an HTML file, the bookmark name should be prefixed by a pound symbol (#). To jump to a named location in another type of file, refer to the application's documentation for the proper naming convention requirements.*

**ADVANCED TOPICS**

5. In the Text To Display box, specify a description for the hyperlink, if desired.

6. Click OK.

**LEARN BY EXAMPLE**

*Link a building elevation to its callouts on a floor plan. Download the files associated with Figure 25-4 from the companion web site, following the instructions on the web site.*

To attach a relative hyperlink to the elevation callout:

1. Open the floor plan drawing.

2. Select the callout labeled 1/A-2, located at the bottom of the plan.

3. From the Insert menu, choose Hyperlink.

4. In the Insert Hyperlink dialog box, click the File button.

5. In the Browse the Web – Select Hyperlink dialog box, select the file 1-A2.DWG, and click Open.

6. In the Type The File Or Web Page Name box, delete the entire path except for the actual filename (1-A2.DWG).

7. In the Text To Display box, type **South Elevation**.

8. Make sure the Use Relative Path For Hyperlink check box is selected, and then click OK.

On your own, repeat Steps 2 through 8 to attach the other elevation drawings you downloaded (2-A2.DWG, 3-A2.DWG, and 4-A2.DWG) to the appropriate callouts on the plan drawing.

*You can copy an existing hyperlink to another object. To do so, right-click the object to which the hyperlink is already attached and then select Hyperlink | Copy Hyperlink from the shortcut menu. Then, from the Edit menu, choose Paste As Hyperlink. AutoCAD prompts you to select objects. Use any selection method to select the objects to which you want to attach the hyperlink, and then press ENTER.*

Once you have attached a hyperlink to an object in a drawing, you can edit that link using the Edit Link dialog box, a dialog box virtually identical to the one you used to create the link. To edit a hyperlink:

1. Start the HYPERLINK command.

2. Select the object to which the hyperlink is attached, and then press ENTER.

3. In the Edit Hyperlink dialog box, make any desired changes to the hyperlink, and then click OK.

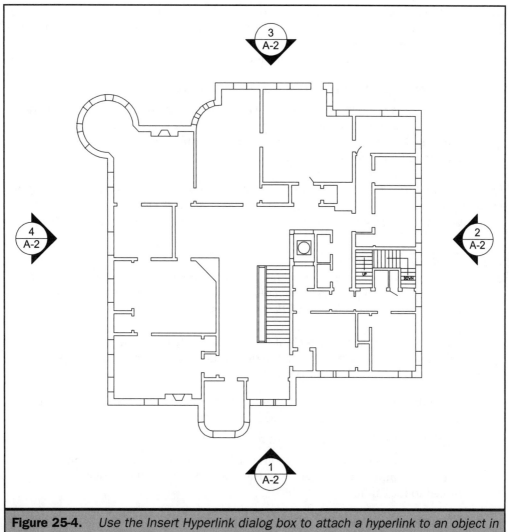

**Figure 25-4.**    *Use the Insert Hyperlink dialog box to attach a hyperlink to an object in a drawing.*

ADVANCED TOPICS

| Tip | *You can also select the object first and then start the HYPERLINK command by clicking the Insert Hyperlink button on the Standard toolbar, selecting Hyperlink from the Insert menu, typing the command, or right-clicking and then choosing Hyperlink | Edit Hyperlink from the shortcut menu.* |

You can also use the Edit Hyperlink dialog box to remove a hyperlink from an object in a drawing. Notice that when you display the dialog box to edit an existing hyperlink, the dialog box contains a Remove Hyperlink button in the lower-left corner. To remove the hyperlink, simply click this button. AutoCAD immediately removes the link and closes the dialog box.

To open a file associated with a hyperlink:

1. Select the object to which the hyperlink is attached.

2. Right-click anywhere in the AutoCAD drawing to display the shortcut menu, and then choose Hyperlink | Open "*name*", where *name* is either the name of the linked file or the hyperlink description specified as the Text To Display in the Insert Hyperlink dialog box.

*You can turn off the display of the hyperlink cursor from the User Preferences tab of the Options dialog box. If you turn off the hyperlink cursor, the Hyperlink shortcut menu becomes unavailable and you will be unable to open files associated with hyperlinks.*

### LEARN BY EXAMPLE
*Use hyperlinks to open other drawings. Continue working with the drawing from the previous example or open the drawing Figure 25-4A that you downloaded from the companion web site (which has all the hyperlinks already inserted).*

As you move the cursor over each of the callouts, notice that the hyperlink cursor appears along with a ToolTip showing the description of the hyperlink. To open the drawing file associated with the hyperlink:

1. Select the callout 3/A-2, the north elevation callout at the top of the drawing, so that it becomes highlighted and its grip appears.

2. Right-click to display the shortcut menu.

3. From the shortcut menu, choose Hyperlink | Open "North Elevation". AutoCAD immediately opens the drawing 3-A2.DWG, the drawing file linked to the callout.

Experiment with opening the other drawings associated with the hyperlinks. Since AutoCAD can open multiple drawings, you can open all the drawings and refer to them as you work on the floor plan.

*Hyperlinks can also be saved with blocks and with objects nested within blocks. Relative hyperlinks within blocks assume the relative base path of the drawing in which they are inserted. If you insert a block containing objects that already have associated hyperlinks, and then attach a hyperlink to that block, when you select an object within the block, you can open, copy, add to favorites, or edit the hyperlink associated with the block and also open the hyperlink associated with the selected object within the block.*

## Attaching, Detaching, and Selecting URLs

You can also attach, detach, and select hyperlink URLs by using the ATTACHURL, DETACHURL, and SELECTURL commands. These commands, which were added to AutoCAD Release 14 as part of the AutoCAD Internet Utilities, remain available in AutoCAD 2002 but are not documented.

The ATTACHURL command attaches a URL to an object or area in the current drawing, thus creating a hyperlink. When you start the ATTACHURL command, AutoCAD prompts you to assign the URL, by either selecting objects or specifying the opposite corners of a rectangular area, and then prompts you for the URL. You must specify the URL by typing the address.

*NOTE: URLs are case-sensitive. AutoCAD does not check the validity of any URLs that you attach. It is up to you to make sure that the address is entered correctly.*

When you attach a URL to one or more objects, AutoCAD attaches the URL to the objects as extended entity data (xdata). If you attach a URL to an area, AutoCAD creates the rectangular area on a special layer named URLLAYER, and then attaches the URL to the rectangle as xdata.

*CAUTION: When attaching URLs to areas, AutoCAD creates the URLLAYER and then attaches the URL string as xdata. You can destroy the hyperlink information if you modify the URLLAYER. You should not freeze, lock, or change the visibility of the URLLAYER or modify any of the objects on that layer. If you turn off the URLLAYER while working on the drawing, be sure to turn it back on before you export the DWF file.*

When should you attach hyperlinks to objects and when should you attach them to areas? Use the following guidelines:

■ Attach hyperlinks to objects when you want to establish a link to several objects that are scattered in several locations in the drawing.

■ Attach hyperlinks to areas when the link encompasses a contiguous area or when there are no objects that can be used as hyperlinks.

*TIP: You can also use the -HYPERLINK command to attach links to areas in the drawing. In a drawing file, only the rectangular perimeter will trigger the link. In a DWF file created from the drawing, simply moving the cursor anywhere within the area will trigger the link.*

To attach a URL to an area:

1. At the command line, type **ATTACHURL** and then press ENTER.

2. Type **A** (for Area) and then press ENTER.

3. Specify the first corner of the rectangle.

4. Specify the opposite corner.

5. Type the URL and then press ENTER.

ADVANCED TOPICS

*NOTE: When you specify the URL, you can enter the URL either as an absolute address (such as http://www.mysite.com/mylink.htm) or as a relative link (such as mylink.htm), which is located relative to the current page.*

You can also use the DETACHURL command to remove a hyperlink from an object or area. When you remove a hyperlink from an area, AutoCAD deletes any associated rectangles from the URLLAYER. When you start the command, AutoCAD prompts you to select objects. You can use any object selection method; AutoCAD ignores any objects that don't have attached URLs.

To detach a URL:

1. At the command line, type **DETACHURL** and then press ENTER.

2. Select the objects or areas from which you want to detach URLs, and then press ENTER.

The SELECTURL command selects all objects and areas in a drawing that have hyperlinks attached to them, and then places them in the current selection set so that they can be edited. If you ever forget which objects and areas in a drawing have hyperlinks, you can also use this command to highlight quickly all the objects with attached URLs.

To select all the URLs in the drawing, at the command line, type **SELECTURL** and then press ENTER.

**Note** *The SHOWURLS Express tool shows all the URLs in a drawing.*

## Creating DWF Files

As you've already learned, AutoCAD can export drawing files in Drawing Web Format (DWF). However, AutoCAD itself can't read DWF files. You should think of DWF files as the electronic version of plot files: they contain an intelligent image of your drawing, not the drawing itself. DWF files can be opened, viewed, and plotted by anyone using one of Autodesk's viewing programs, such as Volo View or Volo View Express. If you publish the DWF files to the Web, anyone can view those DWF files by using the Volo programs or Autodesk's WHIP! browser add-on (which is available as either a plug-in for Netscape Navigator or an ActiveX control for Microsoft Internet Explorer and other Windows programs).

DWF files offer several advantages over other methods of saving and viewing drawings on the Web. DWF files contain two-dimensional vector information, making them more efficient than bitmaps for saving drawings. Unlike a bitmap image, you can use the WHIP! driver or Volo View products to pan and zoom a DWF image quickly. When you zoom in to increase the magnification, the image continues to display as vectors, whereas raster images simply display fatter and fatter pixels.

## Understanding the Differences Between Volo and WHIP!

Beginning in AutoCAD 2002, Autodesk has updated the DWF format such that it is no longer completely compatible with WHIP!, which is being phased out. The new format is provided so that DWF files can be printed with the same level of resolution as the original DWG file. If you intend to view DWF files using WHIP!, you should explicitly save the DWF files in the older format by using the WHIP! 3.1-compatible version of the DWF ePlot driver. When you save DWF files in the new format:

- Raster images are written differently, so that they can be handled better by the Volo View image engine. WHIP! will not display these new raster images.

- The new DWF format supports merge control, which determines how overlapping vectors are viewed and plotted. The new format also supports ISO linetypes and fill patterns. Objects in DWF files may not display properly when viewed with WHIP!.

- The new DWF format provides control over color depth and support for true color plot styles, file size, and tessellation optimization. Objects in the new DWF format may display using different colors when viewed with WHIP!.

While you can explicitly save DWF files in the older format for use with WHIP!, unless you do so, you will need to use the latest version of Volo View or Volo View Express in order to accurately view DWF files. However, the Volo products do not support drag-and-drop for opening or inserting DWG files (described in the next chapter). While creators of web sites incorporating DWF files could configure their pages so that visitors who lacked the WHIP! component could automatically download WHIP!, this capability also doesn't exist with Volo View Express. Instead, you'll need to provide a link to the Volo View Express download page at the Autodesk web site so that visitors can obtain the necessary software to view your DWF files.

When you install AutoCAD 2002, the installation program typically installs two different DWF plot configurations: ePlot (optimized for plotting) and eView (optimized for viewing). In addition, if you use the Publish to Web wizard (described in the next chapter), AutoCAD installs additional DWF plot configurations. You can also use the Add-a-Plotter wizard to add other DWF plot configurations, including DWF Classic (R14 look) and DWF ePlot (WHIP! 3.1-compatible version). It is this latter version you should use in order to create DWF files that are compatible with WHIP!.

DWF files produced by DWF eView (optimized for viewing), DWF Classic (R14 look), and DWF ePlot (WHIP! 3.1-compatible version) are all optimized for viewing. These DWF files have high vector resolution to allow you to zoom very far into the drawings. The resolution depends on paper size and is not expressed in dots per inch (dpi).

DWF files produced by DWF ePlot (optimized for plotting) are optimized for plotting with Volo View. Their resolution is expressed in dpi, independent of paper size. These DWF files have less vector zoom depth but store higher-quality raster images (if raster images are included in the original DWG file) than eView DWF files.

The DWF format is a streaming format, meaning that the DWF image begins to display as data is arriving on your computer from the Web. You can start panning and zooming as soon as you see the image, instead of having to wait until the entire file has been downloaded.

In addition, DWF files are saved and transmitted in compressed form, further reducing download time. Since they contain only two-dimensional vectors, DWF files are also much smaller than the original AutoCAD DWG files.

By saving your drawings in DWF format and publishing them on the Web, users anywhere in the world can view the drawings. You can also control whether users have access to the actual drawing files, by deciding whether to save the original DWG file on the Web, too.

The DWF file does not expose the original AutoCAD DWG file. Because AutoCAD itself can't read the DWF file, viewers of your files can't open the original drawings unless you explicitly make them available. This means that you can publish your drawings to the Web in DWF format, while maintaining the ownership of the intellectual property contained in the DWG file.

When you create hyperlinks in your drawing and then create a DWF file, those hyperlinks are included in the DWF file. You can use hyperlinks to link to other DWF files, AutoCAD drawings, word processor documents, an HTML page, another web site, or any other valid Internet address. When you click the hyperlink in your browser, you immediately jump to that link.

Creating and using DWF files is a very simple three-step process:

1. Add links to your AutoCAD drawing.

2. Plot the drawing as a DWF file.

3. Publish the DWF file to the Web.

After the DWF file is published, anyone can view it by using their web browser and a suitable DWF viewer (such as Volo View Express or WHIP!).

In addition, the DWF format is an open specification that is available free to the general public. The format specification, as well as Autodesk's entire DWF ToolKit, can be downloaded from Autodesk's web site. DWF is becoming a widely accepted standard. As an open specification, other developers can extend the format to accommodate additional needs.

# Understanding What Gets Written to a DWF File

A DWF file is truly an electronic version of a plot file. When you plot the DWF file, you specify settings such as paper size, rotation, scale, and pen assignments, just as you do when plotting to any other device. Only those layers that are visible, actually contain data, and set to plot are written to the DWF file. In addition, the DWF file contains all the linetype, lineweight, and color settings in effect when you plot the drawing to create the DWF file. Once the DWF file is created, you can use the tools contained in the Volo View, Volo View Express, or the WHIP! viewer to turn selected layers on and off, and pan and zoom to limit what is displayed, just as with an AutoCAD drawing.

## Including Raster Images in DWF Files

Any raster images that are attached to the drawing are saved to the resulting DWF file. Because DWF is inherently a vector format, raster images pose a bit of a problem. To preserve the appearance of the drawing and yet minimize the size of the resulting DWF file, raster images are compressed. The new DWF format stores raster data somewhat differently than the older format, in order to provide better image fidelity. This is one of the reasons why DWF files saved in the new format don't display properly when viewed using WHIP!. In addition, raster images are replaced with a purple bounding box when zooming and panning older DWF files in WHIP!.

## Including Named Views in DWF Files

Any named views that were previously saved as part of the DWG file are saved to the resulting DWF file. These views can later be restored when viewing the DWF file using Volo View or WHIP!. For convenience, a view named INITIAL, if not already defined in the drawing, is also saved as part of the DWF file. The INITIAL view corresponds to the current view of the drawing at the time the DWF file was created.

## Including Layers and Blocks in DWF Files

Any layers that are visible in the DWG file when the DWF file is created, except those marked as not to be plotted, are saved to the resulting DWF file. The DWF file also contains the colors, linetypes, and lineweights as determined by the plot settings in effect when you create the DWF file. You can then use Volo View or WHIP! to turn these layers on and off individually. Any layers that are not visible in the DWG file when the DWF file is created, however, are not included.

Individual blocks are not maintained in the resulting DWF file. Instead, all the visible geometry associated with a particular block is saved in the DWF file with the layer on which the block was inserted in the original drawing file. You can therefore turn blocks on and off in the DWF file by turning their layer on and off.

## Including Fonts in DWF Files

When a drawing file contains text created by using standard AutoCAD SHX fonts, the font geometry is converted into simple vectors that mimic the appearance of the text in the resulting DWF file. In this way, the user who is viewing the DWF file does not need to have the SHX fonts loaded on their system. When a drawing contains text that is created by using a Windows TrueType font, however, the text and the actual font are recorded to the resulting DWF file. This has the advantage of retaining the exact appearance of the text, but has the disadvantage of requiring that the TrueType font also be available on the system of the user who is viewing the DWF file. If the associated TrueType font is not loaded on the viewing system, text may appear as gibberish.

*The Electronic Plotting Extension, available only to Autodesk Subscription Members, captures TrueType fonts used in drawings and embeds them into the DWF file so that it can be viewed and plotted exactly as intended.*

## Including the Drawing Name in DWF Files

The original drawing filename is also included in the resulting DWF file. This is done to enable the drag-and-drop capabilities of the WHIP! viewer. You'll learn more about this capability in the next chapter. Even if you save the DWF file with a different filename than the drawing file, the inclusion of the drawing name lets the DWF file keep track of the drawing file from which it was created.

*If you change the name of the drawing file after creating a DWF file, the name contained in the DWF file is not updated.*

The DWG name contained in the DWF file does not include the path. For this reason, if you want to be able to use the WHIP! viewer's drag-and-drop capability, you must save both the DWF file and its DWG file in the same directory. Not storing the path has the advantage, however, of letting you create the files locally and later move them to an Internet server for public access.

**Caution**    *The DWG filename contained in the DWF file is case-sensitive. AutoCAD saves the DWG name with the exact case used to save the drawing file. If you change the case when you copy the files to a UNIX file server (for example, it is common to change filenames to lowercase), drag-and-drop won't work. You can view the source drawing filename by using the About WHIP! function of the WHIP! viewer. You'll learn about this function in the next chapter.*

## Including Drawing Descriptions in DWF Files

The resulting DWF file also contains drawing properties information, such as the date and time the drawing was created and last modified. The DWF file also contains the title, subject, author, and keywords you specify on the Summary tab of the Drawing Properties dialog box. This information can all be displayed when you use the About WHIP! function of the WHIP! viewer and also appears in the summary area of Web pages created using the Publish to Web wizard, described in the next chapter. Volo

View Express only displays the information contained on the General tab of the Drawing Properties dialog box, however.

 *If you define a custom property called copyright, you can also include copyright information in the resulting DWF file. The copyright information only appears in the About WHIP! dialog box.*

# Creating DWF Files

After you complete the drawing and attach any hyperlinks that you want to include in the DWF file, you're ready to create the DWF file. Creating a DWF file is quite simple. You simply plot your drawing using a DWF plotter configuration. AutoCAD 2002 comes with the two preconfigured DWF ePlot PC3 plotter configuration files mentioned earlier—DWF ePlot (optimized for plotting) and DWF eView (optimized for viewing) plus the two older configurations that you can add using the Add-a-Plotter wizard:

- **DWF ePlot (WHIP! 3.1 Compatible version).pc3**   Creates DWF files with a white drawing background (and a paper boundary when plotted from a layout tab)

- **DWF Classic (R14 look).pc3**   Creates DWF files with a black drawing background, similar to the default DWF setting in Release 14

You can also use the Add-a-Plotter wizard to create your own custom DWF plotter configuration. In the Plot Device tab of the Plot or Page Setup dialog box, select the DWF plotter configuration from the Name drop-down list. Then, establish the other plot settings (such as paper size, orientation, and scale), as you would when plotting the drawing using any other type of plotter. When you select a DWF plotter configuration, however, the Plot To File check box is always selected.

To plot a DWF file:

1. Start the PLOT command.

2. In the Plot dialog box, on the Plot Device tab, select the DWF plotter configuration from the Name drop-down list.

3. Under Plot To File, in the File Name box, specify the name for the DWF file.

4. In the Location box, do one of the following:

   - Specify the folder in which you want to create the DWF file.

   - Click the Search The Web button and then use the Browse the Web dialog box to specify the Internet location to which you want to save the DWF file.

5. On the Plot Settings tab, specify the paper size and other parameters, such as drawing orientation, plot scale, and plot area, as necessary.

6. Click OK.

ADVANCED TOPICS

## Controlling DWF Properties

You can control several other properties of DWF files—such as the resolution, format, background color, and inclusion of layer information—by changing the DWF plotter configuration. When you change any of these properties, you can apply those changes to the current plot only or save the changes to the DWF plotter configuration file.

All the changes to DWF properties are made from the DWF Properties dialog box, shown in Figure 25-5. To display this dialog box:

1. Display the Page Setup or Plot dialog box.

2. Select the Plot Device tab.

3. In the Name drop-down list, select the DWF plotter configuration whose properties you want to change, and then click the adjacent Properties button.

4. In the Plotter Configuration Editor dialog box, select the Device and Document Settings tab and then select Custom Properties.

5. Click the Custom Properties button.

After making the desired changes to the DWF properties, click OK to close the DWF Properties dialog box, and then click OK to close the Plotter Configuration Editor. AutoCAD displays a dialog box asking if you want to apply the changes for the current

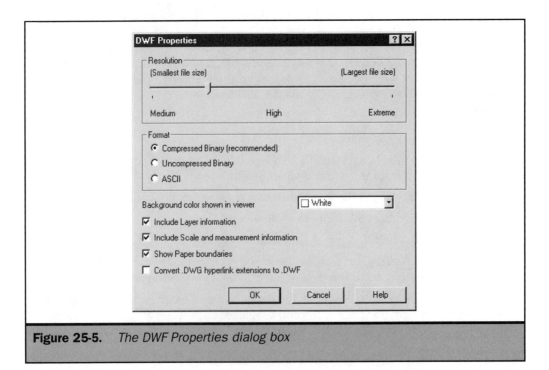

**Figure 25-5.**   *The DWF Properties dialog box*

plot only or save the changes to the current PC3 file. Select the appropriate button, and then click OK.

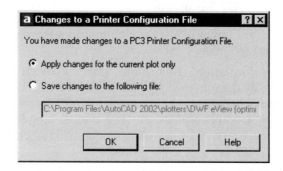

> **Note** To save your changes without altering the existing DWF plotter configurations, in the Plotter Configuration Editor dialog box, click SaveAs to display the Save As dialog box and then specify a new plotter configuration filename in the File Name box. Alternatively, you can use the Add-a-Plotter wizard to create a new DWF plotter configuration and then apply the changes to the new configuration.

## Controlling the Resolution

You can control the resolution of the resulting DWF file by moving the Resolution slider. Higher resolutions produce more precise DWF files, but at the expense of larger file sizes. In most cases, a medium resolution will be sufficient. If your drawing encompasses a large extent, however, increasing the resolution may enable you to zoom in to view smaller details in the resulting DWF file.

## Controlling the Format

The DWF file can be saved in one of three possible formats:

- **Compressed Binary** Produces a compressed binary DWF file, resulting in the smallest DWF file size. This is the default format.

- **Binary** Produces an uncompressed binary DWF file, a format that is not often used.

- **ASCII** Produces an ASCII DWF file that can be modified by using a text editor or external process.

> **Note** In AutoCAD R14, you could use the USERR2 system variable to define what one drawing unit represented. This information was saved as part of the DWF file and appeared in the dialog box title bar when you used the Location function of the WHIP! viewer. This no longer works. Instead, in order to see unit information when using WHIP!, you must save the DWF file in ASCII format and then edit the value for the UNITS field.

ADVANCED TOPICS

## Controlling Other DWF Properties

The controls in the lower portion of the DWF Properties dialog box enable you to control the following other DWF properties:

- **Background Color Shown in Viewer** Determines the background color that is applied to the DWF files.

- **Include Layer Information** Determines whether layer information is included in the DWF file. By default, the DWF file includes all the layers that are visible and plotted, and when viewing the resulting DWF file, users will be able to turn layers on and off. If you clear this check box, the resulting DWF file won't include any layer information, so those viewing the file will not be able to turn layers on and off.

- **Include Scale and Measurement Information** Determines whether scale and measurement information is included in the resulting DWF file.

- **Show Paper Boundaries** Determines whether the paper boundary is visible in DWF files plotted from a layout tab. (Note that the WHIP! viewer does not display paper space boundaries in DWF files.)

- **Convert .DWG Hyperlink Extensions to .DWF** Determines whether hyperlinks to drawing files are included in the resulting DWF file or converted into hyperlinks to other DWF files.

*You can use the BatchPlot Utility, described in Chapter 18, to automate the creation of DWF files from your drawings.*

# Chapter 26

## Leveraging the Internet

In the previous chapter, you learned how AutoCAD interfaces with the Web, how to open and save files across the Internet, and how to embed hyperlinks in AutoCAD drawings. You also learned about Autodesk's Drawing Web Format (DWF) and how to create DWF files. Most of what you've learned so far, however, has been done entirely on your local computer. In this chapter, you'll learn how AutoCAD expands your power by enabling you to collaborate with others across the Web.

This chapter explains how to do the following:

- Send AutoCAD drawings using e-mail
- Pack up and transmit entire drawing sets
- Access the Internet from within AutoCAD
- Share your AutoCAD session with others
- Use Autodesk's viewers
- Create web pages containing DWF files

## Sending AutoCAD Drawings via E-Mail

Even before the World Wide Web became so incredibly popular, many AutoCAD users exchanged drawings by using e-mail, a method that is still often used today. Generally, you send drawings via e-mail when they need to be viewed only by specific recipients, whereas you post drawings and DWF files to the Web when you want to showcase your work or need to collaborate with a larger group of people.

You can send a drawing via e-mail in either of the following two ways:

- By choosing the Send function from AutoCAD's File menu
- By using your e-mail software to create a new message and including the drawing file as an attachment

Both methods include the drawing file as an attachment. Choosing the Send function from AutoCAD automatically starts your default e-mail program (if it is not already active) and creates a new message with the drawing file already attached. When you create the new message in your e-mail software, you are responsible for attaching the drawing file to the message.

> **Note** *To specify the default e-mail program, from the Windows Start menu, choose Settings | Control Panel. In the Control Panel, double-click Internet Options to display the Internet Properties dialog box, and then select the Programs tab. You can then select the e-mail program you want to use by selecting it in the E-Mail pull-down list.*

To e-mail a drawing from within AutoCAD:

1. Open the drawing that you want to send.

2. From the File menu, choose Send.

3. In the e-mail software message dialog box, specify the recipients and subject, and then type the message that will accompany the attached drawing file.

Figure 26-1 shows a typical e-mail message with an attached drawing. In this case, the e-mail software being used is Microsoft Outlook.

When you use AutoCAD's Send function, a new e-mail message is created with the current drawing attached. If you need to send more than one drawing, you'll most likely want to use your e-mail software to create a new message, and then attach all the drawings to the single message.

**Tip**    *Drawing files are often quite large. You can use file-compression programs, such as WinZip, to combine one or more drawing files into a single archive file, and then attach the archive file to your e-mail message. Drawings can be compressed as much as 70 percent, making them much more efficient to transmit. Since WinZip and similar programs use lossless compression, no data is lost. The recipient must also have a copy of the compression utility, to extract the drawing files from the archive file when received.*

**Figure 26-1.**    *It's easy to include a drawing file as an e-mail attachment.*

ADVANCED TOPICS

# Transmitting Files

As you've learned, AutoCAD uses numerous files, such as fonts and external references, that don't actually become part of the drawing file. To move a drawing to another computer or share it with a consultant, you must also copy any of the files that your coworker or consultant doesn't have loaded on his or her system. Previous versions of AutoCAD included an Express Tool called Pack 'n Go, designed to ease this process. This tool has now been significantly enhanced and included as a standard AutoCAD command. The ETRANSMIT command creates a transmittal set of a drawing and all of its related files. To use the command, do one of the following:

- On the Standard toolbar, click eTransmit.
- From the File menu, choose eTransmit.
- At the command line, type **ETRANSMIT** and then press ENTER.

When you start the command, AutoCAD displays the Create Transmittal dialog box, shown in Figure 26-2. This dialog box has three tabs: General, Files, and Report.

The General tab provides controls for specifying the way in which the transmittal set is created. Although there are a lot of controls available, their use is relatively straightforward. In the Notes area, you can enter any notes related to the transmittal set. These notes are included in the transmittal report. You can also specify a template of default notes to be included with all your transmittal sets by creating an ASCII text file called ETRANSMIT.TXT. This file must be saved to a location included in the Support File Search Path option on the Files tab of the Options dialog box.

The Type drop-down list lets you specify the type of transmittal set created. You can choose from the following:

- **Folder**   Creates a transmittal set of uncompressed files in a new or existing folder
- **Self-extracting Executable**   Creates a transmittal set of files as a compressed, self-extracting executable file
- **Zip**   Creates a transmittal set of files as a compressed Zip file

Clicking the Password button opens a Password dialog box in which you can specify a password to protect the transmittal set when it is saved as a compressed file. The recipient of the file must enter the password in order to extract the files.

The Location field specifies the location in which the transmittal set will be created. This drop-down list contains the last ten locations in which transmittal sets were previously created. To specify a new location, click the Browse button and then navigate to the location in which you want to create the transmittal set.

If you select the Convert Drawings To check box, you can specify the format for all the drawings in the transmittal set. Drawings can be saved in either AutoCAD 2000 or R14 format. The four check boxes at the bottom of the General tab do the following:

- **Preserve Directory Structure**    When selected, AutoCAD re-creates the directory structure of all the files in the transmittal set, making it easy to install the files onto another system. When not selected, all the files are created in the target directory.

- **Remove Paths from Xrefs and Images**    When selected, the paths are removed from any externally referenced drawings or images in the transmittal set.

- **Send E-Mail with Transmittal**    When selected, AutoCAD launches the default e-mail application when the transmittal set is created so you can send an e-mail notifying others of the new transmittal set. The e-mail message automatically includes a hyperlink to the transmittal set.

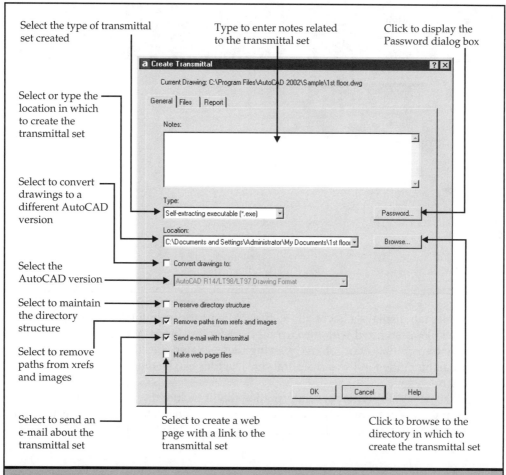

**Figure 26-2.**    *The Create Transmittal dialog box, used for creating a transmittal set of a drawing and all of its related files*

ADVANCED TOPICS

■ **Make Web Page Files**   When selected, AutoCAD generates a web page that includes a bitmap image of the drawing, a link to the transmittal set, and a copy of the transmittal report. The web page is created in the same location as the transmittal set.

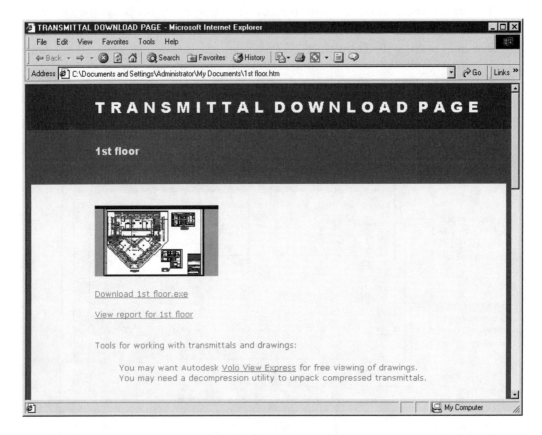

The Files tab displays a list of all the files to be included in the transmittal set. By default, all files associated with the current drawing are listed. You can add additional files to the transmittal set or remove existing files. Related files that are referenced by URLs are not included in the transmittal set. The files can be displayed in either a list view or a tree view.

The Report tab displays report information that is included with the transmittal set. It includes any transmittal notes that you entered on the General tab as well as distribution notes automatically generated by AutoCAD that detail the steps that must be taken for the transmittal set to work properly. For example, if SHX fonts are included in the transmittal, the report includes instructions about where to copy these files so that AutoCAD can detect them on the system on which the transmittal set is installed.

# Accessing the Internet from Within AutoCAD

In addition to launching your e-mail software from within AutoCAD, you can launch your default web browser. Depending on how AutoCAD is configured and how you launch the browser, the browser can display a predefined default web page, or a specific web page on the Autodesk web site. AutoCAD's BROWSER command launches your system's default web browser, as defined in the Windows system registry.

There are numerous ways in which you can launch the browser from within AutoCAD. For example, on the Web toolbar, you can click Browse The Web to display the web page specified in the INETLOCATION system variable. If you type the command, AutoCAD displays a prompt showing the default URL:

```
Enter Web location (URL) <http://www.autodesk.com>:
```

To jump to the default location, simply press ENTER. To start your web browser and jump to a different URL, type the URL and then press ENTER.

You can also use one of several toolbar buttons or menu selections to launch your browser and jump immediately to a specific web page. For example, if you select Product Support On Point A from the Help menu, your browser automatically loads the product support page at Autodesk's Point A web site. Point A, shown in Figure 26-3, is a special web site that provides industry-specific discussion groups, articles, free symbol libraries, and links to product support and updated software downloads. After registering for this free service and setting your preferences (such as your profession), Autodesk Point A displays information personalized just for you. To jump directly to the Point A web site:

■ On the Standard toolbar, click Autodesk Point A.

■ From the Tools menu, choose Autodesk Point A.

The symbol libraries at Point A are particularly useful. You can search for symbols using categories or keywords. As you search, Point A presents a page of preview images. When you find the symbol you want to insert, click the symbol to display it on its own page. You can then download the symbol to your computer or use the drag-and-drop capabilities of WHIP! to insert it into an open AutoCAD drawing. You'll learn more about WHIP!'s drag-and-drop capabilities later in this chapter.

**ADVANCED TOPICS**

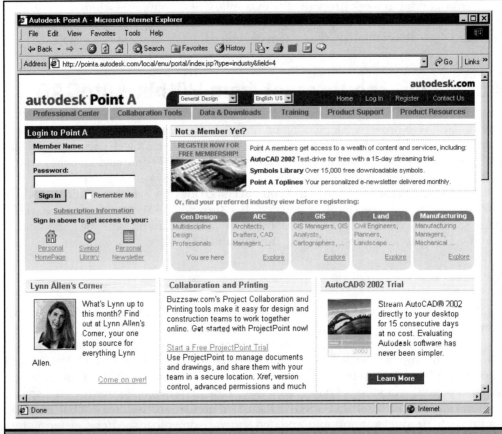

**Figure 26-3.**   Selecting Autodesk Point A from a menu or toolbar displays the Point A home page.

**Caution**   *In order to use drag-and-drop, you must first select a check box at the bottom of the Point A web page indicating that you want to use the WHIP! plug-in for viewing the symbols. However, if you have already installed Volo View Express, selecting this check box will cause WHIP! to be installed on your computer when you view a DWF file on the Point A web site. After downloading and installing WHIP!, your system will use WHIP! for displaying DWF files rather than Volo View Express. You can still use Volo View Express to open DWF files, however, by first starting Volo View Express and then using the Open command to open the DWF file. Since the DWF files in the Point A symbol libraries are saved in the older format, WHIP! has no problem displaying them. But remember that DWF files created using the new format will not display properly when viewed using WHIP!.*

# Using the AutoCAD Today Window

The AutoCAD Today window, shown in Figure 26-4, displays by default whenever you start AutoCAD 2002, and whenever you start a new drawing. You've already learned how to use this window to open drawings, create new drawings, and access the symbol libraries that are installed onto your computer as part of AutoCAD Design Center. This window also provides access to a bulletin board and the Autodesk Point A web site. To display the AutoCAD Today window, do one of the following:

- On the Standard toolbar, click Today.

- From the Tools menu, choose Today.

- At the command line, type **TODAY** and then press ENTER.

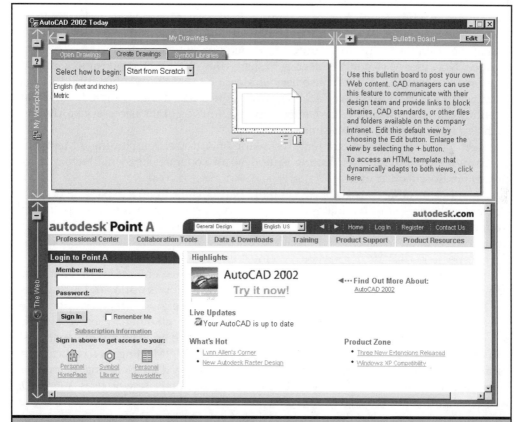

**Figure 26-4.**    *The AutoCAD Today window provides a bulletin board area and direct links to the Autodesk Point A web site.*

ADVANCED TOPICS

The Bulletin Board is designed to serve as a source of communication for your company. Generally, you company's CAD manager should control the information displayed in the Bulletin Board. A special AcTodayMgr utility program provided on the AutoCAD 2002 CD enables the CAD manager to adjust many settings that control the AutoCAD Today window. You'll find detailed information about this utility in Appendix C. If you don't have a CAD manager (or if you are the CAD manager), you can specify a path and filename to the bulletin board file using the Edit button in the upper-right corner of the Bulletin Board area. The default bulletin board file, CADMGR.HTM, is normally located in the AutoCAD 2002 directory.

# Using Meet Now

Meet Now is a powerful tool that enables two or more people to work together in a single AutoCAD session. You can actually share your copy of AutoCAD over the Internet so that while you edit a drawing on your computer, others can watch you work from theirs. You can even let others take control of your AutoCAD session and edit your drawings, even if they don't actually have AutoCAD loaded on their computer.

Meet Now uses Microsoft NetMeeting to share your AutoCAD session among multiple users across a network. You can also use NetMeeting to share any other application on your computer. In addition, NetMeeting provides text-based chat, a shared whiteboard application, and audio and video conferencing (if you've got the necessary hardware, such as a microphone and web camera).

Meet Now provides several features inside of AutoCAD, including a Meet Now toolbar with a shortcut to NetMeeting. Autodesk also operates its own directory server to help you locate other users.

An online meeting consists of two or more participants, up to a maximum of eight. The *host* is the person who initiates the meeting. A *guest* is any person participating in the meeting started by the host. Both the host and guest can share applications, but only one person at a time can control a specific application.

In order to participate in a session, you must first configure NetMeeting. A series of setup dialog boxes step you through this process. You must enter your first and last name and your e-mail address. The rest of the information is optional. During the NetMeeting setup, you can also specify a directory server and directory options. A directory server helps meeting participants connect with each other. When you log on to a directory server, the server registers your presence and adds your name to the list of users connected to it. Then, when using Meet Now or NetMeeting, you can view this list and call someone to participate in your meeting by simply double-clicking their name in the list. Since you are likely to want to connect to someone else who uses AutoCAD, when configuring NetMeeting, you should enter **MEETNOW.AUTODESK.COM** as your directory server, and select the check box labeled Log On To A Directory Server When NetMeeting Starts. If you want your name to appear in the directory server list, be sure to clear the check box labeled Do Not List My Name In The Directory Server.

# Contacting Participants for an Online Meeting

There are two different methods available for contacting participants for an online meeting: using a directory server and contacting someone directly. When you configure NetMeeting, you generally specify the directory server you want to use. By registering your directory server, you are included in a list of users available on that particular server. When NetMeeting is running on your computer, you can be contacted through the specified directory server. If you allowed your name to be included in the list, anyone can contact you by double-clicking your name. If you do not list your name in the directory, however, people can still use the directory server to contact you if they know the e-mail address you specified.

To contact someone directly, you must enter their IP address when calling them using Meet Now or NetMeeting. Every computer connected to the Internet has a unique IP address. For most people, particularly those who connect using a modem, a new IP address is assigned every time they connect, making it next to impossible to contact someone directly. Those who connect to the Internet through a LAN or using a DSL or cable modem have an IP address that doesn't change. You may be able to contact these people directly if you know their IP address, even if they are not connected to a directory server.

ADVANCED TOPICS

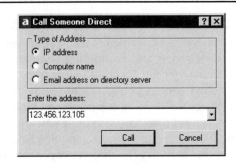

Note that many people with persistent IP addresses, however, may be behind a *firewall*. A firewall serves as a gatekeeper, protecting systems within a company from attack by hackers outside the company. The firewall may disguise the actual IP address of the computer you are trying to reach. If that is the case, you will need to use a directory server in order to contact that participant.

The other information you'll need to provide before you can start using NetMeeting for the first time is your network connection speed. You can then run the Audio Tuning Wizard to configure your audio settings, although this is not required to use NetMeeting. You may want to disable audio if you don't plan to use it.

 *After initially configuring NetMeeting, you can change any of the settings by loading the full NetMeeting application and modifying the program options. To load NetMeeting, click the Windows Start button and choose Programs | Accessories | Communications | NetMeeting. To change the NetMeeting options, from the NetMeeting menu, choose Tools | Options.*

Once you've configured NetMeeting, you can use it or Meet Now to initialize an online meeting. In order to call a guest, the guest must have NetMeeting running on their system so that they can receive your call. If you're going to host a meeting, you should contact the guests in advance, perhaps by sending them e-mail. Let them know when the meeting is going to start (so that they can have NetMeeting running). If you're going to use a directory server, you should let your guests know the name of the directory server so that they can log on to it. Guests should also have their screen configured to the same display resolution (such as 1024×768) as the host. While this isn't a requirement, it helps ensure that all participants see the same thing.

To host an online meeting:

1. Start the NetMeeting session by doing one of the following:

 ■ On the Standard toolbar, click Meet Now.

- From the Tools menu, choose Meet Now.
- At the command line, type **MEETNOW** and then press ENTER.

2. Call a participant by either using a directory server or by calling them directly.

3. The guest you are calling needs to accept the incoming call. (*Note:* You can configure NetMeeting to automatically accept calls.)

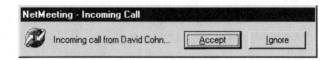

4. Repeat this procedure to add additional guests, up to a maximum of eight.

As soon as you start Meet Now, AutoCAD displays the Online Meeting toolbar. This toolbar provides controls for adding and removing guests, allowing guests to edit your drawing, opening a chat window, using the NetMeeting whiteboard application, and ending the meeting. A drop-down list contains the names of all the guests currently participating in the meeting.

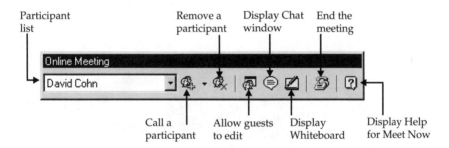

When you first start Meet Now, AutoCAD displays either the Find Someone dialog box (so that you can call a participant using a directory server) or the Call Someone Direct dialog box, depending on which method you used most recently. To use a different method to call someone, cancel the dialog box, click the arrow button adjacent to the Call Participant button on the Online Meeting toolbar, and choose the desired method from the drop-down list.

ADVANCED TOPICS

Once your online meeting has been established, your guests will see everything you do in your AutoCAD session. A guest can leave the session at any time by clicking End Call in the NetMeeting window. As host, you can remove a participant by selecting them in the participant list and clicking Remove Participants. To end the session, click End Meeting.

## Sharing Control of an AutoCAD Session

When you establish a connection to a guest, the guest sees your AutoCAD session but can't manipulate objects within your AutoCAD window, as shown in Figure 26-5. Before a guest can take control, the host must indicate that they are willing to accept control requests. The host does this by clicking the Allow Others To Edit button. When

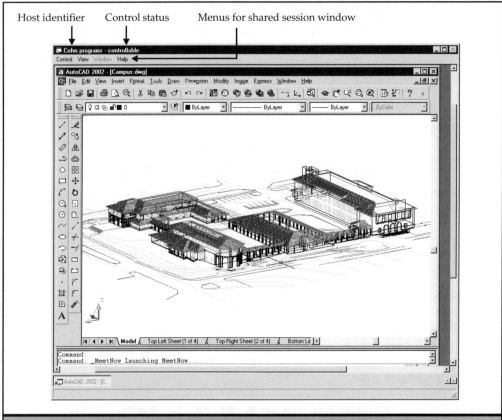

**Figure 26-5.** *A guest sees the AutoCAD session in a special shared session window.*

you click this button, AutoCAD displays a dialog box informing you that you have enabled other users to edit the document.

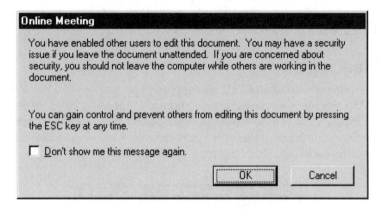

**Caution**    *You should read the message in this dialog box. If you give someone control of your AutoCAD session, they can delete objects and save their changes to the drawing on your computer. They can also gain access to other functions on your computer. Don't ever leave your computer unmonitored when others have control of it. You can regain control of your computer and prevent guests from editing your drawing at any time by pressing the ESC key.*

In order to actually take control of your AutoCAD session, the guest must first request control by selecting Control | Request Control from the menu in the NetMeeting shared session window. As the host, you will see a dialog box informing you that the guest would like to take control of the shared program. You must accept this request before the guest can take control.

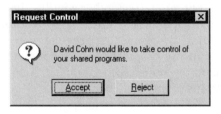

**Note**    *You cannot do any work on your computer while a guest controls your AutoCAD session. If you click your pointing device or press any key on your keyboard, you automatically regain control of your AutoCAD session.*

Once a guest has control, he or she can use AutoCAD and modify the drawing exactly as if AutoCAD were running on their local computer. The guest can relinquish control back to the host, or forward control to another guest. If one guest offers to forward control to another guest, the host is again asked whether it's okay to forward control, and the guest is also asked whether they will accept the forwarded control.

## Using NetMeeting

In addition to sharing your AutoCAD session, you can use the Chat or Whiteboard application. Chat lets meeting participants send each other text-based messages. When you click the Display Chat Window button, you open a Chat window. A similar window also opens on the computers of the other participants. Chat works like other instant messaging systems such as MSN Messenger and AOL Instant Messenger.

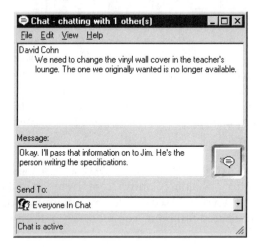

The Whiteboard application provides another window that opens on your computer and those of all the other participants. The Whiteboard window provides tools similar to those found in many "paint" programs. In particular, you can capture the contents of any window or area within any program open on your computer and then use the Whiteboard tools to mark up the static captured image. Again, all of the participants can see and share the contents of the Whiteboard.

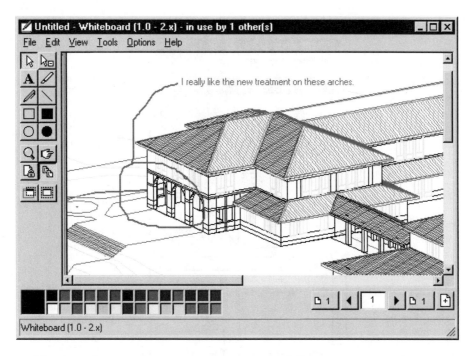

In addition to these tools, you can also open the full NetMeeting application (shown in Figure 26-6). NetMeeting provides several additional capabilities:

- You can communicate with each other using audio and video (if the computers are equipped with the necessary components).
- You can transfer files between participants.
- You can share any program, not just AutoCAD.

**Note**  *Although NetMeeting works relatively well, it can be quite slow, because it operates by sending an image of your shared application via the Internet. Its performance is therefore very much dependent on your connection speed. Using audio or video will further degrade its performance.*

ADVANCED TOPICS

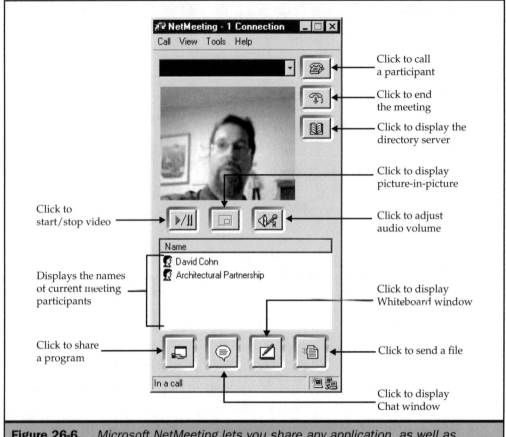

Click to call
a participant

Click to end
the meeting

Click to display the
directory server

Click to display
picture-in-picture

Click to
start/stop video

Click to adjust
audio volume

Displays the names
of current meeting
participants

Click to display
Whiteboard window

Click to share
a program

Click to send a file

Click to display
Chat window

**Figure 26-6.**   *Microsoft NetMeeting lets you share any application, as well as communicate using audio and video.*

# Using Autodesk's Viewers

Over the past five years, Autodesk has created several different technologies for viewing DWF files and other types of files when using your web browser or other Windows applications. Table 26-1 displays the capabilities of these programs.

Both WHIP! and Volo View Express are free and can be downloaded by anyone directly from Autodesk's web site. Volo View Express also comes on the AutoCAD 2002 CD and is normally installed when you install AutoCAD 2002.

*The version of Volo View Express provided on the AutoCAD 2002 CD is an older version of the program that does not support the new default DWF file format created by AutoCAD 2002. Unless you purchase the full Volo View program, you should download the latest version of Volo View Express from Autodesk's web site.*

From Table 26-1, it would appear that Volo View and Volo View Express are the clear winners in terms of capabilities. But even though Autodesk plans to phase out WHIP! and is no longer updating it, WHIP! has several advantages:

■ Only WHIP! currently supports drag-and-drop of DWG content from the DWF file.

■ When displaying a web page containing a DWF file, WHIP! can auto-download to a visitor's system if they don't already have a DWF viewer. Volo View Express doesn't have auto-download capability.

■ The WHIP! driver is quite small (less than 3.5MB). The full Volo View Express program is 27MB. A smaller DWF view-only version is 10MB.

For these reasons, many web sites (including Autodesk's own Point A web site) continue to post DWF files in the older format and to use WHIP! to display those files.

## Getting WHIP!

There are three different ways you can obtain the WHIP! viewer: auto-download when attempting to view a DWF file, auto-download from the AutoCAD web site, or download the WHIP4.EXE installation program.

| Feature | WHIP! | Volo View Express | Volo View |
|---|---|---|---|
| View DWG/DXF | | √ | √ |
| View DWF | √ | √ | √ |
| 3D Orbit/shaded view | | √ | √ |
| File filters | | | √ |
| Print | √ | √ | √ |
| To-scale plotting | | | √ |
| Measure | | | √ |
| Basic sketch and markup | | √ | √ |
| Full ActiveShapes markup | | | √ |
| Object Enabler support | | √ | √ |
| View IPT, IAM, IDW (Autodesk Inventor File Formats) | | √ | √ |

**Table 26-1.**  *Functionality Available in Autodesk's Various Viewing Programs*

ADVANCED TOPICS

*The WHIP! viewer is available only for browsers running under Windows 95/98 and Windows NT. Although it will also run under Windows 2000 and Windows XP, Autodesk does not support it on these platforms.*

The first time you visit a web site and attempt to view a DWF file, if you don't already have a program installed on your system for viewing DWF files and if the web site has been set up to use WHIP!'s auto-download capabilities, you will see a dialog box asking if you want to install the WHIP! plug-in.

To download a copy of the WHIP! viewer, go to the WHIP! home page on the Autodesk web site (**www.autodesk.com/whip**). This page provides links to the WHIP! download page as well as a product overview and a complete online user manual. The WHIP! viewer is available as an ActiveX control for Microsoft Internet Explorer and as a plug-in for Netscape Navigator. When you click a link to download WHIP!, the web site automatically detects which web browser you are using and downloads and installs the proper version.

You can also download the WHIP4.EXE file from the Autodesk web site. This file contains versions of WHIP! for both Internet Explorer and Netscape. After downloading the file, you can run the executable to install WHIP! for either browser. Make sure that your browser is not running. Then, run the WHIP4.EXE program and follow the onscreen instructions. After the installation is complete, you can relaunch your browser and view DWF files.

*Before installing WHIP! by running the WHIP4.EXE program, make sure that your temporary files directory is empty.*

## Viewing DWF Files Using Your Web Browser

Your web browser does not normally recognize Autodesk's DWF file format. That is why you must first install a DWF file viewer before you can display a DWF file in your web browser. After installing either WHIP! or Volo View Express, when you display a DWF file in your browser, or an HTML document that has a DWF file imbedded within it, your browser uses the DWF viewer to display the DWF file.

Whenever you are viewing a DWF file within your browser, you can right-click to display a shortcut menu of DWF viewer commands and then select the desired function from the menu. Although these commands vary slightly depending on whether you are using WHIP! or Volo View Express, both offer many of the same functions. Table 26-2 summarizes the commands available in WHIP! and Volo View Express.

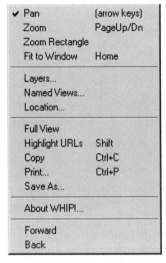

WHIP!

Volo View Express

In addition to these functions, Volo View Express can also display DWG and DXF files and files created using Autodesk Inventor. When viewing DWG files, Volo View Express provides several additional functions:

- **Orbit**   Orbits the camera around the model and also lets you display wireframe or shaded images and parallel or perspective projections, similar to AutoCAD's 3DORBIT command

- **World Plan**   Restores the drawing back to the plan view in the WCS

- **Layouts**   Displays the Layouts dialog box so you can display the model view or any of the drawing's layouts

ADVANCED TOPICS

| Function | WHIP! | Volo View Express |
|---|---|---|
| Pan the DWF image | Pan (default) | Pan |
| Zoom In/Out | Zoom | Zoom \| In/Out |
| Zoom to a rectangle | Zoom Rectangle | Zoom \| Window |
| Zoom previous | n/a | Zoom \| Previous |
| Zoom all | Fit to Window | Zoom \| All |
| Turn layers on/off | Layers | Layers |
| Restore a named view | Named Views | Named Views |
| Display location coordinates | Location | n/a |
| View DWF file in full browser window | Full View | n/a |
| Blink URLs within DWF file | Highlight URLs | n/a |
| Copy DWF file to the clipboard | Copy | n/a |
| Print the DWF file | Print | Print |
| Save the DWF file to your computer | Save As | Save Target As |
| Display information about the DWF file | About WHIP! | n/a |
| Jump forward to a URL | Forward | n/a |
| Jump back to the previous URL | Back | n/a |
| Paste contents of clipboard into DWF image | n/a | Paste |
| Display the DWF page background | n/a | Display Page Background |
| Display DWF in grayscale to improve contrast | n/a | ClearScale |
| Display DWF as black and white only | n/a | Black & White |
| Adjust page parameters before printing | n/a | Page Setup |

**Table 26-2.**   *Comparison of WHIP! and Volo View Express Commands*

While Volo View Express offers lots of additional capabilities, WHIP! provides one valuable function not available in Volo View Express. If the DWG file from which the DWF file was created is saved in the same directory as the DWF file, you can open or insert the DWG file in AutoCAD by dragging-and-dropping the DWF file into AutoCAD. You'll learn more about this function later in this chapter.

## Panning and Zooming

By default, WHIP! always begins in Pan mode, indicated by a hand icon. The hand grabs the screen as you pan the DWF image. To pan, click-and-drag the image. With WHIP!, you can also pan at any time by pressing the arrow keys on the keyboard.

Volo View Express doesn't have a default mode. However, if you are using a wheel-mouse, pressing the wheel activates Pan mode. Otherwise, to pan within the DWF file, you must first right-click to display the shortcut menu, and explicitly select Pan mode. When Pan mode is active in Volo View Express, it works exactly the same as when using WHIP!.

When you choose Zoom mode in WHIP! or Zoom In/Out in Volo View Express, the cursor changes into a magnifying glass icon. To zoom in, click-and-drag toward the top of the screen. To zoom out, click-and-drag toward the bottom. You can also zoom at any time in WHIP! by pressing the PAGEUP and PAGEDOWN keys on the keyboard. You can accomplish the same thing using Volo View Express by rolling the mouse wheel.

The Zoom Rectangle function in WHIP! and the Zoom | Window function in Volo View Express let you zoom in to a rectangular area, similar to AutoCAD's Zoom window function. Click to specify one corner and then drag to specify the opposite corner. To zoom back out to view the entire DWF file, select Fit To Window (WHIP!) or Zoom | All (Volo). Volo View Express also has a Zoom | Previous command, similar to AutoCAD. When viewing a DWF file embedded in an HTML document, you can use WHIP!'s Full View command to view just the DWF file in a full browser window. Volo View Express doesn't provide such a feature.

## Working with Layers

If layer information is included in the DWF file, you can use the Layers command to display a Layers dialog box. This dialog box, which is virtually the same in both WHIP! and Volo View Express, lists all the layers in the DWF file along with their visibility status.

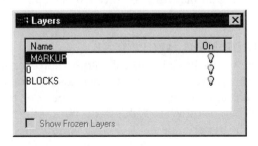

You can selectively turn layers on and off by clicking the adjacent light bulb icon. Pressing the SPACEBAR is the same as clicking the light bulb icon. The Layers dialog box works like most standard Explorer-style dialog boxes. You can select more than one layer and then click the light bulb icon associated with any selected layer to toggle all the selected layers on or off. You can also right-click within this dialog box to display its own shortcut menu, which provides the functions On, Off, Toggle, Select All, and Clear All.

## Working with Views

If the DWF file contains one or more named views, the Named Views option is available. When you choose the Named Views option, both WHIP! and Volo View Express display a Named Views dialog box, which lists all the named views contained in the DWF file.

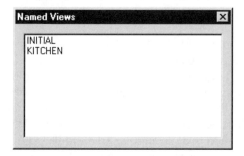

To restore a named view, select it in the list. To close the Named Views dialog box, click the Close button (the X in the upper-right corner). All named views in the original AutoCAD DWG file are saved as part of the DWF file. In addition, every DWF file contains a view named INITIAL, which corresponds to the current view of the original drawing file at the time the DWF file was created.

## Viewing Drawing Coordinates

When you choose the Location option, WHIP! opens a Current Location dialog box, showing the location of the mouse with respect to the DWF file. The coordinates displayed in the dialog box reflect the coordinates in the original DWG file. Although Volo View Express provides similar functionality when running the full application, there is no comparable function when viewing a DWF file within a web page.

## Working with Embedded Hyperlinks

One of the beauties of the Web is its ability to link information enabling you to quickly jump to the linked document. This same capability is also available in DWF files. When you position the mouse pointer over an embedded hyperlink, the pointer changes to a pointing hand icon. When using WHIP!, the status bar (at the bottom of the browser window) also displays the hyperlink's URL. To follow (jump to) the hyperlink when using WHIP!, you merely click once on the link. When using Volo View Express, you must double-click.

If the DWF file contains embedded URLs, WHIP!'s Highlight URLs option causes all the URLs in the file to blink so that you can see where they are located. This function acts as a toggle. To turn off the blinking, choose the Highlight URLs option again (to clear the check mark). You can also highlight embedded URLs by pressing and holding down the SHIFT key. When you release the key, highlighting is turned off. There is no similar function in Volo View Express.

## Printing DWF Files

DWF files can be printed in either of two ways:

- By printing from a DWF viewer
- By printing from your browser

To print just the DWF file, print using the Print function in the shortcut menu. Before printing using Volo View Express, you may want to select the ClearScale or Black & White options to first display the DWF file as either a grayscale or black and white image. When you select the Print function in WHIP!, the Print dialog box provides two check boxes that enable you to force the DWF background to white and the geometry in the DWF image to black.

You can only print to one of the available Windows system printers. When printing from WHIP!, the DWF file will be scaled to fit the paper size. When printing from Volo View Express, you can either scale the DWF image to fit the paper, or retain the paper size as specified in the original AutoCAD drawing and print the drawings across multiple sheets, and then tape those sheets together to create the final drawing.

If you want to print the entire HTML page, including the DWF file, use your browser's Print function instead of the DWF viewer's Print option, by clicking the browser's Print button or selecting File | Print from the browser's menu.

## Saving DWF Files

Whenever you use WHIP! to view a DWF file, you can save that file to your local hard drive. When you choose the Save As option, WHIP! displays a standard Save As dialog box. The Save As Type drop-down list offers three file types:

- **Autodesk Drawing Web Format**   Saves a copy of the DWF file
- **Windows Bitmap**   Saves a copy of the current DWF image as a Windows bitmap (BMP) file
- **AutoCAD Drawing**   Saves a copy of the DWG file used to create the DWF file, provided that the DWG file is actually available

**Note**   *When WHIP! looks for the DWG file, it looks in the same directory in which the DWF file is located. If the DWG file is not saved in that directory, WHIP! displays a dialog box that warns you that the source DWG file can't be found. When you publish DWF files, you can take advantage of this limitation, to prevent unauthorized access to your drawing files.*

ADVANCED TOPICS

When you use Volo View Express to view a DWF file, the Save Target As command only lets you save a copy in DWF format.

## Displaying Additional Information

When you choose About WHIP! in the WHIP! shortcut menu, WHIP! displays a dialog box providing additional information about the DWF file, including the title, author, keywords, and copyright custom property, all as specified in the AutoCAD Properties dialog box of the original drawing file. The dialog box also shows the name of the original AutoCAD DWG file as well as the date and time it was created and last modified. You can also click the WHIP! Home Page button to jump to the WHIP! page on Autodesk's web site.

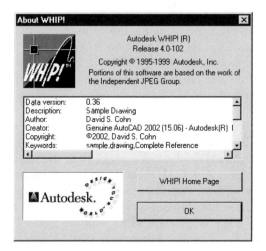

Volo View Express doesn't display any of this data within the browser window. When you use the full Volo View Express application, however, you can display some of the DWF file properties.

*LEARN BY EXAMPLE*
*Use your browser to view a DWF file and practice using the functions of either WHIP! or Volo View Express. You'll find links to sample web pages containing embedded DWF files on the Chapter 26 page of the companion web site.*

To perform this exercise, you should have already installed one of Autodesk's DWF viewers on your computer. The companion web site includes links to the WHIP! and Volo View Express pages on the Autodesk web site so you can download the latest version of either of these viewers.

When you view the sample web pages, practice using the various functions of the WHIP! or Volo View Express viewer. If you are using WHIP!, notice that the viewer starts in Pan mode, indicated by the hand icon. If you're using Volo View Express, right-click and select Pan mode. To pan the image, click-and-drag the mouse.

Right-click and switch to Zoom mode. In Volo View Express, choose Zoom | In/Out. The cursor is now a magnifying glass icon. To increase the magnification, click-and-drag the mouse toward the top of the screen. To decrease the magnification, click-and-drag the mouse toward the bottom of the screen. If your mouse has a wheel, experiment by rolling the wheel to zoom in and out.

To display a view that was saved as part of the DWF file, right-click and select Named Views from the shortcut menu. In the Named Views dialog box, click one of the named views. Notice that the DWF image immediately changes to the saved view. Click INITIAL to return to the original view. To close the Named Views dialog box, click the Close icon (the X) in the upper-right corner of the dialog box.

You can experiment with the other viewer options. Notice that when you move the mouse over some areas of the image, the icon changes to a pointing hand, indicating the presence of a hyperlink. Try jumping to these hyperlinks. If you're using WHIP!, simply click the link. If you're using Volo View Express, double-click the hyperlink. Use your web browser's Back button to return to the original web page.

The original DWG file is also saved in the same directory on the companion web site. If you're using WHIP!, you can right-click and use the Save As function to save a copy of the original drawing file to your hard drive.

## Dragging-and-Dropping Drawings from the Web

As mentioned earlier, when you use WHIP! to view a DWF file or a web page containing an embedded DWF file, if the DWG file used to create the DWF file is located in the same directory as the DWF file, you can either use drag-and-drop to open the drawing in AutoCAD or insert the drawing from the Web into your current drawing. In order to use this functionality, the DWF file needs to be displayed using the WHIP! viewer; this capability is not possible with Volo View Express. You should also have AutoCAD loaded and visible on your monitor.

To open a drawing by using drag-and-drop:

1. Press and hold down the CTRL and SHIFT keys.

2. Click the DWF image and drag it into AutoCAD.

3. Release the mouse button and the CTRL and SHIFT keys.

**Caution**    *Because this feature of the WHIP! viewer currently works only when running AutoCAD in single document mode, in order to be able to open a drawing by using drag-and-drop, you must have one—and only one—drawing currently open in AutoCAD. This restriction doesn't apply, however, when inserting a drawing using drag-and-drop.*

To insert a drawing by using drag-and-drop:

1. Press and hold down the CTRL key.

2. Click the DWF image and drag it into AutoCAD.

3. Release the mouse button and the CTRL key.

ADVANCED TOPICS

The DWG file must be in the same directory as the DWF file for the drag-and-drop feature to function.

*LEARN BY EXAMPLE*
*You can practice using drag-and-drop to open or insert a DWG file. If you are using WHIP! as your DWF viewer, you can use drag-and-drop with any of the sample web pages on the companion web site that contain DWF files.*

To open a drawing file in AutoCAD by using WHIP!'s drag-and-drop capabilities, you need to have both AutoCAD and your browser open and visible. To open a drawing using drag-and-drop, first make sure that there is one and only one drawing currently open in AutoCAD. Then, press and hold down the CTRL and SHIFT keys, click-and-drag the DWF image in your browser into the AutoCAD drawing window, and then release. AutoCAD must first close the current drawing before opening the new one. You may need to save changes before proceeding.

To insert a drawing file as a block, press and hold down the CTRL key, click-and-drag the DWF image into the AutoCAD drawing window, and then release. AutoCAD prompts you to complete the block insertion. You will need to specify the insertion point, scale factors, and rotation angle.

# Publishing Drawings to the Web

When you are ready to publish your DWF files on the Internet, you need to copy them to a web server and then create links so that others will be able to view your work. There are two different methods you can use to accomplish this:

- Create the necessary HTML documents using the appropriate software
- Let AutoCAD create the web pages for you using the Publish to Web wizard

AutoCAD's Publish to Web wizard is certainly the easiest way to create web pages containing DWF files and other content directly from your drawing files. As you will learn later in this chapter, however, creating your own HTML documents provides much more flexibility and originality.

## Using the Publish to Web Wizard

The Publish to Web wizard lets you quickly create DWF files and include them in a predesigned web page, similar to the one shown in Figure 26-7. With the Publish to Web wizard, you can create linked pages containing multiple drawings. You can also use Autodesk's new i-Drop technology to include drag-and-drop links to the original DWG files. I-Drop largely replaces the drag-and-drop capabilities of WHIP! that are missing from Volo View Express. You'll learn more about i-Drop later in this chapter.

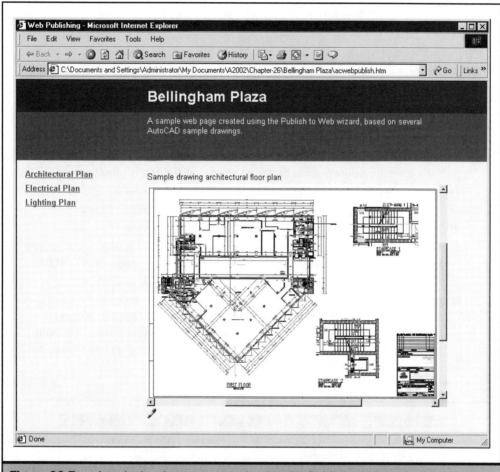

**Figure 26-7.** *A typical web page created using the Publish to Web wizard*

To start the Publish to Web wizard, do one of the following:

■ On the Standard toolbar, click Publish to Web.

■ From the File menu, choose Publish To Web (or choose Tools | Wizards | Publish To Web).

■ At the command line, type **PUBLISHTOWEB** and then press ENTER.

The wizard steps you through the process of creating a new web page. The first page of the wizard asks you if you want to create a new page or edit an existing one.

ADVANCED TOPICS

If you've previously created and saved a web page, you can use the wizard to edit an existing page. To create a new web page using the Publish to Web wizard:

1. On the Begin page, select Create New Web Page, and then click Next.

2. In the Create Web Page screen (shown in Figure 26-8), you must specify the name of the web page. When you create a new web page, the page and all of its configuration files are stored in a directory on your computer for future editing and posting to the Internet. The name of the directory in which these files are created is the same as the name you specify for the web page. You can choose the location (the parent directory) where this new folder is created. You can also provide a written description that will appear at the top of your new web page. When you have finished providing this information, click Next.

3. On the Select Image Type page, you select the type of image files to be created on your web page. The wizard provides three options:

   ■ **DWF**   Creates DWF files. Since the wizard saves DWF files in the newer format, in order for others to view the resulting web page, they will need to have Volo View Express installed on their system.

   ■ **JPEG**   Creates JPEG image files. JPEG files are raster images. You can specify the image size by selecting from an adjacent drop-down list. Although they don't require an additional viewer, the compression mechanism used to reduce their file size makes JPEG files less suitable for large drawing files.

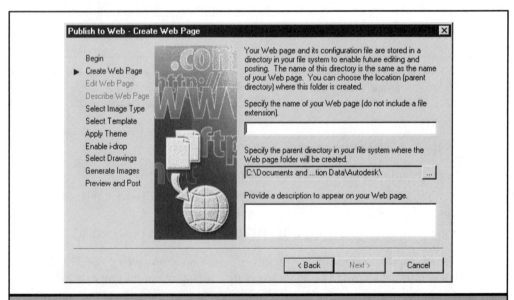

**Figure 26-8.**   *Specify the name, location, and description of your new web page.*

■ **PNG** Creates Portable Network Graphics (PNG) image files. Although these are also raster images, they may be more suitable for larger AutoCAD drawings. Again, you can specify the image size by selecting from the adjacent drop-down list.

After you have specified the image type, click Next.

4. On the Select Template page (shown in Figure 26-9), you select a template to be used to create the web page. The template determines the layout of the images in your web page. The wizard provides four options:

■ **Array of Thumbnails** Creates a page with an array of JPEG thumbnail-size images. Each thumbnail image is linked to a larger version of the image in the format you specified in the previous step.

■ **Array Plus Summary** Creates a page similar to the Array of Thumbnails, but with the addition of summary information about the image the mouse is currently hovering over. The summary information includes data from the Summary tab of AutoCAD's Drawing Properties dialog box.

■ **List of Drawings** Creates a web page containing a list of drawings, along with an image frame. Selecting a drawing from the list updates the image frame to display the image of the selected drawing in the format you specified in the previous step.

■ **List Plus Summary** Creates a page similar to the List of Drawings, but with the addition of summary information about the drawing currently displayed in the image frame.

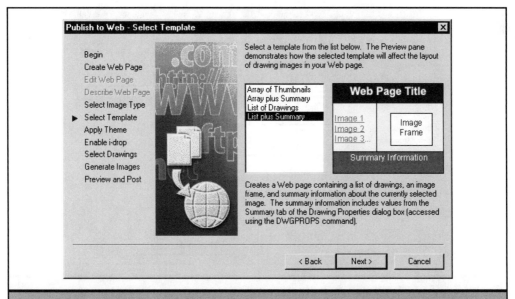

**Figure 26-9.** *Select one of the templates to control the layout of the web page.*

After you select the template, click Next.

5. On the Apply Theme page, you can select a theme from a drop-down list. The theme simply determines the colors and fonts used on the resulting web page, and Autodesk provides seven such themes. As you select a theme from the list, the dialog box displays a preview of its appearance. After you select the desired theme, click Next.

6. On the Enable i-Drop page, you can enable Autodesk's new i-Drop technology on your web page. If you select the Enable i-Drop check box, the resulting web page will include an i-Drop icon. Visitors can then click-and-drag this icon into an open AutoCAD drawing on their system to insert the DWG file associated with the selected drawing as a block. After making your choice, click Next.

**Note** *You cannot use i-Drop to open a drawing file from a web site nor can you insert a drawing file containing external references. You should bind external references to the drawing before using the Publish to Web wizard.*

7. On the Select Drawings page (shown in Figure 26-10), you specify the drawings and layouts you want included in the resulting web page. You can also specify a label and description to be included with each drawing. You can add as many drawings as you wish to the list and change the order of the list. When you are finished adding drawings and arranging the list, click Next.

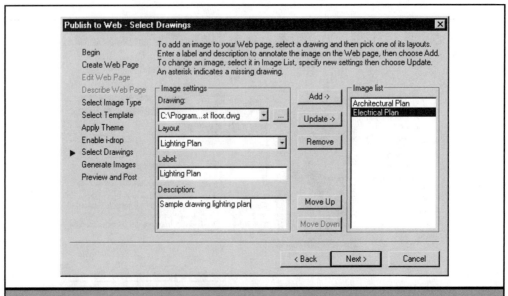

**Figure 26-10.** *Select and organize the drawings you want included on your web page.*

8. On the Generate Images page, you can specify whether you want AutoCAD to regenerate the images for all the drawings you selected or only those images that have changed. Then, click Next. At this point, the wizard actually generates the web page. You can't undo this operation once it has started.

9. After AutoCAD has created all the necessary files, the wizard displays the Preview and Post page. This page provides three buttons:

   ■ **Preview**   Displays the web page in your browser so you can preview it (as shown in Figure 26-7).

   ■ **Post Now**   Displays a Posting Web dialog box, a file dialog box you can use to publish the web page and its related files to the Internet, or to a folder on your computer or local area network. (The most common way to post the files to the Internet is to save them to an FTP site. You may wish to refer to the section in the previous chapter about saving drawings to the Internet.) Depending on the speed of your Internet connection and the number of drawings in your web page, transferring the files can be a lengthy process. After posting the files, AutoCAD displays a message box indicating that the transfer operation was successful.

   ■ **Send Email**   Becomes available after you've posted your web page. When you click this button, AutoCAD launches your e-mail software and prepares a message containing a hyperlink to the new web page. The recipient can click this link to jump directly to the new page.

10. If you wish to make changes to the web page, you can click the Back button and return to any previous step, make the desired changes, and then repeat the steps on the Preview and Post page. If you are satisfied with the new web page, click Finish to close the wizard.

*Remember that you can also use the Publish to Web wizard to edit any web page you previously created using the wizard. When editing an existing web page, the wizard provides tools to help you select the page you want to edit.*

# Creating Your Own HTML Documents

While the Publish to Web wizard is certainly the easiest way to create web pages containing DWF files, it is not very flexible. Because the files created by the wizard make extensive use of JavaScript, they are difficult to modify to match the appearance of other pages on your web site. If you want complete control over the appearance of your web pages and the way the DWF files are integrated into your web site, you'll need to create the HTML documents yourself.

There are two basic methods for making DWF files available in your HTML documents:

■ Create a link to the DWF file from an HTML document

■ Embed the DWF file in the HTML document

## Linking DWF Files to an HTML Document

Linking a DWF file to an HTML document causes the DWF file to appear in the document as a link. When you click the link, the DWF file fills the entire browser window, similar to when you use WHIP!'s Full View option. To add a link to a DWF file, first create the content of your HTML document (using whatever software you normally use to create HTML), and then add the link.

The link to the DWF file is created in the HTML document by using an HREF tag, which is the primary method for specifying any type of file that is not directly viewable by a web browser on a web page.

The HREF tag uses the HTML <A> tag. To link to a DWF file, you add the following tag in your HTML document:

```
<A HREF="DWF-URL">text</A>
```

where *DWF-URL* is the URL of the DWF file, and *text* is the text that actually appears as the link in the HTML document. For example, to provide a link to a DWF file located at http://www.myserver.com/embassy.dwf, with the text Embassy Second Floor Plan displayed as the link in the HTML document, you would use the following HREF tag:

```
<A HREF="http://www.myserver.com/embassy.dwf">Embassy Second Floor Plan</A>
```

When you view the resulting HTML document, the words Embassy Second Floor Plan appear as a hyperlink. Clicking that link causes the browser to display the DWF file.

A URL can include additional information. In the case of DWF files, the options can be used to tell the browser how to display the DWF file. The URL and options take the format

```
http://www.company.com/path/file.dwf?option=value&option=value
```

You can find a complete list of options on the companion web site.

*LEARN BY EXAMPLE*
*Open an HTML document in your web browser to see how easy it is to add a link to a DWF file. You will use one of the sample pages on the companion web site.*

Most people who create web pages learn many HTML tricks by looking at the source code of other web pages. Because HTML documents are stored as ASCII files, you can easily look at how the documents are actually created, by using these steps:

1. In your web browser, display the main Chapter 26 page on the companion web site, and then click Embassy-Link to display a web page containing a link to a DWF file. The page should appear similar to Figure 26-11.

2. From your web browser's View menu, choose Source or Page Source. The source code appears in Windows Notepad (if you're using Microsoft Internet Explorer), as shown in Figure 26-12, or in the Netscape Navigator Source viewer.

Notice that the third line from the bottom contains the HREF tag for the DWF file link. In your web browser, click the link to display the DWF file. You can also click any of the links in the DWF file to view the building elevations. Use your browser's Back button to return to the EMBASSY.HTM file.

## Embedding DWF Files into an HTML Document

Embedding a DWF file into an HTML document takes just a bit more effort than simply including a link, but the results usually are much more appealing. When the resulting HTML page is displayed in a browser, the DWF file appears as part of the page, occupying the space that you specify.

The method used is similar for both WHIP! and the Volo products, the major difference being the classid (which distinguishes various ActiveX controls from each other) and the additional parameters available for Volo View and Volo View Express.

*Remember that Volo View Express can also display DWG and DXF files. You can use methods similar to the ones described in this section to embed drawings and DXF files in your web page, but doing so will prevent them from being visible to those who are using WHIP! to view your web page.*

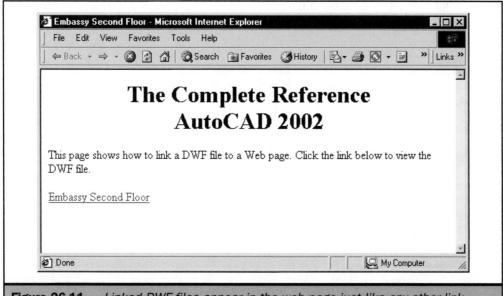

**Figure 26-11.** *Linked DWF files appear in the web page just like any other link.*

ADVANCED TOPICS

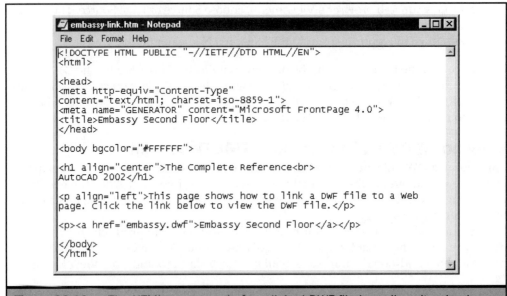

**Figure 26-12.** *The HTML source code for a linked DWF file is really quite simple.*

A typical snippet of HTML used to embed a DWF file into a web page for use with WHIP! might look like the following:

```
<object
id="embassy"
classid="clsid:B2BE75F3-9197-11CF-ABF4-08000996E931"
codebase="ftp://ftp.autodesk.com/pub/whip/whip.cab#version=4,0,42,102"
width="600"
height="400">
<param name="Filename" value="embassy.dwf">
<param name="NamedView" value="STUDY">
<param name="LayersOn" value="WALLS,DOORS,WINDOWS">
<param name="LayersOff" value="LIGHTS">
<param name="UserInterface" value="on">
<param name="BackColor" value="255">
<embed
name="embassy"
pluginspage="http://www.autodesk.com/products/whip"
width="600"
height="400"
src="embassy.dwf"
namedview="STUDY"
layerson="WALLS,DOORS,WINDOWS"
layersoff="LIGHTS"
```

```
userinterface="on"
backcolor="255">
</object>
```

A similar snippet for viewing the DWF file using Volo View Express might look like the following:

```
<object
id="embassy"
classid="clsid:8718C658-8956-11D2-BD21-0060B0A12A50"
width="600"
height="400">
<param name="src" value="embassy.dwf">
<param name="NamedView" value="STUDY">
<param name="LayersOn" value="WALLS,DOORS,WINDOWS">
<param name="LayersOff" value="LIGHTS">
<param name="UserMode" value="PAN">
<param name="BackgroundColor" value="#FFFFFF">
<embed
width="600"
height="400"
src="embassy.dwf"
NamedView="STUDY"
LayersOn="WALLS,DOORS,WINDOWS"
LayersOff="LIGHTS"
UserMode="PAN"
BackgroundColor="#FFFFFF">
</object>
```

Notice the similarity of these two pieces of code. The only major differences are the classid (which is different for the two different viewers), the fact that the DWF file to be embedded is referenced in the Volo View code using a parameter called src rather than the FileName parameter used by WHIP!, a Volo View parameter called UserMode (instead of UserInterface when using WHIP!), and a Volo View! parameter called BackgroundColor, which uses a hexadecimal color definition (instead of WHIP!'s backcolor, which uses a decimal color number).

**Tip**   *Volo View Express will display the DWF file even if you use the WHIP! code, although the parameters that use different names will be ignored. This simple trick has the advantage of enabling the auto-download capability of WHIP! for those visitors to your web page who lack a DWF viewer.*

ADVANCED TOPICS

In both cases, the DWF file is not actually embedded in the HTML document. Rather, you place a reference to the DWF file, similar to the one that simply links to the DWF file, within the HTML file. One of the reasons that this method is a bit more complex is that Microsoft Internet Explorer and Netscape Navigator use different methods to accomplish this. With Internet Explorer, you specify the reference by using the HTML <object> tag. For Navigator, you must use the <embed> tag. Because you want your document to be available to users of either browser, you need to use both methods, effectively specifying the DWF file twice. Navigator ignores the Microsoft tags, and Internet Explorer ignores the ones used by Netscape, which makes your task of accommodating both browsers easier, although a bit more time-consuming.

Notice that the listing begins and ends with the <object> tag, and that Netscape's <embed> tag is actually nested inside the <object> tag. Lots of information is included twice: once for each browser. After you understand what each tag means, embedding a DWF file in an HTML page is quite easy.

**Tip** *Save the preceding HTML code snippets to a separate HTML file. Then, whenever you want to embed a DWF file in an HTML document, you can copy the code into that document, and then modify the tags as needed.*

While some of the tags are required in order to tell the web browser what you want it to display, many of the other tags are optional and are used to modify the way the DWF file is displayed in the browser. For example, you can specify the size of the area in which the DWF file is displayed in the HTML document. You can also turn specific layers on and off or use the UserMode tag to explicitly enable Pan mode so that when the page first loads, Volo View Express behaves the same as WHIP!

Since Autodesk is no longer developing WHIP!, the rest of this section discusses only the tags for Volo View Express. The tags for WHIP! are very similar. A complete reference to all tags used by both viewers can be found on the companion web site.

The following parameters are part of the <object> tag and are used with Microsoft Internet Explorer:

- **id**   A unique identifier that you assign, which can be referenced by JScript or VBScript. Don't confuse this tag with the classid tag.

- **classid**   This identifier is specific to Volo View and should not be changed. The value specified is used to distinguish Volo View from other ActiveX controls.

The following parameters are part of the <object> and <embed> tags and are used with both Microsoft Internet Explorer and Netscape Navigator:

- **width and height**   These two values specify the width and height of the area within the HTML page that is used to display the DWF file. As presented in the previous code listing, these values are specified in pixels, but they can also be represented as a percentage of the available browser window. You can change these values as needed.

The following parameters are specific to Volo View. When using the <object> tag, you must specify param name= and the name of the parameter, and then value= and the value for the parameter. For example:

```
<param name="UserMode" value="PAN">
```

When using the <embed> tag, you only need to specify the parameter name and value. For example:

```
UserMode="PAN"
```

- **src**  Required. The name of the DWF, DWG, or DXF file to be included on the web page.

- **BackgroundColor**  Optional. Specifies the color of the drawing background as a hexadecimal number.

- **BorderStyle**  Optional. Specifies whether the Volo View window has a border (1) or not (0, the default).

- **FontPath**  Optional. Specifies the path where drawing-dependent fonts reside. To specify more than one font path, separate the paths with semicolons.

- **GeometryColor**  Optional. Specifies the color for all geometry in your drawing as a hexadecimal number.

- **HighlightLinks**  Optional. Specifies whether hyperlinks are highlighted (True) or not (False).

- **LayersOff**  Optional. Specifies the name of a layer that you want set to Off (hidden). You must specify a layer already stored in the DWF, DWG, or DXF file. If the specified layer has not been defined in the drawing, the option is ignored. To specify more than one layer, separate the layer names with commas.

- **LayersOn**  Optional. Specifies the name of a layer that you want set to On (shown). Use this option to display layers that were turned off when the DWF, DWG, or DXF file was saved. You must specify a layer already stored in the DWF, DWG, or DXF file. If the specified layer has not been defined in the drawing, the option is ignored. To specify more than one layer, separate the layer names with commas.

- **Layout**  Optional. Specifies an initial layout for your drawing file. You must specify a layout previously saved in the drawing file, or the option is ignored.

- **NamedView**  Optional. Specifies an initial view for your drawing. The initial view can be specified as a named view previously saved in the drawing file, or as drawing coordinates. If a specified named view has not been defined in the drawing file, the option is ignored. If you use a named view, note that the values are case sensitive and that AutoCAD saves named views in uppercase. To specify the initial view using drawing coordinates, use the following format:

```
<param name="NamedView" value="VIEW=(minX,minY)(maxX,maxY)
```

where the *minX,minY* coordinate is the location of the lower-left corner and the *maxX,maxY* coordinate is the location of the upper-right corner. You can use either a named view or a coordinate-defined view to specify the initial view, but not both.

- **PrintBackgroundColor**   Optional. Specifies the background color of the printed drawing as a hexadecimal number.
- **PrintGeometryColor**   Optional. Specifies the color of the geometry of the printed drawing as a hexadecimal number.
- **ProjectionMode**   Optional. Specifies the perspective used in 3D Orbit mode, as either Parallel (default) or Perspective.
- **ShadingMode**   Optional. Specifies a shading option for 3D Orbit mode, as Wireframe, Flat, or Gouraud (default).
- **SupportPath**   Optional. Specifies the path where drawing-dependent files reside. To specify more than one support path, separate the paths with semicolons.
- **UserMode**   Optional. Specifies the initial mode of operation for Volo View as Pan, Zoom, ZoomToRect, Orbit, Sketch, Text, Line, Polyline, Arc, Spline, Rectangle, Circle, Ellipse, Cloud, Note, Callout, MeasureDistance, MeasureMultiDistance, MeasureArea, or Select. (*Note:* Not all of these values are available for Volo View Express.)

Since there is no default mode in Volo View Express as there is in WHIP!, visitors to your web site may not realize that they can right-click to pan, zoom, and perform other operations. You can, however, use the UserMode tag to set a default mode when you first load the DWF file. It is also possible to incorporate Volo View Express interface controls within your web site. This requires a more in-depth understanding of the Volo View application programming interface (API), however, and also requires programming in Visual Basic, C++, or some other language and then incorporating that programming into your web site in the form of scripts. These capabilities are beyond the scope of this book. You can find more information on Autodesk's Volo View web site.

Since there is no auto-download capability when using Volo View Express, you should include a link to the Volo View Express download page on the Autodesk web site. Although the full Volo View Express application is over 25MB, Autodesk has created a DWF-only version of Volo View Express that is approximately 10MB.

*LEARN BY EXAMPLE*
*Open an HTML document in your web browser to see how to embed a DWF file in a web page. You will use one of the sample pages on the companion web site.*

Again, one of the best ways to learn how to create an HTML document with an embedded DWF file is to look at someone else's source code. In your web browser, display the main Chapter 26 page on the companion web site and then click Embassy-Drawing to display a web page containing a DWF file embedded in an HTML document. The page should appear similar to Figure 26-13. You can use the browser's view source function to look at the actual source code.

Since Volo View Express lacks WHIP!'s drag-and-drop capability, you may want to incorporate Autodesk's i-Drop technology on your web pages along with your DWF files. The remainder of this chapter explains how to add i-Drop to your HTML documents.

## Adding i-Drop Content to Your Web Page

Autodesk's new i-Drop technology enables web developers to add drag-and-drop capabilities to their web sites. Unlike WHIP!, which only worked with DWF and DWG files, i-Drop works with all versions of AutoCAD and AutoCAD LT beginning with

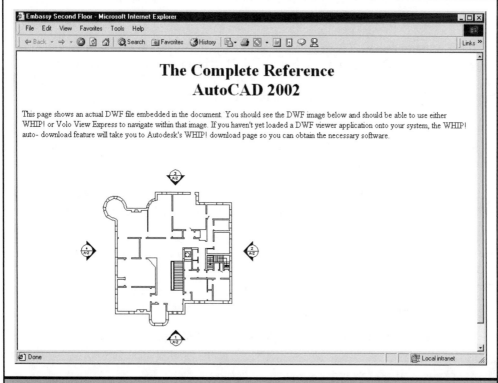

**Figure 26-13.**    *The embedded DWF file can be viewed and navigated directly in the web page.*

ADVANCED TOPICS

2000i, and all applications based on those platforms (such as Architectural Desktop and Mechanical Desktop), as well as Autodesk VIZ and Autodesk Inventor. AutoCAD users can drag-and-drop drawing files as blocks into any open AutoCAD drawing. Autodesk VIZ users can drag-and-drop MAX files into their models. And Inventor users can drag-and-drop parts, assemblies, and drawings into open Inventor files.

> **Caution**   *In order to use i-Drop, you must use Microsoft Internet Explorer as your web browser. i-Drop does not work with Netscape Navigator.*

When you visit a web page that incorporates i-Drop, you can insert a DWG file as a block into an AutoCAD drawing by doing the following:

1. Open the AutoCAD drawing into which you want to insert the block.

2. In your web browser, go to the web page containing the content you want to insert. Notice that when your cursor hovers over i-Drop-enabled content, a special i-Drop cursor appears.

3. Click-and-drag the content into the open AutoCAD drawing.

4. Complete the block insertion as you normally would.

Integrating i-Drop into your web site requires just a bit more coding than embedding a drawing in an HTML document, but not much. The main difference is the inclusion of an XML (eXtensible Markup Language) object file. For each (DWG) object you intend to publish on an HTML page, you must have the following:

- The file or files you want to have transferred (such as a DWG file)

- An image that serves as the trigger for transfer (such as a JPG or GIF 4image)

- The XML object file describing the embedded object

The information and how it is presented is defined in the i-Drop package file. The i-Drop package file is a simple XML file (an ASCII text file) that does two things:

- It describes what you see—the graphic representation to be displayed on the web page to represent the source or trigger of the i-Drop operation (the proxy element)

- It describes what you get—the data files that are to be downloaded to the application once the i-Drop operation is performed

A typical i-Drop package file might look like the following:

```
<package xmlns="x-schema:idrop-schema.xml">
    <proxy defaultsrc="chair.gif">
    </proxy>
    <dataset>
```

```
            <datasrc clipformat="CF_IDROP.dwg">
                <datafile src="chair.dwg"/>
            </datasrc>
        </dataset>
    </package>
```

In the preceding listing, the second line specifies the proxy object, a GIF image file. On the resulting web page, placing the cursor anywhere within the image would initiate the i-Drop operation. The proxy image can be any of a growing number of file types, including bitmaps, DWF, and even animation files such as Shockwave or Metastream.

The code between the dataset tags comprises the data that will be downloaded. The clipformat defines the code as being AutoCAD drawing data, and the line below that specifies the file that will be downloaded, a DWG file.

You can specify several different types of formats, as shown in the following code example:

```
<package xmlns="x-schema:idrop-schema.xml">
    <proxy defaultsrc="chair.gif"></proxy>
    <dataset>
        <datasrc clipformat="CF_IDROP.dwg">
            <datafile src="chair.dwg"/>
        </datasrc>
        <datasrc clipformat="CF_IDROP.max">
            <datafile src="chair.max"/>
            <xreffile src="material.bmp"/>
        </datasrc>
    </dataset>
</package>
```

This listing includes two different clipformats, one for AutoCAD and one for 3D Studio. In addition, notice that the 3D Studio dataset includes both a MAX file and a bitmap used to define a material in that model. You should note that any number of files can be listed. There can be only one datafile, but any number of xreffiles. Thus, the same proxy on a web site could be used to download two different types of data: a 2-D or 3-D block for AutoCAD, and a fully textured 3-D model for 3D Studio.

In addition to the XML object file and the various related files for each object you want to publish, you will also need an XML schema, which is a set of constraints that define the XML tags in the i-Drop package file. The schema is like a legend that tells the browser what to do with your object file. You only need one schema file for your web site, and its contents are always the same. You can download a copy of the schema from **http://idrop.autodesk.com/files/idrop.xdr**.

Once you've got all of these building blocks, the last step is to create the HTML file that gets placed on your web server. The following example includes four files—

CHAIR.DWG, CHAIR.GIF, CHAIR.XML, and IDROP-SCHEMA.XML—plus the HTML file, all located in the same directory.

The following is a typical HTML snippet to include the i-Drop content on a web page:

```
<object
     name="idrop"
     classid="clsid:21E0CB95-1198-4945-A3D2-4BF804295F78"
     width="240"
     height="236">
     <param name="background" value="background.jpg">
     <param name="proxyrect" value="0,0, 240,236">
     <param name="griprect" value="0, 0, 240, 236">
     <param name="package" value="DATA/chair/chair.xml"/>
     <param name="validate" value="1">
     <embed
         width="240"
         height="236"
         background="background.jpg"
         proxyrect="0,0, 240,236"
         griprect="0, 0, 240, 236"
         package="DATA/chair/chair.xml"
         validate="1"
         src="background.jpg"
         name="idrop">
     </embed>
</object>
```

Notice that this snippet of code looks somewhat similar to the code used to embed a DWF file on a web page. It includes an <object> tag and parameters for Microsoft Internet Explorer, and an <embed> tag with an identical set of parameters for Netscape Navigator.

Also notice the inclusion of a file named BACKGROUND.JPG. This is a transparent image file that is used as a *skin* or frame around the CHAIR.GIF file. For simplicity, this file is located in the same directory as the other files.

The following list describes the parameters present in this code snippet:

- **Package** Required. Specifies the XML i-Drop package file URL that in turn describes the display and contents for this instance of the i-Drop control.

- **Background** Optional. Specifies a background skin displayed by the i-Drop control. This can be used to frame the proxy display, so that it is obvious to the user that this is an i-Drop object.

- **ProxyRect** Optional. Specifies a rectangular area in which to display the proxy image. When using a background, you may want to limit the display area of the proxy image or hosted control. This rectangle defines the area where the user can interact with the hosted control.

- **GripRect**  Optional. Specifies the area between the i-Drop control limits and the ProxyRect, which should be smaller.
- **ValidateXml**   Optional. Specifies whether the i-Drop control performs validation. If 1 (on), the control will validate the XML package file, alerting you to possible errors in its structure. If 0 (off, default), no validation is performed.

To insert i-Drop content, you hover over the trigger (for example, the image file) on the web site. The cursor changes to an i-Drop cursor. You click-and-drag the i-Drop cursor into an open AutoCAD drawing. This transfers the XML package file to the application. The application parses the package file, looking for its clipboard format and any files that are referenced. The application then goes back to the server to the referenced data, and those data files are downloaded.

For those visitors who don't yet have the i-Drop ActiveX control on their systems, you can add the control to your web page so that it auto-downloads to the visitor's browser the first time they visit your page. To do this, include the codebase URL along with the classid object in your HTML code, as follows:

```
<object classid="clsid:21E0CB95-1198-4945-A3D2-4BF804295F78"
codebase=http://idrop.autodesk.com/files/idrop.cab>
```

So the HTML snippet would become:

```
<object
name="idrop" classid="clsid:21E0CB95-1198-4945-A3D2-4BF804295F78"
codebase=codebase=http://idrop.autodesk.com/files/idrop.cab
width="240"
height="236">
```

followed by the various parameters. Then, the first time a visitor attempts to view your page that contains i-Drop content, they will have an opportunity to download the i-Drop ActiveX control.

*LEARN BY EXAMPLE*
*Open an HTML document in your web browser to learn how to use i-Drop. You will use one of the sample pages on the companion web site.*

In your web browser, display the main Chapter 26 page on the companion web site and then click i-Drop Examples to display a page containing links to a simple HTML document containing i-Drop content. The page should appear similar to Figure 26-14.

You can use your browser's view source function to look at the actual source code. When you move your mouse over the image of the chair, the i-Drop cursor appears. You can then drag-and-drop a 3-D model of the chair into an open AutoCAD drawing.

You will also find links on the companion web site to download the XML and image files used by this example as well as links to other web sites that use i-Drop content. Feel free to visit these other sites to experiment with the power of i-Drop.

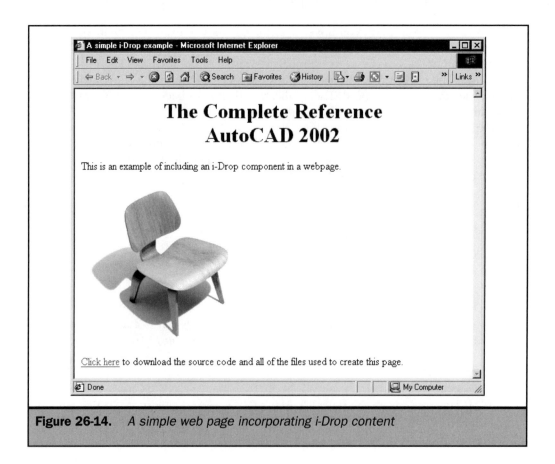

**Figure 26-14.**　*A simple web page incorporating i-Drop content*

# Chapter 27

## Working with External Databases

Y ou've already learned about other methods that you can use to store and retrieve nongraphical information in an AutoCAD drawing. For example, in Chapter 15, you learned how to use attributes to attach text information to blocks and extract that information for use in some other program. You've also seen how AutoCAD can attach extended entity data (xdata) to objects within the drawing. For example, in the previous chapter, you learned that AutoCAD attaches URLs to objects using xdata.

When you attach data by using these methods, the data is stored entirely within the AutoCAD drawing. While this works fine, it is not necessarily the best way to store and retrieve nongraphical information. AutoCAD is, after all, a design and drafting program. Other programs are available—database programs such as Microsoft Access and Oracle— that are much more adept at storing, retrieving, and manipulating data. These programs store data in databases. Although these databases are not part of AutoCAD, you can link them to your AutoCAD drawings to get the best of both possible worlds.

In this chapter, you'll learn about the AutoCAD database connectivity features and how to link AutoCAD to external databases. This chapter explains how to do the following:

- Connect to external databases
- View table data from within AutoCAD
- Work with table data
- Create links to graphical objects
- Create labels from database values
- Search for data and linked objects
- Create selection sets based on data and objects

## Understanding Database Connectivity

The AutoCAD database connectivity feature links AutoCAD to an external database, so that you can work with nongraphical information stored outside of your drawing. The other methods you learned about in previous chapters store nongraphical data within the drawing itself. While any of these methods is fine when dealing with small amounts of data, when working with large amounts of data, AutoCAD's methods other than database connectivity have several drawbacks:

- The data stored in the AutoCAD drawing increases the overall drawing file size, thus reducing performance.
- The data must be extracted to external applications to create reports, thus severing the link between the data and the drawing objects.
- In the case of xdata, AutoCAD can attach a maximum of only 16K of information to each object.

AutoCAD's database connectivity feature does not have any of these limitations. By using AutoCAD's database connectivity feature, you can link objects in an AutoCAD drawing to an external database.

External databases store data in one or more tables. At its most basic, you can think of a table as a spreadsheet or grid consisting of rows and columns. Each row represents a single record in the database, while the columns within each row represent a single field within a database record. Most databases allow you to define relationships between multiple tables. For example, two tables can share a common field, such as a part identification number. Using a relational database, you can then join these two tables by using the common field, so that you don't have to maintain duplicate information in separate tables.

Database links within an AutoCAD drawing are simply pointers to a database table that references data from one or more records in that table. You can also attach labels to your drawing that display data from selected table fields as text objects within your drawing.

## Changes from Previous Versions of AutoCAD

AutoCAD's database connectivity features have changed considerably since R14. Most of the AutoCAD SQL Environment (ASE) terms and commands have been changed. All of the ASE commands have been eliminated and replaced with a single DBCONNECT command that activates database connectivity features. When these features are active, AutoCAD displays the dbConnect Manager, a separate, nonmodal window that is dockable and resizable. The program also displays a dbConnect menu on the main menu bar whenever the dbConnect Manager is available. You'll learn more about these features later in this chapter.

Because AutoCAD stores links in a different format than pre-AutoCAD 2000 versions, you will need to convert your links if you plan to use them with newer versions of AutoCAD. You also need to create a configuration file that points to the data source referenced by the links in your older drawings.

When you open a drawing from an earlier release that contains ASE links, AutoCAD attempts to perform an automatic conversion of the database information by establishing a one-to-one mapping of the data source components. If the automatic link conversion isn't successful, you can use the Link Conversion function to establish a mapping between the old data and the new AutoCAD data sources. You can also use the Link Template Properties dialog box to specify new data source parameters for a selected link template. See the AutoCAD 2002 *User's Guide* for more information about converting ASE links to AutoCAD 2000 format.

If you save an AutoCAD 2000 format drawing containing links to R13 or R14 format, however, AutoCAD 2002 automatically converts the links to the appropriate drawing format. You can't convert AutoCAD 2000 format links to R12 format.

ADVANCED TOPICS

AutoCAD's database connectivity feature provides the following functions:

- An external configuration program that enables AutoCAD to access data contained in external database programs
- A dbConnect Manager in which to associate links, labels, and queries with AutoCAD drawings
- A Data View window that displays records from a database table while using AutoCAD
- A Query Editor in which to construct, execute, and store SQL queries
- A migration tool that converts links and displayable attributes from a previous release of AutoCAD into the AutoCAD 2000 format
- A Link Select operation that creates selection sets based on queries and graphical objects

# Connecting to External Databases

AutoCAD's connectivity feature supports the following external database applications:

- Microsoft Access 97
- dBASE III and V
- Microsoft Excel 97
- Oracle 7.3 and 8
- Paradox 7
- Microsoft Visual FoxPro 6
- SQL Server 6.5 and 7

The descriptions of the procedures that you use to work with external databases are very similar, regardless of the database that you are using. The examples in this chapter are based on using a Microsoft Access 97 database and they utilize sample files provided on the companion web site. You don't need to have Access installed on your system to use these sample files.

**Note** *AutoCAD 2002 also includes a sample drawing and database file that you can use to learn more about the database connectivity features. In addition, you'll find a database connectivity lesson in the AutoCAD 2002 Learning Assistance program that uses another sample drawing. To follow along with the examples on the companion web site, you will need to configure your system to use the example files. To use the sample files that come with AutoCAD, you will need to repeat some of the steps in the following section to readjust your system.*

# Using the ODBC Data Source Administrator

Before you can access an external database from within AutoCAD, you must configure it using the Microsoft ODBC (Open Database Connectivity) and OLE DB program. This is done by using the ODBC Data Source Administrator, accessed from the Windows Control Panel. The configuration process creates a new *data source* that points to a collection of data and provides information about the drivers needed to access it.

**LEARN BY EXAMPLE**
*Use the ODBC Data Source Administrator dialog box to create a new data source. Before proceeding, copy the files ASEMP2002.DWG and ASEMP2002.MDB from the companion web site to your hard drive.*

Use the following steps to create a new data source, which will be used for the other examples in this chapter:

1. Display the Windows Control Panel by selecting Start | Settings | Control Panel.

2. In the Control Panel, double-click Administrative Tools and then double-click Data Sources.

3. In the ODBC Data Source Administrator dialog box, select the User DSN tab, and then click Add to create a new data source.

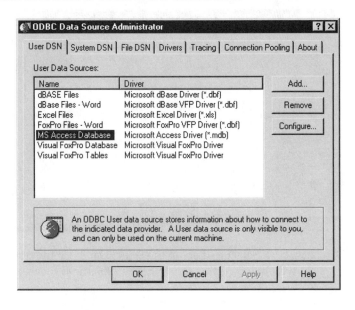

**ADVANCED TOPICS**

4. In the Create New Data Source dialog box, select Microsoft Access Driver (*.mdb), and then click Finish.

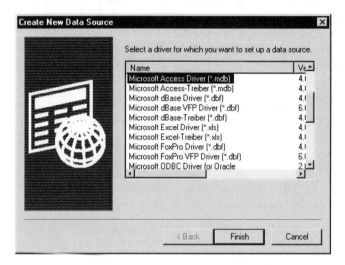

5. In the ODBC Microsoft Access Setup dialog box, in the Data Source Name box, type **mydata**.

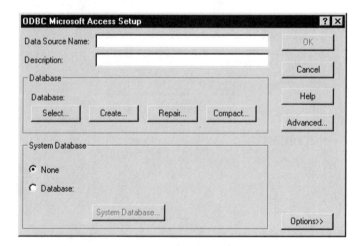

6. Under Database, click Select to display the Select Database file dialog box. Locate the directory in which you copied the ASEMPT2002.MDB file, select the file, and then click OK.

7. Click OK to close the ODBC Microsoft Access Setup dialog box, and then click OK again to close the ODBC Data Source Administrator dialog box.

8. Close the Control Panel.

## Configuring a Data Source

Once you have configured the Microsoft ODBC, you must also configure a data source for use by AutoCAD. AutoCAD 2002 includes several Microsoft Access sample database tables and a preconfigured direct driver (JET_DBSAMPLES.UDL) that you can use to work with these tables.

*LEARN BY EXAMPLE*
*Update the sample configuration file to work with the new data source. Open the drawing*
*ASEMP2002.DWG.*

Use the following steps to update the JET_DBSAMPLES.UDL configuration file so that you can work with the new data source you configured in the previous example:

1. Start AutoCAD's database connectivity feature by doing one of the following:

   ■ From the Tools menu, choose dbConnect.

   ■ At the command line, type **DBCONNECT** and press ENTER, or press CTRL-6.

   AutoCAD adds a new dbConnect pull-down menu to the menu bar.

2. From the dbConnect menu, choose Data Sources | Configure. (*Note:* You can also right-click Data Sources in the dbConnect Manager dialog box and then choose Configure Data Source from the shortcut menu.)

3. In the Configure a Data Source dialog box, under Data Sources, click jet_dbsamples, and then click OK.

ADVANCED TOPICS

4. In the Data Link Properties dialog box, select the Provider tab. Under OLE DB Provider(s), choose Microsoft OLE DB Provider for ODBC Drivers, and then click Next.

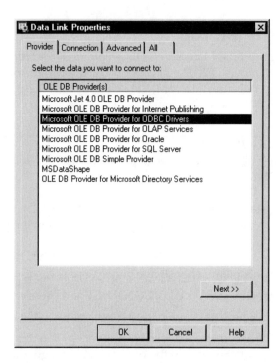

5. On the Connection tab, select Use Data Source Name, and then select mydata from the drop-down list. (If you don't see mydata in the list, click the Refresh button and then try again.)

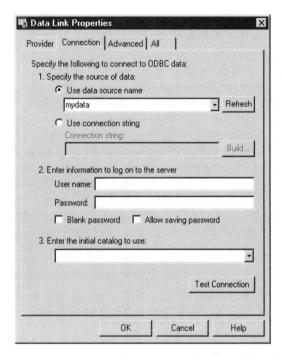

6. Click the Test Connection button to verify that the database connection is functioning.

7. Click OK to close the Microsoft Data Link dialog box, and then click OK again to close the Data Link Properties dialog box.

8. In the dbConnect Manager window, right-click the jet_dbsamples node, and then select Disconnect from the shortcut menu, if that selection is available.

9. Right-click jet_dbsamples again and select Connect from the shortcut menu.

You should now see three database tables in the dbConnect Manager window: Computer, Employee, and Inventory.

ADVANCED TOPICS

# Using the dbConnect Manager

The dbConnect Manager window, shown in Figure 27-1, contains a set of buttons and a tree view. You can use these buttons to view or edit a database table, create or execute a query, create a new link template, or create a new label template.

The tree view contains two sets of nodes: drawings and data sources. The drawing nodes show each open drawing. Within the drawing nodes, you can expand the tree for each drawing to show all the database objects—such as link templates, label templates, and queries—that are associated with the drawing. The data sources node shows all the configured data sources on your system. Each data source represents a database, and within each data source, you can expand the tree to display its tables.

All the various nodes and database objects in the dbConnect Manager have their own shortcut menu options. For example, you can use a shortcut menu to edit a database table. In addition, you can use the buttons to accomplish the following tasks:

- **View Table**   Opens a database table in Read-only mode. This button becomes available when you select a table, link template, or label template.

- **Edit Table**   Opens a database table in Edit mode. This button becomes available when you select a table, link template, or label template.

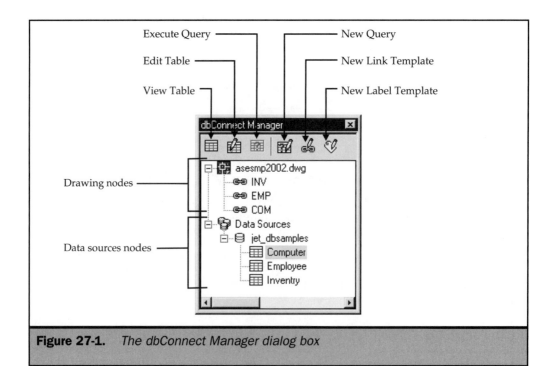

**Figure 27-1.**    *The dbConnect Manager dialog box*

- **Execute Query**   Executes the query selected in the tree view.

- **New Query**   Displays the New Query dialog box, in which you can create a new query, if you select a table or link template; or, displays the Edit Query dialog box, in which you can edit the query, if you select an existing query.

- **New Link Template**   Displays the New Link Template dialog box, in which you can create a new link template, if you select a table; or, displays the Link Template dialog box, in which you can edit the link template, if you select an existing link template.

- **New Label Template**   Displays the New Label Template dialog box, in which you can create a new label template, if you select a table; or, displays the Label Template dialog box, in which you can edit the label template, if you select an existing label template.

> **Note**   *If you right-click a blank area in the dbConnect Manager window, you can enable or disable docking of the window. You can also hide the window. Press CTRL-6 to display it again.*

## Viewing Table Data from Within AutoCAD

Once you have configured a data source, you can access its tables from within AutoCAD by displaying the Data View window. You can open the Data View window in either Read-only or Edit mode.

> **Note**   *When you open a table in Read-only mode, you can't add, delete, or edit any of its records. AutoCAD indicates that the table is read-only by displaying the Data View window with the cell backgrounds shown in the system 3-D objects button color (typically light gray), whereas when the window is open in Edit mode, the cell background is shown using the system window color (typically white).*

To open a table in the dbConnect Manager window, do one of the following:

- Select the table you want and then click View Table (to open the table in Read-only mode) or Edit Table (to open the table in Edit mode).

- Select the table you want, right-click, and then choose either View Table or Edit Table from the shortcut menu.

- Double-click the table to open the table in Edit mode.

> **Note**   *By default, AutoCAD opens tables in Edit mode when you double-click. You can change this default behavior by selecting Open Tables In Read-only Mode under dbConnect on the System tab of the Options dialog box.*

**ADVANCED TOPICS**

Online

EXAMPLES

*LEARN BY EXAMPLE*
*You can also open a table for viewing or editing by selecting the command from the*
*dbConnect menu on the AutoCAD menu bar.*

Open a table for viewing by doing the following:

1. From the dbConnect menu, choose View Data | View External Table.

2. In the Select Data Object dialog box, select Employee and then click OK.

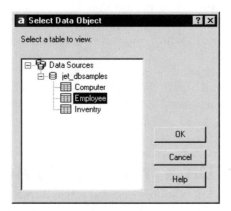

AutoCAD displays the Data View window, similar to the one shown in Figure 27-2.

# Using the Data View Window

The Data View window consists of a *grid* or spreadsheet-like view of the records in the database table, as shown in Figure 27-2. In addition to the grid, the window contains several navigation controls for moving through the record set, as well as its own set of buttons for accessing commands. For example, you can use the scroll bars along the right and bottom of the window to move through the records and fields. Additional buttons let you jump to the first or last record in the table, or move to the previous or next record.

When the Data View window is visible, a Data View menu is also added to the AutoCAD menu bar. As with the dbConnect Manager window, each of the various controls in the Data View window has its own shortcut menu. For example, you can use a shortcut menu to create a link between a record and an AutoCAD graphical object.

Displays either Link!, Create Attached Label, or Create Freestanding Label

View Linked Objects in Drawing

View Linked Records in Data View

AutoView Linked Objects in Drawing

AutoView Linked Records in Data View

Query

Print Data View

Data View and Query Options

Link Template list          Label Template list

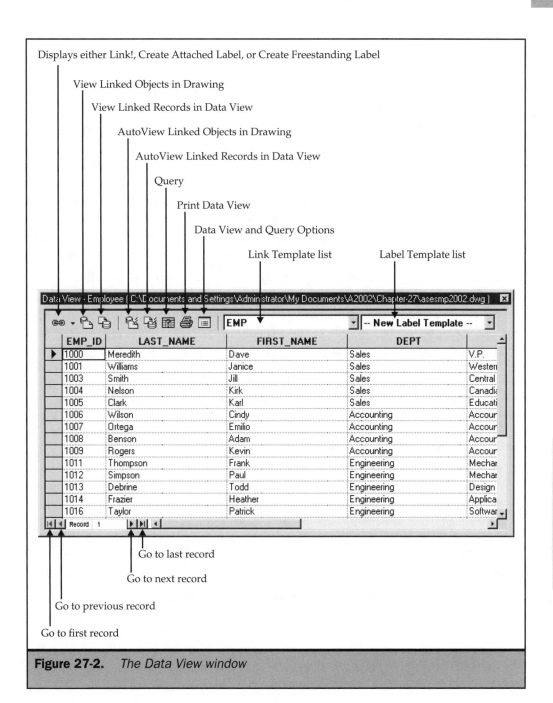

Go to last record

Go to next record

Go to previous record

Go to first record

**Figure 27-2.**    *The Data View window*

ADVANCED TOPICS

In addition, you can use the buttons in the Data View window to accomplish the following tasks:

- **Link!**  Creates a link to one or more graphical objects without creating a corresponding label.

- **Create Attached Label**  Creates a link to one or more graphical objects and creates a corresponding label.

- **Create Freestanding Label**  Creates a freestanding label that isn't associated with a graphical object.

- **View Linked Objects in Drawing**  Selects graphical objects in the current drawing that are linked to the currently selected records in the Data View window.

- **View Linked Records in Data View**  Selects records in the Data View window that are linked to the current selection set of graphical objects in the drawing.

- **AutoView Linked Objects in Drawing**  Displays linked objects in the current drawing automatically as you select records in the Data View window.

- **AutoView Linked Records in Data View**  Displays linked records in the Data View window automatically as you select graphical objects in the current drawing.

- **Query**  Displays the New Query, Query Editor, or Link Select dialog box, depending on the method used to open the Data View window. If the Data View window was opened to view or edit a database table, clicking this button displays the New Query dialog box. If the window was opened to return the results of a query, clicking this button displays the Query Editor dialog box. If the Data View window was opened to return the results of a Link Select operation, clicking this button displays the Link Select dialog box.

- **Print Data View**  Prints the contents of the Data View window to the current Windows system printer.

- **Data View and Query Options**  Displays the Data View and Query Options dialog box, in which you can specify various settings that affect the interaction and display of linked objects in the Data View window and the current AutoCAD drawing.

- **Link Template list**  Lets you select the current link template or create a new one.

- **Label Template list**  Lets you select the current label template or create a new one.

The grid itself contains the following elements:

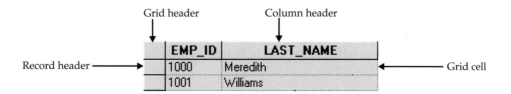

Grid header        Column header

| EMP_ID | LAST_NAME |
|--------|-----------|
| 1000 | Meredith |
| 1001 | Williams |

Record header ───────────▶                                    ◀─────── Grid cell

- **Column header**   Displays the field names. Selecting a column header selects all the records in the column.
- **Record header**   Selecting a record header selects that record.
- **Grid cells**   Displays an individual field. Selecting a grid cell selects that field of the given record.
- **Grid header**   Selecting the grid header selects the entire table.

You can easily adjust the way records are displayed in the Data View window by doing any of the following:

- Move or resize columns
- Hide columns from view
- Freeze columns so they stay in place when scrolling
- Change the alignment of text within columns
- Sort column data
- Change the appearance of fonts within the Data View window

You can move a column by selecting its column header and then dragging it to the desired location. To change the size of a column, select its grid line at the right side of the column header, and then drag it to the left or right to change the width of the column.

To hide a column, right-click its column header and then choose Hide from the shortcut menu. To display hidden columns, right-click any column header and then choose Unhide All from the shortcut menu. You can also freeze a column so that it is relocated to the left side of the Data View window and doesn't scroll as you move the horizontal scroll bar. To freeze a column, right-click its header and choose Freeze from the shortcut menu. To unfreeze columns, right-click any column header and choose Unfreeze All from the shortcut menu.

**Tip**   *You can only freeze contiguous columns. If the columns you want to freeze are not contiguous, first move them, and then freeze them.*

ADVANCED TOPICS

To align the text in one or more columns:

1. In the Data View window, select the columns whose text you want to align.

2. Right-click any column header, choose Align, and then select one of the following options to align the text:

   - **Standard**   Right-aligns numeric fields and left-aligns all other fields
   - **Left**   Left-aligns all the fields
   - **Center**   Center-aligns all the fields
   - **Right**   Right-aligns all the fields

*LEARN BY EXAMPLE*
*Sort the data in a selected column.*

In the dbConnect Manager window, in the data sources node, select the Employee table and then click the View Table button to display the Employee table in the Data View window. The Employee table appears in numerical order based on the employee ID number. Suppose, however, that you want to sort it based on the employees' names. To sort the data in the table:

1. Right-click any column header in the Data View window and choose Sort from the shortcut menu.

2. In the Sort dialog box, under Sort By, select LAST_NAME from the drop-down list.

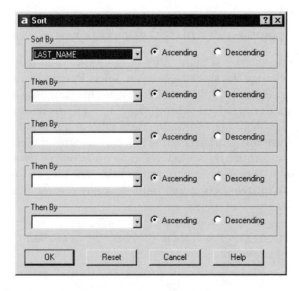

3. Make sure that the adjacent Ascending button is selected, and then click OK.

*When you apply a sort, you can sort on up to four additional fields.*

You can also apply different font settings to the Data View window. The settings you select affect all the text in the window. To change the Data View window font settings:

1. From the Data View menu (on the AutoCAD menu bar), choose Format.

2. In the Format dialog box, specify the font settings you want to use, and then click OK. (*Note:* You can also access this setting by selecting the grid header, so that the entire grid is highlighted, and then right-clicking.)

 *The formatting and sorting you apply within the Data View window is not saved, but rather is discarded when you open a new table or close the current table.*

# Working with Table Data

When you open a table in Edit mode, you can modify the data in the external database by editing the data in any grid cell or adding and deleting records in the database.

*LEARN BY EXAMPLE*
*Use the Data View window to modify data in the external database.*

Suppose Albert Fong's job title changes from Programmer to Application Engineer. You can change that information in the database from within AutoCAD:

1. In the dbConnect Manager window, in the data sources node, double-click the Employee table to open it in the Data View window in Edit mode.

2. In the Data View window, right-click the LAST_NAME column header and then select Find from the shortcut menu.

3. In the Find dialog box, type **Fong**, and then click Find Next.

4. Close the Find dialog box.

5. Scroll the cells until you can see the cell corresponding to Albert Fong's current job title (Programmer).

6. Click to select that cell, and then click again to select the text. Then, type **Application Engineer**.

 *You can also use the Replace function from the Data View menu to do a search and replace of a particular string within a column. Both the Find and Replace functions are limited in their search to the records contained in the same column as the currently selected cell. There is no way to issue a global Find or Replace. In addition, the Find function can only search in one direction (either up or down) from the currently selected cell, while the Replace function can only search down from the currently selected cell.*

ADVANCED TOPICS

To add a new record to the database:

1. In the Data View window, select a record header.
2. Right-click the record header and choose Add New Record from the shortcut menu. AutoCAD immediately adds a new record with blank field values to the end of the current record set.
3. Select a cell in the new record and enter a value.
4. Repeat Step 3 to enter values in additional fields.

**EXAMPLES**

***LEARN BY EXAMPLE***
*You can also simply click inside a blank cell on the last row of the Data View window and start entering new values to add a new record to the database. You can press TAB to move to successive fields. When you reach the last field in a row, AutoCAD automatically adds another new blank record to the database.*

To add another new record to the database:

1. In the Data View window, click the Go To Last Record button.
2. Click in the blank EMP_ID cell on the last row, type **1002,** and the press TAB.
3. In the LAST_NAME cell, type **Simpson**, and then press TAB.
4. Repeat Step 3 for the remaining five blank cells in the new record, entering the following values:

| | |
|---|---|
| FIRST_NAME | **Barton** |
| DEPT | **Sales** |
| TITLE | **Eastern Region Mgr**. |
| ROOM | 103 |
| EXT | 8602 |

5. Press TAB after entering each value, and then press TAB to move to the next row.

To delete a record from the database:

1. In the Data View window, select the record header of the record that you want to delete.
2. Right-click the record header and choose Delete Record from the shortcut menu.

## Saving Your Changes

When you make changes to the records in a database, AutoCAD indicates that the database has been changed by displaying a delta symbol (Δ) in the grid header as well as in the record header of each record that has been edited or added. The changes are not written to the database, however, until you commit them.

To commit the changes that you made in the Data View window to the actual database table, right-click the Data View grid header and choose Commit from the shortcut menu. To restore the original values to the Data View window, right-click the Data View grid header and choose Restore from the shortcut menu.

*If you quit AutoCAD, open a new table, or close the Data View window before committing changes you've made to the database table, AutoCAD automatically commits all changes you've made during the editing session.*

## Creating Links to Graphical Objects

Thus far, you've used AutoCAD's database connectivity functions only to access data in the external database. While that's an interesting capability, you could probably access that data more easily by using the external database program that was used to create the database itself, rather than AutoCAD. Clearly, the true power of database connectivity comes from linking drawing objects to data in the database. For example, you can link an employee to the label representing his or her room number.

You create associations between graphical objects and the database table by creating a link that references one or more records stored in a database table. You can't create links to nongraphical objects, such as layers and linetypes. Links are tightly connected with the graphical objects with which they are associated. If you move or copy a linked object, the link is moved or copied with it. If you delete a linked object, the link is also deleted.

To create an association between the database table records and graphical objects, you must first create a *link template*. Link templates identify the fields in a table that are to be associated with links that share that table. For example, you can create a link template that uses the room number from the Employee database, and then use that template to create links that point to different records in the database table.

Link templates appear in the drawing node within the dbConnect Manager window and can serve as shortcuts that point to the database tables on which they are based. For example, you can double-click a link template to open its table, instead of selecting its table in the data sources node. This can be quite helpful when you have a large number of data sources configured on your system. You can also attach multiple links to a single graphical object by using different link templates. This lets you associate data from more than one database table with a single object.

To create a link template in the current drawing:

1. From the dbConnect menu, choose Templates | New Link Template.

2. In the Select Data Object dialog box, choose the table with which you want to link the data, and then click Continue.

3. In the New Link Template dialog box, enter a name for the new link template in the New Link Template Name box, and then click Continue.

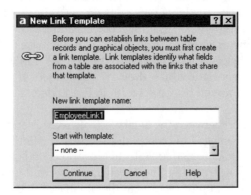

4. In the Link Template dialog box, select one or more key fields by selecting their check boxes. In most cases, you will select one field.

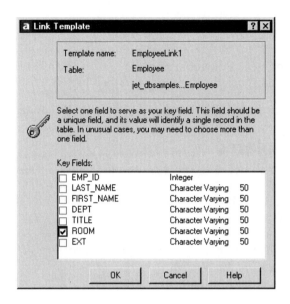

5. Click OK to create the link template.

Once you have created a link template, you can create links between records in the database and graphical objects in the drawing.

*LEARN BY EXAMPLE*

*Create a link between a record in the database and a graphical object in the drawing.*

Our sample drawing and database already have link templates established for linking the records in the database. In addition, several links have already been created. In the previous exercise, however, you added a new employee to the Employee table. You need to add a link from that employee record to his room in the drawing. Before you begin, zoom in on room 103 in the ASEMP2002.DWG sample drawing (shown in Figure 27-3).

1. In the dbConnect Manager window, double-click the EMP link template to open the Employee table. (*Note:* You can also open the Employee table by right-clicking the EMP link template and then selecting Edit Table from the shortcut menu, or by selecting the EMP link template and then clicking the Edit Table button.)

2. Click the record header to select the record you created for Barton Simpson.

3. Make sure that the current mode is Create Links. (In the Data View window, click the arrow adjacent to the Link button and choose Create Links; or from the Data View menu, choose Link and Label Settings | Create Links.)

4. Do one of the following:

   ■ In the Data View window, click Link!.

   ■ From the Data View menu, choose Link!.

The Data View window temporarily disappears and AutoCAD prompts you to select objects.

5. In the drawing, select the room number 103, using any object selection method (for example, just click the number), and then press ENTER.

AutoCAD indicates that the link has been created by highlighting the record in yellow.

## Editing Link Templates

If you need to add an additional key field or need to remove an existing one, you can do so by editing the link template. You may also need to update a link template's properties if any changes are made to the data source that points to it, such as if the database table is renamed or moved.

**Note**
*You can only edit link templates that don't yet have any links defined in the associated drawing. Editing a link template invalidates its links. Before you can edit a link template that already has associated links defined, you must delete all links based on the link template.*

To edit a link template:

1. From the dbConnect menu, choose Templates | Edit Link Template.

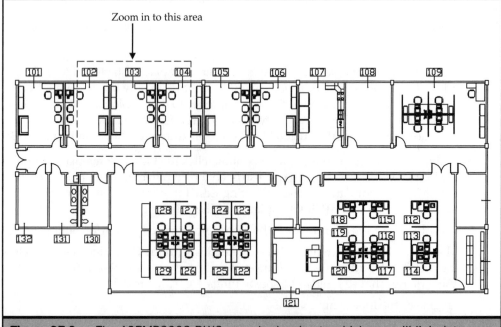

**Figure 27-3.** *The ASEMP2002.DWG sample drawing to which you will link data*

2. In the Select a Database Object dialog box, select the link template that you want to edit, and then click Continue.

3. In the Link Template dialog box, select one or more key fields by selecting their check boxes.

4. Click OK to save the changes to the link template.

To change a link template's properties:

1. From the dbConnect menu, choose Templates | Link Template Properties.

2. In the Select a Database Object dialog box, select the link template whose properties you want to change, and then click Continue.

3. In the Link Template Properties dialog box, update the data source information as necessary, and then click OK.

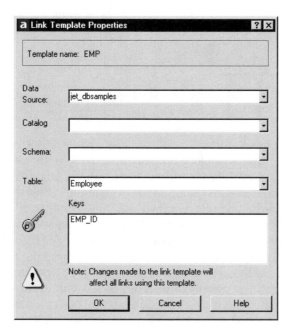

To delete all links based on a link template:

1. From the dbConnect menu, choose Links | Delete Links.

2. In the Select a Database Object dialog box, select the link template whose links you want to delete from the current drawing, and then click OK.

# Viewing Link Data

After you have created links to AutoCAD graphical objects, you can view data associated with them. For example, you can select an employee in the Employee table and locate

## Specifying Link Viewing Settings

The Data View and Query Options dialog box enables you to set a number of viewing options that affect how linked records and linked graphical objects appear whenever a linked object is selected. To display this dialog box, do one of the following:

- In the Data View window, click Data View And Query Options.

- From the Data View menu, choose Options.

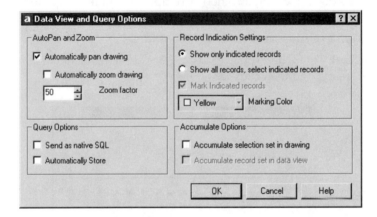

The controls in the AutoPan And Zoom area enable AutoCAD to automatically pan the drawing to display objects that are associated with the current selection set of records in the Data View window. For example, when Automatically Pan Drawing is selected, AutoCAD can pan the drawing to center the graphical objects associated with records you select in the Data View window. If you also select the Automatically Zoom Drawing check box, AutoCAD automatically zooms the drawing so that all the objects associated with the selected record or records are displayed. When this check box is selected, you can also control the zoom factor. This setting determines the size of the extents as a percentage of the drawing area, in the range from 20 to 90 percent. The default value is 50 percent. For example, when AutoCAD automatically pans and zooms to display a graphical object associated with a record, the object appears centered in the drawing, magnified so that the object fills 50 percent of either the height or width of the view, whichever is less.

The Record Indication Settings determine whether all the records or only the subset of records associated with the current selection set of graphical objects are displayed in the Data View window. When Show Only Indicated Records is selected, you can reduce the number of records shown in the Data View window to only those corresponding to graphical objects selected in the drawing. When Show All Records, Select Indicated Records is selected, all the records remain visible in the Data View window. In that case, when you select graphical objects in the drawing, if the Mark

Indicated Records check box is selected, AutoCAD can highlight the corresponding records using the color specified in the Marking Color drop-down list.

The controls in the Accumulate Options area specify whether AutoCAD accumulates selection sets of graphical objects or records in the Data View window or creates new selection sets as additional objects. When the Accumulate Selection Set In Drawing check box is cleared (the default setting), AutoCAD creates a new selection set each time you select a new set of records in the Data View window. Similarly, when the Accumulate Record Set In Data View check box is cleared, the Data View window shows a new set of records each time you create a new selection set of graphical objects in the drawing.

*NOTE:   You'll learn about the controls in the Query Options area later in this chapter.*

that employee's office in the drawing. You can also select an AutoCAD object to which a database record has been linked, and locate that record in the database.

**LEARN BY EXAMPLE**
*Use the Data View tools to quickly locate associated data and graphical objects.*

Before proceeding, display the Data View and Query Options dialog box, select the Automatically Zoom Drawing check box, change the Zoom Factor setting to 20 percent, and then close the dialog box. Then, use the following steps to view graphical objects that are linked to data records:

1. In the dbConnect Manager window, double-click the EMP link to open the Employee table.

2. In the Data View window, select the record header corresponding to an employee (such as Cindy Wilson in Accounting).

3. Do one of the following:

   ■ Right-click the record header and choose View Linked Objects from the shortcut menu.

   ■ In the Data View window, click View Linked Objects In Drawing.

   ■ From the Data View menu, choose View Linked Objects.

Notice that as soon as you select the View Linked Objects function, AutoCAD selects the graphical object associated with the data and pans and zooms the drawing so that the room number corresponding to the employee you selected is centered in the drawing.

You can use a similar function to view database records that are linked to selected graphical objects. Before proceeding, zoom out to the drawing extents, and then zoom in to enlarge the area around rooms 101 through 104 (so that it's easier to select the room

**ADVANCED TOPICS**

numbers). Then, use the following steps to view data that is linked to selected graphical objects:

1. In the dbConnect Manager window, double-click the EMP link to open the Employee table.

2. In the drawing, select the room numbers for rooms 101 and 103 (so that they are highlighted in the drawing).

3. Do one of the following:

   ■ In the Data View window, click View Linked Records In Data View.

   ■ From the Data View menu, choose View Linked Records.

As soon as you select the View Linked Records function, AutoCAD either reduces the records displayed in the Data View window to just those linked to the two room numbers you selected (if the Show Only Indicated Records option is selected in the Data View and Query Options dialog box) or highlights those records using the marking color (if the Show All Records option is selected).

## Automatically Viewing Link Data

You can also instruct AutoCAD to automatically select linked objects in the drawing whenever you select a record, or to automatically select records in the Data View window whenever you select objects in the drawing. Only one of these automatic modes can be active at a time, however.

To automatically view graphical objects linked to records, do one of the following:

■ In the Data View window, click the AutoView Linked Objects In Drawing button (so that it's selected).

■ From the Data View menu, choose AutoView Linked Objects.

Similarly, to automatically view table records that are linked to selected drawing objects, do one of the following:

■ In the Data View window, click the AutoView Linked Records In Data View button (so that it's selected).

■ From the Data View menu, choose AutoView Linked Records.

*LEARN BY EXAMPLE*
*Use the AutoView Linked Objects In Drawing function to automatically display graphical objects linked to selected records.*

Before proceeding, zoom out to the drawing extents. Then, use the following steps to automatically view graphical objects linked to records that you select:

1. In the dbConnect Manager window, double-click the EMP link to open the Employee table.

2. In the Data View window, click the AutoView Linked Objects In Drawing button (so that it's selected).

3. Click a record header.

4. Repeat Step 3.

Notice that as soon as you select a record, AutoCAD automatically selects the room number corresponding to the employee you selected, and automatically pans and zooms the drawing to display that room number at the center of the drawing.

1. In the Data View window, click the AutoView Linked Records In Data View button (so that it's selected).

2. In the drawing, select a room number using any AutoCAD object selection method.

3. Repeat Step 2 to select additional room numbers.

Notice that as you select each room number, depending on how you've set the Data View and Query Options dialog box settings, its corresponding record is either selected in the Data View window or added to the list of records.

**Note**

*By default, when you select an object in AutoCAD, it gets added to the current selection set. When you select records in the Data View window, however, you must press* SHIFT *or* CTRL *as you select records to add them to the current record set. Remember that you can make AutoCAD function in a similar fashion by changing the Use Shift To Add To Selection setting on the Selection tab of the Options dialog box (or by changing the* PICKADD *system variable value). You can also right-click in any record header and select Clear All Marks to reset the Data View window so that no records are selected.*

When you complete this exercise, turn off the AutoView function.

## Editing Link Data

After you have created links, you can update their key field values. For example, if you change the numbering scheme for the employee ID numbers, you can use the Link Manager to enter new key values for a selected link. The key values must reference a record that already exists in the source database table.

To edit a link's key values using the Link Manager:

1. From the dbConnect menu, choose Links | Link Manager.

2. In the AutoCAD drawing, select one linked object.

3. In the Link Manager dialog box, select a link from the Link Template drop-down list.

ADVANCED TOPICS

4. In the Value field, type a new value; or click the button with three dots (...) to display the Column Values list, select the new value, and then click OK.

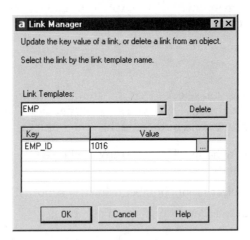

5. Click OK to update the link and close the dialog box.

## Synchronizing Links

You should periodically check the links in your drawing to update or delete broken links. When you start the Synchronize function, AutoCAD first prompts you to select a database link template. The program then checks the links in the drawing that are based on the specified link template, and displays the Synchronize dialog box, showing a list of detected errors.

Some errors, such as a resized column in the source database table, can be fixed directly from the Synchronize dialog box by clicking the Fix button. Other errors, such as links that point to nonexistent records, can only be fixed in the source database table.

To synchronize the links in a drawing:

1. From the dbConnect menu, choose Synchronize.

2. In the Select a Database Object dialog box, select a link template, and then click Continue.

   If no errors are detected, AutoCAD displays a message. If there are errors, they appear in the Errors area of the Synchronize dialog box.

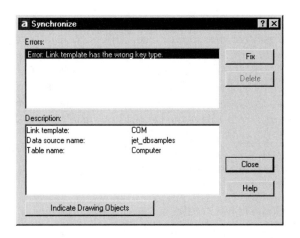

3. In the Synchronize dialog box, select an error and then do one or more of the following:

- Click Indicate Drawing Objects to highlight the linked graphical objects.
- Click Fix to fix the broken link.
- Click Delete to delete the broken link.

4. Repeat Step 3 for any additional errors.

5. Click Close.

*You can also start the Synchronize function from the dbConnect Manager window by right-clicking a link template and selecting Synchronize from the shortcut menu.*

# Exporting Links

You can also export the links contained in your drawing. Because the database doesn't have any knowledge of the links between it and the graphical objects in your drawing, you can use the exported information in your database program to create reports. For example, if you link more than one object in your drawing to the same record in the database table—such as when a single part occurs several times in an assembly—you can export the links from the drawing, thus creating a new file that contains a separate listing for each occurrence of the link. You can then combine this information with the data already in your database, to produce a report that includes the count information. You can export information in the format native to your database, or in comma-delimited or space-delimited format.

The Export Links function exports link data from the drawing. When you start the function, AutoCAD first prompts you to select objects. You can use any object selection method. The command exports link data only for those objects that you select. After you finish selecting objects, if more than one link template exists for the objects that you selected, AutoCAD displays the Select a Database Object dialog box, so that you can select the link template from which to export data. Select the link template, and then click Continue. The program then displays the Export Links dialog box, shown in Figure 27-4, in which you can specify the fields to be included in the output file, as well as the format of the exported data (comma-delimited, space-delimited, or native database format). Specify the filename and the directory in which you want it saved, and then click Save.

To export links from a drawing:

1. From the dbConnect menu, choose Links | Export Links.

2. In the drawing, select the graphical objects whose links you want to export, and then press ENTER.

3. If more than one link template exists for the objects you selected, AutoCAD displays the Select a Database Object dialog box. Select a link template, and then click Continue. If your selection set contains only one link template, the program skips this step.

4. In the Export Links dialog box, select the fields to include by clicking them in the Include Fields list.

5. From the Save As Type drop-down list, choose the type of file you are saving.

6. Specify the filename and the directory in which to save the file, and then click Save.

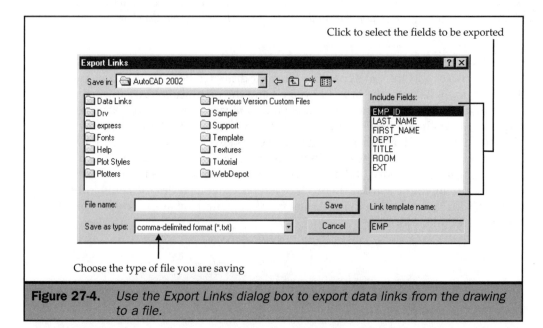

**Figure 27-4.** Use the Export Links dialog box to export data links from the drawing to a file.

# Creating Labels

Although links provide a powerful way for you to associate drawing objects with records in a database, the links are only pointers to the external database table. The information from the database doesn't actually appear in the drawing. When you want data to actually appear in the drawing, you can use labels. Labels enable you to include a visible representation of the external data by placing a multiline text object in the drawing that displays data from selected fields stored in an external database table.

Suppose, for example, that you want to annotate each room in the sample office plan drawing so that each room contains the name of the employee assigned to the room. You can use labels to include that information in the drawing. Labels can be either freestanding or attached to a graphical object. A freestanding label exists in the drawing independent of any other graphical object. Labels that are attached to a graphical object appear with a leader pointing to that object and are tightly connected to the object. If you move the graphical object, the label moves with it. If you copy the object to the Clipboard, the label is also copied. And if you delete the object, the label is also deleted. Regardless of whether the labels are created as freestanding or attached labels, once inserted, you can easily update the labels to reflect any changes that are made to the external data.

To create labels, you must first create a label template. Label templates identify the fields in a table that are to be displayed in the label, and identify how the label text is to be formatted. You create label templates by using the New Label Template function. AutoCAD first prompts you to select a link template to use for the new label template, and to name the new label template. The program then displays a Label Template dialog box, similar to the one shown in Figure 27-5. This dialog box is a modified version of the Multiline Text Editor that you learned about in Chapter 13. The Label Fields and Label Offset tabs are unique to the Label Template dialog box. In it, you specify the field or fields that you want included in the label, by selecting them from the Field drop-down list and then clicking the Add button.

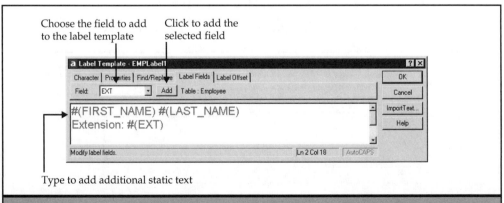

**Figure 27-5.** *Use the Label Template dialog box to select the fields you want to display and control their formatting. Here we see fields added as part of the following Learn by Example exercise.*

The fields you add appear in the dialog box within parentheses, preceded by a pound symbol (#). You can also type to add any additional static text that you want to precede or follow the label. For example, if you want to add the word Extension prior to the employee's telephone extension, you would type **Extension:** in the Label Template dialog box before selecting EXT from the drop-down list.

You can use the controls on the Character and Properties tabs to control the character formatting and properties applied to the label, as you would to format any other multiline text object. You learned about these controls in Chapter 13.

*The controls on the Label Offset tab (shown in Figure 27-6) let you control offset settings that affect the insertion point of labels and, in the case of attached labels, their associated label objects. The Start value specifies the start point for the leader object in relation to the extents of the graphical object to which the label is attached. For example, if you want the leader to be attached to the top left of the object, select Top Left from the Start drop-down list. The Leader Offset values enable you to specify the X and Y offset for the multiline text object in relation to the location of the start of the leader. The Tip Offset values apply to both freestanding and attached labels. In the case of freestanding labels, these values determine the X and Y label offset from the insertion point you specify when placing the label. For attached labels, the Tip Offset values specify the X and Y offset for the beginning of the leader from the point specified by the Start option.*

When you have finished creating the label template, click OK to close the Label Template dialog box. You are then ready to create a link with a label.

*LEARN BY EXAMPLE*
*Create a label template to use for inserting labels.*

Suppose you want to create a label that displays the first and last names and telephone extension of the employees from the Employee table:

1. From the dbConnect menu, choose Templates | New Label Template.

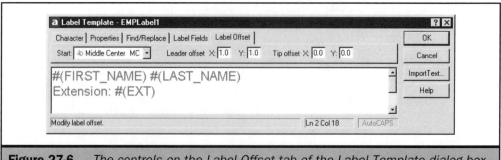

**Figure 27-6.** *The controls on the Label Offset tab of the Label Template dialog box let you control the position of the label.*

2. In the Select a Database Object dialog box, select the EMP link template to use for the new label template, and then click Continue.

3. In the New Label Template dialog box, specify the name for the new label template, and then click Continue. (*Note:* If you wish, you can select an existing label template in the current drawing to use as the starting point for your new label template.) In this case, just click Continue to accept the default label template name.

4. In the Label Template dialog box, from the Field drop-down list, choose FIRST_NAME, and then click Add.

5. Click inside the Label Template dialog box to place the text cursor immediately to the right of the #(FIRST_NAME) label, and then press the SPACEBAR (to include a space between the employee's first and last name).

6. From the Field drop-down list, choose LAST_NAME, and then click Add.

7. Click inside the Label Template dialog box, and then type **Extension:** . (*Note:* Be sure to include a space after the colon.)

8. From the Field drop-down list, choose EXT, and then click Add.

9. Select all the text in the Label Template dialog box, click the Character tab, and then choose a different font (such as Arial) from the Font drop-down list. The resulting label template is shown in Figure 27-5.

10. Click OK to close the Label Template dialog box.

Once you have created a label template, you are ready to create links that display the desired label. Since the label will display data from an existing table, you must open the data table with which the label template and link template are associated. You can then create the links with their associated labels. Although you can create labels that are either freestanding or linked to an object, the process is quite similar.

*LEARN BY EXAMPLE*
*Create a freestanding label linked with a record in a database table.*

Suppose you want to create a label to identify the cubicle for Albert Fong. Begin by zooming in on room 120.

To create a link with a freestanding label:

1. In the dbConnect Manager, double-click the Employee table.

2. In the Data View window, make sure that the EMPLabel1 template is selected in the Select A Label Template drop-down list.

3. Select the row header for Albert Fong's record.

4. Make sure that the current mode is Create Freestanding Labels. (In the Data View window, click the arrow adjacent to the Link button and choose Create Freestanding Labels; or from the Data View menu, choose Link And Label Settings | Create Freestanding Labels.)

5. Do one of the following:

- In the Data View window, click Create Freestanding Label.

- From the Data View menu, choose Link!.

The Data View window temporarily disappears and AutoCAD prompts you to specify a point for the label.

6. In the drawing, specify the point where you want the label displayed.

The label is immediately inserted in your drawing. Because you accepted the default text height when you defined the label template, however, the label appears in the drawing as a small red spot. You could use the DDEDIT command to change the height of the text, but because you are likely to insert additional labels based on the label template, it makes more sense to edit the label template instead.

 *You can also use any other AutoCAD editing command to edit labels. For example, you can use the COPY, MOVE, or SCALE commands to change individual instances of a label. Those changes don't affect or redefine the label template, however.*

 *You should not edit the text of a label by using the DDEDIT command. Any changes you make to the text will be overwritten by the values from the database the next time you reload the labels. You'll learn about updating labels with new database values later in this chapter.*

## Editing Label Templates

After defining a label template, you can edit the template if necessary. For example, you may decide to add additional table fields to the label or to change the font or size used in the label object. In the case of the label you inserted in the previous exercise, you need to increase the font size so that the label is readable.

You can edit label templates by using the Edit Label Template function. Do one of the following:

- From the dbConnect menu, choose Templates | Edit Label Template, select the template you want to edit from the list displayed in the Select a Database Object dialog box, and then click Continue.

- In the dbConnect Manager window, right-click the label template you want to edit, and then select Edit from the shortcut menu. (*Note:* You can also simply double-click the label template.)

AutoCAD displays the same Label Template dialog box you used when you created the label template. Make the desired changes, and then click OK.

Online

EXAMPLES

*LEARN BY EXAMPLE*
*Edit the label template from the previous example so that the label appears with a larger font size.*

When you initially created the label template, AutoCAD assigned the multiline text a text height of 0.2 unit. In order for the text to appear at a more reasonable size in the drawing, you need to increase the text height to approximately 4 units. You can also add additional fields to the label. In this case, you will add the employee's job title.

To edit the label template:

1. In the dbConnect Manager window, right-click the EMPLabel1 label template and choose Edit from the shortcut menu.

2. In the Label Template dialog box, click to the left of the word Extension (to position the text cursor).

3. From the Field drop-down list, select TITLE, and then click Add. The label #(TITLE) should now appear on its own line.

4. Select the Character tab.

5. In the Label Template dialog box, select all of the text (so that it is highlighted). Then, click in the Font Height box, type **4**, and press ENTER.

6. Click OK to close the Label Template dialog box.

The changes you make to the label template are reflected immediately in the drawing. Your drawing should look similar to Figure 27-7.

# Updating Labels with New Database Values

As mentioned previously, if the data in the external database is changed, you can update the labels to reflect those changes. The Reload Labels function updates all the labels associated with a specified label template.

Online

EXAMPLES

*LEARN BY EXAMPLE*
*Update labels by changing their associated data.*

Suppose that you made a mistake, and that Albert Fong's job description should not have been changed from Programmer to Application Engineer. To correct this mistake, you do the following:

1. In the dbConnect Manager window, double-click the Employee table.

2. In the Data View window, locate the record for Albert Fong.

3. In the TITLE column, select the cell corresponding to Albert Fong's job title (which currently says Software Engineer).

ADVANCED TOPICS

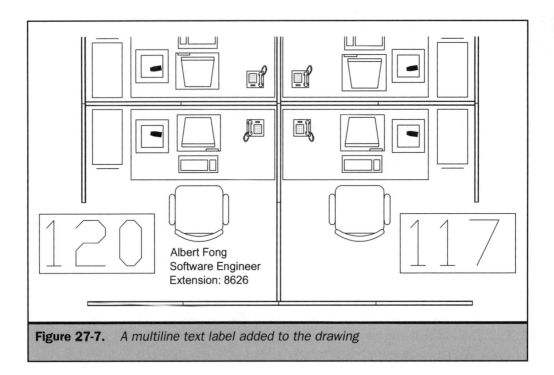

**Figure 27-7.**   *A multiline text label added to the drawing*

4. Select the job title, type **Programmer**, and then click inside another cell.

5. Right-click in the grid header and choose Commit from the shortcut menu.

After you change the data in the database, do the following to update the labels to reflect those changes:

1. From the dbConnect menu, choose Labels | Reload Labels.

2. In the Select a Database Object dialog box, select the label template whose values you want to reload, and then click OK.

*You can also right-click the drawing name in the dbConnect Manager dialog box and then choose Reload Labels from the shortcut menu to reload the data associated with all the label templates in the drawing.*

## Using Shortcut Menu Options for Links and Labels

When you select a linked graphical object or a label, you can right-click to display a shortcut menu containing a number of options for editing links and labels. For example, you can use a shortcut menu to add a label to an existing link. Once you've created a label template, this provides a quick way to add labels to your drawing without having to first locate records in the database table.

Online

EXAMPLES

### LEARN BY EXAMPLE
*Use the Label The Link function to add labels for other employees.*

Since the Employee table already contains links associating employees with their respective room numbers, you can use the shortcut menu to quickly add labels for other employees. Use the following steps to add a label for Maria Sanchez, the mechanical engineer who shares the cubicle adjacent to Albert Fong:

1. In the drawing, click to select the room number 117.

2. Right-click and then choose Label | Label The Link from the shortcut menu.

3. In the Select a Database Object dialog box, select the EMPLabel1 label template, and then click OK. A label for Maria Sanchez immediately appears in the drawing.

4. Press ESC to remove the grip from the room number.

5. Use the MOVE command to relocate the new label so that it doesn't overlap the room number.

   Notice that when you attempt to move the label, it doesn't end up where you placed it. The Label The Link function automatically creates attached labels. Since this isn't what you want in this situation, you can use another shortcut menu function to detach the label from the link.

6. Select the label you just created (so that it is highlighted and its grip is visible).

7. Right-click and then choose Label | Detach Label.
   As soon as you detach the label, the leader disappears and the label moves into the position you originally intended. The label remains associated with the data in the Employee table, and can be updated by reloading the labels, but the label is no longer attached to the link.

On your own, use this method to label the other rooms with the name, job title, and phone extension of the employee assigned to each room.

**Note** *There are over 20 different shortcut menu options for modifying links and labels. For a complete description of all of these options, see the AutoCAD User's Guide.*

---

## Importing and Exporting Link and Label Templates
You can save link and label templates that you create in a drawing to a separate template set file and then import those templates for use in other drawings. This saves time when working in other drawings that are associated with the same database tables, and is also a useful way to develop and share a common set of database tools among several users. When you export a template set, AutoCAD saves all the link and label templates in the current drawing to a template set file (a file with a .DBT file extension).

ADVANCED TOPICS

To export a set of templates from the current drawing:

1. From the dbConnect menu, choose Templates | Export Template Set.

2. In the Export Template Set dialog box, select the directory in which you want to save the template set, specify the name for the template set file, and then click Save.

To import a set of templates into the current drawing:

1. From the dbConnect menu, choose Templates | Import Template Set.

2. In the Import Template Set dialog box, select the template set you want to import, and then click Open.

   If there is a link or label template in the current drawing with the same name as one of the templates being imported, AutoCAD displays an alert dialog box in which you can enter a unique name for the template being imported.

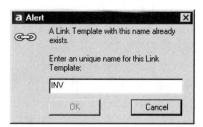

*NOTE:* *You can also import and export template sets from the dbConnect Manager window, by right-clicking the drawing name and choosing Import Template Set or Export Template Set from the shortcut menu.*

# Working with the Query Editor

One of the most powerful aspects of working with a database is the ability to locate a subset of records based on a specific search criteria, or *query*. For example, you might want to locate all the rooms assigned to employees in the Sales department. By using the AutoCAD Query Editor, you can easily construct a query that returns the subset of records or linked graphical objects that you want to find.

The Query Editor, shown in Figure 27-8, consists of a dialog box with four tabs that you can use to develop queries. The dialog box is designed so that it's easy to create a structured query even if you are not familiar with Structured Query Language (SQL). By working with the tabs from left to right, you can create a basic query and then refine

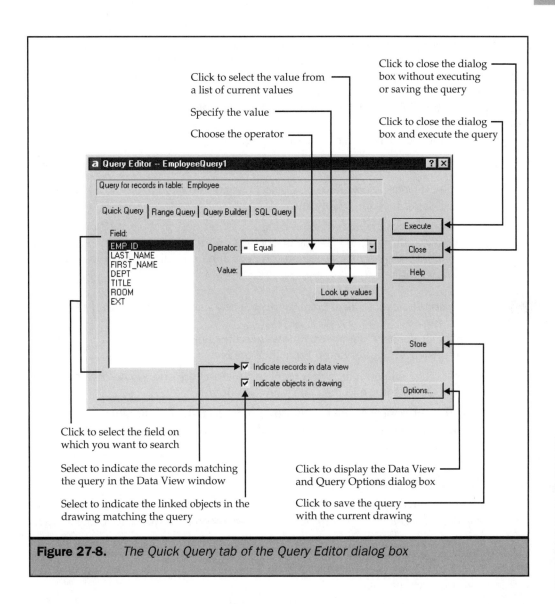

Click to select the value from
a list of current values

Specify the value

Choose the operator

Click to close the dialog
box without executing
or saving the query

Click to close the dialog
box and execute the query

Click to select the field on
which you want to search

Select to indicate the records matching
the query in the Data View window

Select to indicate the linked objects in the
drawing matching the query

Click to display the Data View
and Query Options dialog box

Click to save the query
with the current drawing

**Figure 27-8.**   *The Quick Query tab of the Query Editor dialog box*

it or add additional parameters in subsequent tabs to further narrow your search. For
example, to simply locate all the room numbers assigned to employees in the Sales
department, you could construct the entire query on the Quick Query tab. If you want
to construct a more complex query, such as locating all the room numbers assigned to
employees in the Engineering department whose job title is Programmer, however, you
can use the controls on the Query Builder tab to add the additional conditions.

The tabbed progression of the Query Editor is particularly useful in becoming familiar with the SQL syntax. As you create a query using the first three tabs, the query is actually created as an SQL query. If you select the SQL Query tab, you can see how the query is actually formatted using SQL. As you become proficient in SQL, you can construct your entire query in the SQL Query tab. After constructing a query, you can save it so that you can reuse it later.

To display the Query Editor, do the following:

1. From the dbConnect menu, choose Queries | New Query on an External Table.

2. In the Select Data Object dialog box, select the table on which you want to construct the query, and then click Continue.

3. In the New Query dialog box, enter a name for the query in the New Query Name box. If you wish, you can base the query on a previously saved query. When you have named the query (or accepted the default name), click Continue to display the Query Editor.

**Tip**    *You can also display the Query Editor to create a new query by doing any of the following:*

- In the dbConnect Manager window, select the link or table on which you want to construct the query, and then click the New Query button.

- Open the table on which you want to construct the query, and then click the New Query button in the Data View window.

- In the dbConnect Manager window, right-click the link or table on which you want to construct the query, and then chose New Query from the shortcut menu.

## Using the Quick Query Tab

The Quick Query tab enables you to quickly create a query based on a single field and a single *conditional operator*. For example, you can easily create a query to locate all the rooms assigned to employees in the Sales department.

Conditional operators are constraints such as equal, greater than, or less than. You use these constraints to construct your search. Table 27-1 summarizes query operators available within the Query Editor.

**Note**    *All operators can be applied to both numeric and text-based fields. For example, you can search for all records greater than "n", which would return all records ranging from "na..." through "z..." Note that query searches are case-sensitive: "DOOR" is not the same as "door".*

| Operator | Description |
|---|---|
| = Equal | Returns all records that exactly match the value that you specify. |
| <> Not Equal | Returns all records except those that match the value you specify. |
| > Greater than | Returns all records that exceed the value you specify. |
| >= Greater than or equal | Returns all records that are equal to or exceed the value that you specify. |
| < Less than | Returns all records that are less than the value you specify. |
| <= Less than or equal | Returns all records that are equal to or less than the value you specify. |
| Like | Returns all records that contain the value that you specify. You can use this operator in conjunction with the % wildcard character to return all records containing a particular string. For example, if you query the Employee table for TITLE fields whose value is like **%Mgr%**, AutoCAD returns all records in which the TITLE field contains the letters Mgr within the job title. |
| In | Returns all records that match a set of values that you specify. For example, if you are searching the Employee database for a person, but you're not sure if the last name is spelled Cohn or Cohen, you could use the In operator and provide both spellings in the value (separated by a comma) to return all records spelled either Cohn or Cohen. |
| Is null | Returns all records that don't have values specified for the field you're querying. This operator is useful for locating records in the table that are missing data. |
| Is not null | Returns all records that have values specified for the field that you're querying. This operator is useful for excluding from your query records in the table that are missing data. |

**Table 27-1.** *AutoCAD Query Operators*

*LEARN BY EXAMPLE*
*Use the Quick Query tab of the Query Editor to construct a simple query.*

Suppose that you want to find all the employees in the Sales department. The Quick Query tab makes short work of such a search. Before proceeding, zoom out to the extents of the drawing and press ESC to clear any grips. Then do the following:

1. Open the Query Editor to query the Employee table.

2. On the Quick Query tab, in the Field list, select DEPT.

3. Select the Equal operator from the Operator list.

4. Click the Look Up Values button.

5. In the Column Values dialog box, select Sales, and then click OK.

6. In the Query Editor, make sure that the Indicate Records In Data View and Indicate Objects In Drawing check boxes are both selected.

*Due to a minor bug in AutoCAD, to actually highlight the linked objects in the drawing, you should click the Indicate Objects In Drawing check box once to clear it, and then click it again to select it. In addition, you must first store your query in the current drawing.*

7. Click Store to save the query in the current drawing.

8. Click Execute to execute the query and close the Query Editor dialog box. Notice that a subset of six records that match your query conditions are displayed in the Data View window and that the room numbers corresponding to those employees are selected in the drawing.

Since you saved your query, it appears in the drawing nodes portion of the dbConnect Manager window. You can use this query as the starting point when constructing new queries, or execute the query again later. You can also edit or delete the query from the dbConnect menu or shortcut menus. To return to the Query Editor to refine your query, click the Return To Query button in the Data View window.

**Tip**    *If you select the Automatically Store option in the Data View and Query Options dialog box, you no longer need to click Store to explicitly store a query. Whenever you execute a new query, the query is automatically stored.*

## Using the Range Query Tab

The Range Query tab, shown in Figure 27-9, works much like the Quick Query tab. Instead of specifying an operator, however, you specify the upper and lower ends of a range of values for the query to return. For example, suppose you want to search for all the furniture in the Inventory table in the price range from $60 to $200. You would select the PRICE field in the Field list, specify the From value as 60, and specify the Through value as 200.

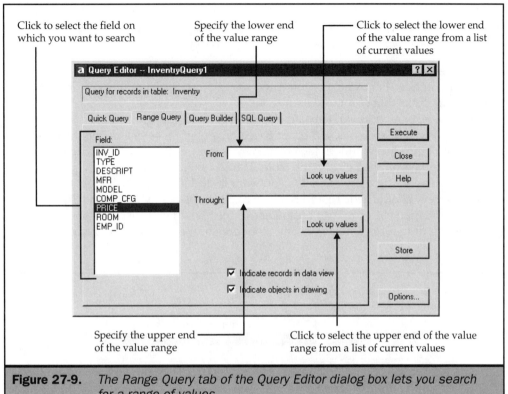

Click to select the field on which you want to search

Specify the lower end of the value range

Click to select the lower end of the value range from a list of current values

Specify the upper end of the value range

Click to select the upper end of the value range from a list of current values

**Figure 27-9.** *The Range Query tab of the Query Editor dialog box lets you search for a range of values.*

# Using the Query Builder Tab

Thus far, all the queries we've constructed were based on a single field. The Query Builder tab, shown in Figure 27-10, lets you create queries based on multiple criteria. For example, you could search for all chairs that cost more than $400. You can also use parentheses to group query criteria, and specify which fields are displayed and how they are sorted when the query is returned in the Data View window.

The Query Builder tab enables you to create compound queries on two or more search criteria by incorporating Boolean operators. To return the list of chairs costing more than $400, the query first assembles a list of chairs and then compares that list to a list of inventory items costing more than $400. Only those items matching both lists are returned by the query. AutoCAD queries support the following Boolean operators:

- **And**    Constructs a query based on multiple criteria, returning a set of records that meet all of the specified criteria; for example, the set of inventory items that are both chairs and cost more than $400.

ADVANCED TOPICS

■ **Or** Constructs a query based on multiple criteria, returning a set of records that meet any of the specified criteria; for example, the set of inventory items that are either chairs or that cost more than $400.

You can use parenthetical grouping within search criteria by bracketing searches within parentheses. For example, you could search for (All chairs that cost more than $400) or (All desks that cost more than $800).

As previously mentioned, you can also specify the fields you want to appear in the returned query, and specify a sort order based on any of those fields. The three lists at the bottom of the Query Builder tab enable you to specify these criteria. The Fields In Table list displays all the fields in the table. The Show Fields list, which is initially empty, is where you specify the fields you want included in the Data View window when the query is returned. To add fields to this list, select the field in the Fields In Table list, and then click the Add button above the Show Fields list. To remove fields from this list, drag it from the list to any other area of the Query Builder tab.

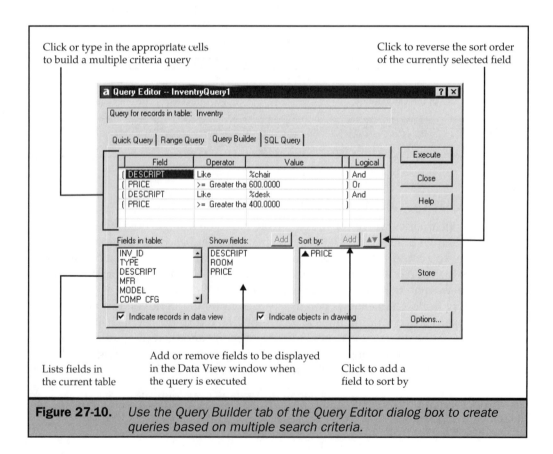

**Figure 27-10.** *Use the Query Builder tab of the Query Editor dialog box to create queries based on multiple search criteria.*

To specify a sort order for the returned query, select the field you want to sort on in the Show Fields list, and then click the Add button above the Sort By list. If the Sort By list contains more than one field, you can change the sort order by selecting a field and dragging it above or below another field in the list. By default, all the fields in the Sort By list are added to the list in ascending sort order, indicated by an up arrow. To change the sort order for any field to a descending sort, select the field and then click the Ascending/Descending Sort button (or double-click the field) to change its symbol to a down arrow. To remove a field from the Sort By list, simply drag it from the list to any area on the Query Builder tab, or select the field and then press the DELETE key. You can also clear fields from the Show Fields or Sort By lists by selecting the field, right-clicking, and choosing Clear Field Name from the shortcut menu.

### LEARN BY EXAMPLE
*Use the Query Builder tab to construct a query with multiple criteria.*

Suppose you want to locate all the chairs that cost more than $600 or all desks that cost $400 or more. Use the following steps to construct this search:

1. Open the Query Editor to query the Inventory table.
2. Click to select the Query Builder tab.
3. On the first row of cells, select DESCRIPT from the drop-down list in the Field cell.
4. Select the Like operator from the drop-down list in the Operator cell.
5. In the Value cell, type **%chair**.
6. Click in the Logical cell to select the And operator.
7. On the second row of cells, select PRICE from the drop-down list in the Field cell.
8. Select the Greater than or equal operator from the drop-down list in the Operator cell.
9. In the Value cell, type **600**.
10. Click in the Logical cell twice to select the Or operator.
11. On the third row of cells, select DESCRIPT from the drop-down list in the Field cell.
12. Select the Like operator from the drop-down list in the Operator cell.
13. In the Value cell, type **%desk**.
14. Click in the Logical cell to select the And operator.
15. On the fourth row of cells, select PRICE from the drop-down list in the Field cell.
16. Select the Greater than or equal operator from the drop-down list in the Operator cell.
17. In the Value cell, type **400**.
18. In the Fields In Table list, select DESCRIPT and then click the Add button above the Show Files list.

ADVANCED TOPICS

19. Repeat Step 18 for the PRICE and ROOM fields.

20. In the Show Fields list, select the PRICE field and then click the Add button above the Sort By list.

21. Click the Store button, and then click Execute.

## Using the SQL Query Tab

The SQL Query tab of the Query Editor dialog box, shown in Figure 27-11, provides you with an area in which you can enter a free-form SQL query, and a set of tools you can use to create your query. You can issue any valid SQL statement that conforms with Microsoft's implementation of the SQL 92 protocol. While you are learning SQL, you may find it helpful to construct queries using the other tabs of the Query Editor and then switch to the SQL Query tab to see how the query actually appears in SQL. For example, the SQL text editor in Figure 27-11 shows the SQL query you constructed in the previous example. (*Note:* The dialog box has been enlarged to display the entire query.)

**Note**    *Unlike the other tabs, you can use the SQL Query tab to perform relational operations on multiple database tables by using the SQL join operator. Because joins and SQL in general are fairly advanced topics, they are not discussed here. Many excellent books on SQL are available, including SQL: The Complete Reference (McGraw-Hill/Osborne). You can also refer to the documentation that came with your database application.*

The following two functions are unique to the SQL Query tab:

■ **Table**    This list displays all the database tables available in the current data source. You can add tables to the SQL query editor by double-clicking them, by selecting them in the list and clicking Add, or by typing their names directly in the SQL text editor area.

■ **Check**    This button enables you to check the SQL query for proper syntax without actually executing the query.

## Working with Stored Queries

If you create a query that you will use more than once, you can store the query with the drawing. Stored queries appear in the dbConnect Manager window under the drawing node of the drawing in which they were created. Once you have stored a query, you can execute, edit, rename, or delete the query, or copy it to another drawing. There are numerous methods to perform these operations. For example, to execute a stored query, you can do any of the following:

■ From the dbConnect menu, choose Queries | Execute Query. In the Select a Database Object dialog box, select the query to execute, and then click OK.

■ In the dbConnect Manager window, select the query to execute, and then either click the Execute Query button or right-click and select Execute from the shortcut menu.

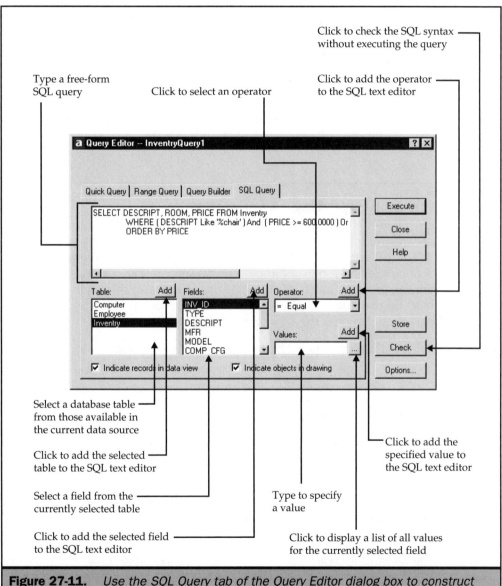

**Figure 27-11.** *Use the SQL Query tab of the Query Editor dialog box to construct SQL queries.*

AutoCAD also makes it easy for you to automatically store all new queries with the current drawing. To do so, display the Data View and Query Options dialog box that you learned about earlier in this chapter. (*Note:* In addition to selecting Options from the Data View menu, you can display this dialog box by clicking the Options button in the Query Editor dialog box any time you are creating or editing a query.) In the Query Options area, select the Automatically Store check box to instruct AutoCAD to always store new queries with the current drawing.

To copy a stored query to another drawing:

1. Open both the drawing containing the stored query and the drawing to which you want to copy the query.

2. In the dbConnect Manager window, select the stored query and drag-and-drop it to the drawing node of the drawing to which you want to copy the query.

## Importing and Exporting SQL Queries

You can save queries you create in a drawing to a separate query set file and then import those queries for use in other drawings. This saves time when working in other drawings that are associated with the same database tables and is also a useful way to develop and share a common set of database tools among several users. When you export a query set, AutoCAD saves all the queries in the current drawing to a query set file (a file with a .DBQ file extension).

To export a set of queries from the current drawing:

1. From the dbConnect menu, choose Queries | Export Query Set.

2. In the Export Query Set dialog box, select the directory in which you want to save the query set, specify the name for the query set file, and then click Save.

To import a set of queries into the current drawing:

1. From the dbConnect menu, choose Queries | Import Query Set.

2. In the Import Query Set dialog box, select the query set you want to import, and then click Open.

   If there is a query in the current drawing with the same name as one of the queries being imported, AutoCAD displays an alert dialog box in which you can enter a unique name for the query being imported.

**NOTE:**   *You can also import and export query sets from the dbConnect Manager window, by right-clicking the drawing name and choosing Import Query Set or Export Query Set from the shortcut menu.*

# Using Link Select to Create Selection Sets

The Link Select function is an advanced implementation of the Query Editor that lets you combine both graphic and data-based selection methods to construct iterative selection sets. When you start the Link Select function, AutoCAD displays the Link Select dialog box, shown in Figure 27-12. You can use the controls in this dialog box to create a selection set consisting of the records or graphical objects meeting a specified search criteria.

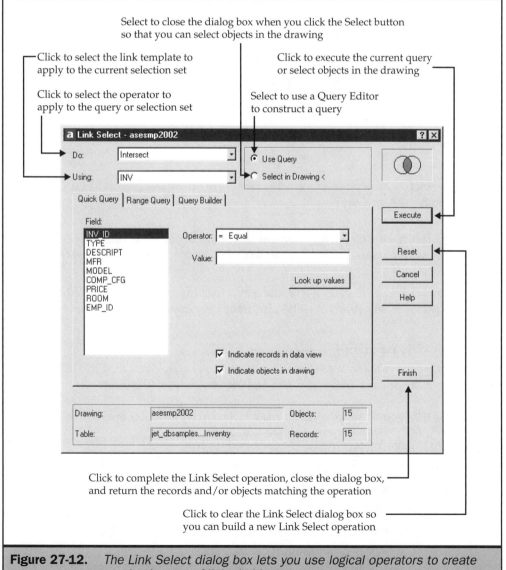

Select to close the dialog box when you click the Select button so that you can select objects in the drawing

Click to select the link template to apply to the current selection set

Click to select the operator to apply to the query or selection set

Click to execute the current query or select objects in the drawing

Select to use a Query Editor to construct a query

Click to complete the Link Select operation, close the dialog box, and return the records and/or objects matching the operation

Click to clear the Link Select dialog box so you can build a new Link Select operation

**Figure 27-12.**   *The Link Select dialog box lets you use logical operators to create a selection set of linked objects.*

ADVANCED TOPICS

To create a selection set based on a database query, select the Use Query button and then use the Query Editor tabs you learned about earlier in this chapter to construct a query. To create a selection set based on linked graphical objects, select the Select In Drawing button. From the Do drop-down list, choose Select and then click Execute to create the first selection set. This selection set is referred to as set A.

You can then choose one of the logical operators from the Do drop-down list and use either graphical or data-based operators to create a second selection set, referred to as set B, which is combined with set A through the logical operator. After applying the logical operator to sets A and B, AutoCAD returns the results. The set matching those results then becomes the new set A, and you can choose another logical operator to establish a new relationship between set A and a new set B to further refine your selection.

The Link Select dialog box supports the following logical operations:

- **Select**  Creates an initial query or selection set (set A) that can be refined through subsequent Link Select operations.

- **Union**  Creates a selection set consisting of the objects or records in both set A and set B.

- **Intersect**  Creates a selection set consisting of only those objects that are members of both set A and set B.

- **Subtract A - B**  Creates a selection set consisting of the results of subtracting the new query or selection set (set B) from the previous selection set (set A).

- **Subtract B - A**  Creates a selection set consisting of the results of subtracting the previous selection set (set A) from the new query or selection set (set B).

*LEARN BY EXAMPLE*
*Use Link Select to create a selection set consisting of objects that meet two selection criteria, one selected graphically, and the other matching a database query.*

Suppose that you want to select all the couches in a particular area of the drawing:

1. From the dbConnect menu, choose Links | Link Select.

2. In the Link Select dialog box, choose Select from the Do drop-down list.

3. Make sure the Use Query button is selected.

4. In the Field list, choose DESCRIPT.

5. From the Operator drop-down list, choose Equal.

6. Click the Look Up Values button.

7. In the Column Values dialog box, select 6×3 couch, and then click OK.

8. Click Execute.

9. Choose Intersect from the Do drop-down list.

10. Click to select the Select In Drawing button.

11. Click Select.

    The Link Select dialog box temporarily disappears, and you are prompted to select objects.

12. Use a selection window to select all of rooms 102, 103, 104, and 105, and then press ENTER.

13. Make sure that both the Indicate Records In Data View and Indicate Objects In Drawing check boxes are selected.

14. Click Finish.

The Data View window returns a list of the four couches in the four rooms you selected. Notice that the four couches are also selected in the drawing.

**Caution**   *If you are performing a Link Select operation on links from more than one database table, you should clear the Indicate Records In Data View check box. Link Select only displays records from the table referenced by the current link template, which can cause misleading results to be returned in the Data View window.*

ADVANCED TOPICS

# Chapter 28

## Customizing AutoCAD Without Programming

One of the factors that has made AutoCAD the most widely used PC-based CAD system is the many ways in which the program can be customized to meet the needs of individual users like you. Many of AutoCAD's features can be modified with virtually no programming whatsoever. The more adventuresome, who want to try their hands at programming, can also create everything from simple macros to complete applications that run inside of AutoCAD. Indeed, more than a thousand companies and individuals have used programming languages, ranging from AutoCAD's own AutoLISP to Visual Basic and C++, to create thousands of custom applications that tailor AutoCAD for the needs of particular professions, such as architecture, mechanical engineering, cartography, and even clothing design. People in practically every profession that requires dimensionally accurate technical drawings have customized AutoCAD in various ways to suit their personal needs.

In this chapter, you will learn how to customize several aspects of AutoCAD without doing any actual programming. All the changes that you make are accomplished either by using easy-to-understand dialog boxes or by making some changes using an ASCII text editor. In the next chapter, you'll learn some of the ways in which you can customize AutoCAD when you are willing to do a bit of programming. In the process, you'll get a brief introduction to several programming languages that you can use in conjunction with AutoCAD.

**Note**   *You can also create custom linetypes, hatch patterns, shapes, and text fonts. You've already learned how to create linetypes in Chapter 8, and how to create hatch patterns in Chapter 12. Because shapes and text fonts are a bit more difficult to create (and because so many good, professionally designed fonts are available already), this book does not cover their creation, although you can use the MKSHAPE Make Shape Express tool to create shapes based on selected entities. If you want to learn more about creating shapes and fonts, you'll find detailed information in AutoCAD's online documentation.*

This chapter explains how to do the following:

- Create and modify command aliases
- Customize AutoCAD's toolbars
- Modify AutoCAD's menus

## Creating Command Aliases

As you've learned in previous chapters, you can start any AutoCAD command by using various different methods. You can click a toolbar button, choose a command from a pull-down menu, or type the command name at the command line. If you're using a digitizer, you may also be able to start a command by clicking a specific area of the tablet. When you type a command at the command line, you often don't need to type the entire command name. Rather, you can type a one-, two-, or three-character abbreviation. For example, you can start the LINE command simply by typing **L** and pressing ENTER. These abbreviations are called *command aliases*.

When you install AutoCAD, numerous command aliases are already defined. As you've read through the instructions in this book for using commands, you've learned several of these aliases. The command aliases are defined in a special file called ACAD.PGP. This file, often referred to as the *AutoCAD program parameters file,* also contains a list of external commands that can be activated from within AutoCAD. The file is located in AutoCAD's support directory (typically \Program Files\AutoCAD 2002\Support\).

The ACAD.PGP file is a simple ASCII file. You can view and modify the contents of this file by using any text editor (such as Windows Notepad or WordPad) or a word processor in text mode. A command alias is defined on a single line, consisting of an abbreviation and a command name. The abbreviation is the alias that you can type to activate the command. The command name can be an AutoCAD command or the command used to start another program (such as an AutoLISP, ADS, ARX, operating system, or graphics display driver command).

Within the ACAD.PGP file, the alias is separated from the command with a comma, and the command itself is preceded by an asterisk. The following listing shows a snippet of the ACAD.PGP file that comes with AutoCAD:

```
;   Sample aliases for AutoCAD commands
;   These examples include most frequently used commands.

3A,         *3DARRAY
3DO,        *3DORBIT
3F,         *3DFACE
3P,         *3DPOLY
A,          *ARC
ADC,        *ADCENTER
AA,         *AREA
AL,         *ALIGN
AP,         *APPLOAD
AR,         *ARRAY
```

Notice that the listing includes blank spaces, so that the columns align for easier readability. You can also include comments on any line within the ACAD.PGP file by typing a semicolon after the command name. AutoCAD ignores anything on the line to the right of the semicolon (including the semicolon itself).

**Note**   *Aliases are limited in that they can only be used to start a command. If the command has several options, you can't specify the options. To start a command and specify options, you need to create a macro using a toolbar, menu, or AutoLISP program.*

**Tip**   *The ALIASEDIT Express tool, described on the companion web site, enables you to create and modify command aliases using an easy-to-use dialog box.*

ADVANCED TOPICS

# Customizing Toolbars

The next step up in AutoCAD customization is the creation of custom toolbars. Like command aliases, toolbars provide an easier way to start commands than typing the entire command name. Toolbars also offer several advantages over aliases. With toolbars, you don't need to take your eyes off the screen or memorize abbreviations. Although AutoCAD already has numerous toolbar buttons, each with its own icon, they are arranged quite logically, and if you forget the purpose of a particular button, you can momentarily pause the arrow cursor over a toolbar button until a ToolTip appears explaining the button's function.

With AutoCAD's toolbar customization tools, you can easily modify existing toolbars or create new ones. All of the toolbar customization begins on the Toolbars tab of the Customize dialog box, shown in Figure 28-1. You've already used this dialog box several times to display toolbars. You also learned how to display toolbars with large buttons and how to enable the display of ToolTips. Notice that in addition to the list of available toolbars and the other controls that you've already used, the dialog box contains a cluster of buttons labeled New, Rename, and Delete. The dialog box also includes three other tabs: Commands, Properties, and Keyboard. The controls on the first three tabs enable you to modify toolbars. The Keyboard tab lets you assign keyboard shortcuts to commands.

To display the Toolbars tab of the Customize dialog box, do one of the following:

- From the View menu, choose Toolbars.
- From the Tools menu, choose Customize | Toolbars.
- At the command line, type **TOOLBAR** (or **TO**) and then press ENTER.
- Right-click any visible toolbar button, and then choose Customize from the shortcut menu.

## Changing a Toolbar Name

When you click the Rename button, AutoCAD displays a Rename Toolbar dialog box, which enables you to define the name of the toolbar. The name is simply the text that appears in the title bar for the toolbar whenever the toolbar is floating.

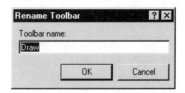

To change the name of any existing toolbar, first select the toolbar in the list on the Toolbars tab of the Customize dialog box and click the Rename button to display the Rename Toolbar dialog box. Then, simply modify the name in the Toolbar Name field and then click OK.

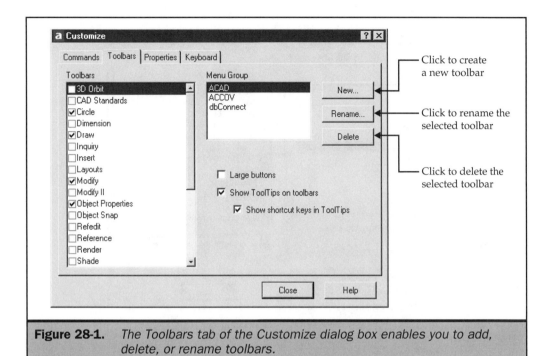

**Figure 28-1.** *The Toolbars tab of the Customize dialog box enables you to add, delete, or rename toolbars.*

# Modifying Existing Toolbars

Although you can easily create your own custom toolbars, AutoCAD already has 26 standard toolbars. Sometimes, rather than create another toolbar, all you really need to do is add a button to one of the existing toolbars. This is probably the easiest way to learn how to customize toolbars.

To modify an existing toolbar, display the Customize dialog box and then switch to the Commands tab, shown in Figure 28-2. When this tab is displayed, you can drag icons from the dialog box to any visible toolbar. You can also move or copy icons from one toolbar to another.

**Note** *You can also create a completely new toolbar by dragging an icon from the dialog box onto any area of the screen other than an existing toolbar. You'll learn about creating new toolbars in the next section.*

## Adding an Icon to a Toolbar

The Commands tab actually contains an icon for every AutoCAD command, even those not included in toolbars. To organize all of those icons, the tab contains two lists. The Categories list on the left arranges all the commands by category. The Commands list on the right displays all the commands in the current category along with its predesigned

ADVANCED TOPICS

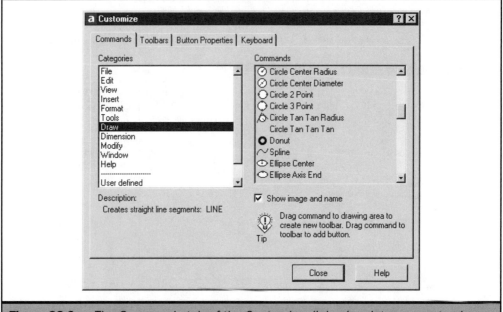

**Figure 28-2.**  *The Commands tab of the Customize dialog box lets you customize existing toolbars and create new ones.*

icon. If a command doesn't have an icon, you'll need to create your own, as explained later in this chapter. To add an icon to a toolbar, first select the proper category from the Categories list, locate the command, and then drag it onto the toolbar.

*LEARN BY EXAMPLE*
*Add a new button to an existing toolbar.*

Many people find the DONUT command quite useful. Unfortunately, the command does not appear on any of the toolbars. However, adding the command to one of the existing toolbars is a simple matter. In this example, you add it to the Draw toolbar:

1. Display the Customize dialog box and switch to the Commands tab.

2. Under Categories, choose Draw.

3. Under Commands, scroll down the list and select Donut, drag it onto the Draw toolbar until it's positioned between the Circle icon and the Spline icon, and then release the mouse button. The Donut icon now appears on the Draw toolbar.

4. Click Close to close the Customize dialog box.

*Notice that some of the toolbar buttons have bars between them to distinguish subgroups of buttons within a toolbar. When the Customize dialog box is open, if you select a button in any visible toolbar, drag it a few pixels away from an adjacent icon, and then release the mouse, a bar appears between the icon you moved and the adjacent icon. Alternately, right-click a button and then choose Begin A Group from the shortcut menu. To remove a bar, simply drag an icon a few pixels toward the bar. You can also right-click the button adjacent to the bar and choose Begin A Group again.*

## Removing an Icon from a Toolbar

You can remove an icon from an existing toolbar as easily as you can add one. The process basically is the reverse:

1. Display the Customize dialog box.
2. In any visible toolbar, select the icon you want to remove, and drag it off of the toolbar (but not onto any other toolbar). AutoCAD displays an alert dialog box asking if you really want to delete the button.
3. Click OK and then click Close to close the Customize dialog box.

# Creating a New Toolbar

Sometimes, you may find that creating a new toolbar containing the commands that you frequently use is more convenient than adding commands to an existing toolbar. You can actually begin to create a new toolbar in one of two different ways:

■ From the Toolbars tab of the Customize dialog box, click New to create a new toolbar. AutoCAD displays the New Toolbar dialog box, in which you specify the name of the toolbar and assign it to one of the loaded menu groups.

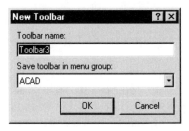

■ From the Commands tab of the Customize dialog box, drag an icon from the dialog box onto any area of the AutoCAD window, except where there is already a toolbar. AutoCAD immediately creates a new toolbar, assigning it the name Toolbar 1 (or Toolbar 2, Toolbar 3, and so on). You can then switch to the Toolbars tab and click the Rename button to assign the toolbar a more appropriate name.

Online

EXAMPLES

*LEARN BY EXAMPLE*
*Create a new toolbar.*

The Circle button on the Draw toolbar starts the CIRCLE command. But, AutoCAD provides five different ways in which you can draw a circle. Use the following steps to create a new toolbar containing all five methods:

1. Display the Toolbars tab of the Customize dialog box.

2. Click New.

3. In the New Toolbar dialog box, under Toolbar Name, type **Circle** and then click OK. AutoCAD creates a new toolbar called Circle. (Look for the new toolbar near the top of the AutoCAD window. It may be difficult to see because it's currently empty.)

Use the following steps to add the five circle buttons to the new Circle toolbar:

1. Switch to the Commands tab.

2. In the Categories list, choose Draw.

3. In the Commands list, locate the button for Circle Center Radius and then drag-and-drop it onto the new Circle toolbar.

4. Repeat Step 3 for the Circle Center Diameter, Circle 2 Point, Circle 3 Point, and Circle Tan Radius buttons. When finished, you should have all five buttons on the new Circle toolbar.

5. Click Close to close the Customize dialog box.

## Adding a Flyout

Flyouts provide a method to help you organize toolbars. A *flyout* is a toolbar that extends out from a particular toolbar button when you click that button and hold down the mouse button. To select a button from a flyout, you continue to hold down the mouse button while pointing to the button that you want. When you release the mouse button, the toolbar button that you selected is activated, and it becomes the default button on the toolbar.

AutoCAD's default toolbars contain a few flyouts, indicated by a small triangle in the lower-right corner of the button. You can easily add your own flyouts. Again, you do this by using the Commands tab of the Customize dialog box.

*LEARN BY EXAMPLE*
*Add a flyout for the Circle toolbar.*

In this example, you replace the Circle button on the Draw toolbar with a flyout button and then link that button to the Circle menu that you just created:

1. Display the Commands tab of the Customize dialog box.

2. In the Categories list, select User Defined.

3. In the Commands list, select User Defined Flyout and then drag-and-drop it onto the Draw toolbar directly over the current Circle button. A new blank flyout button is added to the toolbar adjacent to the existing Circle button.

4. In the Draw toolbar, click the existing Circle button and remove it from the toolbar by dragging-and-dropping it onto an area of the AutoCAD window where there is no toolbar. Confirm that you really want to delete the Circle button.

5. In the Draw toolbar, click the new flyout. AutoCAD displays an alert box informing you that it cannot find a toolbar associated with the flyout. Click OK to close the alert box. Notice that the Customize dialog box has switched to a tab labeled Flyout Properties.

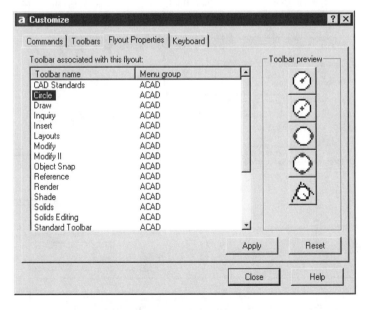

6. In the toolbar list on the Flyout Properties tab, select the new Circle toolbar, and click Close to close the Customize dialog box.

The Circle flyout should now be operational. Test it by clicking the new Circle flyout button on the Draw toolbar. When you click and hold down the mouse button, the flyout should expand to show the entire Circle toolbar.

## Customizing Button Properties

Thus far, all the buttons that you've added to the toolbars have been chosen from those already provided with AutoCAD. In addition, you haven't had to do anything to add the command functions to those buttons, because the commands were also defined along with the buttons. But, you can also create your own buttons and assign to them your own commands, command macros, or even AutoLISP functions. This process consists of two distinct steps: create the custom toolbar button and assign the macro. Assigning the macro is explained first, because creating the button is actually a substep of assigning the button's properties.

You control the command assigned to any toolbar button by using the Button Properties tab of the Customize dialog box, shown in Figure 28-3. To display this version of the Properties tab, select the Properties tab and then click the toolbar button that you want to modify, or right-click the toolbar button any time that the Customize dialog box is open and choose Properties.

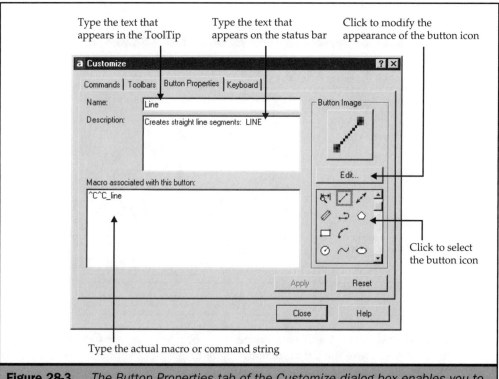

**Figure 28-3.**    *The Button Properties tab of the Customize dialog box enables you to assign macros and customize other button properties.*

The Button Properties tab enables you to control all aspects of the selected toolbar button. The Name field contains the text that appears in the button's ToolTip, while the Description field contains the text that appears in AutoCAD's status bar whenever you pause the mouse over that button. The text in the Macro field is the actual command sequence that AutoCAD executes when you click the button.

**Note** *The Macro field uses some special characters. For example, the caret and letter C combination (^C) cancels any active command (similar to pressing* ESC*). This sequence appears twice, to both cancel an active command and remove any grips that may be displayed. The underscore character appearing in front of a command name translates commands and keywords when you work in languages other than English. A backslash (\\) causes AutoCAD to pause for user input. Within a macro, a series of individual commands can be separated by semicolons, which behave the same as pressing the* ENTER *key. Numerous other characters are used in menu and macro syntax. You can find a complete list in AutoCAD's online* Customization Guide, *included as part of the AutoCAD Help file system.*

The Button Image shows the icon currently assigned to the button. You can modify the icon or design your own by clicking the Edit button. You'll learn how to create your own icons in the next section.

**LEARN BY EXAMPLE**
*Create a new toolbar and custom button to insert the architectural callout symbol that you created in Chapter 15. Open the drawing Figure 28-4 on the companion web site.*

In this example, you create a new toolbar and add a custom button:

1. Display the Customize dialog box.
2. On the Toolbars tab, click New.
3. In the New Toolbar dialog box, under Toolbar Name, type **My Toolbar**, and then click OK.
4. Switch to the Commands tab of the Customize dialog box.
5. In the Categories list, choose User Defined.
6. In the Commands list, select User Defined Button, and then drag-and-drop the blank button onto the new toolbar that you just created.
7. Click the new blank button to display the Button Properties tab of the Customize dialog box.
8. On the Button Properties tab, type **Detail Callout** in the Name field.
9. In the Description field, type **Inserts an architectural detail callout**.
10. In the Macro Associated With This Button box, place the text cursor to the right of the two ^C's already placed there automatically by AutoCAD, type **_-INSERT;DETAIL;\\;;;\\\\**, and then click Apply.

ADVANCED TOPICS

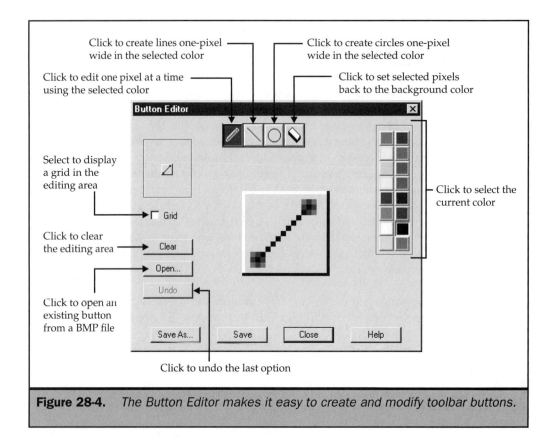

Click to create lines one-pixel wide in the selected color

Click to create circles one-pixel wide in the selected color

Click to edit one pixel at a time using the selected color

Click to set selected pixels back to the background color

Select to display a grid in the editing area

Click to select the current color

Click to clear the editing area

Click to open an existing button from a BMP file

Click to undo the last option

**Figure 28-4.** *The Button Editor makes it easy to create and modify toolbar buttons.*

That completes the steps for creating the new button and assigning the macro to the button. The macro you just created does the following:

- Cancels any active command
- Starts the INSERT command at the command line and inserts the Detail block
- Pauses for you to specify the insertion point and then accepts the default scaling factor and rotation angle
- Pauses while you specify attributes for the detail number and sheet number

 *When you enter the macro string, you need to enter the exact sequence of equivalent keystrokes that you would normally enter if using the command from AutoCAD's command line. You may need to experiment a bit to get this exactly right.*

If you want to, you can close the dialog boxes and try out your new macro button. The button doesn't have an icon, however, making it difficult to select it from among all the other toolbar buttons. Because this isn't one of AutoCAD's standard commands, no predefined button is available. Instead, you need to create your own button icon.

## Creating Custom Button Icons

Toolbar buttons are nothing more than Windows bitmap (BMP) files. Although you can create your own custom button icons by using any paint program or bitmap editor, it is much easier to use the button editor built into AutoCAD. You can display the Button Editor dialog box, shown in Figure 28-4, at any time the Customize dialog box is open by selecting the button you want to modify (so that it appears in the Button Properties tab) and then clicking the Edit button.

The Button Editor displays the current icon, along with a collection of tools for editing and saving the button. You can use these tools to edit the button in the editing area in the middle of the dialog box. While you modify the button in this enlarged view, the image in the upper-left corner of the dialog box displays the button at its actual size. When you finish creating or modifying the button, click the Save or Save As button to save the icon as a BMP file.

*LEARN BY EXAMPLE*
*Create a custom button for the Detail Callout command that you created in the previous exercise.*

Continue the previous exercise. If you already closed the dialog boxes, display the Customize dialog box and then click the blank button in the My Toolbars toolbar that you created earlier. The blank Detail Callout button should appear on the Button Properties tab of the Customize dialog box. Then, follow these steps:

1. On the Button Properties tab of the Customize dialog box, click the Edit button.

2. In the Button Editor dialog box, select the Grid check box (so that a check mark appears in the box and the grid is visible in the editing area).

3. In the Button Editor dialog box, click the circle tool.

4. Click in the center of the editing area and drag the cursor until it is three squares from a corner (measured diagonally). When you release the mouse button, you should see a circle centered in the editing area. If you make a mistake, just click Undo and try again.

5. Click the line tool and then draw a line horizontally across the diameter of the circle.

6. Click a different color in the color palette.

7. Click the pencil tool and then draw symbols to represent the detail number (above the line) and sheet number (below the line).

8. Click Save As. In the Save As dialog box, select AutoCAD's Support directory (or some other directory that you know is in AutoCAD's Support search path), specify a name for your new button icon (AutoCAD automatically adds the .BMP file extension), and click Save.

*Although you could save your new button icon simply by clicking the Save button in the Button Editor dialog box, doing so places the BMP file in whatever directory you started AutoCAD. If that directory is not on AutoCAD's search path, the next time that you start AutoCAD, you'll see smiley faces instead of your custom buttons. In addition, when you use the Save button, AutoCAD assigns the icons generic names, such as ICON0941.BMP. You might inadvertently delete these files, not realizing what they are. For that reason, you should also give the BMP files meaningful names.*

9. In the Button Editor dialog box, click Close.

10. In the Button Properties dialog box, click Apply and then click the Close button.

The new Detail Callout button should now be operational. Test it by clicking the button that you just created on your My Toolbars toolbar. When you click the button, AutoCAD should prompt you for the insertion point. Once you specify the insertion point, AutoCAD should either prompt you for the detail number and sheet number or display an Enter Attributes dialog box.

*The use of the Enter Attributes dialog box is controlled by the ATTDIA system variable, which you learned about in Chapter 15. When you create a macro to insert a block containing attributes, be sure to include enough backslash characters to account for all the attributes, in case the ATTDIA value is 0. When AutoCAD uses an Enter Attributes dialog box, it ignores the extra backslash characters in the macro.*

## Modifying Menus

The command macro that you created in the last set of exercises was your first step in actually programming AutoCAD. In addition to customizing toolbars and command aliases, you can modify AutoCAD's pull-down menus.

AutoCAD actually has six different types of menus:

■ Pull-down and shortcut menus

■ Toolbars

■ Image tile menus

■ Screen menus

■ Tablet menus

■ Button and auxiliary menus

All six types of menus can be defined within a single menu definition file. You can modify an existing menu file, or create your own. Unlike aliases and toolbars (which

## Saving Toolbar Customization

AutoCAD uses several different files for each menu. For example, the original menu template is saved in a file with an .MNU file extension. The menu source file for that menu has the same filename, but is given an .MNS file extension. Similarly, when the menu is compiled, AutoCAD creates another file with the same filename, assigning it an .MNC file extension. (Refer to Appendix E for additional file extensions.)

When you customize toolbars by using the dialog boxes, AutoCAD saves the changes to the menu file with the .MNS file extension. When you edit a menu manually, however, you normally modify the menu file with the .MNU file extension. After you modify an MNU file, if you then load that file, AutoCAD overwrites the MNS file, thus deleting your toolbar customization. You therefore need to take steps to insure that you don't lose your changes. When you are satisfied with your toolbar customization, and before you manually edit the MNU file, you need to copy the MNS file over the MNU file. Alternatively, you can copy just the toolbars section from the MNS file and paste it over the existing toolbar section in the corresponding MNU file.

As with any manual customization you perform on any of AutoCAD's files, you should always make a backup copy of the menu file before you make any changes.

are part of the more generalized menu file), AutoCAD doesn't provide a special tool for modifying menus. Instead, you must use a text editor.

*This section looks at how to customize AutoCAD's pull-down menus. You can use many of the same principles to create custom button, screen, and tablet menus. You can find complete information about creating custom menus in AutoCAD's online* Customization Guide, *available as part of the Help file system.*

# Understanding the Pull-Down Menu Structure

A pull-down menu is the rectangular menu that drops down below a menu bar when the menu bar is selected. These menus are defined in the pull-down menu section of an AutoCAD menu file. The shortcut menus that appear when you right-click your mouse are also defined within the pull-down menu section of an AutoCAD menu file.

Within the menu file, a pull-down menu section is indicated by the section headers ***POP1 through ***POP16. Pull-down menus can also have cascading submenus that are arranged in a hierarchical fashion within the menu file.

*If no POP1 through POP16 section is defined, AutoCAD inserts default File and Edit menus. Any menu sections greater than POP16 and less than POP500 are available to be inserted into the menu bar with the MENULOAD command or by menu swapping. The POP0 section defines the default Object Snap shortcut menu, while sections POP500 through POP999 are used for context-sensitive shortcut menus.*

**ADVANCED TOPICS**

The first line after a pull-down menu is an optional menu name tag. The next line (or the one immediately below the pull-down menu header, if no menu name tag occurs) is the title that appears in the menu bar, which occurs alone on the next line, enclosed in square brackets ([]). Each successive line after the header represents a pull-down menu item.

The first item on each line is the name tag, a string consisting of an alphanumeric, an underscore character, and a unique name that identifies an item within the menu file. Name tags are optional, but they help link pull-down menu items to their associated status line Help. The next item on each line is the text that actually appears on the particular line of the pull-down menu. This text is also enclosed in square brackets. The last item on the line is the menu macro, the AutoCAD command sequence that executes when you click the menu item. The menu items continue until the next pull-down menu section header.

A number of characters have special meaning when included in pull-down menu labels. Some of the most common characters and their meanings are shown in Table 28-1. You'll find a complete listing in AutoCAD's online *Customization Guide*.

One of the best ways to learn how to customize menus is to look at the existing AutoCAD menu. Figure 28-5 shows the Dimension pull-down menu along with the POP8 segment from the default AutoCAD menu definition file, ACAD.MNU, the portion

| Character | Description |
|---|---|
| -- | Displays a separator line in the pull-down or cursor menu. |
| + | Continues a macro onto the next line (when used as the last character on a line). |
| -> | Indicates that a pull-down or cursor menu item has a submenu. |
| <- | Indicates that this item is the last item in a submenu. |
| <-<-... | Indicates that this item is the last item in a submenu, and terminates the parent menu. |
| $ | Enables a pull-down or cursor menu to evaluate a DIESEL expression (when used as the first character). DIESEL, which is an acronym for *Direct Interpretively Evaluated String Expression Language,* is a programming language that can be used in menus, to alter the appearance of menu items. |
| ~ | Grays out a menu item. |
| !. | Marks a menu item with a check mark. |

**Table 28-1.** *Common Characters Used in Menu Files*

| Character | Description |
|-----------|-------------|
| & | Specifies that the letter that follows is a menu accelerator key in a pull-down or cursor menu label. |
| /c | Specifies that the letter that follows appears as the menu accelerator key the first time that it appears in the menu label. |
| \t | Specifies that all label text to the right is pushed to the right side of the menu. |

**Table 28-1.**   *Common Characters Used in Menu Files* (continued)

that defines this pull-down menu. The two have been aligned so that you can better understand how the syntax relates to the menu.

The code ***POP8 marks the start of the menu, the pull-down section header. The next line, **DIMENSION, identifies the pull-down menu name. The third line, ID_MnDimensi [Dime&nsion], creates the text that appears on the menu bar and

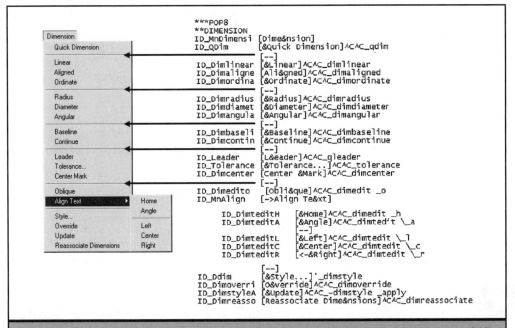

**Figure 28-5.**   *The Dimension pull-down menu along with its code from the AutoCAD menu definition file*

ADVANCED TOPICS

identifies that the letter *n*, the letter following the ampersand, is the accelerator key. Each line containing the code [–] creates the horizontal separator lines in the pull-down menu. Notice that line 22 includes the code [->Align Te&xt]. This marks the start of the Align Text submenu. The next six lines contain the code for that submenu. Notice that the last of those lines contains the code [<-&Right]. This marks the end of the submenu, and the remaining five lines are therefore the portion of the pull-down menu that appears below the submenu.

# Creating a Custom Menu

As you learned in previous chapters, AutoCAD can load more than one menu at a time. For example, when you use the DBCONNECT command, AutoCAD adds a dbConnect menu. These menus load as partial menus—you load just a portion of the menu file—providing access to specific toolbar or pull-down menus without replacing the rest of the main AutoCAD menu.

**Note**   *While you can certainly create custom menus by editing the main AutoCAD menu, doing so requires that you recompile and reload the entire main menu. AutoCAD will do this for you automatically when you load the menu by using the MENU command, but doing so causes AutoCAD to replace any customized toolbars that you may have added via the Customize dialog box. The current discussion of custom menus avoids this potential problem by creating a new menu and loading it as a partial menu. You can also eliminate this problem either by defining new toolbars in a separate menu or by creating them within the ACAD.MNU file, using a text editor. If you do modify any of the AutoCAD menu files, be sure to save a backup copy prior to beginning your customization.*

You load partial menus by using the MENULOAD command. Before you load the menu, however, you need to create it.

**LEARN BY EXAMPLE**
*Create a custom pull-down menu. Open the drawing Figure 28-4 on the companion web site.*

In this exercise, you will use a text editor to create a new pull-down menu, and then load that pull-down menu as a partial menu. Begin by loading a text editor such as Windows Notepad or WordPad. Then, type the menu exactly as shown in the following code. Be sure to press ENTER at the end of each line *and* at the end of the last line. Otherwise, the last menu item won't appear.

**Note**   *You can also find this menu on the companion web site.*

```
***MENUGROUP=MYMENU

***POP1
[My Symbols]
[&Detail Callout]^C^C_-insert;detail;\;;;\\
[D&oor Symbol]^C^C_-insert;door;\;;\
```

Save this file as **MYMENU.MNU**. Be sure to save it as an ASCII text file.

After you create and save the menu file, you can load it into AutoCAD as a partial menu by using the following steps.

1. Do one of the following:

   ■ From the Tools menu, choose Customize | Menus.

   ■ At the command line, type **MENULOAD** and then press ENTER.

2. In the Menu Customization dialog box, click Browse.

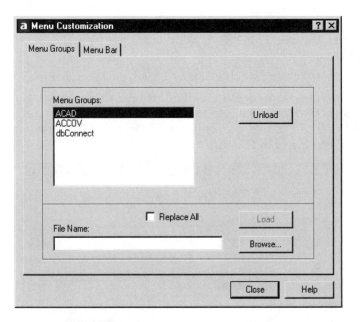

3. In the Select Menu File dialog box, under Files Of Type, select Menu Template (*.mnu). Then, locate the menu file that you just created (MYMENU.MNU), select it, and click Open.

4. In the Menu Customization dialog box, click Load.

**Note**   *AutoCAD displays a warning, telling you that loading a template menu file (MNU file) overwrites and redefines the menu source file (MNS), which results in the loss of any toolbar customization changes that have been made.*

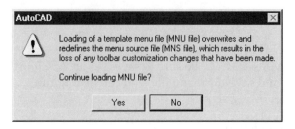

*Only the MNS file associated with the MNU file being loaded is overwritten, however. Because you have not added any customized toolbars to that file, you have none to lose, which is why you created the custom menu in its own file. Had you modified the ACAD.MNU file, you would have lost the toolbar modifications that you made earlier in this chapter.*

5. Click Yes. The menu MYMENU.MNU now appears in the list of available menu groups.

6. In the Menu Customization dialog box, click the Menu Bar tab.

7. Under Menu Group, choose MYMENU from the drop-down list. Under Menus, make sure My Symbols is selected. Then, under Menu Bar, select Format and then click Insert. The pull-down menu My Symbols is inserted into the menu bar between Insert and Format.

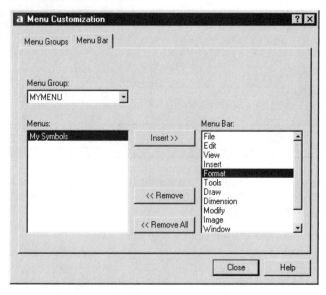

8. Click Close to close the Menu Customization dialog box.

The new menu, My Symbols, should appear on AutoCAD's menu bar between the Insert and Format selections. Try inserting the Detail Callout and Door Symbol by choosing the appropriate menu items.

You can experiment by adding other commands to the menu file and then reloading the partial menu.

*After manually editing your partial menu, remember to unload MYMENU (by selecting the menu in the Menu Groups list on the Menu Groups tab of the Menu Customization dialog box and then clicking the Unload button) and then reload the MNU file. Otherwise, the changes you made won't be loaded.*

## Adding Help Messages to Menu Items

When you highlight an item in a pull-down menu, AutoCAD displays a Help message on the status bar, similar to the messages that are displayed on the status bar when you pause the cursor over a toolbar button. You already learned how you can add these Help messages to your custom toolbar buttons. You can also add these messages to custom menus, but you must again use a text editor to do so.

To add Help messages to your custom menu, you must assign your menu a menu group name. This name appears before any of your pop-up menus and helps AutoCAD identify the individual menus.

In addition to the menu group name, you need to add a name tag for each menu item. Again, this helps AutoCAD identify each item in the menu, so that it can match the command or macro with the proper Help message. Finally, you must add another section to the end of your menu file, called \*\*\*HELPSTRINGS. This section contains all the Help messages, keyed to the menu item's name tag.

### LEARN BY EXAMPLE

*Add Help messages to your custom menu. Open the file MYMENU.MNU that you created in the previous exercise. This file is also found on the companion web site.*

Load the menu file in a text editor such as Windows Notepad or WordPad. Then, edit the menu so that it appears as shown in the following code. (Note that the new items are shown in bold, to help you see where the changes are made.)

```
***MENUGROUP=MYMENU
***POP1
[My Symbols]
```

ADVANCED TOPICS

```
ID_detail    [&Detail Callout]^C^C_-insert;detail;\;;;\\
ID_doorsym  [D&oor Symbol]^C^C_-insert;door;\;;\

***HELPSTRINGS
ID_detail    [Inserts an architectural detail callout]
ID_doorsym  [Inserts a door symbol in plan view]
```

Notice that the name tags are repeated in both the POP1 menu definition section and the HELPSTRINGS section, so that AutoCAD can match the Help message with the appropriate menu item. After you make these changes, save the menu file again. Then, unload the menu and reload it.

**Note**    *Be sure to reload the MNU file, not the MNC or MNS file that were created when AutoCAD compiled the MNU file. Also, note that the name tags are case-sensitive, so you must make sure that they match exactly in the menu definition and HELPSTRINGS sections.*

# Adding Keyboard Shortcuts

Another way in which you can customize AutoCAD is to add shortcut or *accelerator* keys, which, when pressed in combination with the CTRL or SHIFT key, can execute AutoCAD commands or macros. Like the toolbars, pull-down menus, and Help messages, keyboard shortcuts are defined in a special section in the menu file, called ***ACCELERATORS.

You can actually create and modify keyboard shortcuts using the same Customize dialog box you used to create and modify toolbars. The Keyboard tab, shown in Figure 28-6, contains three lists. The Menu Group list contains all the menus currently loaded. The Categories list displays all the menus and toolbars contained in the menu you select in the Menu Group list. The Commands list shows all the commands contained in the menu or toolbar you select in the Categories list.

In addition to the methods you've already learned for displaying the Customize dialog box, you can display the Keyboard tab of the dialog box by choosing Customize | Keyboard from the Tools menu.

When you select a command in the Commands list, if it already has a keyboard shortcut assigned, that shortcut appears in the Current Keys field. You can add a keyboard shortcut for the selected command by clicking in the Press New Shortcut Key field and then pressing the desired key combination.

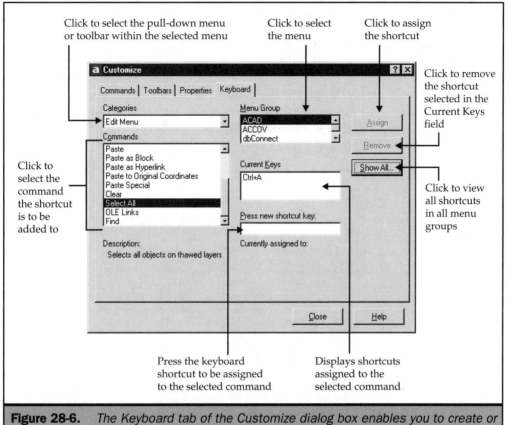

Click to select the pull-down menu or toolbar within the selected menu

Click to select the menu

Click to assign the shortcut

Click to remove the shortcut selected in the Current Keys field

Click to select the command the shortcut is to be added to

Click to view all shortcuts in all menu groups

Press the keyboard shortcut to be assigned to the selected command

Displays shortcuts assigned to the selected command

**Figure 28-6.**   *The Keyboard tab of the Customize dialog box enables you to create or modify keyboard shortcuts.*

You can use the CTRL key, or a combination of the CTRL, ALT, and SHIFT keys, along with most of the keys on the keyboard (including the function keys, numeric keypad keys, INSERT, DELETE, and the arrow keys). If the key combination you specify is already assigned to another command in the currently selected menu group, AutoCAD immediately displays a message directly below the Press New Shortcut Key field. If you don't want to reassign the key combination, press the BACKSPACE key to cancel the keyboard shortcut.

ADVANCED TOPICS

*When you specify a shortcut key assignment, AutoCAD only checks to see if the shortcut is already assigned in the currently selected menu group. While you can assign the same shortcut to commands in different menu groups, to avoid confusion, you shouldn't assign the same shortcut to menu groups that are likely to be loaded at the same time. You can click the Show All button to view a list of all the shortcut key assignments for all the menu groups currently loaded.*

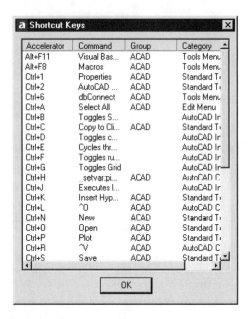

---

### LEARN BY EXAMPLE

*Add a keyboard shortcut to the command you created for inserting the Detail Callout.*

Continue the previous exercise by adding a keyboard shortcut to the command for inserting the Detail Callout. Display the Keyboard tab of the Customize dialog box, and then follow these steps:

1. In the Menu Group list, select MYMENU.

2. In the Categories list, select My Symbols Menu.

3. In the Commands list, select Detail Callout.

4. Click in the Press New Shortcut Key field, and then type CTRL-SHIFT-C (press all three keys simultaneously).

5. Click the Assign button.

6. Click Close to close the Customize dialog box.

The new keyboard shortcut should now be operational. Test it by pressing CTRL-SHIFT-C. AutoCAD should prompt you for the insertion point, just like when you use the toolbar button or the pull-down menu you created earlier.

**Note** *AutoCAD adds the new keyboard shortcuts to the MNS version of your menu file, as it does when you use the Customize dialog box to modify or create toolbars. You should therefore use the same caution when customizing keyboard shortcuts as you would when doing any other menu customization, as described earlier in this chapter. Also use care when assigning keyboard shortcuts, because many are already reserved for Windows or AutoCAD functions. For example, you can't assign any combination using the F1 or ESC keys.*

**Caution** *Keyboard shortcuts do not work unless the Windows Standard Accelerator Keys check box is selected on the User Preferences tab of the Options dialog box.*

**ADVANCED TOPICS**

# Chapter 29

## Some Programming Required

In the previous chapter, you learned some of the ways in which you can customize AutoCAD without having to do any actual programming. In fact, you did begin to learn something about programming when you customized the AutoCAD menus. Reduced to its basics, programming is simply the description of a process or series of steps to be performed by a computer, using a language that can be interpreted by the computer.

Although programs can follow a linear sequence, executing one step after another in the order in which they are included in the program, most programming languages include the means to branch to other portions of the program, based on logical situations. For example, a program may perform one operation if a particular condition exists, and perform a completely different process if that condition does not exist. Many different programming languages are available from which you can choose. Some programming languages use a syntax that is very similar to English. Others utilize a more cryptic syntax that requires a bit more effort to master.

AutoCAD supports several different programming languages and methods. The most basic way to program AutoCAD is to create scripts. An AutoCAD *script* contains a sequence of AutoCAD commands, to be executed one after another, similar to macros used in many other applications. You can also use AutoLISP, a subset of the common LISP programming language, which was invented in the late 1950s for use in artificial intelligence. While AutoLISP programs can be used simply to execute a sequence of commands, AutoLISP also supports logical conditions and branching. AutoLISP is used, both by regular users and by programmers, to create everything from simple macros to complete add-on programs that customize AutoCAD to meet very specific needs. AutoLISP is used by professional programmers to create sophisticated applications, but it can be used by practically anyone.

More-experienced programmers can also customize AutoCAD through its ARX programming environment. ARX, which stands for *AutoCAD Runtime Extension*, is a compiled-language programming environment for developing AutoCAD add-on applications. Unlike AutoLISP, which is an interpreted language (uncompiled) that is built right into AutoCAD, ARX requires the use of a separate programming language, such as C++. Before the C++ ARX programs can be used, they must be compiled and linked with AutoCAD's ARX run-time libraries. Developing ARX applications is well beyond the scope of this book, but you can find more information about creating ARX applications in the *AutoCAD Customization Guide*, part of AutoCAD's online Help system.

AutoCAD also supports Microsoft's *ActiveX Automation* programming interface (previously known as OLE Automation). With ActiveX Automation, users with some programming experience can create add-ons for AutoCAD using other programming environments, such as Microsoft Visual Basic.

This chapter explains how to do the following:

- Create scripts

- Create basic AutoLISP routines
- Integrate other applications with AutoCAD, using ActiveX Automation

## Creating Scripts

An AutoCAD script consists of nothing more than a series of AutoCAD commands, to be executed in succession. These commands usually consist of actual AutoCAD commands, along with the user responses expected by the commands. But scripts can also contain commands in the form of AutoLISP functions, in much the same way that menus can contain AutoLISP functions. Unlike actual AutoLISP functions, however, scripts can't perform conditional branching, nor can they pause for user interaction. AutoCAD cancels any script that tries to pause for you to perform an action. For example, you can use a script to insert a block, but the script must include the block insertion point, scale factors, and rotation angle. The script terminates if AutoCAD has to prompt you for any of this information. The current script also terminates when you invoke another script, or press the ESC or BACKSPACE key.

**Note** *You can get around this limitation by incorporating AutoLISP functions within the script and then using the RESUME command to restart the script from the point at which it was canceled. However, when user interaction is required, simply using AutoLISP instead of scripts usually is better.*

Scripts provide an easy way to create continuously running presentations, such as slide shows. You can also use scripts for many other repetitive tasks, such as turning on and off sets of layers, setting a number of parameters whenever you start a drawing, loading several AutoLISP routines, or performing the same modifications to a batch of drawings. Scripts are saved in an ASCII text file with a .SCR file extension. After you save a script to a script file, you can run the script by using AutoCAD's SCRIPT command.

**Note** *Scripts need to use the command-line version of commands that can be run from either a dialog box or the command line. If a command does not offer a command-line version, the dialog box is displayed. For example, in a script, you should use the -INSERT command rather than the INSERT command.*

Since script files are ASCII text files, you must create them by using a separate text editor, such as Windows Notepad or WordPad. Also, because script files do nothing more than pass a series of *canned* instructions to AutoCAD, the file must contain the precise sequence of steps expected by AutoCAD. Every character in the script file, including blank spaces, is significant. In fact, AutoCAD interprets a blank space as if you had pressed the ENTER key. Therefore, the script must duplicate AutoCAD's command syntax exactly.

ADVANCED TOPICS

*If you type the sequence of commands in AutoCAD, you can switch to the command window (by pressing the F2 key), copy the sequence to the Windows Clipboard, and then paste the sequence into a text editor. Similarly, AutoCAD's Log File feature can be used to retrieve the sequence of steps that was typed into the keyboard. When you construct a sequence of steps for a script, be sure to issue the command-line version of a command by preceding it with a dash (use the format −XREF, for example, rather than XREF). You may also need to turn off the ATTDIA, CMDDIA, and FILEDIA system variables before running a script.*

The only exceptions to this rule involve the use of comments and the inclusion of long filenames in scripts. Any line beginning with a semicolon in a script file is considered a comment, and AutoCAD ignores everything on that line. Unlike menus, however, semicolons can't be substituted for spaces. Scripts fail if AutoCAD encounters a semicolon anywhere other than at the beginning of a line. In addition, because AutoCAD normally interprets spaces within scripts as being the same as pressing the ENTER key, to include a long filename (filename containing spaces) within a script, you must enclose the filename within double quotes.

To run a script:

1. Do one of the following:

   ■ From the Tools menu, choose Run Script.

   ■ At the command line, type **SCRIPT** (or **SCR**) and then press ENTER.

2. In the Select Script File dialog box, select the script file that you want to run, and then click Open. AutoCAD immediately runs the script.

*LEARN BY EXAMPLE*
*Create a simple script and then run the script. Open the drawing Example 29-1 on the companion web site.*

In the first part of this example, you use a text editor to create a simple script. The script that you are creating in the following steps simply draws some text in the current AutoCAD drawing and inserts a block:

1. From the Windows Start button, load Windows Notepad by clicking Start | Programs | Accessories | Notepad.

2. In Windows Notepad, type the following:

```
TEXT 4.5,3 .5 0 AutoCAD 2002
;insert the AutoCAD logo
INSERT LOGO 7.5,1.5 1 1 0
RECTANG 3,1 12,4
ZOOM E
```

Be sure to include one space between each of the commands and values, as shown. Also, be sure to press ENTER at the end of each line, including the last line.

3. In Windows Notepad, choose File | Save. In the Save As dialog box, select the directory in which you want to save the script file, type **MYSCRIPT.SCR** under File Name, and then click Save.

You have just created a simple script. Now, test that script in AutoCAD:

1. In AutoCAD, start the SCRIPT command.

2. In the Select Script File dialog box, select the script file that you just created, and then click Open.

AutoCAD immediately runs your script. All the commands activate very quickly. When the script has finished, you should see the words "AutoCAD 2002" and the Autodesk logo, surrounded by a rectangle.

The script contains the exact same syntax that you would need to type at the command line if you performed the four commands manually. If your script does not run properly, look at AutoCAD's text screen to see where things went wrong, and then reload the script file in your text editor to make corrections. If you still can't find the error, a copy of the script file is on the companion web site.

**Tip** *The UNDO command considers everything performed by a script to be an undo group. Thus, with a single UNDO or U command, you can reverse all the actions performed by the script.*

In addition to all the standard AutoCAD commands, three commands are provided specifically for use with scripts:

- **DELAY** Causes AutoCAD to pause for a specified period of time before continuing with the execution of the next command. DELAY is particularly useful in presentations, such as slide shows, because it pauses the action long enough for the viewer to see each step before it continues to the next.

- **RESUME** Causes an interrupted script to resume from the point at which it was interrupted.

■ **RSCRIPT** Causes the current script to restart from the beginning. Used at the end of a script, this command creates a continuously running presentation.

*When AutoCAD command input comes from a script, AutoCAD assumes that the PICKADD and PICKAUTO system variable values are 1 and 0, respectively. This maintains compatibility of scripts with previous versions of AutoCAD that lacked these variables, and also makes customization easier, because you don't have to check for these values before you execute the script.*

## Invoking a Script when Loading AutoCAD

You can also run a script when you first start AutoCAD, by passing the operating system a command in the following format:

```
acad [drawing] [/t template] [/v view] /b script-file
```

where *drawing* is the name of an existing drawing, *script-file* is the name of the script file that you want to run, and so on. The script file must be the last parameter in the acad program line call. If you pass a drawing filename, the drawing must already exist.

To start AutoCAD in this way, create on the Windows desktop a shortcut to start AutoCAD, and then edit the Target field shortcut properties to include the script. Figure 29-1 shows a typical Shortcut tab whose Target field has been edited in such a way. To create a new shortcut and then display this dialog box, do the following:

1. On the Windows desktop, right-click the existing AutoCAD shortcut.

2. In the shortcut menu, choose Create Shortcut.

3. On the Windows desktop, right-click the new AutoCAD shortcut.

4. In the shortcut menu, choose Properties.

5. In the Properties dialog box, click the Shortcut tab.

6. In the Target field, position the cursor at the end of the line, and then type **/b** followed by the name of the script file that you want to run when you start AutoCAD.

7. Click OK to save your changes and close the Properties dialog box.

*You can use the ScriptPro Utility, part of the AutoCAD Migration Assistance tools, to apply a single script to a list of drawings. You'll find additional information about this and the other Migration Assistance tools in Appendix D.*

## Creating Slide Shows

Slide shows provide a way for you to create and display presentations of your work to an audience. A slide show consists of a series of slides that are then played back by using an AutoCAD command script. Chapter 24 explains how to save an image of the

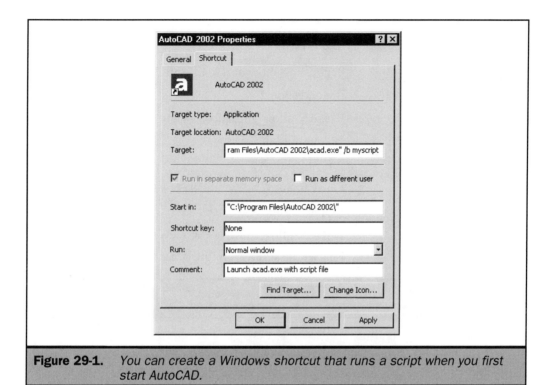

**Figure 29-1.** *You can create a Windows shortcut that runs a script when you first start AutoCAD.*

current AutoCAD viewport as a slide. If you save a series of slides, you can use a script to play back those slides as a presentation.

Although the script for displaying a slide could simply consist of a series of VSLIDE commands, along with the name of each slide, without a built-in delay, those slides would flash onscreen briefly and then be replaced by the next slide in the sequence. This is where the DELAY command comes into play. By inserting a delay between each slide, you can tell AutoCAD to leave the slide onscreen for a long enough period of time for your audience to view it, before loading the next slide.

The speed at which AutoCAD loads the slide is dependent on the speed at which the slide file can be read from your hard drive. This sometimes causes slides to display slowly. AutoCAD provides a neat trick that lets you get around this limitation. You can preload the next slide in the sequence from disk while your audience is viewing the current slide, and then have AutoCAD display that slide. This is accomplished by preceding the name of the slide file with an asterisk (*).

Finally, at the end of the slide show, you can use the RSCRIPT command to restart the presentation from the beginning, thus creating a continuously running presentation. Table 29-1 shows a typical slide show script file, along with an explanation of each line in the file.

**Note** *You can also use scripts to display slides stored in slide libraries, by including the library name and slide name, such as mylibrary(slide1), as you learned in Chapter 24.*

ADVANCED TOPICS

| Program | Explanation |
|---|---|
| VSLIDE SLIDE1 | Begin slide show and load slide 1 |
| VSLIDE *SLIDE2 | Preload slide 2 |
| DELAY 2000 | Pause to allow audience to view slide 1 |
| VSLIDE | Display slide 2 |
| VSLIDE *SLIDE3 | Preload slide 3 |
| DELAY 2000 | Pause to allow audience to view slide 2 |
| VSLIDE | Display slide 3 |
| DELAY 2000 | Pause to allow audience to view slide 3 |
| RSCRIPT | Restart the script from the beginning |

**Table 29-1.**   *A Typical Continuously Running Slide Show Script*

Use the SCRIPT command to start the script. After it starts, the script repeats continuously until you interrupt it, by pressing ESC or BACKSPACE.

*LEARN BY EXAMPLE*
*Create a short slide show and then run the slide show.*

Copy the slide files from the web site onto a folder on your computer. Then, use the following steps to create the slide show:

1. Using a text editor, such as Windows Notepad, create the following script. (*Note:* You may need to change the drive designation and folder name to match the drive and folder on your computer.)

```
VSLIDE d:/folder/slide1
VSLIDE *d:/folder/slide2
DELAY 3000
VSLIDE
VSLIDE *d:/folder/slide3
DELAY 3000
VSLIDE
VSLIDE *d:/folder/slide4
DELAY 3000
VSLIDE
```

```
VSLIDE *d:/folder/slide5
DELAY 3000
VSLIDE
DELAY 3000
RSCRIPT
```

2. Save the script as **SHOW.SCR**. (A copy of this script is also found on the companion web site.)

You can now run the slide show in AutoCAD:

1. In AutoCAD, start the SCRIPT command.

2. In the Select Script File dialog box, select the script file that you just created (SHOW.SCR), and then click Open.

3. The slide show will run continuously. When you are finished viewing the slide show, press ESC or BACKSPACE to interrupt the script.

**Caution**  *Be sure that any of the files called by your script are either in the current subdirectory, on the AutoCAD Support search path, or have their paths included in the script file. Otherwise, AutoCAD won't be able to locate the files and will terminate the script.*

## Introduction to AutoLISP

AutoLISP is a programming language built directly into AutoCAD. With AutoLISP, you can perform arithmetic operations, create variables, execute AutoCAD commands, assign values to system variables, and even create entirely new AutoCAD commands. The power of AutoLISP enables users and independent software developers to provide all sorts of enhancements to the basic AutoCAD program. While a complete AutoLISP programming tutorial is beyond the scope of this book, this section provides you with enough AutoLISP basics to get started.

**Note**  *You'll find a complete guide to AutoLISP included as part of AutoCAD's online Help system.*

AutoLISP is an interpreted language, which means that, instead of being compiled into binary code that only the computer can read, the AutoLISP program source code is read directly and interpreted by the computer when the program is actually run. This means that you can write AutoLISP programs by using any ASCII text editor, and then load and run those programs directly in AutoCAD. You can even type AutoLISP code directly at AutoCAD's command line. As you've already seen, you can also include AutoLISP commands inside menus and scripts. The ability to type AutoLISP commands directly at AutoCAD's command line makes it easy to experiment with AutoLISP.

ADVANCED TOPICS

*AutoCAD also includes Visual LISP, an integrated development environment for developing both interpreted and compiled AutoLISP code. You'll learn more about Visual LISP later in this chapter.*

# Loading AutoLISP Applications

Even if you don't want to do any programming yourself, you can still make use of the many AutoLISP programs available from other sources. In fact, you've already been using many AutoLISP routines in your day-to-day use of AutoCAD. AutoCAD comes with many small AutoLISP programs that enhance its operation.

AutoLISP programs are stored as ASCII text files with the file extension .LSP. You can use a text editor to open any LSP file to determine the function that it performs and how to use the program. Programmers commonly provide a description of the program, along with copyright and author information, as a comment at the beginning of the AutoLISP file. You'll also often find comments throughout the file, explaining the flow of the AutoLISP code. You should follow this same practice with any programs that you create.

Before you can use an AutoLISP application, you must load it into AutoCAD, which you can accomplish by using either the APPLOAD command or the AutoLISP Load function. Loading an AutoLISP program loads the AutoLISP code from the LSP file into your system's memory.

## Using the APPLOAD Command

To load an AutoLISP program by using APPLOAD:

1. Do one of the following:

   ■ From the Tools menu, choose Load Applications or AutoLISP | Load.

   ■ At the command line, type **APPLOAD** (of **AP**) and then press ENTER.

2. In the Load/Unload Applications dialog box (shown in Figure 29-2), locate the file that you want to load, and then click Load. (*Note:* When you load an AutoLISP file, the message box in the lower-left corner of the dialog box indicates that the file was successfully loaded.)

3. Click Close. Depending on how the program was written, it may either begin running or simply be loaded into memory.

 *The Startup Suite is a list of applications that AutoCAD loads each time you start AutoCAD. When you display the Startup Suite dialog box, you can add or delete application files from this list.*

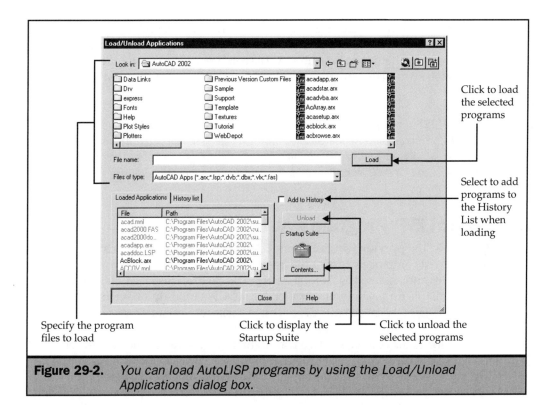

**Figure 29-2.**    *You can load AutoLISP programs by using the Load/Unload Applications dialog box.*

## Loading Programs from the Command Line

To load an AutoLISP program from AutoCAD's command line, you use the AutoLISP Load function. When typed at the command line or included in LSP files, all AutoLISP functions occur inside parentheses. Thus, the Load function appears as

```
(load "filename")
```

where *filename* is the name of the AutoLISP file that you are loading. The filename is enclosed inside double quotes. (The .LSP file extension is not required.) If the AutoLISP file is located in a directory that is specified as part of AutoCAD's search path, AutoCAD locates the file. If the file is not on the search path, you must specify the complete path and filename.

**Note**    *When you specify a complete path, you must use either a forward slash (/) or two backslashes (\\) as the path separator, because a single backslash has a special meaning in AutoLISP (it pauses for user interaction).*

For example, to load the AutoLISP routine MYPROGRAM.LSP, located in the AutoCAD directory, you would type **(load "myprogram")**. If the AutoLISP file is located in a directory that is not on AutoCAD's search path, however, you would have to specify the complete path, such as **(load "c:/jobs/1098/myprogram")**.

**Note** *The rule about enclosing all AutoLISP functions inside parentheses has one exception: If the AutoLISP program includes a custom command function, defined using the Defun AutoLISP function, after you load the AutoLISP file, you can start the custom command function simply by typing the command at AutoCAD's command line, just like any other command. You'll learn how to define and use custom commands in this way later in this chapter.*

## Loading AutoLISP Programs Automatically

As mentioned in the previous section, AutoCAD already uses numerous AutoLISP routines to provide enhancements to the basic program. For example, many of the Express Tools are implemented as AutoLISP programs. In most instances, those programs are already available for use in your copy of AutoCAD. You didn't have to load them to use them, however, because AutoCAD loaded them for you automatically. How did it do that?

AutoCAD loads the contents of two user-definable files automatically—ACAD.LSP and ACAD.MNL (or the MNL file accompanying the AutoCAD menu file that you are currently using)—when you start the program. The program also loads the contents of ACADDOC.LSP and ACAD2000DOC.LSP each time you start or open a new drawing. In addition, if any of these files contains the function S::STARTUP, that function runs immediately as soon as the current drawing is loaded and AutoCAD is fully initialized.

The ACAD.LSP, ACADDOC.LSP, and ACAD2000DOC.LSP files are useful for loading several AutoLISP routines every time that you start AutoCAD or open a new drawing. Each time that you start AutoCAD or start or open a drawing, AutoCAD searches its library path for these files and loads them into memory.

**Note** *If Load acad.lsp With Every Drawing is selected on the System tab of the Options dialog box, AutoCAD reloads the ACAD.LSP file whenever you open a drawing. If this check box is not selected, AutoCAD loads the ACAD.LSP file only when you first start AutoCAD. The ACADDOC.LSP and ACAD2000DOC.LSP files are always reloaded each time you open a new drawing.*

The ACAD.LSP, ACADDOC.LSP, and ACAD2000DOC.LSP files can contain the actual AutoLISP code for one or more routines, or may simply consist of a list of Load functions that load other AutoLISP programs. Using a list of Load functions is actually the better method, because it makes it much easier to modify the ACAD.LSP, ACADDOC.LSP, and ACAD2000DOC.LSP files and the individual programs. For example, a typical ACAD.LSP file might look like this:

```
(load "program1")
(load "program2")
(load "program3")
```

**Note** *Before loading the ACAD.LSP file, AutoCAD loads a special file named ACAD2000.LSP. This file contains AutoLISP functions that are required by AutoCAD. This file actually loads the AutoLISP programs that enable the Express Tools. You should not modify this file in any way.*

The ACAD.MNL file contains any AutoLISP programs that are needed for the proper operation of the associated menu file. For example, if you create a custom menu called MYMENU.MNU, which includes a call to the AutoLISP program MYLINE, rather than include the actual AutoLISP code in the menu file itself, you could include it in the special file MYMENU.MNL. AutoCAD loads this file whenever it loads the menu MYMENU. Again, rather than include the actual code for several AutoLISP programs in the MNL file, the file could simply contain a list of Load functions that, in turn, load the individual AutoLISP programs.

## Handling Load Failures

The AutoLISP Load function actually allows for a second, optional expression, which tells AutoCAD what to do if the AutoLISP file fails to load. A load failure can occur as a result of either an error in the AutoLISP code or AutoCAD's inability to find the specified file. The complete syntax for the Load function is

```
(load "filename" [onfailure])
```

where *onfailure* is an AutoLISP function describing the action that should be taken if the Load function fails. If you want AutoCAD to display a message if any of the programs fail to load that are being loaded by the ACAD.LSP file, you could rewrite the previous ACAD.LSP file as follows:

```
(load "program1" "\nProgram1 didn't load."))
(load "program2" "\nProgram2 didn't load."))
(load "program3" "\nProgram3 didn't load."))
```

If Program2 fails to load for any reason, AutoCAD displays the message "Program 2 didn't load".

ADVANCED TOPICS

### Conserving System Memory

When you load an AutoLISP program by using the Load function, the program is loaded into memory regardless of whether the program is subsequently used. If you load a large number of programs into memory, you may find that AutoCAD runs out of system memory sooner and starts paging to your hard disk. This can slow AutoCAD's performance. Instead of loading all the AutoLISP programs into memory immediately, you can use the Autoload function. Autoload makes the programs available, without loading them into memory. Instead, a program isn't actually loaded into memory until it is used for the first time.

Suppose, for example, that you create an AutoLISP file called MYCMDS that contains three separate AutoLISP commands, CMD1, CMD2, and CMD3. You could replace the Load function in either the ACAD.LSP or MNL file with the following statement:

```
(autoload "MYCMDS" '("CMD1" "CMD2" "CMD3"))
```

AutoCAD makes each of the commands available, but does not load them into memory. The first time that you actually use one of the commands, such as CMD2, AutoCAD automatically loads just that command from the MYCMDS.LSP file.

## Creating an AutoLISP Program

Whenever you type text at AutoCAD's command line, AutoCAD interprets that text by comparing it to an internal list of valid command names. If the word that you type matches a word in that list, AutoCAD performs the task that you specified. When AutoCAD receives AutoLISP code, it passes that code to the built-in AutoLISP interpreter, which in turn evaluates the code. For the code to be evaluated properly, it must adhere to a specific format. All AutoLISP code must be in the following form:

```
(function arguments)
```

where `function` is a valid AutoLISP function, and *arguments* is an optional list of arguments to that function. The entire AutoLISP expression must be contained within a pair of parentheses.

The arguments can be another AutoLISP expression, contained within its own pair of parentheses, and therefore nested inside the first expression, as shown in the following example:

```
(function1 (function2 arguments))
```

The AutoLISP interpreter first evaluates `function2`, the expression nested inside the first function, and then uses the results of that expression as the argument in evaluating `function1`. Depending on the AutoLISP function, the function may have one, two, or more arguments. For example, you can use AutoLISP to multiply two or more numbers:

```
(* 2 3 4)
```

This expression returns 24 as the result. If you type this expression at AutoCAD's command line, AutoCAD also returns the result on the command line, because the expression is not contained in a surrounding expression. While AutoLISP is useful for doing quick mathematical calculations, its real power comes from storing values to user-defined variables, and performing repetitive tasks.

## Defining Variables

You can define variables and assign values to them by using the AutoLISP function Setq. After you assign a value to a variable by using Setq, you can use the variable as the argument for other functions. For example, you could use the following expression to store the value of the previous calculation to the variable A:

```
(setq A (* 2 3 4))
```

The value 24 is thus stored to the variable A. You can then use that variable as the argument in other expressions, without repeating the calculation. For example, the AutoLISP expression (/ A 2) returns the value 12, because it divides the value of A (24) by 2. Variables can store any of the types of information shown in Table 29-2.

## Calling AutoCAD Commands

You can use AutoLISP to run AutoCAD commands, which is accomplished by using the Command function. For example, you could rewrite the door-symbol insertion macro from the previous chapter as an AutoLISP function, as shown in the following example:

```
(command "insert" "door" pause 1 1 pause)
```

This function also illustrates the use of the special AutoLISP variable Pause. Recall that when you create a menu macro, you use a backslash to cause AutoCAD to pause for user interaction. In the case of the door-symbol insertion macro, you specify the insertion point and then, after the block is scaled appropriately, specify the rotation angle. In an AutoLISP function, the Pause variable serves the same purpose.

## Saving System Variables

When you develop AutoLISP routines, you often need to know the state or value of one or more AutoCAD system variables. For example, when you insert a door symbol by using the routine in the previous example, you may want to turn on Orthomode so that the door swing is always set at a 90-degree angle. You can retrieve and set system

| Data Type | Meaning |
|-----------|---------|
| Integer | Whole numbers, which can be positive or negative. Integers have no decimal points, denominators, or exponents. |
| Real | A number containing a decimal point and stored in double-precision floating-point format. Numbers between −1 and 1 must have a leading 0. |
| String | A group of characters enclosed within double quote marks and treated as text. |
| List | A group of related values, separated by spaces and enclosed within parentheses. |
| Selection Set | A group of one or more objects or entities. |
| Entity Name | A numeric label assigned by AutoCAD to objects in the drawing. |
| File Descriptor | A label assigned to files opened by AutoLISP. |

**Table 29-2.**    *Data Types Supported by AutoLISP*

variable values by using the AutoLISP functions Getvar and Setvar, respectively. For example, you could turn on Orthomode by using the following AutoLISP expression:

```
(setvar "orthomode" 1)
```

You can simply change the system variable value as needed, but a better practice is to store the system variable's current value to an AutoLISP variable, change the value as needed, and then restore the original value when your macro is finished.

*LEARN BY EXAMPLE*
*Create an AutoLISP macro that saves the current value of a system variable, performs a command, and then restores the original system variable value. Open the drawing Example 29-2 on the companion web site.*

Using an ASCII text editor, such as Windows Notepad, type the following:

```
(setq ORT (getvar "orthomode"))
(setvar "orthomode" 1)
(command "insert" "door" pause 1 1 pause)
(setvar "orthomode" ORT)
```

This macro does four things. The first line saves the current Orthomode value to the variable ORT. The second line changes the Orthomode value to 1 (Orthomode equals 1). The third line contains the actual insert-door macro. Finally, the fourth line restores Orthomode to whatever value it had prior to the start of the macro.

Save the file using the filename **INSDOOR.LSP**, and be sure to include the .LSP file extension. A copy of this AutoLISP file is also provided on the companion web site.

 *Save the file to a directory on AutoCAD's search path, or be sure to remember the directory in which you save the file. If the file is not in a directory on AutoCAD's search path, you will need to include the path when you load the AutoLISP file.*

In AutoCAD, make sure Orthomode is off (the word ORTHO is not selected on AutoCAD's Status bar). Load the AutoLISP macro by typing **(load "insdoor")** and pressing ENTER. Look at the status bar. The macro has turned on Orthomode, and now prompts you for the insertion point of the door symbol. After you insert the door, the macro turns off Orthomode (its setting before you started the macro).

You can, and should, use this method to save the current values of any system variables whose values you need to alter within any AutoLISP macros that you create.

## Defining Functions

The macro that you created in the previous example works great, but you have no way to restart it without reloading the AutoLISP file. It would be much handier if you could start an AutoLISP macro by typing the name of the macro at AutoCAD's command line, much as you would start any other AutoCAD command. You can do this by assigning the macro a command function name.

A command function is defined by using the AutoLISP function Defun (define function). A custom command takes the following form:

```
(defun c:name () expression)
```

where *name* is the name that you assign to the command function, and *expression* is one or more AutoLISP functions. Note that in this situation, the c: is not a reference to a disk drive, but rather a special prefix used by AutoLISP to denote a command line function.

Once you load an AutoLISP file containing a function created in this way, you can start the function at any time by typing the command function name.

ADVANCED TOPICS

## Defining Local and Global Variables

Notice the pair of parentheses immediately following the command function name. If your function uses variables, you can define these variables as global or local variables by including a list of variables inside these parentheses, separating the global and local variables with a forward-slash character, such as the following:

```
(defun c:myfunction (var1 var2 / var3 var4) expression)
```

Global variables appear before the forward slash and local variables appear after it. Make sure that at least one space is between the slash and the local variables. In this example, var1 and var2 are defined as global variables, while var3 and var4 are local.

Global variables remain in memory and are available for use by other functions. Local variables are used only by the current function, and are then discarded. If you define variables in your function and don't specify whether they are local or global, AutoCAD saves them as global variables.

*LEARN BY EXAMPLE*
*Modify the macro from the previous example so that it can be started from AutoCAD's command line.*

You can easily modify the macro from the previous example, to assign it the command function name INSDOOR. Then, to start the function, you simply type **INSDOOR** at the AutoCAD command line and then press ENTER.

Using an ASCII text editor or Visual LISP, open the AutoLISP file that you created in the previous example, and then add the additional AutoLISP code shown on the first and last lines of the following listing, to change the macro into a command function. The code that you need to add is also shown in bold:

```
(defun c:insdoor (/ ORT)
    (setq ORT (getvar "orthomode"))
    (setvar "orthomode" 1)
    (command "insert" "door" pause 1 1 pause)
    (setvar "orthomode" ORT)
)
```

The code that you added defines INSDOOR as a command function, with the variable ORT defined as a local variable. Notice that the entire macro you created in the previous example is now nested inside the command function as its expression.

**Tip** *Notice, too, that the macro from the previous example is indented. This doesn't affect the AutoLISP code itself. Rather, it is done simply to make the code easier to read. AutoCAD ignores multiple spaces between variables, constants, and function names. Because of the extensive use of parentheses to nest one AutoLISP expression inside of another, indentation is often used to help you see which functions are nested inside of others. Each level of indentation indicates any additional nesting level. Although you can create indentation either by typing spaces or by pressing the TAB key in your text editor, using spaces eliminates any possibility that AutoCAD could incorrectly interpret the code. If you develop your AutoLISP code using Visual LISP, the VLISP editor automatically formats the code for you.*

Save the changes to a new file, using the filename **INSDOOR2.LSP**. (A copy of the INSDOOR2.LSP file is available on the companion web site.) Now, to load the INSDOOR2.LSP AutoLISP file into AutoCAD, type **(load "insdoor2")**, press ENTER, and see what happens. This time, instead of immediately running the macro, AutoCAD displays the command function name

```
C:INSDOOR
```

on the command line. To start the command function, type **INSDOOR** and then press ENTER.

## Creating Keyboard Macros with AutoLISP

In the previous chapter, you learned that you can create aliases for standard AutoCAD commands and for any custom commands that you create. Unfortunately, those aliases can only start commands. If the commands contain options, you can't create an alias for a macro to select the options automatically. For example, you can type the command alias **BR** to start the BREAK command, but you can't create an alias that automatically uses the BREAK command's First point option. Similarly, you can type the command alias **Z** to start the ZOOM command, but you can't create an alias that performs a zoom window.

AutoLISP doesn't have any of these limitations. You can use AutoLISP to create keyboard macros that both start a command and select a specific command option. For example, the following AutoLISP command function starts the BREAK command and then selects the First point option after you select the object that you want to break:

```
(defun C:BRF ()
   (command "break" pause "f" pause pause)
)
```

ADVANCED TOPICS

After you load the AutoLISP file containing this function, type **BRF** and then press ENTER. AutoCAD starts the BREAK command and prompts you to select the object that you want to break. After you select the object, however, instead of using the point that you used to select the object as one of the break points, AutoCAD prompts you for the first and second break points.

Similarly, the following AutoLISP function starts the ZOOM command and automatically selects the Window option:

```
(defun C:ZW ()
   (command "zoom" "w" pause pause)
)
```

Again, once the AutoLISP file containing this function has been loaded into AutoCAD, you can start the function at any time simply by typing **ZW** and pressing ENTER.

**Tip** *You can create many keyboard macros in this fashion. You don't need to save each macro in a separate file, however. A single AutoLISP file can contain multiple command functions. For example, the two functions described here are both saved in the file MYMACROS.LSP, which can be found on the companion web site.*

## Undefining and Redefining AutoCAD Commands

You can use AutoCAD's UNDEFINE command to replace one of AutoCAD's built-in commands with a user-defined command of the same name. For example, suppose you wanted to encourage users in your office to use the BHATCH command instead of the HATCH command. Using an ASCII text editor, create the following AutoLISP function and save it as **HATCH.LSP**. (Note that this file is also provided on the companion web site.)

```
(defun C:HATCH ()
   (princ "\Enter OLDHATCH to use the original HATCH
command.\n")
   (command "BHATCH")
   (princ)
)
(defun C:OLDHATCH ()
   (command ".HATCH")
   (princ)
)
```

Load the HATCH.LSP file into AutoCAD. Then, undefine AutoCAD's normal HATCH command. To undefine an AutoCAD command:

1. At the AutoCAD command line, type **UNDEFINE** and then press ENTER.

2. When you are prompted for the command name, type the name of the AutoCAD command that you want to undefine (in this case **HATCH**) and press ENTER.

Now, when you type **HATCH** and press ENTER, AutoCAD runs the AutoLISP routine that calls the BHATCH command at the command line, instead of the actual HATCH command.

*NOTE: When you undefine a command, you can still start the actual AutoCAD command by including a period as a prefix to the command name. For example, if you undefined the HATCH command and replaced it with the AutoLISP command function called HATCH, you could still start the actual HATCH command, even though it is currently undefined, by typing .HATCH and then pressing ENTER. In fact, the HATCH.LSP AutoLISP routine does exactly that on the seventh line when it issues the AutoLISP command function to call the HATCH command when you type OLDHATCH. You can protect your menus, scripts, and other AutoLISP programs, so that they still work as intended, by including the period prefix in front of all command names. Unfortunately, the period prefix does not work with the RECTANG command.*

To restore the original built-in AutoCAD command, use the REDEFINE command:

1. At the command line, type **REDEFINE** and then press ENTER.

2. When prompted, type the name of the undefined AutoCAD command that you want to restore (in this case **HATCH**) and press ENTER.

# Introducing Visual LISP

AutoCAD includes Visual LISP (VLISP), a powerful extension of the AutoLISP language that enables AutoLISP to interface with objects via the Microsoft ActiveX Automation interface. VLISP enhances AutoLISP's ability to respond to events through the implementation of reactor functions. VLISP also provides a complete, integrated development environment (IDE) that includes an editor, compiler, syntax checker, debugger, and other tools to increase productivity when customizing AutoCAD.

VLISP contains its own set of windows and menus that are distinct from the rest of AutoCAD, as shown in Figure 29-3. AutoCAD must be running, however, whenever you work in VLISP. To start the VLISP IDE, do one of the following:

■ From the Tools menu, choose AutoLISP | Visual LISP Editor.

■ At the command line, type **VLISP** and then press ENTER.

ADVANCED TOPICS

The VLISP screen contains the following elements:

- **Menu**    You can start VLISP commands by choosing menu items. When you highlight an item on a menu, VLISP displays a brief description of the command's function in the status bar area at the bottom of the window.

- **Toolbars**    VLISP contains five toolbars—Standard, Search, Debug, View, and Tools—each containing a distinct set of VLISP commands. Click a button to start a VLISP command.

- **Console window**    You can type AutoLISP commands, similar to the way you do at the AutoCAD command line. You can also issue VLISP commands from this window instead of using the menu or toolbars.

- **Status bar**    Information is displayed on the status bar based on what you are currently doing in VLISP.

In addition, when you open an AutoLISP file in VLISP, it displays within its own text editor window. VLISP can format your code—indenting it and displaying functions and variables in different colors—so that it is easier to read. For example, if you were to load the INSDOOR.LSP function from the previous example, it would appear in VLISP as shown in Figure 29-3.

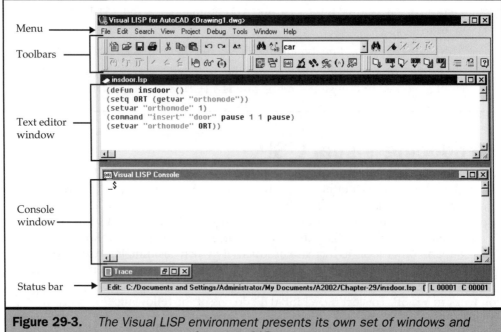

**Figure 29-3.**    *The Visual LISP environment presents its own set of windows and menus separate from AutoCAD.*

You can then test your code in AutoCAD by clicking the Load Active Edit Window button on the Tools toolbar, or by choosing Tools | Load Text In Editor. The focus switches to AutoCAD, and your AutoLISP command is run. After completing the command, the focus returns back to the VLISP window.

 *When you test AutoLISP programs by loading them into AutoCAD, if multiple drawings are currently open, the program is run in the currently active drawing.*

As you've already learned, when you write AutoLISP source code, that code is interpreted—translated into instructions the computer understands—each time you run the AutoLISP program. Interpreted code provides the advantage of being able to make changes and immediately test those changes. Once you are sure that your program is working properly, however, the process that AutoCAD uses to translate that AutoLISP code each time you want to run the program is time-consuming. VLISP provides a compiler that produces executable machine code files from your AutoLISP source code. These executable files have the file extension .FAS. In addition to loading and running faster, compiled code protects your source code, because the source code remains hidden even if you distribute your programs. VLISP also provides features for packaging more complex AutoLISP applications into VLISP executable (VLX) files. VLX files can include additional resource files, such as Visual Basic (VBA) and Dialog Control Language (DCL) files, as well as compiled AutoLISP code.

To compile a single AutoLISP file into an FAS file, you can use the vlisp-compile function. To compile a single AutoLISP file, type the following in the console window:

```
(vlisp-compile 'mode "filename" [out-filename])
```

where *mode* is one of the compilation modes (generally **st** standard build mode for single AutoLISP files), *filename* is the name of the source AutoLISP file, and *out-filename* is the optional name of the compiled output file. If you don't provide an output filename, VLISP uses the same name as the source file, and appends an .FAS file extension.

If the source file is in the AutoCAD support file search path, you don't need to include the path name when specifying the filename. Otherwise, you must specify the complete filename, including the path.

For example, if you copied the INSDOOR.LSP file to AutoCAD's Support directory, to compile the INSDOOR.LSP file, you would type:

```
(vlisp-compile 'st "insdoor.lsp")
```

Visual LISP also includes powerful tools for debugging AutoLISP code. A complete discussion of Visual LISP is beyond the scope of this book. You'll find a complete guide to VLISP, including an online tutorial, in the *Visual LISP Developer's Guide*, part of AutoCAD's Help files.

ADVANCED TOPICS

# Introduction to ActiveX Automation

In addition to all the other programming methods built into AutoCAD, AutoCAD can be customized by using ActiveX Automation, previously known as OLE Automation. To use ActiveX Automation, you must have an application that acts as an Automation controller (such as Visual Basic or Excel). AutoCAD comes with several Visual Basic applications that you can experiment with to learn more about ActiveX Automation. While a complete ActiveX Automation tutorial is beyond the scope of this book, this section provides you with enough ActiveX Automation basics to get started.

**Note** *You can find a complete guide to ActiveX Automation in the* ActiveX and VBA Developer's Guide *and the* ActiveX and VBA Reference, *both of which are included as part of AutoCAD's online documentation. Some of the examples in this section also require that you install AutoCAD's VBA Support, which is an optional component that you can install when you run the Setup program.*

ActiveX Automation provides a way to manipulate AutoCAD both from within and from outside of AutoCAD by exposing various AutoCAD objects. Once these objects are exposed, they can be accessed by different programs and programming environments. By using ActiveX Automation, you can create and manipulate AutoCAD objects from any application that serves as an ActiveX Automation controller. The exposed objects are called *Automation objects*. Automation objects, in turn, expose methods and properties. *Methods* are functions that perform an action on an object. *Properties* are functions that control the state of an object. In ActiveX Automation, AutoCAD acts as the *server*, responding to requests from the *client*, or *Automation controller*, which is the ActiveX Automation application. The client initiates the interaction with the server. The server responds to the commands issued by the client.

## Understanding Objects

An *object* is the main component of any ActiveX Automation application, and each exposed object represents a precise part of AutoCAD. AutoCAD exposes many different types of objects, which can be represented in a hierarchy referred to as the *AutoCAD Object Model*. Objects, in turn, have properties and methods. A *property* describes aspects of the object, such as the center of a circle or the color of a line. A *method* describes actions that can be performed on the object, such as offsetting a circle or moving a line. After you create an object, you can query and edit it based on its properties and methods.

At the topmost level of the Object Model hierarchy is the AutoCAD drawing and the AutoCAD program itself, referred to as the *Application object*. Immediately below the Application object are the Preferences and Document objects. The *Preferences object* provides access to all the settings within AutoCAD's Options dialog box. The *Documents collection* provides access to each of the AutoCAD drawings open in the current session. The *Document object* then provides access to the current AutoCAD drawing. The

Application object also gives you access to application-specific information, such as the application name and version number, and the AutoCAD window size, location, and visibility. You can use this object to start and exit from AutoCAD, as well as to load and unload ARX applications.

**Note** *You can't access AutoLISP routines using ActiveX Automation.*

The Document object is actually the current AutoCAD drawing. It provides access to all the objects in the drawing. Access to graphical objects—such as lines, arcs, and circles—is provided through the ModelSpace and PaperSpace properties. Access to nongraphical objects—such as layer names, linetypes, and text styles—is provided through properties such as Layers, Linetypes, and TextStyles. For ease of access, these properties are grouped as collections. The Document object also provides access to both the Plot object, which provides access to settings in the Plot dialog box, and the Utility object, which provides access to the user-input and conversion functions.

To create a graphical object, you use the appropriate method in the ModelSpace Entities or PaperSpace Entities collection. To modify or query a graphical object, you use the methods or properties of the object itself. Note that visible objects saved as part of a block definition or block insertion are also accessed through the Blocks collection. To create nongraphical objects, you use the Add method of the parent collection object. To modify or query a nongraphical object, you use the methods or properties of the object itself. You can find a diagram of the AutoCAD Object Model in the online *ActiveX and VBA Developers Guide*.

Each object in the hierarchy is permanently linked to its parent object, and all objects originate from a single parent object, called the *root object*. You can access any object by following its links from the root object to the child objects. In addition, all objects have a property called Application that links back to the root object. The root object for AutoCAD is the AutoCAD Application object. All the objects, properties, and methods exposed by Automation objects are described in a *type library*, a special file that stores information about the objects in the Object Model hierarchy. The AutoCAD type library, ACAD.TLB, is located in the main AutoCAD directory. Before using the Automation objects exposed by the application, the application must reference its type library.

## Using Visual Basic for Applications

AutoCAD includes a copy of Microsoft Visual Basic for Applications (VBA). You can use VBA to develop your own applications. VBA is also included with Microsoft Office and is integrated into Microsoft Word and PowerPoint. You can also develop applications by using other programming environments, such as Microsoft Visual Basic. The main difference between developing applications using VBA and Visual Basic is that VBA applications can only be run as embedded applications within AutoCAD, whereas those created using Visual Basic can be compiled into executable programs that can then be run as stand-alone applications.

*To develop your own ActiveX Automation applications by using VBA, you should have a working knowledge of the Visual Basic programming language. This book does not attempt to teach Visual Basic. Many excellent books on Visual Basic are available, including* Visual Basic 6 from the Ground Up *(McGraw-Hill/Osborne, 1998). You can also refer to the Visual Basic for Applications Help file. The examples provided are those that come with AutoCAD.*

You can access VBA and the Visual Basic Editor from AutoCAD's Tools menu. To display the Visual Basic Editor (officially known as the *Visual Basic Interactive Development Environment,* or *VBA IDE*), from the Tools menu, choose Macro | Visual Basic Editor. The Editor is shown in Figure 29-4. You can access other VBA functions, such as loading and unloading projects (ActiveX Automation programs written in VBA and stored with a .DVB file extension), and run a VBA procedure that is already loaded from this menu selection, as well. In addition, AutoCAD offers several commands that let you load and run VBA macros from the command line. For example, VBALOAD loads

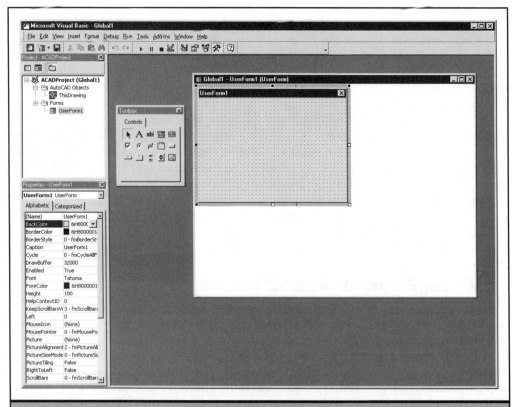

**Figure 29-4.**   *The Visual Basic for Applications Editor (VBA IDE)*

projects into the VBA IDE, and VBAUNLOAD unloads projects from the VBA IDE. The VBARUN command runs a VBA macro from a dialog box, while the VBAIDE command displays the Visual Basic Editor.

**Note**    *The Visual Basic Editor isn't installed if you selected the Compact setup option when installing AutoCAD 2002.*

After you are in the VBA Editor, you can view any of AutoCAD's Automation objects by using the Object Browser, shown in Figure 29-5. To display the Object Browser, from the VBA Editor's View menu, choose Object Browser (or simply press F2). The Object Browser shows a complete list of all the objects available in AutoCAD, along with each object's properties and methods. Objects are shown in the Classes list on the left side of the Browser, and the object's associated properties and methods (or functions) are shown in the Members list on the right.

When you select an object and a method or property, information about that method or property is displayed at the bottom of the Browser window. This information includes a brief description of its purpose. You can view detailed information about the selected method or property by clicking the Help button. Help displays a portion of the online *AutoCAD ActiveX and VBA Reference,* which also includes a code sample for each method or property.

## Creating a Simple VBA Project

The VBA IDE makes it easy to create a user interface and the code that performs an action in AutoCAD. Two separate steps are involved. First, you create a form and use

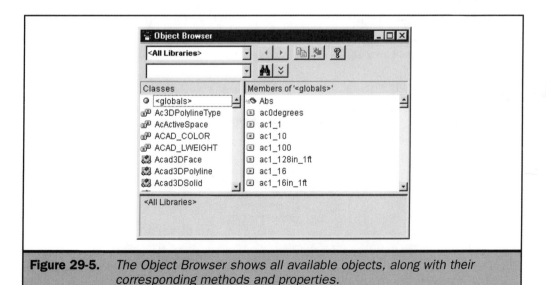

**Figure 29-5.**    *The Object Browser shows all available objects, along with their corresponding methods and properties.*

ADVANCED TOPICS

the VBA IDE tools to create the graphical layout of your project's interface. Visual Basic makes this step incredibly easy, since you don't have to write a single line of code. After you create the interface, you write the actual code that performs the action.

*LEARN BY EXAMPLE*

*Create a simple VBA project that opens a new AutoCAD drawing.*

You begin creating a new VBA project by creating a form. If you are not already in the Microsoft Visual Basic Editor, display it now. In AutoCAD, choose Tools | Macro | Visual Basic Editor.

*If you've already loaded a VBA project, unload it first before starting the VBA IDE. To unload a VBA project, from AutoCAD's Tools menu, choose Macro | VBA Manager to display the VBA Manager dialog box. In the VBA Manager dialog box, select the project in the Projects list, and then click Unload. You can then start the VBA IDE by clicking Visual Basic Editor.*

The VBA Editor should be open. Notice that the ThisDrawing object is highlighted in the Project Explorer.

1. From the Insert menu, choose UserForm. The screen should now look similar to Figure 29-4.

2. In the Toolbox, click the CommandButton tool. Then, click inside the UserForm1 window to place a command button.

3. Click the command button that you just placed, and then drag your cursor to select the word CommandButton1. Type **New Drawing**. The button label should now read New Drawing. You have just created the interface for your VBA project.

*You can resize the button or the dialog box by dragging its handles.*

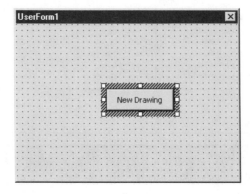

4. Click anywhere in the form window outside the command button and then double-click the button to display the UserForm code window for that button. Notice that the cursor is positioned on a blank line between the lines that begin Private Sub and End Sub. The editor is ready for you to enter the code that will perform the desired action when you click the New Drawing button. Type the following line of code:

```
ThisDrawing.Application.Documents.Add ""
```

This line of code simply starts a new drawing based on the default template drawing. As you type the code, VBA displays valid choices at each step. The code should look similar to Figure 29-6.

5. Choose Insert | Module and then, in the Module1 code window, type the following:

```
Option Explicit
Sub startdrawing()
    UserForm1.Show
```

When you type the line that begins Sub, VBA automatically adds the End Sub line for you.

That's all there is to it. Test the execution of your project by clicking the Run Sub/ User Form button on the VBA IDE toolbar. VBA switches back to AutoCAD, and you should see a dialog box with the New Drawing button that you created. Click the button. AutoCAD opens a new drawing.

 *The VBA application you just created won't work if you've reconfigured AutoCAD for single drawing compatibility mode, since it uses the Documents.Add function. To adapt the code to work in either single-drawing interface (SDI) or multi-drawing interface (MDI) mode, you need to change the UserForm1 code to add some conditional statements to check for the current mode. The complete UserForm1 code appears later in this chapter.*

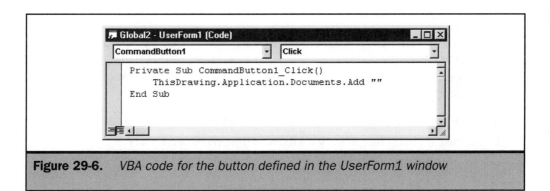

**Figure 29-6.** *VBA code for the button defined in the UserForm1 window*

ADVANCED TOPICS

After the new drawing is open, the dialog box remains onscreen. Notice that you can't do anything in the AutoCAD drawing. VBA dialog boxes are *modal*, which means that while they are active, they have AutoCAD's total focus. To work in AutoCAD, you must close the dialog box. Although you can easily do this by clicking the Close button in the upper-right corner of the dialog box, you can tell VBA to close the dialog box when it is finished, by adding an additional line of code to the New Drawing button, as follows:

1. In AutoCAD, click the Close button to close the dialog box.

2. Press CTRL-TAB if necessary to switch back to the VBA IDE.

3. In the UserForm1 code window, place the cursor on the line of code that you created (the one that begins ThisDrawing) and then press ENTER.

4. Add the following code:

```
Unload Me
```

This line of code simply tells the dialog box to unload itself after the new drawing has been opened.

Now, try your project again by clicking the Run Sub/User Form button. When you click the New Drawing button, AutoCAD starts a new drawing. After the drawing is open, the dialog box closes automatically, and you are returned to the VBA IDE.

To save your project, do one of the following:

- On the toolbar, click Save Global*n*.

- From the File menu, choose Save Global*n*.

- Press CTRL-S.

You are returned to AutoCAD, which displays the Save As dialog box. Specify the directory and filename for your project and then click Save.

 *A copy of this VBA project is also provided as SAMPLE1.DVB on the Chapter 29 page on the companion web site.*

## Loading and Running VBA Projects

After you save a VBA project, you can load and run it from within AutoCAD. In this case, the project is already loaded. Try running the application that you just created. To run a loaded VBA project, do one of the following:

- From the Tools menu, choose Macro | Macros.

- At the command line, type **VBARUN** and then press ENTER.

AutoCAD displays a Macros dialog box, similar to the one shown in Figure 29-7. You can then select any of the valid macros listed in the list, and click Run to run the selected macro.

To load a VBA project, you can use either the VBALOAD command or the VBA Manager dialog box (shown in Figure 29-8). The VBALOAD command displays the

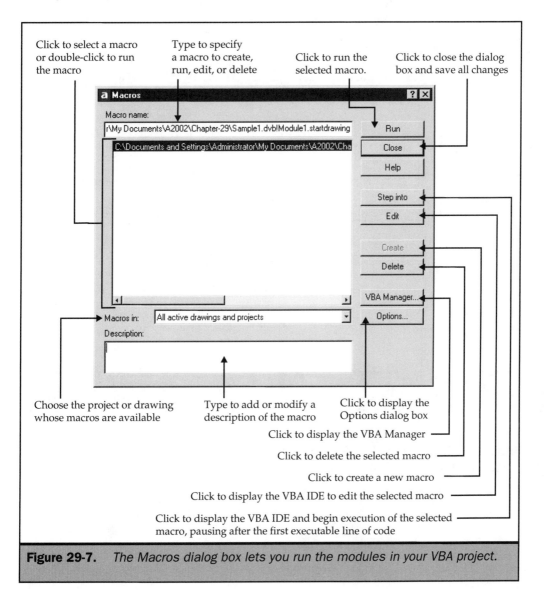

**Figure 29-7.**   *The Macros dialog box lets you run the modules in your VBA project.*

ADVANCED TOPICS

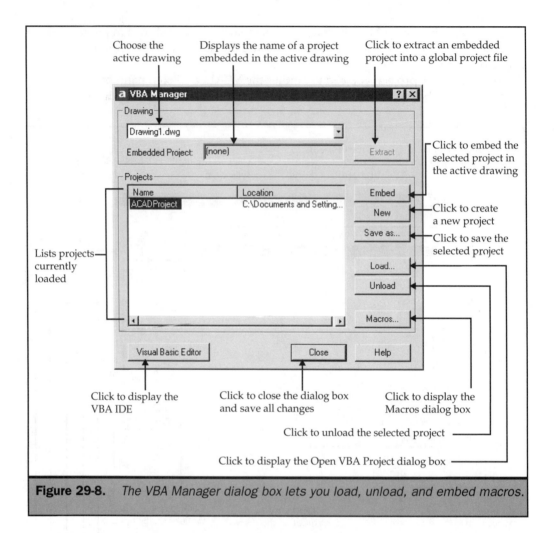

Choose the active drawing

Displays the name of a project embedded in the active drawing

Click to extract an embedded project into a global project file

Click to embed the selected project in the active drawing

Click to create a new project

Click to save the selected project

Lists projects currently loaded

Click to display the VBA IDE

Click to close the dialog box and save all changes

Click to display the Macros dialog box

Click to unload the selected project

Click to display the Open VBA Project dialog box

**Figure 29-8.**    *The VBA Manager dialog box lets you load, unload, and embed macros.*

Open VBA Project dialog box. Select the DVB file you want to load, and then click Open. To display the Open VBA Project dialog box, do one of the following:

■ From the Tools menu, choose Macro | Load Project.

■ At the command line, type **VBALOAD** and then press ENTER.

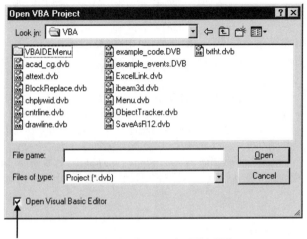

Select to load the project and display it in the VBA IDE

**Note** *When you load a macro, AutoCAD normally displays a dialog box alerting you to the fact that macros can contain viruses. To load the project, click Enable Macros. You can disable the display of this warning by clearing the check box. If you do disable this message, you can turn it back on again from the Options dialog box, displayed by clicking the Options button in the Macros dialog box. Because computer viruses are a real threat, you should always run a full-time virus protection program on your computer to protect your system against infection. In addition, you should periodically update the virus data files used by your antivirus software, to ensure that you're protected from the latest viruses. You can generally download updated data files from your antivirus software provider's web site.*

ADVANCED TOPICS

The VBA Manager dialog box provides even more options for loading and unloading macros. To display this dialog box, do one of the following:

- From the Tools menu, choose Macro | VBA Manager.
- At the command line, type **VBAMAN** and then press ENTER.

**Tip** *As explained earlier, AutoCAD's MDI mode introduces some interesting considerations when writing ActiveX applications. In our previous example, you can deal with the situation in which AutoCAD may be configured for either SDI or MDI mode by including some conditional (If/Then) statements in the UserForm1 code. The following listing shows the complete code, including these conditional statements. A copy of this version of the VBA project is also provided as SAMPLE2.DVB on the Chapter 29 page of the companion web site.*

```
Private Sub CommandButton1_Click()
    Dim sysVarName As String
    Dim sysVarData As Variant
    Dim intSDI As Integer
    ' Check for MDI / SDI
    sysVarName = "SDI"
    sysVarData = ThisDrawing.GetVariable(sysVarName)
    intSDI = CInt(sysVarData)
     ' Starting a new drawing with the default template
    On Error Resume Next
        If intSDI = 0 Then
            ThisDrawing.Application.Documents.Add ""
        Else
            ThisDrawing.New ""
        End If
    Unload Me
End Sub
```

## Sample VBA Projects

Several sample VBA projects are provided with AutoCAD. The files are installed in the AutoCAD 2002/Sample/VBA folder. You can load any of these projects and try them out. You should also feel free to look at the source code for these examples. One of the best ways to learn to program is to look at someone else's code. Here's what you'll find in the sample projects:

- **ACAD_CG.DVB** Demonstrates several AutoCAD functions all run from a single VBA dialog box.

- **ATTEXT.DBV**  Extracts attributes from the ATTRIB.DWG sample drawing (located in the AutoCAD 2002/Sample/ActiveX/ExtAttr folder) and sends them to an Excel spreadsheet. It also creates a chart within the spreadsheet and creates a Word document containing a graph. (*Note:* You must have Excel and Word installed on your system.)

- **BLOCKREPLACE.DVB**  Replaces block references in the current drawing model space.

- **CHPLYWID.DVB**  Uses a dialog box to change the width of all existing polylines in a drawing.

- **CNTRLINE.DVB**  Adds center lines to selected arcs and circles.

- **DRAWLINE.DVB**  Uses a simple dialog box to draw a line between two end points. The point coordinates are hard-coded into the program.

- **EXAMPLE_CODE.DVB**  A VBA project containing code samples for all methods and properties.

- **EXAMPLE_EVENTS.DVB**  A VBA project containing example code for all events.

- **EXCELLINK.DVB**  A demo showing how to pass data between AutoCAD and an external application such as Excel.

- **IBEAM3D.DVB**  Uses a dialog box containing sliders to interactively draw a simple 3-D I-beam.

- **MAP2GLOBE.DWG**  A drawing containing an embedded VBA macro that generates 3-D polylines onto a sphere from 2-D polylines on a map.

- **MENU.DVB**  A sample demonstrating the MenuGroup and MenuBar ActiveX APIs.

- **OBJECTTRACKER.DVB**  A demo showing how to attach persistent information to AutoCAD objects using XRecordData.

- **SAVEASR12.DVB**  Makes a call to a separate application (Convert), which exposes an ActiveX object allowing any Automation client to batch-process AutoCAD 2000 format drawing files to the R12 DWG format.

- **TOWER.DWG**  A drawing containing an embedded VBA macro that builds an electrical power transmission tower and then computes tables of weights and foundation reactions and performs stress analysis on the tower structure.

- **TXTHT.DVB**  Uses a dialog box to change the height of all text and multiline text objects in the current drawing.

All of these projects and drawings are also described in more detail in the README.TXT file located in the AutoCAD 2002/Sample/VBA folder.

ADVANCED TOPICS

# Using ActiveX Automation with Other Applications

As mentioned earlier, you can also use the ActiveX Automation tools built into other applications, such as Microsoft Word, Excel, or PowerPoint, to create programs that link your AutoCAD drawings with other applications. This discussion of ActiveX Automation concludes by looking at a sample macro that comes with AutoCAD.

AutoCAD provides a macro that links an Excel spreadsheet with an AutoCAD drawing. The macro automatically extracts attribute data from the AutoCAD drawing and fills in the cells in the spreadsheet.

To use this macro, load the AutoCAD drawing ATTRIB.DWG, located in the Sample/ActiveX/ExtAttr directory.

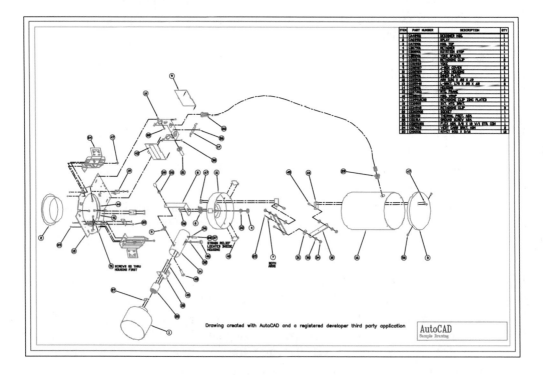

1. From the Tools menu, choose Macro | Load Project.

2. In the Open VBA Project dialog box, load the ATTEXT.DVB project (located in the AutoCAD 2002/Sample/VBA folder).

3. From the Tools menu, choose Macro | Macros.

4. In the Macros dialog box, click Run. AutoCAD displays the Extract Attribute Information dialog box.

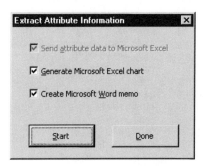

5. Click Start. The macro automatically launches Excel (provided that it is installed on your computer) and extracts the attributes from the drawing to the spreadsheet shown in Figure 29-9. If you selected the Generate Chart check box, the macro also creates an Excel chart showing the quantity of each of the parts in the drawing.

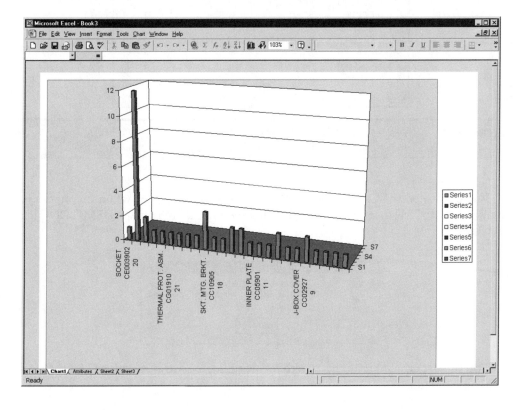

| | A | B | C | D | |
|---|---|---|---|---|---|
| **1** | **1** | **3** | | **4** | **2** |
| 2 | 20 | CE003902 | SOCKET | 1 | |
| 3 | 25 | CH00531 | RIVET #511 X 3/16 | 12 | |
| 4 | 24 | CG17903 | VERT CARR BRKT. ASM | 2 | |
| 5 | 23 | CG1291801 | FLEX ASM. 3/8 X 18 W/1 STR. CON | 1 | |
| 6 | 22 | CG11914 | GROUND SCREW ASM. | 1 | |
| 7 | 21 | CG01910 | THERMAL PROT. ASM. | 1 | |
| 8 | 10 | CC02929 | J-BOX HOUSING | 1 | |
| 9 | 15 | CC07303 | MTG. FRAME | 1 | |
| 10 | 16 | CC08045 | HSG. WRAP | 1 | |
| 11 | 19 | CC14942 | RETAINING CLIP | 3 | |
| 12 | 18 | CC10905 | SKT. MTG. BRKT. | 1 | |
| 13 | 17 | CC09912C02 | RETAINING CLIP ZINC PLATED | 1 | |
| 14 | 13 | CC05942 | L-BRKT. 1.75 X .95 X .62 | 2 | |
| 15 | 12 | CC05931 | ARM 5.56 X .82 X .19 | 2 | |
| 16 | 14 | CC06901 | HOUSING | 1 | |
| 17 | 11 | CC05901 | INNER PLATE | 1 | |
| 18 | 5 | CB08901 | ROTATION STOP | 1 | |
| 19 | 7 | CC00941 | RETAINING CLIP | 2 | |
| 20 | 6 | CB20901 | YOKE SPACER | 1 | |
| 21 | 8 | CC01933 | YOKE | 1 | |
| 22 | 9 | CC02927 | J-BOX COVER | 2 | |
| 23 | 1 | CA48901 | DESIGNER HSG. | 1 | |
| 24 | 2 | CA69901 | SPLAY | 1 | |
| 25 | 3 | CA70901 | HSG. TOP | 1 | |
| 26 | 4 | CB07901 | RETAINER | 1 | |
| 27 | | | | | |

Chart1 **Attributes** Sheet2 Sheet3

**Figure 29-9.** *An Excel spreadsheet created automatically from attributes extracted from an AutoCAD drawing*

If you selected the Create Microsoft Word memo check box, when the macro is complete, AutoCAD displays a dialog box telling you that the file C:\DEMO.DOC has been created.

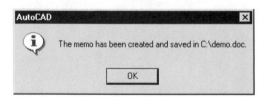

You can experiment with this macro by changing the attribute values and then rerunning the macro.

# The Complete Reference

# Part V

## About the Appendixes Online

The appendixes described in the following pages provide a useful reference to AutoCAD 2002 and many of the additional tools available from Autodesk for use with AutoCAD. These pages describe the contents of the appendixes in general. The entire contents of these appendixes are available online at the companion web site **www.dscohn.com/acad2002**. The web site is password protected. In order to access the site, you will need to enter a user name and password, both of which are printed at the end of this section.

## Appendix A—AutoCAD Command Reference

This appendix contains a complete list of all AutoCAD commands. Although not part of AutoCAD, the list includes the Express Tools as well. The commands are listed alphabetically by command name. To start any of the commands, type the command name at the command line and then press ENTER. If the command name is preceded by an apostrophe ('), the command can be used *transparently*—it can be started while another command is active. When a command can be started by using a keyboard shortcut, such as F1 to start the HELP command, this information is shown as well. Commands from the previous version of AutoCAD that have been eliminated from AutoCAD 2002 are also shown (with a cross reference to the actual command that starts when you type the old command name). Commands that are new to AutoCAD 2002 appear with a special symbol adjacent to the command name.

In addition to the command name, the listing contains:

- The type of command (an actual AutoCAD command, an external command, or an Express tool)

- The toolbar button that is used to start the command (when appropriate) as well as the name of the toolbar on which the command appears

- The menu name and menu selections that are used to start the command (as appropriate)

- A short description of each command

The complete contents of this table are provided on the companion web site.

## Appendix B—System Variable Reference

This appendix contains a complete list of AutoCAD's system variables. System variables store default settings (such as the current elevation), control the appearance of elements in the drawing window (such as the size of the aperture), and determine other aspects of the execution of AutoCAD commands (such as whether text is mirrored when using the MIRROR command). You can fine-tune AutoCAD to meet your specific needs by changing the values of the appropriate system variables. Some system variable settings apply only to the current drawing, others are valid only during the current drawing session, and still others remain as set from that point forward. System variables that are new to AutoCAD 2002 appear with a special symbol adjacent to the variable name.

In addition to the system variable name, the listing contains:

- The name of the AutoCAD command or commands that affect the value of the variable or show its contents
- The variable type (such as integer, real, string, and so on)
- Where the variable is stored (such as the drawing or system registry)
- The variable's default value
- A short description of each variable

The complete contents of this table are provided on the companion web site.

# Appendix C—Installation and Configuration

This appendix contains detailed information about installing and configuring AutoCAD 2002. Before you begin the installation process, you should make sure that your computer system is properly configured and meets the minimum requirements to run AutoCAD. After installing AutoCAD, you can also use information provided in this appendix to customize both your AutoCAD configuration and the operation of the AutoCAD 2002 Today window.

# Appendix D—AutoCAD Migration Assistance

This appendix describes the AutoCAD Migration Assistance tools. Many of these tools are provided on the AutoCAD 2002 CD. Others are available at the Autodesk web site.

# Appendix E—File Types

Most computer programs reserve several file extensions for themselves. The three-letter extension indicates the purpose for the file. For example, all executable program files have an .EXE file extension. Thus, the main AutoCAD program file is named ACAD.EXE. And, as you likely know, AutoCAD drawing files have a .DWG file extension.

AutoCAD uses many file extensions. Some of these file types are specific to AutoCAD; others, like the .EXE file and TrueType fonts (.TTF), are standards used by all programs. In addition, AutoCAD can import and export many other types of files, such as raster-image files saved in BMP or TIF format.

Understanding the purpose for the different files that you're likely to find in your AutoCAD directories can help you better manage your system. For example, depending on how your system is configured, AutoCAD may create backup copies of your drawing files whenever you save a drawing. You may want to delete these BAK files periodically to free space on your hard drive. You can also safely delete ADT, ERR, PLT, SV$, and XLG files when they are no longer needed. You should never delete AutoCAD's temporary working files (AC$) while AutoCAD is being used, but if you find any of these files left on your system when you're not using AutoCAD, they were likely left when AutoCAD was closed improperly, and thus can be deleted.

The complete listing of file types is provided on the companion web site.

# Appendix F—Glossary

This glossary, provided on the companion web site, includes both AutoCAD-specific and general CAD terminology.

# Appendix G—What's On the Companion Web Site

The companion web site—located at **www.dscohn.com/acad2002**—contains links from which you can download all of the files necessary to complete the *Learn by Example* exercises, including all of the drawing and template files used in those exercises.

As you read through the book, you'll encounter instructions about how to use these files. For example, in Chapter 2, the very first *Learn by Example* exercise instructs you to "open the AutoCAD drawing Figure 2-34 from the companion web site." Although you may open these files in AutoCAD directly from the Web, you may find it more useful to first download the files from the web site to your computer, so that your ability to work with the files is not affected by the speed or availability of your Internet connection. When you work through the exercises, you will also need to save the files locally. You will not be able to save your changes back to the web site.

In addition to the drawings and drawing templates for use with the *Learn by Example* exercises, you'll also find other useful files on the companion web site, including:

- Sample linetype library files
- Sample hatch patterns
- Sample menus
- Sample script files to help you learn to create your own macros
- Sample web pages incorporating AutoCAD drawings
- Sample VBA projects to help you get started with ActiveX Automation

■ AutoLISP routines to help you learn to create your own AutoLISP commands

■ A program that automates the creation of custom AutoCAD hatch patterns

■ A program that lets you modify AutoCAD slide library files

The companion web site is organized by chapter. For example, the files used in the exercises for Chapter 20 are linked to the page on the web site dealing with Chapter 20. The full contents of Appendixes A through F are also located on the companion web site. In addition, you'll find lots of useful links to other sites where you can obtain additional AutoCAD utilities, including a link from which you can download a free evaluation copy of AutoCAD 2002.

The companion web site is password protected. In order to access the web site, enter the following URL in the address bar of your web browser:

**www.dscohn.com/acad2002**

When you first attempt to access this page, you will be presented with a dialog box asking you for a user name and password. You can access the web site by entering the following information:

user name: **acad2002**
password: **kelvin8**

If you are also prompted to supply a domain name, you can leave that field blank.

# Index

**Q**

# INTERNATIONAL CONTACT INFORMATION

**AUSTRALIA**
McGraw-Hill Book Company Australia Pty. Ltd.
TEL +61-2-9417-9899
FAX +61-2-9417-5687
http://www.mcgraw-hill.com.au
books-it_sydney@mcgraw-hill.com

**CANADA**
McGraw-Hill Ryerson Ltd.
TEL +905-430-5000
FAX +905-430-5020
http://www.mcgrawhill.ca

**GREECE, MIDDLE EAST,
NORTHERN AFRICA**
McGraw-Hill Hellas
TEL +30-1-656-0990-3-4
FAX +30-1-654-5525

**MEXICO (Also serving Latin America)**
McGraw-Hill Interamericana Editores S.A. de C.V.
TEL +525-117-1583
FAX +525-117-1589
http://www.mcgraw-hill.com.mx
fernando_castellanos@mcgraw-hill.com

**SINGAPORE (Serving Asia)**
McGraw-Hill Book Company
TEL +65-863-1580
FAX +65-862-3354
http://www.mcgraw-hill.com.sg
mghasia@mcgraw-hill.com

**SOUTH AFRICA**
McGraw-Hill South Africa
TEL +27-11-622-7512
FAX +27-11-622-9045
robyn_swanepoel@mcgraw-hill.com

**UNITED KINGDOM & EUROPE
(Excluding Southern Europe)**
McGraw-Hill Education Europe
TEL +44-1-628-502500
FAX +44-1-628-770224
http://www.mcgraw-hill.co.uk
computing_neurope@mcgraw-hill.com

**ALL OTHER INQUIRIES Contact:**
Osborne/McGraw-Hill
TEL +1-510-549-6600
FAX +1-510-883-7600
http://www.osborne.com
omg_international@mcgraw-hill.com